The Paleontology of Gran Barranca
Evolution and Environmental Change through the Middle Cenozoic of Patagonia

Gran Barranca in Patagonia exposes the most complete sequence of middle Cenozoic paleofaunas in South America. It is the only continuous continental fossil record anywhere in the southern hemisphere between 42 and 18 million years ago, when climates at high latitudes transitioned from warm humid to cold dry conditions. Located on a narrow peninsula, surrounded by the southern oceans and close to Antarctica, Gran Barranca was ideally situated to record the biotic response to these climatic changes.

This volume presents the geochronology of the fossil mammal sequence and a compilation of the latest studies of the stratigraphy, sedimentology, mammals, plants, invertebrates, and trace fossils. It is also the first detailed treatment of the vertebrate faunal sequence at Gran Barranca. Based on more than 10 years of fieldwork and study by the contributors, it provides important new evidence about biotic diversity, evolution, and change in the native species. A revised taxonomy allows a re-evaluation of the origination and extinction of herbivorous mammals, marsupials, and xenarthrans, and the earliest occurrence of rodents and primates in southern latitudes.

Academic researchers and advanced students in vertebrate paleontology, geochronology, sedimentology, and paleoprimatology will find a wealth of new information and interpretations about the paleontology of Gran Barranca in this book.

RICHARD H. MADDEN has been a Research Associate at the Duke University Medical Center for the last 20 years, where he assists in the teaching of anatomy in the School of Medicine.

His current research interest is the relationship between climate, earth surface processes, and the geographic and temporal patterns of soil ingestion and tooth wear in mammalian herbivores as these may relate to the evolution of tooth mineral volume.

ALFREDO A. CARLINI is a Research Paleontologist of Consejo Nacional de Investigaciones Científicas y Técnicas (CONICET) and Professor of Comparative Anatomy at the National University of La Plata, Argentina. His research interests focus on the morphological diversity, evolutionary trends, ontogeny, systematics, biostratigraphy, and biogeography of armadillos and living and fossil xenarthrans, with over 100 scientific publications in books and journals.

MARIA GUIOMAR VUCETICH is a Research Palaeontologist of CONICET and Professor of Vertebrate Palaeontology at the National University of La Plata, Argentina, where she has worked since 1971. Her research interests involve the evolutionary history of caviomorph rodents, and she has published nearly 100 scientific articles on this topic.

RICHARD F. KAY is Professor of Evolutionary Anthropology, and Earth and Ocean Sciences at Duke University, North Carolina, where he has worked since 1973. He has edited five books and authored more than 200 research papers on primate paleontology, functional anatomy, adaptations, and phylogenetics. He is an elected Fellow of the American Association for the Advancement of Science (USA).

The Paleontology of Gran Barranca

Evolution and Environmental Change
through the Middle Cenozoic
of Patagonia

EDITED BY

R. H. MADDEN
Duke University, USA

A. A. CARLINI
Museo de La Plata, Argentina

M. G. VUCETICH
Museo de La Plata, Argentina

R. F. KAY
Duke University, USA

CAMBRIDGE
UNIVERSITY PRESS

CAMBRIDGE UNIVERSITY PRESS
Cambridge, New York, Melbourne, Madrid, Cape Town, Singapore,
São Paulo, Delhi, Dubai, Tokyo

Cambridge University Press
The Edinburgh Building, Cambridge CB2 8RU, UK

Published in the United States of America by
Cambridge University Press, New York

www.cambridge.org
Information on this title: www.cambridge.org/9780521872416

First published 2010

Printed in the United Kingdom at the University Press, Cambridge

A catalogue record for this publication is available from the British Library

Library of Congress Cataloguing in Publication Data

The paleontology of Gran Barranca : evolution and environmental change through the middle
Cenozoic of Patagonia / edited by R. H. Madden ... [et al.].
    p.   cm.
  ISBN 978-0-521-87241-6 (Hardback)
  1. Paleontology–Cenozoic.   2. Sequence stratigraphy.   3. Paleontology–Patagonia
(Argentina and Chile)   4. Plant species diversity–Patagonia (Argentina and Chile)
  I. Madden, R. H. (Richard H.)   II. Title.
  QE735.P374 2010
  560′.17809827–dc22

                                                                            2009053545

ISBN 978-0-521-87241-6 Hardback

Contents

Contributors

Abello, María Alejandra
Laboratorio de Sistemática y Biología Evolutiva, Facultad de Ciencias Naturales y Museo, Universidad Nacional de La Plata, Paseo del Bosque s/n, B1900FWA La Plata, Argentina (ANPCyT)

Barreda, Viviana
División Paleobotánica, Museo Argentino de Ciencias Naturales, Av. Ángel Gallardo 470, C1405DJR Buenos Aires, Argentina (CONICET)

Bellosi, Eduardo S.
Departamento de Icnología, Museo Argentino de Ciencias Naturales "Bernardino Rivadavia," Av. Ángel Gallardo 470, C1405DJR Buenos Aires, Argentina (CONICET)

Bond, Mariano
División Paleontología Vertebrados, Facultad de Ciencias Naturales y Museo, Universidad Nacional de La Plata, Paseo del Bosque s/n, B1900FWA La Plata, Argentina (CONICET)

Brea, Mariana
Laboratorio de Paleobotánica, Materi y España s/n, Diamante, E3105XAB, Entre Ríos, Argentina (CICYTTP, CONICET)

Brinkhuis, Henk
Laboratory of Palaeobotany and Palynology, Institute of Environmental Biology, Faculty of Sciences, Utrecht University, Budapestlaan 4, 3584 CD Utrecht, The Netherlands

Candela, Adriana M.
División Paleontología Vertebrados, Facultad de Ciencias Naturales y Museo, Universidad Nacional de La Plata, Paseo del Bosque s/n, B1900FWA La Plata, Argentina (CONICET)

Carlini, Alfredo A.
División Paleontología Vertebrados, Facultad de Ciencias Naturales y Museo, Universidad Nacional de La Plata, Paseo del Bosque s/n, B1900FWA La Plata, Argentina (CONICET)

Cassini, Guillermo H.
División Paleontología Vertebrados, Facultad de Ciencias Naturales y Museo, Universidad Nacional de La Plata, Paseo del Bosque s/n, B1900FWA La Plata, Argentina (ANPCyT)

Chornogubsky, Laura
Sección Paleontología Vertebrados, Museo Argentino de Ciencias Naturales "Bernardino Rivadavia," Av. Angel Gallardo 470, C1405DJR Buenos Aires, Argentina (CONICET)

Ciancio, Martín R.
División Paleontología Vertebrados, Facultad de Ciencias Naturales y Museo, Universidad Nacional de La Plata, Paseo del Bosque s/n, B1900FWA La Plata, Argentina (CONICET)

Czaplewski, Nicholas J.
Oklahoma Museum of Natural History and Department of Zoology, University of Oklahoma, Norman, Oklahoma 73072, USA

Deschamps, Cecilia M.
División Paleontología Vertebrados, Facultad de Ciencias Naturales y Museo, Universidad Nacional de La Plata, Paseo del Bosque s/n, B1900FWA La Plata, Argentina (CIC)

Gelfo, Javier N.
División Paleontología Vertebrados, Facultad de Ciencias Naturales y Museo, Universidad Nacional de La Plata, Paseo del Bosque s/n, B1900FWA La Plata, Argentina (CONICET)

Genise, Jorge F.
Departamento de Icnología, Museo Argentino de Ciencias Naturales "Bernardino Rivadavia," Av. Ángel Gallardo 470, C1405DJR Buenos Aires, Argentina (CONICET)

Geuna, Silvana E.
Departamento de Ciencias Geológicas, Facultad de Ciencias Exactas y Naturales, Universidad de Buenos Aires, Ciudad Universitaria Pab. 2, C1428EHA Buenos Aires, Argentina (CONICET)

Goin, Francisco Javier
División Paleontología Vertebrados, Facultad de Ciencias Naturales y Museo, Universidad Nacional de La Plata, Paseo del Bosque s/n, B1900FWA La Plata, Argentina (CONICET)

González, Mirta G.
Departamento de Icnología, Museo Argentino de Ciencias Naturales, Av. Ángel Gallardo 470, CDJR1405 Buenos Aires, Argentina (CONICET)

Guerstein, G. Raquel
Departamento de Geología, Universidad Nacional del Sur, San Juan 670 (8000) Bahía Blanca, Argentina (CONICET)

Guler, M. Verónica
Departamento de Geología, Universidad Nacional del Sur, San Juan 670 (8000) Bahía Blanca, Argentina (CONICET)

Heizler, Matthew
New Mexico Bureau of Mines and Mineral Resources, New Mexico Tech 801 Leroy Place, Socorro, NM 87801, USA

Josef, Jennifer A.
Department of Geological Sciences, University of South Carolina, Columbia, SC 29208, USA

Kay, Richard F.
Department of Evolutionary Anthropology, Duke University, Durham, NC 27708, USA

Kohn, Matthew J.
Department of Geological Sciences, University of South Carolina, Columbia, SC 29208, USA

Kramarz Alejandro G.
Sección Paleontología Vertebrados, Museo Argentino de Ciencias Naturales "Bernardino Rivadavia," Av. Ángel Gallardo 470, C1405DJR Buenos Aires, Argentina (CONICET)

Laza, José H.
Departamento de Icnología, Museo Argentino de Ciencias Naturales "Bernardino Rivadavia," Av. Ángel Gallardo 470, C1405DJR Buenos Aires, Argentina

López, Guillermo Marcos
División Paleontología Vertebrados, Facultad de Ciencias Naturales y Museo, Universidad Nacional de La Plata, Paseo del Bosque s/n, B1900FWA La Plata, Argentina

Madden, Richard H.
Department of Evolutionary Anthropology, Duke University, Durham, NC 27708, USA

Marenssi, Sergio A.
Instituto Antártico Argentino, Cerrito 1248, C1010AAZ Buenos Aires, Argentina

Miquel, Sergio E.
División Invertebrados, Museo Argentino de Ciencias Naturales "Bernardino Rivadavia," Av. Ángel Gallardo 470, C1405DJR Buenos Aires, Argentina

Palazzesi, Luis
División Paleobotánica, Museo Argentino de Ciencias Naturales "Bernardino Rivadavia," Av. Ángel Gallardo 470, C1405DJR Buenos Aires, Argentina (CONICET)

Pérez, María Encarnación
Museo Paleontológico Egidio Feruglio, Av. Fontana 140, U9100GYO Trelew, Argentina (CONICET)

Prevosti, Francisco J.
División Mastozoología, Museo Argentino de Ciencias Naturales "Bernardino Rivadavia," Av. Ángel Gallardo 470, C1405DJR Buenos Aires, Argentina (CONICET)

Ré, Guillermo H.
Departamento de Ciencias Geológicas, Facultad de Ciencias Exactas y Naturales, Universidad de Buenos Aires, Ciudad Universitaria Pab. 2, C1428EHA Buenos Aires, Argentina

Reguero, Marcelo A.
División Paleontología Vertebrados, Facultad de Ciencias Naturales y Museo, Universidad Nacional de La Plata,

Paseo del Bosque s/n, B1900FWA La Plata, Argentina (CONICET)

Ribeiro, Ana Maria
Museu de Ciências Naturais da Fundação Zoobotánica do Rio Grande do Sul, Av. Salvador França 1427, Cep 90690–000 Porto Alegre, RS, Brazil

Sánchez, M. Victoria
Departamento de Icnología, Museo Argentino de Ciencias Naturales "Bernardino Rivadavia," Av. Ángel Gallardo 470, C1405DJR Buenos Aires, Argentina (ANPCyT)

Scarano, Alejo
División Paleontología Vertebrados, Facultad de Ciencias Naturales y Museo, Universidad Nacional de La Plata, Paseo del Bosque s/n, B1900FWA La Plata, Argentina

Scillato-Yané, Gustavo J.
División Paleontología Vertebrados, Facultad de Ciencias Naturales y Museo, Universidad Nacional de La Plata, Paseo del Bosque s/n, B1900FWA La Plata, Argentina (CONICET)

Vieytes, Emma Carolina
División Zoología Vertebrados, Facultad de Ciencias Naturales y Museo, Universidad Nacional de La Plata, Paseo del Bosque s/n, B1900FWA La Plata, Argentina (CONICET)

Vilas, Juan Francisco
Departamento de Ciencias Geológicas, Facultad de Ciencias Exactas y Naturales, Universidad de Buenos Aires, Ciudad Universitaria Pab. 2, C1428EHA Buenos Aires, Argentina (CONICET)

Vucetich, María Guiomar
División Paleontología Vertebrados, Facultad de Ciencias Naturales y Museo, Universidad Nacional de La Plata, Paseo del Bosque s/n, B1900FWA La Plata, Argentina (CONICET)

Warnaar, Jeroen
Palaeoecology, Laboratory of Palaeobotany and Palynology Institute of Environmental Biology, Faculty of Sciences, Utrecht University, Budapestlaan 4, 3584 CD Utrecht, The Netherlands

Zanazzi, Alessandro
Department of Geological Sciences, University of South Carolina, Columbia, SC 29208, USA

Zucol, Alejandro
Laboratorio de Paleobotánica, Materi y España s/n, Diamante, E3105XAB Entre Ríos, Argentina (CICYTTP, CONICET)

Preface

In *Splendid Isolation*, Simpson described South America as a natural experiment in evolution, where one sees "evolution at work" through the Cenozoic. This experiment, and many of the most dramatic events that shaped the history of the peculiar South American mammal fauna, are recorded in the sediment sequence at Gran Barranca south of Lake Colhue-Huapi in central Patagonia. Gran Barranca is a spectacular sequence of mammalian faunas that serve as the basis for the biostratigraphic and biochronologic sequence of the mammal-bearing middle Cenozoic of South America. To this day, no other sequence with these characteristics has been found.

After its discovery and recognition of its significance around the turn of the twentieth century, over the next 100 years Gran Barranca has been visited by numerous research expeditions, not only from Argentina, but also France, the USA, and other countries. Even after all this effort, our understanding of the middle Cenozoic of South America is still full of uncertainties.

Over the past 15 years, the editors returned repeatedly to Gran Barranca in search of answers to some of the biggest outstanding questions about the mammalian evolution in South America. Duke University paleontologists first visited Gran Barranca at the invitation of Rosendo Pascual in 1990 while in Patagonia working the "Friasian" middle Miocene along the eastern slope of the Andes. Joint Museo de La Plata (MLP)/Duke expeditions started collecting in earnest at Gran Barranca in 1993. At that time, interest centered on the two highest faunal levels, representing the late Oligocene – early Miocene, for their potential content of fossil primates and rodents. The scope of our interests and activities at Gran Barranca changed along with discovery. From 1995 on, MLP/Duke expeditions expanded our work, and we began collecting throughout central Patagonia, sometimes with large field crews, gradually enriching our knowledge of the mammal sequence. At Gran Barranca we undertook detailed study of numerous stratigraphic profiles where we discovered and rediscovered faunal levels, and brought radiometric age control to a comprehensive magnetic polarity stratigraphy. All of this has enabled the elaboration of a new bio- and chronostratigraphic scheme for the middle Cenozoic of South America, and sharpened our understanding of many fundamental questions about mammalian evolutionary history.

The fruits of this research documented by this book could not have been made without a significant and sustained effort in field research at Gran Barranca. Our first expeditions in Patagonia were in the company of the geologist, vulcanologist, and sedimentologist Mario Martin Mazzoni. Mario worked with us in the "Friasian" and at Scarritt Pocket, but his greatest contribution to our work was made at Gran Barranca, where his stimulating ideas were interwoven with a deadpan ironic humor that sustained us through difficult times of conflicting opinion. Mario died prematurely in 1999, and throughout this book he is remembered in many ways.

Many other institutions and individuals have been helpful in the development of this project. Financial support for the field work, laboratory analyses, technical services, and study of museum collections all over the world was provided by the National Science Foundation (grants to R. F. Kay and R. H. Madden), Consejo Nacional de Investigaciones Científicas y Técnicas (CONICET) (grants to M. G. Vucetich), Agencia Nacional de Promoción Científica y Tecnológica (ANPCYT) (grants to M. G. Vucetich and A. A. Carlini), Facultad de Ciencias Naturales y Museo (UNLP) (grants to M. G. Vucetich).

We thank the governments of the Province of Chubut and the Municipality of Sarmiento for permission to undertake fossil collecting at Gran Barranca from 1995 onwards. The authorities of the Museo Paleontológico "Egidio Feruglio" in Trelew allowed access to their collections. For special assistance at Gran Barranca we thank Pan American Energy and the staffs at Cerro Dragón and Valle Hermoso for hospitality, generous collaboration, and many forms of material support during the field work. The Vera family kindly granted permission for us to work on their land.

Many individuals have participated in the Duke University/ Museo de La Plata paleontology expeditions to Gran Barranca. Participants include Richard F. Kay, Richard. H. Madden, Alfredo A. Carlini, Maria Guiomar Vucetich, Alejandra Abello, Alejandra Alcaraz, Ramiro Almagro, Daniel Aquino, Roberto Avila, Judith Babot, Eduardo Bellosi, Valeria Bertoia, Diego Brandoni, Nicolas and Pedro Carlini, Martín Ciancio, Richard Cifelli, Roberto "Tito" Cidale, Valeria Clar, Noelia Corrado, Carlos Dal Molin, Cecilia Deschamps, Georgina Erra, Analía Francia, Patricia García, Germán Gasparini, Javier Gelfo, Jorge Genise, Damián Glaz, Verónica Gomis, Mirta González, Adrián Guillaume, Jennifer Josef, Derek Johnson, Alejandro Kramarz, Cecilia Krmpotic, José H. Laza, Diego Licitra, Jessamyn Markley, Mario M. Mazzoni, Jorge Noriega, María Encarnación Pérez,

Mike Perkins, Sebastián Poljak, François Pujos, Guillermo H. Ré, Marcelo Reguero, Ana María Ribeiro, Alejo Scarano, Viviana Seitz, Esteban Soibelzon, Soledad Magallanes, Caroline Strömberg, Carolina Vieytes, Juan F. Vilas, Mauricio Vinocur, and Alfredo Zurita.

The tiny vertebrate materials studied in this book were found through many hours of diligent and patient hand-picking by Juan Canale, Nico Carlini, Georgina Erra, Verónica Gomis, Clay Madden, Alejandra Medinilla, Sebastián Poljak, Juliana Sterli, Marianela Talevi, Carolina Vieytes, and Danilo Vucetich.

We wish to acknowledge the authorities of the Museo Argentino de Ciencias Naturales "Bernardino Rivadavia" and the Universidad de Buenos Aires for enabling the participation of their scientists in this project. As evident by this book, their contributions have been significant.

We thank the anonymous colleagues who reviewed the chapters.

We want to thank especially Cecilia Deschamps at Museo de La Plata for assisting in the preparation of this volume by editing manuscripts, providing English translations, and supplying constant support, enthusiasm, good humor, and joy.

Irene Lofstrom, Lisa Jones, and Lisa Squires at Duke University provided administrative support.

1 Notes toward a history of vertebrate paleontology at Gran Barranca

Richard H. Madden and Alejo Scarano

Abstract

The turn of the twentieth century began an intense period of paleontological exploration of Patagonia. The field work of Carlos Ameghino in Patagonia, including Gran Barranca, documented in correspondence with his brother Florentino, was central to the discovery of the biostratigraphic sequence of pre-Santacrucian mammalian faunas. The work of the Ameghinos was stimulated by rivalry and benefitted from substantive contributions from contemporaries, notably Andrés Tournouër.

The early field study at Gran Barranca was sustained during the early twentieth century by geologists working in mineral and petroleum exploration in Patagonia, including a noteworthy contribution by Egidio Feruglio. The contributions of paleontologists working for museums of natural history in the United States, Elmer Riggs, Bryan Patterson, and George Simpson, represent another mid-century phase of activity at Gran Barranca. This work yielded exquisite specimens and set a new standard of utility for documentation and stratigraphic resolution. The Second World War and immediate post-war period also saw collecting activity at Gran Barranca by paleontologists associated with Argentine museums, Alejandro Bordas, Alfredo Castellanos, and Galileo Scaglia, work that continued to yield novelty from the exposure.

Rejuvenated by contemporary revolutions in tectonics and geochronology, Rosendo Pascual and his many students and collaborators at the National University and Museum of La Plata initiated multidisciplinary work at Gran Barranca, two noteworthy components of which were the stratigraphy and sedimentology of Luis Spalletti and Mario M. Mazzoni and the geochronology mediated by Larry G. Marshall of the abundant basalts of central Patagonia and Gran Barranca. This work enabled the first refined interpretations of the Patagonian fossil record in light of broader scientific questions about middle Cenozoic earth history.

Continuing this tradition, scientists from the Facultad de Ciencias Naturales y Museo de La Plata, Duke University, the Museo Argentino de Ciencias Naturales, and the Universidad de Buenos Aires, and numerous other institutions, using technical advances in radioisotopic dating and magnetostratigraphy and an intensification and diversification of collecting effort, bring much that is new to an increasingly sophisticated understanding of faunal and floral response to environmental change through the middle Cenozoic.

Resumen

Con el comienzo del siglo XX, arrancó un período de intensiva exploración paleontológica en Patagonia. El trabajo de campo de Carlos Ameghino en Patagonia, y en Gran Barranca, documentada en correspondencia con su hermano Florentino, es punto clave en el descubrimiento de la secuencia bioestratigráfica de las faunas paleomastozoológicas pre-Santacrucenses. El trabajo de los Ameghino vino estimulado por rivales y aprovechó de contribuciones sustantivas de contemporáneos, notablemente Andrés Tournouër.

Trabajos de campo en Gran Barranca fueron sostenidos durante la primera parte del siglo por geólogos involucrados en la exploración minera y petrolera en Patagonia, con una notable contribución de Egidio Feruglio. Las contribuciones de paleontólogos pertenecientes a museos de historia natural en los Estados Unidos, Elmer Riggs, Bryan Patterson, y George Simpson, significan otro período de actividad en Gran Barranca. Este trabajo rindió especímenes exquisitas y sirvió establecer unó nuevo estándar para la documentación de resolución estratigráfica. Durante la Segunda Guerra Mundial y el período inmediatamente posterior continuó la actividad de colección de paleontólogos trabajando para museos argentinos, Alejandro Bordas, Alfredo Castellanos y Galileo Scaglia, trabajo que continuamente rindió novedades del afloramiento.

Junto con las revoluciones en geotectónica y geocronología, Rosendo Pascual y sus estudiantes y colaboradores de la Universidad Nacional y Museo de La Plata comenzaron las primeras investigaciones multidisciplinarias en Gran Barranca, entre las cuales dos componentes notables fueron la estratigrafía y sedimentología de Luis Spalletti y Mario M. Mazzoni y la geocronología de los abundantes basaltos de Patagonia central y Gran Barranca conjugada por Larry G. Marshall. Este trabajo sirvió fundamentar interpretaciones sintéticas acerca del registro paleontológico patagónico en el contexto de preguntas mayores acerca de la historia terrestre en el Cenozoico medio.

Siguiendo esta tradición, científicos de la Facultad de Ciencias Naturales y Museo de La Plata, la Universidad de Duke, el Museo Argentino de Ciencias Naturales, la

The Paleontology of Gran Barranca: Evolution and Environmental Change through the Middle Cenozoic of Patagonia, eds. R. H. Madden, A. A. Carlini, M. G. Vucetich, and R. F. Kay. Published by Cambridge University Press. © Cambridge University Press 2010.

Universidad de Buenos Aires y otras instituciones, empleando avances tecnológicas en datación radiométrica y magnetoestratigrafía, junto con una intensificación y diversificación del esfuerzo de colecta, han podido aportar muchas novedades a la comprensión cada vez más sofisticada de la respuesta de la flora y fauna a los cámbios ambientales del Cenozoico medio.

Introduction

In light of our direct experience at Gran Barranca, we present here some perspectives about the history of discovery of the faunal succession at Gran Barranca, especially as illuminated by our exploration for the intermediate fossil-bearing stratigraphic levels not intensively collected hitherto and developments in field methods. We concentrate here on the establishment of the pre-Santacrucian faunal sequence in Patagonia, a 23-million-year interval of Earth history between about 41 and about 18 Ma, revealed by the exposures at Gran Barranca. The dates of important publications suggest that the overall sequence of evolution of Patagonian mammalian faunas may well have been established on the basis of fossils collected at disparate and geographically isolated localities, ordered into a sequence of evolutionary stages based on ideas about unidirectional progress in evolutionary morphology. Indeed, the clearest and most explicit expression of unidirectional progress in herbivore dental evolution is found in Ameghino (1904), and was published before the stratigraphic superposition at Gran Barranca was depicted in 1906. This order of publication suggests that stage of evolution was the evidence by which the faunal sequence was established before it was confirmed by superposition at Gran Barranca.

Experience leads us to believe otherwise. It seems likely that Gran Barranca was more central to Ameghino's developing ideas about superposition. Nevertheless, one must acknowledge the special problems Gran Barranca presents for the history of Patagonian paleontology. These problems arise from the nature of the fossil record there. Gran Barranca is a cliff and its sediments are poorly consolidated. Weathering by episodic rain and rapid drying by wind of the glass-rich pyroclastic sediments is hard on fossils. Piecing together a faunal sequence based on such fragmentary remains requires sustained collecting over many years. Unless detailed and accurate field notes and sections are kept up to date, and stratigraphic position marked by metal labels affixed by metal wire to steel rebar driven deep into the sediment by sledgehammer, memory is lost and posterity will struggle to build on past work. Notes and even sandwiches blow away in the gusty westerlies. The steep slopes are difficult to prospect, and only the young and intrepid are truly comfortable on Gran Barranca. The energy and fortitude required to stay on the outcrop is

considerable, ascent is strenuous and descent, facilitated by rodados, may be faster but more painful. Gran Barranca is long and high, and the Sarmiento Formation appears monotonous, or not, depending on sunlight. Color is a deceptive guide to the stratigraphy as volcanic glass is subject to subtle changes in tone, varying from pale gray, to greenish, to pinkish, to yellowish, and white with the angle of sunlight and cloud cover. Then too, Gran Barranca, despite its size, is mostly hidden from view. These practical difficulties appear evident throughout the history of exploration and work at Gran Barranca.

Gran Baranca's best-exposed, most accessible, and richest fossil levels are the *Colpodon* and *Notostylops* beds, the highest and lowest beds. The intermediate *Astraponotus* and *Pyrotherium* levels are not nearly as fossil-rich and occur on the steepest slopes. While the *Pyrotherium* beds at Gran Barranca are of limited areal extent, the *Astraponotus* beds and levels between the *Notostylops* and *Pyrotherium* beds present a different problem altogether. These exposures are among the most difficult of all to prospect, occur sporadically between the east and west end of the barranca, have variable lithology that is laterally discontinuous and difficult to characterize, and until recently were not known to be very fossiliferous. Even competent professional collectors have mistaken them for younger and older levels. For these reasons, the history of the discovery of the *Astraponotus* beds at Gran Barranca is most difficult to reconstruct (see Bond and Deschamps this book).

Historical overview

The Ameghinos

In his first contribution to our understanding of the fossil mammals from the "*Pyrotherium*" beds (currently Deseadan), Ameghino (1895) describes numerous new genera and species from the nominal type Deseadan at La Flecha. Judging from the systematic paleontology in that publication, there is no evidence that the faunal sequence at Gran Barranca had been discovered prior to 1895. Soon afterwards, Ameghino (1897a) published a review of his views about the "*Pyrotherium*" beds, and likewise provides no hint of the discovery of either older fossil levels or morphologically more primitive taxa.

Carlos Ameghino is thought to have discovered the first fossil mammals from the barranca south of Lake Colhue-Huapi, during his expedition of 1895–96 (Simpson 1967a), but the supporting evidence for this assertion is problematical. The fossils described by Florentino Ameghino in 1897, and thought by Simpson to have been collected at Gran Barranca, include the type material of 11 taxa typical of the Casamayoran, including *Notostylops murinus*. In the geological appendix to his publication, Florentino states that the formation including *Notopithecus, Pyrotherium,*

Archaeohyrax, and *Didolodus* comprises several levels whose differences have yet to be resolved, alluding to a suspicion that more than one level may have been represented in the collection he described.

Reconstructing the geographic provenance of the fossils discovered by Carlos Ameghino and described by Florentino in 1897 is difficult and establishing stratigraphic provenience is only somewhat easier. More importantly, reconstructing the travel itinerary of Carlos during his many expeditions along the coast of the Gulf of San Jorge and into the interior of Patagonia is especially difficult.

On 6 October 1897, Ameghino (1897a) described fossils collected by Carlos between October 1893 and August 1896, including taxa that in retrospect are of obviously more primitive and of greater antiquity than taxa in the "*Pyrotherium*" fauna. Among other things he proposes the new family "Notopithecidae," including the new genus and species, *Notopithecus adapinus*. While this publication includes numerous descriptions of taxa of Casamayoran age, Ameghino makes no distinction between them as a group from the other more numerous and more derived taxa that are also described. This suggests that his analysis of this fauna had not yet benefitted from the clarity of observed direct stratigraphic superposition.

In that same year, Ameghino (1897b) outlined the broad succession of mammal evolution in Argentina, illustrated with many of the fossils described in 1897a, but the geologic succession does not extend down to resolve levels below the "*Pyrotherium*" beds. Thus, again, the stratigraphic succession had not yet been resolved, although fossils from what we now know are older levels were becoming known to Florentino at this time.

Establishing a sequence between Notostylops *and* Pyrotherium

The first published notice of the existence of a sequence among pre-Santacrucian faunas in Patagonia was by Florentino Ameghino in 1899 (Ameghino 1899). The *Pyrotherium* fauna was well known at the time, based on collections made by Carlos at La Flecha in 1893–94 and at Cabeza Blanca in 1894–95 (Ameghino 1895). Where before he had lumped several groups of fossils of different age, Ameghino (1899) now distinguished the *Pyrotherium* (Deseadan) from the older (Casamayoran) *Notostylops* faunas by taxonomic composition, relative abundances, and some evolutionary differences between primitive and more advanced representatives of Isotemnidae, Notohippidae, and Pyrotheria. Among the contrasts, there are two notable examples of evolutionary change: (1) three derived groups in the *Pyrotherium* fauna (Toxodontia, Leontiniidae, Homalodotheriidae) were claimed to have their ancestry among more primitive isotemnids, and (2) the evolutionary trend to higher tooth crowns (hypsodonty) was first noted among Notohippidae.

Although for the first time the faunas are distinguished by their relative age, there is nothing in this revelation that indicates unequivocally that Carlos had observed the stratigraphic succession at Gran Barranca, as there is more than one place in Patagonia where this particular two-fauna sequence can be observed in superposition or near superposition with a conspicuous discontinuity between them (e.g. Valle Hermoso and the immediate vicinity of Cabeza Blanca).

Sometime between October 1899 (Ameghino 1899) and July 1901 (Ameghino 1901) Ameghino resolved the full sequence of middle Cenozoic faunas in Patagonia to a level of resolution that is essentially modern. (The resolved succession was published in the *Boletín de la Academia Nacional de Ciencias en Córdoba* in 1899, but the date of publication is debated; Simpson [1967b, p. 251] gives it as July 1901.) The faunal sequence included five of the six faunal levels we now recognize at Gran Barranca. Especially important is the intercalation of the beds with *Astraponotus* (Mustersan) between the *Pyrotherium* and *Notostylops* beds and the distinction between the beds with *Astrapothericulus* (early Santacrucian or "Pinturan") and those with *Colpodon* (Colhuehuapian).

The question remains whether this sequence could have been established in any way other than by collecting in direct stratigraphic superposition at Gran Barranca. Could it have been constructed, for example, by inference from evolutionary morphology observed in taxa from assemblages collected at different localities, each with a single fossil-bearing level or at most two superposed horizons? Specifically, with the distinction between the *Notostylops* and *Pyrotherium* faunas resolved (Ameghino 1899), how was the intermediate position of the *Astraponotus* beds (Mustersan) established? And if it required direct superposition, was this sequence established at Gran Barranca?

Carlos' first collection from the Mustersan at Gran Barranca

Bond and Deschamps (this book) review in length the history of the recognition of the Mustersan, and show how the question can only be answered indirectly. Carlos may have collected in the Mustersan at Gran Barranca sometime before 1901 when Ameghino (1901) first mentioned the existence of (1) fossil-bearing beds above the *Notostylops*, (2) the Astraponotéen, and (3) a level immediately above the Astraponotéen but below the *Pyrotherium* beds in 1901. Also, specimen labels and preservation indicate that Carlos Ameghino may well have discovered the locality we call GBV-4 "La Cancha" (the Tinguirirican at Gran Barranca) sometime just prior to the description of its fossils in 1901 (Reguero *et al.* 2003). This sort of evidence bearing on the intermediate levels provides some constraints on the time and activity of Carlos Ameghino at Gran Barranca.

*Carlos' first collection from the Colhuehuapian
at Gran Barranca*

According to Ameghino (1902), Carlos discovered the
"fauna mammalogique des couches à *Colpodon*" in 1898,
during his tenth paleontological expedition between
October 1898 and June 1899. Was this discovery made at
Gran Barranca? In correspondence dated 15 February
1899, Carlos wrote that his 1898–99 expedition was
limited to the vicinity of the Gulf of San Jorge and it is
not clear that his explorations extended as far north as
Gaiman/Trelew or as far west as Gran Barranca, the
nearest localities with "*Colpodon*" faunas. Prior to the
discovery of Gran Barranca by Andrés Tournouër (see
below), Florentino (in a letter dated 18 April 1899) men-
tions that Carlos had discovered very similar fossils "en el
interior del Deseado," but not at Gran Barranca. It wasn't
until six months later when Carlos wrote Florentino the
important letter of 9 October 1899 describing the sequence
of faunas he observed at Gran Barranca, that we know he
had collected the "Colhueapense" fauna at Gran Barranca.
These were fossils corresponding perfectly with those collected by Tournouër. This evidence suggests that Carlos
did not collect in the Colhuehuapian until after Tournouër,
and provides yet another constraint on the time of his
activity at Gran Barranca.

The influence of French paleontology

Albert Gaudry and Henri Gervais

Florentino Ameghino lived in Paris between 1878 and 1881
when he was 24–27 years old, and worked and studied at the
Laboratoire de Paléontologie in Paris with Henri Gervais.
The influence of French paleontology on Ameghino was
significant and sustained over many years. Gaudry's *Les
Enchainements du monde animal dans les temps géologi-
ques: mammifères tertiaires*, the first of a comprehensive
three-volume work, was published in the first year of Ameghino's residence in Paris. At least five subject areas of
Gaudry's inquiry seem to have been particularly influential
on Ameghino's intellectual development and professional
activities: (1) ungulate tooth morphology and homology, (2)
phylogenetic trees, (3) fossil succession, (4) the age of fossil
deposits, and (5) human evolution.

Did the influence and inspiration of Gaudry and Gervais
on Ameghino extend to include field methods for recording
fossil provenance and stratigraphy? Podgorny (2005) claims
that Ameghino "also trained in the observation of fossils,
prehistoric remains, and geological strata." Stratigraphic
profiles had been part of scientific publication in paleontology since the time of Brogniart and Cuvier ("Coupe
général et idéale") as early as 1808 and 1811 (Rudwick
2005) and these published stratigraphic sections are very
much in the style of those made by Ameghino (1906).

Andrés Tournouër

Andrés Tournouër was engaged by Gaudry to collect fossils
from older strata in Argentina, inspired by the scientific
discoveries of Ameghino and the description of new fossil
mammals from Patagonia. Simpson (1984) claimed that
Tournouër made five collecting trips to Patagonia;
Podgorny (2005) claims six, including the last in 1904.
Whatever the total count, two of these are important to the
history of Gran Barranca: (1) Tournouër's first expedition
from November 1898 to April 1899 which overlapped
chronologically with Carlos Ameghino's tenth expedition
between October 1898 and June 1899, and (2) a trip in
February 1903 when Tournouër met the Ameghino brothers
at Cabo Blanco and explored with them along the South
Atlantic coast to the vicinity of Punta Casamayor.

The discoverer of Gran Barranca: Tournouër
or Carlos Ameghino?

On 15 February 1899 Carlos Ameghino wrote a letter to
Florentino from Bahía Camarones describing some of the
results from his previous three-month exploration around
the Gulf of San Jorge. These explorations were of mediocre
yield from the standpoint of paleontology but from the
standpoint of geology were important because they revealed
that heretofore he had confused two distinct faunas as the
Pyrotherium fauna.

Two months later, on 18 April 1899, Florentino wrote
to Carlos from La Plata to the effect that Tournouër had
recently returned to Buenos Aires from an exploration to the
interior of Chubut bringing a collection of fossils that
while not very large, was interesting. Florentino conveyed
that Tournouër mentioned having encountered Carlos at one
point during that expedition, and that the fossils in question
were collected at an outcrop about one hour's horseback
ride from where Carlos and Tournouër had met. Tournouër
mistakenly thought he discovered the *Pyrotherium* fauna
south of Lake Musters, but to Florentino this collection
represented a younger fauna. Among the remains, material
of *Astrapotherium* were abundant, along with a small astrapothere similar to one Carlos had discovered previously in
the interior of Deseado, together with another new astrapothere of gigantic proportions. In addition to these, there
was a notohippid with a diastema between the incisors and
cheek teeth. In these characteristics, this collection appeared to represent a level older than the Santacrucian. This
fauna would eventually become known as the Colhuehuapian
and Tournouër's 1899 collection was the first made from
the type Colhuehuapian at Gran Barranca.

Sometime thereafter, between 18 April and 9 October
1899 Carlos returned to prospect at this locality. After
returning from this trip, Carlos wrote Florentino from Santa
Cruz in which he describes Gran Barranca. In his words
(translated from the original Spanish):

The disentanglement of the *Pyrotherium* fauna, of which I wrote in my letter from Camarones, has been confirmed ... [by] having later discovered an exposure near Colhue-Huapi where the beds are concordant stratigraphically but not paleontologically, as the beds between are nearly devoid of fossils, and between one fauna and another there is a profound difference, always recognizable at first glance. This cliff is the same one where Tournouër collected the fossils he took and that I effectively indicated to him in order to get him off my back, as he asked me to show him a place where he could collect some fossils, as in all his wanderings he had not found anything; but I never imagined that after me he would find what you say he collected there.

Carlos continues his description of Gran Barranca:

The cliff at this place comprises three horizons that correspond to three different faunas: 1) the *Notostylops* fauna, 2) the *Pyrotherum* fauna, very poor and scarce, but typical, as there occurs the pyrothere and astrapotheres with five lower molars, and 3) a fauna that could well be called Colhueapense, and that corresponds perfectly to the Patagonian, as you already discerned from the few fossils that Tournouër took.

By this letter, Carlos admits having returned to the locality Tournouër collected and whose location Florentino had transmitted to him by letter in April, and to have discovered there a sequence of three faunas in stratigraphic superposition, conformable one on the other. In the third of these faunas, which he called "Colhueapense," Carlos claimed to have collected an assemblage that contained all the same taxa collected by Tournouër, and many new ones. Thus, by his own words, Carlos recognized that Tournouër was the first to collect fossil mammals there, and thus to have discovered Gran Barranca.

Of course, it is possible to read Carlos' letter of 9 October 1899 in a different way. First, one could interpret the statement "I never imagined that after me he would find what you say he collected there" to mean that Carlos was surprised Tournouër had managed to find anything at all at Gran Barranca after Carlos had already worked the exposure and high-graded all its fossils, except for the novelty of Tournouër's collection as expressed by Florentino in his letter. Also, the comparison Florentino makes between Tournouër's collection and material collected by Carlos from near the source of the Rio Deseado implies that the only assemblage of similar composition and aspect previously seen by Florentino was the material Carlos collected from the "interior of Deseado," and not at Gran Barranca.

From the description of Tournouër's collection given by Florentino in his letter of 18 April 1899, it appears that Tournouër only collected in the Colhuehuapian at Gran Barranca, and not in lower levels. After Tournouër's discovery of fossils at Gran Barranca, Carlos returned there and discovered a sequence of three faunas.

What supporting material evidence is there to the claim of Tournouër's discovery? The Tournouër Collection from Gran Barranca in Paris appears to be exclusively material belonging to the Colhuehuapian level(s) and was collected before the middle of April 1899. Three documents in the archives of the Laboratoire in Paris describe Tournouër's activities at Gran Barranca in 1899.

(1) The "Compte rendu succinct de voyage de Monsieur A. Tournouër en Patagonie" describes the itinerary of Tournouër's 1899 expedition to Patagonia. This brief narrative on five handwritten pages includes a description of the stratigraphy in the vicinity of Lake Colhue-Huapi. This document is undated and unsigned, but in the handwriting of Tournouër.

(2) A transmittal letter from Tournouër in Buenos Aires to Gaudry of four handwritten pages dated 19 April 1899 that accompanied the shipment of fossils collected at Gran Barranca and conveys what was learned from Florentino Ameghino as to their significance.

(3) An undated and unsigned sketch of a stratigraphic profile of Gran Barranca in the handwriting of Tournouër (Fig. 1.1). On this sketch, there is an annotation "niveau du Colihuapi" above a dashed line that indicates lake base level, analogous to the manner in which sea level is indicated on other published profiles by Tournouër (1903). The profile indicates three fossil levels, with an intervening sterile(?) section. At the top are "gris gris avec couches friables avec astrapotherides." This is how the Colhuehuapian at Gran Barranca was described and characterized by Tournouër in correspondence. Immediately below these stratigraphically are the "argiles rouge a *Pyrotherium*" or red clays with *Pyrotherium*. Below a sterile(?) zone is the "Guaranien

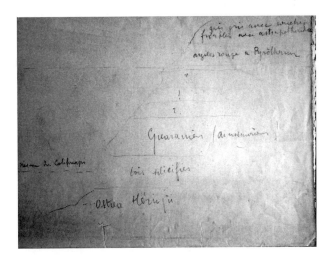

Fig. 1.1. Profile of Gran Barranca sketched by Andrés Tournouër (undated). (Courtesy of the Muséum National d'Histoire Naturelle, Paris.)

(Dinosaurien)" corresponding to Ameghino's *Notostylops* beds. The sketch is reminiscent in orientation and proportions to Ameghino's more detailed profile of 1906. Tournouër sketched the profile in 1899 according to his letter to Gaudry of 19 April 1899.

However, Simpson claims that Carlos Ameghino discovered Gran Barranca. The basis for Simpson's claim (1967a, 1984) is the evidence cited to show that Carlos collected the *Notostylops* fauna at Gran Barranca in 1895–96. Simpson (1967a) wrote that Carlos Ameghino "positively affirmed to me, and collection data and all other evidence are in accordance, that pre-Deseadan mammals were first found in 1895–1896 in this barranca." Furthermore, "[a]ll the pre-Deseadan forms described in 1897 were from there. Both the '*Notostylops* fauna' and the '*Astraponotus* fauna,' our Casamayoran and Mustersan faunas, were first found and (although later) recognized there, and a majority of all Ameghino specimens of both ages are from there … [i]t is the most imposing and important single known fossil mammal locality in South America, and one of the most important in the world. It must also be considered the greatest single discovery of Carlos Ameghino's extraordinary career" (Simpson 1967, p. 64).

The evidence Simpson (1967a) proffers to support his claim comprises the following: (1) all pre-Deseadan taxa described in Ameghino (1897a) were from Gran Barranca and labels on many Casamayoran and Mustersan specimens indicate "Colhue-Huapi" and a few indicate "Colhue-Huapi Sud," (2) that the mixed nature of the *Pyrotherium* fauna described by Ameghino (1895) and resolved in (Ameghino 1897a) was first revealed to Carlos Ameghino in the field by stratigraphic superposition at Gran Barranca, and (3) in interview Carlos claimed to have discovered the barranca in 1895–96.

The claim that Carlos collected the first fossil mammals from the *Notostylops* level at Gran Barranca in 1895–96 is the basis for the claim that Carlos was the first to discover Gran Barranca. But, there is nothing in Carlos' correspondence to Florentino at that time that describes Gran Barranca or the complete sequence there, nor anything that establishes that Carlos revealed the sequence there until October 1899, after Tournouër.

With respect to the 11 Casamayoran taxa described by Ameghino in 1897 and assumed by Simpson to have been collected at Gran Barranca, Simpson found only eight in 1930 (Cifelli 1985). The 93 Ameghino types labeled from "Colhue-Huapi" (Simpson 1967a) represent roughly 46 valid Casamayoran taxa (Cifelli 1985). Of these 46 taxa, Simpson found only 20 at Gran Barranca, despite having prospected and collected over the entire exposure for a month. With respect to labels on specimens in the Ameghino collection Simpson (1967a) claimed that 93 types are labeled from Colhue-Huapi, and that Carlos assured him that most types

not labeled were also from Gran Barranca. While this may be true, most types for valid Casamayoran taxa were described after 1901, after news about Tournouër's discovery was transmitted by letter from Florentino to Carlos, and after Carlos returned to collect there, and after Carlos described Gran Barranca in his letter.

Carlos' claim to have discovered Gran Barranca in 1895–96 was based on recollections made in 1931, 28 years after he was last in Patagonia in 1903.

Reconciliation about who discovered Gran Barranca may be further resolved by recalling that for Simpson, Gran Barranca included exposures and localities extending over 22 kilometers, from near Km 170 of the old Comodoro–Sarmiento railway, all the way to Profile 12 (today known as "Las Flores"). In addition, Simpson extended Gran Barranca to include the exposures at Km 145, the area of Valle Hermoso, and noted that "this is essentially a unit although the exposures are not completely continuous" (Simpson 1948, pp. 212 and 193). Using an expansive definition, it is easy to imagine how Simpson could claim that Carlos discovered Gran Barranca.

The Gran Barranca at the beginning of the twentieth century

In 1903, Tournouër published stratigraphic profiles made at points along the Gulf of San Jorge as far south as the mouth of the Rio Deseado. The five profiles include (1) Casamayor, (2) Punta Nava, (3) the "rive droite du Deseado," (4) the "rive gauche du Deseado," and (5) Florida Negra. Rather surprisingly, Tournouër never published a section for Gran Barranca, although he had made a sketch of it (Fig. 1.1).

The meeting of Tournouër with the Ameghinos in Patagonia in February 1903 is described in a long footnote (Ameghino 1906, p. 135) which provides a brief description of the activities of Florentino, Carlos, and Tournouër around the Gulf of San Jorge in February 1903 when they may have sketched some of the stratigraphic profiles of coastal exposures that were later published by Ameghino (1906, pp. 134–135).

Stylistic similarities between stratigraphic profiles of Tournouër (1903a) and those published by Ameghino (1906) are numerous and include (1) language, (2) depiction of the slope in profile, (3) telegraphic descriptions of lithostratigraphy, (4) special attention to fossil content, (5) strata numbered from base to top using arabic numbers with subordinate lower-case letters, (6) an indication of base level (either sea level or the level of a nearby body of water), and (7) texture infilling to designate lithology. The similarities of convention are so numerous that it seems plausible that the Ameghinos and Tournouër worked on them together.

Tournouër prepared a sketch map showing the location of Gran Barranca and a definitive map of all the localities he visited in Patagonia prior to mid-1903 (before he collected at Cerro Negro). He also prepared a catalog or list of the fossils that served to record the basic succession of three pre-Santacrucian faunas, Casamayoran, Deseadan, and Colhuehuapian. These last items must have been prepared sometime in mid-1903, after almost all Tournouër's field-work was completed, after the fact of superposition at Gran Barranca had been revealed, and after Tournouër and the Ameghino brothers met together on the coast of Patagonia where they confirmed the stratigraphic sequence.

With this basic information about the fossil succession and geologic age, Gaudry reconstructed the broad evolutionary patterns of the Patagonian middle Cenozoic fossil record. Gaudry and Tournouër began to publish the results of their research activities in 1902 and Gaudry began serious study of the South American fossils after he retired the chair of paleontology in 1903. Gaudry's contributions to the evolutionary history of the fossils Tournouër collected were considerable, and include recognition of the geographic isolation of Patagonia and strong homoplastic resemblances in the dentitions of southern ungulates compared with those of the northern continents (Gaudry 1904, 1908).

Working at the same time, Ameghino (1904) established a truly remarkable threefold parallelism among ontogenetic development, phylogenetic descent, and geological succession in the evolutionary history of the upper cheek teeth in the toxodonts, an extinct group of southern ungulates. This example of a threefold parallelism is the first detailed reconstruction of the evolutionary history of hypsodonty among South American Cenozoic mammals. It is difficult to imagine how this threefold parallelism could have been completed without a sequence of fossil taxa in stratigraphic superposition.

Elmer Riggs and the Field Museum
The First Captain Marshall Field Paleontological Expedition to Argentina and Bolivia between 1922 and 1925 yielded many fossil specimens from Gran Barranca. The fossils are important for their quality and completeness, crucial for the determination of more fragmentary material. Elmer S. Riggs and party worked at Gran Barranca between November and December of 1923, with a field crew that included John Bernard Abbott, George F. Sternberg, Jose Strucco, and C. Howard Riggs (see Fig. 15.1 in Kay this book). In a letter to George Gaylord Simpson dated 21 May 1932, Elmer Riggs clarifies how this collection was made: "It may be explained that, in order to hold up the standard of collecting our collectors were given to understand that nothing less than a maxilla or a mandible with a series of teeth was to be regarded as a specimen."

Five sorts of original information about Field Museum of Natural History (FMNH) work at Gran Barranca are available in archives at the FMNH: (1) specimen ticket labels for material collected by E. S. and C. H. Riggs, (2) the field notebooks of G. F. Sternberg and J. B. Abbott, (3) separate

"Record of Collections" made by Sternberg and Abbott, (4) E. S. Riggs' "Private Journal," and (5) glass plate negatives. After their return to Chicago, maps of their travel itinerary, areas collected, and campsites, were drawn for exhibition based on a 1911 map by Alberto Lefrançois. The map indicates the "Horizons with *Colpodon, Pyrotherium, Astraponotus*, and *Notostylops*" as a single unit we recognize today as the Sarmiento Formation. Riggs and party collected from this unit at both Valle Hermoso and Gran Barranca. Later, a list of the fossils "from the Deseado Series (the *Notostylops* Beds and *Pyrotherium* Beds of Ameghino)" including the material from Gran Barranca with genus-level determinations was prepared (with annotations by Bryan Patterson). In the Simpson Archives at the American Philosophical Society is an undated revised list of specimens from the *Notostylops* and *Astraponotus* beds at Gran Barranca collected by the FMNH and temporarily loaned to the American Museum of Natural History (AMNH). This list includes annotations by Patterson and the signatures of Abbott and Patterson.

Riggs (1928) mentioned having made stratigraphic sections during the work in South America, but none were ever published for Patagonian localities. A composite stratigraphic column of the west end of Gran Barranca in the FMNH archives shows profiles of seven numbered stratigraphic units of the Sarmiento Formation in superposition. This section also shows the approximate levels where fossils were collected. Along the right-hand margin are specimen field numbers of Sternberg and along the left-hand margin are field numbers and "names as determined in the museum." As in Ameghino (1906) both Lower and Upper *Notostylops* beds are distinguished, as are the *Astraponotus* beds, the *Colpodon* beds, and the Tehuelche gravels at the top. It is difficult from this composite section to establish more precise stratigraphic or geographic provenance for the fossils collected at Gran Barranca and there is no indication from either the collection or FMNH archives that levels between the Casamayoran and Deseadan were collected.

For example, Riggs and party took around 1200 photographs between 1922 and 1927, and there are 15 black-and-white contact prints from $5' \times 7'$ glass plate negatives of Gran Barranca in a photo album conserved in the FMNH archives. Handwritten captions below the photographs describe each scene. With benefit of hindsight and experience, some of the captions should be amended and a few errors corrected, for example. The "*Astraponotus* beds, Deseado Series" in 71–48921, the "Astrapothere measures" in 72–48923, and the "*Parastrapotherium* Zone" of 73–48929 all correspond with the lower levels of the Colhue-Huapi Member at Profile A. Thus, Riggs used three different terms to describe the beds we consider to be the lower levels of the Colhuehuapian. However, Riggs also used the term

"*Parastrapotherium* Zone" in 74–48931 for the steeply eroding beds of the Upper Puesto Almendra Member at Profile MMZ.

Photographs 73–48928 and 79–48935 are noteworthy as they capture strata of the Vera Member at Profile K. The handwritten caption beneath 79–48935 reads "Upper Measures, Parastrap. Zone" and the photograph is a panoramic view of Profile K looking up from below the level of Simpson's Y Tuff. Photo 73–48928 appears to have been taken from the level of GBV-4 "La Cancha" or just below it. Assuming that Riggs explored along the base of Gran Barranca by vehicle, he had to climb on foot to take this panorama "Overlooking Lake Colhue-Huapi." There are no fossil mammals in the FMNH collection that display the preservation unique to fossils from GBV-4 "La Cancha."

There are 102 specimens in the FMNH collection from Gran Barranca or from localities south of Lake Colhue-Huapi, extending from between Km 143, 145, 163 to 170, and including Valle Hermoso and Gran Barranca. Only 16 are from intermediate levels between the Colhuehuapian and Casamayoran; all of these are said to have been collected in the Deseado Formation and at least half of these (eight, or as many as ten) are material assigned to *Parastrapotherium*.

In correspondence from Simpson to Riggs of 18 April 1932 Simpson admits "I had some tendency to confuse *Astraponotus* and *Pyrotherium* beds in the field at several localities and am still not sure as to which horizon is represented by one or two lots of specimens. In one case, for instance, we have only *Asmodeus* or *Proasmodeus* and these are extremely difficult to separate, although I suppose it can be done when I get down to it." Nevertheless, in another letter written the very next day, Simpson corrects the position of the *Astraponotus* beds on photographs of Gran Barranca that Riggs had labeled (see below).

Egidio Feruglio

The geologist Egidio Feruglio had a little more luck with the *Astraponotus* beds at Gran Barranca than Riggs, and made a small but noteworthy collection of fossils at one locality at Gran Barranca in 1927. A list of the fossils collected by Feruglio at his Locality 31 and now at the Geological and Palaeontological Museum of Padua University in Italy can be found in the Simpson Papers at the American Philosophical Society, along with an original drawing of a stratigraphic section at Gran Barranca stylistically similar to sections Feruglio published in *The Geology of Patagonia* (Feruglio 1949). On the profile where the fossils were found, Feruglio depicts remarkably little stratigraphic detail in the Sarmiento Formation. The lack of stratification at this profile is reminiscent of Profile K, and in particular, the stratigraphic position of the stratum labeled "banco fosilifero f with fósiles No. 31" on the slope

leading to the top of the barranca closely resembles that of GBV-4 "La Cancha." Among Feruglio's material from this locality, Simpson identified four taxa including *Pseudostylops subquadratus*, *Degonia* or *Eohyrax* sp. (frags), ?*Archaeohyrax* sp., and *Propyrotherium* ?*saxum*. According to Simpson, and appropriate to the state of knowledge at the time, these fossil mammals indicated a Mustersan age. Marcelo Reguero (in Reguero *et al.* 2003) was the first to establish that Feruglio had actually discovered the fauna from GBV-4 "La Cancha."

The Simpson Papers at the American Philosophical Society (APS) include three typewritten pages (along with four small pen-and-ink profiles) listing collecting localities (numbered from 1 to 18) in the form of an itinerary, giving directions for access to each. This document reads like an itinerary for a field expedition on which Feruglio made handwritten annotations, dated and signed 30 Noviembre 1930. Someone later, in pencil, made brief additional comments on this document, e.g., no vale, poco interés, no existe, Campamento no. 2, Campamento número 4, etc., suggesting this itinerary was followed (at least in part) and at least some of the sites were visited (see below). The geographic coincidence between this itinerary and the actual itinerary followed by the Scarritt Expedition must be more than simple coincidence.

In addition to this itinerary, another document of six handwritten pages reads "Colección de Mamíferos fósiles de Patagonia del Dr. Egidio Feruglio" and lists and describes the fossil yield from 53 localities. Each entry begins with a number (1, 2, 2bis, 3, 3bis, 3ter, 4, etc. ending with 53), followed by a name, then a description of the geographic situation and major stratigraphic features of each. Following the description of each locality is a date, presumed to be the date the locality was originally worked.

Feruglio took at least some of the fossils with him back to Bologna in Italy in 1932, and in 1934 shipped at least some of them to Simpson in New York, just before fleeing Italy to return to Argentina. The contribution of Feruglio to the success of the Scarritt Patagonia Expeditions was considerable, another example of how a European refugee from fascism provided disinterested support to a US paleontologist (see Madden *et al.* 1997).

In correspondence from Simpson to Feruglio on 3 January 1935, Simpson informs that "other specimens in your collection merit more detailed study: the maxilla of *Pseudostylops* from Locality 31 ... it would be useful to have more details about this locality." Somewhat later, on 12 March 1935, Simpson writes Feruglio to say "we also have some fossils from your Locality 31." To be sure, Simpson did not collect fossils from Feruglio's locality, but Coleman S. Williams collected some important material from near the top of Profile M at a level corresponding to Feruglio's Locality 31.

George Gaylord Simpson and the Scarritt Expeditions
The American Museum of Natural History's (AMNH) Scarritt
Expeditions of 1930–31 and 1932–33 set the standard for
all later fieldwork at Gran Barranca and scientific productivity, although 37 years elapsed between the first
field season at Gran Barranca and the eventual publication (Simpson 1967a) of the last part of his monographic
revision of the older faunas. Two significant parts of the
Scarritt Expedition work at Gran Barranca have never
been published, Patterson's incomplete revision of the
fossils from the Deseadan and Colhuehuapian levels,
and the stratigraphic context of levels above the Barrancan,
especially the Mustersan.

A lasting contribution of the Scarritt Expeditions is the
detailed records of their field activities. Simpson's (GGS)
system of note-taking during the 1930–31 expedition
involved four separate records, all accessible to researchers:
(1) a book of rough notes, the "Rough Book" (at the APS),
(2) a field book of annotated technical notes (at the AMNH),
(3) a photograph log (APS), and (4) a personal diary or
journal (APS). The field books (annotated technical notes)
in the archives of the AMNH had accession numbers for
each specimen added subsequent to the fieldwork. The
1930–31 journal was the basis for *Attending Marvels*
(Simpson 1934).

The "Patagonia, (Rough Book), 1930, G. G. Simpson"
contains rough sketches in pencil or ink with brief notes
labeling features of profiles, lists, sketches, and descriptions
of some photographs, a tally of fossil specimens to measure
progress with collecting effort, brief descriptions of strata,
measured thicknesses, mineral or rock samples, quarry
levels, and highlights of daily activities at Gran Barranca.
The Rough Book sketches include one of Coley's Quarry.
GGS also made sketches of photographs to help memory,
and the Photograph Log also includes a sketch of this same
Mustersan quarry. The notes made on 11 November 1930,
the morning he took photographs, includes a sketch of a
photograph of Coleman "Coley" S. Williams in the vicinity
of GBV-19 "La Cantera" (Fig. 1.2). Another note confirms
that the "Toba del Cocodrilo" is Simpson's Y Tuff.
A separate page of "Analysis" establishes an approximate
comparison of yield for Colhue-Huapi and Valle Hermoso,
made on the basis of field specimen numbers, and establishes the essential facts about the quality of the fossil
record at Gran Barranca. From the *Notostylops* beds,
Simpson tallied 3 skulls, 25 upper and lower jaws, and 80
isolated or more fragmentary specimens, while in the *Astraponotus* beds only 1 skull, 7 jaws, and 35 other specimens,
mostly isolated teeth. Even more sparse were the *Pyrotherium* beds where only 3 jaws and 14 other specimens were
found. By contrast, the *Colpodon* beds yielded 9 skulls,
53 jaws, and 40 other specimens, by far the richest beds
Simpson encountered anywhere in Patagonia, "and rather

Fig. 1.2. Simpson Photo #20 is a detail of Profile A, a photograph
of the slopes around GBV-19 "La Cantera." Coleman "Coley" S.
Williams appears in the middle foreground, with his pick.
The photo is described on the back as "Middle part of Colhue-Huapi
barranca, 11/11/30, Near West End." (Courtesy of the American
Philosophical Society.)

spoiled for us the laborious collecting in the older levels"
(Simpson 19 April 1932 correspondence to E. S. Riggs).

Simpson's "Field books" for 1930–31 and 1934 contain
more fully elaborated themes first noted in the rough book,
with stratigraphic sections neatly drawn to scale, and a numbered entry for every specimen collected with reference to
stratigraphic provenance.[1] Simpson took many photographs
on the expedition but only two are serviceable images of
Gran Barranca.

In the 1930–31 Travel Journal, Simpson described their
activities at Gran Barranca and their collecting efforts in

[1] These exemplary works are now available to researchers through the
AMNH website (http://paleo.amnh.org/notebooks/).

intermediate levels. By 28 October 1930, Simpson and party had finished work in the Colhuehuapian. Simpson writes: "Tomorrow I'll turn the lad loose on the channel beds." On 29 October, he describes the discovery of the crocodilian skull in the "massive lower volcanic ash bed" [= Simpson's Y Tuff]. On 31 October, ever optimistic, Simpson and Justino Hernandez (JH) "took a paseo to examine the barranca farther east and see the full section from Pehuenche (upper dinosaur beds) on through to the *Colpodon*. Did not find any definite fossil layer below our lower tuff, but there must be one." On this same day, JH collected in Bed X at Profile G and on 3 November, GGS and JH ascended Profile G near Cañadón Mazzoni and collected at three levels above Y.

On 4 November, while Coley worked "on a heavy gravel at the presumable base of the *Astraponotus* beds, containing very numerous isolated teeth and a few jaws, all very fragile ... I spent the morning reconoitering in the upper beds, above this gravel and below the rich *Colpodon* level, but found very little except scrap of the great *Parastrapotherium* ... who occupies these beds almost exclusively." Thus, that day, and only that day, Simpson and party prospected the stratigraphic interval that includes GBV-19 "La Cantera," without finding it.

On 5 November they went down to the east end of the barranca, about eight or nine miles away. This was this same day JH and GGS collected the lower levels of Profile K, but didn't ascend up to GBV-4 "La Cancha." Williams returned to his quarry in the *Astraponotus* beds the next day and later that same day, prospected up the guanaco path at Profile J, where he discovered GBV-3 "El Rosado" and the Deseadan and Colhuehuapian levels higher up that same path.

By 14 November 1930, Simpson and party had worked their way eastward to the vicinity of Profiles K and M, and in his journal GGS comments about the difficult nature of the work.

> Where we are now, the badlands seem particularly like a bad dream or a lunar landscape, something completely unearthly. The most fossiliferous zone of the Notostylopense continues hard to prospect. The only really practical scheme we have evolved is to have a troupe of mountain goats trained to wag their tails when they see a fossil – a man on a central hill with a spyglass then sends one of a corps of alpine climbers to the spot with a fish (or whatever delicacy they like, perhaps I'm thinking of seals) for the goat, and tools to extract the fossil. We'll put this in practice as soon as we have trained the goats, after we get them.

Simpson and party collected at GBV-60 "El Nuevo" on the same day Williams climbed up Profile M to collect the *Eomorphippus* material around the unconformity. Coley Williams collected these notohippids only three and a half years after Egidio Feruglio discovered the La Cancha fauna

at Profile K. Friday 21 November was the Scarritt Expedition's last day of work at Gran Barranca.

The extent of Scarritt Expedition exploration of intermediate levels at Gran Barranca can also be reconstructed from handwritten notes entitled "Faunal Lists and Assemblages, Early Tertiary of Patagonia – Scarritt Expedition" (not in the handwriting style of Simpson). The document describes the organization of notes on the biostratigraphy of the collection. The procedure employed was as follows: (1) a folder (now removed except for the label) has been assigned to each locality indicated on the base map (Location #1 = Gran Barranca, according to the numbering used on the map of the Scarritt Expedition among Simpson's reprint collection at the Florida Museum of Natural History in Gainesville), (2) for each profile in a given locality a rough working section is given on a separate page, onto which lithology, thicknesses (not to scale) and field numbers of specimens are listed, and (3) for each fossiliferous horizon in each profile and working section, a separate list of specimens, with their catalog numbers (only SOME of these are identified) and collection notes is given. There is a small note "MTB 2/20/46" at the end of this paragraph and the date may explain why some of the specimens appear to be incorrectly identified. Simpson's work on the biostratigraphy at Gran Barranca was never completed (there are still a number of specimens that remain unidentified), was suspended between 1948 and 1967, and eventually rendered impossible by Simpson's departure from the AMNH.

After returning to New York in 1931, Simpson immediately began work on the new collection, and the 1930s were among the most active decades in the history of Patagonian paleontology. Through mutual consent, Simpson worked on the Mustersan, Casamayoran, and older levels, while Bryan Patterson in Chicago worked on material from the Deseadan and Colhuehuapian levels. Patterson and Simpson shared a rich correspondence over many years. Between 1936 and 1937, much of this correspondence related to finding phylogenetic continuity or connections between the Casamayoran (and older) ungulates and those from Deseadan and younger levels, "the tremendous Casamayoran–Deseadan gap, only partly filled by the Mustersan." Their conversation about the evolutionary transformation between the Casamayoran and Deseadan picked up again in 1946, at which time they had identified two central problems – the relationships of Toxodontidae, Notohippidae, Leontiniidae, and Homalodotheriidae to earlier Isotemnidae between the Casamayoran and Deseadan, and of Hegetotheriidae and Mesotheriidae to other typotherians before the Deseadan (Fig. 1.3). To further this work, Patterson received a Guggenheim Fellowship in 1951 and worked on his revision of the taxonomy of the Ameghino collection at the Museo Argentino de Ciencias Naturales between 1952 and 1955.

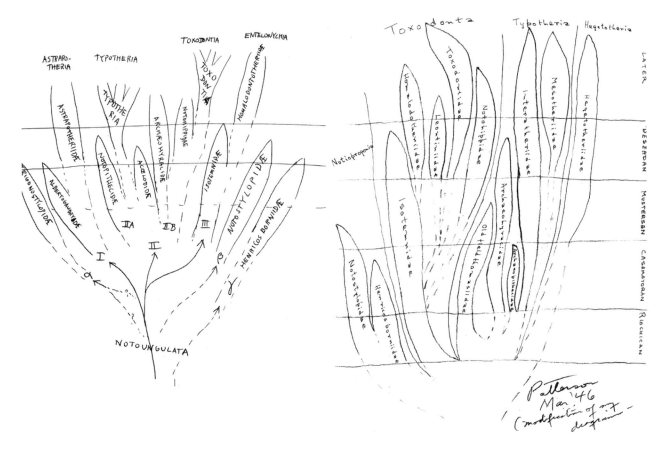

Fig. 1.3. Sketches by G. G. Simpson and B. Patterson representing their views about notoungulate evolution, about 1946. (From correspondence between G. G. Simpson and B. Patterson, Simpson Papers, American Philosophical Society, Philadelphia.)

Simpson never completed work on the Mustersan fossils from Gran Barranca nor the biostratigraphy of the collections made from levels we today consider transitional between Barrancan and Deseadan. Many fossils collected from Mustersan levels at Gran Barranca are not allocated to taxa nor mentioned in his monograph (Simpson 1948, 1967b) and some material was made available to Patterson for study in the mistaken belief it was from the Deseadan. After the Scarritt Expeditions, Simpson never returned to work at Gran Barranca.

Contributions in the second half of the twentieth century

Alfredo Castellanos

The 1930 revolution in Argentina left an imprint on the paleontology of Gran Barranca. Kraglievich, Rusconi, Parodi, and Castellanos were removed from the staff of the Museum of Buenos Aires and took positions in the provinces. Alfredo Castellanos came to reside at the Institute of Physiography of the Universidad del Litoral in Rosario. His work at Gran Barranca in January–February 1945 is remarkable for its documentation, the rock, mineral, and fossil specimens, and the photographs and stratigraphic profile of Cañadon Ameghino.

Accompanied by Federico Henning, Federico Madril, Ing. Lansky, and two soldiers from the 8th Regiment detached to take responsibility for the field vehicles and transport of the camp and fossils, Castellanos installed his first camp between the lake and the barranca from where fossils were collected from 23 January to 7 February. Camp was then moved to Puesto Almendra and then closer to Gran Barranca on 8 February. Castellanos and Henning arrived at Cañadon Ameghino (between Colhue-Huapi West and the West End Basalt) where between 8 and 10 February Castellanos elaborated a stratigraphic section.

In work undertaken over three days, Castellanos distinguished 54 beds or strata by lithology, and grouped these into five units corresponding to (from top to bottom): (1) Trelewense of Kraglievich or Colpodonense of Ameghino, (2) Deseadense of Mercerat or Pyrotheriense of Ameghino, (3) Mustersense of Kraglievich = Astraponotense of Ameghino, (4) Casamayorense of Mercerat = Notostylopense superior of Ameghino, and (5) Colhuehuapiense of Kraglievich = Notostylopense inferior of Ameghino.

The most remarkable discovery of this all too brief work at Gran Barranca, was of a thick lens of sediment of the "Astraponotense" extending from below the basalts to insert itself between the Astraponotense and Casamayorense superior, nearly (but not precisely) like the Vera Member. For reasons that are not clear, Castellanos never returned to Gran Barranca.

Rosendo Pascual et al.

The work of Rosendo Pascual, his colleagues, and students at the Museo and Universidad Nacional de La Plata marks the beginning of modern multidisciplinary research at Gran Barranca. No longer do we speak of the work of individuals. In addition to beginning the renovation and continued improvement in the quality of the fossil record from the Sarmiento Formation (and Gran Barranca) already in Argentine collections and accessible to a larger scientific community, Rosendo's infectious enthusiasm inspired numerous fruitful collaborations. Of these, two collaborations in the late 1970s are particularly important to the history of scientific exploration at Gran Barranca.

Luis Spalletti and Mario Mazzoni of the Centro de Investigaciones Geológicas (CIG) of the Universidad Nacional de La Plata (UNLP) studied and published on the stratigraphy, sedimentology, and genesis of Sarmiento Formation sediments at Gran Barranca (Spalletti and Mazzoni 1977, 1979; Mazzoni 1979, 1985; Franchi and Nullo 1986; Spalletti 1992). Their work continued important traditions in sedimentology (Frenguelli 1925, 1930; Teruggi 1955, 1957), and among other things, led to the nomination of Gran Barranca as the type exposure of the Sarmiento Formation. They brought innovations in petrographic and sedimentological analysis of Sarmiento Formation. That this detailed work has not been further refined nor used more fully in the context of micropaleontology is unfortunate because therein resides clues to provenance, sorting, weathering, erosion, and mixing of fossil-bearing sediments and their influence on microfossil content, mineral phase disparities in radioisotopic dating, and agency in the broad patterns of evolutionary change in vegetation and fauna.

The potassium–argon dating of basalts in stratigraphic relationship to fossil mammal levels was important in bringing the Patagonian record into the emerging global geochronology (Marshall *et al.* 1977). Dates from basalts at Gran Barranca (Marshall *et al.* 1977) originally served to calibrate the duration of the Deseadan–Colhuehuapian hiatus and over the next ten years anchor the age of the Deseadan. While the age of the base of the Deseadan South American Land Mammals Age (SALMA) changed with the interpretation of the stratigraphy at Pico Truncado (Marshall *et al.* 1977; Marshall 1985; Flynn and Swisher 1995), and the age of the top changed with dates from the basalts at Scarritt Pocket (Marshall *et al.* 1986), the radiometric age of the basalts at Gran Barranca has stayed just about the same (Marshall *et al.* 1977, 1986).

Of the faunal sequence revealed by the Sarmiento Formation at Gran Barranca, the essential or core problem remains the faunal evolution between the Barrancan (subage of the Casamayoran) and Deseadan. This is the essence of observed change within and between higher-order faunal units or cycles (Patterson and Pascual 1968; Pascual 1970; Pascual and Odreman-Rivas 1971, 1973, Pascual *et al.* 1985, 1996; Pascual and Ortiz-Jaureguizar 1990, 2007). Changing views about the temporal span of the Deseadan and the duration of the temporal interval between it and the Barrancan, and the imprecision of the temporal boundaries of faunal units interposed between, are reflected in the expansion and contraction of the scope of issues related to this change and discussed in broad faunal syntheses. As the base of the Deseadan SALMA migrated upward in time from the Eocene–Oligocene Transition (Marshall *et al.* 1977) to sometime within the Oligocene (Flynn and Swisher 1995), explanations about the significance of its faunal content expanded to encompass global (Drake Passage, Antarctic ice) and local (diastrophism, grasslands, volcanism) agency. Likewise, the migration of the Barrancan subage from early to late Eocene (Kay *et al.* 1999) brought its faunas into even closer temporal proximity with the Eocene–Oligocene Transition and compressed the temporal duration over which pronounced faunal change must be related to global and regional climate events.

The fossil record from the Sarmiento Formation bears importantly on our understanding of faunal and environmental change during the Prepatagonian faunistic cycle, at the time of transition between the Prepatagonian and Patagonian faunistic cycles (Pascual *et al.* 1996), between the Paleo-Cenozoic and Neo-Cenozoic megacycles (Pascual and Ortiz-Jaureguizar 1990), and within the South American episode (Pascual and Ortiz-Jaureguizar 2007). As this discussion has matured, the basic empirical evidence has remained largely unchanged. This empirical record was first developed around the turn of the century with the discovery of the sequence at Gran Barranca and the description of the faunal content of the Sarmiento Formation (Ameghino 1904, 1906; Gaudry 1904, 1908).

After the interaction among diastrophism, environmental change, and precocious hypsodonty was first enunciated (Patterson and Pascual 1966; Pascual and Odreman-Rivas 1971, 1973), one major new contribution from stratigraphy and sedimentology in the late 1970s (Spalletti and Mazzoni 1977, 1979) opened the door to the possible role of volcanism and/or phytoliths.

During the years between 1934 and 1993, no sustained collecting effort was made at Gran Barranca and there is very little in the published work about the Patagonian fossil mammal record subsequent to the contributions of Spalletti and Mazzoni (1977, 1979) that reflects an indelible and direct influence from studies at or about Gran Barranca. Only after the age of the Barrancan was revised in 1999 (Kay *et al.* 1999) did the daunting task of fieldwork at Gran Barranca, the painstaking and devoted study of new collections of mostly frustratingly fragmentary remains, and the patient and sustained multidisciplinary and international collaboration required to attempt a coherent integration, successfully confront the vagaries of funding and the tragedy of loss, facts of history that conspire against curiosity and the imperatives of modern science.

ACKNOWLEDGEMENTS
During several brief visits to the Laboratoire de Paléontologie of the Muséum National d'Histoire Naturelle in Paris, numerous courtesies and special assistance were provided by Dr. Christian de Muizon and Mlle Claire Sagne. Archival research at the Muséum National benefitted from discussions with Professor Daniel Goujet, Professor Pascal Tassy, and Dr. Marc Godinot.

At the Field Museum of Natural History in Chicago, we wish to thank then curator John J. Flynn for the courtesy of access to the collection and especially wish to acknowledge the conscientious work of William F. Simpson, Collections Manager. In addition, we benefitted from the assistance of Monica Mikulski, Mary R. Carman, and Nina M. Cummings, photo archivist.

At the American Museum of Natural History in New York, we wish to acknowledge our debt to Barbara Mathé, film libarian, the generous spirit of the late Malcolm C. McKenna and the hospitality of Susan Bell. The Osborn Library remains a priceless resource for the historian of paleontology.

A generous Franklin Research Grant from the American Philosophical Society enabled consultation of the George Gaylord Simpson Papers. We wish to thank Valerie-Anne Lutz, Joseph-James Ahern, Eleanor Roach, Ludo Rocher, Mary McDonald, Charles Greifenstein, and Richard Dunn for their many courtesies.

We also thank Dr. Deborah Pacini Hernández of the Center for Latin American Studies of the University of Florida for a Library Travel Grant to enable a visit to the Florida Museum of Natural History in Gainesville where generous access to Simpson's reprint library was facilitated by Dr. Bruce MacFadden and Dr. David Webb.

Alejandro Kramarz at the Museo Argentino de Ciencias Naturales "Bernardino Rivadavia" was very helpful in making available Bryan Patterson's manuscript revisions of the Deseadan and Colhuehuapian material in the Ameghino collection, and otherwise encouraged our inquiry through courteous collegiality.

Many people at the Museo de La Plata gave generously of their time to help us with our work. In particular we acknowledge Rosendo Pascual, Gustavo Scillato-Yané, Eduardo P. Tonni, Mariano Bond, Alfredo Carlini, and Maria Guiomar Vucetich. Guiomar Vucetich, Alfredo Carlini, and Richard Kay read various earlier versions of these notes. Any errors are our own.

REFERENCES

Ameghino, F. 1895. Sur les oiseaux fossiles de Patagonie et la faune mammalogique des couches à *Pyrotherium*. II. Première contribution à la connaissance de la faune mammalogique des couches à *Pyrotherium*. *Boletín del Instituto Geográfico Argentino*, **15**, 603–660. Buenos Aires. [Publication date sometimes given as 1894.]

Ameghino, F. 1897a. Mammifères crétacés de l'Argentine. Deuxième contribution à la connaisasance de la faune mammalogique des couches à *Pyrotherium*. *Boletín del Instituto Geográfico Argentino*, **18**(4–9), 406–521. Buenos Aires.

Ameghino, F. 1897b. *La Argentina a través de las últimas épocas geológicas*. Disertación pronunciada en el acto de la inauguración de la Universidad de La Plata (18 de Abril de 1897). Buenos Aires: Pablo E. Coni e Hijos.

Ameghino, F. 1899. *Sinopsis geológico-paleontológica*, Suplemento (adiciones y correcciones). La Plata.

Ameghino, F. 1901. Notices préliminaires sur des ongulés nouveaux des terrains crétacés de Patagonie. *Boletin de la Academia Nacional de Ciencias en Córdoba*, **16**, 349–426. [Volume dated 1899, but Ameghino refers to it as 1901; "from internal evidence it cannot have been issued before 1900 at the earliest" – Simpson, 1948.] Buenos Aires: Pablo E. Coni e Hijos. [Original dated 1899, but see Simpson 1967b, p. 251.]

Ameghino, F. 1902. Première contribution à la connaissance de la faune mammalogique des couches à Colpodon. *Boletin de la Academia Nacional de Ciencias en Córdoba*, **17**, 71–138.

Ameghino, F. 1904. Recherches de morphologie phylogénétique sur les molaires supérieures des ongulés. *Anales del Museo Nacional de Buenos Aires*, 3(3), 1–541.

Ameghino, F. 1906. Les Formations sédimentaires du Crétacé supérieur en du Tertiare de Patagonie avec un parallèle entre leurs faunes mammalogiques et celles de l'ancien continent. *Anales del Museo Nacional de Buenos Aires*, **8**(3), 1–568.

Cifelli, R. L. 1985. Biostratigraphy of the Casamayoran, Early Eocene, of Patagonia. *American Museum of Natural History Novitates*, **2820**, 1–26.

Feruglio, E. 1949. *Descripción Geológica de la Patagonia*, vol. 2. Buenos Aires: Dirección General de Yacimientos Petrolíferos Fiscales.

Flynn, J. J. and C. C. Swisher III 1995. Cenozoic South American Land Mammal Ages: correlation to global geochronologies. In Berggren, W. A., Kent, D. V., Aubry, M.-P., and Hardenbol, J. (eds.), *Geochronology, Time Scales, and Global Stratigraphic Correlation*, Special Publication no. 54. Tulsa, OK: Society for Sedimentary Geology, pp. 317–333.

Franchi, M. and F. Nullo. 1986. Comentario: Las tobas de Sarmiento en el Macizo de Somoncura. *Revista de la Asociación Geológica Argentina*, **41**(1–2), 219–222.

Frenguelli, J. 1925. Loess y limos pampeanos. *Universidad Nacional de La Plata, Facultad de Ciencias Naturales y Museo de La Plata, Serie Técnica y Didáctica*, **7**, 5–88. [Reprinted in 1955.]

Frenguelli, J. 1930. Partículas de sílice organizada en el loess y en los limos pampeanos: células silificadas de gramineas. *Anales de la Sociedad Científica de Santa Fe*, **2**, 65–109.

Gaudry, A. 1904. Fossiles de Patagonie: Dentition de quelques mammifères. *Mémoires de la Société Géologique de France, Paleontologie*, **12**(I), 1–26.

Gaudry, A. 1908. Fossiles de Patagonie: De l'économie dans la nature. *Annales de Paléontologie* (Paris), **3**, 1–7.

Kay, R. F., R. M. Madden, M. G. Vucetich, A. A. Carlini, M. M. Mazzoni, G. H. Ré, M. Heizler, and H. Sandeman 1999. Revised age of the Casamayoran South American Land Mammals 'Age': climatic and biotic implications. *Proceedings of the National Academy of Sciences USA*, **96**, 13 235–13 240.

Madden, R. H., D. E. Savage, and R. W. Fields 1997. A history of vertebrate paleontology in the Magdalena Valley. In Kay, R. F., Madden, R. H., Cifelli, R. L., and Flynn, J. J. (eds.), *Vertebrate Paleontology in the Neotropics: The Miocene Fauna of La Venta, Colombia*. Washington, DC: Smithsonian Institution Press, pp. 3–11.

Marshall, L. G. 1985. Geochronology and land-mammal biochronology of the transamerican faunal interchange. In Stehli, F. G. and Webb, S. D. (eds.), *The Great American Biotic Interchange*. New York: Plenum Press, pp. 49–85.

Marshall, L. G., R. Pascual, G. H. Curtis, and R. E. Drake 1977. South American geochronology: radiometric time scale for middle to late Tertiary mammal-bearing horizons in Patagonia. *Science*, **195**, 1325–1328.

Marshall, L. G., R. L. Cifelli, R. E. Drake, and G. H. Curtis 1986. Vertebrate paleontology, geology, and geochronology of the Tapera de López and Scarritt Pocket, Chubut Province, Argentina. *Journal of Paleontology*, **60**(4), 920–951.

Mazzoni, M. M. 1979. Contribución al conocimiento petrográfico de la Formación Sarmiento, barranca sur del Lago Colhue-Huapi, Provincia de Chubut. *Revista de la Asociación Argentina de Minerología, Petrología y Sedimentología*, **10**, 33–54.

Mazzoni, M. M. 1985. La Formación Sarmiento y el volcanismo paleógeno. *Revista de la Asociación Geológica Argentina*, **40**(1–2), 60–68.

Pascual, R. 1970. Evolución de comunidades, cambios faunísticos e integraciones biocenóticas de los vertebrados Cenozoicos de Argentina. *Actas IV Congreso Latinoamericano de Zoología, Caracas*, **2**, 991–1088.

Pascual, R. and O. E. Odreman Rivas 1971. Evolución de las comunidades de los vertebrados del Terciario argentino: los aspectos paleozoogeográficos relacionados. *Ameghiniana*, **8**(3–4), 372–412.

Pascual, R. and O. E. Odreman Rivas 1973. Las unidades estratigráficas del Terciario portadores de mamíferos:

su distribución y sus relaciones con los acontecimientos diastróficos. *Actas V Congreso Geológico Argentino*, **3**, 293–338.

Pascual, R. and E. Ortiz-Jaureguizar 1990. Evolving climates and mammal faunas in Cenozoic South America. *Journal of Human Evolution*, **19**, 23–60.

Pascual, R. and E. Ortiz-Jaureguizar 2007. The Gondwanan and South American episodes: two major and unrelated moments in the history of the South American mammals. *Journal of Mammalian Evolution*, **14**, 75–137.

Pascual, R., M. G. Vucetich, G. J. Scillato-Yané, and M. Bond 1985. Main pathways of mammalian diversification in South America. In Stehli, F. G. and Webb S. D. (eds.), *The Great American Biotic Interchange*. New York: Plenum Press, pp. 219–247.

Pascual, R., E. Ortiz-Jaureguizar, and J. L. Prado 1996. Land mammals: paradigm for Cenozoic South American geobiotic evolution. *Münchner Geowissenschaftliche Abhandlungen A*, **30**, 265–319.

Patterson, B. and R. Pascual 1968. The fossil mammal fauna of South America. *Quarterly Review of Biology*, **43**, 409–451.

Podgorny, I. 2005. Bones and devices in the constitution of paleontology in Argentina. *Science In Context*, **18**(2), 249–283.

Reguero, M., D. A. Croft, J. J. Flynn, and A. R. Wyss 2003. Small archaeohyracids (Typotheria, Notoungulata) from Chubut Province, Argentina, and Central Chile; implications for trans-Andean temporal correlation. *Fieldiana, Geology*, n.s., **48**, 1–17.

Riggs, E. S. 1928. Work accomplished by the Field Museum Paleontological Expeditions to South America. *Science*, **67**(1745), 585–587.

Rudwick, M. J. S. 2005. *Bursting the Limits of Time: The Reconstruction of Geohistory in the Age of Revolution*. Chicago, IL: University of Chicago Press.

Simpson, G. G. 1934. *Attending Marvels: A Patagonian Journey*. New York: Macmillan.

Simpson, G. G. 1948. The beginning of the age of mammals in South America. I. *Bulletin of the American Museum of Natural History*, **91**, 1–232.

Simpson, G. G. 1964. Los mamíferos casamayorenses de la Colección Tournouër. *Revista del Museo Argentino de Ciencias Naturales, Paleontología*, **1**, 1–21.

Simpson, G. G. 1967a. The Ameghinos' localities for early Cenozoic mammals in Patagonia. *Bulletin of the Museum of Comparative Zoology*, **136**(4), 63–76.

Simpson, G. G. 1967b. The beginning of the age of mammals in South America. II. *Bulletin of the American Museum of Natural History*, **137**, 1–260.

Simpson, G. G. 1984. *Discoverers of the Lost World*. New Haven, CT: Yale University Press.

Spalletti, L. A. 1992. El loess y el problema de la identificación de las loessitas. *Revista del Museo de La Plata*, n.s., **11** (Geologia, 102), 45–53.

Spalletti, L. A. and M. M. Mazzoni 1977. Sedimentología del Grupo Sarmiento en un perfil ubicado al sudeste del Lago

Colhue-Huapi, Provincia de Chubut. *Revista del Museo de La Plata, Obra del Centenario, Geologia*, **4**, 261–283.

Spalletti, L. A. and M. M. Mazzoni 1979. Estratigrafía de la Formación Sarmiento en la Barranca Sur del Lago Colhue-Huapi, Provincia del Chubut. *Revista de la Asociación Geológica Argentina*, **34**(4), 271–281.

Teruggi, M. E. 1955. Algunas observaciones microscopias sobre vidrio volcánico y ópalo organógeno en sedimentos pampianos. *Notas del Museo, Facultad de Ciencias Naturales y Museo, Universidad Nacional de La Plata*, **18** (Geologia, 66), 17–26.

Teruggi, M. E. 1957. The nature and origin of Argentine loess. *Journal of Sedimentary Petrology*, **27**(3), 322–332.

Tournouër, A. 1903. Note sur la géologie et la paléontologie de la Patagonie. *Bulletin de la Société Géologique de France*, **4**, Série 3, 463–473.

PART I GEOLOGY

2 Physical stratigraphy of the Sarmiento Formation (middle Eocene – lower Miocene) at Gran Barranca, central Patagonia

Eduardo S. Bellosi

Abstract

The middle Eocene to early Miocene pyroclastic Sarmiento Formation is a 319-m thick succession accumulated on loessic (eolian) and fluvial plains, and subordinately in shallow lakes. At the type section three members were originally recognized. Detailed studies along the continuous extent (7 km) of this very well-exposed escarpment, allow the recognition of several stratigraphic discontinuities and lithologic changes, resulting in a new stratigraphic framework with six members. From the base to the top of the section, discontinuities have been numbered from 1 to 10 and classified morphologically and genetically into high-relief erosive unconformities, slightly erosive paraconformities, and non-erosive paleosurfaces. The Gran Barranca Member (middle Eocene) is continuously exposed along the length of the escarpment and includes within it the Discontinuity 1 at the base of the richly fossiliferous marker Bed Y. The Gran Barranca Member is separated by the Paraconformity 2 from the new Rosado Member (late middle Eocene), a paleosol defined in central-eastern profiles. The former Puesto Almendra Member is subdivided by two erosive coplanar unconformities (Discontinuities 5 and 6). Discontinuity 3 marks the base of the Lower Puesto Almendra Member (late Eocene). Discontinuity surface 4 subdivides this member into two units. The new Vera Member (late Eocene to early Oligocene) is delimited at base and top by two deeply erosive unconformities (Discontinuities 5 and 6). As established by Ré *et al.* (Chapter 3, this book), it is chronologically (and physically) interposed between the Lower and Upper Puesto Almendra Members. The Upper Puesto Almendra Member (Oligocene) includes three units separate by slightly erosive surfaces (Discontinuities 8 and 9). The Colhue-Huapi Member (early Miocene) is bounded at the base by the highly erosive and temporally important Unconformity 10. Changing paleotopography in central Patagonia is reconstructed by an assessment of these erosional unconformities.

Resumen

La Formación Sarmiento (Eoceno medio – Mioceno inferior) es una sucesión piroclástica (319 m de espesor) originada en planicies loéssicas (eólicas), fluviales y subordinadamente en lagos someros. En su sección tipo fue inicialmente dividida en tres miembros. Estudios detallados a lo largo de la barranca (7 km), basados sobre discontinuidades estratigráficas y cambios litológicos, determinaron un nuevo armazón estratigráfico con seis miembros. Las discontinuidades fueron numeradas de 1 a 10 y clasificadas por su morfología y origen en: discordancias erosivas de fuerte relieve, paraconcordancias y paleosuperficies no-erosivas. El Miembro Gran Barranca (Eoceno Medio) que aflora con continuidad en toda la barranca, mantiene su definición original e incluye la Discontinuidad 1 en la base del estrato guía fosilífero "Y". La Paraconcordancia 2 lo separa del nuevo Miembro Rosado (Eoceno Medio superior). El Miembro Puesto Almendra se encuentra dividido por la intercalación de dos discordancias coplanares de fuerte relieve erosivo (Discontinuidades 5 y 6) y significativas diferencias en las edades isotópicas y magnéticas. El Miembro Puesto Almendra Inferior (Eoceno Superior), limitado en su base por la Discordancia 3, incluye la Discontinuidad 4 que separa dos unidades internas. El nuevo Miembro Vera (Eoceno Superior – Oligoceno Inferior) está delimitado por las Discordancias 5 y 6. Tal como indican Ré *et al.* (este libro), se intercala cronológica y físicamente entre los Miembros Puesto Almendra Inferior y Superior. El Miembro Puesto Almendra Superior (Oligoceno) se subdivide en tres unidades debido a las Discontinuidades 8 y 9, de menor jerarquía. El Miembro Colhue-Huapi (Mioceno inferior) está limitado en su base por la Discordancia 10, de fuerte relieve y alto valor temporal. El análisis de las discordancias erosivas ha permitido reconstruir los principales cambios paleotopográficos ocurridos en esta región de Patagonia central.

The Paleontology of Gran Barranca: Evolution and Environmental Change through the Middle Cenozoic of Patagonia, eds. R. H. Madden, A. A. Carlini, M. G. Vucetich, and R. F. Kay. Published by Cambridge University Press. © Cambridge University Press 2010.

Introduction

As a consequence of the paleontological collecting work of Carlos Ameghino between 1895 and 1903, the study and

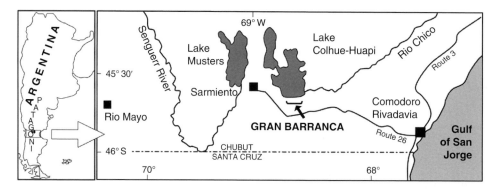

Fig. 2.1. Location map of Gran Barranca and other localities in central Patagonia mentioned in the text.

description of its fossils by Florentino Ameghino (1897 to 1906), the integrated fieldwork and publications of George G. Simpson (between 1930 and 1967) and many others (see Cifelli 1985), Gran Barranca, south of Lake Colhue-Huapi in central Chubut Province (Fig. 2.1), is the most important and well-known mammal-bearing sequence in all of South America. All the mammal fossils from Gran Barranca were collected from the tuffaceous Sarmiento Formation, also called "Tobas mamíferas del Eógeno" by Windhausen (1924), and "Tobas de Sarmiento" by Feruglio (1938, 1949). Stratigraphic, geochronologic, and paleomagnetic data (Ré *et al.* Chapter 3, this book) indicate that the pyroclastic succession outcropping at Gran Barranca is among the most continuous continental stratigraphic records of the middle Cenozoic in the southern hemisphere, although it includes some temporal gaps.

The classic sequence of four mammalian faunas of Gran Barranca, each with characteristic fossils and presumed ages (Casamayoran, Mustersan, Deseadan, and Colhuehuapian), implied the presence of at least three major unconformities within the Sarmiento Formation (Ameghino 1906; Simpson 1940; Marshall *et al.* 1983; Pascual and Ortiz 1990; Legarreta and Uliana 1994). Although the distinctive composition of these faunal units was originally recognized by Ameghino, the idea that these were of vastly different age required the existence of major unconformities representing long temporal hiatuses. Feruglio (1949) and Spalletti and Mazzoni (1979) mention that there are additional lithostratigraphic unconformities unrelated to the mammal record. Spalletti and Mazzoni (1979) identify several "paleosurfaces" and conclude that "most of the Sarmiento time was occupied by intraformational erosive or non-depositional phenomena." The two longest hiatuses were understood to occur between beds of the Casamayoran and Mustersan, and between the Mustersan and Deseadan South American Land Mammals Ages (SALMAs) (Marshall *et al.* 1983; Flynn and Swisher 1995). These hiatuses were thought to have occurred without producing marked or significant erosive topography.

The Sarmiento Formation at Gran Barranca measures 319 meters in total thickness and is exposed continuously along this single escarpment for over 7 kilometers. Most previous studies of the stratigraphy and sedimentology of Gran Barranca concentrated on the westernmost exposures (Spalletti and Mazzoni 1979), although Simpson (1930) recorded substantial lateral variation between the western and eastern extremes.

The aim of recent work reported here is to present a detailed lithostratigraphy for the full extent of the outcrop. The new stratigraphic framework is based on identification and characterization of discontinuity surfaces and changes in lithofacies observed at 11 stratigraphic profiles. Ten discontinuity surfaces can be identified in the Sarmiento Formation at Gran Barranca, some extending the full east-to-west extent of the exposure. Some of these discontinuity surfaces are unconformities, others are sharp contacts that extend laterally and some are undulating erosional surfaces. Most of these discontinuity surfaces are either subparallel, or intersect one another, and reflect a complex sedimentary history with alternating aggradational, erosional, and non-depositional processes.

Cenozoic stratigraphy of central Patagonia

Central Patagonia is broadly coincident geographically with the San Jorge Basin, an extensional intra-plate basin linked to the break-up of Gondwanaland and the opening of the South Atlantic Ocean. Maximum subsidence and accumulation of sediment occurred between the middle Jurassic and late Cretaceous, with the deposition of more than 6 km of mainly continental sediments. Fitzgerald *et al.* (1990) divided the Mesozoic fill into three megasequences, Lonco Trapial, Las Heras, and Chubut groups, representing respectively the early and late rift stages, and subsequent sag stage.

During the Cenozoic, central Patagonia remained active as a foreland ramp and passive margin basin. Latest Cretaceous to middle Miocene marine and continental sediments comprise a distinct 1200-meter thick megasequence

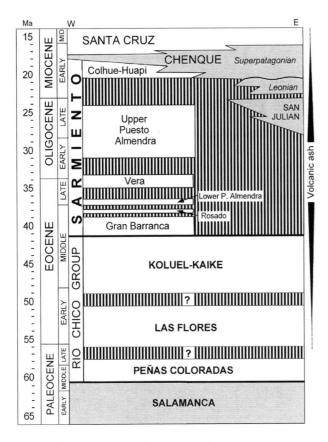

Fig. 2.2. Cenozoic stratigraphy of central Patagonia. Volcanic ash supply, particularly in the San Jorge basin, is shown. Marine units in gray.

that developed on a coastal plain setting (Fig. 2.2). The base of this megasequence has been considered unconformable upon the Chubut Group (Legarreta and Uliana 1994). However, in Cerro Abigarrado, 30 km south of Sarmiento City, the contact is transitional (Feruglio 1929; Bellosi *et al.* 2000). Although this exposure is marginal within the basin, it is assumed that on a regional scale there is no a significant unconformity between the Chubut Group and Salamanca Formation.

Cenozoic infilling of the San Jorge Basin involved alternating phases of epicontinental marine flooding, coastal plain regression and aggradation, and lapses of shelf emersion and fluvial erosion (Legarreta and Uliana 1994; Bellosi 1995; Bond *et al.* 1995), in a regimen of variable extensional subsidence (Fig. 2.2). Continental sequences, such as Rio Chico Group, Sarmiento and Santa Cruz Formations, have a broader geographic distribution than marine units such as Salamanca and Chenque Formations. The Salamanca Formation (upper Maastrichtian to Danian) is a shallow marine to estuarine and coastal swamp succession comprising two depositional sequences (Bellosi *et al.* 2000; Matheos *et al.* 2001). Subsequent regression is recorded by

a sandy and conglomeratic fluvial sequence of late Paleocene age, the Peñas Coloradas Formation, separated from the underlying marine unit by an erosive surface. This redbed sequence includes fossil mammals corresponding to the *Carodnia* association, assigned to the Riochican SALMA (Bond *et al.* 1995). Green and purple lacustrine mudstones and tuffs, corresponding to the Las Flores Formation (late Paleocene – early Eocene), follow and laterally replace the redbed sequence. The Las Flores Formation contains fossil mammals referred to the *Kibenikhoria* faunal zone of Simpson (1940), and assigned to the Itaboraian SALMA by Pascual *et al.* (1994).

Sometime, in the late early to middle Eocene, persistent terrestrial fine-grained volcanic dust/ash sedimentation began in central and northern Patagonia and lasted with some interruptions and change in intensity but with broadly similar characteristics until the early Miocene (Fig. 2.2). The lower part of the Paleogene volcaniclastic succession corresponds to the *Argiles Fissilaires* of Ameghino (1906) or the Koluel-Kaike Formation of Feruglio (1938), a 15- to 55-meter sequence of yellow, red, and gray silicified pyroclastic mudstones, fine tuffs, bentonites, and manganese-rich horizons. Pedogenic modification of these deposits is widespread and intense (Krause and Bellosi 2006). No fossil mammals have been found in this unit (Simpson 1940). The lower boundary of the Koluel-Kaike Formation is concordant with the underlying Las Flores Formation, and the upper contact with the Sarmiento Formation is transitional (Feruglio 1949; Pascual *et al.* 1994). At Las Flores and Cerro Blanco this transition occur in a 10-m thick interval, where reddish and yellowish indurated tuffaceous paleosols of the Koluel-Kaike Formation intercalate with poorly consolidated whitish pyroclastic mudstones of the Gran Barranca Member of the Sarmiento Formation. The detailed internal stratigraphy of the Sarmiento Formation is the subject of the following section of this chapter. During the time of the terrestrial Sarmiento sedimentation, three marine transgressions took place in south and central Patagonia. These are the Julian, Leonian, and Superpatagonian transgressions, collectively known as "Patagonian transgressions" (Ameghino 1906; Feruglio 1949; Bellosi 1995). The first transgressive event, as represented by the San Julián Formation, expanded into the Magallanes Basin in eastern Santa Cruz Province. In the Mazarredo subbasin, the San Julián Formation overlies unconformably beds of the Sarmiento Formation, that yield a fauna of Barrancan age (Cifelli 1985). Pollen grains and dinoflagellate cysts of the San Julián Formation indicate a late Oligocene age (Barreda 1997, Barreda and Palamarczuk 2000).

Subsequent marine episodes invaded a much extensive region of Santa Cruz and Chubut Provinces during the early Miocene (Bellosi 1995). In the San Jorge Basin these transgressions comprise five depositional sequences in the

Chenque Formation (Bellosi 1990; Bellosi and Barreda 1993). The Leonian transgression only flooded the eastern sector of the basin, and extended inland not more than 60 km west of the present coastline (Fig. 2.2). Marine and continental palinomorphs indicate an Aquitanian age (Bellosi and Barreda 1993) and thus agree with a 19.4 Ma radiometric age reported for the Monte León Formation in Santa Cruz Province (Bown and Fleagle 1993). Windhausen (1924) thought that this transgression reworked part of the Sarmiento tuffs, as much as several tens of meters of its total thickness. Sequences 1 and 2 of the Chenque Formation are composed by reworked pyroclastics deposited on a shallow marine shelf (Bellosi 1990) and these unconformably overlie lower beds of the Sarmiento Formation (Casamayoran SALMA), but the precise continental correlatives of these Leonian sequences is still uncertain. Tuffaceous terrestrial deposits of early Miocene age at La Flecha locality, in northeastern Santa Cruz Province, are referred to the Sarmiento Formation and are covered by the Monte León Formation.

The subsequent Superpatagonian transgression (Fig. 2.2) of Burdigalian age, had an even more widespread geographic extent, flooding most of the San Jorge Basin and reaching as far as the Andean foothills in western Patagonia (Bellosi 1995; Barreda and Bellosi 2003). These shallow marine and estuarine deposits comprise the three upper depositional sequences of the Chenque Formation that are correlated with the Colhue-Huapi Member. At Valle Hermoso and Pampa del Castillo, southeast of Gran Barranca, the uppermost beds of the Sarmiento Formation are transitionally covered by upper sequences of the Chenque Formation (Bellosi 1995; Bellosi *et al.* 2002a; Barreda and Bellosi 2003). At the west end of Gran Barranca (in Profiles A and MMZ), Spalletti and Mazzoni (1979) described a thin deposit of olive sandstones above the Colhue-Huapi Member, which they correlated to the marine Patagonia or Chenque Formation. A few kilometers farther to the east (between Profiles L and M), a small and partially covered outcrop of a marine sandy shell bed crops out at the top of Gran Barranca. The stratigraphic relationship of this sandy coquina is unclear, but considering the geographic and topographic location, the most plausible correlation would be with the early Miocene Chenque Formation. The existence of such a discontinuous and thin marine deposit, overlying the Sarmiento Formation, is consistent with a marginal position in the San Jorge Basin. The coastline of the Superpatagonian sea extended very close to the south border of Lake Colhue-Huapi according to Windhausen (1924).

The Neogene infilling of the San Jorge Basin culminated with the Santa Cruz Formation. This non-marine succession is conformable on the Chenque Formation (Feruglio 1936) and accumulated in estuarine, fluvial, and eolian environments during the final stage of the Superpatagonian regression (Bellosi and Jalfin 1996; Bellosi 1998). At Cañadón El Trébol and Estancia Cameron, 60 km east of Gran Barranca, Feruglio (1936, 1949) mentioned mammal remains assigned to the Santacrucian SALMA. Based on isotopic dates and paleomagnetic surveys performed in southern Santa Cruz Province, Marshall *et al.* (1986a) restricted the Santacrucian vertebrate fauna to the interval between 18 and 15 Ma. In the San Jorge Basin, Barreda and Bellosi (2003) suggested that Santa Cruz Formation may be early to middle Miocene in age. Most recently, Bown and Fleagle (1993) and Fleagle *et al.* (1995) reported five Ar/Ar dates between 16.0 and 16.6 Ma for the Santa Cruz Formation at Monte Leon.

In the valley of the Pinturas River, in western central Patagonia, a terrestrial succession correlates in part to the Sarmiento and in part to the Santa Cruz Formation. The Pinturas Formation (late early to middle Miocene) bears two mammal associations (Kramarz and Bellosi 2005). The older "Pinturan" faunal zone also occurs in the uppermost beds of Sarmiento Formation at Gran Barranca (Vucetich *et al.* 2005), and the younger faunal zone corresponds in composition to a typical Santacrucian SALMA association (Ameghino 1906). Thus, the upper section of Colhue-Huapi Member can be correlated to the lower and middle sequences of the Pinturas Formation (Kramarz and Bellosi 2005). Radiometric dates of approximately 18 Ma for both units (Fleagle *et al.* 1995; Re *et al.* Chapter 4, this book) confirm this correlation.

Finally, the youngest unit recognized at Gran Barranca is the "Rodados Patagonicos," a late Miocene–Pleistocene succession of fluvio-glacial and alluvial fan conglomerates (Panza 2002), which exhibits an erosive and unconformable lower contact.

Intraformational stratigraphic discontinuities

The stratigraphy of the Sarmiento Formation at Gran Barranca and in many other localities in central Patagonia has traditionally been based on fossil mammal content. Despite its apparent lithologic homogeneity, several researchers (Windhausen 1924; Feruglio 1949; Spalletti and Mazzoni 1979) have recognized erosion surfaces between successive biostratigraphic units, or "pisos" or "formations" (Casamayor, Musters, Deseado, and Colhue-Huapi). The first of the two most significant of these surfaces was thought by Simpson (1940) to correspond to a time of uplift and erosion after the Mustersan because Deseadan beds are often seen to rest on high relief surfaces. The second of these erosion surfaces was related to a tectonic phase with local faulting and folding after Deseadan sedimentation as the Colhuehuapian beds were deposited in a local depression. Modern sequence stratigraphic analysis also subdivided the Sarmiento succession according to its mammal assemblages (Legarreta and Uliana 1994), implying that faunal changes are coincident with stratal discontinuities in phase with changes in accommodation space and sea-level changes.

Rather than being based on biostratigraphy, the present assessment is based on observations of discontinuity surfaces along the full extent of the exposures at Gran Barranca (Fig. 2.3). Some of these surfaces define lithostratigraphic members and sequence boundaries or sections in the Sarmiento Formation (Figs. 2.4, 2.5). Discontinuities are characterized by their geometry (lateral extent, vertical relief, slope), lithofacies contrast, and mode of origin (Table 2.1). The hierarchy (whether diastems or unconformities) depends on their relationship to other surfaces, the magnitude of erosion and their relative temporal value as inferred from the isotopic dates and paleosol development. In aggradational systems with predominantly eolian or subaerial sedimentation, as is the case with the Sarmiento Formation, some subordinate discontinuities correspond to paleosols. However, we are concerned only with the more significant surfaces, most of which originally developed through erosive processes.

Three types of discontinuity surfaces are distinguished in the Sarmiento Formation at Gran Barranca (Table 2.1).

Type A discontinuities, or pronounced erosive unconformities, are irregular truncating surfaces formed by deep fluvial erosion. These record the removal of a considerable volume of sediment, up to several tens of meters in thickness. Lateral extent varies from hundred to thousand meters. Lithologic change is not always evident across these unconformities, because of frequent facies repetition and similarities in pedogenic features. In places, the steep relief produces vertical and irregular walls, covered by large blocks up to 0.5 m of intraformational material (e.g. soil clasts). The more significant type A unconformities truncate shorter and subordinated discontinuity surfaces, and record the highest temporal gap. They were formed by profound fluvial erosion corresponding to incised fluvial valleys. Unconsolidated fine-grained tephras were rapidly and easily removed by fluvial and eolian processes. Following intervals of rapid aggradation, favored by either intense ash rains and continuous post-eruptive input of reworked material, incised drainage networks would tend to be filled.

Type B discontinuities, or paraconformities, are moderately or slightly erosive low-relief surfaces. These are the most common in the Sarmiento Formation, and extend laterally through hundreds to even thousand meters parallel to bedding. Temporal gaps represented by these surfaces are variable. Often, these surfaces

intersect and indicate a complex history of aggradation (eolian and fluvial), erosive events, and subaerial exposure.

Type C discontinuities, or non-erosive unconformities, are paleosurfaces that coincide with stacked and/or mature paleosols. These discontinuities are characterized by macro- and microscopic pedogenic features and generally thick profiles indicating deep weathering, extended exposure and very slow or no sediment accumulation. The limonitic paleosurfaces in the Gran Barranca and Puesto Almendra Members reported by Spalletti and Mazzoni (1979) do not correspond to this type of discontinuity, but represent poorly developed paleosols. Several authors have stressed the stratigraphic significance and utility of strongly developed paleosols for subdividing continental successions (Hanneman and Wideman 1991; Wright and Marriot 1993; Legarreta and Uliana 1994; Shanley and McCabe 1994; McCarthy and Plint 1998). In particular, Kraus (1999) proposed that the origin of unconformities associated with paleosols might be related to allogenic factors such as regional or global climatic change, sea-level variation, and tectonic movement, operating over timescales of 0.01 to 10 Ma.

Discontinuity 1

Discontinuity 1, a Type B surface (or paraconformity) is a sharp, non-erosive to slightly erosive and somewhat undulating surface at the lower contact of Simpson's Bed Y in the upper portion of the Gran Barranca Member. Bed Y is an easily recognized marker bed with a rather uniform thickness (Figs. 2.3, 2.5). It is a massive pale or whitish pyroclastic mudstone containing abundant fossil vertebrates (Barrancan subage). It rests on a green bentonite and extends unbroken between the western and eastern ends of Gran Barranca (see Simpson 1930; Cifelli 1985). Discontinuity 1 is very continuous, but does not separate contrasting lithologies. Isotopic dates for Simpson's Y Tuff and the immediately adjacent beds, the absence of associated magnetic polarity reversal (Ré *et al.* Chapter 3, this book), the common occurrence of bentonite–mudstone couplets, the geometry and location within the Barrancan Fossil Zone indicates it represents a very short temporal gap (<10 k.y.).

Discontinuity 2

Discontinuity 2 is a Type B surface with a sharp low relief (7 m) contact at the base of Simpson's Bed X. It extends through the central and eastern sectors of Gran Barranca (between Profiles I, J, L, and M). Bed X is a pink, calcareous, indurated horizon, interpreted as a strongly developed pedogenic calcrete (Bellosi *et al.* 2002b), and a new subunit of the Sarmiento Formation, termed Rosado Member (Figs. 2.4, 2.5). The lithologic contrast of Bed X with the

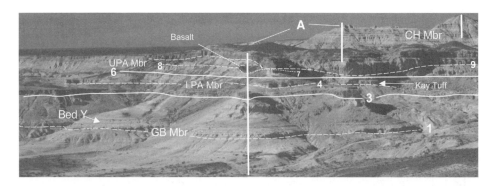

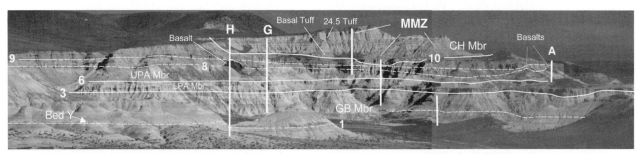

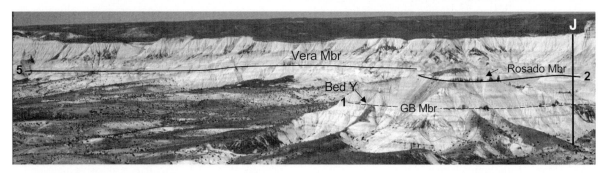

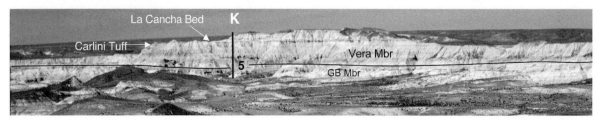

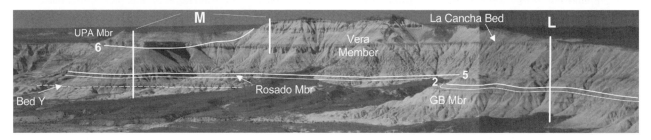

Fig. 2.3. Panoramic pictures of the Gran Barranca showing profiles (vertical lines), discontinuity surfaces (numbers), members, and key beds. East is to the left side of each photo.

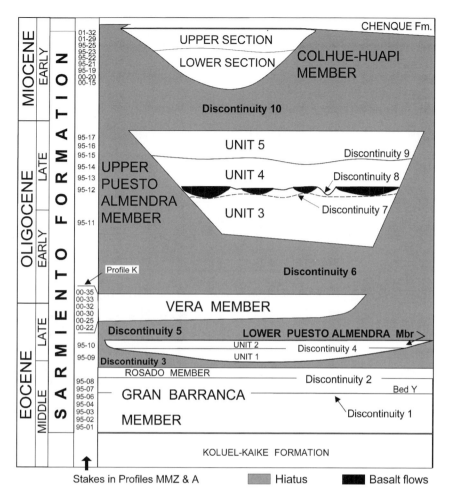

Fig. 2.4. Stratigraphic organization of the Sarmiento Formation at Gran Barranca, showing members, units, and discontinuity surfaces.

underlying poorly consolidated pyroclastic mudstones of the Gran Barranca Member is pronounced. Hypothetically, this surface was originally formed by erosion and later modified by extended pedogenesis. Judging from the stratigraphic relationships, isotopic ages, paleomagnetic data, and thickness between Beds Y and X (Ré *et al.* Chapter 3, this book), the temporal hiatus of this paraconformity is less than 0.6 m.y.

Discontinuity 3

Discontinuity 3 is an unconformity of Type C. It marks the erosional top of the Gran Barranca Member at Profiles A to H and N; and the top of the thick calcareous paleosol of the Rosado Member at Profiles J and M. Unconformity 3 truncates Discontinuity 2, and was eroded by Discontinuity 5 between Profiles J, K, L, and M (Fig. 2.5). This paleosurface is a sharp and undulating contact of moderate relief (10 m). At Profile MMZ it is covered by sandstones and conglomerates of fluvial origin (Spalletti and Mazzoni 1979; Bellosi *et al.* 2002b). The deepest erosion at this surface occurs at Profiles L and M (Fig. 2.4).

The temporal significance of this composite unconformity varies laterally. Given the age of the Rosado and Lower Puesto Almendra members (Ré *et al.* Chapter 3, this book) the hiatus would be small. This is in accordance with the presence of Mustersan fossils in both members. The temporal gap of Unconformity 3 is related to the time of pedogenesis of the moderately to strongly developed Rosado paleosol, estimated in 10^5–10^6 years (Retallack 2001). Thus, Unconformity 3 represents a hiatus about 0.7 m.y. (Ré *et al.* Chapter 4, this book).

Discontinuity 4

Discontinuity 4 is a Type B surface. It is a weakly erosive contact at the base of Bed 95–10 (in Profile MMZ) in the Lower Puesto Almendra Member. It is parallel to Discontinuity 3, defining approximately the top of Simpson (1930) "Lower Channel Series," and can be traced continuously for 1 km between Profiles A and H (Figs. 2.3, 2.5). The deepest erosion occurred at Profile A (8 m). East of Profile H, Discontinuity 5 eroded Discontinuity 4. At Profile MMZ,

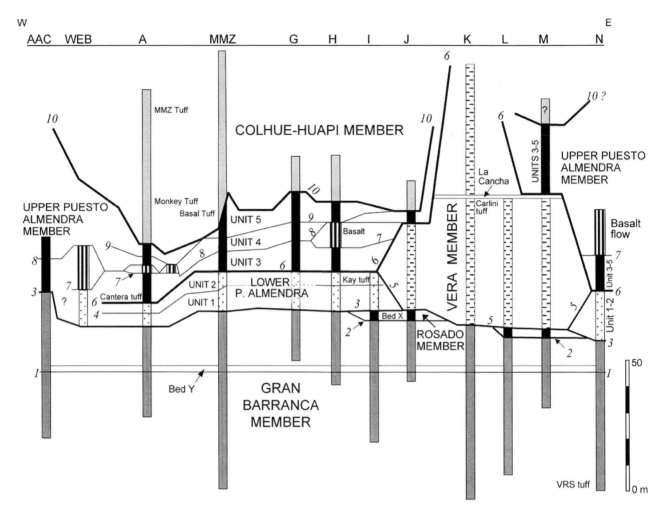

Fig. 2.5. Cross-section of the Gran Barranca (flattened to the top of the marker Bed Y, Gran Barranca Member), showing discontinuity surfaces (numbers in italics) and lithostratigraphic units of the Sarmiento Formation. Horizontal spacing between the Profiles is arbitrary. Lateral extent of Gran Barranca 7 km.

Discontinuity 4 is overlaid by a lenticular coarse breccia of paleosol intraclasts. Given the available isotopic dates (Ré *et al*. this book) and the lack of evidence for erosion or non-deposition, this surface does not appear to mark a significant temporal hiatus.

Discontinuity 5
Discontinuity 5 is a Type A surface. It represents a deep erosive unconformity at the base of the Vera Member (Fig. 2.4), between Profiles J and M. This noteworthy unconformity has removed several subunits of the Sarmiento Formation, including the upper portion of the Gran Barranca Member, and the Rosado and Lower Puesto Almendra Members. Thus, Discontinuity 5 truncates the older discontinuities 2, 3, and 4 in part. The lateral extent of Discontinuity 5 is about 3 km (Fig. 2.5). West of Profile I it was eliminated by a superpositional discontinuity. Discontinuity 5 exhibits a local vertical relief of

between 15 and 20 m. It is deepest at Profile M and shallows west of Profile J. Given its geometry and the absence of the lower section of the Vera Member at Profile N, it is presumed that Discontinuity 5 originated through fluvial erosion, although no coarse lag deposits are observed at its base. The overlying sediments comprise more than 100 m of homogeneous white tephric loessites. The rare and poorly developed paleosols suggests that eolian deposition was continuous. Considering the age of Lower Puesto Almendra and Vera Members (Ré *et al*. Chapter 4, this book), the chronologic extent of Discontinuity 5 is about 1.58 m.y.

Discontinuity 6
Discontinuity 6 is a Type A surface. It is the most conspicuous erosive unconformity in the Sarmiento Formation at Gran Barranca, and in terms of relief, lateral continuity, accompanying change in facies and style of sedimentation,

Table 2.1. *Summary of characteristics of the unconformities (UF) and discontinuity surfaces (DS) of the Sarmiento Formation*

Discontinuity	Type	Lateral extension	Vertical relief	Lithofacies contrast	Relationship to other surfaces	Process	Hierarchy	Temporal gap[a]
D 10	A	Mod–high	27–50 m	Basal conglomerate	Erodes DS 9, parallel to UF 6	Incised valley, pedogenesis	1°	3.48 Ma
D 9	B	Low	5 m	Conglomerate on loessite	Truncates DS 8	Fluvial erosion	3°	Negligible
D 8	B	Moderate	10 m	Basalt breccia	Truncates DS 7, onlaps UF 6	Fluvial erosion after lava flows	3°	Negligible
D 7	B	Low	<5 m	Basalt on variable facies	Onlaps UF 6	Fluvial channel erosion	3°	Negligible
D 6	A	High	105 m	Laterally variable	Truncates DS 5, Onlaps DS 7, 8, 9	Incised valley	1°	2.67 Ma
D 5	A	Moderate	18 m	Loessite on calcrete	Truncates UF 2, 3 and DS 4	Incised valley	1°	1.58 Ma
D 4	B	Moderate	8 m	Breccia on bentonite	Parallel to UF 3	Local fluvial erosion	3°	Negligible
D 3	C	High	10 m	Sandstone on loessite	Truncates UF 2	Fluvial erosion, pedogenesis	2°	0.7 Ma
D 2	B	Low	7 m	Calcrete on loessite	Parallel to DS 1	Slight erosion	2°	<0.1 Ma
D 1	B	Very high	0 m	Loessite on bentonite	Horizontal	Depositional change	3°	Negligible

[a]Temporal gaps according to geochronologic analysis by Re *et al.* (Chapter 4, this book).

and complex relationships with other discontinuities (Figs. 2.4, 2.5). Discontinuity 6 extends the length of Gran Barranca (about 7 km) from Profiles A to N (Fig. 2.3). At the type Profile MMZ, it occurs approximately 9 m above stake 95–10, at the top of paleosols at stake 99–25.5, where it defines the boundary between Lower and Upper Puesto Almendra Members. Discontinuity 6 truncates Discontinuity 5, another high-magnitude surface, and subsequent Discontinuities surfaces 7 and 8 onlap on it. A large volume of intraformational ash was removed by fluvial incision. The total vertical erosional relief is at least the thickness of the Vera Member, about 105 m. The intraformational conglomerates of the Upper Puesto Almendra Member at Profile A would represent the redeposition of these materials. In addition, the stacked paleosols of the Unit 2 at profile MMZ are genetically linked to this unconformity, representing the time of subaerial exposure. The sector where Discontinuity 6 is not present (Profile K) is an interfluve. Discontinuity 6 formed in the early Oligocene (Fig. 2.4) (Ré *et al.* Chapter 4, this book). The erosive episodes represented by the two Unconformities 5 and 6 must be responsible for the less extensive distribution of post-Barrancan to Deseadan beds in Patagonia, as suggested by Ameghino (1906), Windhausen (1924), and Simpson (1940). Considering the age of

Vera and Upper Puesto Almendra Members, Discontinuity 6 has important chronological significance. The temporal gap is about 2.2 m.y. (Ré *et al.* Chapter 4, this book).

Discontinuity 7
Discontinuity 7 is a discontinuous and irregular erosive Type B surface observed at the base of the lenticular basalts at Profiles WEB, A, H, and N (Figs. 2.3, 2.5). It is present in sections dominated by intraformational conglomerates of the Upper Puesto Almendra Member. Discontinuity 7 is concave with a vertical relief of 5 m. Where unprotected between basalt lenses, this surface was later truncated by Discontinuity 8. At Profiles I and M, Discontinuity 7 onlaps Discontinuity 6. The morphology and associated facies suggest that it was formed by fluvial erosion. Similar stratigraphic relationships and geometry are observed in Valle Hermoso (Fig. 2.1), where correlated lenticular basalts also outcrop. Hierarchy and chronologic significance of this surface is not significant.

Discontinuity 8
Discontinuity 8 is a moderately deep erosive Type B surface at the top of the basalts in the Upper Puesto Almendra Member (Fig. 2.4). It extends from Profiles AAC to J

(Fig. 2.5). Vertical relief from the top of the basalts reaches 10 m, exceeding the thickness of the basalts. This indicates that fluvial erosion produced a topographic inversion. Coarse basalt breccias and cross-bedded conglomerates containing large vertebrate remains overlie this surface and extend between the lenticular lavas. Discontinuity 8 erodes Discontinuity 7 and onlaps Discontinuity 6. The temporal value of Discontinuity 8 is not significant.

Discontinuity 9
Discontinuity 9 is a slightly irregular erosive Type B surface at the base of the Upper Channel Series (Simpson 1930) in the Upper Puesto Almendra Member (Fig. 2.4). It can be observed from Profiles A to H, onlapping Discontinuity 6 (Fig. 2.5). In general, vertical relief is not high (5 m), but at two sites between Profiles A and MMZ and at Profile H, this surface is deeper and eroded the tops of the basalts, crossing down Discontinuity 8. At Cerro Blanco, southwest of Gran Barranca (Fig. 2.1), this surface is much deeper with a vertical relief about 25 m. Discontinuity 9 is covered by fluvial deposits (intraformational conglomerates and cross-bedded sandstones), much of them affected by intense pedogenesis. This surface probably does not represent a significant temporal hiatus.

Discontinuity 10
Discontinuity 10 is a first-order erosional unconformity, or a Type A surface, at the base of the Colhue-Huapi Member (Figs. 2.3, 2.4). It cuts down tens of meters into Units 4 and 5 of the Upper Puesto Almendra Member. Discontinuity 10 can be observed as an irregular surface between Profiles AAC and H, and more clearly at Profile MMZ, forming a paleovalley filled by the Colhue-Huapi sediments (Fig. 2.5). At Profile M this unconformity may be present at the basal contact of an unfossiliferous fluvial succession which would therefore be tentatively correlated to the Colhue-Huapi Member. Bellosi et al. (2002a) interpret this unconformity as an incised valley, with a terraced morphology and intense pedogenesis on exposed surfaces. It probably formed inside the older fluvial paleovalley of Discontinuity 6 related to the Upper Puesto Almendra Member. The local orientation of the Colhue-Huapi paleovalley is NNW–SSE. The vertical relief can be directly measured at Profiles A and MMZ, where it is 27 m deep and 150 m wide. However, taking into account the highest point of older units at Profiles K and M, maximum depth is about 50 m. At places, lateral walls of the paleovalley are vertical and show irregular indentations. In deeper portions of the paleovalley, the basal infill consists of conglomerates and cross-bedded sandstones. Large paleosols fragments (up to 60 cm in length), from the Unit 5 of the Upper Puesto Almendra Member, are present on this erosive surface. At local interfluves, this unconformity becomes a paraconformity

(Type B) due to the flat morphology and the subtle facies contrast (paleosols on paleosols).

Discontinuity 10 is correlated with the erosive surfaces underlying the Colhuehuapian beds at Valle Hermoso (Marshall et al. 1986b), and in Rastro del Avestruz and Los Leones in northern Santa Cruz province. It marks an important drop in sea level on the coastal plain of central Patagonia, and in the coeval shallow marine succession of the Chenque Formation where it separates Leonian from Superpatagonian beds (Bellosi 1995; Bellosi et al. 2002a). According to palynological data (Bellosi and Barreda 1993) and the coastal onlap curve (Haq et al. 1987) it would have been produced at about 21 Ma (Legarreta and Uliana 1994, Bellosi 1995). Simpson (1940) believed the pre-Colhuehuapian unconformity represents pronounced and deep erosion but a brief hiatus. Considering the age of Upper Puesto Almendra and Colhue-Huapi Members (Ré et al. Chapter 4, this book), the hiatus represented by Discontinuity 10 is about 3.8 m.y.

Lithostratigraphic division of the Sarmiento Formaton

The new lithostratigraphic framework provided here for the type locality of the Sarmiento Formation departs from the proposal of Spalletti and Mazzoni (1979) and recognizes six members (Fig. 2.4). Maximum vertical thickness in a single profile is about 170 m (Profile MMZ). The cumulative thickness reaches 319 m.

Gran Barranca Member
The Gran Barranca Member and its equivalents is the more extensively exposed unit of the Sarmiento Formation, not only in Gran Barranca (between Profiles AAC to N), but also in most outcrops of the formation in central Patagonia (Simpson 1940). At Gran Barranca, it is a whitish, poorly consolidated, 67-m thick succession of thickly bedded pyroclastic mudstones, bentonites, and poorly developed paleosols (Spalletti and Mazzoni 1979). The base is transitional on the Koluel-Kaike Formation. At Profiles A to H and N, the top of the member is the erosive unconformity of Discontinuity 3 (Fig. 2.5). Abundant fossil mammals recovered from several levels in this member belong to the Barrancan subage (Cifelli 1985). Geochronologic studies elaborated by Re et al. (Chapter 4, this book) indicate a middle Eocene age.

Rosado Member
The Rosado Member is a pink carbonate paleosol outcropping from the central to the eastern end of Gran Barranca (Profiles J, K, M, and N) (Bellosi et al. 2002a; Miquel and Bellosi 2007). This new unit is 7 m thick, and bounded by

Discontinuity 2 at the base and 3 at the top (Figs. 2.3 to 2.5). Vertebrate remains collected from this unit at Profile J are assigned to the Mustersan SALMA. Stratigraphic relationships and geochronologic studies indicate a late middle Eocene age (Ré *et al.* Chapter 4, this book).

Upper and Lower Puesto Almendra Members

The former Puesto Almendra Member of Spalletti and Mazzoni (1979) is here divided into the Upper and Lower Puesto Almendra Members by the unconformities mentioned above. These units are separated by the first-order erosive Discontinuities 5 and 6, which are coplanar from profile I to A (Figs. 2.3, 2.5). To the east, the discontinuities come above and below the Vera Member. Thinner beds, better-defined stratification, more advanced pedogenesis, and higher consolidation are the main characteristics of both members.

The Lower Puesto Almendra Member is 30 m thick and is constituted by two units (Fig. 2.4). Unit 1 is composed of intraformational conglomerates, cross-bedded sandstones, pyroclastic mudstones, bentonites, and moderately developed paleosols (Bellosi this book). Simpson (1930) characterized the lower part of this unit as the "Lower Channel Series," which contains Mustersan mammals. Unit 2 is a thin succession of pyroclastic mudstones and paleosols. Geochronologic studies indicate a late Eocene age (Ré *et al.* Chapter 4, this book).

The Upper Puesto Almendra Member is 46 m thick and is subdivided into three units by erosive surfaces. The lowest or Unit 3 is composed of pyroclastic mudstones, poorly developed paleosols, and lava flows which rest above Discontinuity 7. Unit 4 rests upon Discontinuity 8, and includes intraformational conglomerates, sandstones, pyroclastic mudstones, bentonites, and moderately developed paleosols (Bellosi Chapter 19, this book). Unit 5 develops on Discontinuity 9, and comprises intraformational conglomerates, sandstones, and a thick succession of stacked, moderately to strongly developed paleosols at the upper part. Simpson (1930) identified this last unit as his "Upper Channel Series." Fossil mammals collected from Unit 4 correspond to the Deseadan SALMA. Geochronologic studies indicate an Oligocene age (Ré *et al.* Chapter 4, this book).

Vera Member

This new unit is chronologically intercalated between the Lower and Upper Puesto Almendra Members. It is the thickest member (105 m) but is only developed in the central and eastern part of the Gran Barranca. At Profile MMZ it was totally eroded away (Fig. 2.5). The erosional relief of Discontinuities 5 and 6 defined base and top of the Vera Member (Fig. 2.4). It is a homogeneous succession of massive and poorly bedded pyroclastic mudstones with some very subordinate thin bentonites (Bellosi Chapter 19,

this book). Scarce and weakly developed calcareous paleosols are also present. Weathering into pinnacles is characteristic of the lower part of the Vera Member (Fig. 2.3). The fossiliferous (Tinguirirican mammals) "La Cancha" bed, which occurs in the middle part of the member, is a weakly developed paleosol with a sharp basal contact that does not seem to represent a discontinuity surface. This would otherwise be the only discontinuity within this uniform unit. Geochronologic studies indicate a late Eocene – earliest Oligocene age for this member (Ré *et al.* Chapter 4, this book).

Colhue-Huapi Member

The Colhue-Huapi Member is the uppermost unit of the Sarmiento Formation as defined by Spalletti and Mazzoni (1979) at Profile MMZ (Figs. 2.3 to 2.5). There is an erosive surface of high relief (Discontinuity 10) separating it from the underlying Upper Puesto Almendra Member (Bellosi *et al.* 2002a). The upper contact of the Colhue-Huapi Member is covered at Gran Barranca, but at Valle Hermoso it is a transitional boundary with the shallow marine Chenque Formation (Fig. 2.2). The Colhue-Huapi Member is 64 m thick and includes intraformational conglomerates, pyroclastic mudstones, and weakly to moderately developed paleosols. Mammal remains are very abundant and occur in two well-defined zones. The lower one corresponds to the Colhuehuapian SALMA, and the upper one to the Pinturan mammal assemblage. Geochronologic studies indicate an early Miocene age for this member (Ré *et al.* Chapter 4, this book).

Conclusions

At Gran Barranca, the pyroclastic Sarmiento Formation (middle Eocene – early Miocene) includes ten discontinuity surfaces, categorized into three types: high-relief erosive unconformities, slightly erosive paraconformities, and non-erosive paleosurfaces. These discontinuities provide a stratigraphic framework delimiting six members: Gran Barranca, Rosado, Lower Puesto Almendra, Vera, Upper Puesto Almendra, and Colhue-Huapi.

The new physical stratigraphy for the Sarmiento Formation permits more precise placement of fossil sites, levels, and zones identified at the Gran Barranca. The magnetostratigraphy and radioisotopic dates placed within this framework constrain the temporal magnitude of the hiatuses. The relationships among the unconformities caused by deep erosive events and surfaces truncating one another reflect a complex sedimentary history of alternating aggradation, erosion, and non-deposition. This should serve as a caution to those studying others localities of the Sarmiento Formation across Patagonia that do not exhibit the extent or quality of the exposures at Gran Barranca.

The three upper members are delimited at the base by three major unconformities (Discontinuities 5, 6, and 10). They correspond to the lower surface of deep incised valleys, and record the greatest chronologic gaps. According to Ar/Ar dating and magnetic polarity studies (Ré *et al.* Chapter 3, this book) these unconformities represent temporal gaps between 0.7 and 3.5 Ma, which increase progressively from middle Eocene to Early Miocene. The remaining discontinuities that separate other members or occur within them were formed by subordinated fluvial erosion or prolonged subaerial exposure and pedogenesis.

Flat land surfaces characterized this region of central Patagonia during middle Eocene, but since the late Eocene (Discontinuity 5) local relief was more pronounced. Topographic irregularities attain their maximum relief in the early Oligocene, created by very deep erosion represented by Discontinuity 6. Thus, local topography changed markedly between the Lower Puesto Almendra and Vera Members at Discontinuity 5, when erosion began to have a more marked and lasting imprint on landscape. Perhaps the best example of this irregular relief is observed at Profile MMZ with the Discontinuity 10, a deep and narrow incised fluvial valley.

ACKNOWLEDGEMENTS

I thank Richard H. Madden for the constructive discussions, the photographs and his valuable suggestions, and Richard F. Kay for reviewing and greatly improving the content of this manuscript. This work was supported by grants from NSF-USA (BCS-9318942, DEB-9907985, EAR-0087636, BCS-0090255) to R. F. Kay and R. H. Madden. I am grateful to Pan American Energy for the assistance during field work.

REFERENCES

Ameghino, F. 1906. Les formations sédimentaires du Crétacé supérieur et du Tertiaire de Patagonie. *Anales del Museo Nacional*, **8**, 1–568. Buenos Aires.

Barreda, V. 1997. Palinoestratigrafía de la Formación San Julián en el área de playa La Mina, Oligoceno de la Cuenca Austral. *Ameghiniana*, **34**, 283–294.

Barreda, V. and S. Palamarczuk 2000. Palinoestratigrafía de depósitos del Oligoceno tardío-Mioceno en el área sur del golfo San Jorge, provincia de Santa Cruz, Argentina. *Ameghiniana*, **37**, 103–117.

Barreda, V. and E. Bellosi 2003. Ecosistemas terrestres del Mioceno temprano de la Patagonia central, Argentina: primeros avances. *Revista del Museo Argentino de Ciencias Naturales*, n.s., **5**, 125–134.

Bellosi, E. 1990. Formación Chenque: registro de la transgresión patagoniana (Terciario medio) de la Cuenca San Jorge. *Actas XI Congreso Geológico Argentino*, **2**, 57–60.

Bellosi, E. 1995. Paleogeografía y cambios ambientales de la Patagonia central durante el Terciario medio. *Boletín de Informaciones Petroleras*, **44**, 50–83.

Bellosi, E. 1998. Depósitos progradantes de la Formación Santa Cruz, Mioceno de la Cuenca San Jorge. *Actas VII Reunión Argentina de Sedimentología*, 110–111.

Bellosi, E. and V. Barreda 1993. Secuencias y palinología del Terciario medio en la Cuenca San Jorge, registro de oscilaciones eustáticas en Patagonia. *Actas XII Congreso Geológico Argentino y II Congreso de Exploración de Hidrocarburos*, **1**, 78–86.

Bellosi, E. and G. Jalfin 1996. Sedimentación en la planicie costera santacrucense–superpatagoniana (Mioceno inferior-medio, Cuenca San Jorge). *Actas VI Reunión Argentina de Sedimentología*, 181–186.

Bellosi, E., S. Palamarczuk, V. Barreda, J. Sanagua, and G. Jalfin 2000. Litofacies y palinología del contacto Grupo Chubut: Formación Salamanca en el oeste de la Cuenca Golfo San Jorge, Argentina. *Ameghiniana*, **37**, 45R–46R.

Bellosi, E., M. González, R. Kay, and R. Madden 2002a. El valle inciso colhuehuapense de Patagonia central (Mioceno inferior). *Resumenes IX Reunión Argentina de Sedimentología*, 49.

Bellosi, E., S. Miquel, R. Kay, and R. Madden 2002b. Un paleosuelo mustersense con microgastrópodos terrestres (Charopidae) de la Formación Sarmiento, Eoceno de Patagonia central: significado paleoclimático. *Ameghiniana*, **39**, 465–477.

Berggren, W., D. Kent, C. Swisher, and M. Aubry 1995. A revised Cenozoic geochronology and chronostratigraphy. In Berggren, W. A., Kent, D. V., Aubry, M.-P., and Hardenbol, J. (eds.), *Geochronology, Time Scales and Global Stratigraphic Correlation*, Special Publication no. 54. Tulsa, OK: Society for Sedimentary Geology, pp. 129–212.

Bond, M., A. Carlini, F. Goin, L. Legarreta, J. E. Ortíz, R. Pascual, and M. Uliana 1995. Episodes in South American Land Mammal evolution and sedimentation: testing their apparent concurrence in a Paleocene succession from Central Patagonia. *Actas VI Congreso Argentino de Paleontología y Bioestratigrafía*, 47–58.

Bown, T. and J. Fleagle 1993. Systematics, biostratigraphy and dental evolution of Palaeothentidae, Later Oligocene to Early–Middle Miocene (Deseadean–Santacrucian) caenolestoid marsupials of South America. *Paleontological Society Memoir*, **29**, 1–76.

Cifelli, R. 1985. Biostratigraphy of the Casamayoran, Early Eocene of Patagonia. *American Museum of Natural History Novitates*, **2820**, 1–26.

Feruglio, E. 1929. Apuntes sobre la constitución geológica de la región del Golfo San Jorge. *Boletín de Informaciones Petroleras*, **6**, 925–1025. [Reprint (1995) **44**, 90–123.]

Feruglio, E. 1936. Sobre la presencia del Santacruciano en la Pampa del Castillo (región del Golfo de San Jorge). *Notas Museo La Plata* **1**, (*Geologia* 2), 237–246.

Feruglio, E. 1938. Nomenclatura estratigráfica de la Patagonia y Tierra del Fuego. *Boletín de Informaciones Petroleras*, **171**, 54–67.

Feruglio, E. 1949. *Descripción Geológica de la Patagonia*, vol. 2. Buenos Aires: Dirección General de Yacimientos Petrolíferos Fiscales.

Fleagle, J., T. Bown, C. Swisher, and G. Buckley 1995. Age of the Pinturas and Santa Cruz formations. *Actas VI Congreso Argentino de Paleontología y Bioestratigrafía*, 129–135.

Gradstein, F. M., J. G. Ogg, and A. Smith (eds.) 2004. *A Geological Time Scale 2004*. Cambridge, UK: Cambridge University Press.

Hanneman D. and C. Wideman 1991. Sequence stratigraphy of Cenozoic continental rocks, southwestern Montana. *Geological Society of America Bulletin*, **103**, 1335–1345.

Haq, B., J. Hardenbol, and P. Vail 1987. The chronology of the fluctuating sea level since the Triassic. *Science*, **235**, 1156–1167.

Kramarz, A. and E. Bellosi 2005. Hystricognath rodents from the Pinturas Formation, Early–Middle Miocene of Patagonia, biostratigraphic and paleoenvironmental implications. *Journal of South America Earth Sciences*, **18**, 199–212.

Kraus, M. 1999. Paleosols in clastic sedimentary rocks. *Earth Sciences Reviews*, **47**, 41–70.

Krause, M. and E. Bellosi 2006. Paleosols from the Koluel-Kaike Formation (Lower–Middle Eocene) in south-central Chubut: a preliminary analysis. *IV Congreso Latinoamericano de Sedimentología and 11° Reunión Argentina de Sedimentología*, Abstracts, 125.

Legarreta, L. and M. Uliana 1994. Asociaciones de fósiles y hiatos en el Supracretácico–Neógeno de Patagonia: una perspectiva estratigráfico–secuencial. *Ameghiniana*, **31**, 257–281.

Marshall, L., R. Hoffstetter, and R. Pascual 1983. Mammals and stratigraphy: geochronology of the continental mammal-bearing Tertiary of South America. *Paleovertebrata, Mémoire Extraordinaire*, 1–93.

Marshall, L., R. Drake, G. Curtis, R. Butler, K. Flanagan, and C. Naeser 1986a. Geochronology of type Santacrucian (middle Tertiary) Land Mammal Age, Patagonia, Argentina. *Journal of Geology*, **94**, 449–457.

Marshall, L., R. Cifelli, R. Drake, and G. Curtis 1986. Vertebrate paleontology, geology and geochronology of the Tapera de Lopez and Scarritt Pocket. *Journal of Paleontology*, **60**, 920–951.

Matheos, S., M. Brea, D. Ganuza, and A. Zamuner 2001. Sedimentología y paleoecología del Terciario Inferior en el sur de la provincia del Chubut, Argentina. *Revista de la Asociación Argentina de Sedimentología*, **8**, 93–105.

McCarthy, P. and A. Plint 1998. Recognition of interfluve sequence boundaries: integrating paleopedology and sequence stratigraphy. *Geology*, **26**, 387–390.

Miquel, S. and E. Bellosi 2007. Microgasterópodos terrestres (Charopidae) del Eoceno medio de Gran Barranca (Patagonia Central, Argentina). *Ameghiniana*, **44**, 121–131.

Panza, J. 2002. La cubierta detrítica del Cenozoico superior. *Actas XV Congreso Geológico Argentino*, 259–284.

Pascual, R., A. Carlini, and F. Goin 1994. Paleogene land mammal bearing localities in Central Patagonia, Argentina: field trip guide. *IV Congreso Argentino de Paleontología y Bioestratigrafía*.

Pascual, R. and J. E. Ortíz 1990. Evolving climates and mammal faunas in Cenozoic South America. *Journal of Human Evolution*, **19**, 23–60.

Retallack, G. 2001. *Soils of the Past*, 2nd edn. London: Blackwell Science.

Shanley, K. and P. McCabe 1994. Perspectives on the sequence stratigraphy of continental strata. *American Association of Petroleum Geologists Bulletin*, **78**, 544–568.

Simpson, G. G. 1930. *Scarritt-Patagonian Exped. Field Notes*. New York: American Museum of Natural History. (Unpublished.) Available at http://paleo.amnh.org/notebooks/index.html

Simpson, G. G. 1940. Review of the mammal-bearing Tertiary of South America. *American Philosophical Society Proceedings*, **83**, 649–709.

Spalletti, L. and M. Mazzoni 1979. Estratigrafía de la Formación Sarmiento en la barranca sur del lago Colhue-Huapi, provincia del Chubut. *Revista de Asociación Geológica Argentina*, **34**, 271–281.

Vucetich, M., E. Vieytes, A. Kramarz, and A. Carlini 2005. Caviomorph rodents from Gran Barranca: biostratigraphic and paleoenvironmental contribution. *Resúmenes XVI Congreso Geológico Argentino*, 303.

Windhausen, A. 1924. Líneas generales de la constitución geológica de la región situada al oeste del golfo de San Jorge. *Boletín de la Academia Nacional de Ciencias*, **27**, 167–320.

Wright, V. and S. Marriot 1993. The sequence stratigraphy of fluvial depositional systems: the role of floodplain storage. *Sedimentary Geology*, **86**, 203–210.

3 Paleomagnetism and magnetostratigraphy of Sarmiento Formation (Eocene–Miocene) at Gran Barranca, Chubut, Argentina

Guillermo H. Ré, Silvana E. Geuna, and Juan F. Vilas

Abstract

The cliffs south of Lake Colhue-Huapi consist of pyroclastic sediments of the Sarmiento Formation (Eocene–Miocene), where important paleontological sites have been discovered. Paleomagnetic sampling was performed along five stratigraphic profiles at a sampling interval between sites varying from 2 to 10 meters. Oriented samples from 182 sites (718 specimens) were subjected to demagnetization. Most samples showed just one component of remanent magnetization, which is totally erased by applying linear decay alternating fields (AF) between 15 and 100 mT. Samples with different coercivity usually coexist at the same site.

A magnetostratigraphy was established based on remanent magnetic polarity directions averaged by site as an auxiliary tool in support of stratigraphic guide levels and available radiometric dates. The following magnetic ages were obtained: Colhue-Huapi (CH) Member, ≈18.8 to 20.7 Ma; Upper Puesto Almendra (UPA) Member, 24.2 to ≈30.6 Ma; Vera (VE) Member, 33.3 to 35.0 Ma; Lower Puesto Almendra (LPA) Member, ≈36.6 to 37.3 Ma, Rosado Member, to ≈38 to ≈38.4 Ma, and Gran Barranca (GB) Member, ≈38.4 to 41.7 Ma.

Four paleomagnetic poles for members constituting the Sarmiento Formation were obtained from remanent magnetic directions isolated from higher coercivity samples (>50 mT): Gran Barranca Member: Lat. 86° S, Long. 15° E, N 41, A95 5.4; Vera Member: Lat. 89° S, Long. 262° E, N 11, A95 9.6; Upper Puesto Almendra Member: Lat. 88° S, Long. 81° E, N 10, A95 8.4; Colhue-Huapi Member: Lat. 86° S, Long. 152° E, N 21, A95 6.8. Although statistically indistinguishable from one another, poles from the four members reflect a slight movement northwards of South America during the time the Sarmiento Formation was being deposited.

Resumen

Las barrancas de la margen sur del lago Colhue-Huapi, lugar de afloramiento de la Formación Sarmiento (Eoceno–Mioceno) y de sus principales sitios paleontológicos, fueron objeto de un muestreo paleomagnético. El mismo se realizó a lo largo de 5 perfiles, en los que se establecieron sitios con intervalos variables entre 2 y 10 metros. Muestras de un total de 182 sitios (718 especímenes) fueron sometidas a procedimientos de desmagnetización. La mayor parte de las muestras presentan una sola componente de magnetización remanente, que es eliminada totalmente con campos alternos linealmente decrecientes de entre 15 y 100 mT. Muestras con diferentes fuerzas coercitivas suelen coexistir en un mismo sitio.

Las direcciones de remanencia magnética resultantes fueron utilizadas para establecer una magnetoestratigrafía que, sumada a la presencia de niveles guía y a las edades radimétricas obtenidas en algunos niveles, permitió interpretar las siguientes edades magnéticas: Miembro Colhue-Huapi (CH), desde aproximadamente los 18,8 a los 20,7 Ma; Miembro Puesto Almendra Superior (UPA), entre ~24,2 y 30,6 Ma; Miembro Vera (VE), de 33,3 a 35,0 Ma, Miembro Puesto Almendra Inferior (LPA), desde aproximadamente 36,6 a los 37,3 Ma; Miembro Rosado, aproximadamente entre los 38,0 y los 38,4 Ma y el Miembro Gran Barranca (GB), aproximadamente entre los 38,4 y los 41,5 Ma.

Con las direcciones de remanencia aisladas a partir de las muestras de mayor coercitividad (>50 mT), se calcularon polos paleomagnéticos para los 4 miembros de la Formación Sarmiento: GB Lat. 86° S, Long. 15° E, N 41, A95 5.4; VE Lat. 89° S, Long. 262° E, N 11, A95 9.6; UPA Lat. 88° S, Long. 81° E, N 10, A95 8.4; CH Lat. 86° S, Long. 152° E, N 21, A95 6.8.

Aunque son estadísticamente indistinguibles, los cuatro polos paleomagnéticos reflejan un suave movimiento hacia el norte para la placa Sudamericana, durante el lapso en el cual se depositaba la Formación Sarmiento.

The Paleontology of Gran Barranca: Evolution and Environmental Change through the Middle Cenozoic of Patagonia, eds. R. H. Madden, A. A. Carlini, M. G. Vucetich, and R. F. Kay. Published by Cambridge University Press. © Cambridge University Press 2010.

Introduction

Gran Barranca is the name applied to the cliffs south of Lake Colhue-Huapi, 35 km ESE of Sarmiento, in Chubut Province, Argentina. The rich fossil record, outcrop quality, and long temporal interval of continental deposition make Gran Barranca a classic locality for middle Cenozoic stratigraphy.

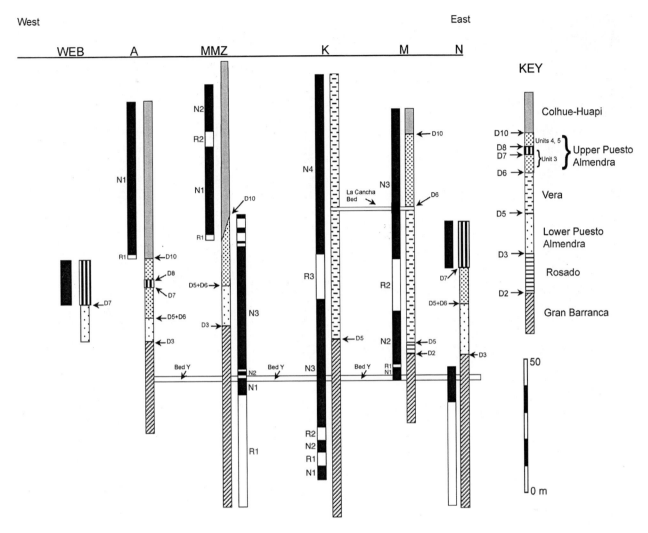

Fig. 3.1. Stratigraphic profiles of the Sarmiento Formation and its principal units at Gran Barranca. Local magnetostratigraphic columns are shown, in black/white as normal/reversed polarity zones. Letters above columns correspond to nomenclature of Simpson (1930), with the addition of Spalletti and Mazzoni's (1979) MMZ profile. Major discontinuities are indicated by D numbers. For a more complete outline of the profiles and discontinuities, see Bellosi (this book, Fig. 2.5).

The first stratigraphic profile of Gran Barranca was published by Florentino Ameghino (1906). G. G. Simpson described the stratigraphy in detail during the Scarritt Patagonian Expeditions in 1930–34. These results were never published, but the original notes are available through the Department of Vertebrate Paleontology, American Museum of Natural History in New York (Simpson 1930).

Simpson analyzed lithological and faunal change at several profiles along the cliff, labeled A to M (Fig. 3.1). This stratigraphic scheme has been followed since that time, and several South American Land Mammal Ages (SAL-MAs) have been described from Gran Barranca following Simpson's scheme. Spalletti and Mazzoni (1979) added Profile MMZ, the thickest and best-studied profile.

The Sarmiento Formation ("*Tobas de Sarmiento*" of Feruglio 1949) comprises most of the outcrop at Gran Barranca. This unit is widely distributed in extra-Andean Patagonia, as it has been described at surface and subsurface from 44° to 48° S, reaching its maximum thickness in the San Jorge Basin (Bellosi and Madden 2005; Bellosi Chapter 2, this book). The extraordinary richness in vertebrate remains, along with the distinctive lithology and uniform color, make the Sarmiento Formation the most representative unit of the continental Cenozoic of Argentina (Bellosi Chapter 2, this book).

The complete sequence at Gran Barranca is composed as follows, from base to top (Spalletti and Mazzoni 1979): Koluel-Kaike Formation, Sarmiento Formation, and Chenque Formation (Bellosi Chapter 2, this book). These outcrops have been the subject of many geological and paleonto-logical studies, the most relevant being Feruglio (1929, 1949), Simpson (1941), Spalletti and Mazzoni (1979), Cifelli (1985), and Kay *et al.* (1999).

The homogeneous lithology, added to the lateral discontinuity of continental deposits and the scarcity of lithostrati-graphic markers prompted a multidisciplinary approach to an understanding of the temporal details of the faunal succession. A further complication is the presence of paleosols and erosional surfaces, evidence of important depositional hiatuses (Kay *et al.* 1996; Bellosi Chapter 19, this book). For this reason, geological and paleontological surveys were supported by paleomagnetic and geochronological studies. The results of our study of paleomagnetism are presented here.

Geological setting

Gran Barranca exposes an important sequence of clastic and pyroclastic sediments, including several levels of Cenozoic mammal remains. The sequence was described by Spalletti and Mazzoni (1979) as modified by Bellosi and Madden (2005) and Bellosi (Chapter 2, this book). The currently recognized sequence is as follows.

The Koluel-Kaike Formation (the Río Chico Formation of Spalletti and Mazzoni 1979) is composed of a succession of continental, multicolored conglomerates, sandstones, tuffs, and limolites, containing amphibians, mammal, and plant remains (Lesta *et al.* 1980). At Gran Barranca, the Koluel-Kaike Formation constitutes the lowest exposed levels and its base is not exposed.

The Sarmiento Formation (Eocene–Miocene) conform-ably overlies the Koluel-Kaike Formation. The unit is sub-horizontal. Spalletti and Mazzoni (1979) reported a total thickness of 165 meters in the sections they measured but the composite thickness is 319 meters (Bellosi Chapter 2, this book). Traditionally known as the Sarmiento Group (Simpson 1941; Pascual and Odreman Rivas 1973; Andreis *et al.* 1975; Spalletti and Mazzoni 1977), Spalletti and Mazzoni (1979) recategorized the unit as a formation. A middle Eocene to early Miocene age was established by Kay *et al.* (1996). The Formation is composed mainly of altered tephra and pyroclastics of dominantly dacitic and calc–alkaline composition, representing the distal deposits of plinian tephra falls derived from a magmatic arc with decreasing activity in the Oligocene (Mazzoni 1985; Bellosi this book). The Sarmiento Formation could be correlated to the "Serie Andesítica" outcropping farther to the northwest (Mazzoni 1985) and interpreted as the product of shallow subduction during the Oligocene (Rapela *et al.* 1983).

Bellosi (Chapter 2, this book) recognizes ten main stratigraphic discontinuities in the Sarmiento Formation. Some of these erosive and non-depositional surfaces are of short duration (possibly less than 50 k.y.) while others represent longer periods up to millions of years of non-deposition or erosion duration. The stratigraphic scheme used in this work is a division into members proposed by Spalletti and Mazzoni (1979), modified by Bellosi (this book) as follows, from base to top.

The Gran Barranca Member (Spalletti and Mazzoni 1979) is the most extensive unit, with a thickness of 67 meters. Its base is transitional on the underlying unit (Koluel-Kaike Formation) and the top is erosive (Discontinuity 2 of Bellosi this book). It is characterized by the predominance of tuffs, with abundant gypsum and siliceous concretions mostly near the base. Pedogenesis is rare and manganese oxides are present as crusts. A layer of regional extent, termed the "Y level" by Simpson (1941) has Discontinuity 1 at its base. The Gran Barranca Member includes three radiometrically dated tuffs ranging in age between 39.1 and 41.7 Ma (Ré *et al.* Chapter 4, this book).

The Rosado Member (Bellosi this book) is a calcrete level, reaching a thickness of 7 meters in Profiles J, M, and MMZ (Bellosi *et al.* 2002b; Miquel and Bellosi 2007). The top is defined by an incised paleosurface (Discontinuity 3). The Rosado Member includes the dated Rosado Tuff at 38.4 Ma (Ré *et al.* Chapter 4, this book).

Spalletti and Mazzoni (1979) identified a Puesto Almendra Member in Profile MMZ near the west end of Gran Barranca, where it reaches a thickness of 80 meters. Towards the east a tongue of sediment is interposed within this member whereas to the west the tongue is absent but a discontinuity (Discontinuity 6) replaces it. This has led Bellosi (this book) to divide the original Puesto Almendra Member into Lower and Upper members, and terms the intervening tongue of sediment the Vera Member. Upper and Lower Puesto Almendra Members shows a predominance of coarse grained intraformational conglomerates, along with tuffs, and abundant evidence of edaphization. Manganese oxides, gypsum, and siliceous concretions are virtually absent. Basalt flows are found in the Upper Member in paleochannels. Discontinuity 6 separates two main intervals. The newly constituted Lower Puesto Almendra (LPA) Member includes the "Lower Channel Series" and part of "Intermediate Tuffs" of Simpson (1941). The age of the LPA unit is constrained by its stratigraphic position above Discontinuity 3 (making it younger than 38 Ma) and its position below Discontinuity 6. The Upper Puesto Almendra Member contains a series of dated basalt flows ranging from 26.3 to 29.2 Ma (Ré *et al.* Chapter 4, this book) and is composed of the remaining "Intermediate Tuffs" plus hardened orange paleosols of the "Upper Channel Series" of Simpson (1941).

The Vera Member (Bellosi and Madden 2005) is well developed in the central and eastern sectors of Gran Barranca (Profiles K, L, M, and N) where it attains 105 m thickness. It represents a continuous accumulation of loess, deposited on Discontinuity 5 (Bellosi this book). Internally it does not show major discontinuities. Discontinuity 6 truncates it laterally and vertically (Bellosi this book). As noted, the Vera Member intercalates between the lower and upper portions of the Puesto Almendra Member. The Vera Member includes one radiometrically dated tuff, the Carlini Tuff, which occurs in a normal polarity zone that helps to constraint its age (Ré *et al*. Chapter 4, this book).

The Colhue-Huapi Member (Spalletti and Mazzoni 1979) is the upper unit of the Sarmiento Formation, reaching 64 m thickness. Its base is marked by Discontinuity 10 separating it from the Upper Puesto Almendra Member (Bellosi *et al*. 2002a; Bellosi Chapter 2, this book). At the top of Profile MMZ, a small amount of marine rock occurs, a possible correlative of the Chenque Formation (Spalletti and Mazzoni 1979; Bellosi Chapter 2, this book). The Colhue-Huapi Member is mainly composed of loessic deposits and subordinate tuffs. Manganese oxides are present as local pigmentation and paleosols show well-developed regional extent. Two sections can be distinguished, the lower corresponding to fluvial infill of an incised valley (low-stand system track), and the upper agradational sequence of loessic and fluvial deposits (Bellosi and Madden 2005; Bellosi this book).

Sampling

Rock samples were collected for paleomagnetic study through four out of five members of the Sarmiento Formation at Gran Barranca; only the Rosado Member calcrete was not sampled. Oriented samples from 205 sites were taken. Most of the samples were drilled with a portable diamond core drill, although block samples were obtained from some sites, from which at least two cores were subsequently subsampled. Orientation was recorded using sun and magnetic compasses. Sampling for radiometric age determination were taken in collaboration with other sampling and paleontologic studies reported elsewhere in this book.

An initial sampling of Profile MMZ (Fig. 3.1) obtained 125 specimens from 25 sites at a sampling interval of 10 meters. The middle of the same profile was later resampled by taking 192 specimens from 45 new sites to obtain a more detailed magnetic record through the Eocene–Oligocene transition. Ten more specimens were taken from two basalt flows outcropping at the west end of the cliff.

The Colhue-Huapi Member was sampled along Profile A (Fig. 3.1) and six specimens were obtained from each of 32 sampled sites with an interval of 2 meters between sites.

Sampling was undertaken in Profiles K and M in an attempt to establish a chronostratigraphy for the Vera Member and the fossil level "La Cancha" (Fig. 3.1). The two profiles are easily correlated because of the presence of both the Y guide level and the "La Cancha" bed.

A total of 48 sites in Profile K were sampled (five specimens per site), separated by an average of 4 meters. It was logistically easier to divide Profile K into three segments with partial superposition between them. The middle section (sites K1 to 14) (see Table 3.1) was sampled along a road some 500 meters to the east of the axis of Profile K; the lower section (sites K15 to 23) was sampled from the base of the cliff along the axis of Profile K; and finally, the upper section (sites K24 to 38) was sampled from a point underlying the "La Cancha" bed at its northern end and proceeding up to the top of the cliff.

Sampling of Profile M was undertaken in 2002 to check the magnetic stratigraphy of the correlative Profile K. Sites MS1 to 13 (Table 3.1) correspond to the upper part; the sample horizons are spaced approximately 3 meters apart, and five specimens were obtained at each site. The lower part, from the level of Simpson's Y Tuff up to the "La Cancha" level, was sampled at 23 sites (five specimens per site) with a sampling interval of 1.5 meters for sites GBMI1 to 8, and of 3 meters for sites GBMI8 to 23.

The most complete section of the lowermost Sarmiento Formation at Gran Barranca is preserved in Profile N (Fig. 3.1). We sampled the lower part of Profile N (about 32 meters thick) and obtained 13 sites (six cores per site) with an average spacing of 2.5 meters. Four additional sites were sampled at four basalt flows at the top of Profile N, where ten specimens per site were taken.

Laboratory procedures

The samples were cut into specimens of 2.2 cm length. At this stage the original sampling was reduced by about 10% as some of the block samples were broken in the process of drilling. Remanent magnetization was measured using a three-axis 2G DC squid cryogenic magnetometer. Alternating field demagnetization was carried out to a maximum of 110 mT (peak) using a static 2G600 demagnetizer attached to the magnetometer. Stepwise thermal demagnetization up to 650 °C was performed using either a two-chambers ASC or Schonstedt TSD-1 oven at the Instituto de Geofísica "Daniel Valencio" (INGEODAV), Universidad de Buenos Aires.

A total of 718 specimens belonging to 182 sites were subjected to stepwise demagnetization, to examine the coercivity and unblocking temperature spectra of the natural remanent magnetization (NRM). Bulk magnetic susceptibility was measured after each thermal step in order to monitor possible magnetic mineral changes, by using a MS2W Bartington susceptometer.

Magnetic behavior of each specimen was analyzed by visual inspection of Zijderveld plots, stereographic projections and intensity demagnetization curves. Magnetic components were determined by using principal component

Table 3.1. *Site mean directions for the Sarmiento Formation at Gran Barranca*

Site	Nl	A/B/C	Unit	Class	Site mean direction				
					N	Dec.	Inc.	α_{95}	K
C32	5	0/0/5	CH	C	–	–	–	–	–
C31	1	0/1/0	CH	B	1	337	−70.6	–	–
C30	5	5/0/0	CH	A	5	24.9	−64.2	15.6	25.0
C29*	5	4/1/0	CH	A	4	342.4	−62.5	32.1	9.1
C28*	3	3/0/0	CH	A	3	1.8	−79.9	91.9	2.9
C27	5	5/0/0	CH	A	5	355.5	−67.6	16.9	21.4
C26	5	0/0/5	CH	C	–	–	–	–	–
C25(OB)	6	1/0/5	CH	A	1	66	−85.1	–	–
C24	5	0/1/4	CH	B	1	347.1	−21.4	–	–
C23	5	0/0/5	CH	C	–	–	–	–	–
C22	5	0/0/5	CH	C	–	–	–	–	–
C21	5	1/1/3	CH	A	1	186.1	67.5	–	–
C20	5	0/4/1	CH	B	4	358.1	−65.2	13.4	48.2
C19	5	0/0/5	CH	C	–	–	–	–	–
C18	4	3/0/1	CH	A	3	31.4	−39.7	27.9	20.5
C17(OB)	5	2/3/0	CH	A	2	305.9	−55.7	–	–
C16*	5	0/4/1	CH	B	4	44.8	−45.5	39.8	6.3
C15	5	2/3/0	CH	A	2	2.8	−55.7	–	–
C14	4	0/0/4	CH	C	–	–	–	–	–
C13	5	2/2/1	CH	A	2	341.6	−55.3	–	–
C12	5	2/1/2	CH	A	2	15.7	−51.4	–	–
C11(OB)	5	2/3/0	CH	A	2	59.2	−54	–	–
C10	5	0/2/5	CH	B	1	136.5	51.4	–	–
C09*	5	4/0/1	CH	A	4	23.7	−45.3	42.5	5.6
C08	4	4/0/0	CH	A	4	318.3	−73.3	6.6	193.9
C07	5	5/0/0	CH	A	5	9.4	−64.1	22.7	12.3
C06*	5	5/0/0	CH	A	5	129.8	−38.3	49.1	3.3
C05(OB)	4	4/0/0	CH	A	4	308.6	−28.7	12.8	52.6
C04	5	0/4/1	CH	B	4	308.9	−48.3	28.3	11.5
C03	4	0/4/0	CH	B	4	16.7	−41.9	25.2	14.2
C02	4	2/0/2	CH	A	2	347.8	−70.1	–	–
C01	4	4/0/0	CH	A	4	185	65.4	7.7	141.9
25	5	3/2/0	CH	A	3	348.6	−60.1	9.3	176.4
24(OB)	5	1/3/1	CH	A	1	67.3	−45.8	–	–
23	4	4/0/0	CH	A	4	163.1	59.1	14.5	40.8
22	7	0/6/1	CH	B	6	26.1	−59.9	10.4	42.7
21	5	2/3/0	CH	A	2	7.1	−56	–	–
20	2	2/0/0	CH	A	2	10.9	−58.4	–	–
19	3	2/0/1	CH	A	2	46.3	−58.3	–	–
18	6	6/0/0	UPA	A	6	359.4	−66.4	8.7	60.1
X48	3	3/0/0	UPA	A	3	352.1	−63.1	9.6	165.3
X47	1	1/0/0	UPA	A	1	351.3	−51.8	–	–
X46	1	1/0/0	UPA	A	1	334.9	−67.5	–	–
X44	1	0/0/1	UPA	C	–	–	–	–	–
17	4	4/0/0	UPA	A	4	178.3	75.6	19.5	23.2
X43	3	2/0/1	UPA	A	2	232.8	60.6	–	–
X41	1	1/0/0	UPA	A	1	184.4	63.9	–	–
X42	2	2/0/0	UPA	A	2	4.8	−59.4	–	–

Table 3.1. (*cont.*)

Site	M	A/B/C	Unit	Class	Site mean direction				
					N	Dec.	Inc.	α_{95}	K
16	3	2/0/1	UPA	A	2	11.8	−34.1	−	−
X40	1	1/0/0	UPA	A	1	3.2	−59	−	−
15	3	0/3/0	UPA	B	3	217.9	−66.1	23.2	29.1
14(OB)	8	7/0/1	UPA	A	7	48.3	−10	4.6	170.8
X30	2	0/0/2	UPA	C	−	−	−	−	−
X29	4	0/1/3	UPA	B	1	31.4	−81.2	−	−
13	6	0/5/1	UPA	B	5	345.9	−60.3	17.2	20.6
X28	2	2/0/0	UPA	A	2	358.7	−68.1	−	−
12	8	0/7/1	UPA	B	7	15.6	−67.7	12.6	23.9
11	3	0/2/1	UPA	B	2	14.2	−61.5	−	−
X24	1	1/0/0	LPA	A	1	20.4	−61.4	−	−
X23	3	3/0/0	LPA	A	3	352.9	−62.7	11.4	118
10	5	4/1/0	LPA	A	4	351.6	−65.5	6	239.3
09	3	3/0/0	GB	A	3	13.3	−50.5	19.3	41.7
X19	1	1/0/0	GB	A	1	34.8	−50.9	−	−
08	5	5/0/0	GB	A	5	338.1	−78.4	17.5	20.1
X13	3	3/0/0	GB	A	3	14.6	−51.3	22.4	31.41
07	6	2/1/3	GB	A	2	221.7	80.2	−	−
X12	3	3/0/0	GB	A	3	1.4	−56.8	18.1	47.5
06	5	0/5/0	GB	B	5	142.6	−15.9	22.8	12.1
X09	7	7/0/0	GB	A	7	310.5	−55.5	12.3	25.0
X08	3	3/0/0	GB	A	3	357.3	−60.1	9.3	176.1
X07	4	4/0/0	GB	A	4	2.1	−61.9	11.4	65.4
05	3	1/2/0	GB	A	1	17.9	−64.6	−	−
X06	4	4/0/0	GB	A	4	302.9	−61.5	10.4	78.8
X04	1	0/1/0	GB	B	1	41.5	−49.3	−	−
X02	1	0/1/0	GB	B	1	343.6	−38	−	−
X01	3	0/3/0	GB	B	3	357.5	−54.6	6.8	334.4
04	5	1/1/3	GB	A	1	148.3	69.3	−	−
03	1	1/0/0	GB	A	1	171.6	64.2	−	−
02(OB)	2	2/0/0	GB	A	2	256.2	85.8	−	−
U6	4	4/0/0	CH	A	4	344	−55.2	18.4	25.8
U5	4	3/1/0	CH	A	3	352.9	−53.1	23.3	29.0
U4*	6	0/4/2	CH	B	4	341.1	−39.4	34.1	8.2
U3	5	2/1/2	CH	A	2	0.5	−48.9	−	−
U2	6	0/4/2	CH	B	4	131	−45.4	23.8	15.9
U1	5	4/0/1	CH	A	4	180.6	54	11.8	61.2
K37	4	0/3/1	V	B	3	356.1	−62.8	14.7	71.1
K36	4	0/2/2	V	B	2	5.3	−59.4	−	−
K35	5	3/2/0	V	A	3	8	−53.4	8	238.8
K34	4	0/2/2	V	B	2	0.4	−60.9	−	−
K33	1	0/1/0	V	B	1	347.1	−47	−	−
K32	3	0/1/2	V	B	1	346.3	−72.4	−	−
K31	3	2/1/0	V	A	2	23.3	−57.6	−	−
K30	3	0/1/2	V	B	1	344.9	−60.8	−	−
K29	2	0/2/0	V	B	2	14.2	−65.4	−	−
K28	5	2/1/2	V	A	2	16.3	−50.6	−	−
K27	1	0/0/1	V	C	−	−	−	−	−

Table 3.1. (*cont.*)

Site	*N*l	A/B/C	Unit	Class	Site mean direction				
					N	Dec.	Inc.	α_{95}	*K*
K26	1	0/0/1	V	C	–	–	–	–	–
K25	1	0/1/0	V	B	1	184.4	66.4	–	–
K24	1	1/0/0	V	A	1	144.9	74.6	–	–
K14	2	0/2/0	V	B	2	166.1	69.7	–	–
K13	4	0/2/2	V	B	2	193.1	62.9	–	–
K12	6	0/3/3	V	B	3	165.3	76.1	8.7	202.6
K11B	4	0/0/4	V	C	–	–	–	–	–
K11	5	0/1/4	V	B	1	185.4	67.1	–	–
K10	5	0/0/5	V	C	–	–	–	–	–
K09	6	6/0/0	V	A	6	342.6	−69.5	8.8	58.5
K08	6	4/2/0	V	A	4	354	−65.6	22.1	18.2
K07	4	3/1/0	GB	A	3	356.6	−60.8	30.2	17.7
K06	6	3/2/1	GB	A	3	9.5	−60.2	29.8	18.2
K05	6	5/1/0	GB	A	5	0.5	−64.5	11.5	45.1
K04D	5	2/0/3	GB	A	2	0.5	−64	–	–
K04C	5	3/0/2	GB	A	3	8.6	−63.4	10.2	147.6
K04B	3	3/0/0	GB	A	3	2.4	−58.4	10.1	151.4
K04	8	6/2/0	GB	A	6	12.1	−56.3	14.4	22.7
K03G	1	1/0/0	GB	A	1	307.4	−52.2	–	–
K03F	1	0/1/0	GB	B	1	97.2	−60.4	–	–
K03E	1	1/0/0	GB	A	1	1.8	−60.9	–	–
K03D	1	1/0/0	GB	A	1	10	−55.5	–	–
K03C	1	1/0/0	GB	A	1	349.3	−50.9	–	–
K03B	1	0/1/0	GB	B	1	352.4	−50	–	–
K03A	1	0/1/0	GB	B	1	23.5	−47.8	–	–
K03	6	2/2/2	GB	A	2	338	−58.9	–	–
K02	4	3/1/0	GB	A	3	347	−63.3	4.4	782.7
K01	4	0/4/0	GB	B	4	346.8	−64.9	24.9	14.5
K23	1	1/0/0	GB	A	1	359.7	−68.8	–	–
K22	6	0/5/1	GB	B	5	2.6	−56.8	10.4	54.6
K21	5	0/2/3	GB	B	2	28	−65.5	–	–
K20	2	0/2/0	GB	B	2	5.5	−63.4	–	–
K19	2	0/2/0	GB	B	2	166.1	57.7	–	–
K18	3	3/0/0	GB	A	3	214.3	66	8.2	227.3
K17	4	1/3/0	GB	A	1	336.1	−63.8	–	–
K16	1	1/0/0	GB	A	1	202.3	66.2	–	–
K15	3	3/0/0	GB	A	3	343.6	−60	4.4	770.7
MS13	5	0/5/0	V	B	5	15.2	−57	11.9	41.9
MS13	1	0/1/0	V	B	1	130.2	72.1	–	–
MS12	4	0/3/1	V	B	3	16.7	−59.3	13.1	89.7
MS11	9	0/7/2	V	B	7	356.3	−60.4	4.4	185.9
MS10	8	0/6/2	V	B	6	1.8	−66.4	7.5	81.2
MS09	8	0/3/5	V	B	3	358.4	−69.4	14.1	77.0
MS08	5	0/1/4	V	B	1	30.9	−70.2	–	–
MS07	5	0/5/0	V	B	5	355.1	−59.1	15	26.8
MS06	8	0/8/0	V	B	8	352.3	−64.8	7.9	50.0
MS05	7	0/6/1	V	B	6	0	−67.8	10.9	39.0
MS04	6	0/0/6	V	C	4	352.4	−56.9	18.5	25.7

Table 3.1. (*cont.*)

Site	N1	A/B/C	Unit	Class	Site mean direction				
					N	Dec.	Inc.	α_{95}	K
MS03	7	0/7/0	V	B	7	346.6	−61.1	15.4	16.4
MS02	2	0/1/1	V	B	1	1.2	−72	–	–
MS01	6	0/4/2	V	B	4	336.4	−56.5	11.7	63.0
MI22	1	0/0/1	V	C	–	–	–	–	–
MI21	1	0/1/0	V	B	1	27.6	−66.9	–	–
MI20	2	0/1/1	V	B	1	37.4	−63.8	–	–
MI18	1	0/1/0	V	B	1	6.7	−52.4	–	–
MI17	1	0/1/0	V	B	1	120.6	68.7	–	–
MI16	2	0/2/0	V	B	2	332.4	−62	–	–
MI15	4	4/0/0	V	A	4	325.4	−67.6	24	15.6
MI13	1	0/1/0	V	B	1	37.1	−46.8	–	–
MI12	3	3/0/0	V	A	3	23.7	−64.6	21.1	35.3
MI11	2	0/0/2	V	C	–	–	–	–	–
MI10	5	0/5/0	V	B	5	10.8	−60.7	14.3	29.7
MI09(OB)	2	2/0/0	GB	A	2	267.8	−64.3	–	–
MI07	2	2/0/0	GB	A	2	338.5	−63.1	–	–
MI06	3	3/0/0	GB	A	3	118.7	60.4	30	18.0
MI05	5	2/1/2	GB	A	2	12.6	−54.2	–	–
MI04	7	0/4/3	GB	B	4	3.2	−46	14	44.2
MI03	5	5/0/0	GB	A	5	342.5	−71	7.3	110.2
MI02	4	4/0/0	GB	A	4	337.9	−64.3	11.3	67.5
MI01	3	2/0/1	GB	A	2	31.5	−53.1	–	–
N13	5	0/4/1	GB	B	4	339.7	−77.9	19.4	23.3
N12	5	0/4/1	GB	B	4	37.9	−71.7	25.8	13.7
N11	2	0/2/0	GB	B	2	344.5	−76.5	–	–
N10	6	0/6/0	GB	B	6	1.5	−60.1	8.8	58.7
N09	4	0/4/0	GB	B	4	0.7	−59.7	15.9	34.4
N08	5	0/4/1	GB	B	4	346.2	−61.5	17.5	28.7
N07	5	0/3/2	GB	B	3	214.2	76.2	35.4	13.2
N06	3	0/3/0	GB	B	3	170.2	68.5	34.8	13.6
N05	4	3/1/0	GB	A	3	160.8	69	7.7	260.5
N04	4	0/2/2	GB	B	2	97.1	74.1	–	–
N03	5	5/0/0	GB	A	5	155	59.7	4.8	252.9
N02	4	3/0/1	GB	A	3	147.9	64.7	16.1	60.0
N01	5	5/0/0	GB	A	5	167.8	67.7	4	367.2
V	8	8/0/0	UPA	A	8	25.6	−71.4	6.6	71.9
B(OB)	8	8/0/0	UPA	A	8	116.6	61.3	7.9	50.7
R1(OB)	4	3/0/1	UPA	A	3	122.8	52.1	17.1	52.9
R2(OB)	5	5/0/0	UPA	A	5	297.3	55.4	5	237.6

Notes: N1: number of specimens submitted to demagnetization procedures.

A/B/C: number of specimens showing Type A, B, or C magnetic behaviour respectively.

Unit: code of member: GB, Gran Barranca; UPA, Upper Puesto Almendra; LPA, Lower Puesto Almendra; VE, Vera; CH, Colhue-Huapi.

Class: quality of site mean direction (A, B, or C; see text).

N: number of specimens used for site mean direction.

α_{95}, K: semiangle of the 95% confidence cone and statistical parameter of site mean direction.

(OB): oblique virtual geomagnetic pole, excluded for paleomagnetic pole calculation.

*excluded from mean because of poor statistical parameters.

analysis (PCA) following Kirschvink (1980). Components were determined with at least four consecutive demagnetization steps and just those defined by a maximum angular deviation (MAD) value lower than 15° were accepted. Programs SuperIAPD (Torsvik *et al.* 2000) and MAG88 (Oviedo 1989) were used, both applying Fisher (1953) statistics.

Results

Magnetic susceptibility of the Sarmiento Formation ranges from 10 to 150×10^{-5} (SI). The NRM varies between 0.5 and 40 mA/m, although it can be higher in some sites, reaching up to 100 mA/m. The variation of NRM intensity is plotted on Fig. 3.2. Although every member is dominated by NRM values below 10 mA/m, with a geometrical mean of 3–4 mA/m, the Vera Member shows a spectrum with characteristic values reaching up to 30 mA/m. The Gran Barranca Member shows very high NRM intensities below the Y Guide Level, in all profiles where it is sampled (MMZ, K, M, and N). Enhancement of NRM could be explained as due to the formation of new

magnetic minerals during the period of exposure of Discontinuity surface 1.

Most of the samples show only one characteristic component of remanence, which in some cases is accompanied by a softer component easily erased after the first steps of demagnetization. Alternating field (AF) and thermal demagnetization procedures allowed classifying samples as follows:

(1) Type A. Magnetic remanence was erased after applying AF peaks higher than 50 mT, usually around 100 mT, with a geometrical mean of 60 mT. Thermal demagnetization of these samples showed unblocking temperatures of 580 °C, with no changes in magnetic susceptibility during the procedure, pointing to (pure) magnetite as the magnetic mineral. There are 295 specimens (~40% of the processed samples) that belong to this group (Fig. 3.3a, b).

(2) Type B. Magnetic remanence showed coercive forces lower than Type A, as AF procedures eliminated the NRM after peaks of 30–50 mT (geometrical mean 35 mT). Thermal demagnetization produced unblocking temperatures varying from 300 to 450 °C (up to

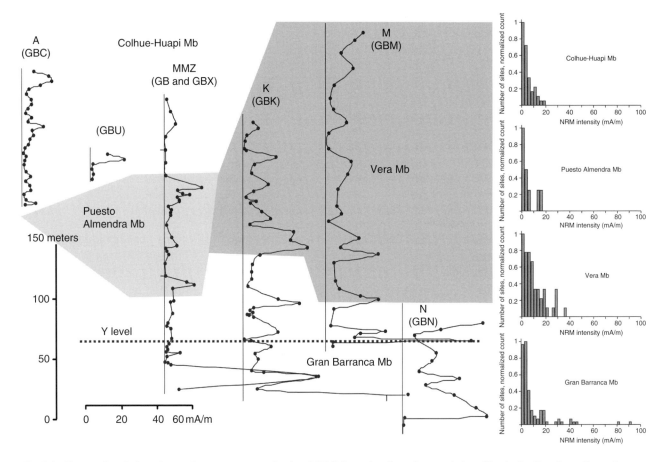

Fig. 3.2. Curves of variation of natural remanent magnetization (NRM) intensity along the sampled profiles in the Sarmiento Formation. To the right, normalized frequency count of NRM intensity for each of the sampled members.

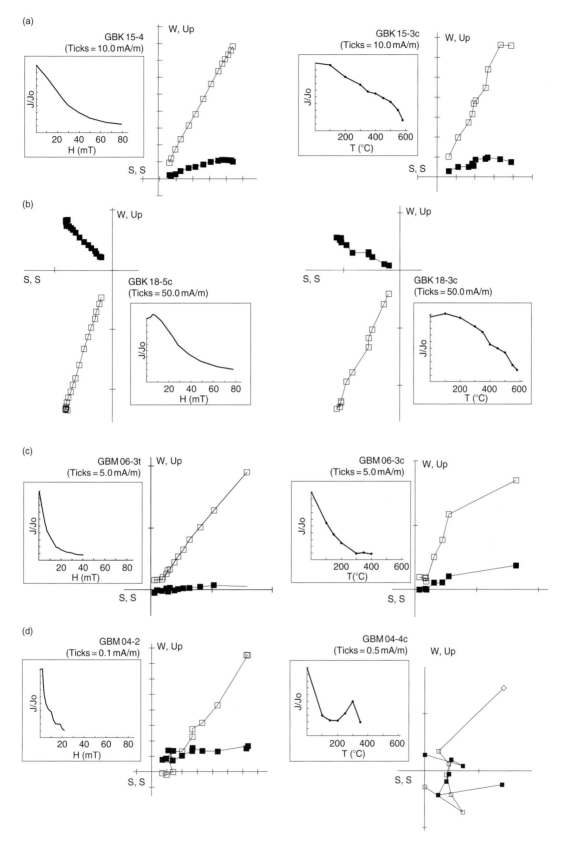

Fig. 3.3. Typical magnetic behavior of Sarmiento Formation samples, illustrated in Zijderveld orthogonal plots and demagnetization curves. The left and right sides show the response to AF and thermal demagnetization, respectively. Open (solid) symbols: vertical (horizontal) projection. (a) Type A behavior, normally magnetized sample; (b) Type A behavior, reversely magnetized sample; (c) Type B behavior; (d) Type C behavior.

550 °C), with no significant changes in the magnetic susceptibility during the procedure (Fig. 3.3c). Unblocking temperatures suggest that (Ti) magnetite is the magnetic carrier in these samples. This includes 255 specimens (~35%).

(3) Type C. Samples of this group became unstable with applied AF peaks lower than 30 mT, and temperatures at around 200 °C. This behavior characterizes 116 specimens (~25%), which were discarded from further analysis (Fig. 3.3d).

Many of the sites contain specimens showing all three kinds of behavior (Table 3.1), indicating that the different behaviors respond to local variation in the nature of the magnetic minerals (including Ti content, degree of oxidation, and effective grain size).

Type A behavior predominates in Colhue-Huapi, Lower and Upper Puesto Almendra, and Gran Barranca Members; sites with enhanced NRM in the Gran Barranca Member also showed this behavior. Conversely, the loessic Vera Member showed lower coercivities, and Type B behavior was more frequently observed (Table 3.1), associated in this case with higher NRM as described before.

As Type A behavior shows the highest coercive force for the magnetic minerals, these specimens are considered the most reliable recorders of the past magnetic field. Consequently, only Type A specimens were used to calculate site mean directions, termed "Class A." For sites where Type A specimens were absent, the site mean direction was calculated using Type B specimens, and termed "Class B."

Site mean directions were grouped by member (see Table 3.1). The only three sampling sites recovered from the Lower Puesto Almendra Member were considered insufficient to provide an independent paleomagnetic pole, and therefore they were combined with sites in the Vera Member, as they represent consecutive time periods.

The four sampled members show site mean directions of positive and negative inclination (i.e. reversed and normal polarity). For every member, Class A paleomagnetic directions of positive and negative inclinations were compared in a reversal test (McFadden and McElhinny 1990). Gran Barranca, Upper and Lower Puesto Almendra, and Colhue-Huapi Members had positive (Class B, C, and B tests, respectively) tests, while Vera Member had an indeterminate result (Fig. 3.4). Four grand mean directions were calculated by combining site directions from each of the sampled members.

Virtual geomagnetic poles (VGP) were calculated from site mean directions. The paleomagnetic pole for each member was calculated as the mean of Class A sites VGPs, with a cutoff angle established by the method of Vandamme (1994). VGPs showing an angular deviation to the mean pole greater than the cutoff were considered oblique

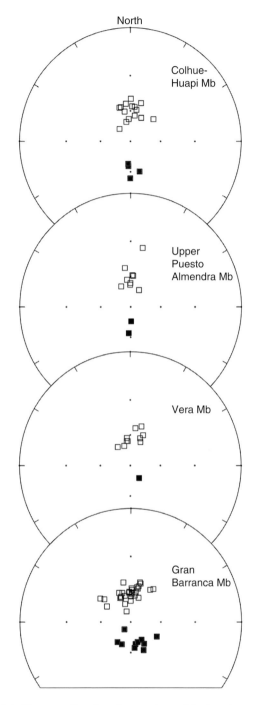

Fig. 3.4. Site mean directions for members of the Sarmiento Formation in stereographic projection. Open/closed symbols: negative/positive inclination.

(see Table 3.1) and were not further considered. The remaining VGPs were averaged to compute the mean paleomagnetic poles (Table 3.2).

We used samples from both Classes A and B for magnetic polarity zonation. Figure 3.1 and 3.5 present the magnetic zones established on this basis for the sampled profiles.

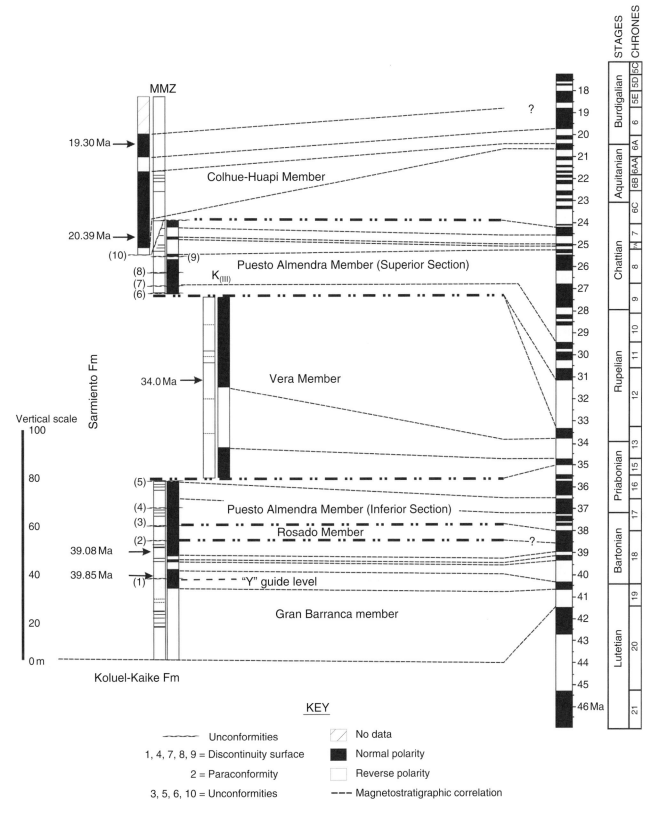

Fig. 3.5. Composite column for Sarmiento Formation in Gran Barranca, with the magnetostratigraphy and its correlation with the geomagnetic polarity timescale of Gradstein *et al.* (2004).

Table 3.2. *Paleomagnetic poles for the Sarmiento Formation at Gran Barranca (45° 42.8′ S, 68° 44.3′ W)*

| Unit | Age (Ma) | N | Site mean direction | | | Paleomagnetic pole | | | | | |
			Dec.	Inc.	α_{95}	Lat. S	Long. E	A_{95}	K	Plat	S
Colhue-Huapi Member	18–20.5	20	3.2	−60.2	5.1	85.8	138.2	6.9	23.6	−41.1	16.7
Upper Puesto Almendra Member	24–30	12	5.0	−63.0	7.5	86.7	205.8	9.6	21.3	−44.5	17.6
Vera Member[a]	33–37	11	2.2	−64.3	6.5	88.6	262.0	9.6	23.4	−46.1	16.7
Gran Barranca Member	37.5–41.5	41	354.1	−63.7	3.8	85.8	14.7	5.4	17.8	−45.4	19.2

Notes: N: number of sites.
α_{95}, A_{95}: semiangle of the 95% confidence cone.
K: Fisher statistical parameter.
Plat: paleolatitude.
S: angular dispersion.
[a]Includes three sites from Lower Puesto Almendra Member.

Discussion

Magnetostratigraphy

The magnetic polarity columns were correlated with the geomagnetic polarity timescale (GPTS) of Gradstein *et al.* (2004). The radiometric age determinations (Ré *et al.* Chapter 4, this book) and the discontinuity surfaces described by Bellosi (Chapter 2, this book) were taken into account.

Figures 3.1 and 3.5 present the correlation of local magnetic polarity stratigraphies for the Sarmiento Formation between profiles and to the GPTS of Gradstein *et al.* (2004). This proposed correlation constrains the age of Sarmiento Formation to the interval 18.8 to 41.7 Ma, with the following ages for the constituent members: Colhue-Huapi Member, ≈18.8 to 20.7 Ma; Upper Puesto Almendra Member, 24.2 to 30.6 Ma; Vera Member, 33.3 to 35.0 Ma; Lower Puesto Almendra (LPA) Member, 36.6 to 37.3 Ma; Rosado Member, 38.0 to ≈38.4 Ma; and Gran Barranca Member, ≈38.4 to 41.7 Ma.

As already noted, the Vera Member is absent at Profile MMZ where it is replaced by an important discontinuity (D6) separating the Puesto Almendra Member into Upper and Lower parts. This interpretation could be better constrained by new data about the regional extent of discontinuities and new radiometric ages.

Paleomagnetic poles

Paleomagnetic poles obtained from Gran Barranca (GB), Upper Puesto Almendra (UPA), Vera (VE), and Colhue-Huapi (CH) Members are virtually coincident within the uncertainty limits (Table 3.2, Fig. 3.6). However, a slight difference can be envisaged among them which denotes a northward displacement for South America during the time elapsed. The GB pole denotes a northward and

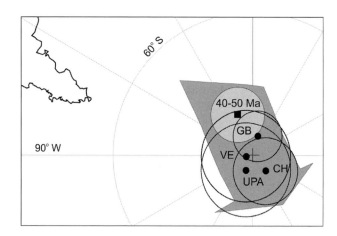

Fig. 3.6. Paleomagnetic poles for members of the Sarmiento Formation, relative to the present position of southern South America. The South American Eocene pole calculated by Somoza (2005) is shown for comparison. 95% confidence cones are depicted. The arrow shows the constant change of the paleomagnetic pole position from ∼50 to ∼18 Ma ago, which reflects the counterclockwise rotation and the northwards displacement of South America during this time.

counterclockwise displacement with respect to Eocene (45 Ma) reference poles for South America (Somoza 2005). The constant northward movement detected by paleomagnetic poles is consistent with the regional tectonic scenario of Somoza (2005). Paleolatitude variation from GB to CH would be approximately 3° as deduced from the observed change in magnetic inclination.

Conclusions

A local magnetostratigraphy was obtained and tentatively correlated to the GPTS based on known depositional discontinuities and available radiometric dates for the

Sarmiento Formation showing that deposition occurred at Gran Barranca between about 18.8 and 41.7 Ma with important discontinuities.

The paleomagnetic poles obtained for the members of Sarmiento Formation describe the expected northward movement of the South American plate during the Cenozoic, with a small component of counterclockwise rotation.

REFERENCES

Ameghino, F. 1906. Les formations sédimentaires du Crétacé superieur et du Tertiaire de Patagonie avec un paralléle entre leurs faunes mammalogiques et celles de l'ancien continent. *Anales del Museo Nacional de Historia Natural*, **15**, 1–568. Buenos Aires.

Andreis, R. R., M. Mazzoni, and L. Spalletti 1975. Estudio estratigráfico y paleoambiental de las sedimentitas terciarias entre Pico Salamanca y Bahía Bustamante, provincia de Chubut, República Argentina. *Revista de la Asociación Geológica Argentina*, **30**(1), 85–103.

Bellosi, E. S. and R. Madden 2005. Estratigrafía física preliminar de las secuencias piroclásticas terrestres de la Formación Sarmiento (Eoceno–Mioceno) en la Gran Barranca, Chubut. *Actas XVI Congreso Geológico Argentino*, **4**, 427–432.

Bellosi, E. S., M. González, R. Kay, and R. Madden 2002a. El valle inciso colhuehuapense de Patagonia central (Mioceno inferior). *IX Reunión Argentina de Sedimentología, Abstracts*, **49**.

Bellosi, E., M. González, R. Kay, and R. Madden 2002b. Un paleosuelo mustersense con microgastrópodos terrestres (Charopidae) de la Formación Sarmiento, Eoceno de Patagonia central: significado paleoclimático. *Ameghiniana*, **39**, 465–477.

Cifelli, R. L. 1985. Biostratigraphy of the Casamayoran, Early Eocene, of Patagonia. *American Museum of Natural History Novitates*, **2820**, 1–26.

Feruglio, E. 1929. Apuntes sobre la constitución geológica de la región del Golfo de San Jorge. *Boletín de Informaciones Petroleras*, **63**, 925–1025. Buenos Aires.

Feruglio, E. 1949. *Descripción Geológica de la Patagonia*, 3 vols. Buenos Aires: Dirección General de Yacimientos Petrolíferos Fiscales.

Fisher, R. A. 1953. Dispersion on a sphere. *Proceedings of the Royal Society of London A*, **217**, 295–305.

Genise, J., E. Bellosi, and M. González 2004. An approach to the description and interpretation of ichnofabrics in palaeosols. In D. McIlroy (ed.), *The Application of Ichnology to Paleonvironmental and Stratigraphic Analysis*, Special Publication no. 228. London: Geological Society of London, pp. 355–382.

Gradstein, F. M., J. Ogg, and A. Smith (eds.) 2004. *A Geological Time Scale 2004*. Cambridge, UK: Cambridge University Press.

Kay, R., R. Madden, M. Mazzoni, and G. Ré 1996. Calibraciones de edades mamífero en la Gran Barranca del lago Colhue-Huapi; provincia del Chubut, Argentina. *Simposio "Paleógeno de América del Sur", abstracts*, **14**.

Kay, R. F., R. H. Madden, M. G. Vucetich, A. A. Carlini, M. M. Mazzoni, G. H. Ré, M. Heizler, and H. Sandeman 1999. Revised geochronology of the Casamayoran South American Land Mammal Age: climatic and biotic implications. *Proceedings of the National Academy of Sciences USA*, **96**, 13 235–13 240.

Kirschvink, J. L. 1980. The least squares line and plane and analysis of paleomagnetic data. *Geophysical Journal of the Royal Astronomical Society*, **62**, 699–718.

Lesta, P., R. Ferello, and G. Chebli 1980. Chubut extraandino. *II Simposio de Geología Regional Argentina, Córdoba, Academia Nacional de Ciencias*, **2**, 1307–1387.

Mazzoni, M. M. 1985. La Formación Sarmiento y el vulcanismo paleógeno. *Revista de la Asociación Geológica Argentina*, **40**, 60–68.

McFadden, P. L. and M. W. McElhinny 1990. Classification of the reversal test in palaeomagnetism. *Geophysical Journal International*, **103**, 725–729.

Miquel, S. E. and E. S. Bellosi 2007. Microgasterópodos terrestres (Charopidae) del Eoceno medio de Gran Barranca (Patagonia Central, Argentina): aporte al conocimiento de los cambios ambientales de la transición Eoceno–Oligoceno. *Ameghiniana*, **44**, 121–131.

Oviedo, E. 1989. Mag88: Un sistema de computación para análisis de datos paleomagnéticos. Ph.D. thesis, Facultad de Ciencias Exactas y Naturales, Universidad de Buenos Aires.

Pascual, R. and O. E. Odreman Rivas 1973. Las unidades estratigráficas terciarias portadoras de mamíferos: su distribución y sus relaciones con los acontecimientos diastróficos. *Actas V Congreso Geológico Argentino*, **3**, 293–338.

Rapela, C., L. A. Spalletti, and J. C. Merodio 1983. Evolución magmática y geotectónica de la "Serie Andesítica" andina (Paleoceno–Eoceno) de la Cordillera Patagónica. *Revista de la Asociación Geológica Argentina*, **38**, 469–484.

Simpson, G. G. 1930. *Scarritt-Patagonia Exped. Field Notes*. New York: American Museum of Natural History (Unpublished.). Available at http://paleo.amnh.org/notebooks/index.html

Simpson, G. G. 1941. The Eogene of Patagonia. *American Museum Novitates* **1120**, 1–15.

Somoza, R. 2005. Polo paleomagnético Eoceno de América del Sur: movimiento hacia el norte en el Cenozoico, apertura del pasaje de Drake y convergencia en el Caribe. *Actas XVI Congreso Geológico Argentino*, **84**.

Spalletti, L. and M. Mazzoni 1977. Sedimentología del Grupo Sarmiento en un perfil ubicado al sudeste del lago Colhue-Huapi, provincia de Chubut. *Obra del Centenario del Museo de La Plata*, **4**, 261–283.

Spalletti, L. A. and M. M. Mazzoni 1979. Estratigrafía de la Formación Sarmiento en la barranca sur del lago Colhue-Huapi, provincia del Chubut. *Revista de la Asociación Geológica Argentina*, **34**, 271–281.

Torsvik, T. H., J. C. Briden, and M. A. Smethurst 2000. *IAPD2000*. hhttp://www.ngu.no/geophysics

Vandamme, D. 1994. A new method to determine paleosecular variation. *Physics of the Earth and Planetary Interiors*, **85**, 131–142.

4 A geochronology for the Sarmiento Formation at Gran Barranca

Guillermo H. Ré, Eduardo S. Bellosi, Matthew Heizler, Juan F. Vilas, Richard H. Madden, Alfredo A. Carlini, Richard F. Kay, and M. Guiomar Vucetich

Abstract

The Sarmiento Formation (middle Eocene – early Miocene) is a terrestrial pyroclastic succession in central Patagonia broadly recognized by its abundant fossil mammals which comprise the standard succession for the South American mammal record. Age calibration of each subdivision of the formation and the internal discontinuities have been controversial and in need of better resolution. Until now, few radiometric dates were available.

Age calibration of the Sarmiento Formation was obtained at the type locality at Gran Barranca in south-central Chubut Province where the most continuous exposures occur. This age calibration integrates information from Ar/Ar dates on plagioclase and volcanic glass mineral separations from ten levels in all six members and five magnetic polarity sequences at different stratigraphic profiles. With this information, ten bounding surfaces with different origins and geometries were correlated along the exposure. The Sarmiento Formation transitionally overlies the Koluel-Kaike Formation. At the top of Gran Barranca is a poorly preserved remnant of the marine Chenque Formation. There are six members of the Sarmiento Formation at Gran Barranca containing seven distinct mammal faunas. The Gran Barranca Member yields Barrancan age fossils and spans the temporal interval from 41.6 to 39.0 Ma or late Lutetian–Bartonian (Chron C19r to C18n.1n). The Rosado Member includes both transitional Barrancan–Mustersan and Mustersan age faunas and spans the interval from about 38.4 to 38.0 Ma or late Bartonian (Chron C18n.1n). The Lower Puesto Almendra Member contains Mustersan age fossils between 37.3 Ma and 36.58 Ma or early Priabonian (Chron C17n). The Vera Member includes Tinguirirican age mammals and spans the interval between 35.0 Ma and 33.3 Ma or Priabonian to early Rupelian (Chron C15n to C13n). Tinguiririan age assemblages within this member are between 33.7 and 33.3 Ma. The Upper Puesto Almendra Member contains a pre-Deseadan age fauna (GBV-19 "La Cantera") and poorly sampled faunules of more typical Deseadan aspect. The member comprises three units and

spans an interval as old as 31.1 Ma to as young as 24.2 Ma or late Rupelian–Chattian (Chron C12n to C7n). GBV-19 is between 31.1 and 29.5 Ma, and the Deseadan assemblages are between 29.2 Ma and 26.3 or younger in age. The Colhue-Huapi Member with Colhuehuapian and Pinturan faunal zones spans the interval between 20.4 Ma and 18.7 Ma or late Aquitanian–Burdigalian (C6An.1r to Chron C6n). The Lower Fossil Zone of the Colhue-Huapi Member containing Colhuehuapian mammals is 20.4 Ma to 20.0 Ma and the Upper Fossil Zone containing Pinturan mammals is between 19.7 and 18.7 Ma.

Primary and secondary pyroclastic sedimentation spans nearly 23 million years in the vicinity of Gran Barranca. During this interval, 8.4 m.y. are not preserved mainly because of significant erosion events and subordinately by non-deposition and soil formation. Temporal hiatuses between members increase from 0.7 m.y. in the middle Eocene to 3.48 m.y. in the early Miocene, probably in response to progressively increasing tectonic and landscape instability.

Resumen

La Formación Sarmiento (Eoceno medio – Mioceno temprano) es una sucesión piroclástica terrestre de Patagonia central, ampliamente reconocida por sus abundantes mamíferos fósiles, que forman la sucesión estándar de los vertebrados cenozoicos sudamericanos. La definición de edades precisas para cada una de sus sub-unidades, así como de posibles discontinuidades y hiatos internos, ha sido un tema discutido y pendiente de resolución desde hace décadas, contándose hasta hoy con muy escasos y restringidos datos de edades absolutas.

Este nuevo ajuste temporal fue elaborado para la localidad tipo de Gran Barranca (centro-sur de la provincia de Chubut), donde se encuentran los afloramientos fosilíferos más extendidos y continuos. Esta calibración geocronológica integró la información procedente de Ar/Ar en plagioclasa y vidrio volcánico de muestras de diez niveles pertenecientes a los seis miembros, cinco perfiles de polaridad magnética y cinco superficies de discontinuidad litoestratigráfica de diferente origen, geometría y jerarquía que fueron correlacionadas a lo largo de la barranca.

La Formación Sarmiento suprayace transicionalmente a la Formación Koluel-Kaike. En el tope de la Gran Barranca hay un remanente pobremente preservado de la Formación

The Paleontology of Gran Barranca: Evolution and Environmental Change through the Middle Cenozoic of Patagonia, eds. R. H. Madden, A. A. Carlini, M. G. Vucetich, and R. F. Kay. Published by Cambridge University Press. © Cambridge University Press 2010.

Chenque, de origen marino. En la localidad de Gran Barranca hay seis miembros de la Formación Sarmiento que contienen siete faunas distintas. El Miembro Gran Barranca contiene fósiles de la Subedad Barranquense y se extiende temporalmente entre los 41.6 y 39.0 Ma; Luteciano tardío – Bartoniano [Crones C19r a C18n.1n]. El Miembro Rosado incluye faunas transicionales Barranquense-Mustersense y Mustersense, y abarca el intervalo de 38.4 a 38.0 Ma; Bartoniano tardío [Cron C18n.1n]. El Miembro Puesto Almendra Inferior contiene fósiles de la SALMA Mustersense y se extiende entre los 37.3 Ma y 36.58; Priaboriano temprano [Chron C17n]. El Miembro Vera contiene mamíferos tinguiririquenses y abarca el intervalo entre 35.0 y 33.3 Ma; Priaboriano – Rupeliano temprano [Crones C15n to C13n]. En este miembro, las localidades tinguiririquenses están entre los 33.7 y 33.3 Ma. El Miembro Puesto Almendra Superior contiene una fauna pre-Deseadense (GBV-19) y fáunulas de aspecto Deseadense más típico, pero pobremente muestreadas. Este Miembro tiene tres unidades y se extiende por un intervalo tan viejo como 31.1 Ma a tan joven como 24.2; Rupeliano tardío – Chatiano [Crones C12n a C7n]. GBV-19 se encuentra entre los 31.1 y 29.5 Ma. Los fósiles deseadenses se encuentran entre los 29.2 y 26.3 Ma, o aún algo más jóvenes. El Miembro Colhue-Huapi, con niveles de faunas del Colhuehuapense y "Pinturense," se extiende por el intervalo entre los 20.4 y 18.7 Ma; Aquitaniano tardío – Burdigaliano [Crones C6An.1r a C6n]. La Zona Fosilífera Inferior, con mamíferos colhuehuapenses, está entre los 20.4 y los 20.0 Ma, mientras que la Zona Fosilífera Superior, con mamíferos del "Pinturense," se encuentra entre los 19.7 y 18.7 Ma.

La sedimentación piroclástica primaria y secundaria se prolongó en la región de Gran Barranca por casi 23 millones de años. Unos 8.4 m.y. de ellos no están registrados, principalmente debido a significativos eventos de erosión y, subordinadamente, por falta de depositación y pedogénesis. Los hiatus temporales entre los distintos miembros se incrementan de 0.7 m.y. en el Eoceno medio a 3.48 m.y. en el Mioceno temprano, probablemente como respuesta a una progresiva inestabilidad tectónica y del paisaje.

Introduction

Gran Barranca exposes the most complete sequence of middle Cenozoic fossil-mammal bearing rocks in South America and the only continuous continental fossil record anywhere in the southern hemisphere through the middle and late Eocene, across the Eocene–Oligocene transition (EOT), and into the early Miocene. The fossil mammal sequence at Gran Barranca is the standard of reference for the middle Cenozoic throughout South America and Western Antarctica. Wherever fossil mammals are collected in South America or Antarctica, comparisons are necessarily made to the standard sequence in Patagonia which in turn relies ultimately on the quality and resolution of the paleontology, stratigraphy, and geochronology at Gran Barranca.

New fossil vertebrate collections are now much larger than ever before available and include new and diverse small mammals collected by both dry and wet screening. The new collections document 49 discrete fossil-mammal levels in the Sarmiento Formation, including levels never before sampled. The fossil content of these levels is described in this book.

This chapter establishes a chronology for the fossil-bearing rocks at Gran Barranca based on radiometric age determinations and magnetic polarity stratigraphy. Until now the geochronology of the Sarmiento Formation at Gran Barranca consisted of only a few $^{40}K/^{40}Ar$ dates on a basalt flow complex high in the section (Marshall *et al.* 1986) and $^{40}Ar/^{39}Ar$ determinations for a single tuff near the base of the section (Kay *et al.* 1999). Although some additional dates for other stratigraphic units were reported without comment by Kay *et al.* (1999), here we report many additional $^{40}Ar/^{39}Ar$ dates and a composite magnetic polarity stratigraphy based on five stratigraphic profiles extending between the west and east end of the continuous exposures at Gran Barranca and build a preliminary geochronologic framework for the Sarmiento Formation at Gran Barranca.

Methodology

During the Scarritt Expeditions of the American Museum of Natural History, George Gaylord Simpson measured and described some of the stratigraphic profiles used here (Simpson 1930; Cifelli 1985). Simpson recorded the position of many of the fossil mammal localities or levels where we have collected. Our geochronology is presented in the context of the physical stratigraphy of the Sarmiento Formation of Bellosi (this book) with reference to the profiles of Simpson and others at Gran Barranca. The new stratigraphy identifies ten temporally significant discontinuities (paleosurfaces, paraconformities, and erosional unconformities), six members and a number of lithostratigraphic subunits in the Sarmiento Formation (Bellosi Chapter 2, this book). The geochronology is organized and presented by lithostratigraphic unit, begininning with the oldest Gran Barranca Member at the base of the Sarmiento Formation and ending with the Colhue-Huapi Member at the top.

Due to their fine-grained texture, bulk plagioclase and glass mineral separations were analyzed using step-heating age spectrum methods. The glass samples were degassed in 8–12 heating steps in a double vacuum resistance furnace. Approximately 3–5 mg splits of plagioclase were analyzed in the laser furnace in 6–12 heating steps. Most step-heated plagioclase samples yielded plateau age spectra for 100% of the ^{39}Ar released. In addition, some plagioclase samples were dated using the total laser fusion method.

Table 4.1. *Preferred ages tuffs for both glass and plagioclase of the Sarmiento Formation*

Tuff Name and Sample	Best Glass Age (error=1 sigma)	Best Plagioclase Age (error=1 sigma)	Tuff Mean Age (N)
Vilas & Re Silicified (VRS) Tuff			
GB 01-9		41.70 (0.38)	
Simpson Y Tuff			39.85 (3)
GB 01-5 (Profile N1)	39.45 (0.24)	40.24 (0.24)	
GB 99-7 (Profile MMZ)		39.87 (0.20)	
Mazzoni Tuff			39.08 (2)
GB 99-1	37.96 (0.34)	40.19 (0.58)	
Rosado Tuff			38.66 (3)
GB 99-4	38.55 (0.11)	39.40 (0.24)	
GB 99-6	38.03 (0.07)		
Kay Tuff			37.045 (2)
GB 99-3	37.21 (0.10)	36.88 (0.19)	
Carlini Tuff			33.995 (2)[*]
GB 02-01	30.15 (0.17)	34.08 (0.25)	
GB 02-02	28.30 (0.23)	31.86 (0.51)	
GB 99-8	33.91 (0.05)		
Big Mammal Tuff			19.75 (3)
RFK-1 (8953-01)	19.13 (0.17)	20.26 (0.70)	
RFK-1 (8953)		19.87 (0.33)	
Basal Colhue-Huapi Tuff			20.39 (3)
GB 01-8 (53459-01)	19.37 (0.08)	21.09 (0.30)	
GB 01-8 (53460)		20.73 (0.26)	
Monkey Tuff			19.81 (5)
RFK-2 (8954)	19.13 (0.16)	19.94 (0.28)	
RFK-2 (8954)		19.54 (0.20)	
GB 99-5	19.12 (0.05)	21.30 (0.28)	
MMZ 24.5 Tuff			19.295 (2)
GB 01-7	18.53 (0.24)	20.06 (0.12)	

Note: *Mean of concordant glass plateau and plagioclase laser fusion ages.

The results of our ^{40}Ar/^{39}Ar dating analyses include step-heating results for glass and plagioclase, laser fusion data for plagioclase are presented in Table 4.1. Laboratory results for the ^{40}Ar/^{39}Ar dates are available directly from MH at the Geochronology Research Laboratory of the New Mexico Bureau of Geology and Mineral Resources. The age interpretations made here depend on the integration of the physical stratigraphy, the ^{40}Ar/^{39}Ar dates and the magnetic polarity stratigraphy. Technical information about magnetic polarity sampling and the results of laboratory analyses are presented in a separate chapter (Ré *et al.* Chapter 3, this book). As reversed polarity sites are relatively rare in the sampled profiles, we critically evaluate the quality of the laboratory results for all reversed polarity sites prior to interpreting the age of the sediments. We organize the

discussion of the geochronology of the Sarmiento Formation with reference to lithostratigraphy rather than fossil content, although noteworthy fossil mammal levels and zones are mentioned where appropriate.

General stratigraphy

Gran Barranca south of the Lake Colhue-Huapi is a continuous escarpment extending some 7 km between two basalt flow complexes, the west and east end basalts. The maximum height of the escarpment is about 200 meters at Profile MMZ near the western end and varies down to about 80 meters at its lowest point between Profiles J and K. The composite stratigraphy of the Sarmiento Formation along this escarpment is based on a series of profiles (Bellosi

Chapter 2, this book) from Profile A at the western end to Profile N at the eastern end. Additional exposures of the Sarmiento Formation occur in smaller isolated areas farther to the west (Profile AAC) and east (Las Flores) of the main escarpment. The stratotype section of the Sarmiento Formation is Profile MMZ near the west end of Gran Barranca (Spalletti and Mazzoni 1979).

In general, the Sarmiento Formation comprises fine-grained primary and reworked volcaniclastic sediments, including pyroclastic mudstones and siltstones, volcaniclastic sandstones, fine tuffs, and bentonitic claystones (Spalletti and Mazzoni 1979; Bellosi this book). These tephric materials originated as distal Plinian or Sub-Plinian pyroclastic ash fall deposits from magmatic arcs in northwestern Patagonia (Mazzoni 1985, Bellosi this book). Several basalt flows intercalate within this pyroclastic succession. Coarser-grained sediments do not occur at Gran Barranca, except as intraformational conglomerates with clasts mostly formed from autochthonous soil fragments and basalt breccias.

The mineral composition of Sarmiento Formation sediments at Gran Barranca is described in detail only for Profile MMZ (Spalletti and Mazzoni 1979). Simpson (1930) described the sediments observed in many additional profiles at Gran Barranca. Cifelli (1985) published new descriptions of the lithostratigraphy for the lower portions of Simpson's profiles, based on Simpson's observations. Simpson measured 11 partial or complete profiles along the continuous escarpment, his A-1, A-2, A-3, G, H, J, K, LB, M, N-1, and N-2 and three additional sections on isolated hills just north of the escarpment (Profiles B, I, and LA). Mazzoni and Spalletti measured two sections at the west end of Gran Barranca, a section described in 1977 (Spalletti and Mazzoni 1977) and a more complete and detailed section that we refer to as Profile MMZ (Spalletti and Mazzoni 1979). We extend measurement of Simpson's Profiles K, L, and M upwards to the top of the escarpment, and continue measurement of Profiles G and H downwards to the base of the exposed section. During the course of magnetostratigraphic work, we remeasured Profile MMZ and Simpson's Profiles A3, K, M, and N-1. Approximate GPS coordinates for the base and top of the profiles and their subdivisions is provided in Table 4.2.

$^{40}Ar/^{39}Ar$ geochronology

The basalts in the Sarmiento Formation at Gran Barranca and elsewhere in Patagonia (Windhausen 1924; Feruglio 1949) that have a direct relationship to fossil mammal levels were the first to be dated by isotopic methods (Marshall *et al.* 1977). While basalts are resistant to erosion and are fairly widespread throughout Patagonia, the episodic nature of basalt production, mostly during a short interval in the Oligocene (Bellosi 1995) has limited their application to the geochronology of mammal-bearing units. The West End Basalt Complex (WEBC) comprises several stacked basalt flows that outcrop immediately west of Profile A. The eastern edge of the WEBC is clearly within the Upper Puesto Almendra Member, a stratigraphic relationship similar to the basalts in Profiles A-2 and H. The East End Basalt at the top of Profile N appears to be a single thick flow, although sampled for magnetic polarity at several different levels.

Table 4.2. *Approximate GPS coordinates for the base and top of profiles of the exposed Sarmiento Formation at Gran Barranca (in decimal degrees South latitude; West longitude)*

Profile	Base	Top
A (Simpson)	45.70588; 68.73935	45.71083; 68.74260
A-1 (Simpson, Colhue-Huapi Mbr)	45.71065; 68.74418	45.71083; 68.74260
G (Simpson)	uncertain	45.71105; 68.74142
I (Simpson)	45.70674; 68.72939	45.70942; 68.73108
MMZ (Mazzoni)	45.70998; 68.73487	45.71024; 68.74400
H (Simpson)	45.71315; 68.73658	45.71358; 68.73788
J (Cifelli 1985 Section II)	45.71430; 68.72855	45.71683; 68.72970
K (Cifelli 1985 Section III)	45.71255; 68.69875	45.71440; 68.69635
LA (Simpson)	45.71953; 68.69304	45.71055; 68.69329
LB (Simpson)	45.72102; 68.69692	45.72000; 68.70110
M (Cifelli 1985 Section V)	45.72242; 68.67763	uncertain
N1 (Simpson)	45.72721; 68.65905	45.72862; 68.65740
N2 (Simpson)	45.72107; 68.65443	45.72363; 68.65593
Las Flores	45.72600; 68.62116	45.72635; 68.63137

With the advent of ^{40}Ar/^{39}Ar dating of airborne tephra, additional levels of the Sarmiento Formation have become suitable for radioisotopic age determination. Tuffaceous sediments of the Sarmiento Formation are widespread in Patagonia (Mazzoni 1985; Franchi and Nullo 1986) and extend geographically and temporally beyond the limits of available basalts. The tuffaceous sediments of the Sarmiento Formation at Gran Barranca have a composition consistent with silicic subduction zone volcanic rocks and a presumptive source area in Neuquen and Rio Negro Provinces (Mazzoni 1979, 1985; Franchi and Nullo 1986).

Despite some complexities in the argon results, both glass and plagioclase provide useful constraints for the interpretation of the local magnetic polarity stratigraphy, the age of members and subunits of the Sarmiento Formation, and the duration of temporal hiatuses identified in the lithostratigraphy. The preferred results of ^{40}Ar/^{39}Ar analyses on Sarmiento Formation tuffs (Table 4.1) are concordant with stratigraphic superposition. Discrepancies between plagioclase and glass ages are found for most of the tuffs in the Sarmiento Formation, with dates for argon consistently younger than those for plagioclase. We attribute part of this discrepancy to argon loss. In some cases, age discrepancies or polymodalities are found in the same mineral and between samples of the same tuff. For example, polymodality is suggested by the plagioclase results for the Monkey Tuff and the Rosado Tuff, and in glass for the Carlini Tuff, the Rosado Tuff, and Simpson's Y Tuff. Thus, sediment reworking and xenocrystic mixing are indicated. Relatively few of the dated tephra are primary air-fall deposits. Most of them show evidence of secondary reworking, such as variable thickness, discontinuity, and inclusion of fragmented and entire clasts, and inclusion of fossils. In general, with the exceptions of the Simpson's Y Tuff, Mazzoni Tuff, and Kay Tuff in the Gran Barranca Member and the MMZ 24.5 Tuff in the Colhue-Huapi Member, all other dated samples at Gran Barranca are from laterally discontinuous local concentrations of tuffaceous material.

The Gran Barranca Member

The Gran Barranca Member is the most widely distributed lithologic unit of the Sarmiento Formation at Gran Barranca. The member is composed of thick, white, massively bedded pyroclastic mudstones, olive-green bentonites, and poorly developed paleosols. The basal contact of the Gran Barranca Member on the Kolhue-Kaike Formation is obscured by the incoherence of the lowermost strata and the low topographic relief of the flatland extending northward from the foot of Gran Barranca. At the Las Flores locality, a distinct exposure 1 km to the east of Profile N-2,

the basal contact is well exposed and transitional (Bellosi Chapter 2, this book). Within this member, Simpson identified a richly fossiliferous tuff called "Bed Y," which is the only continuous marker bed extending unbroken the full length of Gran Barranca. Bed Y rests on Discontinuity 1 (D1). The upper boundary of the Gran Barranca Member is defined by Discontinuity 2. In addition to highly fossiliferous Bed Y, this member contains several other levels with fossil mammals both above and below D1. These assemblages are relatively uniform in composition and form the basis for the recognition of the Barrancan subage of the Casamayoran South American Land Mammal Age (SALMA) (Cifelli 1985). Three pyroclastic mudstones within the Gran Barranca Member have been dated and the ^{40}Ar/^{39}Ar results are discussed here, beginning with the stratigraphically lowest dated level.

VRS or Vilas and Ré Silicified Tuff The lowest pyroclastic mudstone sampled and analyzed from the Sarmiento Formation at Gran Barranca is the VRS Tuff near the base of Profile N-1. The VRS Tuff is below D1 and just overlies the lowest fossil-bearing level at Gran Barranca. A plagioclase laser fusion mean age of 41.70 ± 0.38 Ma was obtained from the VRS Tuff (Table 4.1).

Simpson's Y Tuff Simpson's Y Tuff (also Y Tuff or Bed Y) immediately above D1 has been dated at two different locations along Gran Barranca, at Profile MMZ (sample GB 99–7) and at Profile N-1 (sample GB 01–5). There is close agreement between the plagioclase plateau age (39.87 Ma) and laser fusion mean age (40.24 Ma). Glass results are inconsistent between samples and are younger than the plagioclase results in both samples, suggesting argon loss. The glass plateau age of sample GB 99–7 is much younger than the glass plateau age for GB 01–5. This anomalous result is discounted in the computation of an arithmetic mean age for the Y Tuff. The mean age of 39.85 Ma for Simpson's Y Tuff uses the three preferred results (the GB 01–5 glass plateau and plagioclase laser fusion mean ages plus the GB 99–7 plagioclase plateau age (Table 4.1)).

Mazzoni Tuff The Mazzoni Tuff is a thin dark pyroclastic mudstone just above Simpson's Y Tuff on Profile A-3 and is the "prominent thin hard steely-grey band" of Simpson (1930). A total gas age of 37.7 Ma for glass and a furnace argon total gas age for plagioclase of 37.9 Ma were reported by Kay *et al.* (1999) on samples collected from Profile MMZ. Another sample of the Mazzoni Tuff (GB 99–1) yielded a glass plateau age of 37.96 Ma and a plagioclase plateau age of 40.19 Ma. The arithmetic mean age of the Mazzoni Tuff based on this sample is 39.08 Ma.

The age of the Gran Barranca Member

Magnetic polarity stratigraphies are available for the Gran Barranca Member at four profiles (MMZ, K, M, and N-1).

Of these, the most densely sampled (Profile MMZ) is used to establish the upper age boundary of the Gran Barranca Member. Profile N-1 magnetostratigraphy extends down to the contact with the underlying Koluel-Kaike Formation and is used to establish the age of the base of the Gran Barranca Member, and thus, the age of the base of the Sarmiento Formation at Gran Barranca.

Profile MMZ local normal interval N2, represented by single site (Site X12) is sandwiched between two narrow reversed polarity intervals R2 (Site 06) and R3 (Site 07) likewise supported by single sites. We have some reservation about the behavior of the five cores drilled at Profile MMZ Site 6. Profile MMZ local zone R2 (Site 06) was sampled in Simpson's Y Tuff, a horizon that elsewhere yields only normal polarity (the Y Tuff at Profile K yielded normal polarities for five sites sampled at less than 1 meter intervals). For the age interpretation, we ignore this local reversed polarity zone. Profile K polarity stratigraphy for the Gran Barranca Member yields a complex local polarity sequence. Of these local polarity zones, only R2 is supported by more than one site. Profile K local polarity zones N1, R1, and N2 are supported by single sites, but we are confident about these results. There are no obvious discontinuities and few sharp contacts in the lower part of the Gran Barranca Member at Profile K.

At Profile M magnetic polarity stratigraphy for the member is uniformly normal polarity except for a single site (GBM-03–06 or MI06) stratigraphically just above the Y Tuff. This site is at a stratigraphic level equivalent to Site 7 at Profile MMZ which also yielded reversed polarity, and thus this reverse polarity zone is recognized in the age interpretation. Profile N-1 was sampled at rather coarse density and the analyses for the lower sites reveals a long reverse polarity interval supported by multiple levels.

Judging from the $^{40}Ar/^{39}Ar$ results for the three tuffs and the local magnetic polarity stratigraphy, Chrons 19, 18, and 17 of the GMPTS (Gradstein *et al.* 2004) are the most likely correlatives of the Gran Barranca Member. The simplest interpretation of the age of the Gran Barranca Member begins by noting that at Profile N-1 the VRS Tuff (41.70 Ma) falls in a zone of reversed polarity that we conclude corresponds to Chron 19r calibrated to between 40.671 and 41.590 Ma (Gradstein *et al.* 2004). Simpson's Y Tuff (mean age 39.85 Ma) and the Mazzoni Tuff (mean age 39.08 Ma) both fall in local zones of reversed polarity in Profile MMZ. This suggests that Simpson's Y Tuff identifies Chron C18r at Profile MMZ, and the Mazzoni Tuff may identify Chron C18n.1r. However, Simpson's Y Tuff has been sampled for magnetic polarity stratigraphy at Profiles K, M, and N-1, and in all three profiles, it has normal polarity. This normal polarity is especially significant at Profile K where eight sample sites have been analyzed

(K03, K03A through K03G). In all profiles sampled for magnetostratigraphy, all sites above the Mazzoni Tuff yield normal polarity. Assuming continuous deposition, this normal polarity zone represents Chron C18n.1n and indicates that the age of the top of the Gran Barranca Member is between 38.032 and 38.975 Ma.

In summary of the above, the base of the Gran Barranca Member is within Chron 19r (40.671–41.590 Ma) and its top within Chron C18n.1n (38.032–38.975 Ma). Cifelli's (1985) Barrancan fauna from localities within the Gran Barranca Member falls within the interval between about 41.6 and 38.7 Ma, the age of the Rosado Tuff (see below).

The Rosado and Lower Puesto Almendra Members

The Rosado Member is a 7-m thick pinkish tuffaceous calcic paleosol that outcrops discontinuously between Profiles J and M. Its base is Discontinuity 2, which separates it from the underlying Gran Barranca Member, and its top is the irregular Discontinuity 3. By contrast, the Lower Puesto Almendra Member is a 30-m thick succession of volcaniclastic sandstones, the "Lower Channel Beds" of Simpson (1930), pyroclastic mudstones, bentonites, and paleosols (Bellosi this book) widely exposed west of Profile J and bounded by Discontinuity 3 and Discontinuity 5. One tuff within the Rosado Member and another within the Lower Puesto Almendra Member have been dated.

Rosado Tuff The Rosado Tuff occurs within the Rosado Member just west of the axis of Profile J along a badly eroded and unserviceable road that ascends Gran Barranca. The Rosado Tuff (sample GB 99–4) glass plateau age is somewhat younger than the plagioclase result, an age disparity consistent with argon loss typically observed in Sarmiento Formation tuffs. Sample GB 99–6 yielded a plateau age for glass of 38.03 Ma, close but not within the 2-sigma margin of error of GB 99–4 glass. The arithmetic mean of the $^{40}Ar/^{39}Ar$ results is 38.66 Ma, the age of the Rosado Tuff (Table 4.1).

Kay Tuff The Kay Tuff occurs within the Lower Puesto Almendra Member. It is a thick conspicuous whitish tuff outcropping discontinuously between Profiles A and J. The Kay Tuff rests directly on the surface of a distinct stratum known as Mazzoni Bed 10 in the upper part of the Lower Puesto Almendra Member. In Simpson's Profile A-2, the Kay Tuff corresponds to the "massive grey tuff" (Simpson 1930) about 2 m above a channel bed and Discontinuity 4. The dated sample (GB 99–3) was collected from an exposure about 75 meters east of the axis of Profile A-2 and yielded nearly concordant glass and plagioclase plateau ages of 37.21 and 36.88 Ma, respectively. The arithmetic mean age of the Kay Tuff is 37.05 Ma (Table 4.1).

The age of the Rosado and Lower Puesto Almendra Members

Magnetic polarity stratigraphies are available separately for the Rosado and Lower Puesto Almendra Members. (Ré *et al.* [Chapter 3, this book] report not sampling the Rosado Member at Profile J but they did sample it in Profile M.) All sampled sites in both members have normal polarity. The age of the Rosado Tuff falls within the calibrated age range of Chron C18n.1n from 38.032 to 38.975 Ma (Gradstein *et al.* 2004). Therefore we hypothesize that the Rosado Member represents a portion of the upper part of Chron C18n.1n. The upper Gran Barranca Member represents the lower portion of the same polarity chron. Discontinuity 2 at the contact between the Gran Barranca and Rosado members had little temporal duration. The age of the Kay Tuff (37.05 Ma) falls within C17n.1n from 36.512 to 37.235 Ma (Gradstein *et al.* 2004). Discontinuity 3 at the top of the Rosado Member represents a temporal hiatus related to pedogenesis of the Rosado palaosol and subsequent fluvial erosion. This surface encompasses at least the interval between Chron C18n.1n and Chron C17n.1n, that is, from 37.235 to 38.032 Ma. An estimate of the duration of the temporal hiatus represented by Discontinuity 6 at the top of the Lower Puesto Almendra Member can only be established by considering the age of the overlying Vera Member (see below).

The Rosado and Lower Puesto Almendra members include all levels containing fossils assigned to the Mustersan SALMA. Notably, another locality in the Rosado Member, GBV-60 "El Nuevo," has a distinct faunal composition *intermediate* between temporally adjacent Barrancan and Musteran levels. Chron 18n.1n is a roughly 1-million-year interval. The lower part of this chron is represented by a Barrancan fauna in the Gran Barranca Member. The upper part of the same chron is represented by the Rosado Member, which in its lower parts contains a transitional Barrancan–Mustersan fauna (GBV-60 "El Nuevo"), while the upper part contains a Mustersan fauna (GBV-3 "El Rosado"). We tentatively conclude that the Mustersan SALMA at Gran Barranca must begin towards the end of Chron C18n.1n, slightly older than 38.0 Ma, and can be no younger than the top of C17n.1n at 36.5 Ma.

The Vera Member

The Vera Member is a thick, homogeneous, and poorly stratified succession of pyroclastic mudstones exposed between Profile M and Profile J. This distinctive member is bounded at its base by Discontinuity 5 and at its top by Discontinuity 6. There are no obvious discontinuities within this member. The absence of the Vera Member at Profile MMZ and the obscure contacts at Profile J engender some difficulty interpreting its stratigraphic relationships. The

Vera Member includes two localities at the same stratigraphic level yielding fossil mammals assigned to the Tinguiririean SALMA. These include the faunal assemblage from GBV-4 "La Cancha" and the mammals from GBV-28, a screenwash site (Carlini *et al.* this book; Reguero and Prevosti this book; Goin *et al.* this book).

The only dated rock in the Vera Member is the Carlini Tuff, a pyroclastic mudstone located in the middle part of the member at Profile K. The age of the Vera Member is constrained by $^{40}Ar/^{39}Ar$ dates from this tuff (Table 4.1). Samples of the Carlini Tuff at Profile K come from 1–3 m below the fossil level GBV-4. The "La Cancha" bed (including the dated sample and the fossiliferous bed) is a consolidated pale to pinkish tuffaceous mudstone rich in carbonate and manganese nodules. Numerous fragmentary fossil mammal remains have a characteristic crust or patina of manganese oxide. The Carlini Tuff provides a maximum age for the fossil mammal assemblage at GBV-4 and an age for the base of the local N4 normal polarity interval in Profile K (see below).

Our interpretation of the age of the Carlini Tuff is based on glass ages from samples GB 99–8, GB 02–01, and GB 02–02, and plagioclase results from samples GB 02–01 and GB 02–02. Glass plateau ages range between 28.30 and 33.91 Ma, with no overlap at 2-sigma error. Plagioclase laser fusion results are consistently older than the glass results but show a slightly more constrained age range. The arithmetic mean of the $^{40}Ar/^{39}Ar$ results is 31.66 Ma. The multimodal ages of the $^{40}Ar/^{39}Ar$ results suggest the Carlini Tuff preserves minerals from distinct eruption events, but the age disparities in the $^{40}Ar/^{39}Ar$ results seem inconsistent with the form of the Carlini Tuff in the field, where it appears as a discrete horizon, although of limited lateral extent, and nodular in part suggesting local reworking. The only concordance among the $^{40}Ar/^{39}Ar$ results is found between the best glass plateau age (33.91 Ma) and the plagioclase laser fusions mean age (34.08 Ma) giving a mean age of 34.0 Ma (Table 4.1). An improved constraint on the age of this tuff is suggested by the uniformly normal magnetic polarity of samples from Vera Member sediments adjacent to the "La Cancha" bed and Carlini Tuff. Only one normal interval occurs within the range of dates we have for this tuff. Chron C13n ranges between 33.27 and 33.74 Ma, our preferred age range for the Carlini Tuff.

Age of the Vera Member

Profile K polarity stratigraphy is based on an average sampling interval of 3.78 m for the 131-m portion of the profile sampled. This interval includes the Vera Member in the highest 106 m of the profile. In this interval, there are two normal local polarity zones (N3 and N4) and one reversed zone (R3). Profile M polarity stratigraphy is based on an average sampling interval of 4.13 m for the highest 141 m of the profile above the base of Simpson's Y Tuff. The unconsolidated nature of Vera Member sediments beneath the "La Cancha" level in Profile M obliged a coarser sampling density along this profile. The Vera Member at Profile M includes three local polarity zones, N2 at the base, R2 supported by a single site, and N3 extending through most of the thickness of the member.

The polarity of most sites in both profiles is normal, as is the Carlini Tuff, a finding that is informative given the uncertainty of the ^{40}Ar/^{39}Ar results for the Carlini Tuff and the temporally dominant reversed polarity of Chrons C13, C12, and C11 around the Eocene–Oligocene boundary and into the early Oligocene (Gradstein *et al.* 2004).

The distinctive "La Cancha" bed is a poorly developed paleosol that extends laterally between Profile K and Profile M. As mentioned above, at Profile M, a fluvial channel incises the "La Cancha" bed. The unconformity observed by Simpson at the base of this fluvial channel cannot be traced laterally. A single long normal polarity zone extends below and above the "La Cancha" level and Carlini Tuff and is interpreted to represent the same normal polarity chron, Chron C13n (33.266–33.738 Ma: Gradstein *et al.* 2004), the closest normal polarity interval in the age range of the Carlini Tuff. This correlation and the assumption of nearly continuous deposition imply that the Vera Member represents much of Chron C13 (including all of C13r) and extends down to include some portion of Chron C15n.

As for the age of the base of the Vera Member, the mean age of the Kay Tuff (37.05 Ma) and correlation of the Lower Puesto Almendra Member normal polarity with Chron C17n.1n reveals something about the duration of the temporal hiatus represented by Discontinuity 5. The base of the Vera Member must be between 34.782 and 35.043 Ma and the temporal hiatus between the Lower Puesto Almendra Member and the base of the Vera Member is on the order of 1.5 to 2 m.y. The age of the top of the Vera Member, if only Chron C13n is represented, is no younger than 33.26 Ma. The composition of the fossil assemblage from GBV-4 "La Cancha" suggests a correlation with the Tinguirirican SALMA, which is concordant with our age interpretation, and somewhat older than the minimum age of the type fauna from central Chile.

The Upper Puesto Almendra Member

The Upper Puesto Almendra Member consists of three units: numbers 3, 4, and 5 of Bellosi's scheme (Bellosi this book). The basal Unit 3 rests on Discontinuity 6. Discontinuity 7 within Unit 3 is an erosional surface upon which rests basalt flows. On top of the basalts is another erosional surface (Discontinuity 8) that marks the top of Unit 3 and the base of Unit 4. In turn, the top of Unit 4 is marked by Discontinuity 9 and the top of Unit 5 by Discontinuity 10.

Thus, all the basalts at Gran Barranca occur above Discontinuity 7 within Unit 3 of the Upper Puesto Almendra Member. Marshall *et al.* (1977) reported ^{40}K/^{40}Ar dates for basalt samples collected in 1975 and samples collected in 1981 by Marshall and Drake as reported by Marshall (1985). On the age of these basalts, Marshall concluded "apparently only one flow is represented, in which case the best approximation of its apparent age is the date of 28.8 Ma obtained on the least-weathered sample" (Marshall 1985). The dated basalts correspond to those observed by Simpson in Profiles A, H, and N. "[T]he base of each basalt lies within 2 meters of each other relative to either the 'X' or 'Y' marker tuffs, a relationship which demonstrates that all three basalts occur at the same stratigraphic level relative to these lower marker tuffs" (Marshall *et al.* 1986). Our observations do not confirm Marshall's single-flow hypothesis. We have several dates from these flows and it seems probable they erupted over a period of several million years.

Gran Barranca basalts represent flows with baked zones at their lower contact but not at their tops. The Profile H Basalt, West End Basalt, East End Basalt (in Profile N-2), and the Profile A-2 basalt occur within fluvial deposits of Unit 3 of the Upper Puesto Almendra Member. However, stratigraphic relationships clearly indicate that multiple flows are represented, so relatively brief periods of local erosion must have intervened between them. Erosive discontinuity surfaces also occur at the base and top of the basalts (Bellosi this book). Discontinuity 7, on which most of the basalts rest, is an extensive surface related to fluvial channel erosion. Discontinuity 8 is another irregular erosion surface truncating the top of the lenticular basalts at Profiles A-2 and H, and in some places extending down to the level of the Discontinuity 7 (or possibly lower). Discontinuity 8 is covered by a residual breccia composed of weathered basalt blocks in a white tuffaceous matrix (Bellosi this book).

Four new radioisotopic dates for the basalts are fairly close in age. ^{40}Ar/^{39}Ar plateau ages for two basalt flows, one at the West End (27.87 ± 0.13 Ma) and another at the East End (27.76 ± 0.08 Ma) suggest these represent the same eruption event. The range of ^{40}Ar/^{39}Ar ages for the West End (27.87 ± 0.13, 29.18 ± 0.38 Ma), East End (27.76 ± 0.08 Ma), and Profile H Basalt (26.34 ± 0.32 Ma) point to multiple eruption events with intermittent periods of deposition and erosion. The ∼26.3-Ma Profile H Basalt provides an upper boundary for the age of the basaltic eruption activity at Gran Barranca.

The Upper Puesto Almendra Member has two different fossil-bearing levels. The lower level GBV-19 "La Cantera" occurs near the base of Unit 3 below Discontinuity 7. GBV-19 yields a distinctive assemblage with some Deseadan elements and some elements more reminiscent of the Tinguirirican SALMA. Above D7 in more direct stratigraphic relationship with the basalts are fossil mammals more typical of the Deseadan SALMA. Deseadan age mammals are found in close or even direct association with the basalts and their associated erosional surfaces near the top of Profile J, and in the exposures between Profiles H and A at the west end of Gran Barranca.

Age of the Upper Puesto Almendra Member

The age of the Upper Puesto Almendra Member can be established only in relation to the basalts. The oldest sediments of the Upper Puesto Almendra Member (Unit 3, between Discontinuities 6 and 7) are constrained in age by the lower contact and overlying basalts. The same basalts provide a basal age for Unit 4. The youngest sediments of the Upper Puesto Almendra Member, Unit 5, are constrained in age by the underlying basalts and the age of the overlying Colhue-Huapi Member.

The age of the fossil-bearing sediments of GBV-19 "La Cantera" in Unit 3 between Discontinuity 6 (the contact with the Lower Puesto Almendra Member) and Discontinuity 7 at the base of the basalts hinges on three sorts of information: (1) the stratigraphic relationship of the Kay Tuff in the Lower Puesto Almendra Member to Discontinuity 6 and GBV-19 in Profile A-2, (2) the magnetic polarity stratigraphy of Unit 3, and (3) the age of the basalts.

With respect to the first point, the Vera Member thins west of Profile J and pinches out where the plane of Discontinuities 5 and 6 converge and become coplanar at a level above the Kay Tuff. From this point westward between Profiles I and A-2, the Vera Member cannot be recognized. Between Profiles J and A-2, the Kay Tuff has a discontinuous distribution, but is well exposed below the coplanar Discontinuities 5 and 6 at Profile G and again at Profile A-2. The fossil locality GBV-19 "La Cantera" occurs within a pale grey tuffaceous mudstone in the lowermost part of the Unit 3 of the Upper Puesto Almendra Member, stratigraphically just above Discontinuity 6. Thus, Discontinuity 6 occurs stratigraphically between the Kay Tuff and GBV-19 "La Cantera."

As to the second point, we have five magnetostratigraphic sites in Unit 3 below Discontinuity 7 (the lowest of the units of the Upper Puesto Almendra Member) at Profile MMZ (sites 10.3, 10.5, 11, 12, and 12.7) yield normal polarity. Given our best estimate of the age of the Carlini Tuff in the Vera Member and the overlying basalts in Unit 3 above Discontinuity 7, the Unit 3 normal polarity zone could correspond to either Chron C13n (33.266–33.738 Ma), C12n (30.627–31.116 Ma), C11n.2n (29.853–30.217 Ma), or C11n.1n (29.451–29.740 Ma) following Gradstein *et al.* (2004). Given the apparent magnitude of erosion represented by Discontinuity 6, we consider it unlikely that this is the same normal polarity interval as that sampled by the Vera Member (i.e. Chron C13n). Therefore, the fossil mammals at GBV-19 "La Cantera" (Goin *et al.* this book; Carlini *et al.*

this book; Vucetich *et al.* this book; Lopez *et al.* this book; Ribeiro *et al.* this book) must belong to one of the remaining three chrons of normal polarity. The most precise age we can establish for the fossil-bearing sediments of the lower part of Unit 3 is that they must be younger than 31.1 Ma and older than 29.5 Ma. This interpretation also accords with the ages of the basalts (the third point), all of which are younger than 29.18 Ma.

Upper Unit 3 of the Upper Puesto Almendra Member includes the basalts (29.18–26.34 Ma) above Discontinuity 7 and below Discontinuity 8. Deseadan age mammals occur often along the erosional surfaces distal to the tops of the basalts. If the tops of all the basalts were weathered at the same time, i.e. a single Discontinuity 8, then the Deseadan age assemblages at Gran Barranca must be younger than the youngest basalt. Thus, the Deseadan faunules at Gran Barranca would be younger than about 26.3 Ma. If, on the other hand, there were multiple erosional discontinuities within Unit 3 associated with the tops of several basalt flows of different ages, then some of the Deseadan mammals at Gran Barranca could be slightly younger than 29.2 Ma and some within Unit 4 above Discontinuity 8 could be younger than 26.34 Ma. The beds are not sufficiently well exposed nor the faunas of Deseadan aspect well enough preserved to select among these alternatives.

Unit 5 of the Upper Puesto Almendra Member has no dated rock, so it is constrained only by the age of the underlying Unit 4 basalts and their associated discontinuities and the contact between Unit 5 and the overlying Colhue-Huapi Member. However, the sediments resting on Discontinuity 8 and extending up to Discontinuity 10 at the base of the overlying Colhue-Huapi Member include three normal and two reversed magnetic polarity intervals at Profile MMZ, suggesting this unit extends in age up to Chron 7.

The Colhue-Huapi Member

The Colhue-Huapi Member outcrops only at the west end of Gran Barranca in a single exposure at the top of the escarpment. Two measured stratigraphic profiles extend up through the Colhue-Huapi Member; Profile A-1 at Colhue-Huapi West (CHW) and Profile MMZ at Colhue-Huapi East (CHE). Four tuffs in the Colhue-Huapi Member have been dated. Three of the tuffs occur in the lower section of the Colhue-Huapi Member, distinguished lithologically by more finely stratified low-energy fluvial depositional cycles of intraformational conglomerates, pyroclastic mudstones, and paleosols (Bellosi this book). The fourth and highest of the dated tuffs, the MMZ 24.5 tuff in upper part of the member, is a conspicuous whitish pyroclastic mudstone that forms a distinct laterally continuous stratum of variable thickness. All other tuffs occur as stratiform but laterally discontinuous concentrations of relatively pure fine-grained volcaniclastic material.

Big Mammal Tuff The Big Mammal Tuff occurs on Profile A-1 about 6 meters above the basal Discontinuity 10 of the member and is the stratigraphically lowest of the dated tuffs in the Colhue-Huapi Member. The arithmetic mean age of the Big Mammal Tuff is 19.75 Ma (Table 4.1).

Basal Tuff The Basal Tuff in Profile MMZ at Colhue-Huapi East (CHE) occurs about 10 m above Discontinuity 10 near the axis of the profile. Glass and plagioclase plateau ages for sample GB 01–8 are 19.37 ± 0.08 Ma and 21.09 ± 0.30 Ma (1-sigma error), respectively (Table 4.1). Another sample of the same tuff yielded a plagioclase plateau age of 20.73 ± 0.26 Ma (1-sigma error). The arithmetic mean age of all three determinations is 20.39 Ma (Table 4.1).

Monkey Tuff The Monkey Tuff occurs about 22 m above the base of the member and outcrops in a limited area in close proximity and at the same stratigraphic level (Level C) where most of the fossil primates have been recovered at Colhue-Huapi West (Kay this book). Samples RFK-2 and GB 99–5 yielded an age range between 19.12 and 21.30 Ma. Furnace argon isotopic plateau ages for plagioclase (19.94 and 21.30 Ma) are uniformly older than glass ages (19.13 and 19.12 Ma) for the same samples; however, the glass ages of the two samples are essentially identical, whereas the plagioclase plateau ages differ by about 1.3 m.y. The arithmetic mean of all five determinations (including the laser argon results for RFK-2) is 19.81 Ma (Table 4.1).

MMZ 24.5 Tuff MMZ 24.5 Tuff is a conspicuous 2-m thick impure white pyroclastic mudstone near the top of the northernmost exposures of the Colhue-Huapi Member on Profile MMZ. It occurs about 48.5 m above the base of the member at a level between two discrete biozones; the Lower Fossil Zone with at least 17 distinct fossil-bearing levels of Colhuehuapian age, and the Upper Fossil Zone with two or three levels, a "Pinturan age" correlative. The age of the MMZ 24.5 Tuff falls between the Colhuehuapian and "Pinturan" levels, so it provides age constraints for both ages. The arithmetic mean age of two determinations is 19.30 Ma (Table 4.1).

The age of the Colhue-Huapi Member

Sediments of the Colhue-Huapi Member were sampled for magnetic polarity stratigraphy at sites along two measured sections; Profile A-1 at CHW on the west side of the exposure, and Profile MMZ at CHE on the east side. The Colhue-Huapi Member at Profile A-1 measures 63 m in total thickness, from the basal contact to the top of consolidated sediments of the upper section of this member. The exposures of the Colhue-Huapi Member along Profile MMZ measure about 48.5 m in thickness. The lowest strata of the Colhue-Huapi Member are found above Discontinuity 10 at CHE. The top of the member at CHE is truncated

by erosion and obscured by plant cover. Magnetostratigraphic sampling intervals were spaced along Profile A-1 at an average interval of 2.1 m. This represents a closer sampling interval than along Profile MMZ where sampling achieved an average interval of 4.4 m (Ré *et al.* Chapter 3, this book). There are two normal polarity zones in the Colhue-Huapi Member, with a reversed zone at the base and another near the middle of the member.

Arithmetic mean ages for Colhue-Huapi Member tuffs range between 20.39 and 19.30 Ma, a range of about 0.9 m.y., and well within the 2-sigma errors of the individual estimates. The age range for preferred plagioclase results extends from 21.30 to 19.54 Ma and for glass from 19.37 to 18.53 Ma, with glass ages uniformly younger for all samples and all tuffs.

The temporal interval of the GMPTS (Gradstein *et al.* 2004) between 20.39 and 19.30 Ma (between C6An.1r and C6n) is dominantly normal polarity, but includes two reversed polarity intervals. There is strong evidence for two reversed-polarity intervals in the Colhue-Huapi Member (Ré *et al.* Chapter 3, this book). The lowest sites sampled in both Profiles A-1 (CHW) and MMZ (CHE) yielded reverse polarity with high-quality site mean direction and thus we assume that the base of the long local normal polarity interval N1 of Profile MMZ and the lowest local normal polarity site of Profile A-1 provide a useful time horizon. The second interval of reversed polarity occurs just above the highest fossil mammal level of the Lower Fossil Zone (see below), and was detected with confidence only in Profile MMZ.

Two different interpretations of the age of the Colhue-Huapi Member are possible:

(1) The arithmetic mean age of the Basal Tuff (20.40 Ma) indicates the long normal interval N1 of Profile MMZ (CHE) may represent Chron C6An.2n (20.439–20.709 Ma) and the arithmetic mean age of MMZ 24.5 Tuff (19.30 Ma) assigns the highest normal polarity interval N2 of Profile MMZ to Chron C6n (18.748–19.722 Ma).

(2) The arithmetic mean ages of the Big Mammal (19.75 Ma) and Monkey (19.81 Ma) tuffs fall within the age range of reversed polarity chron C6r (19.722–20.040 Ma). However, the normal polarity of the sediments bracketed by these two tuffs indicates they more likely correspond to C6An.1n (20.040–20.213 Ma) than to C6n (18.748–19.722 Ma). The correlation to C6An.1n agrees better with the plagioclase results for these two tuffs and implies that R1 represents C6An.1r (20.213–20.439 Ma) and R2 represents C6r (19.722–20.040 Ma).

Of the two possibilities, we prefer the second. The first interpretation posits that the interval between 19.722 and 20.439 Ma is not preserved in the Colhue-Huapi Member at Gran Barranca. The second requires no significant temporal hiatus within the member, more in agreement with field

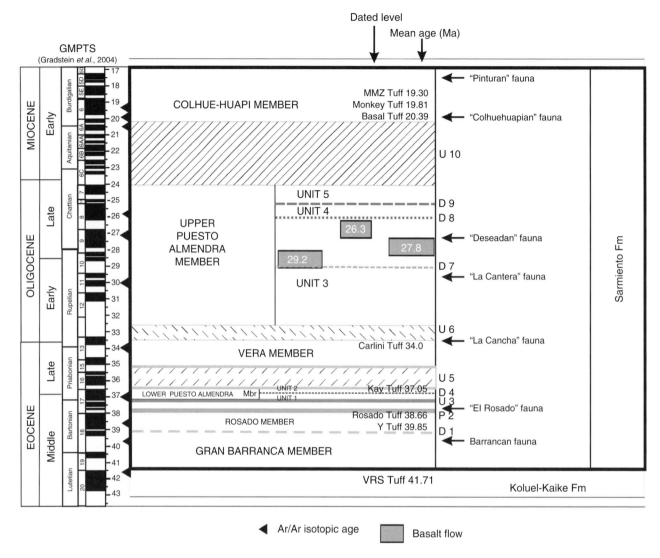

Fig. 4.1. Summary of the geochronology of the Sarmiento Formation at Gran Barranca, indicating temporal hiatuses and stratigraphic discontinuities.

observations of depositional continuity. An assignment of the lower normal polarity interval of the Colhue-Huapi Member to Chron C6An.1n and the upper normal polarity interval to Chron C6n seems to best fit the available evidence.

The Colhue-Huapi Member has many fossil bearing levels. More than 16 of them in the Lower Section are grouped together into the Lower Fossil Zone. All occur in the lower interval of normal polarity. The highest levels at the top of the Upper Section in Profile A-1 comprise the Upper Fossil Zone in the upper normal polarity interval. The fauna of the Lower Fossil Zone is the type fauna for the Colhuehuapian SALMA, and that of the Upper Fossil Zone is correlated with the Pinturan (see Kramarz and Bellosi 2005; Kramarz *et al.* this book). The Colhuehuapian

SALMA at Gran Barranca is interpreted to represent 20.0 and 20.2 Ma (Chron C6An.1n). The Upper Fossil Zone has a more derived fauna (Kramarz *et al.* this book) and would be within C6n, between 18.7 and 19.7 Ma in age.

Summary and implications

The geochronology of the Sarmiento Formation at Gran Barranca, including its temporal hiatuses and stratigraphic discontinuities, is summarized in Fig. 4.1.

Gran Barranca Member The base of the Gran Barranca Member is within Chron 19r (40.671–41.590 Ma) and its top within Chron C18n.1n (38.032–38.975 Ma). Chrons 18 and 19 encompass the latest Lutetian and most of the Bartonian (Gradstein *et al.* 2004). This interval of the late middle Eocene that includes the Middle Eocene Climate Optimum (MECO), a stable isotope anomaly of transient high pCO_2 levels in the Indian (Kerguelen Plateau) and Atlantic (Maud Rise) sectors of the Southern Ocean (Bohaty and Zachos 2003; Luterbacher *et al.* 2004; Jovane *et al.* 2007). The Barrancan mammal fauna at Gran Barranca (Cifelli 1985) would thus fall within the interval between about 41.7 and 39.0 Ma.

Rosado and Lower Puesto Almendra Members The Rosado and Lower Puesto Almendra Members have a temporal duration equal to or less than the interval from near the top of Chron C18n.1n to the top of C17n.1n, that is, from slightly older than 38.0 to younger than 36.5 Ma. The Rosado Member contains a fauna transitional between the Barrancan and Mustersan SALMAs as well as a slightly younger Mustersan fauna. Faunas of Mustersan aspect also occur in the Lower Puesto Almendra Mamber. We tentatively conclude that the Mustersan SALMA at Gran Barranca may have a temporal duration equal to or less than the interval from near the top of Chron C18n.1n to the top of C17n.1n, that is, from slightly older than 38.0 to younger than 36.5 Ma. The Rosado and Lower Puesto Almendra members represent part of the late Bartonian (late middle Eocene) and Priabonian (late Eocene). The Mustersan SALMA temporal interval at Gran Barranca occurs at the beginning of a marked decline in atmospheric carbon dioxide concentrations following the MECO (Pagani *et al.* 2005) (although not documented from South Atlantic DSDP/ODP cores 511 and 513) and part of the post-MECO cooling of both surface and intermediate waters documented in the Atlantic and Indian sectors of the Southern Ocean (Bohaty and Zachos 2003). Discontinuity 3 between the Rosado Member and Lower Puesto Almendra Member represents a temporal hiatus encompassing at least the interval from 37.235 to 38.032 Ma.

Vera Member The base of the Vera Member is between 34.8 and 35.0 Ma. The age of the top of the Vera Member is no younger than 33.3 Ma. A temporal hiatus between the Lower Puesto Almendra Member and the base of the Vera Member is on the order of 1.5 to 2 m.y. The age of the Tinguiririican fauna from GBV-4 "La Cancha" is between 33.3 and 33.7 Ma. The marked increase in sedimentation rate observed between C13r and C13n in the Vera Member corresponds to the oxygen isotope excursion of the Eocene–Oligocene transition into the Oi-1 glaciation of C13n as understood in the Southern Ocean deep sea record (Zachos *et al.* 1996; Salamy and Zachos 1999). At the western end of Gran Barranca, nearly all of the Vera Member was removed by erosion associated with Discontinuity 6, the erosional unconformity corresponding to the Oi-1 glaciation, with its inferred consequences for intensification of Southern Ocean circulation, atmospheric circulation over Patagonia, and continental shelf exposure off

Patagonia. Correlation of the upper part of the Vera Member to C13n would attribute their stratigraphy and sedimentology to gradually increasing rates of eolian sediment deposition through C13r, followed by accelerating deposition into C13n, and eventually by still further increases in wind intensity at the erosional unconformity represented by Discontinuity 6 at the top of the Vera Member.

Upper Puesto Almendra Member The Upper Puesto Almendra Member is composed of three units. The lowest Unit 3 below Discontinuity 8 at the top of the basalts and above Discontinuity 6 has normal polarity. Given our best estimate of the age of the Vera Member and the age of the basalts, the Unit 3 normal polarity zone below the basalts and Discontinuity 7 corresponds to Chron C12n, C11n.2n, or C11n.1n. Thus, lower Unit 3 and the contained fossil mammals at GBV-19 "La Cantera" are between 31.1 and 29.5 Ma in age. The upper part of Unit 3 could be as young as 26.3 Ma. Units 4 and 5 are not dated. The top of the Upper Puesto Almendra Member must be younger than the age of the basalts at the top of Unit 3, and older than the age of the base of the overlying Colhue-Huapi Member (20.4 Ma). It is not established whether the Deseadan faunules at Gran Barranca are found in upper Unit 3 or at the base of Unit 4, or both. Deseadan assemblages at Gran Barranca could be as old as 29.2 Ma and as young a 26.3 Ma. Unit 3 sediments represent the beginning of the long 5-million-year period of sustained low atmospheric carbon dioxide concentrations that lasted through much of the Oligocene (DeConto *et al.* 2008), the temporal interval extending into the time when Units 4 and 5 were deposited.

Colhue-Huapi Member There are two normal polarity zones in the Colhue-Huapi Member, with a reversed zone at the base and another near the middle of the member. The reversed polarity interval at the base of the Member is C6An.1r (20.2–20.4 Ma). The lowest normal polarity interval represents Chron C6An.1n (20.0–20.2 Ma). The fauna of the Colhuehuapian SALMA at Gran Barranca is contained within this lower normal interval. The age of MMZ 24.5 Tuff (19.29 Ma) places the highest normal polarity interval to Chron C6n (18.748–19.722 Ma). The fauna of this interval has been correlated with that of the lower faunal levels in the vicinity of the Rio Pinturas. This temporal interval corresponds to the beginning of the Miocene Climate Optimum (Zachos *et al.* 2004).

ACKNOWLEDGEMENTS

The paleomagnetics laboratory was supported by grant UBACyT to Juan F. Vilas. Field research was supported by US National Science Foundation grants EAR-0087636, BCS-0090255, and DEB-9907985 to Richard F. Kay and Richard H. Madden.

REFERENCES

Bellosi, E. S. 1995. Paleogeografía y cambios ambientales de la Patagonia central durante el Terciario medio. *Boletín de Informaciones Petroleras*, **44**, 50–83.

Bellosi, E. S. and R. Madden 2005. Estratigrafía física preliminar de las secuencias piroclásticas terrestres de la Formación Sarmiento (Eoceno–Mioceno) en la Gran Barranca, Chubut. *Actas XVI Congreso Geológico Argentino*, **4**, 427–432.

Bohaty, S. M. and J. C. Zachos 2003. Significant Southern Ocean warming event in the late middle Eocene. *Geology*, **31**, 1017–1020.

Cifelli, R. L. 1985. Biostratigraphy of the Casamayoran, Early Eocene, of Patagonia. *American Museum of Natural History Novitates*, **2820**, 1–16.

DeConto, R. M., D. Pollard, P. A. Wilson, H. Pälike, C. H. Lear, and M. Pagani 2008. Thresholds for Cenozoic bipolar glaciation. *Nature*, **455**, 652–655.

Feruglio, E. 1949. *Descripción Geologica de la Patagonia*, 3 vols. Buenos Aires: Dirección General de Yacimientos Petrolíferos Fiscales.

Franchi, M. and F. Nullo 1986. Las tobas de Sarmiento en el macizo de Somuncura. *Revista Asociación Geológica Argentina*, **41**, 218–222.

Gradstein, F. M., J. G. Ogg, and A. Smith (eds.) 2004. *A Geological Time Scale 2004*. Cambridge, UK: Cambridge University Press.

Jovane, L., F. Florindo, R. Coccioni, J. Dinarès-Turell, A. Marsili, S. Monechi, A. P. Roberts, and M. Sprovieri 2007. The middle Eocene climatic optimum event in the Contessa Highway section, Umbrian Apennines, Italy. *Geological Society of America Bulletin*, **119**, 413–427.

Kay, R. F., R. H. Madden, M. Guiomar Vucetich, A. A. Carlini, M. M. Mazzoni, and G. H. Ré 1996. Revised geochronology of the Casamayoran South American Land Mammal Age: climatic and biotic implications. *Proceedings of the National Academy of Sciences USA*, **96**, 13 235–13 240.

Luterbacher, H. P., J. R. Ali, H. Brinkhuis, F. M. Gradstein, J. J. Hooker, S. Monechi, J. G. Ogg, J. Powell, U. Röhl, A. Sanfilippo, and B. Schmitz 2004. The paleogene period. In Gradstein, F. M., Ogg, J. G., and Smith, A. (eds.), *A Geological Time Scale 2004*, Cambridge, UK: Cambridge University Press, pp. 384–408.

Marshall, L. G. 1985. Geochronology and land-mammal biochronology of the transamerican faunal interchange. In Stehli, E. G. and Webb, S. D. (eds.), *The Great American Biotic Interchange*. New York: Plenum, pp. 49–85.

Marshall, L. G., R. Pascual, G. H. Curtis, and R. E. Drake 1977. South American geochronology: radiometric time scale for middle to late Tertiary mammal-bearing horizonts in Patagonia. *Science*, **195**, 1325–1328.

Marshall, L. G., R. L. Cifelli, R. E. Drake, and G. H. Curtis 1986. Vertebrate paleontology, geology, and geochronology of the Tapera de Lopez and Scarritt Pocket, Chubut Province, Argentina. *Journal of Paleontology*, **60**, 920–951.

Mazzoni, M. M. 1979. Contribución al conocimiento petrográfico de la Formación Sarmiento, barranca sur del Lago Colhué Huapi, Provincia del Chubut. *Revista de la Asociación de Minerologia, Petrología y Sedimentologia, Buenos Aires*, **10**(3–4), 33–54.

Mazzoni, M. M. 1985. La Formación Sarmiento y el vulcanismo paleogeno. *Revista de la Asociación Geológica Argentina*, **40**(1–2), 60–68.

Miller, K. G., M. A. Kominz, J. V. Browning, J. D. Wright, G. S. Mountain, M. E. Katz, P. J. Sugarman, B. S. Cramer, N. Christie-Blick, and S. F. Pekar 2005. The Phanerozoic record of global sea-level change. *Science*, **310**, 1293–1298.

Pagani, M., J. C. Zachos, K. H. Freeman, B. Tipple, and S. Bohaty 2005. Marked decline in atmospheric carbon dioxide concentrations during the Paleogene. *Science*, **309**, 600–602.

Salamy, K. A. and J. C. Zachos 1999. Late Eocene–early Oligocene climate change on Southern Ocean fertility: inferences from sediment accumulation and stable isotope data. *Palaeogeography, Palaeoclimatology, Palaeoecology*, **145**, 79–93.

Simpson, G. G. 1930. *Scarritt-Patagonian Exped. Field Notes*. New York: American Museum of Natural History. (Unpublished.) Available at http://paleo.amnh.org/notebooks/index.html

Spalletti, L. A. and M. M. Mazzoni 1977. Sedimentologia del Grupo Sarmiento en un perfil ubicado al sudeste del Lago Colhué Huapi, Provincia del Chubut. *Revista Museo de La Plata (Obra del Centenario)*, **4**, 261–283.

Spalletti, L. A. and M. M. Mazzoni 1979. Estratigrafia de la Formación Sarmiento en la barranca sur del Lago Colhué Huapi, Provincia del Chubut. *Revista de la Asociación Geológica Argentina*, **34**(4), 271–281.

Swisher, C. C. III, L. Dingus, and R. F. Butler 1993. $^{40}Ar/^{39}Ar$ dating and magnetostratigraphic correlation of the terrestrial Cretaceous–Paleogene boundary and Puercan Mammal Age, Hell Creek–Tullock formations, eastern Montana. *Canadian Journal of Earth Sciences*, **30**, 1981–1996.

Windhausen, A. 1924. Líneas generales de la constitución geológica de la región situada al oeste del Golfo de San Jorge. *Boletín de la Academia Nacional de Ciencias*, **27**, 167–320.

Zachos, J. C., T. M. Quinn, and K. A. Salamy 1996. High-resolution (104 years) deep-sea foraminiferal stable isotope records of the Eocene–Oligocene climate transition. *Paleoceanography*, **11**, 251–266.

Zachos, J., M. Pagani, L. Sloan, E. Thomas, and K. Billups 2001. Trends, rhythms, and aberrations in global climate 65 Ma to present. *Science*, **292**, 686–693.

PART II   SYSTEMATIC PALEONTOLOGY

5 Middle Eocene – Oligocene gastropods of the Sarmiento Formation, central Patagonia

Sergio E. Miquel and Eduardo S. Bellosi

Abstract

Diverse fossil gastropods are preserved in the lower and middle sections of the continental Sarmiento Formation in Chubut Province, Argentina. The assemblage is dominated by land specimens assigned to five families (15 genera). A new genus and species of Charopidae, *Colhueconus simpsoni*, is described. The presence of *Austroborus* sp. (Strophocheilidae) and *Plagiodontes* sp. (Odontostomidae) is noted for the first time. Stratigraphic distribution and ecological affinities with related living specimens allow us to infer that warm–temperate and humid environments prevailed in the late middle Eocene (Casamayoran SALMA), with species of Strophocheilidae, Orthalicidae, and Chilinidae. In the Mustersan SALMA (middle–late Eocene) a significant increase in diversity is observed, mostly dominated by Charopidae, which indicates cooler conditions in accordance with the global climatic trend. The absence of gastropods in Tinguirirican SALMA deposits could be related to the enhanced cooling and drying at the terminal Eocene – early Oligocene. The fauna preserved in Deseadan SALMA beds (late Oligocene) is similar but less diverse than the Casamayoran fauna, indicating a return to probably locally wetter and warmer conditions.

Resumen

Los depósitos continentales de las secciones inferior y media de la Formación Sarmiento, en diversas localidades de Chubut (Argentina), preservan una variada fauna de gasterópodos mayormente terrestres, integrada por 5 familias (15 géneros). Se describe un nuevo género y especie de Charopidae: *Colhueconus simpsoni*, y se citan por primera vez *Austroborus* sp. (Strophocheilidae) y *Plagiodontes* sp. (Odontostomidae). La distribución estratigráfica y las afinidades ecológicas con formas emparentadas actuales sugieren para el Eoceno medio (Edad mamífero Casamayorense) ambientes templado-cálidos y húmedos con especies de Strophocheilidae, Orthalicidae y Chilinidae. El marcado aumento de diversidad en estratos más jóvenes (Mustersense), donde predominan los carópidos, denota condiciones más frías hacia el Eoceno medio–tardío, en concordancia con la tendencia climática

global. La ausencia de gasterópodos en niveles del Tinguirririquense se vincularía con el aumento del enfriamiento y desecación ocurrido en el Eoceno terminal – Oligoceno temprano. La malacofauna asociada a estratos Deseadenses del Oligoceno tardío, aunque menos diversa, es similar a la del Casamayorense, sugiriendo un retorno a condiciones más húmedas y cálidas, posiblemente de carácter local.

Introduction

Knowledge of fossil gastropods in the middle Cenozoic Sarmiento Formation dates back to the beginning of the twentieth century with the first Strophocheilidae collected by Carlos Ameghino and described by von Ihering (1904). Later, the Scarritt Patagonian Expedition collected Chilinidae and Orthalicidae specimens that were analyzed by Parodiz (1946, 1949a, 1963). Lastly, detailed studies using screen-washing techniques allow us to recognize new and very diverse microgastropods of the family Charopidae (Bellosi *et al.* 2002; Miquel and Bellosi 2004, 2007).

The aim of this contribution is to review the gastropod faunas preserved in the Sarmiento Formation and to discuss their paleoenvironmental implications, particularly for the highly dominant land snails, as most of them can be compared to living taxa. This study includes specimens collected by one of authors (E. S. B.) at Gran Barranca (Chubut) and the revision of the historical collections of museums of Argentina. Distribution maps of extant land snails are based on the collections of the Department of Invertebrates of the Museo Argentino de Ciencias Naturales "Bernardino Rivadavia" and the Museo de La Plata.

The Sarmiento Formation is a continental succession widely exposed in central and northern Patagonia and well known by the abundance of fossil mammals. It records a time span of approximately 23 m.y. (middle Eocene – early Miocene), a time in which significant climate and biotic change occurred in Patagonia. Despite numerous studies of its fossil content, regional lithostratigraphic characterization of the Sarmiento Formation is still in infancy because lateral changes and internal discontinuity surfaces make correlation difficult. The original mammal assemblages recognized by Ameghino (1906), with recent age calibration of the type locality of Gran Barranca (Ré *et al.* Chapter 4, this

The Paleontology of Gran Barranca: Evolution and Environmental Change through the Middle Cenozoic of Patagonia, eds. R. H. Madden, A. A. Carlini, M. G. Vucetich, and R. F. Kay. Published by Cambridge University Press. © Cambridge University Press 2010.

Table 5.1. *Geographic and stratigraphic distribution of fossil gastropod species, materials studied and main references from the Sarmiento Formation*

Taxa	Localities	Age (SALMA)	Materials studied	Main references
Chilina stenostylops Parodiz, 1963	Chubut (Cañadón Vaca, Cañadón Hondo)	Middle Eocene (Casamayoran)	CMNH 73401. Holotype. CMNH 73402. Paratypes (2). MACN-Pi 4856	Parodiz, 1969
Strophocheilus chubutensis Ihering, 1904	Chubut (Cañadón Blanco, Mallín Blanco, Laguna de la Bombilla)	Middle Eocene (Casamayoran) to late Oligocene (Deseadan)	MACN-Pi 559. Lectotype and paralectotypes (2) (mislaid); MACN-Pi 560. Paralectotypes (2) (mislaid); MACN-Pi 466; 4858; 4862. MLP 399; 4086; 10550	Parodiz, 1996
Megalobulimus hauthali (Ihering, 1904)	Chubut (Cañadón Blanco, Colhue-Huapi, Solano Bay, Musters Lake, Guacho Hill, Bombilla Lake, Aguada La Escondida, La Herrería)	Middle to late Eocene (Casamayoran and Mustersan SALMAs). Probably late Oligocene (Deseadean)	MACN-Pi 557. Lectotype and paralectotypes (5). MACN-Pi 558; 4859. MLP 2695–2696–2697; 4084; 4085; 7449; 7618; 10123; 31515	Parodiz, 1996
Austroborus sp.	Chubut (Cañadón Pelado)	?Eocene	MACN-Pi 4855	This chapter
Thaumastus patagonicus Parodiz, 1946	Chubut (Cañadón Hondo, Cañadón Blanco, Mallín Blanco)	Middle to late Eocene (Casamayoran and Mustersan). Probably late Oligocene (Deseadan)	MACN-Pi 4521 ?mislaid; 4860; 4861	Parodiz, 1969
Paleobulimulus eocenicus Parodiz, 1949a	Chubut (Mallín Blanco)	Late Eocene (Mustersan)	MACN-Pi. W/no.?mislaid	Parodiz, 1969
Plagiodontes sp.	Chubut (Cañadón Hondo)	Paleogene	MACN-Pi 4857	This chapter
Gyrocochlea? sp. cf. *Gyrocochlea? mirabilis* Hylton Scott, 1968	Chubut (Gran Barranca)	Late Eocene (Mustersan)	MACN-Pi 4627; 4633	Bellosi *et al.* 2002
Lilloiconcha sp. cf. *Lilloiconcha gordurasensis* (Thiele, 1927)	Chubut (Gran Barranca)	Late Eocene (Mustersan)	CNP-PIIc 191; 192	Miquel & Bellosi, 2007
Radiodiscus sp. cf. *Radiodiscus riochicoensis* Crawford, 1939	Chubut (Gran Barranca)	Late Eocene (Mustersan)	CNP-PIIc 203; 204; 205	Miquel & Bellosi, 2007
Rotadiscus sp. cf. *Rotadiscus amancaezensis* (Hidalgo, 1869)	Chubut (Gran Barranca)	Late Eocene (Mustersan)	CNP-PIIc 206	Miquel & Bellosi, 2007
Stephanoda? mazzonii Miquel & Bellosi, 2004	Chubut (Gran Barranca)	Late Eocene (Mustersan)	MACN-Pi 4631 (Holotype); MACN-Pi 4678 (Paratype). CNP-PIIc 193; 194; 195	Miquel & Bellosi, 2007

Table 5.1. (*cont.*)

Taxa	Localities	Age (SALMA)	Materials studied	Main references
Stephadiscus sp. cf. *Stephadiscus lyratus* Couthouy in Gould, 1846	Chubut (Gran Barranca)	Late Eocene (Mustersan)	CNP-PIIc 196; 197	Bellosi *et al.* 2002
Stephadiscus sp. cf. *Stephadiscus celinae* (Hylton Scott, 1969)	Chubut (Gran Barranca)	Late Eocene (Mustersan)	CNP-PIIc 196; 197	Miquel & Bellosi, 2007
Stephadiscus sp.	Chubut (Gran Barranca)	Late Eocene (Mustersan)	CNP-PIIc 198; 199	Miquel & Bellosi, 2007
Zilchogyra sp. 1	Chubut (Gran Barranca)	Late Eocene (Mustersan)	MACN-Pi 4629	Bellosi *et al.* 2002
Zilchogyra sp. 2	Chubut (Gran Barranca)	Late Eocene (Mustersan)	MACN-Pi 4630	Bellosi *et al.* 2002
Colhueconus simpsoni gen. et sp. nov.	Chubut (Gran Barranca)	Late Eocene (Mustersan)	CNP PIIc 232. Type	This chapter

Notes: CMNH, Carnegie Museum of Natural History; CNP-PIIc, Centro Nacional Patagónico, Paleontología de Invertebrados e Icknología; MACN-Pi, División Paleoinvertebrados, Museo Argentino de Ciencias Naturales "Bernardino Rivadavia"; MLP, Museo de La Plata.

book) are used to correlate the isolated localities where gastropods are found. Thus, the gastropod-bearing beds are assigned to South American Land Mammal Ages (SALMAs).

The Sarmiento Formation comprises primary and secondary pyroclastic facies representing a variable spectrum of sedimentary settings and processes, including fine ash falls, paleosols, fluvial channel fill, and ponded plains (Bellosi this book). Gastropods generally occur in terrestrial facies such as paleosols and massive pyroclastic mudstones (Andreis 1977; Bellosi *et al.* 2002) interpreted as loessites (Spalleti and Mazzoni 1979). Because of this characteristic sedimentology and the fine detail of the shell sculpture that is preserved (particularly in the terrestrial snails of the family Charopidae) these assemblages appear to represent lag accumulations. Such accumulations are compatible with the alkaline–neutral and calcareous soils preserved in the Sarmiento Formation at Gran Barranca (Bellosi and González this book), as in more acidic conditions the shells would otherwise have been quickly dissolved.

In this chapter the occurrences of two genera of land snails are cited for the first time and a new genus of

Charopidae is described. Table 5.1 lists the gastropods recorded in the Paleogene of Chubut.

Systematic paleontology

Abbreviations used are: CMNH (Carnegie Museum of Natural History), CNP-PIIc (Centro Nacional Patagónico, Paleontología de Invertebrados e Icnología), MACN-Pi (División Paleoinvertebrados, Museo Argentino de Ciencias Naturales "Bernardino Rivadavia", MLP (Museo de La Plata), SALMA (South American Land Mammal Age). The units are in millimeters (mm).

Class **GASTROPODA** Cuvier 1797
Subclass **PULMONATA** Cuvier 1817
Order **STYLOMMATOPHORA** Schmidt 1855
Family **STROPHOCHEILIDAE** Pilsbry 1902
Austroborus Parodiz 1949c

Type species *Bulimus lutescens* (King and Broderip 1831).

Modern occurrence Central area of Argentina and Uruguay (Fernández 1973).

Age range Eocene – Recent.

Fig. 5.1. (1), *Strophocheilus chubutensis* Ihering, 1904: MLP 4.086. (2), *Austroborus* sp.: MACN-Pi 4855. (3), *Plagiodontes* sp. MACN-Pi 4857. (4), *Colhueconus simpsoni* gen. et sp. nov. CNP-PIIc 232: Type. Apertural view. (5), *Colhueconus simpsoni* gen. et sp. nov. CNP-PIIc 232: Type. Detail of first whorls. (6), *Colhueconus simpsoni* gen. et sp. nov. CNP-PIIc 232: Type. Detail of protoconch.

Remarks First mention for the Sarmiento Formation.

Austroborus sp.

Fig. 5.1: 1, 2.

Description Shell small, globose, 18 mm, last whorl prominent, short spire, aperture large (almost 60% of total length), suture impressed.

Age ?Eocene.

Remarks The specimen is fragmentary, with apex lost and labral area incomplete. The cast is similar to species of this endemic genus of Argentina and Uruguay.

Repository MACN-Pi 4855. Cañadón Pelado, Chubut. ?Eocene.

Collector Suero, 1948. 1 specimen.

Family **Odontostomidae** Zilch, 1960

Plagiodontes Doering 1875 [1877]

Type species *Helix dentata* Wood 1828.

Modern occurrence North and central Argentina; Uruguay.

Age Paleogene – Recent.

Remarks First record for the Sarmiento Formation.

Plagiodontes sp.

Fig. 5.1: 3.

Description Specimen fragmentary of 5 – approximately – convex whorls, with curved axial ribs (35 on half whorl), suture marked, protoconch and last whorl lost; specimen a little compressed, height 15.0 mm, diameter maximum 12.0 mm.

Remarks The specimen is similar to species of *Plagiodontes* with axial ribs in teleoconch.

Occurrence Cañadón Hondo, Chubut.

Age Paleogene.

Repository MACN-Pi 4857. Cañadón Hondo, Chubut. Collector: Scarritt Patagonian Expedition. 1 specimen.

Family **Charopidae** Hutton 1884

Colhueconus simpsoni gen. et sp. nov.

Fig. 5.1: 4–6.

Diagnosis Shell trochoid, 1.52 mm, teleoconch with at least 4 slightly convex whorls and retractives axial ribs, approximately 55 ribs on last whorl.

Description Shell very small, trochoid; with conspicuous retractives axial ribs; protoconch smooth (maybe by abrasion), with at least 1¼ whorls; teleoconch with at least 4 lightly convex whorls, suture slightly impressed; with 22 to 25 ribs in one millimeter; approximately 55 ribs on last whorl.

Observation The apertural area and protoconch is lost.

Type locality El Rosado Member, Gran Barranca south of Lake Colhue-Huapi, Chubut, Argentina.

Dimensions Maximum diameter 1.52 mm, height 1.42 mm.

Derivation of names Colhue: from the Lake Colhue-Huapi; *conus*: from Latin *conus* because of its shape; *simpsoni*: in honor of George Gaylord Simpson (1902–84), who studied the Cenozoic rocks and fossils of Patagonia.

Comparisons *Colhueconus* differs from *Zilchogyra costellata* (d'Orbigny 1835) (type species of genus, from the Río de la Plata region in Argentina and Uruguay) and *Z. hyltonscottae* Weyrauch 1965, from Tucumán, because these are bigger (4.5 and 2.3 mm, respectively), have a suborbicular shell, flat spire and canaliculated suture. It differs from *Z. michaelseni* (Strebel 1907), a Patagonian species, and *Radioconus bactricolus* (Guppy 1868) (type species of genus, inhabit the Antilles and Venezuela), in that the latter two are bigger (5.0 and 3.0 mm, respectively), with a trochoid and flat shell, more convex whorls, and canaliculated sutures. It differs from *Lilloiconcha superba* (Thiele 1927) (type species of genus, ranging from Argentina, Brazil, and Colombia) in being

Table 5.2. *Geographic distribution of living and stratigraphic distribution of fossil gastropod genera of the Sarmiento Formation*

Taxa	South America (excluding Argentina–Chile)	Argentina–Chile: N of 39° S	Patagonia	Casamayoran	Mustersan	Deseadan
Chilinidae						
Chilina	*	*	*	*		
Strophocheilidae						
Strophocheilus	*			*	*	*
Megalobulimus	*	*		*	*	?
Austroborus	*	*		?	?	
Orthalicidae						
Thaumastus	*			*	*	?
Paleobulimulus					*	
Odontostomidae						
Plagiodontes	*	*				
Charopidae						
Gyrocochlea?			*		*	
Lilloiconcha	*	*	*		*	
Radiodiscus	*	*	*		*	
Rotadiscus	*	*	*		*	
Stephanoda			*		*	
Stephadiscus			*		*	
Zilchogyra	*	*	*		*	
Colhueconus gen. et sp. nov.			*		*	

Note: Unknown stratigraphic position.

bigger (3.0 mm), having very convex whorls, and a sculpture of strong, almost straight axial ribs.

Occurrence Only known from its type locality.

Age Middle – late Eocene.

Repository CNP-PIIc 232. Type, 1 specimen.

Discussion and conclusions

In this chapter a new genus and species of Charopidae, *Colhueconus simpsoni*, is described. The revision of the collections of the Museo Argentino de Ciencias Naturales "Bernardino Rivadavia" and the Museo de La Plata allows us to add some new specimens of genera previously published for the Sarmiento Formation (*Strophocheilus, Megalobulimus, Thaumastus*), besides noting for the first time two genera (*Austroborus* and *Plagiodontes*).

Five families of gastropods are preserved in the lower and middle sections of the Sarmiento Formation. The stratigraphic distribution of these taxa (Table 5.2) presents a few uncertainties arising from older collections. The Casamayoran deposits (middle Eocene) include *Chilina*, the only aquatic genus; *Strophocheilus, Megalobulimus*,

Thaumastus, and probably *Austroborus*. In the subsequent Mustersan interval (middle – late Eocene) no significant changes are noted in the composition of the large-sized gastropod fauna, but a highly diverse assemblage of small land snails Charopidae is found exclusive in these strata. However, microgastropods and macrogastropods have not been found in the same locality. The frequency of fossil snails decreases significantly in younger deposits, presumably because of climate cooling and drying after the middle Eocene (Zachos *et al.* 2001). No specimens were recovered from Tinguirirican (latest Eocene – early Oligocene) or Colhuehuapian (early Miocene) beds, whereas in the Deseadan interval (late Oligocene – early Miocene) *Strophocheilus* and probably *Megalobulimus* and *Thaumastus* are present.

As most of gastropods of the Sarmiento Formation have closely related living representatives in South America, it is possible to infer something about paleoenvironmental conditions. Chilinidae live in temperate to cold lakes and rivers (Castellanos and Gaillard 1981). The present distribution of the three most frequent families of terrestrial gastropods from the Sarmiento Formation

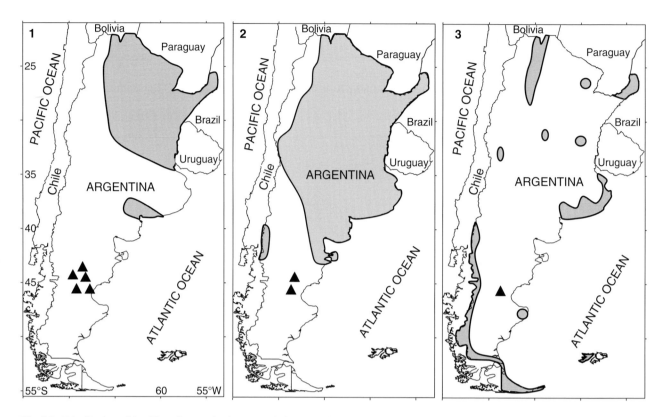

Fig. 5.2. Distribution of families of extant land gastropods in Argentina and their fossils in the Sarmiento Formation.
(1) Strophocheilidae; (2) Orthalicidae; (3) Charopidae.

(i.e. Strophocheilidae, Orthalicidae, and Charopidae) is
shown in Fig. 5.2. The selected areas in these maps have
contrasting climatic and biotic conditions, from tropical or
cool–temperate forests to semi-arid steppes. *Thaumastus* is
not present in Argentina, but in Venezuela, Ecuador, Peru,
Brazil, and Bolivia where it is found in seasonal forests
from tropical to equatorial latitudes (Breure 1979). *Strophocheilus* can occur in sites that are densely covered with
vegetation as in tropical southeastern Brazil (Bequaert
1948; Leme 1973). *Paleobulimulus* is closely related to
the extant *Bulimulus* which lives (as does *Megalobulimus*)
today in humid to subhumid and warm to temperate areas of
America (Bequaert 1948; Breure 1979). *Megalobulimus*
suggests seasonal precipitation because it buries deeply
for aestivation (Bequaert 1948). The environmental significance of genera of the dominant family Charopidae was
recently discussed by Miquel and Bellosi (2007), who
inferred a prevailing cold–temperate and humid habitat
for this fauna.

In summary, warm–temperate and humid conditions are
suggested by Casamayoran gastropods. Mustersan gastropods indicate a relatively colder environment. The contrasting ecological characteristics suggested by the larger-sized
genera may be an artifact of age uncertainties, given that

both large- and small-sized snails never occur together in
the same locality. This climate change is in general agreement with global cooling in the middle – late Eocene
(Bohaty and Zachos 2003). This trend might also explain
the absence of gastropods in Tinguiririran age beds of the
Sarmiento Formation of the early Oligocene (Kay *et al.*
1999; Ré *et al.* Chapter 3, this book). Finally, the faunal
composition of Deseadan strata denotes a return to warmer
conditions, as previously reported, suggested by the palynology of late Oligocene marine deposits of central Patagonia
(Barreda 1997; Barreda and Bellosi 2003).

ACKNOWLEDGEMENTS
We thank to Dr. H. Camacho, Dr. A. Riccardi, Lic.
A. Tablado, and Dr. C. Ituarte for making available the
collections of Departments of Paleoinvertebrates and Inver-
tebrates at the Museo Argentino de Ciencias Naturales
"Bernardino Rivadavia" and the Museo de La Plata; to
Ameghiniana for authorizing the reproduction of photo-
graphs "5a" and "5b"; to Dr. S. Martínez for his construct-
ive criticism. This work was supported by grants from
NSF-USA (BCS-9318942, DEB-9907985, EAR-0087636,
BCS-0090255) to R. F. Kay and R. H. Madden.

REFERENCES

Ameghino, F. 1906. Les formations sédimentaires du Crétacé Superieur et Tertiaire de Patagonie. *Anales del Museo Nacional de Buenos Aires*, **8**(3), 1–568.

Andreis, R. R. 1977. Geología del Área de Cañadón Hondo, Depto. Escalante, provincia del Chubut, República Argentina. *Revista del Museo de La Plata, "Obra del Centenario", Geología*, **4**, 77–102.

Barreda, V. D. 1997. Palinoestratigrafía de la Formación San Julián en el área de Playa La Mina, Oligoceno de la Cuenca Austral. *Ameghiniana*, **34**, 283–294.

Barreda, V. D. and E. S. Bellosi 2003. Ecosistemas terrestres del Mioceno temprano de la Patagonia central: primeros avances. *Revista del Museo Argentino de Ciencias Naurales*, n.s., **5**, 125–134.

Bellosi, E. S., S. E. Miquel, R. F. Kay, and R. H. Madden 2002. Un paleosuelo mustersense con micrograstrópodos terrestres (Charopidae) de la Formación Sarmiento, Eoceno de Patagonia central: significado paleoclimático. *Ameghiniana*, **39**, 465–477.

Bequaert, J. C. 1948. Monograph of the Strophocheilidae, a Neotropical family of terrestrial mollusks. *Bulletin of the Museum of Comparative Zoology*, **100**, 1–210.

Bohaty, S. M. and J. C. Zachos 2003. Significant southern ocean warming event in the late middle Eocene. *Geology*, **31**, 1017–1020.

Breure, A. S. H. 1979. Systematics, phylogeny and zoogeography of Bulimulinae (Mollusca). *Zoologische Verhandelingen Leiden*, **168**, 3–215.

Castellanos, Z. A. de and M. C. Gaillard 1981. Chilinidae. In *Fauna de Agua Dulce de la República Argentina*, vol. 15. Buenos Aires: Fundación para la Educación, la Ciencia y la Cultura, pp. 23–44.

Crawford, G. I. 1939. Notes on *Stephanoda patagonica* (Suter) and the genus *Radiodiscus*, with a new name for *R. patagonicus* Pilsbry. *Nautilus*, **52**(4), 115–117.

Doering, A. 1875 [1877]. Apuntes sobre la fauna de moluscos de la República Argentina (continuación). *Periódico Zoológico*, **2**(4), 219–258.

d'Orbigny, A. 1835. Sinopsis terrestrium et fluviatilium molluscorum, in suo per Americam meridionalem itinere. *Magasin de Zoologie*, **5**(61), 1–44.

Fernández, D. 1973. *Catálogo de la Malacofauna Terrestre Argentina*. La Plata: Comisión de Investigaciones Científicas.

Gould, A. A. 1846. Descriptions of the shells collected by the United States Exploring Expedition under Captain Wilkes. *Proceedings of the Boston Society of Natural History*, **2**, 165–167.

Guppy, R. J. L. 1868. On the terrestrial Mollusca of Dominica and Grenada; with an account of some new species from Trinidad. *Annals and Magazine of Natural History*, **1**, 429–442.

Hidalgo, J. G. 1869. Description d'espèces nouvelles. *Journal de Conchyliologie*, **9**, 410–413.

Hylton Scott, M. I. 1968. Endodóntidos neotropicales III. *Neotropica*, **14**(45), 99–102.

Hylton Scott, M. I. 1969. Endodóntidos neotropicales IV (Moll. Pulm.). *Neotropica*, **15**(47), 59–63.

Ihering, H. von 1904. Nuevas observaciones sobre moluscos cretáceos y terciarios de Patagonia. *Revista del Museo de La Plata*, **11**, 227–244.

Ihering, H. von 1907. Les Mollusques fossiles du Tertiaire et du Crétacé supérieur de l'Argentine. *Anales del Museo Nacional de Buenos Aires*, **14**, 1–611.

Ihering, H. von 1914. Catálogo de Moluscos cretáceos e Terciarios da Argentina da colecção do author. *Notas Preliminares do Museu Paulista*, **1**, 1–148.

Kay, R. F., R. H. Madden, M. G. Vucetich, A. A. Carlini, M. M. Mazzoni, G. H. Ré, M. Heizler, and H. Sandeman 1999. Revised geochronology of the Casamayoran South America Land Mammal Age: climatic and biotic implications. *Proceedings of the National Academy of Sciences USA*, **96**, 13235–13240.

King, P. P. and W. J. Broderip 1831. Description of the Cirripedia, Conchifera and Mollusca. *Zoological Journal*, **5**, 332–349.

Leme, J. L. M. 1973. Anatomy and systematics of the Neotropical Strophocheilidae (Gastropoda, Pulmonata) with the description of a new family. *Arquivos de Zoologia*, **23**, 295–336.

Miquel, S. E. and E. S. Bellosi 2004. Un nuevo microgastrópodo terrestre (Charopidae) del Eoceno de Patagonia central, Argentina. *Ameghiniana*, **41**, 111–1114.

Miquel, S. E. and E. S. Bellosi 2007. Microgasterópodos terrestres (Charopidae) del Eoceno medio de Gran Barranca (Patagonia Central, Argentina) y su relación con los cambios ambientales de la transición Eoceno–Oligoceno. *Ameghiniana*, **44**, 121–131.

Parodiz, J. J. 1946. Bulimulinae fósiles de la Argentina. *Notas del Museo de La Plata, Paleontología*, **11**, 301–310.

Parodiz, J. J. 1949a. Un nuevo gastrópodo terrestre del Eoceno de Patagonia. *Physis (Buenos Aires)*, **20**, 174–179.

Parodiz, J. J. 1949b. Notas sobre "*Strophocheilus*" fósiles de Argentina. *Physis (Buenos Aires)*, **20**, 180–184.

Parodiz, J. J. 1949c. *Austroborus* n. nom. pro *Microborus* Pilsbry, 1926. *Physis (Buenos Aires)*, **20**, 189–190.

Parodiz, J. J. 1963. New fresh-water mollusca from the Eogene of Chile and Patagonia. *Nautilus*, **76**, 145–147.

Parodiz, J. J. 1969. The Tertiary non-marine Mollusca of South America. *Annals of the Carnegie Museum of Natural History*, **40**, 1–242.

Parodiz, J. J. 1996. The taxa of fossil mollusca introduced by Hermann von Ihering. *Annals of the Carnegie Museum of Natural History*, **65**, 183–296.

Spalleti, L. A. and M. M. Mazzoni 1979. Estratigrafía de la Formación Sarmiento en la barranca sur del lago Colhue-Huapi, provincia del Chubut. *Revista de la Asociación Geológica Argentina*, **34**, 271–281.

Strebel, H. 1907. Beiträge zur Kenntnis der Molluskenfauna der Magalhaen-Provinz. V. *Zoologischer Jahrbucher*,

Abteilung für Systematik, Okologie und Geografie der Tiere, **25**, 79–196.

Thiele, J. 1927. Über einige brasilianische Landschnecken. *Abhandlungen der Senckenberggischen Naturforschenden Gesellschaft*, **40**, 307–399.

Weyrauch, W. K. 1965. Neue und verkannte Endodontiden aus Südamerika. *Archiv für Molluskenkunde*, **94**(3/4), 121–134.

Wood, W. 1828. *Index testaceologicus*, Supplement i–iv, 1–59. London.

Zachos, J., M. Pagani, L. Sloan, E. Thomas, and K. Billups 2001. Trends, rhythms, and aberrations in global climate 65 Ma to present. *Science*, **292**, 686–694.

Zilch, A. 1959–60. Gastropoda, Euthyneura, In O. H. Schindewolf, *Handbuch der Paläozoologie* **6**(2) i–xii, 1–834. Berlin.

6 Middle Tertiary marsupials from central Patagonia (early Oligocene of Gran Barranca): understanding South America's *Grande Coupure*

Francisco Javier Goin, María Alejandra Abello, and Laura Chornogubsky

Abstract

We describe two new marsupial assemblages from GBV-4 "La Cancha" and GBV-19 "La Cantera," localities at Gran Barranca south of Lake Colhue-Huapi in central Patagonia (Sarmiento Department, Chubut Province, Argentina). These remarkable new faunas (around 400 specimens referable to 24 species, 18 genera, 14 families, and 5 orders) correspond to a time interval previously unknown for central Patagonia: the post-Mustersan, pre-Deseadan, early Oligocene. A Tinguiririan age (SALMA) can be inferred for the La Cancha levels and fauna (18 species), based on its similarity with the marsupials of Tinguririca in Chile. In turn, the La Cantera association (10 species) represents a younger, pre-Deseadan age. The La Cancha marsupials represent the most dramatic faunal turnover in South American marsupials (and probably of all other mammalian clades in the continent as well) during the Cenozoic Era, an event here termed the Patagonian Hinge ("Bisagra Patagónica"). This turnover coincides with a sudden drop in global temperatures at latest Eocene – early Oligocene time. Some of the features that characterize this turnover among marsupials are the last records of Polydolopiformes and bonapartherioid Bonapartheriiformes, the beginning of the Argyrolagoidea (and of hypsodonty in marsupials), a rapid radiation of paucituberculatans, and the development of gigantism among borhyaenid sparassodonts and the last polydolopines.

Resumen

Se describen dos nuevas asociaciones de marsupiales exhumadas en La Cancha y La Cantera, sobre la Barranca Sur del lago Colhue-Huapi, en Patagonia Central (Departamento de Sarmiento, Provincia del Chubut, República Argentina). Estas nuevas y notables faunas (unos 400 especímenes referibles a 24 especies, 18 géneros, 14 familias y cinco órdenes) se corresponden con un lapso temporal previamente desconocido para esta región: el Oligoceno

Temprano (post-Mustersense—pre-Deseadense). Se infiere una edad Tinguiririquense para los niveles y fauna de La Cancha (18 especies), en base a la asociación faunística exhumada, notablemente similar a la de niveles tinguiririquenses de Chile central. Por su parte, la asociación de La Cantera (con 10 especies) parece representar una edad más joven, pre-Deseadense. Los marsupiales de La Cancha son representativos del recambio faunístico más dramático ocurrido entre los marsupiales sudamericanos (y probablemente también en el resto de los mamíferos) durante la Era Cenozoica. Este evento, aquí denominado "Bisagra Patagónica", ocurrió en coincidencia con una súbita caída de las temperaturas globales hacia el Oligoceno más temprano. Algunos de los procesos que caracterizaron este recambio son: último registro de Polydolopimorphia Polydolopiformes y Bonapartheriiformes Bonapartherioidea, y primeros registros de Argyrolagoidea (e inicios de la hipsodoncia en los marsupiales); radiación rápida de los Paucituberculata y el desarrollo de procesos de gigantismo en los últimos polidolopinos y en algunos Sparassodonta.

Introduction

As revealed by recent work (Goin and Candela 2004; Goin *et al.* 2007a) the application of matrix screen-washing and sorting techniques to fossil-bearing levels is leading to a revolution in our knowledge of extinct, small-sized South American mammals. This is especially true for marsupials, as a large majority of species (living and extinct) range from minute to medium-sized (e.g. Gordon 2004). Ten years of fossil prospecting at Gran Barranca in central Patagonia, Argentina (see Zucol and Brea 2005, Fig. 1), including screen-washing and picking efforts led by paleontologists from Duke University and the Museo de La Plata, have resulted in an impressive amount of new specimens, part of which are described here.

The new material described in this chapter were collected from La Cancha levels and the La Cantera locality (see below), in the middle section of the Sarmiento Formation in

The Paleontology of Gran Barranca: Evolution and Environmental Change through the Middle Cenozoic of Patagonia, eds. R. H. Madden, A. A. Carlini, M. G. Vucetich, and R. F. Kay. Published by Cambridge University Press. © Cambridge University Press 2010.

Gran Barranca. Recent advances in the understanding of the geology and paleontology of the middle Cenozoic at Gran Barranca have been provided by Ré *et al.* (2005), Zucol and Brea (2005), Madden *et al.* (2005a, 2005b), Vucetich *et al.* (2005), López *et al.* (2005), Carlini *et al.* (2005a, 2005b), Bellosi (2005), and Bellosi and Madden (2005) and by Bellosi (this book) and Ré *et al.* (Chapter 4, this book). The La Cancha levels are exposed at two localities corresponding to Profiles K and M of eastern Gran Barranca (Ré *et al.* 2005, Fig. 2; Ré *et al.* Chapter 3, this book), situated within the Vera Member of the Sarrmiento Formation. Based upon a combination of radiometric age of the Carlini Tuff and paleomagnetics, the base of the Vera Member is no older than 35.0 Ma and no younger than 33.26 Ma. The La Cancha faunal level and the Carlini Tuff falls within a single normal polarity zone interpreted as Chron C13n (33.266–33.738 Ma: Gradstein *et al.* 2004) (Ré *et al.* Chapter 4, this book). Referring to these levels, Carlini *et al.* (2005b) suggested that "the age of the base of a Tinguiririrican Land Mammal Age may be more precisely constrained at Gran Barraca" because the age of the Chilean Tinguiririrican can be constrained only to between ∼31.5 and 37.5 Ma (Flynn *et al.* 2003). In turn, the La Cantera locality (GBV-19) in Unit 3 of the Upper Puesto Almendra Member is below those that yield typical Deseadan age mammals. The age of Unit 3 and the fossils from GBV-19 "La Cantera" is between 31.1 and 29.5 Ma (Ré *et al.* Chapter 4, this book). Updated stratigraphy and sedimentology reviews of the Sarmiento Formation are provided by Bellosi (this book) and Ré *et al.* (Chapter 3, this book).

Marsupials from La Cantera and La Cancha comprise two completely new assemblages, outstanding for two reasons: (1) they reveal a previously unknown diversity of taxa and adaptive types among mid-Cenozoic South American marsupials, and (2) they give decisive information on the extensive faunal turnover that took place across the Eocene–Oligocene boundary. The boundary is at 33.9 ± 0.1 Ma, such that the La Cancha levels could be as little as 100 k.y., and no more than 1 million years younger than the Eocene–Oligocene boundary. The La Cantera fauna is 2.2 to 4.4 m.y. younger than the boundary. This faunal turnover could represent a direct consequence of the global cooling that took place at the earliest Oligocene, a thermal event that signalled the beginning of the climate conditions known as the "icehouse world." A preliminary account of the La Cancha marsupials was given by Goin and Candela (1997).

Due to the limitations of space, we present here the new marsupial faunas (including taxon diagnoses, types, and referred specimens), together with a brief comment on their significance. Detailed descriptions, as well as a more thorough discussion of the bearing of these faunas to marsupial evolution in South America during the early Oligocene, are to be published in a future work. This study does not exhaust the richness of levels and faunas from Gran Barranca. Other marsupial associations from different levels of this same area are being studied and published elsewhere, including the Col-huehuapian South American Land Mammal Age (SALMA) marsupials (Goin *et al.* 2007a) and the Itaboran SALMA taxa from the Las Flores formation. Finally, an updated review of the Barrancan (= Casamayoran *pars*) and Mustersan marsupials from Gran Barranca is still forthcoming.

This work is dedicated to Florentino Ameghino and Rosendo Pascual, successive fathers of vertebrate paleontology in Argentina, and senior analysts of South American faunal turnovers.

Materials and abbreviations

All materials here described, around 400 specimens consisting mostly of isolated teeth, belong to the collections of the Sección Paleontología Vertebrados of the Museo Paleontológico "Egidio Feruglio" in Trelew (MPEF-PV). Abbreviations used are: Ma, *megannum*, 1 million years in the radioisotopic timescale; EOB, Eocene–Oligocene Boundary. Dental nomenclature follows Goin *et al.* (2003). In the description or argyrolagoid material, we follow Hershkovitz (1971) for the nomenclature of neomorphic cusps. Our use of these terms does not imply homologies with similar structures in eutherian mammals. To emphasize this, we put quotes around terms that refer to new structures described here (e.g., "epiconule," "epiconular shelf," "ectostylid," and "entostylid"). Dental measurements are given in Appendix 6.1. L, length; W, width. In lower molars, width corresponds to talonid width. Measurements are in millimeters (mm).

In the following section, specimens are referred either to the La Cancha or to the La Cantera levels/localities, each associated with the following geographic, stratigraphic, and biochronologic information:

(1) *La Cancha* The La Cancha bed or level outcrops between profiles K and M at Gran Barranca south of Lake Colhue-Huapi in Sarmiento Department, Chubut Province (Central Patagonia), Argentina; 45° 42′ S; 68° 44′ W. Vera Member, Sarmiento Formation. Earliest Oligocene (Tinguiririrican SALMA).

(2) *La Cantera* GBV-19 "La Cantera" is a locality at Profile A (west of Profile MMZ) at Gran Barranca south of Lake Colhue-Huapi in Sarmiento Department, Chubut Province (Central Patagonia), Argentina; 45° 42′ S; 68° 44′ W. Unit 3, Upper Puesto Almendra Member, Sarmiento Formation. Early Oligocene. Maps, profiles, and geological descriptions of these levels and localities are provided by Bellosi (this book) and Ré *et al.* (Chapter 3, this book).

Systematics

Infraclass **METATHERIA** Huxley 1880
Supercohort **MARSUPIALIA** Illiger 1811
Order **DIDELPHIMORPHIA** Gill 1872
Superfamily **PERADECTOIDEA** Crochet 1979

Note We follow Goin (2007) who provisionally included the Peradectoidea (Peradectidae, Caroloameghiniidae, and Mayulestidae) within the Didelphimorphia. In this concept, *Alphadon* and allies (Alphadontidae *sensu* Kielan-Jaworowska *et al.* 2004) are explicitly excluded from the Didelphimorphia. Thus, we retain the concept of Cohort Alphadelphia (Marshall *et al.* 1990) in a more restricted sense (i.e. excluding Peradectidae and Caroloameghiniidae).

Family **CAROLOAMEGHINIIDAE**
Ameghino 1901
Canchadelphys, new genus
Etymology *Cancha-*, from La Cancha, the type level/locality, and *delphys*, Greek for "womb," common ending of generic names for "opossum-like" marsupials. Gender is feminine.
Type species *Canchadelphys cristata*, new species
Included species The type only.
Diagnosis As for the type species.
Occurrence Early Oligocene, southern South America.

Canchadelphys cristata, new species
Etymology From the Latin *cristatus*, "crested"; after the more crested molar pattern of the new taxon, as opposed to the bunoid condition of molars in all other caroloameghiniid species.
Type MPEF-PV 4460, a left M?2 (Fig. 6.1A).
Hypodigm The type and MPEF-PV 4461, a right M?3, and MPEF-PV 4242, a right M?1.
Occurrence La Cancha.
Measurements See Appendix 6.1.
Diagnosis *Canchadelphys cristata* differs from all other caroloameghiniids in its much better developed crests; upper molars with more reduced StD, which is anteriorly placed in relation to the metacone; the postmetacrista joins posteriorly to a crest projecting posteriorly from StD; differs from *Caroloameghinia* in that the anterior cingulum formed by the preparaconular crest is wider, the postmetaconular crest is proportionally longer and ends labially upwards, the trigon basin is anteroposteriorly longer, and that enamel molars are less "wrinkled."
Remarks The type and specimen MPEF-PV 4461 were already mentioned (as *?Caroloameghinia*) by Goin (2007) in a review of the Caroloameghiniidae. *Canchadelphys cristata* represents the last occurrence of a carolameghiniid marsupial in the fossil record. Most notable of *Canchadelphys* is the development of crests out of the more bunoid, crestless pattern of other caroloameghiniids.

SUPERFAMILY indet.
Family **STERNBERGIIDAE** McKenna and Bell 1997
Genus and species indet.
Referred specimens MPEF-PV 4459, left M4; MPEF-PV 4457, right mx.
Occurrence La Cancha.
Measurements See Appendix 6.1.
Remarks Specimens are too scanty to allow a taxonomic allocation closer than Sternbergiidae. The lower molar resembles those of the Itaboraian *Itaboraidelphys camposi*; however, the paracristid is more trenchant, and the La Cancha specimens are smaller.

Order **SPARASSODONTA** Ameghino 1894
Family **HATHLIACYNIDAE** Ameghino 1894
Genus and species **indet**.
Referred specimens MPEF-PV 4345, a right M?1 lacking the protocone (Fig. 6.1B, C).
Occurrence La Cancha.
Measurements Length 4.95
Description MPEF-PV 4345 has a reduced stylar shelf, a shallow ectoflexus, a moderately developed StB (which is paired with the larger paracone), StC absent, and a small and labiolingually compressed StD located quite posteriorly, near the distal end of postmetacrista. The parastyle is quite reduced; the paracone is only slightly smaller than the metacone, and the postmetacrista is well developed.
Remarks We tentatively include this specimen within the Hathliacynidae, though we note that a thorough review of early sparassodonts is needed. Specimen 4345 seems to be more specialized for a carnivorous diet than species of *Patene*, from the Paleocene–Eocene of South America.

?HATHLIACYNIDAE
Referred specimens MPEF-PV 4344, a right ?lower premolar.
Occurrence La Cancha
Measurements Length 5.74, width 3.68
Remarks Because of its larger size, MPEF-PV 4344 obviously does not belong to the same taxon as MPEF-PV 4345. It is referred to the Sparassodonta, but whether it represents a hathliacynid or a small borhyaenid remains unclear.

Family **BORHYAENIDAE** Ameghino 1894
Genus *Pharsophorus* Ameghino 1897
Pharsophorus cf. *P. lacerans* Ameghino 1897
Referred specimens MPEF-PV 4190, an almost complete left dentary with c–m4 (Fig. 6.1D).

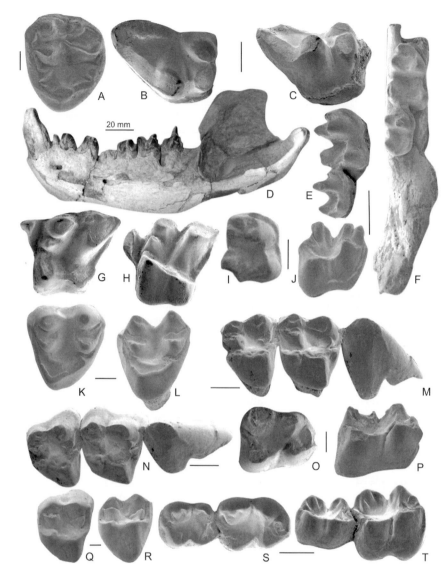

Fig. 6.1. (A) *Canchadelphys cristata*, gen. et sp. nov. (Caroloameghiniidae). MPEF-PV 4460 (type), a left M?2 in occlusal view. (B–C), Hathliacynidae, gen. et sp. indet. MPEF-PV 4345, a right M?1 lacking the protocone, in occlusal (B) and lingual (C) views. (D) *Pharsophorus* cf. *P. lacerans* Ameghino (Borhyaenidae). MPEF-PV 4190, an almost complete left dentary with c–m4 in labial view. (E–H) *Evolestes* sp. MPEF-PV 3804, fragment of right dentary with m3–4, in labial (E) and occlusal (F) views. (G–H) MPEF-PV 3849, left M1 in occlusal (G) and lingual (H) views. (I–L) Pichipilidae, gen. et sp. indet.; MPEF-PV 3828 (detail), right m2 in occlusal (I) and labial (J) views; MPEF-PV 3890, left M2 or M3 in occlusal (K) and occlusal-lingual (L) views. (M–P) *Pilchenia antiqua*, sp. nov. (Palaeothentidae); MPEF-PV 4490 (type), a right maxillary with alveoli and roots for the M3 and complete P3–M2 in occlusal–lingual (M) and occlusal (N) views; MPEF-PV 4235, right m2 in occlusal (O) and labial (P) views. (Q–T) *Pilchenia intermedia* sp. nov. (Palaeothentidae); MPEF-PV (type) 3836, a right M2 in occlusal (Q) and lingual (R) views; MPEF-PV 3898, fragment of right dentary with m2–3 in occlusal (S) and labial (T) views.

Occurrence La Cantera.

Measurements c, length 17.04, width 12.2; p1, length 7.38, width 3.4; p2, length 11.31, width 7.2; p3, length 12, width 8.85; m1, length 12.83, width 8.24; m2, length 13.72, width 9.05; m3, length 15.34, width 9.81; m4, length 16.74, width 6.92.

Remarks Differences with the type of *P. lacerans* are only minor and mostly restricted to the proportions of p2–3 (stouter in the La Cantera specimen). A review of all specimens currently assigned to *P. lacerans* could prove the full assigment to this species of specimen 4190.

Order **PAUCITUBERCULATA** Ameghino 1894
Family **?CAENOLESTIDAE** Trouessart 1898
Genus *Evolestes* Goin, Sánchez-Villagra,
Abello & Kay 2007
***Evolestes* sp**.
Referred specimens MPEF-PV numbers 3804, fragment of right dentary with m3–4 (Fig. 6.1E, F); 3849, left M1 (Fig. 6.1G, H); 3906, fragment of left M1; 3913, right M2; 3864, right M2; 3918, right M2; 4152, right M2; 3827, left M2; 3835, left M3; 3850, left M3; 3813, right M3; 3834, right M4; 3830, left m2; 3811, right m3; 3812, trigonid of a left m1; 3857, talonid of right m2; 3866, trigonid of right m2; 3846, left m3; 3844, fragment of left dentary with m2 and partial of m3.
Occurrence GBV-19 La Cantera.
Measurements See Appendix 6.1.
Description Molars sharply decrease in size from M/m1–2 to M/m3–4. M1–2 are subquadrangular in occlusal view, whereas M3–4 are subtriangular (the metaconule is much more developed and lingually expanded in M1–2). In M1–2 the paracone is smaller than the metacone; the metacone is twinned to StD, whereas the paracone is placed lingually and slightly anterior to StB; StD is much smaller than StB; both cusps are labiolingually compressed; the postmetaconular crest extends labially to the metastylar corner, shaping a "cingulum-like" structure. The centrocrista is open: the premetacrista and the postparacrista meet the lingual slopes of StD and StB, respectively. In M1–3 a parastylar cusp (StA?) can be observed, especially in M1. In the lower molars, the entoconid is labiolingually compressed; the hypoconid is labially salient; the para- and metaconids are moderately twinned (though set more apart than in other paucituberculatans). The paracristid is not straight: there is a ~90° angle at the meeting of the postparacristid and the preprotocristid. The hypoconulid is very small and antero-posteriorly compressed. Anterobasal cinguli are short and narrow.
Remarks The recently described *Evolestes* (Goin *et al.* 2007b), from the Deseadan beds of Salla, Bolivia is notable in being generalized as compared to all other known caenolestoid paucituberculatans. Specimens here described are obviously referable to this genus; better-preserved specimens of *Evolestes* from La Cantera may allow the recognition of a new species. Up to now, caenolestids are regarded as the most generalized members of the order Paucituberculata. The molar morphology of *Evolestes* sp. preserves features which are intermediate between those of the generalized didelphimorphian pattern and those of Neogene caenolestids. For instance, in the lower molars the paraconid and the metaconid are relatively close to each other (as compared to didelphimorphians); however the cusps are not twinned as in living caenolestids. Also, the entoconid is large and labiolingually compressed, but is not as large as in caenolestids and it also lacks a sharp preentocristid. In the upper molars, the StB, even though large, is not proportionaly enormous as in other paucituberculatans; the para- and metacone are still distinct and separated from the StB and StD. Finally, the StD is only moderately sized (very large in M1–2 of other paucituberculatans). In short, *Evolestes* may prove to be morphologicaly basal to the entire order Paucituberculata.

Family **PICHIPILIDAE** Marshall 1980, new rank
Genus and species **indet**.
Referred specimens MPEF-PV 3828, right dentary fragment with alveoli for p3–m1 and complete m2 (detail of m2 in Fig. 6.1I, J); 3910, right m?3; MPEF-PV 3848, left M2 or M3; MPEF-PV 3890, left M2 or M3 (Fig. 6.1K, L).
Occurrence GBV-19 La Cantera.
Remarks A series of features suggest that these specimens belong to pichipilid Paucituberculata. The lower molars have a wide talonid basin; the entoconid is laterally compressed, and has a short preentocristid which is oriented towards the talonid basin; there is a well-developed postentocristid; the metaconid is anteriorly placed with respect to the protoconid, and is subequal in size to the paraconid; there is a deep metacristid; the protoconid and metaconid have posterior crests running vertically to each cusp; there is a long anterobasal cingulum, extending posteriorly to the base of the hypoconid. There are a few differences between the lower molars: in MPEF-PV 3910 the hypoconid is less labially salient than in MPEF-PV 3828; although broken, it is evident that the entoconid was large; also different from MPEF-PV 3828, the paraconid is slightly smaller with respect to the metaconid; also in this molar, there is a short anterobasal cingulum. Upper molars are subtriangular in outline, as the metaconule is small; the para- and metacone are well developed, though smaller than StB and StD. As in other pichipilids, both the protocone and the metaconule are located in a relatively anterior position with respect to the stylar cusps. Because of their sizes (upper molars are larger), upper and lower molars do not belong to the same species.

?PICHIPILIDAE
Referred specimens MPEF-PV 4410, right mx; MPEF-PV 4402 left mx; MPEF-PV 4494, fragment of right dentary with m3–4.

Occurrence La Cancha.

Remarks All three specimens share at least two pauci-tuberculatan features: proportionally large, labially salient hypoconids, and large, laterally compressed, and crested entoconids. Also, the metaconid is anteriorly placed relative to the protoconid at least in MPEF-PV 4402 and 4410. On the other hand, all three specimens lack the vertical posterior crests to the metaconid and protoconid, otherwise ubiquitous among pichipilids.

Family **PALAEOTHENTIDAE** Sinclair 1906
Genus *Pilchenia* Ameghino 1903
[= *Palaeothentes* Patterson and Marshall 1978]
Type species *Pilchenia lucina* Ameghino 1903
Extended diagnosis To the diagnostic features mentioned by Marshall (1978, p. 82), we add the following, relative to the upper dentition: in M1–3, para- and metacone proportionally larger and more separated from StB and StD, respectively, than in other known palaeothentids; metacone is not anteriorly displaced relative to the StD.
Occurrence Early–late Oligocene, southern South America.

Pilchenia antiqua, new species
Etymology From the Latin *antiquus*, old.
Type MPEF-PV 4490, a right maxillary with alveoli and roots for the M3 and complete P3–M2 (Fig. 6.1M, N).
Hypodigm The type and MPEF-PV numbers 4486, fragment of right maxillary with M2–3; 4482, fragment of left M1; 4483, fragment of left M1; 4484, right M2; 4148, fragment of right M2; 4479, right M3; 4481, left M3; 4480, left M3; 4478, right M3; 4474, left M4; 4129, fragment of right dentary with p3, anterior root of m1, and posterior alveolus of m1 partially preserved; 4126, left p3; 4489, right m1; 4235, right m2 (Fig. 6.1O, P); 4470, left m2; 4488, partial right m2; 4469, right m3; 4465, right m3; 4468, right m3; 4467, right m3; 4471, left m3; 4466, left m3; 4464, fragment of right m3; 4462, right m4; 4463, right m4.
Occurrence La Cancha.
Measurements See Appendix 6.1.
Diagnosis Differs from *P. lucina* in being smaller; P3 is proportionally larger and with a vestigial anterobasal cusp (larger in *P. lucina*); the labial face of M1–2 is higher; StB in M2 is much higher than StD; metaconule smaller and lower relative to the trigon basin; the premetaconular crest does not end labially in a cusp; more rapid decrease in size from M/m1 to M/m4; m1 has a shorter trigonid and a proportionally larger paraconid; the posterior crest of the entoconid in m1–3 does not connect with the postcristid; in m2 the cristid obliqua ends anteriorly at the metacristid

notch; paraconid in m2–3 is more labial relative to the metaconid. Differs from *P. intermedia* n. sp. in being smaller, metaconule in M1–2 is slightly lower relative to the trigon basin; metaconule in M1 is proportionally smaller; trigonid in m2 is narrower; the cristida obliqua in m2 ends anteriorly at the metacristid notch.
Remarks *Pilchenia antiqua* constitutes the oldest record of an undoubted palaeothentid marsupial. *Sasawatsu mahaynaq* Goin and Candela 2004, from the ?Mustersan SALMA of eastern Peru, was tentatively referred to "cf. Family Palaeothentidae" (Goin and Candela 2004; see the discussion on the taxonomic position of *Sasawatsu* and on uncertainties regarding the age of the Peruvian assemblage). *Pilchenia antiqua* adds significant knowledge on the dental morphology characteristic of this genus. Species of *Pilchenia* show several features that are generalized among the Palaeothentidae: proportionally large paracone and metacone in M1–4, and, in *P. antiqua* and *P. intermedia*, the metaconule is low regarding the trigon basin. *Pilchenia lucina* and *P. antiqua* also have relative large paraconids in m1, while in m2–3 all species of this genus have a distinct paraconid. Finally, another generalized feature shared by *P. antiqua* and *P. intermedia* is the absence of any connection between the postcristid and the posterior crest of the entoconid in m2–3.

Pilchenia intermedia, new species
Etymology From the Latin *intermedius*, intermediate; representatives of this species are intermediate in size and morphology between the type species of the genus and *P. antiqua*.
Type MPEF-PV 3836, a right M2 (Fig. 6.1Q–T).
Hypodigm The type and MPEF-PV numbers 3869, fragment of left M1; 3893, fragment of left M1; 3878, fragment of right M1; 3798, right M1; 3825, right M1; 3855, fragment of right M1; 3902, fragment of left M1; 3887, right M2; 3829, right M2; 3897, fragment of right M2; 3903, left M3; 3853, right M3; 3891, right M3; 3803, fragment of right M3; 3919, fragment of right M3; 3859, left M4; 3852, left M4; 3831, right M4; 3922, fragment of right dentary with p3; 3845, fragment of left dentary with alveoli for p2–m1; 3898, fragment of right dentary with m2–3 (Fig. 6.1S, T); 3900, left p3; 3901, right p3; 3894, left p3; 3814, talonid of right m1; 3842, talonid of left m1; 3892, talonid of right m2; 3873, right m2; 3889, fragment of right m2; 3802, fragment of left m2; 3886, left m3; 3905, left m3; 3805, left m3; 3895, right m3; 3904, fragment of right m3; 3868, right m4; 3815, left m4.
Tentatively referred specimen MPEF-PV 3920, right P3.

Occurrence GBV-19 La Cantera.

Measurements See Appendix 6.1.

Diagnosis Differs from *P. antiqua* in being larger, metaconule in M1–2 is slightly higher relative to the trigon basin; metaconule in M1 is proportionally larger; trigonid in m2 is wider; the cristid obliqua in m2 does not end anteriorly at the metacristid notch. Differs from *P. lucina* in being somewhat smaller; molar size decreases more rapidly from M/m1 to M/m4; upper molars have the metaconule less lingually salient, the premetaconular crest lacks a cusp at its distal end.

PALAEOTHENTIDAE indet.

Referred specimens MPEF-PV 3838, right m1 trigonid; MPEF-PV 3843, right m2.

Occurrence GBV-19 La Cantera.

Remarks MPEF-PV 3843, similar in size to *Carlothentes chubutensis* (Ameghino 1897), is clearly larger than MPEF-PV 3838, so that they do not belong to the same taxon. MPEF-PV 3843 cannot be referred to *Pilchenia*, as the hypoconulid is large and dorsoventrally compressed, the paraconid is vestigial, and the postparacristid is not perpendicular to the molar axis but instead oblique to it (a feature that characterizes more modern palaeothentids). MPEF-PV 3838 is too worn and fragmentary to attempt any generic referral.

Order **MICROBIOTHERIA** Ameghino 1889
Family **MICROBIOTHERIIDAE** Ameghino 1887
Genus *Eomicrobiotherium* Marshall 1982a
Eomicrobiotherium matutinum, new species

Etymology From the Latin *matutinus*, early, from the morning; implying that the new species is the oldest known for the genus.

Type MPEF-PV 4369, right m1 (Fig. 6.2A, B).

Hypodigm The type and MPEF-PV numbers 4370, right m2; 4352, left m2; 4350, trigonid of right m2; 4373, right m3; 4232, right M1; 4374, left M2; 3872, left M3.

Tentatively referred specimen MPEF-PV 4357, talonid of left mx.

Occurrence All specimens are from La Cancha, except MPEF-PV 3872, recovered from GBV-19 La Cantera.

Measurements See Appendix 6.1.

Diagnosis Differs from the type species of the genus, *E. gaudryi* (Simpson 1964), in being larger, the paraconid in m1 is more anteriorly placed, the metaconid is placed not so posterior relative to the protoconid, and the hypoconid is more labially salient.

Clenia Ameghino 1904
Clenia brevis, new species

Etymology From the Latin *brevis*, short; in reference to the shorter talonid of the m1 in the new species.

Type MPEF-PV 4368, left m1 (Fig. 6.2C, D).

Hypodigm The type and MPEF-PV numbers 4353, left m1; 4364, left m1; 4349 right m1; 4361, right m1; 4358, left m2; 4354, left m2; 4363, right m2; 4355, right m3; 4360, right m3; fragment of left maxilla with M2–3; 4146, fragment of right maxilla with M2–3; 4238, left M1; 4372, right M2.

Tentatively referred specimens MPEF-PV 4346, right mx; MPEF-PV 4356, right mx.

Occurrence La Cancha.

Measurements See Appendix 6.1.

Diagnosis Differs from the type species of the genus, *Clenia minuscula*, in the following features: molars are wider and more robust; talonid is shorter in m1; metaconid of m2 is less posteriorly placed.

Remarks In the last thorough review of representatives of the Microbiotheriidae, Marshall (1982a) synonymized *Clenia* Ameghino 1904 and *Oligobiotherium* Ameghino 1904, with *Microbiotherium* Ameghino 1887. We do not concur with his conclusions. The La Cancha specimens add new features diagnostic of *Clenia* as a distinct genus of microbiotheriid marsupials. A current review of Colhuehuapian marsupials (Goin *et al.* 2007a) adds more information on *Clenia*'s full generic status.

Genus *Microbiotherium* Ameghino 1887
Microbiotherium sp.

Referred specimens MPEF-PV numbers 4365, left mx; 4348, left mx; 4366, fragment of left dentary with p3–m1 (Fig. 6.2F, G); 4371, right Mx; 4362, right M1 (Fig. 6.2E); 4367, right Mx; 4359, left M?3; 4135, left M?3.

Occurrence La Cancha.

Measurements See Appendix 6.1.

Remarks We provisionally assign these specimens to *Microbiotherium* sp. As in the La Cancha specimens, those referable to species of *Microbiotherium* have lower molars with the metaconid not so posteriorly displaced as in other genera (e.g. *Oligobiotherium* Ameghino 1902, *Clenia* Ameghino 1904, and *Pachybiotherium* Ameghino 1902), the paraconid is not reduced, and the hypoconulid is closer to the entoconid; in turn, upper molars have a more or less distinguishable StC in the already reduced stylar shelf.

Order **POLYDOLOPIMORPHIA** Marshall 1987
Suborder **HATCHERIFORMES** Case, Goin, and Woodburne 2005
Family **GLASBIIDAE** Clemens 1966
Periakros, new genus

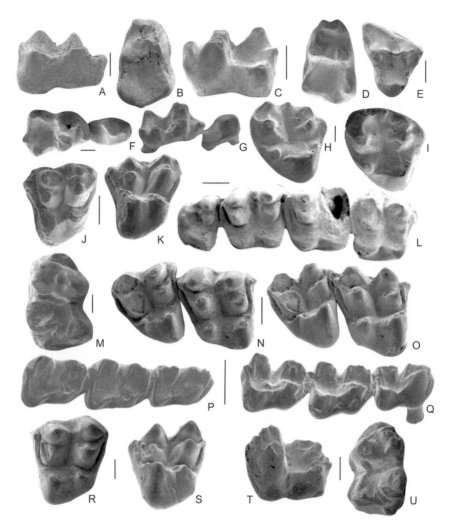

Fig. 6.2. (A–B) *Eomicrobiotherium matutinum* sp. nov. (Microbiotheriidae); MPEF-PV 4369 (type), right m1 in labial (A) and occlusal (B) views. (C–D) *Clenia brevis* sp. nov. (Microbiotheriidae); MPEF-PV 4368 (type), left m1 in labial (C) and occlusal (D) views. (E–G) *Microbiotherium* sp. (Microbiotheriidae); MPEF-PV 4362, right M1 in occlusal (E) view; MPEF-PV 4366, fragment of left dentary with p3–m1 in occlusal (F) and lingual (G) views. (H–I) *Periakros ambiguus* gen. et sp. nov. (Glasbiidae); MPEF-PV 4498 (type), right Mx in occlusal–lingual (H), and occlusal (I) views. (J–K) *Rosendolops ebaios* sp. nov. (Rosendolopidae); MPEF-PV 4449 (type), left M?3 in occlusal (J) and lingual (K) views. (L–M) *Hondonadia praecipitia* sp. nov. (Rosendolopidae, n. fam.); MPEF-PV 4144 (type), a right maxillary with complete M1–4 in occlusal view (L); MPEF-PV 4420, right m3 in occlusal view (M). (N–Q) *Hondonadia pumila* sp. nov. (Rosendolopidae); MPEF-PV 4239 (type), fragment of right maxillary with M2–3 in occlusal (N) and occlusal–lingual (O) views; MPEF-PV 4406 (detail), m1–3 in occlusal (P) and occlusolabial (Q) views. (R–S) *Hondonadia parca* sp. nov. (Rosendolopidae); MPEF-PV 3823 (type), left M2 in occlusal (R) and lingual (S) views. (T–U) *Hondonadia* cf. *H. fierroensis* Flynn and Wyss 1999 (Rosendolopidae); MPEF-PV 4388, left mx in labial (T) and occlusal (U) views.

Etymology From the Greek *peri*, "around," and *akros*, "point," "tip" (the type specimen of the type species is characterized by a set of cusps surrounding the trigon basin). Gender is masculine.

Type species *Periakros ambiguus* n. sp.

Included species Only the type species.

Occurrence Early Oligocene, South America.

Diagnosis As for the type and only known species.

Periakros ambiguus, new species

Etymology From the Latin *ambiguus*, "doubtful," "uncertain"; acknowledging the uncertain affinities of the type species among the Glasbiidae.

Type MPEF-PV 4498, right Mx (Fig. 6.2H, I)

Hypodigm The type only.

Tentatively referred specimens MPEF-PV numbers 4496, fragment of right Mx; 4495, fragment of left maxillary with two Mx, 4493, fragment of Mx lacking the protocone.

Occurrence La Cancha.

Measurements See Appendix 6.1.

Diagnosis Upper molars with very wide trigon basin; centrocrista absent (as in *Palangania*), and reduced paraconule; well-developed postmetaconular crest (even more than in *Glasbius*); the lingual slope of the metacone develops a short crest oriented towards the trigon basin; StC present.

Remarks Having a reduced stylar shelf, and twinned para- and metacone with StB and StD respectively, *Periakros ambiguus* can be confidently referred to the Polydolopimorphia. Two features of the type (and only known) specimen preclude us from referring it to the Bonapartheriiformes: StC and paraconule are still present, and the metacone is larger than the paracone. Also, it cannot be referred to the Polydolopiformes: the para- and metaconule are not lingually aligned with the protocone, and the StC is not lingually invasive (see Goin *et al.* 2003). All these generalized features are present in the Hatcheriformes, the basalmost group of Polydolopimorphia. However, *Periakros* shows several derived features that make it difficult to state its affinities with other hatcheriforms. Even though it shares a closer morphology in the upper molars with the Glasbiidae (*Glasbius* and *Palangania*), its molar cusps are not bunoid but acute, and some crests are well developed. Other differences with *Glasbius* are the lack of centrocrista (probably merged in the inner slopes of the paracone and the metacone), and an unreduced paraconule (reduced in *Glasbius*, vestigial or absent in *Palangania*). A remarkable feature of *Periakros ambiguus* is the presence of a crest on the anterolingual slope of the metacone. We hypothesized that this crest is homologous to the lingual edge of the metacone in the generalized tribosphenic molar. It is noteworthy that in *Glasbius* this same edge is also somewhat sharp and ending basally, not at the metaconule, but (as in *Periakros*) at the trigon basin.

Suborder **BONAPARTHERIIFORMES**
Pascual 1981
Superfamily **BONAPARTHERIOIDEA** Goin
and Candela 2004
Family **ROSENDOLOPIDAE**, new

Type genus *Rosendolops* Goin and Candela 1996.

Included genera *Rosendolops* and *Hondonadia* Goin and Candela 1998.

Occurrence Late Eocene – early Oligocene, Patagonia.

Diagnosis Rosendolopids differ from all other bonapartherioids in the following combination of features: upper and lower molars with relatively acute cusps and

trenchant cristae rather than bunoid; upper molars with unreduced paracone and metacone; preparacrista and postmetacrista present and moderately developed; premetaconular crest meets labially the trigon basin; labiolingually expanded trigon basin; lower molars with salient hypoconids, enlarged entoconids, and twinned para- and metaconids at the lingual face of the trigonid.

Remarks In a recent rearrangement of the major groups of Polydolopimorphia, Goin and Candela (2004) stated that "*Rosendolops* and *Hondonadia* [belong] to and as yet undetermined family of Bonapartherioidea." They recognized the affinities between *Hondonadia* and *Rosendolops*, and opted not to regard *Rosendolops* as a prepidolpid (as originally stated by Goin and Candela 1996). Here we go further on this same line of reasoning and formally propose a new family of Bonapartherioidea: the Rosendolopidae. Rosendolopids are unique among polydolopimorphians in their sharp cusps and crests, a feature that could either represent the generalized condition for bonapartherioids or, alternatively, a secondarily acquired condition.

At first sight, rosendolopids have an upper and lower molar morphology similar to that of basal paucituberculatan marsupials. However, we are confident that rosendolopids belong with polydolopimorphians on the basis of several key characters: (1) no paucituberculatan has the trigonid of m1 with paired, closely twinned paraconid and metaconid (instead, the metaconid is well posteriorly placed, even regarding the protoconid); (2) basal paucituberculatans have strongly reduced M/m4, something lacking in rosendolopids or other Polydolopimorphia; (3) on upper molars of paucituberculatans there is a sharp decrease in size of StD relative to StB, from M1 to M4, while in rosendolopids, as in other polydolopimorphians, StB and StD are subequal in size; (4) the metaconule in paucituberculatans is hypoconelike only in M1–2, whereas in M3–4 it is much more reduced; on the contrary, in rosendolopids, as well as in most other bonapartheriform polydolopimorphians (prepidolopids being an exception), the metaconule is very enlarged in M3 and M4).

Genus *Rosendolops* Goin and Candela 1996

Type species *Rosendolops primigenium* Goin and Candela 1996.

Included species The type and *R. ebaios*, n. sp.

Occurrence Late Eocene – early Oligocene, Patagonia.

Diagnosis See Goin and Candela (1996).

Rosendolops ebaios, new species

Etymology From the Greek *ebaios*, "small," "little."

Type MPEF-PV 4449, left M?3 (Fig. 6.2J, K).

Hypodigm The type and MPEF-PV numbers 4477, left M?1; 4446, right M?2; 4476, right M?2.

Tentatively referred specimen MPEF-PV 3917, left mx.
Occurrence All specimens are from La Cancha, except 3917, recovered from GBV-19 La Cantera.
Measurements See Appendix 6.1.
Diagnosis Differs from *Rosendolops primigenium* in its smaller size; upper molars have a less-developed metaconule, postmetaconular crest is present, and there is a deeper ectoflexus.

Rosendolops sp.

Referred specimens MPEF-PV 4487, left Mx; MPEF-PV 4497, left Mx.
Occurrence La Cancha.
Measurements See Appendix 6.1.
Remarks MPEF-PV 4487 is significantly larger than MPEF-PV 4497. In all other confrontable features they are very similar. MPEF-PV 4497 has a better developed anterior and posterior cinguli.

Genus *Hondonadia* Goin and Candela 1998
[= *Pascualdelphys* Flynn and Wyss 1999, p. 543]

Type species *Hondonadia feruglioi* Goin and Candela 1998.
Included species The type species and *H. fierroensis* (Flynn and Wyss 1999), *H. pittmanae* Goin and Candela 2004, *H. praecipitia* n. sp., *H. pumila* n. sp., and *H. parca* n. sp.
Occurrence Late Eocene – early Oligocene, South America.
New diagnosis (cf. Goin and Candela 1998, p. 81). Differs from *Rosendolops* in the following features: upper molars with well-developed, and lingually expanded, metaconule; the labiolingual expansion of the trigon basin is extreme; StB, paracone, and protocone aligned and parallel to a row formed by the StD, metacone, and metaconule; the cusp row formed by the StD, metacone, and metaconule is slightly labial to the anterior cusp line.
Remarks *Pascualdelphys* Flynn and Wyss 1999 is a junior synonym of *Hondonadia* Goin and Candela 1998. The type (and, apparently, only known specimen of *Pascualdelphys fierroensis*) is a right dentary with "fragments and mold impressions (filled with resin during preparation)" (Flynn and Wyss 1999). As a consequence of this peculiar type of preservation, several details of the molar morphology at the lingual side are unknown. Specimens here assigned to *Hondonadia* represent complete upper and lower molar series, and lead to a much better understanding of the dental structures of this genus. The generic identity of *H. fierroensis* with the *Hondonadia* species here described is warranted by the following features: very high, and closely

appressed paraconid and metaconid which, in worn molars, appear as a single cusp (as is happens in the type of *H. fierroensis*); these cusps are much higher than the protocone; high and laterally compressed entoconid; labially salient hypoconid; short and poorly developed anterobasal cinguli (well developed in *H. parca* n. sp.). These features, as well as those of the upper molars, confirm that *Hondonadia* is not a didelphimorphian but a bonapartherioid polydolopimorphian, allied to, but more derived than, *Rosendolops*.

Rosendolpids reached their climax in the early Oligocene. In fact, most taxa from La Cantera and La Cancha mark a radical morphological departure from the ubiquitous bunoid morphologies of previous (Paleocene–Eocene) polydolopimorphian lineages. Their molar morphology suggests a stronger lateral (ectental) component during mastication, suggesting at least some vegetarian component in their diet.

Not all species of *Hondonadia* are represented by upper and lower molar series (upper molars are unknown in *H. fierroensis*, while lower molars are not known for *H. feruglioi* and *H. pittmanae*). In addition, *Hondonadia* species show a whole mosaic of evolution in many features. For these reasons we have made comparative diagnoses, showing the specific variation to each one of these features in each one of the diagnoses here presented.

Hondonadia praecipitia, new species
Etymology From the Latin *praecipitium*, "cliff," in reference to the Gran Barranca southern cliffs.
Type MPEF-PV 4144, a right maxillary with complete M1–4 (Fig. 6.2L).
Hypodigm The type and MPEF-PV numbers 4383, right maxillary with M2–3; 4384, a left maxillary with M1–2; 4435, left M1; 4442, left M1; 4386, left M1; 4151, left M1; 4429, right M2; 4438, right M2; 4441, left M2; 4437, right M2; 4434, left M3; 4436, left M3; 4132, fragment of right dentary with roots of p2–3 and complete, but worn, m1; 4233, left m1; 4423, left m2; 4418, right m2; 4229, fragment of right m2; 4417, right m2; 4385, left m2; 4250, left m2; 4420, right m3 (Fig. 6.2M); 4419, fragment of right mx.
Referred specimens MPEF-PV numbers 4440, fragment of Mx; 4427; fragment of Mx; 4428, fragment of Mx; 4430, fragment of Mx; 4432, fragment of Mx; 4228; fragment Mx; 4431, fragment of Mx; 4433, fragment of Mx; 4415, mx; 4422, mx; 4414, mx; 4425, mx; 4426, mx; 4413, mx; 4421, mx.
Occurrence La Cancha.
Measurements See Appendix 6.1.
Diagnosis Largest species of the genus; molars do not significantly decrease in size from M/m2 to M/m3

(similar to *H. fierroensis*; different from *H. pumila* and *H. parca*, were there is a noticeable decrease in M/m3 size); the postmetaconular crest forms a wide posterior cingulum in M1–3 (less developed in all other species with known upper molars); the M1–2 metacone is almost at the same level than the paracone, though slightly placed labially (in *H. pumila* and *H. parca* the metacone is moderately displaced labially, while in *H. pittmanae* this condition is extreme); the metaconule is large and reaches lingually the protocone level (as in *H. feruglioi*; in *H. pumila* and *H. parca* the metaconule is less developed and does not reach the protocone level); the trigon basin extends labially up to the labial face of the molar (as in *H. feruglioi*; in *H. pumila*, *H. parca*, and *H. pittmanae* it does not reach the labial face); the StD in M1 is not posteriorly displaced respect to the metacone (in *H. pumila* and *H. parca* the StD it is posteriorly displaced); the parastylar edge of M1 is less salient anteriorly (as in *H. pumila*, different from *parca* and *H. pittmanae*, which have more salient parastyle); in m2–3, the paraconid and metaconid are almost completely fused (in *H. fierroensis*, *H. pumila*, and *H. parca*, the paraconid and metaconid are very close to each other but not fused); anterobasal cinguli in m1–3 are very short and reduced (similar to *H. fierroensis*; in *H. pumila* cinguli are moderately developed, while in *H. parca* are well developed); the entoconid is large, spire-like, and has a well-developed lingual crest (in *H. pumila* and cf. *H. fierroensis* the entoconid is moderately developed and laterally compressed; in *H. parca* it is extremely developed and laterally compressed); the hypoconulid is well developed and posteriorly projected (in *H. fierroensis* and *H. pumila* it is reduced, while in *H. parca* is moderately developed); hypoconids are labially salient (similar to *H. fierroensis* and *H. pumila*; in *H. parca* the hypoconids are not so labially salient).

Hondonadia pumila, new species

Etymology From the Latin *pumilus*, "dwarfish," "little"; in reference to the small size of this species as compared to other species of this genus present at Gran Barranca.

Type MPEF-PV 4239, fragment of right maxillary with M2–3 (Fig. 6.2N–O).

Hypodigm The type and MPEF-PV numbers 4247, left M1; 4448, left M1; 4450, right M1; 4447, left M2; 4444, right M2; 4445, right M2; 4127, right M2; 4454, left M2; 4451, right M2; 4455, right M2; 4406, fragment of a right dentary with alveoli for p2–3 and m1–3 (Fig. 6.2P, Q); 4397, right m1; 4411, left m2; 4408, left m2; 4399, right m2; 4409, right m2; 4393, left m3; 4401, left m3; 4403, left m3; 4390, right m3; 4452, right M2; 4453, right M2; 4234, left m2; 4395, left m?3; 4491, fragment of right M1; 4475, left M3; 4473, right M3.

Tentatively referred specimens MPEF-PV numbers 4404, fragment of left dentary with m1–2; 4405, fragment of right dentary with m1–2; 4407, fragment of left dentary with mx; 4391, left mx; 4394, left mx; 4400, fragment of left mx; 4393, left mx; 4387, fragment of right mx.

Occurrence La Cancha.

Measurements See Appendix 6.1.

Diagnosis Small size (similar to *H. pittmanae*); M/m3 is significantly smaller than M/m2 (similar to *H. parca* and differently to *H. praecipitia* and *H. fierroensis*, with less differences in size); the postmetaconular crest is narrow (similar to *H. parca* and *H. pittmanae*; in *H. praecipitia* is very wide, whereas in *H. feruglioi* is moderately developed); the parastyle is not anteriorly salient (similar to *H. praecipitia*; in *H. parca* and *H. pittmanae* is very salient); relative to the paracone, the metacone in M1–2 is moderately displaced labially (similar to *H. parca* and unlike *H. praecipitia* and *H. feruglioi* where the metacone is almost not displaced labially, in *H. pittmanae* the metacone is extremely displaced labially); the metaconule does not reach lingually the protocone level (similar to *H. parca*, unlike *H. praecipitia* and *H. feruglioi*, where the metaconule reaches lingually the protocone level); the labial expansion of the trigone basin is less developed than in *H. praecipitia* and *H. feruglioi* (similarly expanded than in *H. parca*); StD in M1 is well displaced posteriorly relative to the metacone (similar to *H. parca*, unlike *H. praecipitia* were StD is not displaced); in the lower molars, the para- and metaconid are closely set but not fused (similar to *H. fierroensis* and *H. parca*; in *H. praecipitia* both cusps are almost completely fused); anterobasal cinguli are moderately short (in *H. fierroensis* and *H. praecipitia* they are extremely short, whereas in *H. parca* they are well developed); the entoconid is large and laterally compressed (in *H. praecipitia* is spire-like, whereas in *H. parca* is proportionally huge and less laterally compressed); the hypoconulid is reduced (similar to *H. fierroensis*; in *H. praecipitia* is large; in *H. parca* is moderately developed); the hypoconid is labially salient (similar to *H. fierroensis* and *H. praecipitia*; in *H. parca* is less labially salient).

Hondonadia cf. *H. pumila*

Referred specimens MPEF-PV numbers 3861, left M1; 3915, fragment of right M2; 3809, right m2; 3826, right mx.

Occurrence GBV-19 La Cantera.

Measurements See Appendix 6.1.

Remarks These specimens are slightly larger than those of *H. pumila*. Lower molars have the paraconid lightly more labially placed than those of *H. pumila* from La Cancha; the anterobasal cingulum is somewhat more developed than that of *H. pumila*. These differences, plus the provenience of the referred specimens, may prove that the La Cantera species is different from *H. pumila*. However, and given the scanty materials here presented, we opt to regard them as *Hondonadia* cf. *H. pumila*.

Hondonadia parca, new species,

Etymology From the Latin *parcus*, "frugal," "scanty"; in reference to the few specimens referable to this species, as compared to the other species of this genus coming from Gran Barranca.

Type MPEF-PV 3823, left M2 (Fig. 6.2R, S).

Hypodigm The type and MPEF-PV numbers 3884, right M1; 3863, left M2; 3808, right M2; 3856, right M2; 3885, left M2; 3865, right M3; 3911, left M4; 3399, fragment of right dentary with m2–3 and roots of m4; 3883, right m1; 3916, left m2; 3896, right m2.

Tentatively referred specimens MPEF-PV numbers 3914, fragment of Mx; 3881, fragment of Mx; 3841, fragment of Mx; 3840 fragment of mx; 3847, fragment of mx.

Occurrence GBV-19 La Cantera.

Measurements See Appendix 6.1.

Diagnosis Moderately sized (similar to *H. feruglioi* and *H. fierroensis*; smaller than *H. praecipitia*; larger than *H. pumila* and *H. pittmanae*); M/m 3 is significantly smaller than M/m2 (similar to *H. pumila*; in *H. praecipitia* and *H. fierroensis* there is no sharp distinction in size between molars); very narrow posterior cingulum formed by the labial expansion of the postmetaconular crest (similar to *H. pittmanae* and *H. pumila*; in *H. praecipitia* and *H. feruglioi* cinguli are wider); anteriorly salient parastyle (as in *H. pittmanae*; in *H. praecipitia*, *H. pumila*, and *H. feruglioi* the parastyle is less salient); the metacone in M1–2 is slightly displaced labially respect to the paracone (as in *H. pumila*; in *H. feruglioi* and *H. praecipitia* the metacone is less displaced, while in *H. pittmanae* it is very much displaced); the enlarged metaconule does not reach lingually the protocone level (as in *H. pumila*; in *H. praecipitia* and *H. feruglioi* the metaconule does reach lingually the protocone level); the labially expanded trigon basin does not reach the labial face of the upper molars (as in *H. pittmanae* and *H. pumila*; in *H. praecipitia* and *H. feruglioi* the expansion meets the labial face); M1's StD is posteriorly placed regarding the metacone position (as in

pumila; in *praecipitia* the StD is not displaced); m2–3 with para- and metaconid closely appressed to each other but not fused (similar to *H. pumila* and *H. fierroensis*; in *H. praecipitia* both cusps are almost fused); anterobasal cingulum well developed (narrower in *H. pumila*, vestigial in *H. praecipitia* and *H. fierroensis*); very large entoconid which is laterally compressed (in *H. fierroensis* and *H. pumila* is proportionally not as large, while in *H. praecipitia* is large but spire-like); moderately developed hypoconulid (reduced in *H. pumila* and *H. fierroensis*; large in *H. praecipitia*); hypoconid not very salient labially (very salient in all other species with known lower molars).

Remarks In some features, *H. parca* seems to represent the generalized condition for the whole genus: lower molars have a well-developed anterobasal cingulum, and the hypoconid is not as labially salient as in the remaining species. In the upper molars the metaconule is proportionally smaller (relative to the total molar size).

Hondonadia cf. *H. fierroensis* Flynn and Wyss 1999

Referred specimens MPEF-PV numbers 4412, left mx; 4388, left mx (Fig. 6.2T, U); 4439, left Mx.

Occurrence La Cancha.

Remarks The lower molars here referred to are only slightly larger than the type of *H. fierroensis*. Additionally, their hypoconids are more labially salient than those of the Chilean species. In all other features these specimens are very close to *H. fierroensis*; however, it should be noted that the type (and only known specimen of this species) shows the lower molars heavily worn, this precludes us to go further than our tentative assignation.

Hondonadia sp.

Referred specimens MPEF-PV numbers 4416, left mx; 4142, right mx; 4424, fragment of Mx; 4446, right Mx; 4485, right Mx.

Occurrence La Cancha.

Measurements See Appendix 6.1.

Remarks The specimens listed here, although referable to *Hondonadia*, are too worn or fragmentary to attempt any specific assignation.

Rosendolopidae indet.

Referred specimens MPEF-PV numbers 4477, left Mx; 4476, right Mx; 4446, right Mx.

Occurrence La Cancha.

Measurements See Appendix 6.1.

Remarks These three molars are, in their molar morphology, closer to *Rosendolops* than to *Hondonadia*.

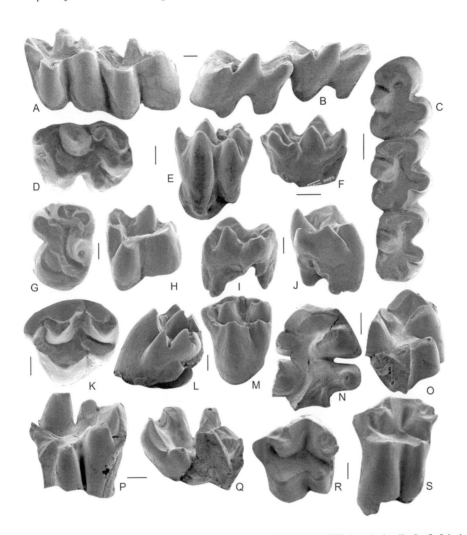

Fig. 6.3. (A–M) *Klohnia major* sp. nov. (Argyrolagoidea, fam. indet.); (A–C) MPEF-PV 4279 (type), detail of m2–3 in labial (A) and lingual (B) views, and m1–3 in occlusal (C) view; (D–F) MPEF-PV 4288, right m2 in occlusal (D), labial (E), and lingual (F) views; (G–J) MPEF-PV 4291, left m3 in occlusal (G), labial (H), lingual (I), and anterior (J); (K–M) MPEF-PV 4199, left M2 in occlusal (K), posterior (L), and lingual (M) views. (N–S) *Epiklohnia verticalis* gen. et sp. nov. (Argyrolagoidea, fam. indet.); MPEF-PV 3837 (type), right m1 lacking the hypoconid, in occlusal (N), posterior (O), labial (P), and lingual (Q) views; MPEF-PV 3839, left Mx in occlusal (R) and lingual (S) views.

However, specimens are too worn or fragmentary to go further than the present assignation.

Superfamily **ARGYROLAGOIDEA** Ameghino 1904
Family **INDETERMINATE**
Genus *Klohnia* Flynn and Wyss 1999
Type species *Klohnia charrieri* Flynn and Wyss 1999.
Included species The type and *K. major* n. sp.
Occurrence Early Oligocene, southern South America.

Klohnia major, new species
Etymology From the Latin *major*, "greater," as the new species is larger than the type species of the genus.

Type MPEF-PV 4279, left dentary with roots of p3 and complete m1–3 (Fig. 6.3A–C).
Hypodigm The type and MPEF-PV numbers 4217, fragment of left dentary with m1–3; 4224, fragment of left dentary with m1–3; 4283, fragment of right dentary with m2–3; 4133, right p3; 3788, left ?dp3; 3795, right m1; 4280, left m1; 4271; left m1; 4285, right m1; 4287, left m1; 4267, left m2; 4281, left m2; 4284, right m2; 4288, right m2 (Fig. 6.3D–F); 4223, left m2; 4254, right m2; 4215, left m2; 3787, right m3; 4291, left m3 (Fig. 6.3G–J); 3789, left m3; 4276, right m3; 4261, left m3; 4147, left m3; 4272, right m3; 4282, right m3; 4273, left m3; 4149, right m3; 4263, left m3; 4212, left m3; 3790, left m3; 3785, fragment of right dentary with m?2; 4225, mx; 4257,

mx; 4213, right mx; 4216, left mx; 3784, mx; 3796, left mx; 3793, mx; 4170, mx; 4138, mx; 3786, mx; 4256, left mx; 4219, right mx; 3794, left mx; 4269, right mx; 4214, right mx; 3792, right mx; 4277, fragment of left maxilla with M1–2; 4251, fragment of right maxilla with M2–3; 4173, fragment of right maxilla with M1–2; 4136, left P3; 4137, P3; 4139, left P3; 4145, P3; 4259, upper incisor; 4134, left P3; 4150, right M3; 4270, right M3; 4197, right M3; 4286, left M3; 4175, right M3; 4264, right M3; 4178, left M3; 4184, left M3; 4180, left M3; 4174, left M3; 4199, left M2 (Fig. 6.3K–M); 4201, right M1; 4186, right M2; 4292, right M1 or M2; 4185, left M1 or M2; 4143, left M1 or M2; 4179, left M1 or M2; 4153, right M1; 4262, right M1 or M2; 4265, left M1; 4252, right M1 or M2; 4193, right M2; 4289, left M1 or M2; 4196, right M1; 4293, left M1; 4205, left M1; 4202, right M1; 4204, right M1 or M2; 4177, left M2; 4275, right M2; 4194, left M2; 4290, left M2; 4176, right M1 or M2; 4192, right M1 or M2; 4260, left M1 or M2; 4183, right M1; 4253, right M1; 4258, mx; 3791, mx.

Occurrence La Cancha.

Measurements See Appendix 6.1.

Diagnosis Differs from *K. charrieri* in being around 40% larger; the tip of p3 is more labiolingually compressed; trigonid of m1 is wider; ectoflexid of m2 is deeper and more open outwards; talonid of m3 proportionately narrower; upper molars are less buccolingually compressed; the metaconule is set less lingually (i.e. it is not at the same level than the protocone); StA is larger, at least in M2.

Remarks Among upper molars, the morphology of M3 is easily recognizable because of its size proportions: the posterior lobe is longer than that of M1–2. However, differences between M1 and M2 are very subtle, being extremely hard to recognize the proper molar locus in isolated molars. None of the preserved maxillae bear unworn M1 or M2, which makes it difficult recognizing these molars between each other. These problems aside, it is noteworthy that the unworn specimens of *K. major* give unquestionable evidence of the argyrolagoid nature of *Klohnia*, as well as on the cusps homologies of upper and lower molars. The unilateral hypsodonty developed by *Klohnia* molars is striking: the labial face of the lower molars and the lingual face of the upper molars are four times higher than their respective opposite faces. Even though *Klohnia* molars are clearly rooted and have a distinct neck, it is obvious that this genus anticipates the protohypsodont/hypsodont conditions of the other argyrolagoids (*Proargyrolagus*, *Argyrolagus*, and *Patagonia*).

Klohnia cf. *K. major*

Referred specimens MPEF-PV numbers 4131, fragment of left dentary with alveoli for i1–p2, and complet p3–m2; 4171, left m1; 4274, right m1; 4222, left m2; 4221, right m3; 4226, right m3; 4278, fragment of right maxilla with P3–M1; 4198, right M1; 4141, right M1; 4266, left M1; 4200, left M1; 4181, right M2; 4204, right M2; 4172, right M2; 4203, right M3; 4187, left M3; 4195, left M3; 4210, right mx; 3797, right mx; 4218, left mx.

Occurrence La Cancha.

Measurements See Appendix 6.1.

Remarks Specimens listed here are somewhat larger and more robust than those of the hypodigm of *Klohnia major*. Even though we cannot discount the possibility that they belong to a new, larger species than *K. major*, these observed differences could be also due to intraspecific variation. Lacking better and more complete materials, we regard them as *Klohnia* cf. *K. major*.

Epiklohnia, new genus

Etymology From the Greek *epi*, "upon," "beside," "over," "after"; in recognition of the the derived condition of the new genus with respect to its most closely related taxon, *Klohnia*.

Type (and only known) species *Epiklohnia verticalis*, new species.

Occurrence Early Oligocene, South America.

Diagnosis As for the type and only known species.

Epiklohnia verticalis, new species,

Etymology From the Latin *verticalis*, "vertical"; in reference to the vertical orientation of the anterior, posterior, labial, and lingual faces of the crown in the new species.

Type MPEF-PV 3837, right m1 lacking the hypoconid (Fig. 6.3N–Q).

Hypodigm The type and MPEF-PV numbers 3888, fragment of left m2; 3800, fragment of left m3; 3874, left mx talonid; 3839 left Mx (Fig. 6.3R, S); 3799, fragment of left Mx; 3821, left Mx; 3860, fragment of left Mx.

Occurrence GBV-19 La Cantera.

Measurements See Appendix 6.1.

Diagnosis Genus and species similar and most probably related to *Klohnia*, but with the following differences: large anterolabial conule instead of parastyle, hyperdevelopment and individualization of the "ectostylid"; molars are more anteroposteriorly compressed; lower molars have the hypoconid proportionally smaller and more anteriorly set in relation to the entoconid; in the upper molars, the entoflexus is deeper.

Remarks *Epiklohnia* is even more derived than *Klohnia* in its sharp, vertical faces and the extreme anteroposterior compression of the trigon. This compression is expressed in the trigonid as a loph-like structure. The upper and lower molar outlines are similar in being bilobed.

Praedens, new genus
Etymology From the Latin *praedium*, "plot of ground" (implying "field," or *cancha* in Spanish; for the La Cancha locality), and *dens* or *dentis*, "tooth." Gender is neuter.
Type (and only known) species *Praedens aberrans* n. sp.
Occurrence Early Oligocene, South America.
Diagnosis As for the type and only known species.

Praedens aberrans, new species,
Etymology From the Latin *aberrans*, "wandering"; acknowledging the extremely derived molar pattern of the type species.
Type MPEF-PV 4243, left Mx (Fig. 6.4A–C).
Hypodigm The type and MPEF-PV numbers 4381, right mx (Figs. 6.4D–F); 4382, right mx; 4206, right mx; 4209, right mx (Fig. 6.4G–J); 4380, left mx; 4377, right Mx; 4376, left Mx.
Tentatively referred specimens MPEF-PV numbers 4378, right Mx; 4379, Mx; 4244, Mx.
Occurrence La Cancha.
Measurements See Appendix 6.1.
Diagnosis Differs from all other argyrolagoids in the following combination of features: very small size; upper and lower molars are brachydont; lower molars with the labial face higher than the lingual one; hypoconid anteriorly displaced regarding the entoconid; well-developed hypoconulid; trigonids with a single major cusp (the protoconid) which projects the postprotocrista posteriorly and downwards toward a neocusp ("ectostylid"), which is located anterior to the hypoconid; the metaconid is very small and locates posterolingual to the protoconid; the paraconid is vestigial or absent; upper molars almost lack a trigon basin; the protocone is labially displaced; there is no preprotocrista but instead a strong, conspicuous postprotocrista that forms a cingulum-like structure posterior to the metacone; StB is not facing the paracone but placed posterolabially to it, even more than in other argyrolagoids.
Remarks *Praedens aberrans* is one of the most distinct metatherians so far known; its tiny size should not preclude the significance of its numerous derived features. Some of them relate *P. aberrans* with *Wamradolops tsullodon*, from the ?Mustersan of the Peruvian Amazonia: labially displaced protocone, strong posterior cingulum basal to the protocone. Other features ally it with *Klohnia*, from the Tinguirirican SALMA of

Chile and Argentina (vertical postprotocristid, presence of a neocuspid anterior to the hypoconid). Actually, it seems to be intermediate between Bonapartherioidea and Argyrolagoidea bonapartheriforms. However, due to its highly derived feature of bearing distinct neocusps, and because of the radical reorientation of some of its crests, we regard *Praedens aberrans* as basal to all mesodont and hypsodont argyrolagoids.

Neocusps are a striking feature of *Praedens*. For instance, there is an "epiconular shelf" anterior and lingual to the parastyle; this neostructure is placed near the crown base in a middle point between the labial and lingual faces of the tooth. The epiconular shelf ends anteriorly in a neocusp, the "epiconule." In lower molars both crests of the metacristid (i.e. the postprotocristid and the postmetacristid) do not meet in a notch but instead the postprotocristid turns down and then towards the labial face of the crown, until it meets the "ectostylid," a neocusp that is placed anterior to the hypoconid. This "ectostylid" is homologous to the large cusp that is placed anterior to the hypoconid in *Klohnia* and *Epiklohnia* n. gen. This set of neostructures, unique among metatherians, is even beter developed in argyrolagids. At present, it is imposible to determine wether groeberids also followed this pattern (most all known molars of *Groeberia* are heavily worn).

Suborder **POLYDOLOPIFORMES** Kinman 1994
Family **POLYDOLOPIDAE** Ameghino 1897
Kramadolops, new genus
[= *Polydolops* (part) Odreman Rivas 1978, Marshall 1982b, Flynn and Wyss 1999, Flynn and Wyss 2004]
Etymology From the Greek *kramatos*, "mixture," and *-dolops*, common ending for polydolopine generic names. The name *Kramadolops* implies that representatives of this genus have a mixture of features previously characterizing other polydolopine genera. Gender is masculine.
Type species *Kramadolops mayoi* (Odreman Rivas 1978).
Included species The type species and: *K. abanicoi* (Flynn and Wyss 1999); *K. mckennai* (Flynn and Wyss 2004); *K. fissuratus* n. sp.; and *K. maximus* n. sp.
Occurrence Late Eocene – early Oligocene, Patagonia.
Diagnosis Differs from other polydolopines in the following combination of features: large size, molars subequal in length and width (i.e. they do not decrease rapidly in size from m1 to m3); upper molars without accessory labial cuspules; upper and lower molars (except m3) divided into two lobes (anterior and posterior) due to deep labial and lingual flexa/flexids; P2 considerably larger than P3; p3 large, with

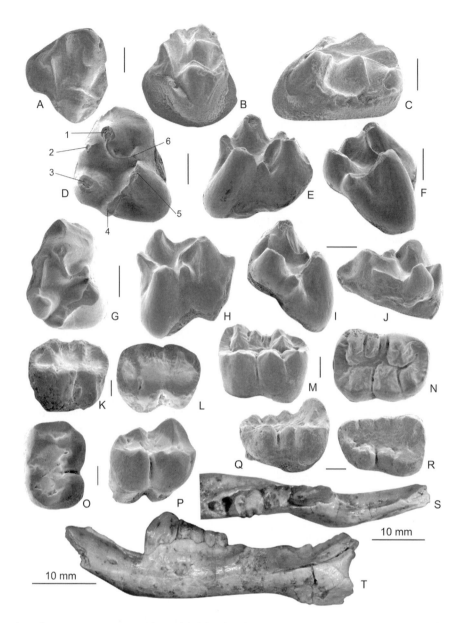

Fig. 6.4. (A–J) *Praedens aberrans* gen. et sp. nov. (Argyrolagoidea, fam. indet.); (A–C) MPEF-PV 4243 (type), left Mx in occlusal (A), occlusal–lingual (B), and posterior (C) views; (D–F) MPEF-PV 4381, right mx in occlusal (D), labial (E), and posterior (F) views; references to D: (1) Protoconid, (2) metaconid, (3) entoconid, (4) hypoconulid, (5) hypoconid, (6) neocusp ("ectostylid"); (G–J) MPEF-PV 4209, right mx in occlusal (G), labial (H), posterior (I), and lingual (J). (K–R) *Kramadolops fissuratus* sp. nov. (Polydolopidae), MPEF-PV 4342 (type), right M1 in lingual (K) and occlusal–lingual (L) views; (M–N) MPEF-PV 4343, left M2 in lingual (M) and occlusal (N) views; (O–P) MPEF-PV 4339, right m2 in occlusal (O) and labial (P) views; (Q–R) MPEF-PV 4340, right m3 in labial (Q) and occlusal (R) views. (S–T) *Kramadolops maximus* sp. nov. (Polydolopidae); MPEF-PV 4500 (type), a left dentary with p3–m2 in occlusal (S) and labial (T) views.

asymmetrical (in lateral view) anterior and posterior crests (the posterior crest is shorter and more horizontally set).

Discussion Representatives of *Kramadolops* constitute the last occurrence of polydolopiform marsupials in South America. Their known biochron ranges from the late Eocene (Mustersan SALMA; *Kramadolops mayoi*) to the early Oligocene (pre-Deseadan of La Cantera; *K. maximus* n. sp.). As implied by its name, species of *Kramadolops* display a mosaic of features, some of them resembling *Eudolops* (large size, simplified upper and lower molars without accessory cuspules, large M3) and some other resembling *Polydolops* (large p3 which is asymmetrical in lateral view and strongly ribbed). Species of *Kramadolops* include some of the best-known remains for the whole family: almost complete dentaries for *K. abanicoi* (Flynn and Wyss 1999) and *K. maximus* n. sp., and the first known cranial remain for a polydolopid, the type of *K. mckennai* (Flynn and Wyss 2004). Our interpretation of the generalized dental formula of *Kramadolops* is as follows: I3/i1, C1/c1, P2/p2, M3/m3 (but see Marshall 1982b, Flynn and Wyss 1999, Flynn and Wyss 2004). *Kramadolops maximus* constitutes an extreme pattern for the whole family, both in terms of size (very large) and in its simplification of the dental formula (i1, c0, p1, m3). This last feature is convergent with that of *Eudolops* species (see Marshall 1982b for a review of the genus). The basic molar pattern of *Kramadolops* is already evident in *K. mayoi*, the oldest species of the genus: lower molars show a simplified cusp number, strong labial and lingual cusps, and a deep ectoflexid dividing the m1–2 in two lobes, anterior and posterior. Upper molars are not known in *K. mayoi*, but in all other species of the genus (*K. mckennai*, *K. fissuratus* n. sp., and *K. maximus* n. sp.), upper molars are also divided in anterior and posterior lobes by a deep entoflexus.

Kramadolops fissuratus, new species,

Etymology From the latin *fissura*, "break," "cleft"; in reference to the deep flexures that divide the molars in two lobes.

Type MPEF-PV 4342, right M1 (Fig. 6.4K, L).

Hypodigm The type and MPEF-PV numbers 4249, left P3; 4327, right P3; 4343, left M2 (Fig. 6.4M, N); 4246, right P2; 4298, right P2; 4301, fragment of right P3; 4302, fragment of P2; 4317, fragment of M1; 4329, left M3; 4338, right m2; 4339, right m2 (Fig. 6.4O, P); 4341, left m2; 4328, fragment of left m2; 4335, left m2; 4315, right m3; 4336, right m3; 4340, right m3 (Fig. 6.4Q, R); 4303, fragment of a right p3.

Occurrence La Cancha.

Measurements See Appendix 6.1.

Diagnosis Differs from all other species of the genus in that the upper and lower molars have more distinct anterior and posterior lobes (flexa/flexids are deeper). Differs from *K. mckennai* and *K. maximus* n. sp. in that the posterior lobe of M1 is considerably larger than the anterior one, and in having a small cusp just lingual to the ectoflexus. Differs from *K. mckennai* in that the posterolabial cusps of M1 are closer to each other, and in that the posteriormost cusp is almost undistinguishable from the posterior edge of the crown. The p3 is, proportionally, more laterally compressed than in *K. mayoi* and *K. abanicoi*, but wider than in *K. maximus*. Differs from *K. abanicoi* in that the p3 is more asymmetrical in lateral view (i.e., the posterior crest is subhorizontal).

Kramadolops maximus, new species,

Etymology From the latin *maximus*, "greatest."

Type MPEF-PV 4500, a left dentary with p3–m2 (Fig. 6.4S, T).

Hypodigm The type and MPEF-PV 4501, left M1.

Occurrence GBV-19 La Cantera.

Measurements See Appendix 6.1.

Diagnosis Largest species of *Kramadolops*; differs from other species of *Kramadolops* in having a deeper ectoflexus in M1; fused posterolabial cusps in M1; lower dental formula: i1, c0, p1, m3 (i.e. the posteriormost incisors, canine, p1 and p2 are absent); p3 large, very long, and laterally compressed; extremely long, diastema between i1 and p3; the dentary is lower and more elongated than in other species of the genus.

Discussion

Diversity

Table 6.1 summarizes the taxonomic results of this work for both La Cancha and La Cantera marsupial assemblages. There are no fewer than 24 marsupial species referable to 18 genera, 14 families, and 5 orders. The numbers of referred specimens are unevenly distributed among taxa; almost 80% of the entire analyzed sample (400 specimens) are referable to species of just five genera: *Klohnia* (102 specimens, or 28% of the sample), *Hondonadia* (103 specimens, 25.75%), *Pilchenia* (64 specimens, 16%), *Kramadolops* (19 specimens, 4.75%), and *Evolestes* sp. (19 specimens, 4.75%). Individuals referable to species of *Klohnia* and *Hondonadia* account for more than half of the entire sample.

The scarcity of the Didelphimorphia is noteworthy, which otherwise compose the majority of taxa in late Neogene and Recent South American marsupial assemblages. In turn, the extreme diversity of polydolopimorphians (Hatcheriformes, Bonapartheriiformes, and Polydolopiformes), seems to reach a peak (and for many of them, an end) within the Tinguirirican fauna of La Cancha. Microbiotherians are represented by three microbiotheriid species (one of them unnamed), while the Sparassodonta (with one borhyaenid and one hathlyacinid) have a poor record as compared to their Miocene diversity. Finally, among paucituberculatans, the La Cancha record suggests a very rapid radiation of all lineages just at the Eocene–Oligocene boundary. Older Paleogene associations show a much larger diversity of didelphimorphian groups (e.g. Protodidelphidae, Derorhynchidae, Sternbergiidae), almost no paucituberculatans, and a much higher diversity of polydolopines. In turn, more modern (Neogene) marsupial faunas show an increased diversity of Sparassodonta, Microbiotheria, all clades of Paucituberculata, only argyrolagoid polydolopimorphians,

Table 6.1 *Marsupial taxa from La Cancha (■) and La Cantera (□) localities*

Order Didelphimorphia Gill 1872	
Superfamily Peradectoidea Crochet 1979	
Family Caroloameghiniidae Ameghino 1901	
Canchadelphys, new genus	
Canchadelphys cristata, new species	■
Superfamily indeterminate	
Family Sternbergiidae Szalay 1994	
Genus and species indet.	■
Order Sparassodonta Ameghino 1894	
Family Hathliacynidae Ameghino 1894	
Genus and species indet.	■
Family Borhyaenidae Ameghino 1894	
Genus *Pharsophorus* Ameghino 1897	
Pharsophorus cf. *P. lacerans*	□
Order Paucituberculata Ameghino 1894	
Family ?Caenolestidae Trouessart 1898	
Evolestes sp.	□
Family Pichipilidae Marshall 1980, new rank	
Genus and species indet.	■ □
Family Palaeothentidae Sinclair 1906	
Genus *Pilchenia* Ameghino 1903	
Pilchenia antiqua, new species	■
Pilchenia intermedia, new species	□
Order Microbiotheria Ameghino 1889	
Family Microbiotheriidae Ameghino 1887	
Genus *Eomicrobiotherium* Marshall 1982	
Eomicrobiotherium matutinum, new sp.	■ □
Genus *Clenia* Ameghino 1904	
Clenia brevis, new species	■
Genus *Microbiotherium* Ameghino 1887	
Microbiotherium sp.	■
Order Polydolopimorphia Marshall 1987	
Suborder Hatcheriformes Case, Goin and Woodburne 2005	
Family Glasbiidae Clemens 1966	
Periakros, new genus	
Periakros ambiguus, new species	■
Suborder Bonapartheriiformes Pascual 1981	
Superfamily Bonapartherioidea Pascual 1981	
Family Rosendolopidae, new family	
Genus *Rosendolops* Goin and Candela 1996	
Rosendolops ebaios, new species	■ □
Rosendolops sp.	■
Genus *Hondonadia* Goin and Candela 1998	
Hondonadia praecipitia, new species	■
Hondonadia pumila, new species	■
Hondonadia cf. *H. pumila*	□
Hondonadia parca, new species	□
Hondonadia cf. *H. fierroensis*	■

Table 6.1 (*cont.*)

Superfamily Argyrolagoidea	
Family indet.	
Genus *Klohnia* Flynn and Wyss 1999	
Klohnia major, new species	■
Klohnia cf. *K. major*	■
Epiklohnia, new genus	
Epiklohnia verticalis, new species	□
Praedens, new genus	
Praedens aberrans, new species	■
Suborder Polydolopiformes Ameghino 1897	
Family Polydolopidae Ameghino 1897	
Kramadolops, new genus	
Kramadolops fissuratus, new species	■
Kramadolops maximus, new species	□

and several lineages of mostly carnivorous didelphimorphians (Goin *et al.* 2007a).

In short, the La Cancha fauna (and the Tinguirirican marsupials in general) constitute an unusual association that marks the end of one era for marsupial evolution and the beginning of a new one.

Age

The original Tinguirirican SALMA fauna of central Chile included three marsupial taxa: *Klohnia charrieri*, *Hondonadia* (= *Pascualdelphys*) *fierroensis*, and *Kramadolops* (= *Polydolops* partim) *abanicoi* (Flynn and Wyss 1999; Flynn *et al.* 2003). All three genera are represented in the La Cancha assemblage, while two of them (*Hondonadia* and *Kramadolops*) are also present in the La Cantera fauna. Our results agree with previous analyses (Carlini *et al.* 2005a; López *et al.* 2005) in two respects:

(1) The La Cancha fauna is referable to the Tinguirirican SALMA. Following previous studies, Flynn and Swisher (1995) stated that the Chilean Tinguirirican Fauna is as young as 31.5–32 Ma, and as old as 36–37 Ma. Subsequently Flynn *et al.* (2003: 229) stated that, even though the Tinguirirican age "potentially spans a range as large as 31–37.5 Ma or more, various lines of evidence hint that this SALMA is probably of short duration (possibly less than 2 m.y.)." More recently Flynn and Wyss (2004) estimated the Tinguirirican SALMA at *c.* 31–32 Ma. In turn, Vucetich *et al.* (2004) supported an age of *c.* 33.4 Ma for the La Cancha levels. The new evidence suggests a date of 33.3 to 33.7 Ma for La Cancha (Ré *et al.* Chapter 4, this book). Because of this difference in estimated ages, Carlini *et al.* (2005b) suggested that the La Cancha levels and fauna could represent a basal Tinguirirican age. On the basis of

marsupials alone, we cannot test the hypothesis that the La Cancha fauna is basal to the Tinguirirican SALMA as marsupials from the type locality are still rare. In turn, the La Cantera locality in Unit 3 of the Upper Puesto Almendra Member of the Sarmiento Formation, is below levels that yield typical Deseadan-aged mammals. The age of this unit and the fossils at GBV-19 "La Cantera" is between 31.1 and 29.5 Ma (Ré *et al.* Chapter 4, this book).

(2) The La Cantera fauna is younger than that of La Cancha. Whereas Carlini *et al.* (2005a) favored a pre-Deseadan age for the La Cantera fauna, López *et al.* (2005) suggested it could be referred to the Deseadan or an immediately older age, intermediate between the Tinguirirican and the Deseadan. On the basis of the marsupial association, we cannot be conclusive. Some evidence, however, suggests that the La Cantera marsupials may be pre-Deseadan in age: (a) the microbiotheriid *Eomicrobiotherium matutinum* and the rosendolopid *Rosendolops ebaios* are present both at La Cantera and La Cancha; (b) in both La Cancha and La Cantera there are closely related species: *Pilchenia antiqua* in La Cancha, *P. intermedia* in La Cantera; *Hondonadia praecipitia*, *H. pumila*, and *H.* cf. *H. fierroensis* in La Cancha, and *H.* cf. *H. pumila* and *H. parca* in La Cantera; *Kramadolops fissuratus* in La Cancha, *K. maximus* in La Cantera; (c) *Epiklohnia*, from La Cantera, seems to be a direct derivative of *Klohnia* (see above) from La Cancha. On the other hand, (d) with only three exceptions (the borhyaenid *Pharsophorus*, the palaeothentid *Pilchenia*, and the ?caenolestid *Evolestes*), no other marsupial taxa are shared with the Deseadan SALMA. Deseadan marsupials include the following genera: Argyrolagidae: *Proargyrolagus*; Abderitidae: *Parabderites*; Caenolestidae: *Pseudhalmariphus* and *Evolestes*; Palaeothentidae:

Acdestodon, Carlothentes, Pilchenia, and *Palaeothentes;* Borhyaenidae: *Pharsophorus;* Hathliacynidae: *Notogale, Sallacyon,* and *Andinogale;* Proborhyaenidae: *Paraborhyaena* and *Proborhyaena* (modified from Pascual *et al.* 1996). In short, though not conclusive, the marsupial association of La Cantera suggests a pre-Deseadan age, intermediate in time between the Tinguiririran SALMA and the Deseadan SALMA.

Vucetich *et al.* (2004) hypothesized that the La Cantera fauna occurs between 34 and 30 Ma. Taking in account that (a) the La Cancha fauna has been estimated at *c.* 33.4 (Vucetich *et al.* 2004), (b) our own estimate suggests a time span of *c.* 31–33.4 Ma for the Tinguiririran SALMA, (c) the La Cantera fauna seems to be post-Tinguiririran and pre-Deseadan in age, and (d) Deseadan levels at Central Patagonia have been estimated at between 27 and 29 Ma (Flynn and Swisher 1995), we hypothesize a tentative age of *c.* 29–31 Ma for the deposition of the La Cantera sediments (Fig. 6.5). This line of reasoning is supported by the geochronology presented by Ré *et al.* (Chapter 4, this book) who interpret the age of Unit 3 and the fossils at GBV-19 "La Cantera" to between 31.1 and 29.5 Ma.

Marsupials and the early Oligocene faunal turnover

The marsupial associations described in this chapter constitute, both in taxonomic and ecological terms, a radical departure from any other Paleogene fauna known up to now. (However, it should be noted that latest Eocene, Mustersan SALMA marsupials are as yet poorly known; a review of the Mustersan marsupial association recently recovered from the El Rosado Member at Gran Barranca will probably add much information about the marsupial faunal changes across the Eocene–Oligocene transition.) Among the most important features of these early Oligocene (La Cancha + La Cantera) assemblages, we note the last record of: (1) Caroloameghiniidae (and peradectoids in general), with *Canchadelphys cristata;* (2) sternbergiid didelphimorphians; (3) hatcheriform polydolopimorphians (*Periakros ambiguus*); (4) all bonapartherioid Bonapartheriiformes (rosendolopids); (5) all polydolopiform polydolopimorphians (*Kramadolops*). On the other hand, (6) argyrolagoid Bonapartheriiformes make their first appearance in the fossil record with *Klohnia, Epiklohnia,* and *Praedens;* (7) the Paucituberculata underwent a rapid diversification, with ?Caenolestidae, Pichipilidae, and Palaeothentidae already represented; (8) large, modern borhyaenids are represented by *Pharsophorus* from La Cantera.

Paleocene and Eocene marsupials were characterized by the dominance of frugivorous, omnivorous, or insectivorous types, or a combination of these; many of the early Oligocene marsupials show a definite trend toward more herbivorous, and probably granivorous, feeding habits. A majority of specimens from La Cancha and La Cantera (53.75%)

are represented by *Hondonadia* spp. and *Klohnia* spp. with molar morphologies indicative of some ectental (*Hondonadia*), or anteroposterior (*Klohnia*) component in mastication. Moreover, representatives of both genera show unilateral hypsodonty in upper and lower molars (the labial crown of the lower molars, and the lingual crown of the upper ones, are much higher than their counterparts). Most probably, these molar features relate to feeding habits that included either more abrasive items, or distinct vegetal structures such as seeds, hard fruits (or fruits covered by hard pericarpia), or a combination of these. This aspect is probably the most distinctive feature of the La Cancha fauna as compared to other (older) Paleogene associations. Future research on this topic should stress the new floral assemblages contemporary with the sharp, global cooling event that took place at the Eocene–Oligocene boundary (EOB).

The newly described specimens of *Praedens, Klohnia* (both from La Cancha), and *Epiklohnia* (from GBV-19 La Cantera) show the acquisition of neomorphic structures to the generalized tribosphenic pattern of marsupial molars. These novelties imply a degree of specialization probably unique among metatherians, living or extinct. They led to the development of a rodent-like molar pattern of many Argyrolagoidea. *Praedens aberrans* shows, in the lower molars, distinct "ectostylid" and "entostylid" neocusps, while in the upper molars there is an "epiconule" and an "epiconular shelf." These novel structures are more distinctly developed in *Klohnia,* and, specially, in *Epiklohnia.* Both *Klohnia* and *Epiklohnia* show unilateral hypsodonty in upper and lower molars, as well as a thick tooth enamel. Mastication in these taxa must have had a more anteroposterior component than in most other marsupials, with typically orthal occlusal movements. These rodent-like adaptations among South American marsupials suggests a rapid response towards more cool, open and, probably, more arid environments, which opened a new series of adaptive zones by the early Oligocene. Further evolution of argyrolagoid marsupials led to a consolidation of few, though distinct, dental morphotypes (e.g. those represented by *Proargyrolagus, Argyrolagus,* or *Patagonia*), some of which persisted throughout the Neogene. An interesting question is what may have transpired to this incipient argyrolagoid radiation had caviomorph rodents not arrived in southern South America by La Cantera times.

Size increase occurred in the La Cantera polydolopids and borhyaenids. This body size enlargement has been also observed in several native ungulates from these localities and levels (López *et al.* 2005). *Kramadolops maximus* is the last and largest polydolopid. *Kramadolops fissuratus, K. abanicoi,* and *K. mckennai* are also larger when compared with the first representative of the genus, *K. mayoi,* from the Mustersan SALMA. In turn, *Kramadolops* includes, together with *Eudolops* all of the largest species

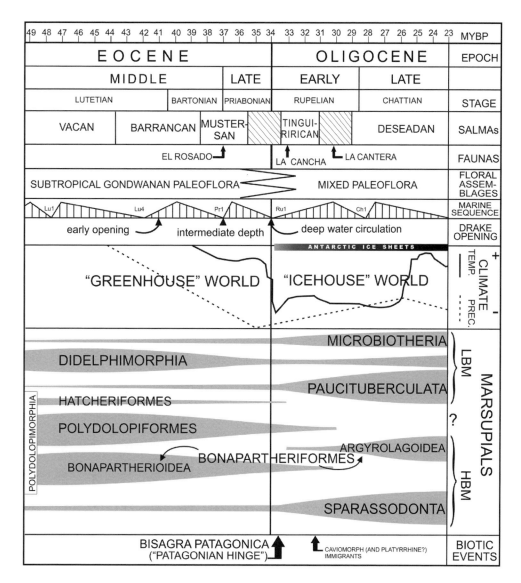

Fig. 6.5. Summary of the biotic and abiotic events discussed in this work. The time series (Mybp, epochs, and stages), as well as the eustatic marine sequence, follows Gradstein *et al.* (2004). Timing of SALMAs is partially derived from Carlini *et al.* (2005b) and Madden *et al.* (2005a). The time spans for the Tinguiririran SALMA, as well as the estimated age for the La Cantera locality, are discussed in this work. El Rosado, La Cancha, and La Cantera are the three faunas/localities mentioned in this work. Floral assemblages, as well as the precipitation curve (dashed line in "Climate"), follow Hinojosa (2005). Temperature curve (thick line in "Climate") follows Zachos *et al.* (2001). Successive stages in the opening of the Drake Passage, expressed as depth of circulating water, were taken from Scher and Martin (2006). In "Marsupials," the thickness of the different bars corresponding to the orders and suborders are schematic, and they do not strictly express taxon numbers. "LBM" and "HBM" signify that the taxa enclosed in each one of them had inferred diets compatible with low and high basal metabolic rates, respectively, in eutherians. The timing of the immigration events involving caviomorph rodents and platyrrhine monkeys is tentative. *Bisagra patagónica* ("Patagonian Hinge") is the biotic event here described.

of polydolopids. *Pharsophorus* from La Cantera is a very large borhyaenoid; coevolutionary procesess with medium to large placental herbivores should be expected in the evolution of this predator lineage. Body size enlargement has been correlated with decreasing temperatures in other temporal sequences, as well as with variations in latitude and altitude (Bergmann's Rule). This correlation can be

seen in several mammalian orders and families (e.g. phalangerid and macropodid marsupials, soricid insectivorans: Ashton *et al.* 2000). In the case of *Kramadolops*, a size increase from the Mustersan SALMA to the pre-Deseadan (La Cantera), passing through the Tinguiririran SALMA, seems to correlate with progressively cooler environments following Bergmann's Rule. If this is the case, Cope's Rule applies too, as a special case of Bergmann's. The former asserts that in time organisms tend to enlarge their size, as can be evidenced in *Kramadolops* species.

It is highly speculative to evaluate the extent of marsupial extinctions by the early Tinguiririan SALMA due to the scarce marsupial record for earlier ages within the Paleogene. However, it can be roughly estimated that the La Cancha marsupials represent 50% fewer species than those of Itaboraian SALMA assemblages. Itaboraian marsupials represent the most diverse fossil assemblage known up to now, and they were contemporary with the peak in warm temperatures that transpired during the late Paleocene and early Eocene. Unpublished data by F. Goin and A. Forasiepi suggest that no fewer than 35 species have been recovered from the Las Flores Formation (Central Patagonia; Itaboraian SALMA). A recently recorded marsupial association from Paso del Sapo (northeastern Chubut province; early–middle Eocene) led to the recognition of around 30 species (Goin *et al.* 2005). These numbers are almost double the number of taxa in the marsupial assemblage of La Cancha (18 species). Post-Itaboraian, pre-Tinguiririan marsupial assemblages are still scanty, mostly due to the lack of sampling of the small-sized marsupials; in fact, borhyaenids and polydolopids (which include some of the largest South American marsupials) comprise a majority of the taxa known for the Casamayoran and Mustersan SALMAs. It is interesting to note that Itaboraian and Tinguiririan SALMA marsupials from Central Patagonia are known from very well-sampled sites that included screen-washing of sediments, whilst Casamayoran and Mustersan fossils are known up to now only from surface prospecting. We suspect that Casamayoran and Mustersan mammals included many more marsupial species than those presently known. If such was the case, the extinction rate of the Tinguiririan marsupials of La Cancha may prove to be significant in the context of Paleogene marsupial associations.

As hypothesized below, we regard the marsupial turnover (both in lineages and in adaptive types) evidenced at the La Cancha fauna, as part of a more general, biotic change resulting from the global cooling process that transpired during the latest Eocene – earliest Oligocene (Fig. 6.5). A related question is whether marsupials are reliable indicators of temperature changes, or not. The answer is yes, at least when temperature changes refer to cooling events. There is a clear correlation between marsupial occurrence and environmental temperature, as can be demonstrated from the distribution of extant South American marsupials. In explaining the relationship between the latitudinal gradient of South American marsupial distribution (marsupials decrease significantly southwards in species numbers), Birney and Monjeau (2003) stated that the mean minimum extreme temperature is the best parameter correlative to that

distribution. Mean annual temperature is also correlated with this distribution, over other physical parameters (such as continental area, biome diversity, or mean annual precipitation). Similarly, they stated that thermal range is the best predictor of the proportion of carnivorous and frugivorous marsupials in a stepwise regression model: "there is no additional variable to add significant explanation to the single thermal range model" (Birney and Monjeau 2003: 303). In short, low temperatures constitute a primary limiting factor in marsupial evolution. Reasons that explain this correlation between marsupial distribution and temperature are based on marsupial physiology and on its reproductive and developmental strategies. Marsupials are highly immature at birth, as compared to placentals: (1) energy metabolism, as assessed by standard metabolic rates, is initially at very low level; it only increases in proportion to body size; (2) the young pouch-dwelling marsupial is not able to fully regulate its body temperature until weaning. Only "[a]t about the time it reaches a 'mammalian' level of energy metabolism, it also develops the ability to sense low environmental temperatures and respond by increasing thermogenesis" (Hulbert 1988: 160; see also Hunsaker 1977). This implies that marsupial pouch young are more vulnerable to changes in environmental temperatures. A second argument has been settled by McNab, who stated that "the absence of high basal rate in marsupials reflects the absence of a correlation of rate of reproduction with basal rate, a correlation present in eutherians" (McNab 2005: 183). This difference in energy expenditure between marsupials and eutherians leads to a restricted tolerance of marsupials to cold environments. It is interesting to note that, as revealed by Itaboraian and Paso del Sapo mammals, marsupials comprise more than 50% of mammalian species by the late Paleocene – early Eocene Thermal Maximum. In contrast, the La Cancha marsupials comprise clearly less than 50% of the mammalian association.

The Patagonian Hinge (*Bisagra Patagónica*) event
The global cooling at the EOB was one of the most dramatic thermal events that exerted environmental pressure over Cenozoic marine and terrestrial biotas, all over the world (Merico *et al.* 2008). "The opening of Southern Ocean gateways was critical to the formation of the Antarctic Circumpolar Current and may have led to Cenozoic global cooling and Antarctic glaciation" (Livermore *et al.* 2004: 797). This glacial event corresponds to the first major expansion of Antarctic ice in the Cenozoic, and coincides with a significant positive shift in the oxygen isotope value of marine carbonates (the Oi-1 event) of Miller *et al.* (1991). Evidence of substantial ice-sheets in Antarctica were first found in the east of this continent, until Ivany *et al.* (2006) reported on a regionally extensive West Antarctica ice-sheet already by the EOB, extending up to the

Antarctic Peninsula. Scher and Martin (2006) summarized estimated ages for the opening of the Drake Passage: early opening at *c.* 41 Ma (shallow waters), intermediate depths at *c.* 37 Ma, and deep water circulation by *c.* 34 Ma. Accordingly, a strong Antarctic Circumpolar Current was already occurring at least at 32 Ma (Lagabrielle *et al.* 2009). There is some dispute about whether the Shackleton Fracture Zone could have blocked the deep-water oceanic gateway until the early Miocene, thus slowing its effect as a motor of global cooling. Recent work, however, demonstrates that a deep-water connection between the Pacific and Atlantic Oceans was probably established soon after spreading began in Drake Passage during the earliest Oligocene. "Hence, thermal isolation of Antarctica by the development of the Antarctic Convergence and onset of the Antarctic Circumpolar Current remains a prime candidate for the underlying mechanism driving global changes in circulation and climate at the Eocene–Oligocene boundary" (Livermore *et al.* 2004: 800). Central Patagonia is, and was, a peninsular region standing some 2000 km north of the Drake area. Thus, the climate consequences of the establishment of the Antarctic Circumpolar Current must have been important. Our hypothesis is that the La Cancha assemblage directly reflects global cooling initiated by these tectonic and oceanic events.

There is a basic coincidence between the timing of the geophysical parameters occurring near the EOB, their paleoclimatic consequences, and floral and faunal evidence, all suggesting significant cooling by this time in Patagonia. Paleobotanical evidence favors a sharp decrease in temperatures at the EOB. The Middle Ñirihuau Flora (*c.* 35 Ma) from Western Patagonia has been characterized as an Antarctic Paleoflora; its mean temperature values (14.5–15.3 °C) as inferred from the floral assemblage correspond to the Cenozoic's minimum records (Hinojosa 2005). Also, a sharp decrease in mean precipitation is also recorded after the late Eocene, reaching minimum values of 573 mm of annual rainfall by the EOB. Subsequent to the EOB and up to the early Miocene, southern South America's paleofloras have a strong Austral-Antarctic component together with less conspicuous, Neotropical and Pantropical elements. This "Mixed Paleoflora" (Romero 1978) developed under colder and drier conditions than those of the late Eocene (Hinojosa 2005). Accordingly, and also referring to Patagonian environments, Barreda and Palazzesi (2005) state: "Latest Eocene–Early Oligocene fossil floras indicate the development of temperate forests along with a reduction of floristic diversity ... Communities were dominated by Nothofagaceae and Podocarpaceae without megathermal elements" (see also Quattrocchio 2006; Barreda and Palazzesi this book).

As shown above, the early Oligocene marsupial assemblage from La Cancha represents a significant faunal turnover with respect to previous Paleogene marsupial assemblages. Something similar can be said of placental

faunas, even though such turnover may have not been so dramatic (López *et al.* 2005; but see Carlini *et al.* 2005a). Pascual *et al.* (1996) argued in favor of a major shift in the faunal successions of South America beginning with the late Oligocene Deseadan SALMA mammalian associations. In light of the new findings here reported, we suggest that the La Cancha fauna, rather than the Deseadan fauna, initiates the major change leading to the modernization of South American mammals (i.e. the "Patagonian faunistic cycle" of Pascual *et al.* 1996). A similar conclusion was reached by Flynn *et al.* (2003) when analyzing the Tinguiririran fauna of Central Chile (see below).

We propose to name the evolutionary event expressed by the major taxonomic and ecologic shift in land-mammal faunas of this part of the South American continent at the EOB the Patagonian Hinge (*Bisagra Patagónica*). We regard the Hinge as equally significant as the European *Grand Coupure* (Stehlin 1909) or the Central Asian "Mongolian Remodeling" (Meng and McKenna 1998). All these events were triggered by global cooling at the EOB (Zachos *et al.* 2001). This cooling event led to the development of more open and arid environments in both northern and southern hemispheres. This climatic–environmental change also accounts for the faunal turnover in marine and continental biotas, of which the *Grand Coupure*, Mongolian Remodeling, and Patagonian Hinge are regional examples among land-mammal faunas. A comparison of these regional phenomena suggests close parallelisms between the Mongolian Remodeling and the Patagonian Hinge. Reflecting on the former event, Meng and McKenna (1998) stated: "Our results reveal a distinctive pattern of faunal turnovers that is reflected by taxonomic differentiation, species appearances and extinctions, reorganization of faunal compositions, changes of species body sizes, and evolution of rodent dentitions." Provided a terminological change, from "rodent dentitions" to "rodent-like dentitions," this characterization accurately applies to the La Cancha marsupials, and, we suspect, to all other mammals from La Cancha as well (Carlini *et al.* 2005a; López *et al.* 2005). It is interesting to note that, contrary to what happened in the *Grand Coupure* event, those of the Patagonian Hinge and Mongolian Remodeling did not involve immigration events. Thus, "changes apparently occur across whole faunas, not just in potentially competing lineages" (Meng and McKenna 1998). The arrival of caviomorph rodents to central Patagonia was a younger event, the oldest recorded at La Cantera times (Vucetich *et al.* 2005). It is also probable, but there is yet no evidence, that the first immigrant platyarhine primates were present in central Patagonia by immediately post-Tinguiririran times.

Summing up: "[f]rom Early Paleocene to Late Pleistocene, the overall trend of environmental change in SSA [Southern South America] moved from more equable to less equable conditions. More specifically, climates changed from warm, wet, and non-seasonal, to cold, dry, and seasonal. Concomitantly, biomes moved from tropical forest to steppes, across a sequence constituted by subtropical forests, woodland savanna, park-savanna, and grassland savanna" (Ortiz-Jaureguizar and Cladera 2006, p. 524). As stated above, we regard the Patagonian Hinge event as the most notable marker, the "hinge" moment of this shift.

La Cancha's bearing on macroevolution

The early Oligocene faunal turnover evidenced by the La Cancha and the La Cantera mammals seems to be in sharp contrast with what happened in North America. Summing up his, and other authors', conclusions on the topic, Prothero and Emry (1996, p. 664) stated that: "[t]he earliest Oligocene climatic event ... was almost ignored by land mammals" (see also Prothero 2004). However, it should be noted that (1) "[t]he Oligocene mammalian record is biased toward that of the eastern flank of the Rocky Mountains ... and adjacent plains" (Woodburne 2004, p. 330), and (2) actually, *there was* some kind of biotic turnover in North America, as revealed by the development of dry, open shrublands, as well as by the first record of ursids, nimravids, diceratherine rhinos, and extinction of cilindrodont rodents during the Orellan NALMA (earliest Oligocene; see fig. 8.1 in Woodburne 2004, and statements in pp. 330–331). When specific areas of North America are analyzed, megafloral assemblages also show abrupt, successive turnovers beginning at the EOB (Retallack *et al.* 2004). Finally, it may be the case that the faunal patterns verified in North America during the early Oligocene are not a good mirror for South America's faunal events. In turn, Central Asian, European, and South American land mammal faunas show significant turnovers that the North American succession may not reflect to such an extent.

In their analysis of the significance of Central Chile's Tinguiririran Fauna, Flynn *et al.* (2003, pp. 229–230) concluded that "the most dramatic shift in Cenozoic South American paleoecology and paleoenvironment occurred between the Mustersan and Tinguiririran SALMAs"; i.e. immediately after the Eocene–Oligocene transition. A radically different view has been more recently advanced by Kohn *et al.* (2004, p. 621), who suggested that, at least for southern South America, "the paradigm of major global cooling at the Eocene–Oligocene transition is largely false." As exposed above, our own results sharply differ with this last interpretation. As far as marsupials are concerned, the early Tinguiririran SALMA, as represented by the La Cancha Fauna of Gran Barranca, indicates that by this time there

had occurred the most significant, Cenozoic faunal turnover the southern hemisphere's continental record. The *Bisagra Patagónica* was a major evolutionary change in marsupial history, and suggests that global average temperature may have been, at least for marsupials, the most important trigger for both radiation and extinction events. At this point, several interesting hypotheses for future testing can be derived: (1) marsupial origins and initial radiation were triggered by the beginning of the "greenhouse world" conditions – roughly, by the medial Cretaceous as attested by the fossil record (Kielan-Jaworowska *et al.* 2004); (2) South American, and probably Antarctic and Australian, marsupial radiations reached their climaxes by the early Eocene Climatic Optimum; (3) the *Bisagra Patagónica* implied, at least for southern South American marsupials, a change in diets and adaptive niches which, in eutherian mammals, are correlated with high basal metabolic rates (i.e. herbivory, carnivory) (Fig. 6.5).

A final consideration is concerned with two alternative mechanisms frequently advocated in explaining evolutionary events: Red Queen vs. Court Jester. Briefly, Red Queen hypotheses maintain that biotic interactions are the most important drivers of evolutionary change, whereas Court Jester hypotheses regard physical–environmental perturbations, such as climate change, as most important (Barnosky 2001). As revealed by the La Cancha and La Cantera marsupials, our inclination towards a Court Jester explanation for the *Bisagra Patagónica* event needs no further elaboration.

ACKNOWLEDGEMENTS
We thank the editors of this volume for inviting us to participate. Richard Kay, Richard Madden, Alfredo Carlini, and Maria Guiomar Vucetich lead a decade of field prospecting at Gran Barranca; nearly all specimens here studied were collected by them or their students. Analía Forasiepi helped much with the determination of borhyaenid material. Discussions with Richard Madden, Alfredo Carlini, and Guiomar Vucetich were helpful and enlightening on several topics covered by this work. Michael Woodburne and Marcelo Sánchez-Villagra made useful critical reviews of the original manuscript. Marcela Tomeo helped with the figures and manuscript preparation. F. Goin thanks the Alexander von Humboldt Foundation and CONICET (PIP 5621) for their support. Field research was supported by US National Science Foundation grants EAR-0087636, BCS-0090255, and DEB-9907985 to Richard F. Kay and Richard H. Madden.

Appendix 6.1   Dental measurements (in mm) of the La Cancha and La Cantera marsupials; an asterisk (*) means that the measurement is approximate

Upper molars

Taxon	Specimen (MPEF-PV)	LP3	WP3	LM1	WM1	LM2	WM2	LM3	WM3	LM4	WM4	LMx	WMx
Canchadelphys cristata, n. gen. et sp.	4242							2.32	2.4				
Sternbergiidae, gen. et sp. indet.	4459									1.2	2.12		
Evolestes sp.	3849			1.8	1.52								
Evolestes sp.	3906				1.72								
Evolestes sp.	3913					1.8	1.8						
Evolestes sp.	3864					1.72	1.8						
Evolestes sp.	3918					1.76	1.84						
Evolestes sp.	4152					1.68	1.72						
Evolestes sp.	3827					1.68	1.68						
Evolestes sp.	3835							1.4	1.48				
Evolestes sp.	3850							1.4	1.44				
Evolestes sp.	3813							1.44	1.56				
Evolestes sp.	3834									1.12	1.56		
Pilchenia antiqua, n. sp.	4490	2.8	1.92	2.4	2.68	1.96	2.8						
Pilchenia antiqua, n. sp.	4486					2.04	2.56*	1.48*	1.88*				
Pilchenia antiqua, n. sp.	4483			2.4*									
Pilchenia antiqua, n. sp.	4148					2.04							
Pilchenia antiqua, n. sp.	4484					2.08	2.84						
Pilchenia antiqua, n. sp.	4479							1.48	1.8				
Pilchenia antiqua, n. sp.	4478							1.56	2				
Pilchenia antiqua, n. sp.	4480							1.44	1.84				
Pilchenia antiqua, n. sp.	4481							1.52	1.84				
Pilchenia antiqua, n. sp.	4474			2.92						1.16	1.2		
Pilchenia intermedia, n. sp.	3825					2.52	3.32						
Pilchenia intermedia, n. sp.	3836					2.4	3						
Pilchenia intermedia, n. sp.	3829					2.4	3.2						
Pilchenia intermedia, n. sp.	3887					2.4							
Pilchenia intermedia, n. sp.	3897												
Pilchenia intermedia, n. sp.	3831									1.68	1.72		
Pilchenia intermedia, n. sp.	3859									1.64	1.8		
Pilchenia intermedia, n. sp.	3852									1.56	1.6		
Pilchenia intermedia, n. sp.	3891							1.96	2.44				
Pilchenia intermedia, n. sp.	3853								2.28				
Pilchenia intermedia, n. sp.	3903							1.88	2.28				
Pilchenia intermedia, n. sp.	3803							1.92					
Pilchenia intermedia, n. sp.	3919							1.92					

Species	No.	1	2	3	4	5	6	7	8
Clenia brevis, n. sp.	4146		1.84	2.16	1.8	2.04			
Clenia brevis, n. sp.	4375		1.84	2	1.84	2.16			
Clenia brevis, n. sp.	4238	2.2	1.72	2.16					
Clenia brevis, n. sp.	4372				2.36*	2.6*			
Eomicrobiotherium matutinum, n. sp.	3872	2.04*							
Eomicrobiotherium matutinum, n. sp.	4374	2.52*	2.08	2.52					
Eomicrobiotherium matutinum, n. sp.	4232	1.72							
Microbiotherium sp.	4362	1.64							
Microbiotherium sp.	4135				1.44*	1.8*			
Microbiotherium sp.	4359				1.56				
Microbiotherium sp.	4371							1.6	1.84
Microbiotherium sp.	4367								1.88
Rosendolops sp.	4487							2.64	2.8
Rosendolops sp.	4497							1.8	2.12
Rosendolops ebaios, n. sp.	4449	1.2			1.44	1.48			
Rosendolops ebaios, n. sp.	4477	1.44	1.24	1.48					
Rosendolops ebaios, n. sp.	4446		1.24	1.56					
Rosendolops ebaios, n. sp.	4476								
Hondonadia praecipitia, n. sp.	4144	2.8		2.72	2.72	2.56			
Hondonadia praecipitia, n. sp.	4383	2.6	2.84	2.8*	2.6*	2.76*			
Hondonadia praecipitia, n. sp.	4384	2.96	2.8	2.96					
Hondonadia praecipitia, n. sp.	4435	2.8							
Hondonadia praecipitia, n. sp.	4386	2.92							
Hondonadia praecipitia, n. sp.	4442	2.68							
Hondonadia praecipitia, n. sp.	4151	2.8							
Hondonadia praecipitia, n. sp.	4429		3.04	3.04					
Hondonadia praecipitia, n. sp.	4438		3	3					
Hondonadia praecipitia, n. sp.	4441		2.72*	2.88			1.8		
Hondonadia praecipitia, n. sp.	4437		3*	3.12*			2.2		
Hondonadia praecipitia, n. sp.	4434				2.76	3			
Hondonadia praecipitia, n. sp.	4436				2.76	2.92			
Hondonadia praecipitia, n. sp.	4428								2.96
Hondonadia praecipitia, n. sp.	4427								2.44
Hondonadia praecipitia, n. sp.	4433							2.4	3
Hondonadia pumila, n. sp.	4239	1.52	1.56	1.6	1.44	1.52			
Hondonadia pumila, n. sp.	4247	1.48							
Hondonadia pumila, n. sp.	4448	1.4							
Hondonadia pumila, n. sp.	4450	1.48							
Hondonadia pumila, n. sp.	4454	1.52*	1.64	1.76					
Hondonadia pumila, n. sp.	4452	1.72	1.6	1.8					

(cont.)

Taxon	Specimen (MPEF-PV)	LP3	WP3	LM1	WM1	LM2	WM2	LM3	WM3	LM4	WM4	LMx	WMx
Hondonadia pumila, n. sp.	4445					1.52*	1.72						
Hondonadia pumila, n. sp.	4444					1.4*	1.68						
Hondonadia pumila, n. sp.	4451					1.68	1.84						
Hondonadia pumila, n. sp.	4127					1.4*	1.8						
Hondonadia pumila, n. sp.	4447						1.68						
Hondonadia pumila, n. sp.	4455					1.64	1.92						
Hondonadia pumila, n. sp.	4453					1.4*	1.84*						
Hondonadia pumila, n. sp.	4475							1.44*	1.64*				
Hondonadia pumila, n. sp.	4473							1.4	1.6				
Hondonadia parca, n. sp.	3823			2.28		2.08	2.4						
Hondonadia parca, n. sp.	3884				1.76								
Hondonadia parca, n. sp.	3863					2.04	2.2						
Hondonadia parca, n. sp.	3808					2.2	2.4						
Hondonadia parca, n. sp.	3856					2	2.48						
Hondonadia parca, n. sp.	3885					2.2	2.32*						
Hondonadia parca, n. sp.	3865							1.56	1.84				
Hondonadia parca, n. sp.	3911									1.48	1.72		
Hondonadia parca, n. sp.	3914											1.6	2.2*
Hondonadia parca, n. sp.	3881											1.64*	1.88
Hondonadia parca, n. sp.	3841												
Hondonadia cf. *H. fierroensis*	4439											2.2*	2.6*
Hondonadia sp.	4424											2.6*	2*
Hondonadia sp.	4485											2.04	2*
Hondonadia cf. *H. pumila*	3861			1.8	1.72								
Hondonadia cf. *H. pumila*	3915												
Klohnia major, n. sp.	4173			2.6	2.6	2.52	2.48						
Klohnia major, n. sp.	4251					2.4	2.56	2.6	2.2				
Klohnia major, n. sp.	4277			2.28*	2.32*	2.16	2.28						
Klohnia major, n. sp.	4150							3.16	2.2				
Klohnia major, n. sp.	4270							3.12	2				
Klohnia major, n. sp.	4197							3.2	2.2				
Klohnia major, n. sp.	4175							3*	2.36*				
Klohnia major, n. sp.	4286							3.52	2.84				
Klohnia major, n. sp.	4264							3.2	2*				
Klohnia major, n. sp.	4184							3.2*	2.44*				
Klohnia major, n. sp.	4178							3.2	2.36				
Klohnia major, n. sp.	4174							3.2*	2*				

Species	No.								
Klohnia major, n. sp.	4180							2.8	2.4*
Klohnia major, n. sp.	4199					2.8	1.8		
Klohnia major, n. sp.	4186					2.6	2.2		
Klohnia major, n. sp.	4193					2.48*	2.2		
Klohnia major, n. sp.	4177					2.6	1.8		
Klohnia major, n. sp.	4275					2.6	2.2		
Klohnia major, n. sp.	4194					2.72*	2.8		
Klohnia major, n. sp.	4290					2.64	2.4		
Klohnia major, n. sp.	4201			2.76*	2.32				
Klohnia major, n. sp.	4265			2.6	2.08				
Klohnia major, n. sp.	4196			2.88	2.4				
Klohnia major, n. sp.	4293			2.6	2.08				
Klohnia major, n. sp.	4205			2.8	2.4				
Klohnia major, n. sp.	4202			2.84	2				
Klohnia major, n. sp.	4183			2.64*	2*				
Klohnia major, n. sp.	4253			2.48	2.8				
Klohnia major, n. sp.	4134	1.6	1.2						
Klohnia major, n. sp.	4136	1.6	1.2						
Klohnia major, n. sp.	4145	1.4	1.2						
Klohnia major, n. sp.	4137								
Klohnia major, n. sp.	4139	1.6	1.48						
Klohnia major, n. sp.	4207	1.44	1.16						
Klohnia major, n. sp.	4185							2.4*	2.44*
Klohnia major, n. sp.	4176							3.12	2.48
Klohnia major, n. sp.	4153			2.76	2.2				
Klohnia major, n. sp.	4204							3	2
Klohnia major, n. sp.	4292							2.88	1.8
Klohnia major, n. sp.	4289							2.68	2.64
Klohnia cf. K. major	4141			2.8*	2.6*				
Klohnia cf. K. major	4198			2.8*	2.2				
Klohnia cf. K. major	4172					2.8*	2.88		
Klohnia cf. K. major	4200			3.28	2.28				
Klohnia cf. K. major	4181					2.76*	2.56*		
Klohnia cf. K. major	4204					2.88*	2.84		
Klohnia cf. K. major	4203							3.04	2.88
Klohnia cf. K. major	4195							3.08	2.32
Klohnia cf. K. major	4278	1.36	1.12	2.76	2.56				
Epiklohnia verticalis n. gen. et sp.	3821					2.8		3*	
Epiklohnia verticalis n. gen. et sp.	3829						2.2		2.8
Kramadolops fissuratus n. gen. et sp.	4249	4.63	2.89						

(cont.)

98

Taxon	Specimen (MPEF-PV)	LP3	WP3	LM1	WM1	LM2	WM2	LM3	WM3	LM4	WM4	LMx	WMx
Kramadolops fissuratus n. gen. et sp.	4327	4.13	2.4										
Kramadolops fissuratus n. gen. et sp.	4342			5.62	3.55								
Kramadolops fissuratus n. gen. et sp.	4329							4.38*	3.63*				
Kramadolops fissuratus n. gen. et sp.	4343					4.38	3.55						
Pichipilidae gen et sp. indet.	3848											1.88	2.08
Pichipilidae gen et sp. indet.	3890											1.92	2.08
Lower molars													
Pichipilidae, gen. et sp. indet.	4457											1.88	1.12
Evolestes sp.	3804							1.48	1.08	1.2	0.7		
Evolestes sp.	3830					1.72	1.2						
Evolestes sp.	3811							1.36	0.84				
Evolestes sp.	3812												
Evolestes sp.	3857						1						
Evolestes sp.	3866												
Evolestes sp.	3846							1.56	1.08				
Evolestes sp.	3844					1.68	1.24	1.4*	0.92*				
Pilchenia antiqua, n. sp.	4129	2	1.5										
Pilchenia antiqua, n. sp.	4489			3.4	2.2								
Pilchenia antiqua, n. sp.	4126	2	1.6										
Pilchenia antiqua, n. sp.	4470					2.32	1.92						
Pilchenia antiqua, n. sp.	4235					2.24	2						
Pilchenia antiqua, n. sp.	4488					2.28*							
Pilchenia antiqua, n. sp.	4468							1.8	1.52				
Pilchenia antiqua, n. sp.	4465							1.4*	1.52				
Pilchenia antiqua, n. sp.	4464								1.4*				
Pilchenia antiqua, n. sp.	4466							1.88	1.52				
Pilchenia antiqua, n. sp.	4467							1.88	1.52				
Pilchenia antiqua, n. sp.	4471							1.92	1.48				
Pilchenia antiqua, n. sp.	4469							1.92	1.56				
Pilchenia antiqua, n. sp.	4462									1.48	1.2		
Pilchenia antiqua, n. sp.	4463									1.4	1.1		
Pilchenia intermedia, n. sp.	3892						1.76						
Pilchenia intermedia, n. sp.	3873					2.68							
Pilchenia intermedia, n. sp.	3905							2.2	1.72				
Pilchenia intermedia, n. sp.	3886							2.08	1.6				
Pilchenia intermedia, n. sp.	3805							2.28	1.68				
Pilchenia intermedia, n. sp.	3895							2.28	1.76				

Taxon	No.							
Pilchenia intermedia, n. sp.	3904							
Pilchenia intermedia, n. sp.	3868		1.84	1.3				1.76*	1.16*
Pilchenia intermedia, n. sp.	3815		1.8	1.4				1.84*	1.24*
Pilchenia intermedia, n. sp.	3898		2.64	2.12	2.2	1.76			
Pilchenia intermedia, n. sp.	3814		2.4						
Pilchenia intermedia, n. sp.	3842		2.3						
Pilchenia intermedia, n. sp.	3894	2	1.8						
Pilchenia intermedia, n. sp.	3900	2	1.8						
Pilchenia intermedia, n. sp.	3901	3	1.9						
Pilchenia intermedia, n. sp.	3922	2	1.9						
Clenia brevis, n. sp.	4368		1.64	1.3					
Clenia brevis, n. sp.	4346								
Clenia brevis, n. sp.	4356		1.84	1.2					
Clenia brevis, n. sp.	4353								
Clenia brevis, n. sp.	4360		1.96	1.24					
Clenia brevis, n. sp.	4364		1.8	1.3					
Clenia brevis, n. sp.	4355		1.96	1.28					
Clenia brevis, n. sp.	4363				1.96*	1.32*			
Clenia brevis, n. sp.	4358				1.8*	1.12*			
Clenia brevis, n. sp.	4361		1.88	1.28					
Clenia brevis, n. sp.	4354		1.76	1.2					
Clenia brevis, n. sp.	4349		1.76	1.2			1.44*		
Eomicrobiotherium matutinum, n. sp.	4350						1.48		
Eomicrobiotherium matutinum, n. sp.	4357								
Eomicrobiotherium matutinum, n. sp.	4373		2.28	1.5	1.96	1.32			
Eomicrobiotherium matutinum, n. sp.	4369				2.12	1.4			
Eomicrobiotherium matutinum, n. sp.	4352				2.12	1.52			
Eomicrobiotherium matutinum, n. sp.	4370								
Microbiotherium sp.	4366	1	1	0.7	1.64				
Microbiotherium sp.	4365		1.64					1.12*	
Microbiotherium sp.	4348		1.6					0.96*	0.96
Rosendolops ebaios, n. sp.	3917								
Hondonadia praecipitia, n. sp.	4132		2.4	1.8					
Hondonadia praecipitia, n. sp.	4233		2.56	2					
Hondonadia praecipitia, n. sp.	4423				2.84*				
Hondonadia praecipitia, n. sp.	4229								
Hondonadia praecipitia, n. sp.	4418				3				
Hondonadia praecipitia, n. sp.	4417				2.76*	2.04			
Hondonadia praecipitia, n. sp.	4385				2.96	2.08*			
Hondonadia praecipitia, n. sp.	4419							1.8*	

(cont.)

Taxon	Specimen (MPEF-PV)	LP3	WP3	LM1	WM1	LM2	WM2	LM3	WM3	LM4	WM4	LMx	WMx
Hondonadia praecipitia, n. sp.	4420							2.8	1.8				
Hondonadia praecipitia, n. sp.	4250					3*	2.2*						
Hondonadia praecipitia, n. sp.	4415												1.6*
Hondonadia praecipitia, n. sp.	4422												
Hondonadia praecipitia, n. sp.	4414												
Hondonadia praecipitia, n. sp.	4413												1.88
Hondonadia praecipitia, n. sp.	4426												2
Hondonadia praecipitia, n. sp.	4425												2
Hondonadia praecipitia, n. sp.	4421												1.96*
Hondonadia pumila, n. sp.	4406			1.48	1.1	1.52	1.32	1.56	1.32*				
Hondonadia pumila, n. sp.	4397			1.52	1.1								
Hondonadia pumila, n. sp.	4411					1.6	1.2						
Hondonadia pumila, n. sp.	4408					1.48	1.04						
Hondonadia pumila, n. sp.	4234					1.84	1.36						
Hondonadia pumila, n. sp.	4409					1.64	1.24						
Hondonadia pumila, n. sp.	4399					1.6	1.2						
Hondonadia pumila, n. sp.	4395							1.68*	1.12*				
Hondonadia pumila, n. sp.	4403							1.56	1.16				
Hondonadia pumila, n. sp.	4390							1.6	1.28				
Hondonadia pumila, n. sp.	4401							1.48	1.12				
Hondonadia pumila, n. sp.	4393							1.52	1.16				
Hondonadia pumila, n. sp.	4404			1.24	1	1.2*	1.04*						
Hondonadia pumila, n. sp.	4405			1.2*	1.1	1.28	1.2						
Hondonadia pumila, n. sp.	4407											1.56	
Hondonadia pumila, n. sp.	4391											1.52	1.12*
Hondonadia pumila, n. sp.	4394											1.6	1.28
Hondonadia pumila, n. sp.	4400												1*
Hondonadia pumila, n. sp.	4387												1.16
Hondonadia pumila, n. sp.	4392											1.56*	
Hondonadia parca, n. sp.	3883			2.04*	1.24*								
Hondonadia parca, n. sp.	3399					2.24	1.44	2	1.4				
Hondonadia parca, n. sp.	3916					2.12	1.64						
Hondonadia parca, n. sp.	3896					2.12	1.84						
Hondonadia parca, n. sp.	3840												1.36*
Hondonadia parca, n. sp.	3847												
Hondonadia cf. *H. fierroensis*	4412											2.16	1.4
Hondonadia cf. *H. fierroensis*	4388											2.12	1.56

Taxon	Specimen									
Hondonadia sp.	4416								2.48	2
Hondonadia sp.	4142								2.4*	1.72*
Hondonadia cf. *H. pumila*	3809								1.56	1.2
Hondonadia cf. *H. pumila*	3826				1.68	1.28				
Klohnia major, n. sp.	4224			2.6	1.4*	2.4*	2*	2.8	2*		
Klohnia major, n. sp.	4217			2.48	1.9	2.6	2.16	2.68	2.2		
Klohnia major, n. sp.	4279				2.3	2.64	2.44	2.6	2.4		
Klohnia major, n. sp.	3785					2.36	2			2.24*	2*
Klohnia major, n. sp.	4283							2.72	2.08		
Klohnia major, n. sp.	3787							2.32	1.88		
Klohnia major, n. sp.	4291							2.16	1.72		
Klohnia major, n. sp.	3789							2.48	1.88*		
Klohnia major, n. sp.	4276							2.68	2.24		
Klohnia major, n. sp.	4261								2.2		
Klohnia major, n. sp.	4147							2.48	2.24		
Klohnia major, n. sp.	4272							2.88	2.28		
Klohnia major, n. sp.	4282							2.8	2.4		
Klohnia major, n. sp.	4273							2.16	2.24		
Klohnia major, n. sp.	4149							2.48	1.92		
Klohnia major, n. sp.	4263							2.6	2.2		
Klohnia major, n. sp.	4212							2.8	2.36		
Klohnia major, n. sp.	3790							2.2	1.8		
Klohnia major, n. sp.	4267					2.52*	2.08*				
Klohnia major, n. sp.	4281					2.8	2*				
Klohnia major, n. sp.	4284					2.88*	2				
Klohnia major, n. sp.	4288					2.84	1.84				
Klohnia major, n. sp.	4223					2.6	2.28				
Klohnia major, n. sp.	4254						2.16				
Klohnia major, n. sp.	4215					2.8*	2.4*				
Klohnia major, n. sp.	3795			2.44	1.4						
Klohnia major, n. sp.	4280			2.8*	2						
Klohnia major, n. sp.	4271			2.72	1.6						
Klohnia major, n. sp.	4285			2.92	1.7						
Klohnia major, n. sp.	4287			2.4*	1.8						
Klohnia major, n. sp.	4616									2.48*	1.88*
Klohnia major, n. sp.	3792									2.48*	2
Klohnia major, n. sp.	3796									2.2	1.72
Klohnia major, n. sp.	4214									2.8	2.36
Klohnia major, n. sp.	4258									2.32*	1.68
Klohnia cf. *K. major*	4131	1	0.6	2.4	2	2.64	2.08				

(cont.)

Taxon	Specimen (MPEF-PV)	LP3	WP3	LM1	WM1	LM2	WM2	LM3	WM3	LM4	WM4	LMx	WMx
Klohnia cf. *K. major*	4171			3.2	1.6								
Klohnia cf. *K. major*	4221							3	2.4				
Klohnia cf. *K. major*	4226							2.68	2.32				
Klohnia cf. *K. major*	4222					2.96	2.36						
Epiklohnia verticalis, n. gen. et sp.	3837			2.36*	2.4*								
Kramadolops fissuratus, n. gen. et sp.	4339					4.63	3.06						
Kramadolops fissuratus, n. gen. et sp.	4338					4.13*	3.22*						
Kramadolops fissuratus, n. gen. et sp.	4335					4.87*	3.39*						
Kramadolops fissuratus, n. gen. et sp.	4340							4.54	3.14				
Kramadolops fissuratus, n. gen. et sp.	4315							4.54*	2.81*				
Kramadolops fissuratus, n. gen. et sp.	4336							4.38					
Kramadolops fissuratus, n. gen. et sp.	4303			3.72*									
Kramadolops fissuratus, n. gen. et sp.	4166	1.6*	0.83*										
Kramadolops fissuratus, n. gen. et sp.	4267	2	0.8										
Pichipilidae, gen. et sp. indet.	3828					1.24	1.12						
Pichipilidae, gen. et sp. indet.	3910							1.44	1				
?Pichipilidae, gen. et sp. indet.	4410											1.72	1.32
?Pichipilidae, gen. et sp. indet.	4402											1.68*	1.24
?Pichipilidae, gen. et sp. indet.	4494							1.2*	0.8	1	0.6		

REFERENCES

Ameghino, F. 1887. Enumeración sistemática de las especies de mamíferos fósiles coleccionados por Carlos Ameghino en los terrenos eocenos de la Patagonia austral y depositados en el Museo de La Plata. *Boletín del Museo de La Plata*, **1**, 1–26.

Ameghino, F. 1889. Contribución al conocimiento de los mamíferos fósiles de la República Argentina, obra escrita bajo los auspicios de la Academia Nacional de Ciencias de la República Argentina para presentarla a la Exposición Universal de Paris de 1889. *Actas de la Academia Nacional de Ciencias, Córdoba*, **6**, 1–1027.

Ameghino, F. 1894. Enumération synoptique des espèces de mammifères fossiles des formations éocènes de Patagonie. *Boletín de la Academia Nacional de Ciencias, Córdoba*, **13**, 259–452.

Ameghino, F. 1897. Mammifères crétacés de l'Argentine: deuxième contribution á la connaissance de la faune mammalogique des couches à Pyrotherium. *Boletín del Instituto Geográfico Argentino*, **18**, 406–521.

Ameghino, F. 1901. Notices préliminaires sur de ongulés nouveaux de terrains crétacés de Patagonie. *Boletín de la Academia Nacional de Ciencias, Córdoba*, **16**, 349–426.

Ameghino, F. 1902. Premièr contribution à la connaissance de la faune mammalogique des couches à Colpodon. *Boletín de la Academia Nacional de Ciencias, Córdoba*, **17**, 71–141.

Ameghino, F. 1903. Los Diprotodontes del orden de los Plagiaulacoideos y el origen de los roedores y de los Polimastodontes. *Anales del Museo Nacional de Buenos Aires*, **1**, 81–192.

Ameghino, F. 1904. Nuevas especies de mamíferos cretáceos y terciarios de la República Argentina. *Anales de la Sociedad Cientifica Argentina*, **56**, 193–208 (1903); **57**, 162–75, 327–341 (1904); **58**, 35–71, 182–192, 225–291 (1904).

Ashton, K. G., M. C. Tracy, and A. de Queiroz 2000. Is Bergmann's Rule valid for mammals? *American Naturalist*, **156**, 390–415.

Barnosky, A. D. 2001. Distinguishing the effects of the Red Queen and Court Jester on Miocene mammal evolution in the Northern Rocky Mountains. *Journal of Vertebrate Paleontology*, **21**, 172–185.

Barreda, V. and L. Palazzesi 2005. Patagonian vegetation turnovers during the Paleogene–early Neogene: origin of arid-adapted floras. *Actas XVI Congreso Geológico Argentino*, **4**, 445–446.

Bellosi, E. S. 2005. Sedimentology and depositional setting of the Sarmiento Formation (Eocene–Miocene) in Central Patagonia. *Actas XVI Congreso Geológico Argentino*, **4**, 439–440.

Bellosi, E. S. and R. H. Madden 2005. Estratigrafía física preliminar de las secuencias piroclásticas terrestres de la Formación Sarmiento (Eoceno–Mioceno) en la Gran Barranca, Chubut. *Actas XVI Congreso Geológico Argentino*, **4**, 427–432.

Bown, T. M. and J. G. Fleagle 1993. Systematics, bioestratigraphy, and dental evolution of the Palaeothentidae, later Oligocene to early–middle Miocene (Deseadan–Santacrucian) caenolestoid marsupials of South America. *Journal of Paleontology*, **67**, 1–76.

Birney, E. C. and J. A. Monjeau 2003. Latitudinal variation in South American marsupial biology. In Jones, M., Dickman, C., and Archer, M. (eds.) *Predators with Pouches: The Biology of Carnivorous Marsupials*. Collingwood, NSW: CSIRO, pp. 297–317.

Carlini, A. A., M. Ciancio, and G. J. Scillato-Yané 2005a. Los Xenarthra de Gran Barranca: Más 20 Ma de Historia. *Actas XVI Congreso Geológico Argentino*, **4**, 419–424.

Carlini, A. A., R. H. Madden, M. G. Vucetich, M. Bond, G. López, M. Reguero, and A. Scarano 2005b. Mammalian biostratigraphy and biochronology at Gran Barranca: the standard reference section for the continental middle Cenozoic of South America. *Actas XVI Congreso Geoló gico Argentino*, **4**, 425–426.

Case, J. A., F. J. Goin, and M. O. Woodburne 2005. "South American" Marsupials from the Late Cretaceous of North America and the Origin of Marsupial Cohorts. *Journal of Mammalian Evolution*, **11**, 223–255.

Clemens, W. A. 1966. Fossil mammals of the Type Lance Formation, Wyoming. Part II: Marsupialia. *University of California Publications in Geological Science*, **62**, 1–122.

Crochet, J. -Y. 1979. Les marsupiaux fossiles d'Europe. *Proceedings VII Réunion Annuelle des Sciences de la Terre, Lyon*, **137**.

Flynn, J. J. and C. C. Swisher III 1995. Chronology of the Cenozoic South American Land Mammal Ages. In Berggren, W. A., Kent, D. V., Aubry, M. P., and Hardenbol, J. (eds.), *Geochronology, Time-Scales, and Global Stratigraphic Correlation*, Special Publication no. 54. Tulsa, OK: Society for Sedimentary Geology, pp. 317–33.

Flynn, J. J. and A. R. Wyss 1999. New marsupials from the Eocene–Oligocene transition of the Andean Main Range, Chile. *Journal of Vertebrate Paleontology*, **19**, 533–549.

Flynn, J. J. and A. R. Wyss 2004. A polydolopine marsupial skull from the Cachapoal Valley, Andean Main Range, Chile. *Bulletin of the American Museum of Natural History*, **285**, 80–92.

Flynn, J. J., A. R. Wyss, D. A. Croft, and R. Charrier 2003. The Tinguiririca Fauna, Chile: biochronology, paleoecology, biogeography, and a new earliest Oligocene South American Land Mammal 'Age'. *Palaeogeography, Palaeoclimatology, Palaeoecology*, **195**, 229–259.

Gill, T. 1872. Arrangement of the families of mammals with analytical tables. *Smithsonian Miscellaneons Collections*, **11**, 1–98.

Goin, F. J. 2007. A review of the Caroloameghiniidae, Paleogene South American "primate-like" marsupials (Peradectia). In Kalthoff, D., Martin, T., and Möors, T. (eds.), *Festband für Herrn Professor Wighart v. Koenigswald anlässlich seines 65. Geburtstages*, Stuttgart: Schweizerbart'sche Verlagsbuchhandlung, pp. 57–67.

Goin, F. J. and A. M. Candela 1996. A new Early Eocene Polydolopimorphian (Mammalia, Marsupialia) from Patagonia. *Journal of Vertebrate Paleontology*, **16**, 292–296.

Goin, F. J. and A. M. Candela 1997. New Patagonian marsupials from Ameghino's "Astraponoteén plus supérieure" (post-Mustersan/pre-Deseadan age). *Journal of Vertebrate Paleontology*, **17** (Suppl. to No.3), unpublished abstract for replacement presentation at annual meeting.

Goin, F. J. and A. M. Candela 1998. Dos nuevos marsupiales "Pseudodiprotodontes" del Paleógeno de Patagonia. In Casadío, S. (ed.), *Paleógeno de América del Sur y de la Península Antártica*. Buenos Aires: Asociación Paleontológica Argentina, pp. 79–84.

Goin, F. J., and A. M. Candela 2004. New Paleogene marsupials from the Amazonian basin, Southeastern Perú. In Campbell, K. E. Jr. (ed.), *The Paleogene Mammalian Fauna of Santa Rosa, Amazonian Perú*. Los Angeles, CA: Natural History Museum of Los Angeles County, pp. 15–60.

Goin, F. J., A. M. Candela, and C. de Muizon 2003. The affinities of *Roberthoffstetteria nationalgeographica* (Marsupialia) and the origin of the polydolopine molar pattern. *Journal of Vertebrate Paleontology*, **23**, 869–876.

Goin, F. J., M. F. Tejedor, M. O. Woodburne, L. Chornogubsky, G. Martin, and E. Aragón 2005. Marsupiales paleógenos de la región de Paso del Sapo, Chubut, Argentina. *Ameghiniana*, **42**, 30R.

Goin, F. J., M. A. Abello, E. Bellosi, R. Kay, R. H. Madden, and A. A. Carlini 2007a. Los Metatheria sudamericanos de comienzos del Neógeno (Mioceno temprano, Edad-mamífero Colhuehuapense). I. Introducción, Didelphimorphia y Sparassodonta. *Ameghiniana*, **44**, 29–71.

Goin, F. J., M. Sánchez-Villagra, M. A. Abello, and A. M. Candela 2007b. A new generalized paucituberculatan marsupial from the Oligocene of Bolivia, and the origin of 'shrew-like' opossums. *Palaeontology*, **50**, 1267–1276.

Gordon, C. L. 2004. A first look at estimating body size in dentally conservative marsupials. *Journal of Mammalian Evolution*, **10**, 1–21.

Gradstein, F., J. Ogg, and A. Smith (eds.) 2004. *A Geological Time Scale 2004*. Cambridge, UK: Cambridge University Press.

Haq, B. U., J. Hardenbol, and P. R. Vail 1987. Chronology of fluctuating sea levels since the Triassic. *Science*, **235**, 1156–1166.

Hershkovitz, P. 1971. Basic crown patterns and cusp homologies of mammalian teeth. In Dahlberg, A. A. (ed.), *Dental Morphology and Evolution*. Chicago, IL: University of Chicago Press, pp. 95–150.

Hinojosa, L. F. 2005. Cambios climáticos y vegetacionales inferidos a partir de paleofloras cenozoicas del sur de Sudamérica. *Revista Chilena Historia Natural*, **32**, 95–115.

Hulbert, A. J. 1988. Metabolism and the development of endothermy. In Tyndale-Biscoe, C. H. and Janssens, P. A. (eds.), *The Developing Marsupial*. Berlin: Springer-Verlag, pp. 148–161.

Hunsaker, D. I I 1977. Ecology of New World Marsupials. In Hunsaker, D. (ed.), *The Biology of Marsupials*. New York: Academic Press, pp. 95–156.

Huxley, T. H. 1880. On the application of the laws of evolution to the arrangement of the Vertebrata and more particularly of the Mammalia. *Proceedings of the Zoological Society of London*, **1880**, 649–662.

Illiger, C. 1811. *Prodromus systematics mammalium et avium additis terminis zoographicis utrudque classis*. Berlin: C. Salfeld.

Ivany, L. C., S. Van Simaeys, E. W. Domack, and S. D. Samson 2006. Evidence for an earliest Oligocene ice sheet on the Antarctic Peninsula. *Geology*, **34**, 377–380.

Kielan-Jaworowska, Z., R. L. Cifelli, and Z. -X. Luo 2004. *Mammals from the Age of Dinosaurs: Origins, Evolution, and Structure*. New York: Columbia University Press.

Kohn, M. J., J. A. Josef, R. H. Madden, R. Kay, M. G. Vucetich, and A. A. Carlini 2004. Climate stability across the Eocene–Oligocene transition, southern Argentina. *Geology*, **32**, 621–624.

Lagabrielle, Y., Y. Goddéris, Y. Donnadieu, J. Malavieille, and M. Suarez 2009. The tectonic history of Drake Passage and its possible impacts on global climate. *Earth and Planetary Science Letters*, **279**, 197–211.

Livermore, R., G. Eagles, P. Morris, and A. Maldonado 2004. Shackleton Fracture Zone: no barrier to early circumpolar ocean circulation. *Geology*, **32**, 797–800.

Loomis, F. B. 1914. *The Deseadan Formation of Patagonia*. New Haven, CT: Concord Press.

López, G., M. Bond, M. Reguero, J. Gelfo, and A. Kramarz 2005. Los ungulados del Eoceno-Oligoceno de la Gran Barranca, Chubut. *Actas XVI Congreso Geológico Argentino*, **4**, 415–418.

Madden, R. H., E. Bellosi, A. A. Carlini, M. Heizler, J. J. Vilas, G. H. Ré, R. F. Kay, and M. G. Vucetich 2005a. Geochronology of the Sarmiento Formation at Gran Barranca and elsewhere in Patagonia: calibrating Middle Cenozoic mammal evolution in South America. *Actas XVI Congreso Geológico Argentino*, **4**, 411–412.

Madden, R. H., A. A. Carlini, M. G. Vucetich, and R. Kay 2005b. The Paleontology of Gran Barranca: evolution and environmental change through the middle Cenozoic of Patagonia. *Actas XVI Congreso Geológico Argentino*, **4**, 409–410.

Marshall, L. G. 1978. Evolution of the Borhyaenidae, extinct South American predaceous marsupials. *University of California Publications in Geological Science*, **117**, 1–89.

Marshall, L. G. 1980. Systematics of the South American marsupial family Caenolestidae. *Fieldiana, Geology*, **5**, 1–145.

Marshall, L. G. 1982a. Systematics of the South American marsupial family Microbiotheriidae. *Fieldiana, Geology*, **10**, 1–75.

Marshall, L. G. 1982b. Systematics of the extinct South American marsupial family Polydolopidae. *Fieldiana, Geology*, **12**, 1–106.

Marshall, L. G. 1987. Systematics of Itaboraian (middle Paleocene) age "opossum-like" marsupials from the limestone quarry at Sao José de Itoboraí, Brazil. In Archer, M. (ed.), *Possums and Opossums: Studies in*

Evolution. Sydney, NSW: Royal Zoological Society of New South Wales, pp. 91–160.

Marshall, L. G. and R. Pascual 1977. Nuevos marsupiales Caenolestidae del "Piso Notohippidense" (SW de Santa Cruz, Patagonia) de Ameghino: sus aportaciones a la cronología y evolución de las comunidades de mamíferos sudamericanos. *Publicaciones Museo Municipal de Ciencias Naturales Mar del Plata*, **2**, 91–122.

McKenna, M. C. and S. K. Bell 1997. *Classification of Mammals above the Species Level*. New York: Columbia University Press.

McNab, B. K. 2005. Uniformity in the basal metabolic rate of marsupials: its causes and consequences. *Revista Chilena de Historia Natural*, **78**, 183–198.

Merico, A., T. Tyrrell, and P. A. Wilson 2008. Eocene/Oligocene ocean de-acidification linked to Antarctic glaciation by sea-level fall. *Nature*, **452**, 979–983.

Meng, J. and M. C. McKenna 1998. Faunal turnovers of Paleogene mammals from the Mongolian Plateau. *Nature*, **394**, 364–367.

Miller, K. G., J. D. Wright, and R. G. Fairbanks 1991. Unlocking the ice house: Oligocene–Miocene oxygen isotopes, eustasy, and margin erosion. *Journal of Geophysical Research*, **96**, 6829–6848.

Odreman Rivas, O. E. 1978. Sobre la presencia de un Polydolopidae (Mammalia, Marsupialia) en capas de Edad Mustersense (Eoceno Medio) de Patagonia. *Obra Centeniales Museo de La Plata*, **5**, 29–38.

Ortiz-Jaureguizar, E. and G. A. Cladera 2006. Paleoenvironmental evolution of southern South America during the Cenozoic. *Journal of Arid Environments*, **66**, 498–532.

Pascual, R. 1981. Adiciones al conocimiento de *Bonapartherium hinakusijum* (Marsupialia, Bonapartheriidae) del Eoceno temprano del Noroeste argentino. *Anais II Congreso Latino-Americano Paleontología, Porto Alegre*, **2**, 507–520.

Pascual, R., E. Ortiz Jaureguizar, and J. L. Prado 1996. Land mammals: paradigm for Cenozoic South American geobiotic evolution. *Münchner Geowissenschaftliche Abhandlungen*, **30**, 265–319.

Patterson, B. and L. G. Marshall 1978. The Deseadan Early Oligocene Marsupialia of South America. *Fieldiana, Geology*, **41**, 37–100.

Prothero, D. R. 2004. Did impacts, volcanic eruptions, or climate change affect mammalian evolution? *Palaeogeography, Palaeoclimatology, Palaeoecology*, **214**, 283–294.

Prothero, D. R. and R. J. Emry 1996. Summary. In Prothero, D. R. and Emry, R. J. (eds.), *The Terrestrial Eocene–Oligocene Transition in North America*. New York: Cambridge University Press, pp. 664–683.

Quattrocchio, M. 2006. Palynology and palaeocommunities of the Paleogene of Argentina. *Revista Brasiliense Paleontología*, **9**, 101–108.

Ré, G. H., R. H. Madden, M. Heizler, J. F. Vilas, M. E. Rodríguez, R. F. Kay, and A. A. Carlini 2005. Polaridad magnética preliminar de las sedimentitas de la Formación Sarmiento (Gran Barranca del Lago Colhue Huapi, Chubut, Argentina). *Actas XVI Congreso Geológico Argentino*, **4**, 387–394.

Retallack, G. J., W. N. Orr, D. R. Prothero, R. A. Duncan, P. R. Kester, and C. P. Ambers 2004. Eocene–Oligocene extinction and paleoclimatic change near Eugene, Oregon. *Bulletin of the Geological Society of America*, **116**, 817–839.

Romero, E. J. 1978. Paleoecología y paleofitogeografía de las tafofloras del Cenofítico de Argentina y áreas vecinas. *Ameghiniana*, **15**, 209–227.

Scher, H. D. and E. E. Martin 2006. Timing and climatic consequences of the opening of Drake Passage. *Science*, **312**, 428–430.

Simpson, G. G. 1964. Los mamíferos Casamayorenses de la Colección Tournouër. *Revista de Museo Argentino de Ciencias Naturales, Paleontología*, **1**, 1–21.

Sinclair, W. J. 1906. Mammalia of the Santa Cruz Beds: Marsupialia. In *Reports of the Princeton University Expeditions of Patagonia 1896–1899*, vol. **4**, Part III. Princeton, NJ: Princeton University Press, pp. 333–460.

Stehlin, H. G. 1909. Remarques sur les faunules de mammifères des couches éocènes et oligocènes du Basin de Paris. *Bulletin de la Société de Géologie de France*, **9**, 488–520.

Trouessart, E. L. 1898. *Catalogus mammalium tam viventium quam fossilium*, 2nd edn. Berlin: Friedländer und Sohn.

Vucetich, M. G., A. A. Carlini, R. H. Madden, R. F. Kay, and E. C. Vieytes 2004. Nuevos hallazgos entre los más antiguos roedores de América del Sur: una dispersión post-transición Eoceno–Oligoceno. *Ameghiniana*, **41** (Supl.), 24R–25R.

Vucetich, M. G., E. C. Vieytes, A. Kramarz, and A. A. Carlini 2005. Los roedores caviomorfos de Gran Barranca: aportes bioestratigráficos y paleoambientales. *Actas XVI Congreso Geológico Argentino*, **4**, 413–414.

Woodburne, M. O. 2004. Global events and the North American mammalian biochronology. In Woodburne, M. O. (ed.), *Late Cretaceous and Cenozoic Mammals of North America: Biostratigraphy and Geochronology*. New York: Columbia University Press, pp. 315–343.

Zachos, J., M. Pagani, L. Sloan, E. Thomas, and K. Billups 2001. Trends, rhythms, and aberrations in global climate 65 Ma to present. *Science*, **292**, 686–693.

Zucol, A. F. and M. Brea 2005. El registro fitolítico de los sedimentos cenozoicos de la localidad de Gran Barranca: su aporte a la reconstrucción paleoecológica. *Actas XVI Congreso Geológico Argentino*, **4**, 395–402.

7 Middle Eocene – Early Miocene Dasypodidae (Xenarthra) of southern South America: faunal succession at Gran Barranca – biostratigraphy and paleoecology

Alfredo A. Carlini, Martín R. Ciancio,
and G. J. Scillato-Yané

Abstract

Most biostratigraphic sequences in Patagonia and elsewhere in South America have been based on the evolutionary stage and taxonomic representation of "ungulates" (archaic endemic herbivores or southern ungulates) and/or marsupials. Recent collections of microfossils made at Gran Barranca, Patagonia, allow us to assess critically the value of Cingulata (mainly the Dasypodidae) and cingulate assemblages in mammalian biostratigraphy. The numerous and rich fossil localities and levels at Gran Barranca provide the most stratigraphically complete sampling of Cingulata available anywhere in South America. Starting with the "Barrancan" (Eocene) level at the base of the sequence and ending with the "Pansantacrucian" (Miocene) beds at the top of the Sarmiento Formation, dasypodid assemblages demonstrate a very close relationship with sea-surface temperature variation through the late Paleogene. We have studied seven successive faunas with a very rich dasypodid record. While the astegotheriine Dasypodinae decrease in diversity through the late Paleocene – late Eocene interval, the stegotheriine Dasypodinae, and especially the Euphractinae, increase in diversity through the middle Eocene and into the early Oligocene, until euphractines became dominant in the late Eocene. Cingulate assemblages permit speculation that between the upper Casamayoran (Cifelli's "Barrancan," around 39–41.6 Ma) and the typical Mustersan (about 38.7 to 37.0 Ma) a distinct faunal assemblage may represent a new subdivision of the mammalian sequence in Patagonia.

Resumen

La mayor parte de las secuencias bioestratigráficas en Patagonia y en otros lugares de Sudamérica ha estado
basada en los estados evolutivos y la representación taxonómica de los "ungulados" (antiguos herbívoros endémicos o ungulados sudamericanos) y/o marsupiales. Recientes colecciones de microfósiles realizadas en Gran Barranca, Patagonia, nos permitieron evaluar el valor de los Cingulata (principalmente los Dasypodidae) y de las asociaciones de cingulados en el contexto de la bioestratigrafía de mamíferos. Las numerosas y ricas localidades y niveles fosilíferos de Gran Barranca proveen la más completa muestra de Cingulata en una secuencia estratigráfica disponible en cualquier lugar de Sudamérica. Comenzando con el nivel "Barranquense" (Eoceno) en la base de la secuencia y finalizando con las capas "Pansantacrucianas" (Mioceno) en el tope de la Formación Sarmiento, los ensambles de dasipódidos muestran una relación muy estrecha con las variaciones de las temperaturas marinas superficiales a través del Paleógeno tardío. Se estudiaron 7 faunas sucesivas con un registro de dasipódidos muy rico. Mientras los Dasypodinae Astegotheriini disminuyen en diversidad a través del intervalo Paleoceno tardío-Eoceno tardío, los Dasypodinae Stegotheriini y especialmente los Euphractinae aumentan en diversidad hacia el Eoceno medio-Oligoceno temprano, hasta que los eufractinos se vuelven dominantes en el Eoceno tardío. Las agrupaciones de Cingulata nos permiten especular que entre el Casamayorense tardío ("Barranquense" de Cifelli, 39–41.6 Ma) y el típico Mustersense (*ca.* 38.7 to 37.0 Ma), hay una fauna distinta que podría representar una nueva subdivisión de la secuencia mamífera en Patagonia.

Introduction

The remains of dasypodids collected at Gran Barranca and other exposures of the Sarmiento Formation along the cliffs south of Lake Colhue-Huapi (Bellosi this book) since the

The Paleontology of Gran Barranca: Evolution and Environmental Change through the Middle Cenozoic of Patagonia, eds. R. H. Madden, A. A. Carlini, M. G. Vucetich, and R. F. Kay. Published by Cambridge University Press. © Cambridge University Press 2010.

end of the nineteenth century form a conspicuous group of taxa notably diverse and abundant. They were found in different levels of the Paleogene and earliest Neogene. They are mainly isolated or associated osteoderms from the carapace, a few of them are associated or even associated with fragments of skeleton. The study of the Dasypodidae from the Paleogene – early Miocene of the Argentine Patagonia (including those from Gran Barranca) may be historically separated into three stages. The first studies were those of Florentino Ameghino. Since before the end of the nineteenth century, remains had been collected by his brother Carlos in southern Patagonia. These were published (Ameghino 1902a, 1902b) in works focused mainly on the recognition of new genera and species from his "Couches à *Colpodon*" (Colhuehuapian, early Miocene: Ameghino 1902a). However, when he studied the remains found in his "Couches à *Notostylops*" (Casamayoran, middle Eocene), "Couches à *Astraponotus*" (Mustersan, latest Eocene), and "Couches à *Pyrotherium*" (Deseadan, Oligocene), the number of named taxa became rather excessive, as in several cases he established new taxa on the basis of osteoderms from non-homologous regions of the carapace (Ameghino 1902b). At first he identified the geographic localities, although the location of these was never precise, given the lack of detailed cartography of the area. But by 1902, he was precise neither with respect to the locality nor age of the material. Despite this imprecision, many of his new species are undoubtedly valid and contribute importantly to our present knowledge of the group. In fact, many of these species are revalidated in this chapter, after the study of a large amount of additional material collected during our field trips to Patagonia. After the work of the Ameghino brothers at the end of the nineteenth century, there were no significant contributions concerning the Patagonian dasypodids for another half century until Simpson (1948) revised the Casamayoran and Mustersan faunas of Patagonia based on new collections made by him and those collected by Ameghino.

Of the dasypodid taxa described by Ameghino (1887, 1894, 1897, 1902a, 1902b) and revised by Simpson (1948), 14 species definitely come from Gran Barranca (Ciancio and Carlini 2008), whereas the remaining species were referred to an age (e.g. Notostylopéen) rather than to a locality. New collections at Gran Barranca made by a collaborative research between Duke University, USA and Museo de La Plata, Argentina, permit a comprehensive revision of the Dasypodidae for the middle Eocene – early Miocene, with more precise stratigraphic control. We revisit those taxa described by Ameghino (1902a, 1902b), revalidate some species that Simpson (1948) had placed in synonymy, and describe new taxa. The new collections from Gran Barranca were made with strict temporal and stratigraphic control (Ré *et al.* Chapter 4, this book; Bellosi

this book) in which at least eight successive faunas are recognized, seven of which have Dasypodidae. The revision, redescription and nomination of taxa here reported, together with the evolutionary and geochronological pattern, will certainly allow the correlation of many Paleogene Patagonian faunas of which almost nothing is known, except for a tentative assignment of specimens. Currently, many type specimens of the species described by Ameghino are missing, and consequently, the validity of these names is dubious (Ciancio and Carlini 2008). However, in this chapter we assign new materials to these names following certain characters mentioned in the original descriptions.

Abbreviations

AMNH, American Museum of Natural History; FMNH, Field Museum of Natural History; MACN A, Museo Argentino de Ciencias Naturales "Bernardino Rivadavia" (Argentina), Colección Ameghino; MLP, Museo de La Plata; MPEF PV, Museo Paleontológico Egidio Feruglio Paleontología Vertebrados (Trelew, Chubut Province, Argentina); SGOPV, Museo Nacional de Historia Natural, Santiago (Chile), Paleontología de Vertebrados; *s.l.*, *sensu lato*.

Descriptions

The definition of dasypodid species is based on the morphological variation of osteoderms in specimens that preserve large articulated parts, or large groups of associated osteoderms (e.g. SGOPV 3122, SGOPV 2841, MLP 61-VIII-3–272, MLP 61-VIII-3–271, MLP 61-VIII-3–273, AMNH 28668, AMNH 29036, FMNH P13199, FMNH P12069, among others). In this way, each species recognized as valid, encompasses a wide range of morphological variation. In addition, the intraspecific variation observed in the carapaces of living species (e.g. *Chaetophractus villosus*, *Zaedyus pichiy*, *Dasypus hybridus*, *D. novemcinctus*) has also been considered when defining the limits of the species. For morphological descriptions we use those osteoderms that belong to the medial region of the dorsal shield in order to compare osteoderms from homologous regions of the carapace. We do not describe new taxa represented by poorly represented remains.

Systematic paleontology

Subfamily **DASYPODINAE** Gray 1821
Tribe **Astegotheriini** Ameghino 1906
Genus *Stegosimpsonia* Vizcaíno 1994

Type species *Stegosimpsonia chubutana* (Ameghino 1902b).

Age and distribution Paso del Sapo, Chubut Province, Argentina, lower Eocene?, "Riochican SALMA" (Tejedor *et al.* 2009); Sarmiento Formation,

A.A. *Carlini* et al.

DASYPODINAE
ASTEGOTHERIINI

STEGOTHERIINI

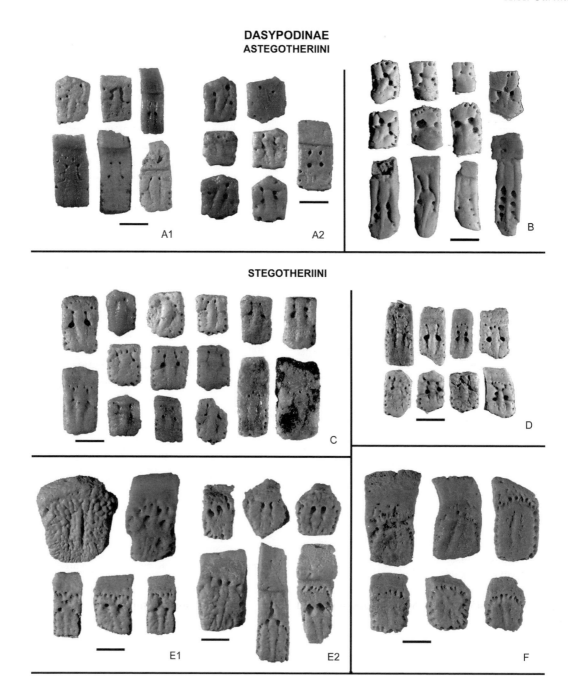

Fig. 7.1. (A) *Stegosimpsonia chubutana* (A1, MPEF PV 7660B; A2, MPEF PV 5425); (B) *Pseudostegotherium glangeaudi* (MPEF PV 5451A); (C) Stegotheriini gen. nov. *A* sp. nov. *a* (MPEF PV 7635C); (D) Stegotheriini gen. nov. *A* sp. nov. *b* (MPEF PV 7378F); (E) *Stegotherium variegatum* (E1, MPEF PV 5456; E2, MPEF PV 6474); (F) *Stegotherium tessellatum* (MPEF PV 7315D). Scale bar 5 mm.

Gran Barranca, Chubut Province, Argentina, Barrancan Subage, Casamayoran SALMA.
Stegosimpsonia chubutana (Ameghino 1902b)
Fig. 7.1A1, A2.
Syntype MACN A-10438, osteoderms of the carapace.

Age and distribution Sarmiento Formation, Gran Barranca, Chubut Province, Argentina, Barrancan Subage, Casamayoran SALMA.
Assigned materials MPEF PV: 5425, 6344, 6398, 6465, 6516, 7234A, 7660B and AMNH: 28745, 28851; groups of isolated osteoderms.

Comments This species is widely distributed in the Gran Barranca Member of the Sarmiento Formation. The main feature of this species is the lageniform (bottled-shaped) central figure (to isodiametric in the fixed osteoderms) surrounded by numerous small pits including the posterior margin (Vizcaíno 1994). In the syntype and in the numerous osteoderms we have recovered, the number of perforations is frequently less than 15. We have collected new Stegotheriini that extend by some 20 Ma the history of the tribe (Carlini *et al.* 2004a). Stegotheriini are recognized mainly by distinctive large perforations (among other characters), suggesting that the amount and position of the perforations is diagnostic. At this time, and in view of the new diversity reported for the phylogenetic history of the Dasypodinae, it is quite probably that the Stegotheriini were derived from such forms as *Stegosimpsonia*. The size of the osteoderms are 6–8 mm long, 6–7 mm wide, 2.5 mm thick for fixed and 14–16 mm long, 5–6 mm wide, 2 mm thick for moveable osteoderms.

Genus *Pseudostegotherium* Ameghino 1902a
Type species *Pseudostegotherium glangeaudi* Ameghino 1902a
Age and distribution Sarmiento Formation, Gran Barranca, Chubut Province, Argentina, Barrancan Subage, Colhuehuapian SALMA.
Pseudostegotherium glangeaudi Ameghino 1902a
Fig. 7.1B.
Holotype MACN A-12679, osteoderms of the carapace and left mandibular fragment with two teeth.
Age and distribution As for the genus by monotypy.
Assigned materials MPEF PV: 6377B, 6384B, 6416B, 6504, 6519B, 6532A, 6659, 7056B, 7153B, 7220B, 7334C, 5451A, 5458, 5462; groups of isolated osteoderms. MPEF PV 6530; associated osteoderms.
Comments This species has no clear relationships with previous forms of Astegotheriini, whereas its characters are quite far from the general pattern of the tribe. However, it seems to be related to *Nanoastegotherium prostatum*, described by Carlini *et al.* (1997) for the middle Miocene of Colombia. Both taxa are very small dasypodids with large perforations on the surface of the osteoderms (fixed osteoderm meassurements are 4–7 mm length, 3–4.5 mm width for *N. postatum* and 8–10 mm length, 4.5–6.5 mm width for *P. glangeaudi*).

Tribe **Stegotheriini** Ameghino 1889
Gen. nov. *A*
Sp. nov. *A a*
Fig. 7.1C.

Age and distribution Sarmiento Formation, Gran Barranca, Chubut Province, Argentina, GBV-60 "El Nuevo" and GBV-3 "El Rosado" (Mustersan SALMA) faunas.
Assigned material MPEF PV: 7376A, 7377C, 7378E, 7379B, 7486E, 7633D, 7635C, 7636B, 7658B, 5427, 7845C, 5432; groups of isolated osteoderms.
Description Size similar to *Stegosimpsonia*; the osteoderms have a more rugose surface than *Stegosimpsonia chubutana*, are thicker than those of Astegotheriini in general, and the lateral contact surfaces among osteoderms are even and planar.

Fixed osteoderms are quadrangular (8–9 mm length, 6–7 mm width and 3–3.5 mm thick); they possess a lageniform central figure with a medial keel wider and higher than that of *S. chubutana*. Anteriorly, there is one or no anterior peripheral figure; laterally a couple of peripheral figures are found adjacent to the neck of the lageniform figure. On the furrow surrounding the neck of the main figure, there are two foramina at each side. The posterior part of the central figure at the lateral and posterior margins is surrounded by small foramina, more numerous than in *S. chubutana*. Additionally, there may be foramina on the furrows separating the peripheral figures.

Mobile osteoderms are 14–16 mm long and 6–5 mm wide, with a lageniform central figure and a pair of anterior lateral figures. The central figure has a median keel, with the anterior third elevated above the posterior third. Pits on the exposed surface and posterior margins are equivalent to those of the fixed osteoderms. Anterior the central figure there are four to six pits in semicircle or aligned transversally.

Sp. nov. *A b*
Fig. 7.1D.
Age and distribution Sarmiento Formation, Gran Barranca, Chubut Province, Argentina, GBV-60 "El Nuevo" and GBV-3 "El Rosado" (Mustersan SALMA) faunas.
Assigned materials MPEF PV: 7378F, 7379C, 7635D, 5428, 6590B, 6598, 7845F; groups of isolated osteoderms.
Description Small size, smaller than sp. nov. *A a*. Surface of the osteoderms with less verticle relief than in sp. nov. *A a*, and relatively thick and with the lateral surfaces in contact between osteoderms even and planar.

Fixed osteoderms are quadrangular (6–7 mm length and 5 mm width), with a lageniform central figure surrounded by numerous (16–18) and small foramina that are smaller than those surrounding the central surface of *Stegosimpsonia*. The central figure has a short neck and a narrow, well-defined medial keel. A pair of figures are present anterolateral to the central figure, on both sides of the neck of the lageniform figure. The pits at both sides of the neck of the lageniform figure are larger. There may be pits on the furrows separating the peripheral figures.

Mobile osteoderms are rectangular (9 mm length and 4 mm width); these have also a central lageniform figure, and at both sides of the neck a pair of elongate anterolateral figures. In the middle of the central figure there is a well-developed median keel. The entire central figure is surrounded by numerous foramina. Those surrounding the figure lateroposteriorly and posteriorly are small and very close to each other, unlike in *Stegosimpsonia* (in which they are more separated and somewhat larger). Those at both sides of the neck of the lageniform figure are larger (the posterior pair, is oblique). Anterior to the central figure there is a row of pits in a semicircle.

Genus *Stegotherium* Ameghino 1887
Type species *Stegotherium tessellatum* Ameghino 1887
Referred species The type; *S. variegatum* Ameghino 1902a; *S. caroloameghinoi* Fernícola and Vizcaíno 2008; *S. pascuali* Fernícola and Vizcaíno 2008; and *S. tauberi* González and Scillato-Yané 2008.
Age and distribution Cliffs of the Santa Cruz River, Santa Cruz Province, Santa Cruz Formation, latest early Miocene, Santacrucian *s.l.* SALMA (Scott 1903–05); Sarmiento Formation, Gran Barranca, Chubut Province, Argentina, Colhuehuapian and "Pinturan" SALMAs.

Stegotherium variegatum Ameghino 1902a
Fig. 7.1E1, E2.
Syntype Carapace osteoderms, MACN A-12680 (see comments below).
Age and distribution Sarmiento Formation, Gran Barranca, Chubut Province, Argentina, Colhuehuapian and "Pinturan" SALMAs.
Assigned materials MPEF PV: 6377D, 6484A, 6416A, 6429, 6442A, 6451, 6474, 6485, 6519A, 6523, 6529B, 6541, 6647A,, 6656B, 7039, 7056A, 7670C, 7102, 7130B, 7146, 7153A, 7188, 7197, 7220A, 7303, 7334B, 7842, 5452, 5456, 5460, 5465, 5969A, 5960B, 5876A, 5604A, 6883, 6902, 6930, 8118A, and 8122F; groups of isolated osteoderms. MPEF PV: 6656A, 5967; groups of associated osteoderms.
Comments See Kramarz *et al.* (this book).

Stegotherium tessellatum Ameghino 1887
Fig. 7.1F.

Lectotype MACN-A 781, single osteoderm of dorsal shield (according Fernícola and Vizcaíno 2008).
Age and distribution Cliffs of the Santa Cruz River, Santa Cruz Province, Santa Cruz Formation, Middle Miocene, Santacrucian *s.l.* SALMA (Scott 1903–05); Sarmiento Formation, Gran Barranca, Chubut Province, Argentina, Colhuehuapian and "Pinturan" SALMAs.
Assigned materials MPEF PV: 7315D, 8118B, 8122A; groups of isolated osteoderms.
Comments See Kramarz *et al.* (this book).
General comments about stegotheriines

(1) *A a* sp. nov. occurs in the Casamayoran SALMA (in the "Barrancan" and "Vacan" subages; see Carlini *et al.* 2002a, 2004a, and 2005). It is quite difficult to say whether with this species we are dealing with *Stegosimpsonia* or a stegotheriin. Both groups differ in size, the development of the mean keel, the foramina anterior to the central figure in mobile osteoderms, the size and number of pilipherous pits, etc. According to the studied remains, there are two hypotheses for the *Stegosimpsonia*–Stegotheriini problem: (a) *Stegosimpsonia* is a primitive Stegotheriini; or (b) *Stegosimpsonia* is a derived Astegotheriini.

(2) The few remains of stegotheriines found in GBV-4 La Cancha (early Oligocene, Tinguirirican SALMA), MPEF PV 6784C (one isolated osteoderm) and GBV-19 La Cantera (late early Oligocene, Deseadan? SALMA), MPEF PV 5443 (one isolated osteoderm) are not enough to diagnose a new species but do confirm the presence of stegotheriines in Patagonia during this time interval. They are also present during the late Oligocene in typically Deseadan localities (e.g. the type Deseadan locality, La Flecha, in Santa Cruz Province).

(3) There have been two recent revisions of the genus *Stegotherium* (Fernícola and Vizcaíno 2008; González and Scillato-Yané 2008), in which new species and emended diagnoses for the already known taxa have been proposed. Particularly Fernícola and Vizcaíno (2008) propose two new species for the genus, based mainly in the remains that were part of the syntype of *S. variegatum*, quite abundant in the Colhuehuapian of Gran Barranca. However, and having in mind the morphological variation observed in new associated remains found in this locality, we are cautious about the validity of these two new taxa and prefer grouping the materials within *Stegotherium variegatum*, until we can evaluate the actual specific variation.

EUPHRACTINAE
"UTAETINI"

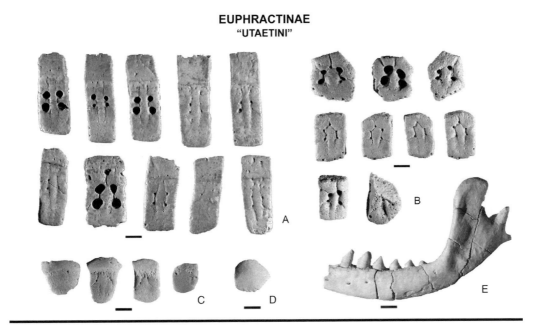

Fig. 7.2. *Utaetus buccatus* (MPEF PV 5426): (A) movable osteoderms; (B) fixed osteoderms; (C) tail osteoderms; (D) cephalic osteoderm; (E) left mandible. Scale bar 5 mm.

Subfamily **EUPHRACTINAE** Winge 1923
Tribe **"Utaetini"** Simpson 1945

We provisionally keep the name of the tribe Utaetini (to incude *Utaetus*) but it probably has to be considered Euphractini with retained primitive characters such as having areas of the carapace with large pits, the presence of enamel on the teeth and other features (see Simpson 1948), which led Simpson (1945) to interpret them as members of a different tribe.

Genus *Utaetus* Ameghino 1902b

Type species *Utaetus buccatus* Ameghino 1902b
Referred species The type, *U. argos* Ameghino 1902b, *U. laxus* Ameghino 1902b, and *?U. deustus* Ameghino 1902b.
Age and distribution Patagonia, Argentina, middle – late Eocene, Casamayoran (and ?Mustersan *sensu* Simpson 1948) SALMAs.

Utaetus buccatus Ameghino 1902b
Fig. 7.2.

Syntype MACN A-11622A, 10 osteoderms of the dorsal shield, left calcaneus and one metapodial.
Age and distribution Sarmiento Formation, Gran Barranca, Chubut Province, Argentina, middle Eocene, Casamayoran SALMA (Barrancan Subage).
Assigned materials AMNH: 28668 (associated remains that Simpson 1948 used to designate a neotype of *Utaetus buccatus*), 28481. MPEF PV: 6339B,

6468A, 6548, 6818, 7660A, 5424; groups of isolated osteoderms. MPEF PV 5426 associated remains.
Comments *Utaetus buccatus* is common in Barrancan levels, and among the few taxa collected in Gran Barranca with associated remains (AMNH 28668 and MPEF PV5426). AMNH 28668 has been studied by Simpson (1948), who proposed it as the neotype and was the basis for a proposed a new tribe (Utaetini: Simpson 1945). Moreover, Simpson (1948) based on the study of this specimen, synonymized 10 species of Ameghino with *Utaetus buccatus* (*U. argos, U. laxus, Posteutatus indentatus, P. scabridus, P. indemnis, Parutaetus chicoensis, P. clusus, P. signatus, Orthutaetus crenulatus,* and *O. clavatus*; all described by Ameghino in 1902b). However, based on comparative studies with the most complete specimens, we do not consider as valid all of Simpson's actions (see below, and Ciancio and Carlini 2008). The size of the fixed osteoderms are 14 mm long, 9–13 mm wide, 4–6 mm thick; moveable ones are 20–22 mm long, 7–9 mm wide, 3 mm thick.

?Utaetus deustus Ameghino 1902b
Syntype Lost.
Age and distribution Gran Barranca, Lake Colhue-Huapi, Chubut Province, Argentina, Sarmiento Formation, Middle Eocene, upper levels with *Notostylops* (Ameghino 1902b), late Eocene, "El Nuevo" fauna.
Assigned material MPEF PV 7376C.

Comments This species is represented only by a single fragment of a mobile osteoderm. Based on the original description of Ameghino (1902b: 60) for *Utaetus deustus*, "... *Plaques à surface externe rugueuse et avec sutures dentées quoique non parfaites. Les plaques mobiles ont le corps avec les bords latéraux relévés en forme de crêtes longitudinales, l'espace intermédiaire entre les deux crêtes étant creusé et au centre de ce deux, il y a une carène longitudinale médiane pas trop haute ... ces plaques ont de 3 à 4 centimètres de longueur et 1,5 centimètres de large ...*", MPEF PV 7376C is the single specimen that may be trustfully assigned to this species. Ameghino described this taxon for the Notostylopéen plus supérieur, south of Colhue-Huapi Lake (Simpson 1948). We found this species in Gran Barranca, in the "El Nuevo" level. Like Ameghino (1902b), we cannot determine whether this species belongs to the genus *Utaetus*, but certainly it does not belong to any other known species.

Tribe **Eutatini** Bordas 1933
Genus *Meteutatus* Ameghino 1902b

Type species *Meteutatus lagenaformis* (Ameghino 1897)

Referred species The type; *M. concavus* Ameghino 1902b; *M. rigidus* Ameghino 1902b; *M. attonsus* Ameghino 1902b; *M. percarinatus* Ameghino 1902b; *M. lucidus* Ameghino 1902b; and *M. anthinus* Ameghino 1902b.

Age and distribution Patagonia, Argentina, middle Eocene – late Oligocene, Casamayoran (Barrancan Subage) to Deseadan SALMAs (Ameghino 1902b). "La Cancha" and "La Cantera" faunas, early Oligocene, Tinguirirican, and Deseadan *s.l.* SALMAs.

Diagnosis (modified from Carlini *et al.* 2009) Medium-sized, between *Euphractus sexcinctus* and *Chaetophractus villosus*. According to Ameghino (1902b, p. 54), molars are rectangular in shape, the crown section a little longer than wide, with the lingual side (in lowers) convex and the labial bilobed by a vertical furrow, instead of being somewhat elliptic as in *Stenotatus, Doellotatus,* and *Eutatus* or subtriangular as in *Proeutatus*; teeth implanted with their longer axis longitudinal, and not oblique as in *Eutatus* and *Proeutatus*; more posterior molars decrease progressively in size and are only formed by a homogeneous mass of dentine with the occlusal surface lacking the protruding central keel of hard dentine observed in molars of *Eutatus* and *Proeutatus*. Osteoderms of the carapace are thick, rugose, and with areas in contact with other osteoderms,

covered surface with denticulate well-developed projections. Fixed osteoderms are rectangular with a lageniform central figure, with a neck at the middle portion that widens posteriorly, reaching the laterals and therefore occupying the whole posterior portion of the exposed surface, unlike in *Proeutatus,* in which the posterior rami of the furrow limiting the lageniform central figure continue parallel to the lateral margins of the osteoderm to end at the posterior margin. The central area of the central figure has a well-developed median keel, and is rounded in cross-section. Anteriorly, there are one or two well-developed figures, and at both sides of the neck of the central figure, a pair of lateral figures. Figures are convex and rounded in shape; furrows surrounding them are narrow and shallow. The posterior margin of the external surface of the osteoderms has a furrow occupying the entire width and widely open to the surface. This furrow is straight, and divided by septa. The piliferous system is prominent, with large channels produced by the septa division of the mentioned furrow. Behind the furrow, a posterior edge develops vertically, at the end of the osteoderm. Mobile osteoderms have a lageniform central figure that begins at the anterior portion of the exposed surface and, as in fixed osteoderms, reaches the lateral margins of the osteoderm, occupying the entire posterior area (the difference from *Proeutatus* is equivalent to that mentioned for fixed osteoderms). The central figure has a median keel similar to that of the fixed osteoderms. At both sides of the neck, there is a pair of lateral figures. The piliferous system is developed as in the fixed osteoderms.

Meteutatus aff. *M. lagenaformis* (Ameghino 1902b)
Fig. 7.3A.

Assigned materials MPEF PV: 6600B, 7440A, 7737B, 7835B, 6636, 5441; isolated osteoderms.

Age and distribution "La Cantera" fauna, early Oligocene, Deseadan *s.l.* SALMA.

Description The remains are not enough to determine that they belong with certainty to *Meteutatus lagenaformis*. We have only eight fixed osteoderms, six of which are broken, and one broken mobile osteoderm. The fixed osteoderms are thick, rugose, and similar in size to those of the type of *M. lagenaformis* (20 mm long, 14 mm wide, and 7.5 mm thick). The fixed osteoderms have a lageniform central figure with a median keel low and rounded in section. There are one or two peripheral anterior figures and a pair of lateral peripheral figures. Contact surfaces occur by denticular projections between osteoderms. Posteriorly, there is a furrow with three septa inside that divide it in four cavities.

EUPHRACTINAE
EUTATINI

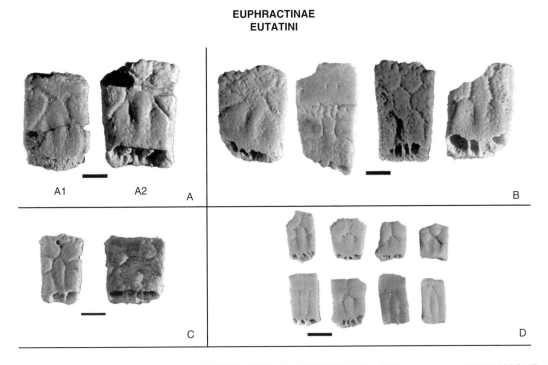

Fig. 7.3. (A) *Meteutatus* aff. *M. lagenaformis* (A1, MPEF PV 7737B; A2, MPEF PV 5441); (B) *Proeutatus* sp. (MPEF PV 5447); (C) *Meteutatus* sp. (MPEF PV 5438); (D) *Stenotatus* cf. *S. patagonicus* (MPEF PV 7315B). Scale bar 5 mm.

Meteutatus sp.
Fig. 7.3C.

Assigned materials MPEF PV: 6412A, 5438, 7673C; isolated osteoderms.

Age and distribution "La Cancha" fauna, early Oligocene, Tinguiririran SALMA.

Description We have several osteoderms that according to the external morphology belong to *Meteutatus*. However, we cannot assign them to a species. They are smaller than the other species of the genus, and typical of fixed osteoderms of the central area of the pelvic shield is 15.5 mm long, 9.5 mm wide, and 4 mm thick.

Genus *Barrancatatus* nov. gen.

Type species *Meteutatus rigidus* Ameghino 1902b

Etymology "*Barranca-*" from Gran Barranca, reflecting the type and referred material's origin, "*-tatus*" to indicate a member of the Eutatini.

Referred species The type, *Meteutatus tinguirirquensis* Carlini *et al.* 2009 and *Barrancatatus maddeni* sp. nov.

Age and distribution Sarmiento Formation at Gran Barranca, Chubut Province, Argentina; "La Cancha" and "La Cantera" faunas, early Oligocene, Tinguiririran and Deseadan *s.l.* SALMAs; also Andean Main Range, Chile, Abanico (= Coya-Machalí) Formation; early Oligocene, Tinguiririran SALMA.

Diagnosis Medium-sized, between *Zaedyus* and *Euphractus sexcinctus*. Osteoderms of the carapace are thick, with very rugose surfaces; denticulate projections provide the contact margins between adjacent osteoderms. Exposed surfaces of the fixed osteoderms have a lageniform central figure, with a well-developed median keel, but distally acute in section or scarcely developed. At both sides of the narrowest portion of the lageniform figure (neck) there are two lateral figures, and anterior to the central figures, one to three anterior figures are well developed. All the figures are planar and angular in shape. On the posterior margin of each osteoderm there is a straight and well-developed furrow, and inside this furrow, many small and thin septa limiting thin conducits (suggesting abundant pilosity) (Fig. 7.4E2). Behind this furrow, a vertical edge forms the posterior margin of the osteoderm. In some areas of the pelvic buckler some osteoderms have accessory figures intercalated among those mentioned. Mobile osteoderms are thinner than fixed ones, with less defined ornamentation, and a mean keel with variable development; as in fixed osteoderms, there is a lageniform central figure which, when it widens, delimits two lateral figures. On the transitional zone between the covered area and the exposed surface of the osteoderm, several rugosities are limited by deep furrows. The piliferous

EUPHRACTINAE
EUTATINI

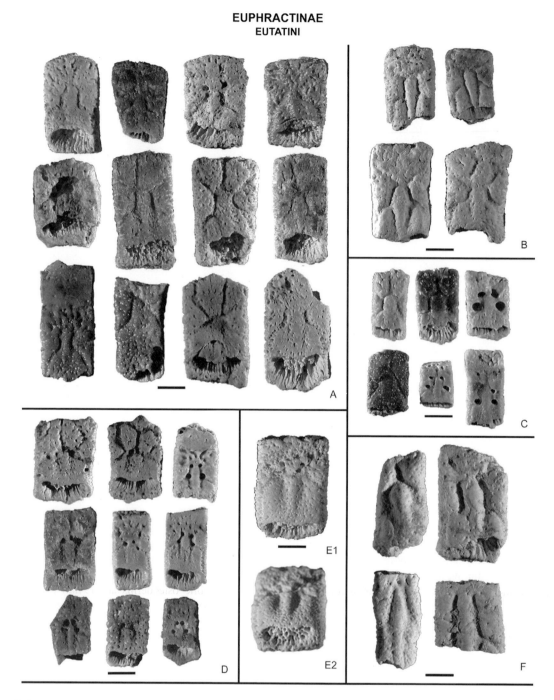

Fig. 7.4. (A) *Barrancatatus rigidus* (MPEF PV 6300B); (B) *Sadypus* aff. S. *confluens* (MPEF PV 6301D); (C) *Sadypus minutus* (syntype, MPEF PV 6301C); (D) *Barrancatatus maddeni* (syntype, MPEF PV 6299B); (E) *Barrancatatus* sp. (E1, MPEF PV 6600C; E2 the same osteoderm, showing the morphology of the piliferous system); (F) *Sadypus tortuosus* (MPEF PV 6602, associated). Scale bar 5 mm.

system of mobile osteoderms is restricted to the posterior margin and with an equivalent development.

Barrancatatus rigidus new combination
(Ameghino 1902b)
Fig. 7.4A.

Syntype MACN A-10958, osteoderms of the carapace.
Assigned material MPEF PV: 6299A, 6300B, 6301E, 6784E, 6645, 6676, 7267A, 7477B, 7673A; groups of isolated osteoderms.
Age and distribution Material of known provinence: Sarmiento Formation at Gran Barranca, Chubut

Province, Argentina, "La Cancha" fauna, early Oligocene, Tinguiririran SALMA.

Diagnosis (modified from Ameghino 1902): Similar size to *Euphractus sexcinctus*. Surface of osteoderms rugose and lateral contact areas among osteoderms completely covered with serrated projections. Fixed osteoderms are rectangular, thick and larger than those of *Barrancatatus tinguirirquensis* (17–23 mm long, 10–12 mm wide, and 5–6 mm thick). Central figure lageniform, the neck generally continues posteriorly in a wide low keel. Two anterior peripheral figures, and one pair of laterals. Some fixed osteoderms have accessory figures. Piliferous system similar to the one described for the genus, i.e. with a straight furrow and numerous bony septa, thin and arranged crosswise.

Mobile osteoderms with a lageniform central figure, without a median keel. At both sides of the central figure there are two elongate, lateral, peripheral figures. Remains of mobile osteoderms are 9 mm wide and 3.5 to 4.5 mm thick; length could not be precisely determined because they are frequently broken distally, but they would have been approximately 22 mm long.

Comments Probably Ameghino would have not been aware of the peculiar morphology of the piliferous zone at the posterior margin of the osteoderm. Certainly this was due to the thick cover of manganese oxide that hides the characters of the material he studied, which is the type of this species. Hundreds of osteoderms collected in GBV-4 La Cancha are also covered by a thick cover of manganese oxide. However, after removing this deposit of manganese, we could observe their characteristics.

Another species described by Ameghino (1902b, p. 55), *Meteutatus attonsus*, could also belong to this genus, but so far we consider it as *nomen vanum*, because the type is lost. The specimen referred by Simpson (1948) as the type (MACN A-10957) does not match with the material housed today in the Ameghino collection of the MACN, and, unfortunately, while the original description of the species states that its size is similar to *M. lagenaformis* it later says that osteoderms reach 30 mm length (which would be 45% larger than those of *M. lagenaformis*).

Barrancatatus maddeni sp. nov.
Fig. 7.4D.
Syntype MPEF PV 6299B, several isolated osteoderms.
Etymology "*maddeni*" for our friend and colleague Richard H. Madden.
Referred material MPEF PV: 6300A, 6301B, 6674, 6784F, 7267C, 7477C, 7673G, 5436; groups of isolated osteoderms.

Age and distribution Gran Barranca of Lake Colhue-Huapi, Chubut Province, Argentina, Sarmiento Formation, "La Cancha" fauna, early Oligocene, Tinguiririran SALMA.

Diagnosis Smaller than *Barrancatatus rigidus* and *B. tinguiririquensis*. Fixed osteoderms are 14 mm long, 9–10 mm wide, and 4.5 mm thick. The surface of the osteoderms is rugose, and the lateral surfaces in contact among osteoderms with smooth denticular projections. The posterodorsal margin of osteoderms is widely open (this posterior opening is related to a strong development of the pilosity). The furrow is divided into two large cavities by one or two thick septa; inside each cavity there is a complex partition formed by numerous, crossed, bony laminae limiting small canals (as described in the genus). Fixed osteoderms are quadrangular, very rugose, and with a lageniform central figure with a short neck; in the more quadrangular osteoderms this figure becomes semicircular, as in *B. rigidus*. The central figure occupies the posterior third of the osteoderm and has a well-developed medial keel, which is wide anteriorly and narrows abruptly at the posterior margin of the osteoderm. This keel is more conspicuous than that of *B. rigidus* and blunt in cross-section, not distally acute as in *B. tinguiririquensis*. Foramina are grouped in a semicircle on the most anterior portion of the furrow that surrounds the central figure. These foramina are small and generally four in number, but some osteoderms may bear five, and those of the marginal sectors of the pelvic shield may bear only two or three. Four main peripheral figures surround the central figure, two smaller, anterolateral, triangular-shaped ones and two larger ones, anteriorly positioned and polygonal. Small accessory figures are intercalated anterior to these.

Mobile osteoderms are 19 mm long, 9 mm wide, and 3 mm thick. In semi-mobile and mobile osteoderms the central figure has a short anterior neck that determines a lageniform shape. It has a central keel similar to that of the fixed osteoderms. On the anterior portion, before the beginning of the overlapping area there is a rugose area with furrows. At both sides of the most anterior portion of the central figure there are two anterolateral figures. On the exposed side of these osteoderms, four foramina are placed at both sides of the neck of the central figure. The piliferous system is similar to that of fixed osteoderms.

Barrancatatus sp.
Fig. 7.4E1, E2.
Assigned materials MPEF PV: 7001B, 6600C; isolated osteoderms.

Age and distribution "La Cantera" fauna, early Oligocene, Deseadan *s.l.* SALMA.

Description These are two fixed osteoderms (19 mm long, 14 mm wide, and 6 mm thick), with a central figure occupying almost the whole surface of the scute. Anterior to this figure, there are two lateral figures and an anterior one not well defined and relatively smaller. On the posterior margin of the osteoderm there is a straight and well-developed furrow, and inside this furrow, many little and thin septa limiting thin conducts (as described in the genus; see Fig. 7.4E2).

Genus *Sadypus* Ameghino 1902b

Type species *Sadypus confluens* Ameghino 1902b
Referred species The type; *S. ascendens* Ameghino 1902b; *S. nepotulus* Ameghino 1902b; *S. minutus* sp. nov.; and *Anutaetus tortuosus* Ameghino 1902b.
Age and distribution Patagonia, Argentina, late Eocene, Mustersan SALMA (Ameghino 1902b); southern cliff of Lake Colhue-Huapi, Chubut Province, Argentina, Sarmiento Formation, "La Cancha" and "La Cantera" faunas, early Oligocene, Tinguirian and Deseadan *s.l.* SALMAs.
Diagnosis (modified from Ameghino 1902b) Size between extant *Chaetophractus villosus* and *Euphractus sexcinctus*. Osteoderms rectangular to quadrangular. Lateral contact surfaces among osteoderms completely covered with denticulate projections. Ornamentation of osteoderms conspicuous, figures very convex, and furrows delimiting figures wide. Fixed osteoderms with a large, lageniform central figure occupying two-thirds of the exposed surface, and a central keel. The keel may be narrow or wide, rounded in section and higher than peripheral figures. Surrounding the central figure there are one or two anterior figures, and two laterals at each side of the neck of the central figure. The posterior margin of the osteoderms have a large transverse semicircular furrow, widely open to the surface and posterior border of osteoderms. This furrow is occupied by numerous thin, intercrossed septa, which limit small ducts that tend to be convergent anteriorly and centrally. Posteriorly, this opening is limited by an oblique edge. This posterior morphology of the osteoderm suggests a large development of pilosity. The surface of the osteoderms rises posteriorly, but distally, on the area of the furrow, it goes down. The margin of this posterior end is not straight, but semicircular; hence the wide zone where hairs emerge is partially covered in dorsal view. Mobile osteoderms are equally thick and with lageniform central figure that begins at the anterior portion of the exposed surface and, as in fixed osteoderms, reaches the lateral margins and occupies the

whole posterior area. The central figure has a median keel as in fixed osteoderms. The piliferous system is equivalent to that of fixed osteoderms.

Comments The study of this genus is complex for the following reasons.

(1) Ameghino (1902b) proposed the genus *Sadypus*, with three species *S. confluens*, *S. ascendens* (both for the Mustersan SALMA), and *S. nepotulus* (for the Deseadan SALMA). Simpson (1948) considered in his revision that *Sadypus* was a junior synonym of *Meteutatus* Ameghino 1902, while recognizing the new combinations as valid species: *M. confluens*, *M. ascendens*, and *M. nepotulus* (although this latter was not considered by Simpson, as his paper was only about the Casamayoran and Mustersan SALMAs). Definitively we do not agree with this position.

(2) The type material of each species of the genus has been lost, so the basis for assessing the validity of the genus are: (a) the original diagnoses of Ameghino, (b) the brief diagnoses given by Simpson (1948) for his new combinations, and (c) a drawing of Simpson of a single fixed osteoderm of the type material of *S. confluens*.

Despite these difficulties, we propose to revalidate *Sadypus* since the description of Ameghino matches with the materials we assign to it, which in turn, clearly are distinct from the characteristics of *Meteutatus*. At this point, the most significant character of the description of Ameghino (1902b) is his reference to the kind, amount, and distribution of the piliferous areas: "Système pilifère du bord postérieur enormement développé, représenté par une grande vacuité qui occupe toute la largeur du bord postérieur de la plaque et une partie considérable de la face supérieure" (p. 64). Concerning the *Sadypus* species described by Ameghino, we consider *S. confluens* as a valid species because there are referred materials that can be used in comparisons (the drawing of the type and one osteoderm referred by Simpson). However, the other species, *S. nepotulus* and *S ascendens*, are considered as *nomina dubia*, because we cannot identify the type materials, and the descriptions are insufficient to assign the remains collected at Gran Barranca.

Sadypus aff. *S. confluens* Ameghino 1902b
Fig. 7.4B.
Holotype ?MACN A-10954, osteoderm from the carapace (see Ciancio and Carlini 2008).
Assigned material MPEF PV 6301D, MPEF PV 7267B; groups of isolated osteoderms.
Age and distribution Gran Baranca south of Lake Colhue-Huapi, Chubut Province, Argentina, Sarmiento

Formation, "La Cancha" fauna, early Oligocene, Tinguirirican SALMA.

Description Only a few fixed osteoderms of the dorsal carapace are available. Their mean measurements are 13 mm long, 8.5 mm wide, and 4.5 mm thick. The development of the piliferous system is equivalent to that described for the genus. The sculpture consists of a lageniform central figure with a narrow and high median keel, an anterior peripheral figure, and a pair of lateral figures. All figures are convex and conspicuous, and the furrows that separate them are wide.

Comments Under the exposed circumstances, it is difficult to establish whether these osteoderms belong to this species or not, since the type material is lost and there is only one drawing (not detailed; see Simpson 1948) of the single osteoderm that could belong to the type of the species. Consequently, we based the assignment on the original description of Ameghino (1902b, p. 64). Although Ameghino (1902b) reported *S. confluens* from the "Astraponotéen," this species has not been found among the materials at "El Rosado," nor in other Mustersan localities.

Sadypus minutus sp. nov.
Fig. 7.4C.

Syntype MPEF PV 6301C, several isolated osteoderms.
Etymology From Latin "*minutus*," very small, according to its size among Eutatini.
Referred material MPEF PV: 6299C, 6784D; groups of isolated osteoderms.
Age and distribution Gran Barranca, Chubut Province, Argentina, Sarmiento Formation, "La Cancha" fauna, early Oligocene, Tinguirirican SALMA.
Diagnosis Small-sized, similar to *Zaedyus*. Typical fixed osteoderms are quadrangular, 11 mm long, 9 mm wide, and 4.5 mm thick. Osteoderms from the carapace have a smooth surface with punctuations, lateral areas show contacts among adjacent osteoderms with slight rugosities. The osteoderms have a well-developed piliferous system formed by one transverse furrow that occupies the width of the osteoderm. This furrow is narrower than that of other *Sadypus* species, open posteriorly, not occupying a large part of the exposed surface. Internally, this opening is divided by numerous dorsoventral septal partitions that bifurcate forming numerous small canals. Figures are convex.

Fixed osteoderms have a lageniform central figure with a wide neck. When this figure opens, the divergent branches project toward the posterior vertices of the osteoderm. In front of the central figure there are two anterior peripheral polygonal figures, and at both sides one pair of elongated lateral peripheral figures.

On the anterior portion of the lageniform figure, there are generally four large circular foramina, but sometimes there may be a fifth anterior to the others. Some fixed osteoderms are a little narrower and do not bear perforations on the exposed surface. Probably the quadrangular osteoderms would have been at the middle of the pelvic shield, and those without perforations, at the lateral margins (as in *Utaetus buccatus*).

Semi-mobile osteoderms are 14 mm long, 7 mm wide, and 3 mm thick, with similar sculpture to fixed osteoderms. On the anterior portion of the osteoderm there is a short overlapping region, beside a transverse even depression. As in fixed osteoderms, four large perforations are seen on the exposed surface. The piliferous system is similar to that of fixed osteoderms.

Sadypus tortuosus new combination
(Ameghino 1902b)
Fig. 7.4F.

Holotype MACN A-10446, a single osteoderm.
Assigned material MPEF PV: 6352A, 6600A, 5780A, 5963B, 7001A; groups of isolated osteoderms. MPEF PV: 6602, 6603, 7780; groups of associated osteoderms.
Age and distribution Gran Barranca of Lake Colhue-Huapi, Chubut Province, Argentina, Sarmiento Formation, "La Cantera" fauna, early Oligocene, Deseadan *s.l.* SALMA.
Description Osteoderms of medium size, similar to *Euphractus sexcinctus*. Osteoderms of the carapace are thick, rugose, and larger than *Sadypus confluens*. Fixed osteoderms are rectangular (18–19 mm long, 12 mm wide, and 5 mm thick), with a lageniform central figure with a wide neck, extended backwards. The median keel is wider and higher than that of *S. confluens*. Surrounding the central figure, there is an anterior peripheral figure and a pair of elongate, lateral peripheral figures. All the figures are very convex (more so than in *S. confluens*) and the central one stands out from the rest. Furrows limiting figures are deep and wide. The piliferous system is equivalent to the one described for the genus. Mobile osteoderms are approximately 20–25 mm long, 8–12 mm wide, and 4 mm thick, and also have a lageniform central figure with long neck and a well-developed median keel. Two anterior peripheral lateral figures are present alongside the narrowest part of the central figure. Figures are very convex and the furrows limiting them are deep and wide. The transition zone between the overlapping portion and the posterior part is concave and has shallow longitudinal furrows. The piliferous system is equivalent to that of fixed osteoderms.

Comments The presence of associated osteoderms of this taxon allowed us to see that the osteoderm holotype of *Anutaetus tortuosus* (see Ciancio and Carlini 2008) matches with those marginal of the associated specimens (MPEF PV 6602, MPEF PV 6603, and MPEF PV 7780). Consequently, we propose this new combination, given that the characters that distinguish one genus from the other are easily encompassed within intraspecific variation. The type material of *A. tortuosus*, according to Ameghino, comes from the Pyrotherium beds at Colhue-Huapi (see Ciancio and Carlini 2008), although it is not clear whether the remains come from a locality at Gran Barranca because there is no known Deseadan cingulate fauna there.

Genus *Proeutatus* Ameghino 1891

Type species *Proeutatus oenophorus* (Ameghino 1887)
Assigned species The type; *P. lagena* (Ameghino 1887); *P. deleo* (Ameghino 1891); *P. carinatus* (Ameghino 1891); *P. postpuntum* Ameghino 1902a; *P. robustus* Scott 1903.
Age and distribution Cliffs of the Santa Cruz River, Santa Cruz Province, Santa Cruz Formation, late early Miocene, Santacrucian *s.l.* SALMA (Scott, 1903–05); Gran Barranca, Lake Colhue-Huapi, Chubut Province, Argentina, Sarmiento Formation, early Miocene, Colhuehuapian and "Pinturan" SALMAs; Collón-Curá Formation, middle Miocene, "Friasian" SALMA (Roth 1899).
Comments It is difficult to establish clear differences among the species of *Proeutatus* on the basis of isolated osteoderms from the carapace. We collected several osteoderms of this genus in Colhuehuapian and Pinturan levels but we cannot assign them to a species. Ameghino (1902a), when describing the Colpodonéen fauna (= Colhuehuapian), on the basis of isolated osteoderms, nominated the species *P. postpuntum* and assigned other materials to *P.* aff. *lagena*. The type material of *P. postpuntum* is lost; however, according to the description of Ameghino, it has certain distinctive characters that we do not find in the osteoderms we collected.

Proeutatus sp.
Fig. 7.3B.

Assigned material MPEF PV: 6376, 6529A, 6657, 7069, 7076A, 7114, 7130A, 7156A, 7211A, 7305, 7334A, 5447, 5455, 5463, 7314A, 7315A, 8122D, 7321, 5960A, 5876E, 5604C, 6757C, 6898, 7026, 7031; groups of isolated osteoderms.
Age and distribution Cliffs of the Santa Cruz River, Santa Cruz Province, Santa Cruz Formation, late early Miocene, Santacrucian *s.l.* SALMA (Scott 1903–05);

Gran Barranca of Lake Colhue-Huapi, Chubut Province, Argentina, Sarmiento Formation, early Miocene, Colhuehuapian and "Pinturan" SALMAs.
Comments See Kramarz *et al.* (this book).

Genus *Stenotatus* Ameghino 1891

Type species *Stenotatus patagonicus* (Ameghino 1887)
Assigned species The type; *S. hesternus* (Ameghino 1889); *S. ornatus* (Ameghino 1897); *S. centralis* (Ameghino 1897); and *S. planus* (Scillato-Yané and Carlini 1998).
Age and distribution Patagonia, Argentina, "Couches à *Pyrotherium*" (Ameghino 1897); cliffs of the Santa Cruz River, Santa Cruz Province, Santa Cruz Formation, late early Miocene, Santacrucian *s.l.* SALMA (Scott 1903–05); valley of Collón-Curá River, Neuquén Province, Collón-Curá Formation, middle Miocene, "Friasian" SALMA (Roth 1899); Cerro Boleadoras Formation, middle Miocene (Scillato-Yané and Carlini 1998); Chucal Formation, northern Chile, late early Miocene, Santacrucian SALMA (Croft *et al.* 2007); Gran Barranca, Lake Colhue-Huapi, Chubut Province, Argentina, Sarmiento Formation, early Miocene, Colhuehuapian and "Pinturan" SALMAs.

Stenotatus cf. *S. patagonicus* (Ameghino 1887)
Fig. 7.3D.

Syntype Lost.
Age and distribution Cliffs of the Santa Cruz River, Santa Cruz Province, Santa Cruz Formation, late early Miocene, Santacrucian *s.l.* SALMA (Scott 1903–05); Gran Barranca, Lake Colhue-Huapi, Chubut Province, Argentina, Sarmiento Formation, early Miocene, "Pinturan" SALMA.
Assigned materials MPEF PV: 7315B, 8121, 8122C; groups of isolated osteoderms.
Comments Morphological characters of the osteoderms match with those of *S. patagonicus*; hence, the assignment of this species – which is always smaller (8–11 mm length, 5–7 mm width), even considering the variation of size recorded in the materials assigned to *S. patagonicus* – is dubious (see see Kramarz *et al.* this book).

Tribe **Euphractini** Winge 1923
Genus *Parutaetus* Ameghino 1902b

Type species *Parutaetus chicoensis* Ameghino 1902b
Referred species The type; *P. clusus* Ameghino 1902b; *P. signatus* Ameghino 1902b; and *P. chilensis* Carlini *et al.* 2009.
Age and distribution Gran Barranca, Lake Colhue-Huapi, Chubut Province, Argentina, Sarmiento Formation, upper part of levels with *Notostylops*

EUPHRACTINAE
EUPHRACTINI

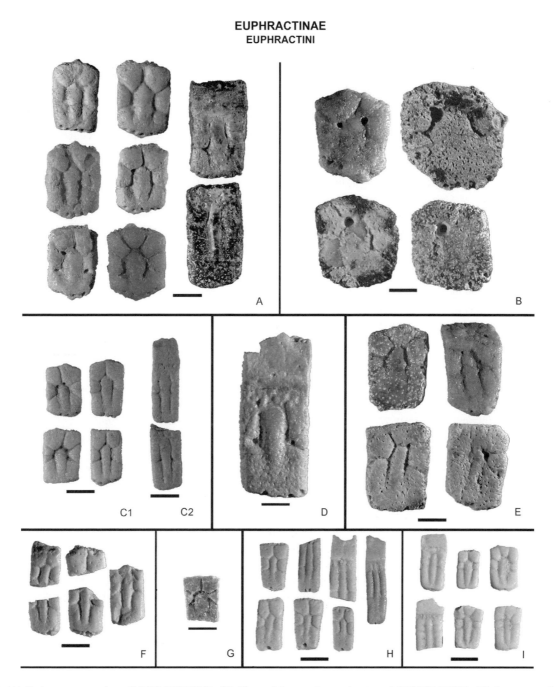

Fig. 7.5. (A) *Orthutaetus crenulatus* (MPEF PV 7633B); (B) *Mazzoniphractus ingens* (syntype, MPEF PV 7632B); (C) *Parutaetus chicoensis* (C1, MPEF PV 7633C; C2, MPEF PV 7658D); (D) *Archaeutatus* aff. *A. malaspinensis* (MPEF PV 7798); (E) *Anteutatus lenis* (MPEF PV 7632A); (F) ?*Pachyzaedyus* sp. (MPEF PV 7779); (G) *Parutaetus chilensis* (MPEF PV 6299F); (H) *Prozaedyus humilis* (MPEF PV 5446); (I) *Prozaedyus* sp. (MPEF PV 7315C). Scale bar 5 mm.

(Ameghino 1902b); "El Nuevo" and "El Rosado" faunas, late Eocene; "La Cancha" fauna, early Oligocene; Mustersan and Tinguiririran SALMAs; Andean Main Range, Chile, Abanico (= Coya-Machalí) Formation; early Oligocene, Tinguiririran SALMA (Carlini *et al.* 2009).

Diagnosis See Carlini *et al.* (2009).

Parutaetus chicoensis Ameghino 1902b
Fig. 7.5C1, C2.
Syntype MACN-A 11617, osteoderms of the carapace.

Referred material MPEF PV: 7376B, 7377B, 7378D, 7486B, 7633C, 7635B, 7658D, 5429, 6368B, 6589, 6597, 7845D; groups of isolated osteoderms.

Age and distribution Gran Barranca, Lake Colhue-Huapi, Chubut Province, Argentina, Sarmiento Formation, levels with *Notostylops* (Ameghino 1902b); Late Eocene, "El Nuevo" and "El Rosado" faunas, Mustersan SALMA.

Diagnosis (modified from Ameghino 1902b) Small-sized, similar to extant *Zaedyus*. Fixed osteoderms are 10 mm long, 6 mm wide, and 3 mm thick, larger than those of *P. chilensis*. Osteoderms are smooth, with the contact areas between adjacent osteoderms concave in section. Their surface is mainly smooth with scattered small denticular projections with rounded points. The fixed osteoderms have a central figure with a wide anterior third, narrowing backwards, reaching almost to the posterior margin of the osteoderm. The central figure is surrounded by a pair of anterior polygonal figures, two triangular anterolaterals, and two elongate posterolaterals. Some osteoderms show a third anterior accessory figure, intercalated between the two typical anterior figures. All the figures are convex. At the intersection of furrows of different figures, there are two to five conspicuous foramina, unlike in *P. chilensis,* which has two or three. On some osteoderms these foramina are large and form a semicircular furrow. This is likely related to the presence of glandular cavities as in the pelvic shield of some species of living Euphractines. The piliferous system is restricted to the posterior margin and poorly developed, forming a few small foramina (probably a single hair per foramen). The number of these foramina varies between three and six (generally four), unlike the osteoderms of *P. chilensis* in which there are only two. Mobile osteoderms are 14 mm long, 5 mm wide, and 2.5 mm thick. Their external surfaces have two longitudinal furrows that define a central figure and two laterals. Each lateral figure is divided by a shallow transverse furrow on the anterior third, which defines an anterior lateral figure and a longer posterior one. As in fixed osteoderms, a pair of foramina are found at the intersection of the furrows that delimit the figures. In mobile osteoderms there may be two to four piliferous foramina, but most commonly three.

Comments Ameghino (1902b) proposed the genus *Parutaetus* with three species, two from the *Notostylopéen* (= Casamayoran), *P. chicoensis* and *P. signatus*, and one from the *Notostylopéen plus superieur* (also regarded as Casamayoran), *P. clusus*. Simpson (1948) placed the three species as synonyms of *Utaetus buccatus*, but based on our revision of more

material, we disagree with Simpson's assessment and return to Ameghino's concept, except in the sense that we think *P. signatus* is an invalid junior synonym of *P. chicoensis* because the characters given for each in the original descriptions are encompassed within intraspecific variation (especially on the basis of the remains of a carapace from Chile).

Among the groups of osteoderms assigned to this species, there are instances where pairs of osteoderms are laterally fused (MPEF PV 7635B), making a single figure formed by the lateral figures of these two osteoderms. This is evidence for the presence of epidermal horny scales shared by two adjacent osteoderms as in the carapace of *P. chilensis* (see Carlini *et al.* 2009).

Although the name seems to refer to Río Chico, one of the numerous localities in which Carlos Ameghino collected materials, the type materials of this species were found at Gran Barranca (Simpson 1967). No materials have been found in Barrancan levels of Gran Barranca that may be referred to this species; however, we find it abundantly in the "El Nuevo" and "El Rosado" levels.

Parutaetus chilensis Carlini *et al.* 2009
Fig. 7.5G.

Holotype SGOPV 3122, a large portion of the carapace including several fixed and movable bands, and a significant portion of the pelvic shield.

Referred material MPEF PV: 6299F, 5435; isolated osteoderms.

Age and distribution Gran Barranca, Lake Colhue-Huapi, Chubut Province, Argentina, Sarmiento Formation, "La Cancha" fauna, early Oligocene; Andean Main Range, Chile, Abanico (= Coya-Machalí) Formation; early Oligocene, Tinguiririran SALMA.

Comments There are only a few osteoderms, which are slightly different from those of the type material and could be within the intraspecific variation (measurements are 8.5 mm length, 6.5 mm width, and 2.5 mm thick). The presence of this species in these levels supports the correlation with the Tinguiririran of Chile. Remains coming from "La Cantera" (MPEF PV 7343, MPEF PV 5442) might also be referred to this taxon, but they are still too scarce to confirm this assignment.

Genus *Orthutaetus* Ameghino 1902b
Type species *Orthutaetus crenulatus* Ameghino 1902b
Referred species The type and *O. clavatus* (synonym of *U. buccatus*) Ameghino 1902b
Age and distribution Gran Barranca, Lake Colhue-Huapi, Chubut Province, Argentina, Sarmiento

Formation, upper levels with *Notostylops* (Ameghino 1902b); late Eocene, "El Nuevo" fauna.

Diagnosis (modified from Ameghino 1902b) Medium-sized, similar to extant *Euphractus*. The carapace osteoderms are thick and rugose. Osteoderm figures convex; the contact areas between adjacent osteoderms have numerous denticular projections, as in *Meteutatus*, not smooth as in *Utaetus*. The piliferous system is restricted to the posterior margin and formed by small circular foramina. Fixed osteoderms are quadrangular to rectangular and thick (12–14 mm long, 8–11 mm wide, and 4–5 mm thick). The central figure is wide and elongate occupying approximately half the osteoderm, and not reaching its posterior margin; two anterior peripheral polygonal figures have their posterior vertex posteriorly oriented; there are two quadrangular, middle peripheral figures and one large U-shaped posterior figure that occupies the whole posterior margin of the osteoderm. The posterior part of the central figure penetrates between the rami of the U-shaped figure. In the more quadrangular fixed osteoderms there is a triangular anterior figure intercalated between the aforementioned anterior figures. Furrows separating figures are deep and narrow. There are three to seven piliferous foramina (depending on the width of the osteoderms, the widest having more foramina). Typical mobile osteoderms are 23 mm long, 10 mm wide, and 4.5 mm thick, with a wide central figure that does not reach the posterior margin. Furrows delimiting this central figure form two lateral figures that are divided into anterior and posterior laterals by a transverse furrow. The furrow separating the central figure from the peripheral ones is lost on the posterior part of the osteoderm, where the posterior lateral figures merge with the median figure. All the figures are convex, as in the fixed osteoderms. The piliferous system is less developed than in the fixed osteoderms, consisting on a pair of foramina placed at the corners of the posterior margin.

Orthutaetus crenulatus Ameghino 1902b
Fig. 7.5A.

Syntype Lost.
Assigned material MPEF PV: 7376F, 7377D, 7378D, 7379D, 7486D, 7632C, 7633B, 7658C.
Age and distribution As for the genus, by monotypy.
Diagnosis As for the genus, by monotypy.
Comments This genus was erected by Ameghino (1902b); later, Simpson (1948) synonymized it to *U. buccatus*. We resurect the genus *Orthutaetus* and the species *O. crenulatus*, on the basis of numerous isolated osteoderms whose characters match the description of Ameghino (see 1902b, p. 63). Ameghino

nominated *O. clavatus* for the "Partie inférieure des couches à *Notostylops*," and *O. crenulatus* for the "Partie supérieure des couches à *Notostylops*." We found no taxa assignable to the first species, hence, we keep it as a synonym of *U. buccatus*, because according to the original description (Ameghino 1902b, p. 63) it could fit with the variation observed in this species. Simpson (1948) and Mones (1986) reported MACN A-10430 to be the type specimen but the specimen with this catalog number is a tardigrade astragalus. Consequently, it is necessary to create a neotype and make a new diagnosis for this species. This species is found in Gran Barranca, in the "El Nuevo" level but not at GBV-3 "El Rosado."

Genus *Anteutatus* Ameghino 1902b
Type species *Anteutatus lenis* Ameghino 1902b
Referred species The type and *A. laevus* Ameghino 1902b (see comments below).
Age and distribution Gran Barranca, Lake Colhue-Huapi, Chubut Province, Argentina, Sarmiento Formation, upper part of levels with *Notostylops* – Astraponotéen (Ameghino 1902b); Late Eocene, "El Nuevo" and "El Rosado" (Mustersan SALMA) faunas.

Anteutatus lenis Ameghino 1902b
Fig. 7.5E.

Syntype MACN A-11621, osteoderms from the carapace.
Assigned material MPEF PV: 7379A, 7486C, 7632A, 7633E, 6588, 6847A; groups of isolated osteoderms.
Age and distribution Gran Barranca, Lake Colhue-Huapi, Chubut Province, Argentina, Sarmiento Formation, upper part of levels with *Notostylops* (Ameghino 1902b); late Eocene, "El Nuevo" and "El Rosado" faunas (Mustersan SALMA).
Diagnosis (modified from Ameghino 1902b) Large-sized, similar to the extant *Euphractus sexcinctus*. Osteoderms from the carapace are large, with rugose surface and proportionally thin (14–16 mm long, 10–11 mm wide, and 3–5 mm thick). Lateral contact surfaces between osteoderms are planar or slightly concave, with an even surface or with a few denticular projections, as is the case in *Utaetus*, but unlike *Meteutatus*. Fixed osteoderms have a lageniform central figure with a low medial keel reaching almost the posterior margin of the osteoderm (as in *Meteutatus, Barrancatatus*, and *Sadypus*). Surrounding this figure there are two anterior peripheral figures and two lateral peripheral figures placed at both sides of the neck of the central figure. All figures are flat and low. Piliferous perforations of the posterior margin are few (two to four) or none, and they are small (in contrast

to *Meteutatus*) and subcircular. Fixed osteoderms are 14–16 mm long, 10–11 mm wide, and 3–5 mm thick. Mobile osteoderms also have a lageniform central figure with a low medial keel that reaches almost the posterior margin of the osteoderm. The neck of the lageniform figure is short; where this figure widens, it delimits two lateral peripheral figures. The transition area between the tecla and the main portion of the osteoderm has rugosities and furrows. At each one of the lateral margins there is a row of very small perforations. The piliferous system is similar to that of mobile osteoderms. Mobile osteoderms are 26 mm long and 11 mm wide.

Comments Simpson (1948) found no evidence to separate this genus from *Utaetus* and synonymized it. We validate the original designation of Ameghino because its distinctive morphology and keep this species within *Anteutatus*. We consider the other species of the genus, *A. laevus*, as *nomen dubium* because the type is lost and Ameghino's original description is insufficient to designate a neotype.

Genus *Archaeutatus* Ameghino 1902b
Type species *Archaeutatus malaspinensis* Ameghino 1902b
Age and distribution Patagonia, Argentina, "Couches à *Pyrotherium*" (Ameghino 1902b); Gran Barranca, Lake Colhue-Huapi, Chubut Province, Argentina, Sarmiento Formation, early Oligocene, "La Cantera" fauna, Deseadan *s.l.*

Archaeutatus aff. *A. malaspinensis* Ameghino 1902b
Fig. 7.5D.
Syntype MACN-A 10440, osteoderms from the carapace.
Assigned material MPEF PV 7798, MPEF PV 5444; isolated osteoderms.
Age and distribution As for the genus, by monotypy.
Description Remains consist of a mobile and a damaged fixed osteoderm. The description of Ameghino is based on fixed osteoderms of the pelvic girdle, but in the group of osteoderms belonging to the type there is a mobile osteoderm, broken at its proximal end, that was used to assign the new specimens. The new mobile osteoderm has a large lageniform figure, which occupies the whole posterior part, and narrows anteriorly. At both sides of this narrow portion there are two lateral figures. It has two circular foramina in the furrow limiting the central figure, placed where it widens to reach the lateral margins. On the posterior margin there are five small piliferous perforations. This osteoderm is 28 mm long, 12 mm wide, and

5 mm thick. This osteoderm is larger than the type material.

Genus *Mazzoniphractus* nov. gen.
Type species *Mazzoniphractus ingens* nov. sp.
Etymology "*Mazzoni-*" in honor of Mario Martín Mazzoni, one of first geologists who studied the main sequence of Gran Barranca in detail, and "*-phractus*" to indicate a member of the Euphractini.
Age and distribution Gran Barranca, Lake Colhue-Huapi, Chubut Province, Argentina, Sarmiento Formation, late Eocene, "El Nuevo" and "El Rosado" (Mustersan SALMA) faunas.
Diagnosis Osteoderms are larger and thicker than those of *Anteutatus lenis*. Fixed osteoderms are quadrangular (15–18 mm long, 13–17 mm wide, and 6–8 mm thick), with very rugose surfaces. The contact areas with adjacent osteoderms are completely covered by denticular projections and cavities, as in *Meteutatus*, but not as in *Anteutatus*. The lageniform central figure is very wide, short-necked, with no median keel, unlike those in *Anteutatus*. Four peripheral figures are arranged in a semicircle at the anterior end of the central figure. The accessory figures are circular in shape and convex, not planar as those of *Anteutatus*. On the posterior margin of the osteoderm there are two to four small and circular piliferous foramina. At the exposed surface there are one or two large circular foramina.

Mazzoniphractus ingens nov. sp.
Fig. 7.5B.
Syntype MPEF PV: 7632B.
Etymology From Latin *ingens* (= large, huge), because its size among the Euphractini.
Referred material MPEF PV: 6579.
Age and distribution As for the genus, by monotypy.
Diagnosis As for the genus, by monotypy.

Genus *Prozaedyus* Ameghino 1891
Type species *Prozaedyus proximus* (Ameghino 1887)
Referred species The type; *P. exilis* (Ameghino 1887); *P. impressus* Ameghino 1897; *P. planus* Ameghino 1897; *P. humilis* Ameghino 1902a; and *P. tenuissimus*, Ameghino 1902b.
Age and distribution Patagonia, Argentina, "Couches à *Pyrotherium*" (Ameghino 1897); cliffs of the Santa Cruz River, Santa Cruz Province, Santa Cruz Formation, early Miocene, Santacrucian *s.l.* SALMA (Scott 1903–05); valley of Collón-Curá River, Neuquén Province, Collón-Curá Formation, middle Miocene, "Friasian" SALMA (Roth 1899); Cerro Bandera Formation, Neuquén, early Miocene, Colhuehuapian SALMA (Kramarz *et al.* 2005);

Choquecota Formation, Oruro, Bolivia, middle Miocene, Colloncuran SALMA (Marshall and Sempere 1991); Gran Barranca, Lake Colhue-Huapi, Chubut Province, Argentina, Sarmiento Formation, late Oligocene – early Miocene, Colhuehuapian and "Pinturan" SALMAs.

Prozaedyus humilis Ameghino 1902a
Fig. 7.5H.

Syntype Lost.

Assigned material MPEF PV: 6377A, 6442B, 6519C, 6532B, 7076B, 7130C, 7156B, 7211B, 5446, 5453, 5464, 5969C, 5973, 5960D, 5876B, 6757D; groups of isolated osteoderms.

Age and distribution: Gran Barranca, Lake Colhue-Huapi, Chubut Province, Argentina, Sarmiento Formation, early Miocene; Colhuehuapian SALMA.

Diagnosis (modified from Ameghino 1902b) Similar or slightly smaller in size than *Prozaedyus proximus*. Osteoderms are smooth, with the contact areas between adjacent osteoderms straight in section and completely covered with denticulate projections. Fixed osteoderms are small (7–9 mm long, 4–5.5 mm wide, and 2.5–3 mm thick), with a narrow and elongate central figure that does not reach the posterior margin. Adjacent to the central figure there are two anterior peripheral figures, and two lateral peripheral figures. Lateral figures are divided by a transverse furrow, which defines an anterior lateral figure and a longer posterior one. Fixed osteoderms have a single large piliferous pit at the middle of the posterior border. Mobile osteoderms are narrow and long (14–16 mm long, 4–5 mm wide, and 1.5–2 mm thick), with an elongate narrow and globose central figure that does not reach the posterior margin. There are two lateral figures parallel to the central one and divided by a transverse furrow at the middle into two portions. The posterior margin has two small piliferous pits placed at the middle of the exposed surface.

Prozaedyus sp.
Fig. 7.5I.

Assigned materials MPEF PV: 7315C, 8119, 8122B; groups of isolated osteoderms.

Age and distribution Gran Barranca, Lake Colhue-Huapi, Chubut Province, Argentina, Sarmiento Formation, early Miocene; "Pinturan" SALMA.

Comments See Kramarz *et al.* (this book).

?*Pachyzaedyus* sp.
Fig. 7.5F.

Assigned material MPEF PV 7779, associated osteoderms.

Description We describe a group of associated osteoderms. They are all broken, but collectively they show some distinctive characters. Osteoderms are small, approximately 11 mm long, 6 mm wide, and 4 mm thick. A central elongate figure occupies the posterior third of fixed osteoderms. This central figure is a little narrower posteriorly. Six peripheral figures surround the central figures, two large and polygonal figures are situated anteriorly, two small figures are postioned anterolaterally, and two elongate figures are positioned posterolaterally. The contact surfaces between osteoderms are somewhat concave and have numerous small projections. The piliferous system is restricted to the posterior margin and formed by five to six small piliferous foramina grouped in the middle of the posterior margin. These foramina are elongate dorsoventrally.

Other species recorded in Gran Barranca

Ameghino (1902a, 1902b) mentioned other dasypodid species as coming from Gran Barranca. However, and despite the intensive work made in this sequence, we have found no record of them. In some cases, these species are recorded in other localities of this age, and in others, the type specimen is lost and no materials of the new or old collections could be assigned to these taxa.

(1) *Prostegotherium notostylopianum* Ameghino 1902b and *P. astrifer* Ameghino 1902b. These two species described by Ameghino for Gran Barranca are not represented in the thousands of osteoderms collected in our successive field trips to Gran Barranca. Moreover, we have not found them among the materials from Gran Barranca housed in other collections both in Argentina and abroad. However, these supposed two species (which could be a single one: Carlini *et al.* 2002a) are abundant in Paso del Sapo (lower Eocene, "Riochican SALMA") (Carlini *et al.* 2002a; Tejedor *et al.* 2009). Hence, although we previously reported these two species among the taxa described for Gran Barranca (Carlini *et al.* 2005), we now believe there must have been a mistake in the locality reported in the original catalogs, and that these were never collected there.

(2) *Posteutatus indentatus* Ameghino 1902b and *P. indemnis* Ameghino 1902b. Both described by Ameghino from Casamayoran levels of Great Barranca (see Ciancio and Carlini 2008). This situation is similar to that just discussed: we do not find these two species in any new collections. With the evidence provided by the type material, *Posteutatus indentatus* could be a synonym of *Utaetus buccatus* (as proposed by Simpson 1948), whereas the second (*P. indemnis*) is a valid taxon but we are not sure that it comes from Gran Barranca.

(3) *Meteutatus percarinatus* Ameghino 1902b. Ameghino (1902b) described this taxon for the *Notostylopéen plus supérieur*, and Simpson (1948) pointed out that these remains come from the south of Lake Colhue-Huapi. However, according to the original descriptions (the type MACN A-10453 is lost) no material in the collections from Gran Barranca (either our recent collections or those made by the middle of the twentieth century) may be assigned to this species. It is noteworthy that no undoubted member of the Eutatini was found in any of the faunas older than "La Cancha" in Gran Barranca; in error, we recorded a few supposed eutatin osteoderms from GBV-60 "El Nuevo" and GBV-3 "El Rosado" (Carlini *et al.* 2005).

(4) *Stenotatus centralis* (Ameghino 1902a) and *Proeutatus pospunctum* (Ameghino 1902a). Holotypes of these species are lost. In the Ameghino collection at MACN there are fossils assigned to this species (MACN A-10447 and MACN A-10448), but they do not match with the original descriptions (Ameghino 1902a, pp. 134–135). In recent collections from Gran Barranca we have not found osteoderms that could be referred to these species; in particular, we have not found a single specimen referable to *Stenotatus* in Colhuehuapian levels.

The faunas and their diversity

Among the numerous materials collected (see Ré *et al.* Chapter 3, this book; Madden *et al.* this book) we recognize the following assemblages of Dasypodidae (Table 7.1).

Barrancan fauna

The levels in which these remains were collected are very well represented in Gran Barranca, from east to west, closely related to the guide level Simpson "Y" Tuff. Often in the paleontological literature, references to upper *Notostylops* beds or late Casamayoran faunas have been made, in contrast to lower *Notostylops* beds or early Casamayoran, suggesting faunal differences. Cifelli (1985) formally proposed the Barrancan and Vacan subages for the Casamayoran, with the type faunas at Gran Barranca and Cañadón Vaca, respectively.

All the remains collected in Gran Barranca from Casamayoran levels belong to the Barrancan Subage, since we have not found any lower fossiliferous level bearing identifiable vertebrate materials. The Barrancan fauna at Gran Barranca is between 41.6 and 39.0 Ma (Ré *et al.* Chapter 4, this book). We do not include in this analysis taxa cited by authors for the "upper Casamayoran" outside of Gran Barranca because their provenance data needs to be revised.

Among the collected material, one species of Dasypodinae and one of Euphractinae are recognized here: (a) Dasypodinae Astegotheriini, *Stegosimpsonia chubutana*;

(b) Euphractinae "Utaetini", *Utaetus buccatus*. *Prostegotherium notostylopianum* Ameghino 1902b and *Prostegotherium astrifer* Ameghino 1902a have been frequently cited for the Barrancan of Gran Barranca (e.g. Vizcaíno 1994). However, the large collections made during the last 20 years at Gran Barranca have not yielded a single osteoderm of either of these species. Possibly, when material of these taxa was collected, the osteoderms were recorded from a geographic locality that may not be the provenance *sensu stricto*. To date, these last two species are only known from Paso del Sapo (Chubut) in levels that are even older than those of the Vacan Subage (see Carlini *et al.* 2002b; Tejedor *et al.* 2009).

The Barrancan fauna has compositional differences from the Vacan at the type locality (Carlini *et al.* 2002a, 2002b, 2004a, 2005), with the Dasypodinae species being different, and the Eupractinae are absent in the Vacan but present in the Barrancan.

"El Nuevo" fauna

The fossiliferous level and locality (GBV-60) that yielded the "El Nuevo" fauna is not formally recognized as a different faunal unit, age, or subage. It occurs stratigraphically immediately above levels that yield the Barrancan fauna. The "El Nuevo" level is estimated to be between 38.0 to 39.0 Ma old (Ré *et al.* Chapter 4, this book). Compositional differences in the dasypodid fauna can be identified by comparison with underlying Barrancan levels. The following taxa are recognized for this level: (a) Dasypodinae Stegotheriini, two species, both also present in the overlying "El Rosado" fauna but only one of which is present in the Barrancan; (b) Euphractinae "Utaetini", a species of *Utaetus* (*U. deustus*) distinct from the one in the Barrancan; (c) four species of Euphractinae Euphractini, *Parutaetus chicoensis, Anteutatus lenis, Orthutaetus crenulatus,* and *Mazzoniphractus ingens* – all new compared with the Barrancan. Additionally, "El Nuevo" contains no known Astegotheriini.

"El Rosado" fauna

The fossil-bearing level at GBV-3 "El Rosado" yields an assemblage that belongs to the Mustersan SALMA *sensu* Bond and Deschamps (this book). The date of this fauna is from 38.7 to 37.0 Ma (Ré *et al.* Chapter 4, this book). The fauna in general is poorly represented, unlike older levels; however, the following Dasypodidae can be recognized: the same two Stegotheriini that appear in "El Nuevo" and three Euphractini (*Parutaetus chicoensis, Anteutatus lenis,* and *Mazzoniphractus ingens*) that also appear in "El Nuevo." The only difference with respect to the assemblage from GBV-60 "El Nuevo" is the decreased diversity of Euphractini (absence of *Orthutaetus crenulatus*). However, the fossils recovered from GBV-3 "El Rosado" are

Table 7.1. *Dasypodids recorded at Gran Barranca*

	Barrancan	"El Nuevo"	"El Rosado"	"La Cancha"	"La Cantera"	Colhuehuapian	"Pinturan"
DASYPODINAE							
ASTEGOTHERIINI							
Stegosimpsonia chubutana	X						
Pseudostegotherium glangeaudi						X	
STEGOTHERIINI							
Stegotheriini gen. *A* sp. *a*	?	X	X				
Stegotheriini gen. *A* sp. *b*		X	X				
Stegotheriini indet.				X	X		
Stegotherium variegatum						X	X
Stegotherium tessellatum							X
EUPHRACTINAE							
"UTAETINI"							
Utaetus buccatus	X						
?Utaetus deustus		X					
EUTATINI							
Barrancatatus rigidus				X			
Barrancatatus maddeni				X			
Barrancatatus sp.					X		
Sadypus aff. *S. confluens*				X			
Sadypus minutus				X			
Sadypus tortuosus					X		
Meteutatus aff. *M. lagenaformis*					X		
Meteutatus sp.				X			
Proeutatus sp.						X	X
Stenotatus cf. *S. patagonicus*							X
EUPHRACTINI							
Parutaetus chicoensis		X	X				
Parutaetus chilensis				X	?		
Orthutaetus crenulatus		X					
Anteutatus lenis		X	X				
Mazzoniphractus ingens		X	X				
Archaeutatus aff. *A. malaspinensis*					X		
?Pachyzaedyus sp.					X		
Prozaedyus humilis						X	
Prozaedyus sp.							X

less abundant, so this may prove to be an artifact of specimen number.

"La Cancha" fauna

The fauna from this level seems to be equivalent to the Tinguirirican (Flynn *et al.* 2003). At Gran Barranca we estimate the age of the Tinguiririan fauna from GBV-4 "La Cancha" to between 33.3 and 33.7 Ma (Ré *et al.* Chapter 4, this book). Dasypodidae are very well represented, with high diversity and abundance. Compared with older levels at Gran Barranca, there is a reduction from two to one representatives of Stegotheriini indet., and Euphractini decrease in diversity from three genera and species to one, *Parutaetus chilensis*. Most striking is the appearance of five species of Eutatini, *Barrancatatus rigidus*, *Barrancatatus maddeni*, *Sadypus* aff. *S. confluens*, *Sadypus minutus*, and *Meteutatus* sp. Eutatins are not recorded in older beds at Gran Barranca. In addition to a reduced diversity and abundance of Stegotheriini and Euphractini, and the first appearance, notable diversity and abundance of Eutatini, we also note the generally larger size, and pilosity developed by the Dasypodidae in general at this level.

"La Cantera" fauna

This fauna is younger than that of GBV-4 "La Cancha" (Ré *et al.* Chapter 4, this book; Vucetich *et al.* this book), estimated at between 31.1 and 29.5 Ma, but older than the poorly documented Deseadan fauna of Gran Barranca between about 29 and 26 Ma. While the assemblage from GBV-19 "La Cantera" includes only a few xenarthran taxa, their representation is important within the context of the different "pan-Deseadan" faunas known today (e.g. that of La Curandera in Chubut Province). Taxa from GBV-19 include one Stegotheriini indet., two species of Euphractini (?*Pachyzaedyus* sp. and *Archeutatus* aff. *A. malaspinensis),* and three species of Eutatini (*Meteutatus* aff. *M. lagenformis*, *Sadypus tortuosus*, and *Barrancatatus* sp.). The compositional change with respect to the underlying level involves a decrease in eutatine diversity (from three genera and five species to the same three genera but only three species) and the relatively large size of one species of Euphractini.

Colhuehuapian fauna

The Colhuehuapian fauna here studied was collected in the classic levels at Gran Barranca dated between 20.4 and 20.0 Ma (Ré *et al.* Chapter 4, this book). The following taxa are recognized: one Astegotheriini (*Pseudostegotherium glangeaudi*), one Stegotheriini (*Stegotherium variegatum*), one Euphractini (*Prozaedyus humilis*), and one Eutatini (*Proeutatus* sp.). It is noteworthy that between the Colhuahuapian fauna and the older La Cantera fauna there is a hiatus of approximately 9 to 10 Ma, a temporal hiatus equivalent to the interval encompassing all five older faunas (approximately 11 Ma). In the interval before the Colhuehuapian would be included almost all the Oligocene sites (except for "La Cancha" and "La Cantera") that are commonly referred to the Deseadan *s.l.* The main differences between the Colhuehuapian Dasypodidae compared with those from GBV-19 "La Cantera" are a reappearance of the Astegotheriini in the record at Gran Barranca, and a decrease of the Euphractinae diversity, although one taxon appears that will be frequent in the Pansantacrucian (i.e. *Proeutatus*).

"Pinturan" fauna

This fauna from the uppermost levels of the Gran Barranca sequence (Kramarz *et al.* this book) dates to between 19.7

and 18.7 Ma (Ré *et al.* Chapter 4, this book), somewhat older than the fauna from near the Río Pinturas. The following taxa are recognized in the assemblage: two species of Stegotheriini (*Stegotherium variegatum* and *Stegotherium tessellatum*), one Euphractini (*Prozaedyus* sp.), two Eutatini (*Proeutatus* sp. and *Stenotatus* cf. *S. patagonicus*) as described in Kramarz *et al.* (this book). Although the exposures are poor and the material not abundant, the following changes are noted with respect to the underlying Colhuehuapian: (1) the absence of Astegotheriini, (2) presence of two species of Stegotheriini (one surviving from the Colhuehuapian), (3) one Euphractini that is more derived than the taxon from the Colhuehuapian, and (4) the first record of *Stenotatus* at Gran Barranca.

Discussion

The succession of xenarthran faunas at Gran Barranca can be discussed from several related perspectives; systematics, biostratigraphy, paleoenvironments, and paleoclimate. The relative diversity of each xenarthran group recorded in each of these faunas suggests a positive correlation with marine temperature change (Fig. 7.6). Certainly, the "peninsular" effect on Patagonia implies that temperature change in marine water will be more directly and immediately reflected on inland climate than at latitudes farther north. The main changes observed in the diversity of Dasypodidae are discussed below.

During the Barrancan, there are fewer Dasypodinae species compared with older faunas such as Cañadón Vaca, and even more so, compared with Laguna Fría (Carlini *et al.* 2002b, 2002c; Tejedor *et al.* 2009). From a paleoclimatic standpoint, the presence of Euphractinae may be significant because today they frequent temperate areas and are scarcely represented in the tropical zone. Plausibly, the abundance of Euphractinae in these sediments may be related to the gradual decrease in temperature after the early Eocene. This change in the composition of dasypodid diversity initiates the subsequent prevalence of Euphractinae over Dasypodinae at Gran Barranca and other Patagonian localities.

The Eocene–Oligocene transition is documented in Gran Barranca (Scarano *et al.* 2004; Zucol *et al.* 2004; Ré *et al.* Chapter 4, this book). Concerning the xenarthrans, toward the end of the Eocene a decrease both in diversity and relative abundance of the Astegotheriini is evident. Immediately after the Eocene–Oligocene transition, in the early Oligocene (represented at Gran Barranca by La Cancha), the diversity and representation of Euphractinae Eutatini increased notably. These medium- to large-sized cingulates show a pronounced development of pilosity. Coincidently, other Cingulata that are larger than those in older levels also appear (e.g. *Machlydotherium* and Glyptatelinae). All these

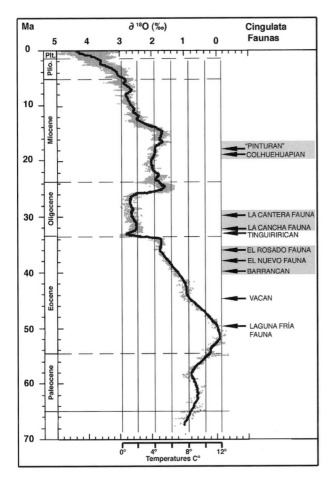

Fig. 7.6. Approximate temporal placement of Cenozoic dasypodid faunas discussed in the text, in relation to Cenozoic marine temperatures (Zachos *et al.* 2001) Darker shading indicates the faunas at Gran Barranca. (Figure modified from Carlini *et al.* 2005.)

Because of the large collections we have made at Gran Barranca, the record of Stegotheriini, previously only known from the Colhuehuapian–Colluncuran interval (early to middle Miocene), has been greatly extended. Now, there is an almost continuous record of this tribe starting with the oldest levels at Gran Barranca, i.e. beginning in the later middle Eocene (Carlini *et al.* 2004a). In addition, the evolutionary transition from taxa referred to the Astegotheriini to those assignable to Stegotheriini is probably recorded among the recognized taxa.

In the fauna from GBV-19 "La Cantera," although taxa are few, the record of large-sized species is outstanding. The one species of Euphractini is larger than any other Euphractini from Gran Barranca, both in older and younger units. Also, the La Cantera Glyptatelinae are comparable to the gigantic *Clypeotherium magnum* Scillato-Yané 1977 from the Deseadan at "El Pajarito" in central Chubut. Another interesting point is that this fauna allows the correlation among assemblages of Cingulata from other Patagonian localities, and now within a geochronological framework. To date, we have not found any typical Deseadan dasypodids at Gran Barranca.

Xenarthans present at GBV-60 "El Nuevo" suggest an assemblage that is different from, and transitional between the Barrancan and typical Mustersan assemblages from localities outside Gran Barranca. However, the "El Nuevo" dasypodid fauna has few diferences with that from GBV-3 "El Rosado," either taxonomically or in terms of the relative abundance of taxa. Definition about which of these is closest is still controversial, since the record of Mustersan-like faunas at Gran Barranca is scant, but certainly, this is different from any of the faunas already known.

At the top of the Gran Barranca sequence, there is a distinct cingulate assemblage, different from that of the underlying Colhuehuapian, and which in turn, is older than the typical Santacrucian fauna (Kramarz *et al.* this book).

forms may be associated with more herbivorous diets and more open environments (Carlini *et al.* 2004b).

In the early Miocene Colhuehuapian fauna some peculiar taxa are recorded. A small Euphractini with little pilosity, a very small species of Astegotheriini, and one species of the genus *Stegotherium* are also very abundant. The species of *Stegotherium* is particularly important as it is among the most specialized Dasypodidae with a strictly insectivorous diet (as inferred from the skull morphology of *S. tessellatum* in Scott 1903–05). These peculiarities suggests a warmer climate during the Colhuehuapian.

A coincidence between change in xenarthran faunas and marine temperature is noteworthy. When the changing pattern of marine temperature is compared with change in the successive faunas of Dasypodidae, there is an association between characteristics of these faunas and the temperature inferred for each biostratigraphic level or interval (Fig. 7.6).

ACKNOWLEDGEMENTS
We are indebted to J.J. Flynn (American Museum of Natural History), W. Simpson (Field Museum of Natural History, Chicago), M. Reguero (Museo de La Plata), and A. Kramarz (Museo Argentino de Ciencias Naturales "Bernardino Rivadavia") for allowing access to collections under their care; to the Vera family for permitting us to work at Gran Barranca; and to Pan American Energy for its continuous support of our fieldwork. This research was supported by US National Science Foundation (to R. F. Kay and R. H. Madden), and PICT-R 074, PICT 1860 and UNLP-FCNYM N-514 (to AAC).

REFERENCES

Ameghino, F. 1887. Enumeración sistemática de las especies de mamíferos fósiles coleccionados por Carlos Ameghino en los terrenos eocenos de Patagonia austral y depositados en el museo de La Plata. *Boletín del Museo de La Plata*, **1**, 1–26.

Ameghino, F. 1889. Contribución al conocimiento de los mamíferos fósiles de la República Argentina. *Actas de la Academia Nacional de Ciencias en Córdoba*, **6**, 1–1027.

Ameghino, F. 1891. Nuevos restos de mamíferos fósiles descubiertos por Carlos Ameghino en el Eoceno inferior de la Patagonia actual. Especies nuevas, adiciones y correcciones. *Revista Argentina de Historia Natural*, **1**(5), 289–328.

Ameghino, F. 1894. Première contribution à la connaissance de la faune mammalogique des couches à *Pyrotherium*. *Boletín del Instituto Geográfico Argentino*, **15**, 603–660.

Ameghino, F. 1897. Les Mammifères crétacés de l'Argentine: deuxième contribution à la connaissance de la faune mammalogique des couches à *Pyrotherium*. *Boletín del Instituto Geográfico Argentino*, **18**, 406–521.

Ameghino, F. 1902a. Première contribution à la connaissance de la faune mammalogique des couches à *Colpodon*. *Boletín de la Academia Nacional de Ciencias en Córdoba*, **17**, 71–138.

Ameghino, F. 1902b. Notices préliminaires sur des mammifères nouveaux des terrains crétacés de Patagonie. *Boletín de la Academia Nacional de Ciencias en Córdoba*, **17**, 5–70.

Ameghino, F. 1906. Les formations sédimentaires du crétacé supérieur et du tertiaire de Patagonie avec un parallèle entre leurs faunes mammalogiques et celle de l'ancien continent. *Anales del Museo Nacional de Buenos Aires*, **15**, 1–568.

Bordas, A. F. 1933. Notas sobre los Eutatinae: nueva subfamilia extinguida de Dasypodidae. *Anales del Museo Nacional de Historia Natural de Buenos Aires*, **37**, 583–614.

Carlini, A. A., S. F. Vizcaíno, and G. J. Scillato-Yané 1997. Armored xenarthrans: a unique taxonomic and ecologic assemblage. In Kay, R. F., Madden, R. H., Cifelli, R. L., and Flynn, J. J. (eds.), *Vertebrate Palaeontology in the Neotropics: The Miocene Fauna of La Venta, Colombia*. Washington, DC: Smithsonian Institution Press, pp. 213–226.

Carlini, A. A., G. J. Scillato-Yané, R. H. Madden, M. Ciancio, and E. Soibelzon 2002a. Los Dasypodidae (Mammalia, Xenarthra) del Eoceno. III. Las especies del Casamayorense de la Barranca Sur del lago Colhué Huapi, S de Chubut (Argentina): el establecimiento de los Euphractinae. *Resumenes I Congreso Latinoamericano de Paleontología de Vertebrados*, 24–25.

Carlini, A. A., G. J. Scillato-Yané, R. H. Madden, E.Soibelzon, and M. Ciancio 2002b. Los Dasypodidae (Mammalia, Xenarthra) del Eoceno. II. El conjunto de especies del Casamayorense de Cañadón Vaca, SE de Chubut (Argentina) y su relación con los que le suceden. *Resumenes I Congreso Latinoamericano de Paleontología de Vertebrados*, 24.

Carlini, A. A., G. J. Scillato-Yané, F. J. Goin, and F. Praderio 2002c. Los Dasypodidae (Mammalia, Xenarthra) del Eoceno. I. El registro en Paso del Sapo, NO de Chubut (Argentina): exclusivamente Astegotheriini. *Resumenes I Congreso Latinoamericano de Paleontología de Vertebrados*, 23.

Carlini, A. A., M. Ciancio, and G. J. Scillato-Yané 2004a. La tribu Stegotheriini (Xenarthra, Dasypodidae), 20 Ma más de registro paleógeno. *Ameghiniana*, **41**(4) (Supl.), 39R.

Carlini, A. A., M. Ciancio, and G. J. Scillato-Yané 2004b. La transición Eoceno–Oligoceno y su manifestación en la diversidad de los Cingulata (Mammalia, Xenarthra): Inferencias paleoecológicas. *Ameghiniana*, **41**(4) (Supl.) 23–24R.

Carlini, A. A., M. Ciancio, and G. J. Scillato-Yané 2005. Los Xenarthra de Gran Barranca, más de 20 Ma de historia. *Actas XVI Congreso Geológico Argentino*, **4**, 419–424.

Carlini, A. A., M. R. Ciancio, J. J. Flynn,G. J. Scillato-Yané, and A. R. Wyss 2009. The phylogenetic and biostratigraphic significance of new armadillos (Mammalia, Xenarthra, Dasypodidae, Euphractinae) from the Tinguirirican (Early Oligocene) of Chile. *Journal of Systematic Palaeontology*, doi:10.1017/S1477201909002740.

Ciancio, M. R and A. A. Carlini 2008. Identificación de Ejemplares Tipo de Dasypodidae (Mammalia, Xenarthra) del Paleógeno de Argentina. *Revista del Museo Argentino de Ciencias Naturales*, n.s., **10**(2), 221–237.

Cifelli, R. L. 1985. Biostratigraphy of the Casamayoran, Early Eocene, of Patagonia. *American Museum of National History Novitates*, **2820**, 1–26.

Fernicola, J. C. and S. F. Vizcaíno 2008. Revisión del género *Stegotherium* Ameghino, 1887 (Mammalia, Xenarthra, Dasypodidae). *Ameghiniana*, **45**(2), 321–332.

Flynn, J. J., A. R. Wyss, D. A. Croft, and R. Charrier 2003. The Tinguiririca Fauna, Chile: biochronology, paleoecology, biogeography, and a new earliest Oligocene South American Land Mammal 'Age'. *Palaeogeography, Palaeoclimatology, Palaeoecology*, **195**, 229–259.

González, L. R. and G. J. Scillato-Yané 2008. Una nueva especie de *Stegotherium* Ameghino (Xenarthra, Dasypodidae, Stegotheriini) del Mioceno de la provincia de Santa Cruz (Argentina). *Ameghiniana*, **45**(4), 641–648.

Gray, J. E. 1821. *Catalogue of Carnivorous, Pachydermatous, and Edentate Mammalia in the British Museum*. London: British Museum.

Kramarz, A., A. Garrido, A. Forasiepi, M. Bond, and C. Tambussi 2005. Estratigrafía y vertebrados (Aves y Mammalia) de la Formación Cerro Bandera, Mioceno Temprano de la Provincia del Neuquén, Argentina. *Revista Geológica de Chile*, **32**(2), 273–291.

Marshall, L. G. and L. Sempere 1991. The Eocene to Pleistocene vertebrates of Bolivia and their stratigraphic context: a review. In Suárez-Soruco, R. (ed.), *Fósiles y Facies de Bolivia*, vol. 1, *Vertebrados. Revista Técnica de Yacimientos Petrolís Fiscales de Bolivia*, **12**, 631–652.

Mones, A. 1986. Paleovertebrata sudamericana: catálogo sistemático de los vertebrados fósiles de América del

Sur. I. Lista preliminar y bibliográfica. *Courier Forschungs-Institut Senckemberg*, **82**, 1–625.

Roth, S. 1899. Apuntes sobre la geología y paleontología de los territorios de Río Negro y Neuquén (diciembre de 1895 a junio de 1896). *Revista del Museo de La Plata*, **9**, 141–197.

Scarano, A. C., M. Reguero, and A. A. Carlini 2004. Evolución de la hipsodoncia en los Interatheriidae de Gran Barranca, una evaluación antes y después de la transición Eoceno–Oligoceno. *Ameghiniana*, **41**(4) (Supl.), 24R.

Scillato-Yané, G. J. and A. A. Carlini 1998. Nuevos Xenarthra del Friasense (Mioceno medio) de Argentina. *Studia Geologica Salmanticensia*, **34**, 43–67.

Scillato-Yané, G. J. 1977. Sur quelques Glyptodontidae nouveaux (Mammalia, Edentata) du Déséadien (Oligocène inférieur) de Patagonie (Argentine). *Bulletin du Muséum National D'Histoire Naturelle, Sciences de la Terre*, **64**, 249–262.

Scott, W. B. 1903–05. Mammalia of the Santa Cruz Beds. Volume V, Paleontology. Part I, Edentata. In Scott, W. B. (ed.), *Reports of the Princeton Expedition to Patagonia 1896–99*. Stuttgart: E. Schweizerbart'sche Verlagshandlung (E. Nägele), pp. 1–364.

Simpson, G. G. 1945. The principles of classification and a classification of Mammals. *Bulletin of the American Museum of Natural History*, **85**, 1–350.

Simpson, G. G. 1948. The beginning of the Age of Mammals in South America. I. *Bulletin of the American Museum of Natural History*, **91**, 1–232.

Simpson, G. G. 1967. The beginning of the Age of Mammals in South America. II. *Bulletin of the American Museum of Natural History*, **137**, 1–259.

Tejedor, M. F., F. J. Goin, J. N. Gelfo, G. López, M. Bond, A. A. Carlini, G. J. Scillato-Yané, M. O. Woodburne, L. Chornogubsky, E. Aragón, M. Reguero, N. Czaplewski, S. Vincon, G. Martin, and M. Ciancio. 2009. New early Eocene mammalian fauna from western Patagonia, Argentina. *American Museum of Natural History Novitates*, **3638**, 1–43.

Vizcaíno, S. F. 1994. Sistemática y anatomía de los Astegotheriini Ameghino, 1906 (nuevo rango) (Xenarthra, Dasypodidae, Dasypodinae). *Ameghiniana*, **31**(1), 3–13.

Winge, H. 1923. *Pattedyr-Slaegter. Volume I, Monotremata, Marsupialia, Insectivora, Chiroptera, Edentata*. Copenhagen: Hagarup.

Zachos, J., M. Pagani, L. Sloan, E. Thomas, and K. Billups, 2001. Trends, rhythms, and aberrations in global climate 65 Ma to present. *Science*, **292**, 686–693.

Zucol, A. F., M. Brea, R. Madden, and E. Bellosi 2004. Análisis fitolítico de la transición Eoceno–Oligoceno en el perfil tipo de la Formación Sarmiento (Gran Barranca), Chubut. *Ameghiniana*, **41**(4) (Supl.), 25R.

8 The "condylarth" Didolodontidae from Gran Barranca: history of the bunodont South American mammals up to the Eocene–Oligocene transition

Javier N. Gelfo

Abstract

The "Condylarthra" have been traditionally viewed as a group of primitive eutherian mammals postulated to be the stem group for most extant and extinct ungulates. The Didolodontidae were endemic condylarths, small to medium-sized ungulates with bunodont dentition that were well represented in the Paleogene mammal communities. In this chapter, the systematics of the Didolodontidae are summarized, with a focus on the fossil record of the Gran Barranca locality. New didolodontid remains and a new species are described and illustrated. A phylogenetic analysis is performed to test the position of the new taxon. The analysis shows the Kollpaniinae and the Didolodontidae as monophyletic sister groups. Like other middle Eocene didolodontids, *D. magnus* sp. nov. developed larger size, in coincidence with climate cooling. The morphology of didolodontids suggests they filled a unique and distinctive ecological role among native archaic ungulates, one that seems to have been susceptible to environmental change during the Eocene–Oligocene transition.

Resumen

Los "Condylarthra" han sido vistos tradicionalmente como un grupo de primitivos mamíferos eutéricos postulados como el "stem group" de muchos ungulados extintos y vivientes. Los Didolodontidae fueron ungulados endémicos, de talla pequeña a mediana, con denticiones bunodontes, bien representados en las comunidades de mamíferos paleógenos. En el presente capítulo, la sistemática de los didolodontidos es compendiada, enfatizando el registro fósil de la localidad de Gran Barranca. Nuevos restos y una nueva especie son descriptos e ilustrados. Para evaluar la posición del nuevo taxón se ha realizado un análisis filogenético. El análisis recupera a los Kollpaniinae y a los Didolodontidae como grupos hermanos

monofiléticos. Tal como se observa en otros didolodóntidos del Eoceno medio; *D. magnus* sp. nov. desarrolla un mayor tamaño, en coincidencia con el enfriamiento global. La morfología de los didolodóntidos permite inferir que ocupaban un rol ecológico distintivo y único entre los ungulados nativos del Paleógeno de América del Sur, un rol que parece haber sido sumamente susceptible a los cambios climáticos ocurridos durante la transición Eoceno–Oligoceno.

Introduction

The Didolodontidae are a problematic group of strictly bunodont mammals recorded exclusively in South America. Their fossil record extends from the Selandian (middle Paleocene – Peligran SALMA) until the Priabonian (late Eocene – Mustersan SALMA), with one taxon doubtfully referred to this family recorded in the late Oligocene (Deseadan SALMA) of Bolivia (Gelfo 2006).

Originally, Ameghino (1897) described the new taxon *Didolodus multicuspis*, the first of a long list referred by him to the Condylarthra and included in Holarctic families. He also distinguished three native families for "condylarths": Selenoconidae, which he interpreted as the link between Phenacodontidae and Meniscotheridae (Ameghino 1902a), Pantostylopidae, and Catathleidae (Ameghino 1906). However, most of the "condylarths" described by Ameghino were later placed in different groups of litopterns, notoungulates (Simpson 1948), and even as Phocidae (Muizon and Bond 1982). The families Selenoconidae and Pantostylopidae were regarded as junior synonyms of the notoungulate family Henricosborniidae, and the Catathleidae were identified as Periptychidae (Simpson 1934). Osborn (1910) identified the genera *Didolodus*, *Notoprotogonia*, *Lambdaconus*, and *Proectocion* as Condylarthra *incertae sedis*, and Scott (1913) placed them in the Didolodidae, emended to the correct derivative Didolodontidae by Simpson (1937).

The Paleontology of Gran Barranca: Evolution and Environmental Change through the Middle Cenozoic of Patagonia, eds. R. H. Madden, A. A. Carlini, M. G. Vucetich, and R. F. Kay. Published by Cambridge University Press. © Cambridge University Press 2010.

The Didolodontidae diagnosis was mostly based on tarsal characters, with only one dental trait (Cifelli 1983a, 1993). However, none of the postcranial elements assigned to didolodontids had been found unquestionably in direct association with dental elements; instead, they had been reassociated based on their relative abundance and relative size (Cifelli 1983b). The probably didolodontid tarsal elements were described as having "phenacodontid" structure, and a new family, the Protolipternidae, was created to include the smallest litopterns, characterized by a "condylarth" dental structure and derived litoptern tarsal elements. Soria (2001) argued against the reassociation of these tarsal elements to the Didolodontidae. Nevertheless, he accepted the *Protolipterna* reassociation, but considered invalid the Protolipternidae, arguing that all didolodontids should have a litoptern-like tarsal (Soria 2001). Neither this conflict, nor the questioned severance of the Sparnotheriodontidae from the Litopterna and their assignment to the "Condylarthra," will be unambiguously resolved until complete or associated remains are found. These problems are reflected in the few cladistics analyses available for this group (Muizon and Cifelli 2000; Gelfo 2006). In the present chapter the knowledge of the didolodontids recorded at Gran Barranca is updated following Gelfo (2004, 2006) and a new species of *Didolodus* is described. A phylogenetic analysis is performed to test the position of the new taxon. Lastly, the Paleogene history of South American bunodont ungulates is reviewed and analyzed.

Material and methods

Institutional abbreviations

AMNH, American Museum of Natural History, New York, USA; MACN, Museo Argentino de Ciencias Naturales "Bernardino Rivadavia," Buenos Aires, Argentina; MMP, Museo Municipal de Mar del Plata "Lorenzo Scaglia," Mar del Plata, Argentina; MPEF-PV, Museo Paleontológico "Egidio Feruglio," Trelew, Argentina.

Phylogenetic analysis

The data matrix has 18 taxa and 33 characters. The analysis was performed with TNT (Goloboff *et al.* 2003). The ingroup included 12 didolodontids, and the outgroup five Mioclaenidae, Kollpaniinae, and *Protoungulatum donae*. The characters were modified from Muizon and Cifelli (2000) and Gelfo (2004, 2006). All characters (Appendix 8.1) were treated as unordered and equally weighted. Characters 11, 12, and 23 were considered as polymorphic.

Systematic paleontology

Order **PANAMERIUNGULATA** Muizon and Cifelli 2000
Family **DIDOLODONTIDAE** Scott 1913

[Didolodidae Scott 1913, p. 489. Bunolitopternidae Schlosser 1923, p. 525; Abel 1928]

Diagnosis South American ungulates with complete dentition, brachydont bunodont teeth. Molar cusps with wide base, never slender, lophoid or selenodont in structure. Paracone and metacone without the labial folds seen in Proterotheriidae. P4 molarized, without developed hypocone, but with paraconule and metaconule usually present, except in *Salladolodus*. In contrast with the Kollpaniinae, molars with well-developed hypocone, which is often present in the M3, at least as a protuberance of the postcingulum. Lower dentition: p4 with a small paraconid slightly differentiated from the paracristid; m1–2 with talonid wider than the trigonid.

Included genera *Asmithwoodwardia*, *Didolodus*, *Escribania*, *Ernestokokenia*, *Lamegoia*, *Paulacoutoia*, *Paulogervaisia*, *Raulvaccia*, *Salladolodus*, and *Xesmodon*. *Nomen dubium*: *Megacrodon*; *species inquirenda*: "*Lambdaconus*" *alius*.

Genus *Didolodus* Ameghino 1897

[*Didolodus*: Roth 1927, p. 200. (typographic mistake). *Lonchoconus*: Ameghino 1901, p. 379. *Nephacodus*: Ameghino 1902, p. 19. *Cephanodus*: Ameghino 1902, p. 25. *Argyrolambda*: Ameghino 1904a, vol. 57, p. 338.]

Type species *Didolodus multicuspis* Ameghino 1897
Diagnosis I?/3, C?/1, P4/4, M3/3: upper molars with well-developed protostyle over the precingulum, mesiolingual to the paraconule and mesiolabial to the protocone; premetaconular crista absent; centrocrista projected to the labial side, contacting a robust metastyle; labial cingulum usually interrupted at mesostyle level; hypocone well developed in M3.

Didolodus multicuspis Ameghino 1897

[*Didolodus crassicuspis*: Ameghino 1901, p. 376. *Lonchoconus lanceolatus*: Ameghino 1901, p. 379. *Didolodus colligatus*: Ameghino 1902a, p. 18. *Cephanodus colligatus*: Ameghino 1902b, p. 25, fig. 12. *Nephacodus latigonus*: Ameghino 1902a, p. 19. *Didolodus dispar*: Ameghino 1904a, vol. 57, p. 333. *Argyrolambda conidens*: Ameghino 1904a, vol. 57 p. 338. *Argyrolambda conulifera*: Ameghino 1904b, p. 123, fig. 140. *Didolodus multicuspis*: Roth 1927, p. 247 (typographic mistake). *Didolodus latigonus*: Simpson 1948, pp. 103, 105. *Didolodus conidens*: Simpson 1967, p. 10]

Types of the synonyms *Didolodus crassicuspis*: MACN 10689, right jaw fragment with p2–m3. *Lonchoconus lanceolatus*: MACN 10730, isolated left M1. *Cephanodus colligatus*: MACN 10736 isolated right m3 and left p3–4 and m3. *Nephacodus* *latigonus*: MACN 10725 isolated left m2. *Didolodus dispar*: MACN 10733, isolated right M2. *Argyrolambda conidens*: MACN A55–8, isolated right M1.

Type MACN 10690 left maxillary fragment with P2–M3, and alveoli of P1 and C.

Diagnosis I?/3 C1/1 P4/4 C3/3; size larger than *D. minor* but smaller than *D. magnus* sp. nov.; upper molars with labial cingulum well developed but interrupted at the mesostyle; centrocrista strong and projected toward mesostyle; M1–2 quadrangular in outline with hypocone distal to the protocone; M3 with trapezoid outline because of the distolingual position of hypocone; lower molars with strong cristid obliqua associated to a heavy centroconid; paracristid with accesory cusps mesial to the protocone.

Occurrence The only reference given by Ameghino (1897, 1901, 1902a, 1904a, 1904b) for the types of *D. multicuspis, D. crassicuspis,* and *D. dispar* was "*Notostylops* beds," a provenance that was referred to the Casamayoran by Simpson (1948). These remains were recorded at the Gran Barranca Member of the Sarmiento Formation, Casamayoran SALMA, Barrancan subage (middle Eocene). Particularly, MPEF 7992 described here came from Simpson's "Y" level at Profile Carlini.

Description A partial left dentary with p4–m3, MPEF 7992, is one of the most complete jaw material found to date (Fig. 8.1A, B, C). The slender dentary bone is broken on the lingual side from the p4 to the m2 exposing the roots of these teeth. The p4 trigonid is higher than the talonid, while the protoconid and metaconid are subequal in size. The paraconid is a lower elevation on the mesial side. A low and strong precingulid runs from the labial side of the tooth up to the mesiolingual base of the protoconid. The talonid basin is mesiodistally short, and divided into lingual and labial sides by a central hypoconid. The m1 is more heavy erased by wear, but the paraconid can still be identified twinned to the metaconid. In the molars the trigonid area is completely reduced, fully occupied by the bases of protoconid and metaconid. The paracristid is short, distally concave, and associated with a small low cusp on the mesiolabial side, particularly evident on m2–3. The protocristid is not so bent as the paracristid; it runs distolingually due to the distal placement of the metaconid. The talonid basin is almost filled by the bulbous structure of the cusps and represented by a lingually open deep furrow. The hypoconid is the largest cusp, and projects a low and rounded cristid oblique to the labial side of the metaconid. A strong centroconid is present over this crista. The entoconid is small, located only slightly distally with respect to the hypoconid and very close to the

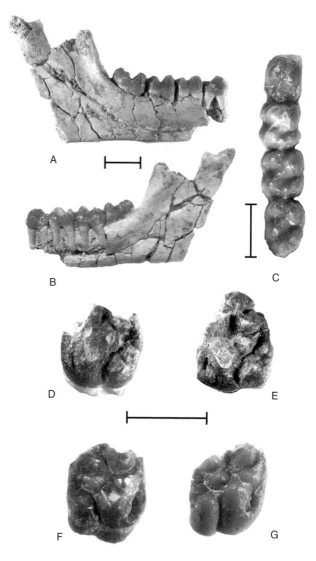

Fig. 8.1. *Didolodus multicuspis* MPEF 7992: (A) labial view, (B) lingual view, (C) occlusal view of dental series. Occlusal views of *Didolodus magnus* sp. nov.: (D) MPEF 7858, (E) MPEF 7857. *Paulogervaisia inusta*: (F) MPEF 6587a, (G) MPEF 6587b. Scale bar 10 mm aprox.

hypoconulid. The hypoconulid is clearly separated from the entoconid by a deep furrow, but related to the hypoconid by a well-developed hypocristid. In the first two molars, the postcingulid is short; in distal view, it descends from the hypoconulid to the distal wall of the hypoconid and the entoconid. In contrast to previously described remains (Simpson 1948, Gelfo 2006) a short postcingulid is present distolingually to the m3, as a reduced shelf. The precingulid is well developed and in the m2 runs over the lingual side of the trigonid, encircling the base of the protoconid. The m3 is mesiodistally longer, with the hypoconulid larger than the entoconid and clearly separated from the rest of the talonid cusps.

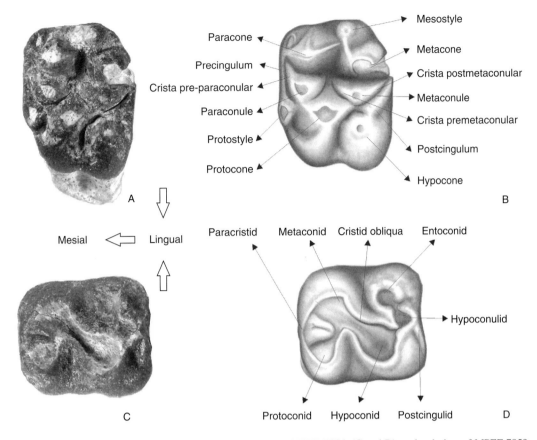

Fig. 8.2. *Didolodus magnus* sp. nov.: (A and B) occlusal view of the type MPEF 7856, (C and D) occlusal view of MPEF 7859.

As a consequence the talonid basin of m3 is also larger than in m1–2.

Remarks Considering Soria's (2001) observations, there is some doubt about the provenance of *Didolodus* types. But, despite the uncertainty surrounding the type locality, all other known remains of *Didolodus multicuspis* were recovered from the Sarmiento Formation, Casamayoran SALMA, Barrancan Subage (middle Eocene). Particularly MPEF 7992 here described was found Simpson's "Y" level at Profile Carlini.

Didolodus minor Simpson 1948

Type AMNH 28473 left M1

Diagnosis I?/3 C1/1 P4/4 C3/3; smaller-sized didolodontid, upper molars narrower than in *D. multicuspis*; premolars strongly molarized, P2 with well developed precingulid, strong metacone almost as large as the paracone, and similar is size to the paracone of P-4.

Occurrence The remains here described came from the south margin of Lake Colhue-Huapi, in the Gran Barranca Member of the Sarmiento Formation, Casamayoran SALMA, Barrancan Subage (middle Eocene). Is important to note that the type material did not come from the Sarmiento Formation but

from Cañadón Vaca, another Patagonian locality, and belongs to the older Vacan Subage of the Casamayoran.

Remarks Simpson (1948) defined this species taking into consideration five remains discovered at Cañadón Vaca which differ from *D. multicuspis* in size and proportions, while considering that these could probably represent a geographic race or a temporal mutation. Later, when he studied the MMP 696 and MMP 138 specimens that came from the Gran Barranca he rejected this view (Simspon 1964, 1967a). In fact both species were coeval and co-occurring in several localities of the Barrancan Subage in Chubut Province.

Didolodus magnus sp. nov.
Figs. 8.1D, E and 8.2A–D.

Etymology "*Magnus*" meaning "large" in Latin, because is the largest species of the genus.

Type MPEF 7856: isolated left M2? with the lingual root.

Hypodigm The type, MPEF 7857: left fragment of M2?, MPEF 7858: isolated left M3? with lingual root, ?MPEF 7859: isolated left m2.

Diagnosis I?/3 C1/1 P4/4 C3/3; didolodontid larger than *D. multicuspis* but somewhat smaller than *Paulogervaisia inusta*. In contrast to the other species of *Didolodus*, the labial cingulum is clearly reduced to a thin rim at the base of the paracone and metacone.

Occurrence The remains described here were collected at the south margin of Lake Colhue-Huapi, from the Gran Barrancan Member of the Sarmiento Formation. All the specimens assigned to this new species are from the "El Nuevo" local fauna (GBV-60) which seems to represent a post-Barrancan, pre-Mustersan age.

Description The M2? is rectangular in outline, with labiolingually oriented major axis, and much larger than any other known tooth ever assigned to *Didolodus*. The thin precingulum runs from the more labial side of the tooth through the base of the protocone. The protostyle is almost as large as the paraconule, and closer to the latter than to the protocone. The parastyle is reduced to a cuspule at the intersection between preparacrista and precingulum. The strong reduction of the labial cingulum contrasts with what can be seen in *D. minor* or *D. multicuspis* in which is not reduced but interrupted at the mesostyle. The paracone and metacone are subequal in size, with the former located somewhat more labially than the latter. The centrocrista is projected toward the mesostyle, but the premetacrista is wider than the postparacrista. The conules are smaller in size than the main cusps, and are connected to the protocone by a strong and short pre- and postprotocrista respectively. The outline of the paraconule is more distinct from the cristae than that of the metaconule. The preparaconular crista runs labially but without interrupting the continuity of the precingulum. Both postparaconular and premetaconular cristae are poorly developed. The postmetaconular crista is longer than the premetaconular one and directed labially. The hypocone is subequal in size to the protocone and joined to it by a short and bulky entocrista. The hypocone is well separated from the protocone. The presence of a protuberance between the lingual bases of hypocone and protocone is similar to the condition seen in some remains of *D. minor*. The only remain assigned here to an M3 is badly preserved, with the labial side of the tooth broken. It is much larger than the M2 and more than twice the size of the M3 of *Didolodus multicuspis*. The lingual root is strong and divided by a vertical furrow into one portion below the protocone and a second one related to the hypocone. The metaconule of the M3 differs from the one in the M2 by its more circular outline and lesser development of the cristae.

A worn left m2 could belong to this taxon according to its size, but this assignation requires confirmation.

Genus *Paulogervaisia* Ameghino 1901
[*Lambdaconus* Ameghino 1901, p. 376.]
Type species *Paulogervaisia inusta* Ameghino 1901
Diagnosis I?/? C?/? P4?/4? M3/3; large-sized didolodontid; M3 as wide as M2 and metacone more lingual than paracone. Postmetaconular crista extended distally to the postcingulum. Centrocrista only slightly bent labially. Mesostyle smaller than in *Didolodus*. Lower molars with entoconid as large as the hypoconulid in the m3.

Remarks *Paulogervaisia* was originally considered by Ameghino as a primitive Proboscidea that allowed a close relationship with the Condylarthra. Simpson (1948) argued that it was a Didolodontidae and distinguished it from *Didolodus*. Van Valen (1978) regarded *Paulogervaisia* as a junior synonym of *Didolodus* without justification, and was followed by McKenna and Bell (1997) but not by Cifelli (1983a).

Paulogervaisia inusta Ameghino 1901
[*Paulogervaisia mamma* (Ameghino 1901) Simpson 1948, p. 107; *Lambdaconus mamma* Ameghino 1901, p. 376.]
Type MACN 10664 isolated right m3, deeply erased by wear.
Types of the synonyms *Lambdaconus mamma*: MACN 10719 right jaw fragment with p3–m2 and maxillary fragment with M2–3 and an isolated right M2.
Diagnosis The same as for the genus.
Occurrence All the remains were collected at the south margin of Lake Colhue-Huapi. Type and types of the synonyms from the Great Barrancan Member of the Sarmiento Formation, Casamayoran SALMA, Barrancan Subage (middle Eocene). MPEF 6587a–b described here, came from GBV-3 "El Rosado" in the El Rosado Member of the Sarmiento formation, Mustersan SALMA.
Description An isolated right m2 MPEF 7621 is similar to the m2 of MACN 10719. Two isolated right upper molars that could be referred with doubts to *P. inusta*. MPEF 6587a, an M2 with the mesiolabial side broken (Fig. 8.1F) which is square in outline, with straight and almost parallel precingulum and postcingulum. In contrast to the M2 of the set MACN 10719, there is no protostyle and only an enlargement of the precingulum mesial to the preprotocrista can be observed. The labial edge of the paracone is broken, but this cusp was the largest and somewhat more labially located than the metacone. The postcingulum is straight and extends from the distal side of the hypocone through the distolabial side of the metacone. A partially broken cuspidated cingulum is present labial to the centrocrista. The metacone is similar

in size to the paraconule and metaconule. In contrast to MACN 10719 no mesostyle is present. No postparaconular crista is present. The preparaconular and postmetaconular cristae extend mesiolabially and distally respectively. The hypocone is only slightly lingual with respect to the protocone and clearly separated from it. A strong entocrista relates both lingual cusps.

MPEF 6587b is an isolated right M2 without wear but broken at the labial side of the metacone (Fig. 8.1G). It resembles MPEF 6587a, but with some important differences. The enamel is crinkly or wrinkled at the base of the cusps and at the top of the crest. The mesostyle is present. There are two cusps associated with the postcingulum, the hypocone which is more lingually located than the protocone, and a massive but smaller cusp distolabial to the metaconule. The premetaconular crista is a low edge of the metaconule which contact the metacone.

Remarks *Paulogervaisia inusta* and *Lambdaconus mamma* (= *P. mamma*) were published in the same work (Ameghino 1901), but the latter name is preferable for the synonym because it is the type of the genus (Gelfo 2004b, 2006). MPEF 6587a–b expands the biochron of *P. inusta* up to the Mustersan. However, these isolated teeth could also be related to *Xesmodon langi* Roth 1899, a known larger Mustersan species, which contrastingly, comes from Gran Hondonada and Cerro Humo (*sensu* Bond pers. comm.; Roth locality Cerro Humo is Colhue-Huapi Norte).

Genus ***Ernestokokenia*** Ameghino 1901
[*Notoprotogonia* Ameghino 1904a, vol. 57, p. 336; *Protogonia* Ameghino 1906, p. 467; *Protogonia (Euprotogonia)* Cope, Gaudry 1904, p. 9; *Notoprogonia* Scott 1913, p. 489; 1937, p. 490; *Enneoconus* Ameghino 1901 p. 378; *Euneoconus* Roth 1927, p. 249 (typographic mistake).]
Type species *Ernestokokenia nitida* Ameghino 1901
Diagnosis Upper molars wider than mesiodistally long, mesostyle if present very small and not interrupting the continuity of the labial cingulum as in *Didolodus*. Usually with a cusp over the postcingulum situated labial to the hypocone; paraconule and metaconule of small to medium size compared to the labial cusps. First and second lower molars with talonid wider than the trigonid; paraconid absent or if present, much smaller than the metaconid and coalescent to the base of the latter.
Occurrence The genus *Ernestokokenia* has been found in several Chubut localities, and its stratigraphic provenance corresponds to Las Flores and Koluel-Kaike formations (Rio Chico Group). Also from the Gran Barranca Member of the Sarmiento

Formation at Gran Barranca. This taxon is represented by diverse species in the Itaboraian, Riochican, and Casamayoran (Vacan and Barrancan Subages) SALMAs. An isolated tooth was reported also from the Geste Formation at Antofagasta de la Sierra in Catamarca Province (López 1997).
Remarks Simpson (1948) considered that *Notoprotogonia* created by Ameghino (1901) corresponded to the upper molar of *Ernestokokenia*. Similarly, *Enneoconus parvidens* Ameghino 1901, only known from an isolated right M3, corresponds to the upper dentition of *Ernestokokenia nitida* (Gelfo 2006).

Despite Simpson's (1948) highlighting of the similarities between *Ernestokokenia* and *Asmithwoodwardia*, he considered both taxa as valid and as members of Didolodontidae. More recently, several authors have considered *Ernestokokenia* as invalid (Van Valen 1978; McKenna and Bell 1997), but their proposals have not been followed by other authors (Cifelli 1983a; Muizon and Cifelli 2000; Soria 2001; Gelfo 2006). In fact, *Asmithwoodwardia* differs from *Ernestokokenia* not only by its smaller size, but by several qualitative characters (Gelfo and Tejedor 2004). Of the four valid species of *Ernestokokenia* (Table 8.1) (Gelfo 2006), only *E. nitida* and *E. patagonica* have been recorded at Gran Barranca.
Ernestokokenia nitida Ameghino 1901
[*Ernestokokenia marginata* Ameghino 1901, p. 380; *Enneoconus parvidens* Ameghino 1901, p. 378.]
Type species MACN 10735 left m2 and distal remain of left m3 talonid.

Types of the synonyms *E. marginata*: MACN 10722, isolated left m1. *Enneoconus parvidens*: MACN 10726, isolated right M3.
Diagnosis I?/? C?/? P4/4 M3/3. Middle-sized didolodontid compared with other species of the genus, larger than *E. yirunhor* and somewhat smaller than *E. chaishoer*. Upper molars with preparaconular crista projected between parastyle and paracone. Lower molars with short entocristid, hypoconid large invading the talonid basin adjacent the hypoconulid. Entoconid of m1–2 larger than the hypoconulid. Hypoconulid of the m3 contacting the hypoconid with the hypocristid, and with a short postcristid that does not contact the entoconid.
Occurrence No precise reference of the type locality was given by Ameghino (1901). Simpson (1948) referred several remains from the south margin of Lake Colhue-Huapi to this taxon, and assumed that the type locality is the Gran Barranca Member of the Sarmiento Formation. Also from this locality are all the materials in the Tournouër collection (MNHN CAS) labeled as Cerro Negro. The type of the

Table 8.1. *Comparison of Didolodus, Ernestokokenia, and Paulogervaisia and their species*

Character	Didolodus	Ernestokokenia	Paulogervaisia
Centrocrista	Short and bent labially	Mesiodistally straight, not bent labially (but see remarks for *E. nitida*)	Slightly bending labially
Mesostyle	Strong and interrupting the labial cingulum	Absent	Weak
Hypocone	Strong, with a wide and lingually projected base	Subequal to the protocone	Subequal to the protocone
Cristid obliqua	Short, associated with a well developed centroconid	Without centroconid	Without centroconid (due to wear, inferred from the reduced base of the cristida)

Character	*D. minor*	*D. multicuspis*	*D. magnus*	*E. nitida*	*E. chaishoer**	*E. yirunhor**	*E. patagonica*	*P. inusta*
Relative size	Smallest species	Smaller than *D. magnus*	Larger than any *Ernestokokenia*	Smaller than *E. chaishoer*	Larger species	Smallest species	Subequal to *E. nitida*	Somewhat larger than *D. magnus*
Upper teeth	P2–3 with metacone; Molars with labial cingulum; No lingual cingulum	P2–3 without metacone; Molars with well-developed labial cingulum; No lingual cingulum	Premolars unknown; Molars with labial cingulum reduced to a faint rim; No lingual cingulum	Molars with labial cingulum crenulated; No lingual cingulum	Molars with labial cingulum well developed; No lingual cingulum	Molars with labial cingulum reduced to a faint rim; No lingual cingulum	Molars with labial cingulum well developed; Lingual cingulum around the base of protocone	Molars with labial cingulum absent labial to the metacone; No lingual cingulum
Lower teeth	Molars with paraconid close to the metaconid	Molars with paraconid close to the metaconid	Molars without paraconid (due to wear, inferred from the reduced size of the trigonid)	Molars without paraconid	Molars with paraconid close to the metaconid	Molars with paraconid close to the metaconid	Lower dentiton unknown	Molars without paraconid (due to wear, inferred from the reduced size of the trigonid)

Note: *Not recorded at Gran Barranca.

synonym *Enneoconus parvidens* came from the locality known as "Oeste de Río Chico" (West of Río Chico) near Cañadón Vaca (Simpson 1967b). All the remains are from the Casamayoran SALMA, Barrancan Subage (middle Eocene).

Remarks *Ernestokokenia marginata* and *E. nitida* were referred respectively as the m1 and an m2 of the same species (Simpson 1948). *Enneoconus parvidens*, only known from an isolated left M3 (MACN 10725), was considered as a junior synonym of *E. nitida* on the basis of the perfect fit of these teeth in direct occlusion (Gelfo 2006).

Ernestokokenia patagonica Ameghino 1901
[*Euprotogonia patagonica* Ameghino 1901, p. 375;
Notoprotogonia patagonica Ameghino 1904a, p. 336;
Protogonia (Euprotogonia) patagonica Gaudry 1904,
p. 8; *Ernestokokenia trigonalis* Simpson 1948, p. 111;
Euprotogonia trigonalis Ameghino 1901, p. 375;
Notoprotogonia trigonalis Ameghino 1904a, p. 336.]
   Type species MACN 10687 isolated right M2.
Types of the synonyms *Ernestokokenia trigonalis*, MACN 10688 right M1.
Diagnosis Medium-sized didolodontid with well-developed labial and lingual cingulum. Base of the protocone surrounded by the lingual cingulum. Premetaconular and postparaconular crista present. Preparaconular crista contacting the precingulum at the parastyle. Postmetaconular crista reduced. No mesostyle developed on the labial edge.
Occurrence The few known remains are from Gran Barrancan Member of the Sarmiento Formation. Casamayoran SALMA, Barrancan Subage (middle Eocene).
Remarks Simpson (1948) recognized these species as new combinations of *Ernestokokenia*. *Ernestokokenia trigonalis* was later considered as a junior synonym of *E. patagonica* (Gelfo 2006).

Discussion

Cladistic analysis

The phylogenetic analysis resulted in two most parsimonious trees of 77 steps, the strict consensus of which is 78 steps long (Fig. 8.3). In contrast to previous analyses (Muizon and Cifelli 2000; Gelfo 2004) in which the Kollpaniinae appeared as paraphyletic, the consensus recovered them as monophyletic and supported by two derived characters: 23 (1) and 25 (1). The Didolodontidae are also monophyletic and supported by three characters: 10 (1), 12 (2), and 18 (1). All the didolodontids share a paraconule and metaconule located closer to a middle position respect to the lingual and labial edges (10), and the subequal size of the paracone and metacone in the M2 (18).

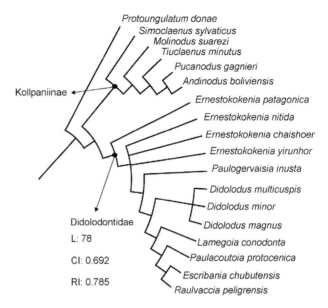

Fig. 8.3. Strict consensus of two most parsimonious trees (L:77) using the implicit enumeration option of TNT software.

In contrast, the development of the postmetaconular crista associated to the postcingulum (12) seems to be variable. Differing from Cifelli (1983a, 1993), in this analysis the presence of hypocone in the M3 is not a derived character, even when is present in all the Didolodontidae considered here, in contrast to the Kollpaniinae. The *Ernestokokenia* species are paraphyletic, probably due to the large amount of missing characters. *Paulogervaisia* appears as a sister group of two clades, one of them an unresolved polytomy with the *Didolodus* species (Fig. 8.3), supported by six synapomorphies: 5 (2), 7 (2), 14 (0), 28 (1), 29 (1), and 30 (0). The new taxon described here is thus upheld as part of the genus. The last clade joins the Itaboraian taxa *Lamegoia* and *Paulocoutoia* consecutively as sister taxa of *Escribania* and *Raulvaccia*. To sum up, this phylogenetic analysis succeeds in the recognition of the Kollpaniinae and the Didolodontidae as monophyletic groups, and also supports the assignation of the new species to *Didolodus*; but fails to obtain good resolution for the *Ernestokokenia* species.

The bunodont mammals up to the Eocene–Oligocene transition

During the Paleogene, several South American ungulates (e.g. Litopterna and Notoungulata) while retaining the primitive brachydont molars, developed some differences that could be related with trophic roles. Prima facie browsers and omnivorous ungulates can be distinguished by having lophed and bunodont tooth structure respectively. The vast majority of litopterns and notoungulates developed a lophed molar pattern, with high-relief cusps extended into cutting ridges, and most tended to increase their crown height. In contrast, the primitive Kollpaniinae pattern was specialized in didolodontids by rounding the cusps, adding new ones (e.g. hypocone, mesostyle, and protostyle) and developing low-relief cusps. Despite the low diversity of Litopterna Protolipternidae from the early Eocene (Cifelli 1983a; Gelfo 2006), didolodontids were the most diverse among the strictly bunodont ungulates in the Paleogene of South America. However, in addition to ungulates, a wide range of bunodont marsupials were also present (Case *et al.* 2005).

Despite the gaps in the South American fossil record a general trend for bunodont ungulates may be traced through the Paleogene. There were two maxima of bunodont ungulate diversity. The first one coincides with the most pronounced Cenozoic warming trend, which peaked with the early Eocene Climatic Optimum (Zachos *et al.* 2001). The Itaboraian SALMA, which seems to match a late Paleocene – early Eocene span, comprises at least two Protolipternidae (*Miguelsoria* and *Protolipterna*) and five Didolodontidae. Particularly in Patagonia, this climate optimum coincides with the occurrence of rainforest floras, including angiosperms with warm–temperate affinity (Barreda and Palazzesi 2005), hence, it is possible to infer the availability of large variety of resources for bunodont omnivorous mammals. Mainly small-sized didolodontids were recorded during this phase. In fact, even though large-sized taxa are known from Peligran (e.g. *Escribania chubutensis*) and Itaboraian (e.g. *Lamegoia conodonta*) SALMAs, the most frequent bunodont ungulates in the warm Paleocene and early Eocene were small, "rat-like" in size.

Larger-sized didolodontids dominated the Casamayoran and Mustersan SALMAs, coincidently with the trend toward cooler conditions that occurred from the early–middle Eocene to the late Eocene (Zachos *et al.* 2001). The Barrancan Subage of the Casamayoran also coincides with the second high-diversity peak of bunodont ungulates, which included seven species of didolodontids. Among them, at least two lineages, namely *Didolodus* and *Ernestokokenia*, independently evolved a molar structure comparable to that of the North American "condylarths" *Phenacodus* and *Tetraclaenodon* respectively.

The diversity of didolodontid taxa decreased drastically during the Mustersan to only two taxa (i.e. *Xesmodon langi* and cf. *Paulogervaisa*), and no didolodontids are known for the Tinguiririca SALMA. Apart from *Salladolodus* from the Deseadan of Bolivia, a taxon doubtfully assigned to this family, no didolodontids are recorded after the early Oligocene.

A worldwide decline of "archaic ungulates" occurred after the middle Eocene, following the global drop in temperature due to changes in ocean circulation, and the development of cold bottom-water currents from the poles (Zachos *et al.* 2001). Bunodont ungulates seem to have been particularly affected by major extinctions during the middle and late Eocene in North and South America (Janis 2000), and at the Eocene–Oligocene transition (EOT) in Europe (Hooker *et al.* 2004). After the EOT, the most successful bunodont ungulate groups to date have been the artiodactyl families Tayassuidae and Suidae, which occur in faunas from both tropical and temperate habitats. However, these taxa did not reach South America until the Pliocene (peccaries) and historical times (pigs) (Stehli and Webb 1985). Thus, after the extinction of Didolodontidae and before the entrance of artiodactyls, the only strictly bunodont ungulates recorded in South America were the Proterotheriidae Megadolodinae, a scarce group of litopterns represented by only two taxa from the Miocene of Colombia (Cifelli and Guerrero 1997) and Venezuela (Carlini *et al.* 2006).

ACKNOWLEDGEMENTS

I am grateful to Guiomar Vucetich, Alfredo Carlini, and Richard Madden for inviting me to participate in this book. I also thank Cecilia Morgan who improved the English, Guillermo Lopez and Mariano Bond for the very productive discussions, and Agustín Viñas who drew the figures. Hans Thewissen made comments on an earlier version of this chapter and Richard Kay edited this chapter. All of these scientists do not necessarily agree with the contents of this work, for which I am solely responsible.

Appendix 8.1 Description of characters

0 postprotocrista of P3 (0) P3 with incipient or small protocone and non-expanded distal border, (1) small to medium sized protocone present and postprotocrista expanded posteriorly; **1** metacone of P4 (0) absent, (1) small projection of the postparacrista; **2** outline of P4 (0) triangular, (1) quadrangular; **3** hypocone on M2 (0) absent, (1) small lingual cusp in the postcingulum, (2) large cusp distal or distolingual to the protocone (3) large cusp apprised to the protocone; **4** hypocone in M3 (0) absent, (1) present; **5** labial cingulum M1–2 (0) reduced to a labial rim with no separation of the bases of paracone and metacone, (1) interrupted labially to the paracone, (2) interrupted labially to the mesostyle, (3) well developed mesiodistally; **6** labial cingulum M3 (0) absent, (1) reduced to a labial rim with no separation of the bases of paracone and metacone, (2) interrupted labially to the paracone, (3) interrupted labially to the mesostyle, (4) well developed mesiodistally, (5) interrupted labial to the metacone; **7** mesostyle (0) absent, (1) small, (2) large; **8** parastyle–stylocone relationship (0) separated, (1) fused; **9** size of the conules (0) small to medium, (1) large; **10** position of the conules (0) closer to the labial side of the protocone, (1) in a middle position; **11** preparacrista (0) bent labially, (1) projected mesially, (2) bent lingually; **12** crista postmetaconular (0) in contact with the postcingulum distally, (1) in contact with the postcingulum at the metastyle, (2) reduced or absent; **13** crista preparaconular (0) not related with the precingulum, (1) in contact with the precingulum at the parastyle, (2) projected between the parastyle and paracone, (3) reduced or absent;

14 crista premetaconular (0) present, (1) absent; **15** crista postparaconular (0) present, (1) absent; **16** protostyle (0) absent, (1) small, (2) large; **17** protocone (0) not mesio-distally expanded, (1) mesio-distally expanded, (2) not expanded but with a wide base; **18** paracone and metacone (0) paracone larger and higher than the metacone, (1) both cusps subequal; **19** mandibular symphysis (0) not fused, (1) ankylosed; **20** p4 metaconid (0) small and appressed to the protoconid, (1) enlarged and separated from the protoconid; **21** p4 talonid (0) formed by a simple cusp, (1) presence of hypoconulid in labial position and incipient talonid; **22** paracristid (0) not distally arched (1) distally arched; **23** posterior slope of the metaconid (0) not inflated or invading the talonid, (1) strongly inflated, invading the talonid basin; **24** m1–2 entoconid and hypoconulid relation (0) distantly separated, (1) almost fused, (2) near but not fused; **25** m1–2 hypoconid (0) comprises the labial half or less of the talonid, (1) large extending also in the lingual half of the talonid; **26** m1–2 entoconid size, (0) smaller than the hypoconulid, (1) subequal, (2) larger than the hypoconulid; **27** m3 entoconid (0) smaller than the hypoconulid, (1) subequal, (2) larger than the hypoconulid; **28** m1–2 cristid obliqua (0) strong, (1) with centroconid, (2) reduced or almost absent; **29** m3 cristid obliqua (0) strong, (1) with centroconid, (2) reduced or almost absent; **30** centrocrista (0) labially projected, (1) mesiodistally straight, (2) bent labially; **31** M2 precingulum (0) more or less straight (1) strongly distally concave; **32** P2–3 metacone (0) absent, (1) smaller than paracone, (2) subequal to the paracone.

Appendix 8.2   Data matrix

Taxa	Characters 0–9	10–19	20–29	30–32
Protungulatum donae	0001034000	0011??0000	0000202?00	100
Molinodus suarezi	0000031010	0113110100	1011110000	10?
Tiuclaenus minutus	0000001010	0103100110	0011110022	100
Pucanodus gagnieri	0000001010	0103100100	001[01]111122	100
Simoclaenus sylvaticus	???0034010	0111100100	1011012100	10?
Andinodus boliviensis	??????????	??????????	??11111022	???
Escribania chubutensis	???3135011	110111121?	??11001000	11?
Lamegoia conodonta	1112101011	110311221?	1110001?0?	100
Paulacoutoia protocenica	1112110011	1103111210	1110012000	100
Didolodus multicuspis	1112123211	11[02]0012211	1110001011	001
D. minor	1112?23211	11[02]00122??	1110001?11	002
D. magnus	???212?211	120001221?	??10001???	00?
Paillogervaisia inusta	???21??110	120011221?	1110???100	20?
Ernestokokenia nitida	???2112010	1[12]2211221?	??10212000	10?
E. chaishoer	????21?010	11201122??	??10201000	10?
E. yirunhor	????1?2010	11201102??	??10201?0?	10?
E. patagonica	???2?3?010	112100121?	??????????	10?
Raulvaccia peligrensis	???3?1?011	12010112??	??10001?0	11?

140

REFERENCES

Abel, O. 1928. Unterordnung: Litopterna. In Weber, M. (ed.), *Die Säugetiere*, 2nd edn, vol. 2. Jena: Gustav Fisher, pp. 695–700.

Ameghino, F. 1897. Mammifères crétacés de l' Argentine: deuxième contribution à la connaissance de la faune mammalogique des couches à Pyrotherioum. *Boletín Instituto Geográfico Argentino Buenos Aires*, **18**, 406–429, 431–521.

Ameghino, F. 1901. Notices préliminaires sur des ongulés nouveaux des terrains crétacés de Patagonie. *Boletín de la Academia Nacional de Ciencias en Córdoba*, **16**, 350–426.

Ameghino, F. 1902a. Notices préliminaires sur des mammifères nouveaux des terrains crétacés de Patagonie. *Boletín de la Academia Nacional de Ciencias en Córdoba*, **17**, 5–70.

Ameghino, F. 1902b. Líneas filogenéticas de los proboscídeos. *Anales del Museo Nacional de Buenos Aires*, **8**, 19–43.

Ameghino, F. 1904a. Nuevas especies de mamíferos cretáceos y terciarios de la República Argentina. *Anales de la Sociedad Científica Argentina*, **56**, 193–209–8; **57**, 162–175, 327–341; **58**, 35–41, 56–71, 182–192, 225–240, 241–291.

Ameghino, F. 1904b. Recherches de morphologie phylogénétique sur les molaires supérieures des ongulés. *Anales del Museo Nacional de Buenos Aires*, **9**, 1–541.

Barreda, V. and L. Palazzesi 2005. Patagonian vegetation turnovers during the Paleogene-early Neogene: origin of arid-adapted floras. *Actas XVI Congreso geológico Argentino*, 445–446.

Carlini, A. A., J. N. Gelfo, and R. Sánchez 2006. First record of the strange Megadolodinae (Mammalia, Litopterna, Protherotheriidae). *Journal of Systematic Paleontology*, **4**, 279–284.

Case J., F. J. Goin, and M. Woodburne 2005. "South American" marsupials in the Late Cretaceous of North America and the Origin of Marsupial Cohorts. *Journal of Mammalian Evolution*, **12**, 461–494.

Cifelli, R. L. 1983a. The origin and affinities of the South American Condylarthra and early Tertiary Litopterna (Mammalia). *American Museum of Natural History Novitates*, **2772**, 1–49.

Cifelli, R. L. 1983b. Eutherian tarsals from the late Paleocene of Brazil. *American Museum of Natural History Novitates*, **2761**, 1–31.

Cifelli, R. L. 1985. Biostratigraphy of the Casamayoran, Early Eocene, of Patagonia. *American Museum of Natural History Novitates*, **2820**, 1–26.

Cifelli, R. L. 1993. The Phylogeny of the native South American Ungulates. In Szalay, F. S., Novacek, M. J., and McKenna, M. (eds.), *Mammal Phylogeny: Placentals*. Springer-Verlag: New York, pp. 95–216.

Cifelli, R. and J. Guerrero 1997. Litopterns. In Kay, R., Madden, R. H., Cifelli, R. L., and Flynn, J. J. (eds.), *Vertebrate Paleontology in the Neotropics: The Miocene Fauna of La Venta, Colombia*. Washington, DC: Smithsonian Institution Press, pp. 289–302.

Gaudry, A. 1904. Fossiles de Patagonie: dentition de quelque mammifères. *Mémoires de la Societé Géologique de France*, **12**(31), 1–43.

Gelfo, J. N. 2004a. A new South American mioclaenid (Mammalia Ungulatomorpha) from the Tertiary of Patagonia, Argentina. *Ameghiniana*, **41**, 475–484.

Gelfo, J. N. 2004b. The validity of genus *Paulogervaisia* (Mammalia Didolodontidae) from the Eocene of Patagonia, Argentina. *Ameghiniana*, **41**, 12R–13R.

Gelfo, J. N. 2006. Los Didolodontidae (Mammalia: Ungulatomorpha) del Terciario sudamericano: sistemática, origen y evolución. Ph.D. thesis, La Plata University.

Gelfo, J. N. and M. Tejedor 2004. Implicancias sistemáticas de nuevos restos de *Asmithwoodwardia subtrigona* Ameghino (Mammalia: Liptopterna?) del Paleógeno de Patagonia. *Ameghiniana*, **41**, 48R.

Goloboff, P., J. Farris, and K. Nixon 1999. *TNT: Tree Analysis using New Technology*. www.zmuc.dk/public/phylogeny/tnt.

Hooker J. J., M. E. Collinson, and N. P. Sille 2004. Eocene–Oligocene mammalian faunal turnover in the Hampshire Basin, UK: calibration to the global time scale and the major cooling event. *Journal of the Geological Society of London*, **161**, 161–172.

Janis, C. M. 2000. Patterns in the evolution of herbivory in large terrestrial mammals: the Paleogene of North America. In Sues, H. D. and Labanderia, C. (eds.), *Origin and Evolution of Herbivory in Terrestrial Vertebrates*. Cambridge, UK: Cambridge University Press, pp. 168–221.

López, G. M. 1997. Paleogene faunal assemblage from Antofagasta de La Sierra (Catamarca Province, Argentina). *Palaeovertebrata*, **26**, 61–81.

McKenna, M. C. and S. K. Bell 1997. *Classification of Mammals above the Species Level*. New York: Columbia University Press.

Muizon, C. de and M. Bond 1982. Les Phocidae (Mammalia) miocènes de le formation Paraná (Entre Ríos, Argentine). *Bulletin du Muséum National d'Histoire Naturelle, Sciences de la Terre*, **4**(3–4), 165–207.

Muizon, C. de and R. L. Cifelli 2000. The "condylarths" (archaic Ungulata, Mammalia) from the early Palaeocene of Tiupampa (Bolivia): implications on the origin of the South American ungulates. *Geodiversitas*, **22**, 47–150.

Osborn, H. F. 1910. *The Age of Mammals in Europe, Asia and North America*. New York: Macmillan.

Roth, S. 1899. Aviso preliminar sobre mamíferos mesozoicos encontrados en Patagonia. *Revista del Museo de La Plata*, **9**, 381–388.

Roth, S. 1927. La diferenciación del sistema dentario en los ungulados, notoungulados y primates. *Revista del Museo de La Plata*, **30**, 172–255.

Scott, W. B. 1913. *A History of Land Mammals in the Western Hemisphere*. New York: Macmillan.

Schlosser, M. 1923. 5. *Klasse: Mammalia – Säugetiere*. In Broili, F. and Sctilosser, M. (eds.) *Grundzüge der Paläontologie (Paläozoologie)*, vol. II, *Vertebrata*, 4th edn. Berlin: Oldenbourg, pp. 402–689.

Soria, M. F. 2001. Los Proterotheridae (Litopterna, Mammalia): sistemática, origen y filogenia. *Monografía Museo Argentino de Ciencias Naturales*, **1**, 1–167.

Simpson, G. G. 1934. Provisional classification of extinct South American hoofed mammals. *American Museum of Natural History Novitates*, **775**, 1–29.

Simpson, G. G. 1935. Descriptions of the oldest known South American mammals from Río Chico formation. *American Museum of Natural History Novitates*, **793**, 1–25.

Simpson, G. G. 1948. The beginning of the Age of Mammals in South America. I. *Bulletin of the American Museum of Natural History*, **91**, 1–232.

Simpson, G. G. 1964. Los Mamíferos Casamayorenses de la Colección Tournoüer. *Revista de Museo Argentino Ciencias Naturales, Paleontología*, **1**, 1–21.

Simpson, G. G. 1967a. The beginning of the Age of Mammals in South America. II. *Bulletin of the American Museum of Natural History*, **137**, 1–259.

Simpson, G. G. 1967b. The Ameghinos' localities for early Cenozoic mammals in Patagonia. *Bulletin of the Museum of Comparative Zoology*, **136**(4), 63–76.

Stehli, S. and D. Webb 1985. *The Great American Biotic Interchange*. New York: Plenum.

Van Valen, L. 1978. The beginning of the Age of Mammals. *Evolutionary Theory*, **4**, 45–80.

Zachos, J., M. Pagani, L. Sloan, E. Thomas, and K. Billups 2001. Trends, rhythms, and aberrations in global climate 65 Ma to present. *Science*, **292**, 686.

9 The Notohippidae (Mammalia, Notoungulata) from Gran Barranca: preliminary considerations

Guillermo M. López, Ana Maria Ribeiro, and Mariano Bond

Abstract

The Notohippidae include notoungulates from Casamayoran to the "*Piso Notohippidense*" of the early Santacrucian. The oldest forms have brachydont to mesodont dentitions, while later forms attain hypsodonty and some taxa from the late Oligocene to middle Miocene evolved a thick coating of external cement on their teeth. The greatest diversity of taxa has been found in the Deseadan localities of La Flecha, Cabeza Blanca, and Scarritt Pocket in Argentine Patagonia. New material of Notohippidae has been recovered from several stratigraphic levels of different ages at Gran Barranca. GBV-3 "El Rosado" (Rosado Member) from the Mustersan SALMA has yielded material referred to *Puelia* with some doubt. GBV-4 "La Cancha" (Vera Member) of Tinguiririan SALMA has yielded fragmentary remains that suggest considerable diversity, including material referred to *Puelia plicata*, *Puelia* sp., *Eomorphippus obscurus*, *Eomorphippus pascuali*, and *Eomorphippus* sp. GBV-19 "La Cantera" (Unit 3, Upper Puesto Almendra Member) of pre-Deseadan aspect has yielded two new notohippids: *Patagonhippus canterensis* n. gen. et sp. and *Patagonhippus dukei* n. gen. et sp. Both species are closely related to well-known taxa from La Flecha and Cabeza Blanca, but are more primitive morphologically. Finally in the Lower Fossil Zone of the Colhue-Huapi Member, of Colhuehuapian SALMA age, occurs *Argyrohippus* cf. *A. boulei*.

Resumen

Los Notoungulata Notohippidae se registran desde niveles referibles a la SALMA Casamayorense hasta el "Piso Notohippidense" de la SALMA Santacrucense. Los Notohippidae basales presentan dentición brachydonta-mesodonta, mientras que las formas terminales son claramente hypsodonctas y pastadoras. Algunos taxones del Oligoceno tardío-Mioceno desarrollan una gruesa cobertura de cemento en sus dientes. La mayor diversidad ha sido registrada en las localidades deseadenses

de La Flecha, Cabeza Blanca y Scarritt Pocket (Patagonia, Argentina). Nuevos materiales de esta familia han sido recuperados de varios niveles estratigráficos de Gran Barranca. En el nivel GBV-3 "EL Rosado" (Miembro Rosado, SALMA Mustersense) se registra *Puelia?*. El GBV-4 "La Cancha" (Miembro Vera, SALMA Tinguiririquense) presenta una considerable diversidad, incluyendo materiales referidos a *Puelia plicata*, *Puelia* sp., *Eomorphippus obscurus*, *Eomorphippus pascuali* y *Eomorphippus* sp. En el nivel GBV-19 "La Cantera" (Unit 3, Miembro Puesto Almendra Superior, Deseadense temprano?) se registran dos nuevos Notohippidae: *Patagonhippus canterensis* n. gen. et sp. y *Patagonhippus dukei* n. gen. et sp. Ambas especies están relacionadas con las registradas en las localidades de La Flecha y Cabeza Blanca, pero morfológicamente son más generalizadas. Finalmente en la Zona Fosilífera Inferior del Miembro Colhue-Huapi (SALMA Colhuehuapense) se registra *Argyrohippus* cf. *A. boulei*.

Introduction

The Notohippidae are medium-sized notoungulates with a distinctive cheek tooth occlusal morphology of "toxodontoid" pattern (see for example Simpson 1932) but differing from the Toxodontidae in the retention of all upper incisors and without developing "tusk-like" I2 and I3. The notohippids range in age from Casamayoran to the "*Piso Notohippidense*" of the early Santacrucian South American Land Mammal Age (SALMA). The oldest taxa in the family have brachydont dentitions and later forms attain hypsodonty. Also some specialized later taxa display a thick layer of external cementum around the perimeter of the crown. Judging broadly by cheek tooth crown height, some early notohippids may have been browsers but most have been interpreted as open-country grazers based on the hypsodonty and cementum. The best-known notohippids are from the Deseadan SALMA of Patagonia. Recent discoveries at Gran Barranca warrant a review of the taxonomy of the Patagonian members of the family to achieve a better understanding of their phylogenetics and adaptive diversity.

The Paleontology of Gran Barranca: Evolution and Environmental Change through the Middle Cenozoic of Patagonia, eds. R. H. Madden, A. A. Carlini, M. G. Vucetich, and R. F. Kay. Published by Cambridge University Press. © Cambridge University Press 2010.

Eocene notohippids (i.e. *Pampahippus*, *Plexotemnus*, *Puelia*, *Eomorphippus*, and *Trimerostephanus*) are the least well-known members of the family and generally comprise the smaller Toxodontia in paleofaunas where they occur (Bond and López 1993). The Patagonian *Plexotemnus*, *Puelia*, and *Eomorphippus* have had a long and complex taxonomic history (Simpson 1967). The family attained peak diversity in the Deseadan (i.e. *Morphippus*, *Rhynchippus*, *Interhippus*, *Nesohippus*, *Eurygenium*, *Coresodon*, *Pascualihippus*, and *Moqueguahippus*) when their range extended from Patagonian Argentina to Bolivia, Brazil, and Peru. The family quickly declined thereafter such that by the early Miocene Colhuehuapian SALMA only *Argyrohippus* is known. The last surviving representative, the rather poorly known *Notohippus*, was found in the "Notohippidense" fauna at Karaiken in western Patagonia before the turn of the twentieth century (Ameghino 1891; Ribeiro and Bond 1999). In those taxa where the postcranium is known, notohippids were quadrupedal terrestrial herbivores displaying some specializations for cursorial locomotion (Shockey 1997a, 1997b).

Notohippid dental morphology is generally described as being similar to that of the horse, that is, with an arcade of incisors and canines that form a functionally unified homomorphic "cropping" battery set in a smoothly rounded anterior rostrum, and unilaterally hypsodont cheek teeth with a complex pattern of enamel folds and fossettes variably obliterated by wear. In pre-Miocene taxa the dentition is complete and closed, but the anterior teeth become separated from the cheek teeth by a diastema and a thick layer of external cementum envelops the crowns in taxa from the Colhuehuapian SALMA (*Argyrohippus*) and "Piso Notohippidense" (*Notohippus*). A newly described notohippid from the Deseadan SALMA of southern Peru also displays external cementum on the crown (Shockey *et al.* 2006).

Ameghino (1906) included Notohippidae in Hippoidea and recognized 14 genera and many species among the South American branches of the superfamily. Loomis (1914) abandoned the family term Notohippidae and proposed the family Rhynchippidae (suborder Toxodontia, order Notoungulata) for the Deseadan genera *Rhynchippus*, *Morphippus*, and *Eurygenium*. He included in Nesodontidae the Deseadan taxa *Coresodon*, *Interhippus*, and *Nesohippus*.

Simpson (1932) followed Loomis (1914) and recognized the family Rhynchippidae, but differed with Ameghino (1897) on the relationship between *Rhynchippus* and *Notohippus* by including *Argyrohippus* together with *Notohippus* in another family, the Notohippidae. By contrast, Patterson (1934) did not recognize Rhynchippidae, and returned the genera that Loomis (1914) included in Nesodontidae to the Notohippidae.

Simpson (1945) proposed two subfamilies Notohippinae and Rhynchippinae, and included *Interhippus*, *Nesohippus*,

Argyrohippus, *Stilhippus*, *Perhippidium*, and *Notohippus* in the former and *Pseudostylops*, *Morphippus*, *Rhynchippus*, and *Eurygenium* in the latter.

Patterson simplified the taxonomy of the Deseadan and Colhuehuapian notohippids in his unpublished revision and synonymy of the Ameghino collection (Patterson 1952). Simpson (1967) and Patterson (in Simpson 1967) further simplified the taxonomy of the oldest members of the family by recognizing *Pseudostylops* to be a synonym of *Eomorphippus*, and *Interhippus* a synonym of *Coresodon*. In the case of *Nesohippus*, Simpson (1967, p. 180) believed it to have been listed by Ameghino (1906) mistakenly "because no Mustersan species was ever described, there are in the Mustersan specimens in the Ameghino Collection labeled as of this genus, and there is no other mention of its occurrence in the Mustersan."

When Bond and López (1993) described *Pampahippus arenalesi* from Lumbrera Formation (Salta Province, northwestern Argentina), the isotemnid-like taxa *Plexotemnus* and *Puelia* were included in Notohippidae. Other Paleogene genera (e.g. *Coelostylodon* and *Trimerostephanus*) were regarded as notohippids. This proposal was discussed by Shockey (1997b).

The record of notohippids outside Argentina is less well known. In Brazil, Soria and Alvarenga (1989) described *Rhynchippus brasiliensis* and a Notohippidae indet. from the Oligocene Tremembé Formation in the Taubaté Basin of São Paulo State. Paula-Couto (1981, 1982) described *Notohippus* sp. and *Purperia cribatidens* from localities 48 and 28, respectively, in the Miocene Solimões Formation along the Juruá River in Acre State. Paula-Couto's specimens are considered Notohippidae indet. and Leontiniidae respectively by Ribeiro and Bond (2000).

Shockey (1997a, 1997b) described *Pascualihippus* from the Deseadan of Salla, Bolivia, whose rostral morphology seems to suggest a close relationship with the origin of the Toxodontidae. Based on a phylogenetic analysis he inserted toxodontids within the paraphyletic Notohippidae. This clade also has a close relationship with the Leontiniidae. *Pampahippus*, *Plexotemnus*, and *Puelia* are considered to be Notohippidae *sensu lato* when Leontiniidae are included. These results differ from Cifelli (1993) who viewed Oligocene to Miocene notohippids as comprising a monophyletic clade. Shockey (1997a, 1997b) described two other taxa from Salla: *Eurygenium pacegnum* and *Rhynchippus brasiliensis*.

Shockey *et al.* (2004) described a very small m1 (or m2) (smaller than *Rhynchippus pumilus*) as cf. Notohippidae. The specimen is from the west bank of the Juruá River near its confluence with the Rio Breu, in Ucayali Department, Peru. Recently, Shockey *et al.* (2006) described *Moqueguahippus glycisma* from southern Peru, thus increasing the geographic distribution of this group during the Deseadan.

Table 9.1. *Notohippidae species reported at Gran Barranca from the literature before this chapter*

Species	Current status[a]	SALMA[b]
Argyrohippus boulei Ameghino 1901	*Argyrohippus boulei* (Patterson 1952, *in litteris*)	Co
Argyrohippus fraterculus Ameghino 1901	*Argyrohippus boulei* (Patterson 1952, *in litteris*)	Co
Eomorphippus obscurus Ameghino 1901	*Eomorphippus obscurus* (Simpson 1967)	M
?Eomorphippus pascuali Simpson 1967	*?Eomorphippus pascuali* (Simpson 1967)	M
Interhippus deflexus Ameghino 1902	*Nomen dubium* (Simpson 1967)	?T
Perhippidion tetragonidens Ameghino 1904	*Argyrohippus boulei* (Patterson 1952, *in litteris*)	Co
Plexotemnus complicatissimus Simpson 1970	*Plexotemnus complicatissimus* (Bond and López 1993)	C
Pseudhippus tournoueri Ameghino 1902	*Argyrohippus boulei* (Patterson 1952, *in litteris*)	Co
Pseudostylops subquadratus Ameghino 1901	*Eomorphippus obscurus* (Simpson 1967)	M
Stilhippus deterioratus Ameghino 1904	*Argyrohippus boulei* (Patterson 1952, *in litteris*)	Co
Trimerostephanus coalitus Ameghino 1901	*Trimerostephanus coalitus* (Bond and López, 1993)	M
	Patagonhippus canterensis n. gen. et sp.	?De
	Patagonhippus dukei n. gen. et sp.	?De
	Puelia plicata, this book	T
	Puelia sp., this book	T

Notes: [a]Bibliographical references in parentheses.
[b]C, Casamayoran; De, Deseadan; Co, Colhuehuapian; M, Mustersan; T, Tinguiririan.

The greatest diversity of notohippid taxa is found in the Deseadan of Argentine Patagonia at localities such as La Flecha, Cabeza Blanca, Scarritt Pocket, and Gran Barranca (see Table 9.1). Here we describe new material of Notohippidae from the Sarmiento Formation at Gran Barranca. This material was collected at GBV-3 "El Rosado", GBV-4 "La Cancha", GBV-19 "La Cantera", and from levels of the Colhue-Huapi Member.

The Sarmiento Formation (middle Eocene – early Miocene) at Gran Barranca is the most complete continental sequence of middle Cenozoic mammal-bearing sedimentary rocks anywhere in South America, and serves as the standard reference section for this interval. Based on the stratigraphic occurrences of index and guide taxa, at least six different faunal zones can been differentiated (Carlini *et al.* 2005; Bellosi this book).

Three levels at Gran Barranca are of major importance for this work: localities GBV-3 "El Rosado" and GBV-4 "La Cancha," and GBV-19 "La Cantera." GBV-4 lies stratigraphically above GBV-3 and is separated from it by several stratigraphic discontinuities (Bellosi this book). In turn GBV-19 is younger than GBV-4 based on radiometric, paleomagnetic, and stratigraphic evidence (Ré *et al.* Chapter 3, this book; Bellosi this book). Locality GBV-3 is referred to the Mustersan SALMA on the basis of the composition of its native ungulates (Madden *et al.* this book). The locality contains some notoungulate taxa characteristic of the Mustersan SALMA but also has taxa not found in the Mustersan but characteristic of the Tinguiririan SALMA (e.g. *Archaeotypotherium propheticus*, *Protarchaeohyrax gracilis*, and *Proargyrohyrax* sp.) and also

includes other toxodontian families of a Deseadan age (López *et al.* 2005). The marsupial fauna of GBV-4 closely resembles that of the Tinguiririran (Goin *et al.* this book) and the absence of rodents is significant (Vucetich *et al.* this book). Overlying the Tinguiririran level is GBV-19. GBV-19 contains a mix of Deseadan and pre-Deseadan elements (Goin *et al.* this book; Carlini *et al.* this book; Vucetich *et al.* this book; see also Bond *et al.* 2004), befitting its intermediate stratigraphic and temporal position in the Gran Barranca sequence.

Abbreviations

Specimens of notohippids were studied in the following collections (given with their abbreviations used in the text): FMNH, Field Museum of Natural History, Chicago, USA; MACN, Museo Argentino de Ciencias Naturales "Bernardino Rivadavia," Buenos Aires; MLP, Museo de La Plata, La Plata; MNHN-Bol, Museo Nacional de Historia Natural de Bolivia; MPEF-PV, Museo Paleontológico "Egidio Feruglio," Trelew; UNPSJB-PV, Universidad Nacional de la Patagonia "San Juan Bosco". Dental terminology follows the convention of using upper-case letters for the upper incisors, canines, premolars, and molars, lower-case for the lower series.

Systematic paleontology

Order **NOTOUNGULATA** Roth 1903
Suborder **TOXODONTIA** Owen 1853
Family **NOTOHIPPIDAE** Ameghino 1894
Genus ***Puelia*** Roth 1902
Puelia plicata Roth 1902

Material MPEF-PV 6215, upper molars, MPEF-PV 6186, M2.
Locality Gran Barranca, GBV-4 "La Cancha" level, Vera Member, Sarmiento Formation.

Puelia sp.

Material MPEF-PV 6170, right and left upper molars.
Locality Gran Barranca, GBV-4 "La Cancha" level.
***Puelia?* sp.**
Material and localities MPEF-PV 6713, jugal and maxilla fragments, MPEF-PV 6853, left upper M1 from Gran Barranca, GBV-3 "El Rosado," Rosado Member, Sarmiento Formation. MPEF-PV 5732, upper molar, MPEF-PV 6180, P4 and M1, MPEF-PV 6200, right lower p3, MPEF-PV 6206, left upper M1 from Gran Barranca, GBV-4 "La Cancha" level.
Comments To date, *Puelia* has been reported from the Mustersan localities of Cerro del Humo, Sierra Talquino, Gran Hondonada, and Laguna del Mate. The materials here presented are the first record of the genus at Gran Barranca, and the discovery of more complete remains may permit recognition of a new species, as well as the extension of the stratigraphic distribution of the genus to the Tinguiririran SALMA. Some even more fragmentary material from GBV-3 El Rosado (MPEF-PV 6713 and MPEF-PV 6853) may be related to *Puelia?* sp.

The problem of the systematic arrangement of *Puelia* is complex. Bond and López (1993) recognize this taxon as a form related to *Pampahippus* and *Plexotemnus*, two Casamayoran genera that they considered basal Notohippidae (as does Shockey 1997b). Simpson (1967) was well acquainted with the similarities that his "*Acoelohyrax*" group (including what we are calling now *Plexotemnus* and *Puelia*) had with the earlier Notohippidae such as *Eomorphippus*. Confronted with the possibility that his "*Acoelohyrax*" group could be either an early offshoot of the Notohippidae or Isotemnidae evolving in parallel, Simpson (1967) thought their many resemblances possibly but unlikely convergent and stated that "the difference between the two interpretations is rather formal and not very important" a somewhat surprising statement given the phylogenetic consequences. Bond and López (1993) evaluating the characters mentioned by Simpson (1967) and other not known to him for *Plexotemnus* (such as the incisor region) concluded that *Pampahippus*, *Plexotemnus*, and *Puelia*, could be clearly segregated from the Isotemnidae and considered early representatives of the Notohippidae.

Bond and López (1993) recognized three valid species within genus *Puelia*, namely *P. plicata* Roth 1902, the larger-sized *P. coarctatus* Ameghino 1901, and *P. sigma* Ameghino 1901. However, the validity of the latter species remains uncertain, and it could be a junior synonym of *P. plicata*. Several specimens at GBV-4 "La Cancha" are referred to *Puelia*. MPEF 6215 is a left M1 or M2 that is referred to *Puelia plicata*, and represents the youngest record for this species, formerly known only from the Mustersan SALMA. The type and other referred specimens come from other Mustersan localities (e.g. Cerro del Humo, in Chubut: Simpson 1967). Some fragmentary upper and lower molars (i.e. MPEF-PV 6170) also from GBV-4 are referred to *Puelia* sp.

Genus *Eomorphippus* Ameghino 1901
Eomorphippus obscurus Ameghino 1901
Fig. 9.1A–B.

Material MPEF-PV 6154, isolated upper and lower premolars and molars; MPEF-PV 6188, left lower m3; MPEF-PV 6197, isolated lower incisor; MPEF-PV 6791, isolated upper incisor; MPEF-PV 7945, left dentary fragment with p4–m3.
Locality Gran Barranca, GBV-4 "La Cancha" level.
***Eomorphippus* sp.**
Material MPEF-PV 6217, isolated upper premolars and molars.
Locality Gran Barranca, GBV-4 "La Cancha" level.
Comments The genus *Eomorphippus* presents a suite of characters including mesodonty and the absence of external cementum that place it in an intermediate position phylogenetically between the Casamayoran–Mustersan taxa on the one hand (e.g. *Pampahippus*, *Plexotemnus*, *Puelia*) and those of the Deseadan on the other hand (e.g. *Rhynchippus, Morphippus*). This intermediate phylogenetic position agrees with stratigraphic superposition at Gran Barranca, between the Mustersan levels with *Puelia* and pre-Deseadan (GBV-19) and Deseadan levels with more progressive notohippids. *Eomorphippus* occurs exclusively in Tinguiririran SALMA levels of Patagonia (Gran Barranca and Cañadón Blanco) and Chile. In addition to materials collected by Simpson in Profile M at Gran Barranca and Feruglio from the same bed in Profile K, we identify new materials (e.g. a dentary fragment, upper and lower incisors, as well as isolated molars) from GBV-4 in Profile K at Gran Barranca. The GBV-4 is a manganese-rich locality and the fossils are covered with manganese oxides. Much of the material described by Ameghino from "une partie supérieure" of the *Astraponoteen* (see Reguero *et al.* 2004) is also covered with manganese (e.g. *Pseudopachyrucos foliiformis* Ameghino 1901 and *Interhippus deflexus* Ameghino 1904).

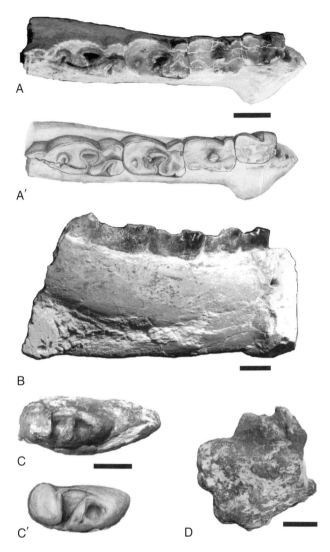

Fig. 9.1. (A, A′, B) *Eomorphippus obscurus* Ameghino 1901; MPEF-PV 7945, left dentary fragment with p4–m3 in occlusal and lingual views. (C, C′, D) *Eomorphippus pascuali* Simpson 1967; MPEF-PV 6243, right m2 isolated in occlusal and labial views. Scale bar 1 cm.

Eomorphippus pascuali Simpson 1967
Material MPEF-PV 6159, right upper M1–2 and left upper M3, MPEF-PV 6243, right m2.
Locality Gran Barranca, GBV-4 "La Cancha" level.
Comments Simpson (1967, plate 41, figures 13–15) described a new species of Notohippidae collected by Coley Williams from the fluvial beds from near the top of his Profile M, and referred with doubt to the genus *Eomorphippus* as *E.? pascuali*, characterized among other things by its small size when compared to *E. obscurus*. *Eomorphippus pascuali* is clearly different morphologically from *E. obscurus*

and as Simpson (1967) opined, it could represent a new genus. Given our state of knowledge of this taxon, we maintain it as *E. pascuali*. *Eomorphippus pascuali* is larger than the two known species of *Puelia*, with higher upper molars crowns and deeper lingual clefts reaching nearly to the bases of the molar crowns. In these characters *E. pascuali* is more progressive than *Puelia*. In this light, the synonymy of *E. pascuali* to *Puelia* (Wyss *et al.* 2005) cannot be accepted. In the new collections made at Gran Barranca there are some upper molars that can be referred to *E. pascuali*, and an m3 (MPEF-PV 6243, Fig. 9.1C–D) very similar in size to the m3 in the lower jaw assigned by Simpson (1967) to *E. pascuali*. These crowns are lower than those of *E. obscurus*. The new material establishes the presence of *E. pascuali* at GBV-4 "La Cancha" level at Profile K. They also serve to show that this taxon represents a notohippid distinct and more progressive than *Puelia* and also distinct, but lower crowned, than *E. obscurus*. Pending revision of all Tinguirirican notohippids we prefer to maintain *E. pascuali* as a distinct taxon until its generic assignment can be fully resolved.

Genus *Patagonhippus* n. gen.
Type species *Patagonhippus canterensis* n. sp.
Referred species *Patagonhippus canterensis* n. sp. and *Patagonhippus dukei* n. sp.
Etymology From Patagonia, the geographic area where this taxon was found, and *hippus*, a suffix commonly used for notohippid genera.
Diagnosis Notohippidae with teeth lacking cement, as in all the Deseadan notohippids with the exception of *Argyrohippus praecox*, *Eurygenium latirostris*, and *Moqueguahippus glycisma*. Subtriangular upper premolars differ from the molarized premolars of *Pascualihippus*, *Rhynchippus*, and *Eurygenium*. Premolars and molars with paracone column and parastyle more pronounced than in *Rhynchippus*, *Coresodon*, and *Eurygenium* and with a straight ectoloph. Mesial cingulum in upper premolars as in *Rhynchippus* but not *Eurygenium*. M1–2 with a groove in the mesiolingual corner of the protocone that originates as a mesial expansion as in *Eurygenium*. M1–2 with first crista in the lingual ectoloph and central valley open lingually as distinct from *Pascualihippus*. M3 with central valley open distally similar to *Eurygenium*. Lower molars with deep groove on the lingual face and without posterior fossettid in the entolophid as in *Eurygenium*, *Morphippus*, and *Moqueguahippus* and different from *Rhynchippus* and *Coresodon*.

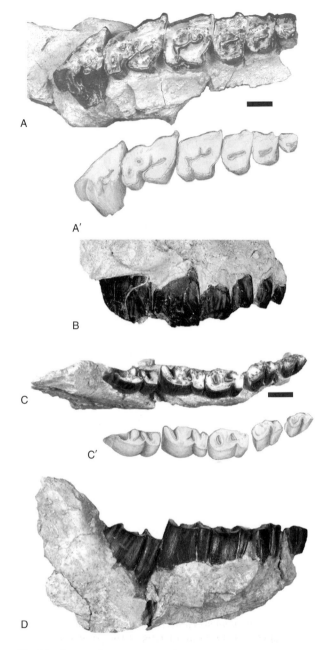

Fig. 9.2. *Patagonhippus canterensis* n. gen. et n. sp. (A–B) MPEF-PV 7087, right maxilla fragment with P2-M3 in occlusal and labial views. (C–D) MPEF-PV 6264, right dentary with p3–m3 in occlusal and labial views. Scale bar 1 cm.

Geographic and chronologic distribution Argentina, early? Deseadan SALMA.

Patagonhippus canterensis n. sp.
Fig. 9.2.
Holotype MPEF-PV 7087, maxilla fragments with right P2–M3 and left M2–3.

Referred material MPEF-PV 6264, right dentary with p3–m3 and left dentary with canine, p1 and m3; MPEF-PV 5775, left P2–M1; MPEF-PV 6112, left lower m2–3; MPEF-PV 6120, incisor; MPEF-PV 6136, left lower p4; MPEF-PV 6138, upper right P1 or P2; MPEF-PV 6239, left upper molar; MPEF-PV 6247, left lower p2; MPEF-PV 6255, dentary fragment with p4–m2; MPEF-PV 6272, right upper M1; MPEF-PV 6617, right dentary fragment with dp4, m1–2 and left dentary fragment with m1; MPEF-PV 6811, premolar; MPEF-PV 7501, P4–M2, MPEF-PV 7698, dentary fragment with m1; MPEF-PV 7720, left dentary with m2–3; MPEF-PV 7721, teeth; MPEF-PV 7729, mandible; MPEF-PV 7740, symphysis with teeth; MPEF-PV 7742, dentary fragment with teeth; MPEF-PV 7743, dentary fragment with teeth; MPEF-PV 7746, mandible; MPEF-PV 7762, teeth; MPEF-PV 7775, right maxilla with M1; MPEF-PV 7776, upper incisor; MPEF-PV 7777, dentary fragment with tooth; MPEF-PV 7834, teeth.

Type locality GBV-19 "La Cantera," Unit 3, Upper Puesto Almendra Member, Sarmiento Formation at Gran Barranca, Chubut Province, Argentina.

Etymology In reference to the La Cantera locality at Gran Barranca.

Diagnosis Similar in size to *Morphippus imbricatus* and 10% smaller than *Rhynchippus equinus*. P2 half the size of P3 (unlike *Rhynchippus equinus, Pascualihippus boliviensis, Eurygenium latirostris,* and *E. pacegnum*) and with protocone less developed than the hypocone. P3–4 subtriangular in outline; compared with other Deseadan species (*Pascualihippus boliviensis, Eurygenium latirostris,* and *E. pacegnum*), the upper premolars and molars are less hypsodont and have a more persistent mesial cingulum. Upper molars central valley opens lingually and the lingual groove is deep and persists into advanced stages of wear. Lower molars without posterior fossettid in the entolophid, as in *E. pacegnum* and *M. imbricatus* but differs from these species in having a deeper groove between metalophid and entolophid, and entolophid and hypolophid.

Patagonhippus dukei n. sp.
Fig. 9.3.
Holotype MPEF 6127, right dentary fragment with canine, p1–4, m1–3.

Type locality GBV-19 "La Cantera," Unit 3, Upper Puesto Almendra Member, Profile A, Sarmiento Formation at Gran Barranca, Chubut Province, Argentina.

Etymology For Duke University, North Carolina, in recognition of the contributions of its faculty to the knowledge of the Patagonian vertebrate fossils.

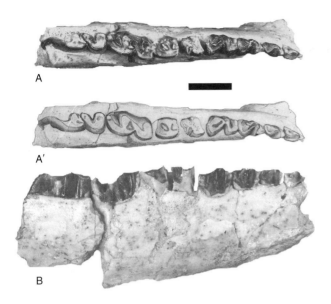

Fig. 9.3. *Patagonhippus dukei* n. gen. et n. sp. (A–B) MPEF-PV 6127, right dentary fragment with canine, p1–m3 in occlusal and lingual (reversed) views. Scale bar 1 cm.

Diagnosis Morphologically similar to *Patagonhippus canterensis* n. sp., but 50% smaller in size. Size comparable to *Rhynchippus pumilus*, but differs from this species because lacks a posterior fossettid in the entolophid.

Comments and discussion As commented previously, Deseadan notohippids include *Rhynchippus, Eurygenium, Morphippus, Coresodon, Argyrohippus praecox, Pascualihippus,* and *Moqueguahippus*. The first five taxa are known from the Deseadan of Patagonia and occur in localities with typical Deseadan faunas (e.g. Cabeza Blanca, La Flecha), whereas the latter are from tropical areas (Peru and Bolivia). The type material of *Morphippus imbricatus* (MACN A 52–76), *Coresodon scalpridens* (MACN A 52–1), and *Argyrohippus praecox* (FMNH P13334) comes from La Flecha, while the type material of the two species of *Rhynchippus* (i.e. *R. equinus* MACN A 52–31 and *R. pumilus* MACN A 52–61) comes from Cabeza Blanca. The type locality of *Eurygenium latirostris* (MACN A 52–70) is unknown (probably Cabeza Blanca or La Flecha). *Patagonhippus* from the La Cantera level (early? Deseadan) at Gran Barranca represents the first record of a new notohippid genus for Patagonia since the time of Ameghino.

When compared with taxa from the Mustersan and Tinguirirican, *Patagonhippus canterensis* n. gen. et n. sp. and *P. dukei* n. gen. et n. sp. share the absence of a posterior fossettid, but have a more hypsodont dentition.

Patagonhippus canterensis and *P. dukei* differ in many respects from Deseadan taxa *Rhynchippus* and *Coresodon* (occlusal morphology), and *Argyrohippus* (e.g. absence of cement), and have a closer affinity with *Morphippus* and *Eurygenium*.

Six species were proposed for the genus *Morphippus* (*M. complicatus* Ameghino 1897, *M. hypselodus* Ameghino 1897, *M. fraternus* Ameghino 1901, *M. imbricatus* Ameghino 1897, *M. quadrilobus* Ameghino 1901, and *M. corrugatus* Ameghino 1899). Loomis (1914) considered the first five species as *M. imbricatus*, but Patterson (1952) in his unpublished revision recognized *M. imbricatus* (only the specimen MACN A 52–76) as valid species; he considered *M. quadrilobus* to be a toxodontid; and he placed *M. complicatus, M. hypselodus, M. fraternus,* and *M. imbricatus* (specimen MACN A 52–59) into synonymy with *Rhynchippus equinus* Ameghino 1897; Patterson does not mention *M. corrugatus*. Examination of the type material (MACN A 52–602) of the latter shows it to be a toxodontid.

Patagonhippus canterensis is similar in size to *M. imbricatus* (MACN A 52–76, right dentary with all teeth). The two resemble one another in having lower molars without a posterior fossettid on the entolophid, but differ in the depth of the groove between the entolophid and hypolophid; *Patagonhippus dukei* also differs from *M. imbricatus* by the smaller size (50%).

Also in the Patagonia Deseadan, two species were proposed for the genus *Eurygenium* (*E. latirostris* Ameghino 1894 and *E. normalis* Ameghino 1897). Patterson (1952) recognized *E. latirostris* as valid but he sank *E. normalis*, based on MACN A 52–85, a left maxilla fragment with P4–M3, into *Morphippus imbricatus* but Marani and Dozo (2008) retain *E. normalis* as a smaller species of *Eurygenium* because it possesses premolar characters of that genus. More recently Shockey (1997b) described a new species of *Eurygenium, E. pacegnum*, from the Deseadan of Bolivia.

Patagonhippus canterensis shares with *E. latirostris* (MACN A 52–71 and UNPSJB-PV-60) the first crista on the lingual ectoloph and with *E. pacegnum* (MNHN-Bol-V-003643) the absence of posterior fossettid on the entolophid, but differs from both species in having a mesial premolar cingulum, in lacking cementum on the mesial faces of the cheek teeth and in having deeper grooves between metalophid and entolophid, and entolophid and hypolophid. *Patagonhippus dukei* has the diagnostic *Patagonhippus* characters mentioned above and is 50% smaller than *E. latirostris* and *E. pacegnum*.

Others two species of Deseadan notohippids are recognized outside Patagonia: *Pascualihippus boliviensis* Shockey 1997a from Salla, Bolivia, and *Moqueguahippus glycisma* Shockey *et al.* 2006, from the upper Moquegua Formation, Peru. Both new species differ from *P. boliviensis* in having subtriangular premolars, P2 and P3 having the first crista in the lingual ectoloph and lacking a posterior fossettid on the entolophid. The two new species *P. canterensis* and *P. dukei* share the posterior fossettid in the entolophid with *M. glycisma*, but differ in absence of cementum, present in *M. glycisma*.

Genus ***Argyrohippus*** Ameghino 1901
Argyrohippus cf. ***A. boulei*** Ameghino 1901
Material MPEF-PV 5388, mandible with right and left i1–3, c, p2–m3; MPEF-PV 5922, fragment of upper molar and left lower m1; MPEF-PV 7325, left lower molar.
Locality Lower Fossil Zone, Colhue-Huapi Member of the Sarmiento Formation at Gran Barranca, Chubut Province, Argentina. Colhuehuapian SALMA.
Comments Three species have been described for *Argyrohippus*: *A. boulei* Ameghino 1901 and *A. fraterculus* Ameghino 1901 from the Colhuehuapian levels at Gran Barranca, and *A. praecox* Patterson 1935 from the Deseadan at La. Flecha. Patterson (1952) considered the two Colhuehuapian species to be conspecific with *A. boulei* the nominal species. These species are based on fragmentary remains, and new material MPEF-PV 5388 contributes a better knowledge of this taxon.

Bioestratigraphy and paleoecology

Some notohippid taxa at Gran Barranca cannot be referred with certainty to any known species. This suggests that either the stratigraphic levels are not equivalent in age with localities in Patagonia yielding previously known taxa, or that the evolutionary radiation of notohippids in this interval was characterized by morphological experimentation and rapid transformation.

As with equids, the major trends observed in the history of notohippids are towards increasing height of the cheek teeth from brachydont to hypsodont, but notohippids were more conservative in podial morphology, and never attained the extreme reduction in the number of digits seen in horses (Loomis 1914; Shockey 1997b).

An extended, continuous and rich fossiliferous sequence of notohippids at Gran Barranca is an ideal opportunity to follow the evolutionary increase in crown height. In the low levels of this stratigraphic sequence ("El Rosado") the notohippids are represented by remains assigned to *Puelia?*, which has low-crowned cheek teeth. "La Cancha" level has yielded remains that suggest a high diversity of brachydont

notohippid taxa (*Puelia plicata* and *Puelia* sp.) along with others that are somewhat more high-crowned or mesodont: *Eomorphippus obscurus* and *Eomorphippus pascuali*. Mesodont taxa occur further up in the sequence. "La Cantera" level has yielded two new mesodont notohippids (*Patagonhippus canterensis* and *P. dukei*), and the Lower Fossil Zone of the Colhue-Huapi Member, has the mesodont taxon *Argyrohippus* cf. *A. boulei*.

Although *Moqueguahippus glycisma*, a Deseadan notohippid from southern Peru, has a layer of external cementum surrounding the crowns, in the sequence at Gran Barranca this feature does not appears before the Colhuehuapian levels (*Argyrohippus*).

These changes in tooth morphology and limb structure are accompanied by an increase in body size during notohippid evolution. These changes have been proposed as a consequence of the appearance and development of grassland biomes at a time of major environmental change that occurred during the late Eocene through the late Oligocene interval in South America (Pascual and Ortiz Jaureguizar 1990). *Eomorphippus obscurus* was among the earliest notoungulates to developed high-crowned teeth (Patterson and Pascual 1972). Some more advanced forms present external cement, which possibly made these animals more suited to foraging on plants like grasses that contain large amounts of silica phytoliths or to diets containing more grit in general. Notohippidae show a reduction of the lateral digits (loss of digit I and reduction of digit V), having tetradactyl but functionally tridactyl feet (Shockey 1997b). Other modifications in the body and tarsus, indicate that they probably had been animals of cursorial habits, possibly living in more open areas.

ACKNOWLEDGEMENTS
The authors are grateful to the editors of this book for the invitation to contribute to this work. This study was greatly facilitated by access to specimens of collections at the Museo de La Plata, the Museo Argentino de Ciencias Naturales "Bernardino Rivadavia," and the Museo Paleontológico "Egidio Feruglio" (courtesy of Marcelo Reguero, Alejandro Kramarz, and Eduardo Ruigómez, respectively). We also thank Richard Madden and Jorge Ferigolo for discussion and revision of the text; Richard Kay for his critical review of the final manuscript; Guiomar Vucetich and Cecília Dechamps for the aid with information on the material of Gran Barranca; and Rejane Rosa (MCN/Fundação Zoobotânica do Rio Grande do Sul) for the drawings. Field work was supported by grants from the US National Science Foundation (EAR- 0087636, BCS-0090255, and DEB-9907985) to Richard F. Kay and Richard H. Madden.

REFERENCES

Ameghino, F. 1891. Caracteres diagnósticos de cincuenta especies nuevas de mamíferos fósiles argentinos. *Revista Argentina de Historia Natural*, **1**, 129–167.

Ameghino, F. 1897. Mammifères crétacés de l'Argentine: deuxième contribution à la connaissance de la fauna mammalogique de couches à Pyrotherium. *Boletin del Instituto Geográfico Argentino*, **18**, 406–521.

Ameghino, F. 1899. *Sinopsis geológico–paleontológica: Suplemento (Adiciones y correcciones)*. La plata: Imprenta La Libertad.

Ameghino, F. 1901. Notices préliminaires sur des ongulés nouveaux des terrains crétacés de Patagonie. *Boletín de la Academia Nacional de Ciencias en Córdoba*, **16**, 349–426.

Ameghino, F. 1906. Les formations sédimentaires du Crétacé supérieur et du Tertiaire de Patagonie avec un parallèle entre leurs faunes mammalogiques et celles de l'ancien continent. *Anales del Museo Nacional de Buenos Aires*, **3**, 1–568.

Bond, M. and G. M. López 1993. El primer Notohippidae (Mammalia, Notoungulata) de la Formación Lumbrera (Grupo Salta) del Noroeste Argentino: consideraciones sobre la sistemática de la Familia Notohippidae. *Ameghiniana*, **30**, 59–68.

Bond, M., G. M. López, R. H. Madden, M. A. Reguero, and A. Scarano 2004. Los Ungulados no mienten. *Ameghiniana*, **41**(4)(Supl.), 36–37R.

Carlini, A. A., R. H. Madden, M. G. Vucetich, M. Bond, G. M. López, M. A. Reguero, and A. Scarano 2005. Mammalian biostratigraphy and biochronology at Gran Barranca: the standard reference section for the continental middle Cenozoic of South America. *Actas XVI Congreso Geológico Argentino*, **4**, 425–426.

Cifelli, R. L. 1993. The phylogeny of the native South American ungulates. In Szalay, F. S., Novacek, M. J., and McKenna, M. C. (eds.), *Mammal Phylogeny*, vol. 2. New York: Springer-Verlag, pp. 195–216.

Loomis, F. B. 1914. *The Deseado Formation of Patagonia*. Concord, NH: Rumford Press.

López, G. M., M. Bond, M. A. Reguero, J. N. Gelfo, and A. Kramarz 2005. Los ungulados del Eoceno–Oligoceno de la Gran Barranca, Chubut. *Actas XVI Congreso Geológico Argentino*, **4**, 415–418.

Marani, H. and M. T. Dozo 2008. El cráneo más completo de *Eurygenium latirostris* Ameghino, 1895 (Mammalia, Notoungulata), um Notohippidae del Deseadense (Oligoceno Tardío) de la Patagonia, Argentina. *Ameghiniana*, **45**, 619–626.

Pascual, R. and E. Ortíz Jaureguizar 1990. Evolving climates and mammal fauna in Cenozoic South America. *Journal of Human Evolution*, **19**, 23–60.

Patterson, B. 1934. Upper premolar–molar structure in the Notoungulata with notes on taxonomy. *Geological Series of the Field Museum of Natural History*, **6**, 91–111.

Patterson, B. 1935. A new *Argyrohippus* from the Deseado beds of patagonia. *Geological Series of the Field Museum of Natural History*, **6**, 161–166.

Patterson, B. 1952. *Catálogo de los mamíferos del Deseadiano y Colhuehuapiano*. Buenos Aires: Informe Seción Paleozoología Vertebrados Museo Argentino de Ciencias Naturales "Bernardino Rivadavia." (Unpublished.).

Patterson, B. and R. Pascual 1972. The fossil mammal fauna of South America. In Keast, A., Erk, F. C., and Glass, B. (eds.), *Evolution, Mammals, and Southern Continents*. Albany, NY: State University of New York Press, pp. 247–309.

Paula Couto, C. de 1981. Fossil mammals from the Cenozoic of Acre Brazil. IV. Notoungulata Notohippidae and Toxodontidae Nesodontinae. *Anais II Congreso Latinoamericano de Paleontologia, Porto Alegre*, **2**, 461–477.

Paula Couto, C. de 1982. *Purperia*, a new name for *Megahippus* Paula Couto, 1981. *Iheringia*, **7**, 69–70.

Reguero, M. A., G. M. López, M. Bond, and R. H. Madden 2004. Faunas de Edad Tinguiririquense (Eoceno tardío–Oligoceno temprano) en Patagonia. *Ameghiniana*, **41**(4)(Supl.), 24R.

Ribeiro, A. M. and M. Bond 1999. Novos materiais de *Notohippus toxodontoides* Ameghino, 1891 (Notohippidae, Notoungulata) do Piso Notohippidense, SW da Província de Santa Cruz, Argentina. *Ameghiniana*, **36**(4)(Supl.), 19–20R.

Ribeiro, A. M. and M. Bond 2000. New data about the Notoungulata (Leontiniidae and Notohippidae) from Tertiary of Acre State, Brazil. *Revista Universidade Guarulhos*, **5**(6), 47–53.

Shockey, B. J. 1997a. Toxodontia of Salla, Bolivia (Late Oligocene): Taxonomy, systematics, and functional morphology. Ph.D. dissertation, University of Florida.

Shockey, B. J. 1997b. Two new notoungulates (Family Notohippidae) from the Salla Beds of Bolivia (Deseadan: Late Oligocene): systematics and functional morphology. *Journal of Vertebrate Paleontology*, **17**, 584–599.

Shockey, B. J., R. Hitz, and M. Bond 2004. Paleogene notoungulates from the Amazon Basin of Peru. In Campbell, K. E. Jr. (ed.), *The Paleogene Mammalian Fauna of Santa Rosa, Amazonian Peru*. Los Angeles, CA: Natural History Museum of Los Angeles County, pp. 61–69.

Shockey, B. J., R. Salas, R. Quispe, A. Flores, E. J. Sargis, J. Acosta, A. Pino, N. J. Jarica, and M. Urbina 2006. Discovery of Deseadan fossils in the Upper Moquegua Formation (Late Oligocene–?Early Miocene) of Southern Peru. *Journal of Vertebrate Paleontology*, **26**, 205–208.

Simpson, G. G. 1932. New or little-known ungulates from the Pyrotherium and Colpodon beds of Patagonia. *American Museum Novitates*, **576**, 1–13.

Simpson, G. G. 1945. The principles of classification and a classification of mammals. *Bulletin of the American Museum of Natural History*, **85**, 1–350.

Simpson, G. G. 1967. The beginning of the Age of Mammals in South America. II. *Bulletin of the American Museum of Natural History*, **137**, 1–259.

Soria, M. F. and H. Alvarenga 1989. Nuevos restos de mamíferos de la Cuenca de Taubaté, Estado de São Paulo, Brasil. *Anais Academia Brasileira de Ciências*, **61**, 157–175.

Wyss, A. R., J. J. Flynn, and D. A. Croft 2005. New notohippids (Notoungulata, Eutheria) from the central Chilean Andes. *Journal of Vertebrate Paleontology*, **25**, 132A.

10 Rodent-like notoungulates (Typotheria) from Gran Barranca, Chubut Province, Argentina: phylogeny and systematics

Marcelo A. Reguero and Francisco J. Prevosti

Abstract

Rodent-like notoungulates, traditionally included in suborders Typotheria and Hegetotheria, experienced an important evolutionary radiation in South America. This evolution is well documented in the fossil record at Gran Barranca in Patagonia, Argentina, where 14 taxa occur. The phylogenetic position of Typotheria and Hegetotheria is poorly understood and has become controversial in recent years. This study traces the origin and early evolution of the rodent-like notoungulates (Typotheria and Hegetotheria) through an analysis of phylogenetic relationships among genera that are well known cranially and dentally. For this we use parsimony analysis on 51 taxa using 78 morphological characters. The results of the phylogenetic analysis suggest the abandonment of the concept of Hegetotheria as a separate suborder of Notoungulata, and expand the concept of Typotheria to include the more advanced families of rodent-like notoungulates: Mesotheriidae, Archaeohyracidae, and Hegetotheriidae (the two latter originally allocated in Hegetotheria). Included within this clade is *Campanorco inauguralis* Bond *et al.* 1984, the sister species of Archaeohyracidae + Hegetotheriidae + Mesotheriidae. Oldfieldthomasiidae and Archaeopithecidae, previously considered typotheres, are removed and form the sister clade to Typotheria. By our definition Typotheria is a more inclusive monophyletic group containing the most recent common ancestor of *Notopithecus* and *Mesotherium* and all of its descendants, including *Campanorco inauguralis*, archaeohyracids, and hegetotheriids. The implications of this phylogenetic analysis for the early evolutionary history of rodent-like typotherians are explored further in a separate chapter of this book.

Resumen

Los notoungulados rodentiformes, tradicionalmente incluidos en dos subórdenes, Typotheria y Hegetotheria, experimentaron una gran radiación evolutiva en América
del Sur. Este proceso está bien documentado en el registro fósil de la Gran Barranca, Patagonia, Argentina. La posición filogenética de los Typotheria y Hegetotheria ha sido pobremente entendida, pero muy controvertida en años recientes. En este estudio se explora el origen y evolución de los tipoterios rodentiformes, a través de un análisis filogenético de los géneros mejor conocidos morfológicamente, con 51 taxones y 78 caracteres morfológicos. Las relaciones filogenéticas entre las familias de estos notoungulados son hipotetizadas en este trabajo. Los resultados del análisis filogenético sugieren abandonar el concepto Hegetotheria como un suborden separado, y expandir el concepto de Typotheria para incluir a las familias más avanzadas de notoungulados rodentiformes: Interatheriidae, Mesotheriidae, Archaeohyracidae y Hegetotheriidae (estas dos últimas originalmente incluidas en Hegetotheria). También se incluye dentro de este clado a *Campanorco inauguralis* Bond *et al.* 1984 que representa la especie hermana de los Archaeohyracidae + Hegetotheriidae + Mesotheriidae. También sugieren remover a Oldfieldthomasiidae y Archaeopithecidae, previamente considerados tipoterios, para formar el clado hermano de Typotheria. Typotheria es el grupo monofilético más inclusivo definido como el clado que contiene al ancestro común más reciente de *Notopithecus* y *Mesotherium* y todos sus descendientes incluyendo *Campanorco inauguralis*, Archaeohyracidae y Hegetotheriidae. Las implicancias de estos resultados en la historia evolutiva temprana de los tipoterios rodentiformes son exploradas en un capítulo aparte en este libro.

Introduction

During the early and middle Cenozoic, notoungulates were the most successful and diverse ungulate group living in South America (Patterson and Pascual 1972; Reig 1981; Cifelli 1985, 1993). They experienced a broad adaptive radiation during the Eocene–Oligocene transition (EOT) diversifying into small rodent- (Interatheriidae) and rabbit-like (Hegetotheriidae) forms and medium to large-sized tapir- (Isotemnidae) and rhino-like (Toxodontidae) forms. The two suborders Typotheria and Hegetotheria, hypsodont

The Paleontology of Gran Barranca: Evolution and Environmental Change through the Middle Cenozoic of Patagonia, eds. R. H. Madden, A. A. Carlini, M. G. Vucetich, and R. F. Kay. Published by Cambridge University Press. © Cambridge University Press 2010.

rodent- and rabbit-like notoungulates, became taxonomically diverse and numerically abundant in the late Eocene.

The rodent-like notoungulates were allocated by Simpson (1967) in two suborders, Typotheria and Hegetotheria. In 1893, Karl Zittel proposed the name Typotheria as a subgroup of Rodentia to include the families "Protypotheriidae" (now Interatheriidae) and "Typotheriidae" (now Mesotheriidae). Later, Ameghino (1906) included the Hegetotheriidae within Typotheria. In 1913, Scott expanded the concept of Typotheria to include "Notopithecidae" (= Notopithecinae), Archaeopithecidae, and Archaeohyracidae. Simpson (1945, p. 238) proposed a separate origin for Hegetotheriidae, and removed it from Typotheria arguing that "the typothere-like aspect of the hegetotheres is superficial and is apparently due to the rodent-like habitus, independently acquired, a trend that appeared also in other notoungulates of surely non-typothere origin (see, *e.g.* Patterson 1936)" and erected a new suborder Hegetotheria to include it. Finally, Simpson (1967), in his last classification of Typotheria, followed Patterson (1934a, 1934b) and placed "Trachytheridae" (= Trachytheriinae) in the Mesotheriidae and "Notopithecidae" (= Notopithecinae) in the Interatheriidae. In the same work, Simpson grouped within Typotheria the families Oldfieldthomasiidae, Archaeopithecidae, Interatheriidae, and Mesotheriidae. The suborder Hegetotheria (Simpson 1945, 1967) presently includes the families Hegetotheriidae and Archaeohyracidae. The phylogenetic positions of Typotheria and Hegetotheria have been poorly understood and have become highly controversial in recent years and only a few phylogenetic studies of Typotheria and Hegetotheria relationships have been published (Cifelli 1993; Reguero 1994, 1999; Reguero *et al.* 1996; Hitz 1997; Croft 2000). Cifelli (1993) suggested the abandonment of the concept of Hegetotheria as a separate suborder of Notoungulata.

Starting from their earliest appearance in Patagonia in the *Kibenikhoria* faunal zone (late Paleocene), Typotheria show evolutionary tendencies toward higher-crowned incisors and cheek teeth, somewhat earlier than contemporary Toxodontia. Despite their late Paleocene first appearance, it was not until the late Eocene that members of this suborder attained high-crowned incisors and cheek teeth and began assuming the "rodent-like habitus" (*sensu* Simpson 1967).

Cifelli (1993) tentatively grouped six families within Typotheria: Oldfieldthomasiidae, Archaeopithecidae, Archaeohyracidae, Hegetotheriidae, Interatheriidae, and Mesotheriidae. All these notoungulates share cranial and dental morphology that includes: (1) I1 somewhat enlarged; (2) anterior root of the zygomatic arch expanded and flattened; (3) upper molar pattern with posterior crista joining the crochet and the anterior crista joining the protoloph or antecrochet, forming the typical "face" pattern of fossettes; and (4) lower molars with trigonid and talonid

divided by a deep labial sulcus. These features are also shared with the monotypic family Campanorcidae (*Campanorco inauguralis*) from the early Eocene of the northwest of Argentina and we regard it here as a primitive family of Typotheria. This grouping seems to be natural but has never been defined phylogenetically.

The relationships of Archaeopithecidae and Oldfieldthomasiidae, two groups considered by Simpson (1967) to be typotheres, are also problematical; they share only a few of the rodent-like characters, and differ from the remaining rodent-like families in, for example, the morphology of the arcade and the anterior dentitions. Therefore, their inclusion in this grouping remains questionable.

Gran Barranca contains the most complete sequence (Sarmiento Formation) of Paleogene continental sediments known in South America. A refined local biostratigraphy allows assignment of most mammal-bearing horizons to specific biostratigraphic zones. The most noteworthy period of evolutionary change in Typotheria occurred during the late Eocene to early Oligocene and it is well documented in the Sarmiento Formation at Gran Barranca. Large samples of specimens of Typotheria made over the past ten years are available for study from stratigraphicallly controlled localities throughout this sequence ranging from the late middle Eocene (about 41 Ma) to the early Miocene (about 18.5 Ma).

Abbreviations

AMNH, American Museum of Natural History, New York; FLMNH, Florida Museum of Natural History, Gainesville, Florida; FMNH, Field Museum of Natural History, Chicago; HI, Hypsodonty Index; MACN, Museo Argentino de Ciencias Naturales "Bernardino Rivadavia"; MLP, Museo de La Plata, Argentina; MNHN, Vertebrate Paleontology Collections, Museum Nacional de Historia Natural, La Paz, Bolivia; MPEF, Museo Provincial "Egidio Feruglio," Trelew, Argentina; SGOPV, Museo Nacional de Historia Natural, Santiago, Chile; UNSJB, Universidad Nacional "San Juan Bosco," Comodoro Rivadavia, Argentina. SALMA, South American Land Mammal Age. Upper tooth loci are indicated by upper-case letters (e.g. I1, P2, M1) and lower tooth loci by lower-case letters (e.g. i1, p2, m1); deciduous teeth are indicated by a "d/D" preceding the tooth position. Abbreviations of localities' names are given in Appendix 10.3.

Phylogenetic analysis

This chapter does not intend to deal with the species-level taxonomy of all taxa included in the group, much of which has been revised recently by several authors (Cerdeño and Bond 1999; Croft 2000; Hitz *et al.* 2000; Cerdeño and Montalvo 2001; Reguero *et al.* 2002, 2003, 2007;

Croft *et al.* 2003; Reguero and Castro 2004; Croft and Anaya 2006). Instead, we limit our scope to an analysis of relationships among genera, and we generally only treat taxa that are well known cranially and dentally. The following taxa were not included in this study: *Antepithecus brachystephanus* Ameghino 1901; *Transpithecus obtentus* Ameghino 1901; *Eohyrax rusticus* Ameghino 1901; *Eohyrax isotemnoides* Ameghino 1904; *Eohyrax praerusticus* Ameghino 1902; *?Eohyrax platyodus* Ameghino 1904; *Pseudhyrax strangulatus* (Ameghino 1901); *Pseudopachyrucos foliiformis* Ameghino 1901; *Proargyrohyrax curanderensis* Hitz *et al.* 2000; *Argyrohyrax acuticostatus* Ameghino 1901.

Character polarities were determined using three notoungulate species as outgroups: *Simpsonotus praecursor* (Henricosborniidae), *Notostylops murinus* (Notostylopidae), *Pleurostylodon modicus* (Isotemnidae) (Simpson 1948, 1967, Cifelli 1993). *Simpsonotus praecursor* Pascual *et al.* 1978 is a basal notoungulate (Henricosborniidae) from the Paleocene Mealla Formation of Jujuy Province, northwestern Argentina. It is the most complete henricosborniid known. *Pleurostylodon modicus* Ameghino 1897 is a primitive isotemnid from the Barrancan of Chubut Province. *Notostylops murinus* Ameghino 1897 is a primitive isotemnid also from the Barrancan of Chubut. Appendix 10.1 lists definitions of characters used in this analysis, and Appendix 10.2 shows the data matrix.

Morphological data were assessed through a review of the literature (e.g. Cifelli 1993; Reguero 1994, 1999; Hitz 1997; Cerdeño and Bond 1999; Croft 2000; Hitz *et al.* 2000; Cerdeño and Montalvo 2001; Croft *et al.* 2003; Reguero *et al.* 2003; Reguero and Castro 2004; Croft and Anaya 2006) and study of original specimens and casts from different institutions and museums (AMNH, FLMNH, FMNH, MACN, MLP, MNHN, MPEF, SGOPV, and UNSJB).

The cladistic analysis was rooted with *Simpsonotus praecursor* (Henricosborniidae) because it is the geologically oldest notoungulate close to the ancestry of all later notoungulates, and well known from both upper and lower dentitions.

Methods

We performed a parsimony analysis on 51 taxa (Appendix 10.1) using 78 morphological characters (75 of which are cladistically informative). The data matrix is presented in Appendix 10.2. For the analysis we assumed maximum parsimony under equal and implied weights (Goloboff 1993, 1995, 1997) and used the computer program TNT 1.0 (Goloboff *et al.* 2003a). The character analysis follows the "composite code" criterion (see Lee and Bryant 1999; Strong and Lipscomb 1999). We conducted heuristic searches with Tree Bisection Reconnection (TBR) using 1000 random addition sequences and saving 10 trees

per round. After these, a new TBR search was made from previously obtained trees, saving the new trees. Zero branch lengths were collapsed with rule 3 (minimum length = 0; see Coddington and Scharff 1994). We explored a wide range of concavity values (k 1–100) during the implied weights searches and utilized a stability test (as implemented by Ramírez 2003; see also Lopardo 2005) to identify which parameter (or range of parameters) are more efficient. This test performs a jackknife (500 pseudo-replicates) that randomly deletes 30% of the original characters and compares the percent of correct and false recovered nodes (expressed as a proportion of total recovered nodes in the original or resampled consensus, respectively) in the strict consensus obtained from the pseudo-replicates, under equal weights and different k values. Correct and false nodes are those present or absent in the original strict consensus search under equal weight of different k values. The significance of the observed differences between each concavity constant values (k) and equal weights was established with the non-parametric Mann–Whitney U-test ($p < 0.05$). The non-ambiguous information present in the topology found with the better parameters was synthesized with strict consensus trees (Bremer 1990). The listed synapomorphies are common to all these trees. Branch support was estimated by absolute group frequencies and frequency group differences (1000 pseudo-replicates) obtained with Symmetrical Resampling (Goloboff *et al.* 2003b). The potential effect of missing entries on the recovered topology was measured with the Missing Entry Replacement Data Analysis (MERDA) with 1000 replications (see Norell and Wheeler 2003), using the frequencies of original nodes in the replications. The correlation between MERDA values and branch supports was established with R, the Spearman Rank Correlation coefficient.

Results

The stability test shows that $k4$–$k100$ recovered more correct nodes than equal weights and $k1$–$k3$, and those differences are statistically significant ($p < 0.05$). The significant difference in percentage of false recovered nodes is between equal weights and implied weights. Equal weight presents lower error, but as was mentioned by Lopardo (2005) this appears to be an artifact because its strict consensus is less resolved than those obtained with implied weights.

With $k4$–$k100$ we found 13 topologies that differ only in the position of *P. sculptum*, *P. schiaffinoi*, *S. altiplanense*, *A. paucidens*, *A. chucalensis*, *M. choquecotense*, *E. lehmannitschei*, *E. superans*, and *P. achirense*, which form three polytomies in the consensus tree (Fig. 10.1). The consensus shows that most nodes have good support (see Fig. 10.1), especially Typotheria (Node 4), Interatheriidae (Node 7), Typotherioidea (Node 6), Hegetotheriidae (Node 21), Pachyrukhinae (Node 23), and Mesotheriinae

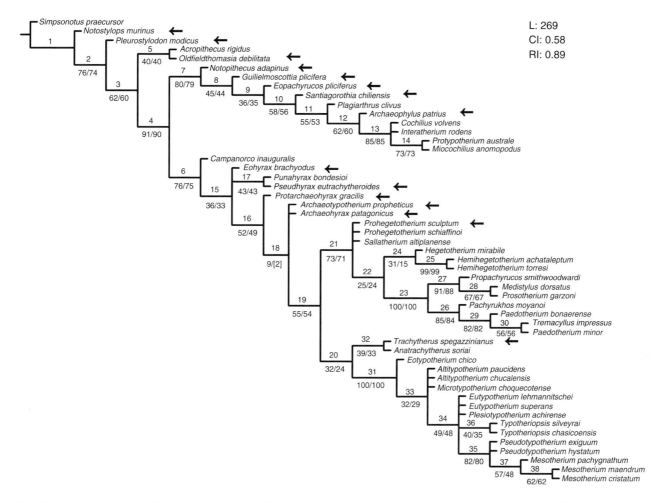

Fig. 10.1. Interrelationships of South American rodent-like notoungulates and other selected notoungulates. The arrows indicate the species recorded at Gran Barranca. Abbreviations: L, length; CI, consistency index; RI, retention index. The numbers below the branches are absolute frequencies and frequency differences from the symmetrical resampling. The numbers above the line correspond to the node's number in the text.

(Node 31). Mesotheriidae (Node 20) and Trachytheriinae (Node 32) have low but positive values (Fig. 10.1).

The MERDA values (frequencies of the Fig. 10.1 nodes recovered) show that Typotheria (Node 4), Hegetotheriidae (Node 21), Pachyrukhinae (Node 23), and Mesotheriinae (Node 31) are robust (values above 75%) to the potential effect of the missing entries. On the other hand, Typotherioidea (Node 6), Interatheriidae (Node 7), and Mesotheriidae (Node 20) are more subject to missing entries, with values around 40–50%. The MERDA values show a good correlation (R: 0.72, $p < 0.0001$) with support values (group frequencies and frequency group differences). Some nodes (e.g. Node 6 and 7; see Fig. 10.1) have high supports and moderate MERDA values.

Based on this analysis (Fig. 10.1) we recognize the following monophyletic groups/taxa.

Unnamed taxon: *Acropithecus rigidus* + *Oldfieldthomasia debilitata* **(Node 5)**

This clade is diagnosed by one unequivocal synapomorphy (Fig. 10.1): relative size of upper incisors [C2] subequal.

Typotheria (Interatheriidae + *Campanorco inauguralis* **+ "Archaeohyracidae" + Hegetotheriidae + Mesotheriidae) (Node 4)**

This monophyletic group is diagnosed by five unequivocal synapomorphies (Fig. 10.1): (1) I1 with angle between the mediolateral face of the I1 and the sagittal plane [C7] approx. 40°; (2) P3–4/p3–4 [C13] completely molarized; (3) M1–2 [C16] quadrangular; (4) ratio m2/m3 [C40] = ½; (5) rostrum [C47] long and tall.

Based on the topology obtained in this analysis, the species classically referred to as typotheres and hegetotheres share an exclusive common ancestor, and thus form an evolutionary entity worthy of naming. Following Cifelli's criteria we name this clade Typotheria, the clade that includes all the hypsodont rodent-like notoungulate families. One of the many plausible ways to translate this name (as traditionally used: Simpson 1945) into phylogenetic taxonomy would be to define Typotheria as the clade stemming the most common recent ancestor (MRCA) of the earliest diverging member *Notopithecus adapinus* plus all of its descendants (a node-based definition *sensu* de Queiroz and Gauthier 1990).

Interatheriidae (Node 7)
This clade is diagnosed by five unequivocal synapomorphies (Fig. 10.1): (1) paracone/metastyle of the P3–4 [C12] deep; (2) metacristid [C34] very well developed; (3) postmetacristid on p3–4 [C35] present; (4) jugal [C54] reduced, between the maxilla and the squamosal; (5) descending process of the maxilla [C55] moderately developed.

As generally conceived, Interatheriidae is the most basal member of the group, concordant with its stratigraphic occurrence. Interatheriidae is the sister taxon of the clade that includes *C. inauguralis* + "Archaeohyracidae" + Hegetotheriidae + Mesotheriidae (Node 82, Fig. 10.1).

Typotherioidea *sensu* Reguero and Castro (2004) (*Campanorco inauguralis* + "Archaeohyracidae" + Hegetotheriidae + Mesotheriidae) (Node 6)
This clade is diagnosed by five unequivocal synapomorphies (Fig. 10.1): (1) I1 [C4] obliquely implanted and not procumbent; (2) I1 [C5] subtriangular, pointed distally, with mesial sulcus; (3) metastyle on M3 [C10] present; (4) zygomatic plate [C52] little developed; (5) suborbital fossa [C60] poorly defined. Typotherioidea is defined as the clade stemming from the MRCA of the earliest diverging member *Campanorco inauguralis* plus all of its descendants. The present analysis confirms that the Archaeohyracidae are a paraphyletic group whose species represent successive outgroup taxa to a monophyletic clade Hegetotheriidae + Mesotheriidae. Previous analyses of Cifelli (1993), Hitz (1997), and Croft (2000) reached the same conclusion.

Unnamed taxon: Hegetotheriidae + Mesotheriidae (Node 19)
This clade is diagnosed by five unequivocal synapomorphies (Fig. 10.1): (1) lower molars [C36] without fossettids; (2) paracristid [C37] absent; (3) rostrum [C47] short and tall; (4) anterior rostral notch (premaxillae) [C62] forming obtuse angle; (5) maxillar fossa [C72] large and shallow.

Hegetotheriidae ("Hegetotheriinae" + Pachyrukhinae) (Node 21)
This clade is diagnosed by six unequivocal synapomorphies (Fig. 10.1): (1) metastyle on M3 [C10] absent; (2) P4 [C14] subtriangular, short, without central fossette; (3) M1–2 (16) ovoid; absence of lobes on M1–2 [C17]; (4) ratio of i1/i2 [C24] = 1.5; (5) metacristid [C34] absent; (6) anterior and posterior sides of M1 middle lobe [C64] absent. Hegetotheriidae could potentially be defined as the MRCA of *Prohegetotherium* (the most basal member of the clade) and *Paedotherium* (or any other hegetotheriid) and all of its descendants.

Pachyrukhinae (Node 23)
This clade is diagnosed by eight unequivocal synapomorphies (Fig. 10.1): (1) angle between the mediolateral face of the I1 and the sagittal plane [C7] <45°; (2) m3 [C43] with third lobe triangular and with the same size than the second; (3) diastemata (upper and lower) [C45] little developed; (4) upper molar imbrication [C46] between 1.25 and 1.50; (5) mastoid bulla [C56] large, bulbous, expanded dorsally and medially, easily visible; (6) shape of the molar trigonid [C58] triangular; (7) suborbital fossa [C60] well developed; (8) sagittal crest [C71] absent. Pachyrukhinae could potentially be defined as the MRCA of *Propachyrucos* and *Paedotherium* and all of its descendants.

Mesotheriidae (Trachytheriinae + Mesotheriinae) (Node 20)
This clade is diagnosed by nine unequivocal synapomorphies (Fig. 10.1): (1) metastylid on M3 [C21] present; (2) section of i1 [C27] rounded with little or no lingual sulcus; (3) postorbital process [C59] long and transverse; (4) occipital notch [C61] present; (5) postpalatal notch [C63] wide and removed from M3; (6) lingual exposure of M1 middle lobe [C65] little or none; (7) position of carotid foramen [C75] at anterior margin of bulla; (8) vertical septum in auditory bulla [C76] absent; (9) hamulus process of pterygoid [C77] present.

Trachytheriinae (Node 32)
This clade is diagnosed by two unequivocal synapomorphies (Fig. 10.1): (1) p3–4/p3–4 [C13] little molarized; (2) lower molars [C36] with only one fossettid (central).

Annotated systematic list

Order **NOTOUNGULATA** Roth 1903
Suborder **TYPOTHERIA** Zittel 1893
Comments Simpson's (1967, p. 15) definition remains a reasonable description of typotheres.
Family **INTERATHERIIDAE** Ameghino 1887

Protypotheridae Ameghino 1891; Notopithecidae Ameghino 1897

Subfamily **NOTOPITHECINAE** Ameghino 1897
(Simpson 1945 new rank)
Genus *Notopithecus* Ameghino 1897
Adpithecus Ameghino 1901; *Infrapithecus* (*partim*) Ameghino 1901; *Pseudadiantus* (*partim*) Ameghino 1901; *Patriarchippus* (*partim*) Ameghino 1901
Type species *Notopithecus adapinus* Ameghino 1897.
Range of m1 length 3.8–3.9 mm.
Included species and distribution *N. adapinus* (RDL, LGF): *Notopithecus* sp. (AGB, BJP, CB2, CEB, CGU, CÑL, CÑV, CU3, EP2, CA2, CB2, GBT, NUE, LGF, LBP).

Genus *Antepithecus* Ameghino 1901
Infrapithecus (*partim*) Ameghino 1901; *Pseudadiantus* (*partim*) Ameghino 1901; *Patriarchippus* (*partim*) Ameghino 1901
Type species *Antepithecus brachystephanus* Ameghino 1901
Range of m1 length 4.1–5.2 mm.
Included species and distribution *A. brachystephanus* (ROS); *Antepithecus* sp. (NUE, GBT, FLO).

Genus *Transpithecus* Ameghino 1901
Type species *Transpithecus obtentus* Ameghino 1901
Length of m1 4.3 mm.
Included species and distribution *T. obtentus* (GBT); *Transpithecus* sp. (NUE, CÑH).

Genus *Guilielmoscottia* Ameghino 1901
Fig. 10.2A.
Type species *Guilielmoscottia plicifera* Ameghino 1901
Range of M1 length 5.0–5.4 mm.
Included species and distribution *G. plicifera* (CDH, LGH, STQ); *Guilielmoscottia* sp. (ROS, CAN).

Family **INTERATHERIINAE** Simpson 1945
Genus *Eopachyrucos* Ameghino 1901
Type species *Eopachyrucos pliciformis* Ameghino 1901
Range of m1 length 4.1–8.7 mm.
Included species and distribution *E. pliciformis* (LCN, CÑB, RBY, CAN); *E. ranchoverdensis* (FYB); *Eopachyrucos* sp. (ROS).

Genus *Santiagorothia* Hitz *et al.* 2000
Fig. 10.2B, E.
Type species *Santiagorothia chiliensis* Hitz *et al.* 2000
Range of m1 length 4.96–6.93 mm.
Included species and distribution *S. chiliensis* (CÑB, RBY, CU2, LLB, LCN, CAN, NTN).

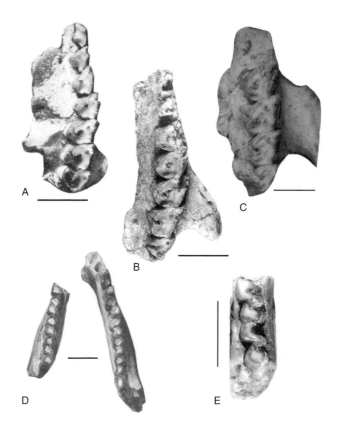

Fig. 10.2. (A) *Guilielmoscottia plicifera* Ameghino, type, MACN 10898, partial left maxilla with P1–M3; (B) *Santiagorothia chiliensis* Hitz *et al.*, MLP 91-IX-5–15, left maxillary fragment with P2–M3; (C) *Proargyrohyrax curanderensis* Hitz *et al.*, holotype, MLP 61-VIII-3–27, left maxillary fragment with P1–M3; (D) *Plagiarthrus clivus* Ameghino, type, MACN A52–474, partial left and right mandible with p3-m2 and p2-m3 respectively; (E) *Santiagorothia chiliensis* Hitz *et al.*, MLP 91-IX-5–14a, partial left mandibular fragment with m1–2. Scale bar 1 cm.

Genus *Proargyrohyrax* Hitz *et al.* 2000
Fig. 10.2C.
Type species *Proargyrohyrax curanderensis* Hitz *et al.* 2000
Range of m1 length 7.3–7.6 mm.
Included species and distribution *P. curanderensis* (CU2); *Proargyrohyrax* sp. (CAN).

Superfamily **TYPOTHERIOIDEA** Reguero and Castro 2004
Comments Typotherioidea is defined as the clade stemming the MRCA of *Campanorco inauguralis* and plus all of its sister taxa.

Family **"ARCHAEOHYRACIDAE"** Ameghino 1897
Genus *Eohyrax* Ameghino 1901
Type species *Eohyrax rusticus* Ameghino 1901
Range of m1 length 6.7–8.3 mm.

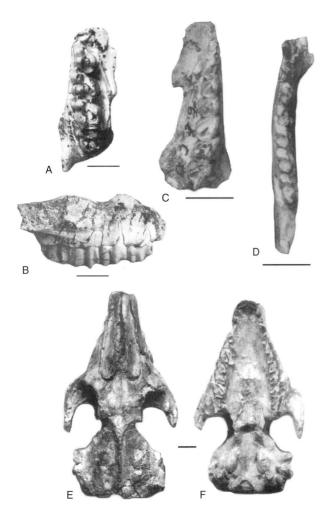

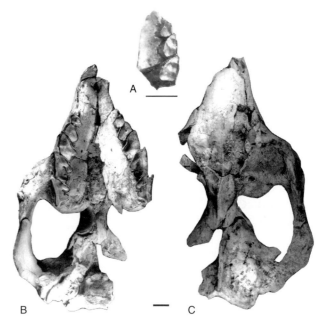

Fig. 10.4. (A) *Prohegetotherium sculptum* Ameghino, lectotype, MACN A52–443, left maxillary fragment with P1–3; (B, C) *Trachytherus spegazzinianus* Ameghino, UNPSJB PV-112, incomplete skull, occlusal and dorsal views respectively. Scale bar 1 cm.

Fig. 10.3. (A, B) *Pseudhyrax eutrachytheroides* Ameghino, type, MACN A11622, left maxillary fragment with dM3–4, M1–2 occlusal and labial views respectively; (C) *Protarchaeohyrax gracilis* (Roth), holotype, MLP 12–1522, fragment of left maxilla with P1–M3; (D) *Protarchaeohyrax gracilis* (Roth), holotype, MLP 12–1518, left mandibular fragment with p1–m2; (E, F) *Archaeohyrax patagonicus* Ameghino, holotype, MACN A52–617, skull in dorsal and occlusal view respectively. Scale bar 1 cm.

Type species *Pseudhyrax eutrachytheroides* Ameghino 1901
Range of m1 length 7.6–9.5 mm.
Included species and distribution *P. eutrachytheroides* (CDH, CBL, NTN); *P. strangulatus* (CDH, ROS, NTN).

Included species and distribution *E. rusticus* (GBY, VAP); *E. isotemnoides* (GBY, CSM); *E. praerusticus* (BJP); *Eohyrax* sp. (NUE).
Comments Simpson (1967) viewed archaeohyracids as spanning the latest Riochican through Deseadan, but he noted that their occurrence in the Riochican (late Paleocene) is doubtful. They are first definitively known from Barrancan (middle Eocene) faunas, with all currently described species from this temporal interval being referred to *Eohyrax*.

Genus ***Protarchaeohyrax*** Reguero *et al.* 2003
Fig. 10.3C, D.
Eohegetotherium Ameghino 1901; *Archaeohyrax* (*partim*) Ameghino 1897; *Bryanpattersonia* (*partim*) Simpson 1967
Type species *Archaeohyrax gracilis* Roth 1903
Range of m1 length 4.2–8.0 mm.
Included species and distribution *P. gracilis* (CBL, LCN, CAN, PER, NTN); *P. intermedium* (ERT); *P. minor* (LCN, CBL); *Protarchaeohyrax* sp. (SAL).

Genus ***Pseudhyrax*** Ameghino 1901
Fig. 10.3A, B.
Degonia Roth 1902; *Pseudopithecus* Roth 1902; *Rankelia* Roth 1902

Genus ***Archaeotypotherium*** Roth 1903
Archaeohyrax (*partim*) Ameghino 1897; *Eomorphippus* (*partim*) Ameghino 1901; *Bryanpattersonia* (*partim*) Simpson 1967
Type species *Archaeotypotherium propheticus* (Ameghino 1897)
Range of m1 length 6.9–10.7 mm.
Included species and distribution *A. propheticus* (CÑB, CAN, LCN); *A. tinguiriricaense* (NTN, PEF); *A. pattersoni* (NTN).

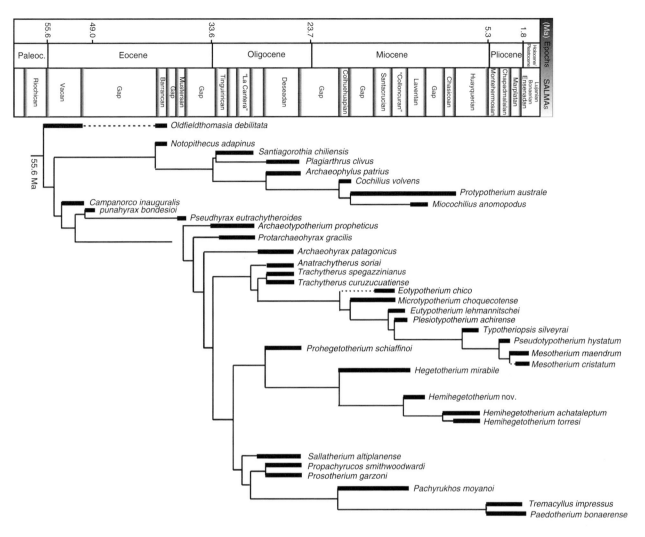

Fig. 10.5. Phylogeny of the Typotheria and the relative stratigraphic position of the taxa.

Genus ***Archaeohyrax*** Ameghino 1897
Fig. 10.3E, F.
Type species *Archaeohyrax patagonicus* Ameghino 1897
Range of m1 length 7.8–10.2 mm.
Included species and distribution *A. patagonicus* (BDD, CBL, LFL, CAS1, EP1, PTR, SPK, LCN, RBY, QFI); *A. suniensis* (SAL).

Family **HEGETOTHERIIDAE** Ameghino 1894
Subfamily **"HEGETOTHERIINAE"** Ameghino 1894
Comments The monophyly of Hegetotheriinae is currently an open question; although some hegetotheriines appear to share a derived configuration of distal tibiofibular fusion (Croft *et al.* 2004) craniodental data indicate they may instead constitute a paraphyletic collection of basal hegetotheriids (Croft and Anaya 2004, 2006). The phylogenetic analysis of

hegetotheriid relationships presented below favors the latter interpretation. The monophyly of Pachyrukhinae, in contrast, has not been questioned (Cifelli 1993; Cerdeño and Bond 1998). Croft and Anaya (2004, 2006) demonstrated that one of the characters previously used to diagnose Pachyrukhinae – a sharply trilobed m3 – is no longer unique to members of this subfamily. Accordingly, it is omitted from the above diagnosis.

Genus ***Prohegetotherium*** Ameghino 1897
Fig. 10.4A.
Ethegotherium Simpson, Minoprio, and Patterson 1962; *Propachyrucos* (*partim*) Ameghino 1897
Type species *Prohegetotherium sculptum* Ameghino 1897
Range of m1 length 5.8–8.8 mm.
Included species and distribution *P. sculptum* (CB1, SPK, CA1, RDZ, LLM, QFI); *P. schiaffinoi*

(SAL, FYB, DVL or MAÑ, CHJ, PER, CON); *P. shumwayi* (CB1); *Prohegetotherium* sp. (LCN).

Family **MESOTHERIIDAE** Alston 1876
Typotheriidae Lydekker 1886 (including Trachytheriidae Ameghino 1894; Eutrachytheriidae Ameghino 1897)
Comments See above-mentioned phylogenetic definition of this clade.

Subfamily **Trachytheriinae** Ameghino 1894
(Simpson 1945 new rank)
Genus ***Trachytherus*** Ameghino 1889
Fig. 10.4B, C.
Eutrachytherus Ameghino 1897; *Proedrium* Ameghino 1895; *Proedrium* Ameghino 1897; *Isoproedrium* Ameghino 1904; *Coresodon* (*partim*) Ameghino 1901
Type species *Trachytherus spegazzinianus* Ameghino 1889
Range of m1 length 15.4–19.4 mm.
Included species and distribution *T. spegazzinianus* (CB1, LFL, LDM, EP1, QFI, CA1, NEQ, SAL); *T. curuzucuatiense* (FYB, PER); *T. subandinus* (RPL); *T. alloxus* (SAL); *T.? mendocensis* (DVL or MAÑ, see Cerdeño *et al.* 2008); *Trachytherus* sp. (LCN, CMO).

Conclusions

A notable feature of this analysis is the congruence between the branching order of the cladogram and the relative stratigraphic position of the taxa (Fig. 10.5). In no instance does the proximal outgroup to a clade have a first appearance later than any of its ingroup taxa. As many of the derived character of Typotheria are likely associated with the consumption of more abrasives (e.g. hypsodonty, simplified occlusal pattern of the cheek teeth, hypertrophied I1), the tight correlation between cladogenesis and morphologic evolution suggests that dietary selection strongly influenced typotherian evolution. The most robustly supported major clade in the analysis is that defined by Node 4 and makes its first appearance in the late Eocene.

ACKNOWLEDGEMENTS
We thank the editors for the invitation to participate in this volume and for many stimulating discussions and ideas. To Pablo Goloboff in helping us to use the script the MERDA in TNT, and to Martín Ramírez in helping us to use the script of the stability analysis for TNT, written by him. We are grateful to A. Scarano for sharing information on Interatheriidae. We also thank Consejo Nacional de Investigaciones Científicas y Técnicas (CONICET) for permanent research support.

Appendix 10.1 List of characters utilized in the phylogenetic analysis

0. Cristae on upper molars: 0: absent; 1: present.
1. Lingual ridge on lower premolars: 0: absent; 1: present.
2. Relative size of upper incisors. 0: subequal; 1: I1 enlarged.
3. Upper molar crown pattern: 0: cristae unattached or joining crochet; 1: crista 1 joins the protoloph/antecrochet, forming the "face" pattern of enclosed fossettes.
4. I1 implantation: 0: vertically implanted; 1: obliquely implanted and not procumbent; 1: obliquely implanted and procumbent; 2: more transverse implantation and procumbent.
5. I1 form: 0: oval; 1: subtriangular, pointed distally, with mesial sulcus width; 2: rounded, with lingual sulcus; 3: wide and compressed labiolingually with two lingual sulci; 4: subtriangular, pointed distally, with two lingual sulci.
6. I1: 0: with continuous enamel; 1: with labial enamel only.
7. Angle between the mediolateral face of the I1 and the sagittal plane: 0: $<35°$; 1: $40°$; 2: $<45°$; 3: $=45°$; 4: $>45°$ and $<49°$; 5: $=50°$; 6: $=90°$.
8. I2–3-C/i2–3-c: 0: developed; 1: reduced, at times absent in the adult; 2: always absent.
9. Enamel on teeth: 0: continuous around margins; 1: reduced or lacking at the antero- and posterolabial angles of molar; 2: confined to the labial and lingual faces.
10. Metastyle on M3: 0: absent; 1: present.
11. Relative size of P2: 0: at least 80% length of P3; 1: less than 75% of P3; 2: absent.
12. P3–4 paracone/metastyle: 0: shallow; 1: deep.
13. P3–4/p3–4: 0: not molarized; 1: little molarized; 2: completely molarized.
14. P4 shape: 0: subtriangular, short, with central fossette; 1: subtriangular, short, without central fossette; 2: bilobed, with short and poorly defined lingual placation; 3: bilobed, with well-developed and patent lingual placation.
15. Relationship of canine and anterior premolars: 0: all lie directly in line with rest of toothrow; 1: root of C lingual to P1 and p1 lingual to p2; 2: canine absent.
16. M1–2 shape: 0: triangular; 1: quadrangular; 2: bilobed; 3: ovoid; 4: trilobed; 5: rectangular.
17. Presence of lobes on M1–2: 0: absence; 1: median lobe enclosed and vestigial in the adult; 2: median lobe developed and prominent.
18. Mandibular symphysis: 0: wide and long; 1: narrow and short.
19. M3 shape: 0: subtriangular, with fossette; 1: trilobed with little differentiated middle lobe; 2: with short middle lobe, surrounded by the other two lobes; 3: with the middle lobe less enclosed but still shorter than the others; 4: ovoid, without fossete.
20. Upper molars in adults with: 1: three fossettes; 2: one central fossette; 3: without fossettes.
21. Metastylid on M3: 0: absent; 1: present.
22. Mesostyle: 0: present; 1: absent.
23. m2 proportions (length): 0: between 1.60 and 2.30; 1: >2.30; 2: <1.60.
24. Ratio of i1/i2: 0: 1; 1: $=1.5$; 2: between >1.5 and <1.8; 3: between 2.00 and $3.00 > 3.00$.
25. Orientation of deciduous canine and p1: 0: vertical; 1: procumbent.
26. i1–2 shape: 0: cylindrical; 1: lingually bifid; 2: bicolumnar.
27. Section of i1: 0: subcylindrical; 1: rounded with little or no lingual sulcus; 2: subtriangular with smooth sulcus; 3: trapezoidal with well-demarcated lingual sulcus elliptical with smooth sulcus.
28. C/c: 0: hypertrophied; 1: not hypertrophied; 2: absent.
29. P1/p1: 0: present, not reduced; 1: reduced; 2: absent.
30. Angle formed by the ectoloph and the distal face of the M1: 0: greater than 90°; 1: approximately equal to 90°; 2: less than 90°.
31. P3 shape: 0: subquadrangular; 1: subtriangular; 2: absent.
32. Trigonid/talonid of p3: 0: subequal; 1: trigonid $>$ talonid; 2: absent.
33. Maxillary process: 0: absent; 1: present as spur; 2: present as salient, straight and laminar wall.
34. Metacristid on molars: 0: poorly developed; 1: very well developed; 2: absent.
35. Postmetacristid on p3–4: 0: absent; 1: present.
36. Lower molars: 0: with only one fossettid (central); 1: with fossettids, central fosettid + postlingual fossettid; 2: without fossettids.
37. Paracristid on lower molars: 0: prominent; 1: somewhat reduced; 2: absent.

38. Entoconid on lower molars: 0: little expanded; 1: very well expanded; 2: wholly incorporated to the talonid.
39. Accessory cuspid on lower molars: 0: absent; 1: present between the entoconid and hypoconulid.
40. Ratio m2/m3: 0: 1/3; 1: 1/2; 2: 1.
41. Relative size of P2: 0: at least 80% length of P3; 1: less than 75% length of P3; 2: absent.
42. Molars with single deep lingual cleft: 0: absent; 1: present.
43. m3: 0: third lobe absent; 1: third lobe rounded and smaller than the second; 2: third lobe triangular and with the same size as the second.
44. Imbrication of the nasals: 0: absent; 1: with maxillae and premaxillae; 2: with maxillae, premaxillae, and frontals.
45. Diastemata (upper and lower): 0: absent; 1: little developed; 2: posterolaterally divergent; 3: parallel or gently convergent; 4: very convergent.
46. Upper molar imbrication: 0: < 1.25; 1: between 1.25 and 1.50; 2: > 1.50.
47. Rostrum: 0: elongated and low; 1: long and tall; 2: short and tall; 3: short, rodent-like.
48. Lengthening of premaxillae: 0: absent; 1: poorly defined; 1: very pronounced.
49. Paraoccipital process: 0: short; 1: long.
50. Maxillary apophysis: 0: absent; 1: present and prominent; 2: present and laminar.
51. Postorbital constriction: 0: very developed; 1: little developed.
52. Zygomatic plate: 0: absent or hardly developed; 1: little developed; 2: very developed.
53. Root of zygomatic arch: 0: posteriorly directed from the level of M1 or M2; 1: perpendicular to M1; 2: perpendicular to M1 with biconcave edge.
54. Jugal: 0: not reduced; 1: reduced, between the maxillary and the squamosal.
55. Descending process of the maxilla: 0: absent; 1: moderately developed; 1: very developed and long; 2: prominent and short.
56. Mastoid bulla: 0: not expanded or easily visible in dorsal view of skull; 1: large, bulbous, expanded dorsally and medially, easily visible.

57. Feet paraxonic, with lateral digits reduced: 0: absent; 1: present.
58. Shape of the molar trigonid: 0: quadrangular; 1: triangular; 2: rounded.
59. Postorbital process: 0: short; 1: long and transverse; 2: long, directed posteriorly.
60. Suborbital fossa: 0: absent; 1: poorly defined; 2: well developed.
61. Occipital notch 0: absent; 1: present.
62. Anterior rostral notch (premaxillae): 0: very smoothly concave; 1: forming obtuse angle; 2: forming acute angle; 3: tall, wide, U-shaped.
63. Postpalatal notch: 0: narrow, deep, and removed from M3; 1: wide and removed from M3; 2: deep to the level of M3.
64. Anterior and posterior sides of M1 middle lobe: 0: absent; 1: lingually convergent; 2: subparallel; 3: lingually divergent.
65. Lingual exposure of M1 middle lobe: 0: absent; 1: little or none; 2: extensive.
66. Post-incisive depressions: 0: absent; 1: present.
67. Size of the incisive foramina: 0: restricted to maxillae; 1: extending beyond premaxillae.
68. Metapodial distal keel: 0: incomplete; 1: complete.
69. Configuration of tibia and fibula: 0: unfused; 1: short, broad distal fusion; 2: long, narrow distal fusion.
70. Dental cementum: 0: absent; 1: present as a thin layer; 2: present as thick layer.
71. Sagittal crest: 0: low, narrow ridge extending from posterior orbit to occiput; 1: low, Y-shaped ridge bounded by deep grooves for temporalis; 2: absent.
72. Maxillary fossa: 0: small and shallow; 1: large and shallow; 2: well expanded and deep.
73. Hypsodonty Index: 0: < 1; 1: > 1.
74. Horizontal septum in auditory bulla: 0: absent; 1: present.
75. Position of carotid foramen: 0: at anterior margin of bulla; 1: between bulla and basicranium, well forward of the foramen.
76. Vertical septum in auditory bulla: 0: absent; 1: present.
77. Hamular process of pterygoid: 0: present; 1: absent.

Appendix 10.2   Matrix for the phylogenetic analysis

Characters

```
                                   0000000000111111111122222222223333333333444444444455555555556666666666777777777 7
                                   0123456789012345678901234567890123456789012345678901234567890123456789012345 67
```

Taxa	Characters
Simpsonotus praecursor	00000000000000000000000000000010000000000000000?0000?00000000??00000???
Notostylops murinus	001000000200101100010200000012020000011000200000000000??00000003
Pleurostylodon modicus	1100000000001100000010000000000000000002000000000000000??00001000
Acropithecus rigidus	10010000000010000000000100010000000000000000000000?10000000??00000???
Oldfieldthomasia debilitata	100100000000010000000100000010000000000000000000000000?0000000000
Notopithecus adapinus	10110001000012001000000100000011000010000001102?000000000??00000000
Guilielmoscottia plicifera	10?10?0?000012001000?0010000?710000010000?110000010?0??????0??0?1????
Eopachyrucos pliciferus	10?1???00?00?01?0?0?0010000???0001?710000100010?0??????0??0??1????
Santiagorothia chiliensis	101100010000120011000010000100010101100200000001102000000010000??10010000
Plagiarthrus clivus	101100010000120021000010000010001100200110020000000??10020000
Archaeophylus patrius	101?????00012002100000010???710000??2102200??0010?0??2??????
Cochilius volvens	1011000012120002000100000020022001000001101000000100000010020000
Interatherium rodens	10110001000012120200000100010000020002201000000120000000001000001010040000
Protypotherium australe	101100010000121050000200100010200200000000110100000010000001000000100030000
Miocochilius anomopodus	1011000100001210500020100101000102000220100000011010000001000000100030000
Campanorco inauguralis	1?1111011010020010?000?0?00?2?710000?0?0?0102000010000?200110000002?00000000
Eohyrax brachyodus	10?1?????0010020010000010???710000?0010100?00?00?20?????????0??2?2???????
Punahyrax bondesioi	10?1?????01?020?11?000010???710000?0011200??000?0?????????0??2?2???????
Pseudhyrax eutrachytheroides	10111101001002001100001102?710000102?1110002001?0?????????2?0???0??1????
Protarchaeohyrax gracilis	101111011002001100101000000010100000001010000000?00100000?00110000?10020111
Archaeotypotherium propheticus	10111101100200110001000001000010000000101000001010000?2?1?0?????1?0???2?111
Archaeohyrax patagonicus	1011110111100200111010010010010000000010201001000010000?0010000??10030111
Prohegetotherium smithwoodwardi	1011211100003020142010110002100212022201201120011100010102010000?0011214011
Medistylus dorsatus	1?11211220003020?42002????210?1??????0101001201121012100000??202010000??12140111
Prosotherium garzoni	1011211220003020142010110002100020222010100101021012101200010101001120111
Pachyrukhos moyanoi	101121122100021030142010110021200220222010020011200212000100002121601 11
Tremacyllus impressus	101121122100021030142010110021200220220201020010201001012150111
Paedotherium minor	1011211221000210301420101100212002202220110201020100000121150111
Paedotherium bonaerense	1011211221000210301420101100212002202220110201020100000121601 11
Prohegetotherium sculptum	10?1????0100021030042010100112002202220100?200???????2?00???1?12????
Prohegetotherium schiaffinoi	10112111100002130140201011001100020222010000002000120000?0?201010000??111120111
Sallatherium altiplanense	101121111000021030142010110011001100020222010000022001200000?20101000007?11130111
Hegetotherium mirabile	1011211110002130140201011002110002022201000020011200011130111
```

164
```

```
Hemihegetotherium          10112111110102113014201010110021100002022201100000200012000 0?201010000?0111240111
  achataleptum

Hemihegetotherium torresi      10112111110102113014201010110021100002022201100?0020?0120?????????0000?0111?40111
Trachytherus spegazzinianus    10111101110010111100011110000111000000022010001002000010030021211110000101300000
Anatrachytherus soriai         ?0???????01??01??0??0??0??0110110?0??0?000022?0?2??0?0?????????1?2????1??3????
Altitypotherium paucidens      101124012202124213211001042020000222012000210322000210?20?1??0??20023?????
Altitypotherium chucalensis    1011?401221202124212211001042201200022201200021022000210?20?2?0?20023????
Eotypotherium  chico           1011?10?2211202024212211001012200200020002220120002?0?20?210?2?1?????2022????
Microtypotherium               10112101220212421221122??12201200022201200022201200210?3??20021?20?1?0??20230???
  choquecotense

Eutypotherium lehmannitschei   101122022211202022442212211001022201200022201200210?1?0?200020?11?112??0020023????
Eutypotherium superans         101122022212021242132112??????220012000222012000220021?0?20000?07112112000?220230000
Typotheriopsis silveyrai       1011??322120212421221112????220012000222012000220031100?200020?0221202200?20230000
Typotheriopsis chasicoensis    101122032212021242124213211011022201200022201200223110020000?0121?02200?20230000
Plesiotypotherium achirense    101122052212021242132110010222012000222012000220311002100002201112200?20240000
Pseudotypotherium exiguum      1011??????2212023242132111210?2220120002220120002?23?20?210?2?01???12???????2004????
Pseudotypotherium hystatum     1011??712212023242132111?????2220120002220120002?23110020000001?120?200???2004????
Mesotherium pachygnathum       10113201221202324213211113102220120002220120002231100210000020137230000200240000
Mesotherium maendrum           101133012212023242132111310322012000222012000220232321000000201223200002200240000
Mesotherium cristatum          1011320122120232421321113102220120002220120002201320000002013232000000200240000
```

Appendix 10.3   Localities listed in the systematic paleontology

Code	Locality name	State/country	SALMA	Stratigraphy	Age
AGB	Aguada Batistín	Chubut/Argentina	Barrancan	Sarmiento Fm.	middle Eocene
BDD	Bajada del Diablo	Chubut/Argentina	Deseadan	Sarmiento Fm.	late Oligocene
BJP	Bajo Palangana	Chubut/Argentina	Riochican	Koluel-Kaike Fm.	late Paleocene
CA1	Las Cascadas	Chubut/Argentina	Deseadan	Sarmiento Fm.	late Oligocene
CA2	Las Cascadas	Chubut/Argentina	Barrancan	Sarmiento Fm.	middle Eocene
CAN	La Cancha (GBV-4), Gran Barranca	Chubut/Argentina	Tinguirirican?	Sarmiento Fm.	early Oligocene
CB1	Cabeza Blanca	Chubut/Argentina	Deseadan	Sarmiento Fm.	late Oligocene
CB2	Cabeza Blanca	Chubut/Argentina	Barrancan	Sarmiento Fm.	middle Eocene
CDH	Cerro del Humo	Chubut/Argentina	Mustersan	Sarmiento Fm.	late Eocene
CEB	Cerro Blanco	Chubut/Argentina	Barrancan	Sarmiento Fm.	middle Eocene
CGU	Cerro Guacho	Chubut/Argentina	Barrancan	Sarmiento Fm.	middle Eocene
CHJ	Estancia La Matilde, Chajarí	Entre Ríos/ Argentina	Deseadan	Fray Bentos Fm.	late Oligocene
CMO	Cerro Mono, Pan de Azúcar	Moquegua/Peru	Deseadan	Upper Moquegua Fm.	late Oligocene
CÑB	Cañadón Blanco	Chubut/Argentina	Tinguirirican	Sarmiento Fm.	early Oligocene
CÑL	Cañadón Lobo/Cañadón Tournouer	Santa Cruz/ Argentina	Barrancan	Sarmiento Fm.	middle Eocene
CÑV	Cañadón Vaca	Chubut/Argentina	Vacan	Sarmiento Fm.	?early Eocene
CON	Colón	Entre Ríos/ Argentina	Deseadan	Fray Bentos Fm.	late Oligocene
CSM	Punta Casamayor	Santa Cruz/ Argentina	Barrancan	Sarmiento Fm.	middle Eocene
CU2	La Curandera/Lomas Blancas	Chubut/Argentina	Tinguirirican	Sarmiento Fm.	early Oligocene
CU3	La Curandera	Chubut/Argentina	Barrancan	Sarmiento Fm.	middle Eocene
DVL	Divisadero Largo	Mendoza/ Argentina	Divisaderan	Divisadero Largo Fm.	early Eocene?
EP1	El Pajarito	Chubut/Argentina	Deseadan	Sarmiento Fm.	late Oligocene
EP2	El Pajarito	Chubut/Argentina	Barrancan	Sarmiento Fm.	middle Eocene
ERT	East Ridge Tinguiririca	Región IV/Chile	Tinguirirican	Abanico (Coya-Machalí) Fm.	early Oligocene
FLO	Las Flores	Chubut/Argentina	Itaboraian	Las Flores Fm.	middle–late Paleocene
FBY	Fray Bentos	Fray Bentos/ Uruguay	Deseadan	Fray Bentos Fm.	late Oligocene
GBT	Y Tuff, Gran Barranca	Chubut/Argentina	Barrancan	Sarmiento Fm.	middle Eocene
GBY	Above Y, Gran Barranca	Chubut/Argentina	Barrancan	Sarmiento Fm.	middle Eocene

(*cont.*)

Code	Locality name	State/country	SALMA	Stratigraphy	Age
LCN	La Cantera (GBV-19), Gran Barranca	Chubut/Argentina	Tinguirirican?	Sarmiento Fm.	?early Oligocene
LDB	Laguna de la Bombilla	Chubut/Argentina	Deseadan	Sarmiento Fm.	late Oligocene
LDM	Laguna del Mate/ Laguna Grande	Chubut/Argentina	Deseadan	Sarmiento Fm.	late Oligocene
LFL	La Flecha	Santa Cruz/ Argentina	Deseadan	Sarmiento Fm.	late Oligocene
LGF	Laguna Fría, Paso del Sapo	Chubut/Argentina	Vacan?	"Tufolitas de Laguna del Hunco"	early? Eocene
LGH	La Gran Hondonada	Chubut/Argentina	Mustersan	Sarmiento Fm.	late Eocene
LLM	Laguna Los Machos	Chubut/Argentina	Deseadan	Sarmiento Fm.	late Oligocene
MAÑ	Mariño	Mendoza/ Argentina	Santacrucian?	Mariño Fm.	Miocene
NEQ	Neuquén	Neuquén/ Argentina	Deseadan	Unnamed fm.	late Oligocene
NTN	North Tinguiririca	Región IV/Chile	Tinguirirican	Abanico (Coya-Machalí) Fm.	early Oligocene
NUE	El Nuevo (GBV-60), Gran Barranca	Chubut/Argentina	Barrancan?	Sarmiento Fm.	middle Eocene
PEF	Portezuelo El Fierro	Región IV/Chile	Tinguirirican	Abanico (Coya-Machalí) Fm.	early Oligocene
PER	Arroyo María Grande, Perugorria	Corrientes/ Argentina	Deseadan	Fray Bentos Fm.	late Oligocene
PTR	Pico Truncado	Santa Cruz/ Argentina	Deseadan	Sarmiento Fm.	late Oligocene
QFI	Quebrada Fiera	Mendoza/ Argentina	Deseadan	Agua de la Piedra Fm.	late Oligocene
RBY	Rocas Bayas	Río Negro/ Argentina	Tinguirirican	Unnamed fm.	early Oligocene
RCO	Río Corrientes	Corrientes/ Argentina	Deseadan	Fray Bentos Fm.	late Oligocene
RDL	Rinconada de los López	Chubut/Argentina	Barrancan	Sarmiento Fm.	middle Eocene
RDZ	Rincón del Zampal	Chubut/Argentina	Deseadan	Sarmiento Fm.	late Oligocene
RLL	Rinconada de los López	Chubut/Argentina	Tinguirirican	Sarmiento Fm.	early Oligocene
ROS	El Rosado (GBV-3), Gran Barranca	Chubut/Argentina	Musteran	Sarmiento Fm.	late Eocene
RPL	Río Pluma	Cochabamba/ Bolivia	Deseadan	Unnamed fm.	late Oligocene
SAL	Salla	La Paz/Bolivia	Deseadan	"Estratos de Salla"	late Oligocene
SPK	Scarritt Pocket	Chubut/Argentina	Deseadan	Sarmiento Fm.	late Oligocene
STQ	Sierra Talquino	Chubut/Argentina	Mustersan	Sarmiento Fm.	late Eocene
VAP	Valle de Punilla	Córdoba/ Argentina	Mustersan	Unnamed fm.	late Eocene
VRS	VRS Tuff, below Y, Gran Barranca	Chubut/Argentina	Barrancan	Sarmiento Fm.	middle Eocene

REFERENCES

Alston, E. R. 1876. On the classification of the order Glires. *Proceedings of the Zoological Society of London*, **1876**, 61–98.

Ameghino, F. 1887. Enumeración sistemática de las especies de mamíferos fósiles coleccionados por Carlos Ameghino en terrenos eocenos de la Patagonia austral y depositados en el Museo La Plata. *Boletín del Museo de La Plata*, **1**, 1–26.

Ameghino, F. 1889. Contribución al conocimiento de los mamíferos fósiles de la República Argentina. *Actas de la Academia Nacional de Ciencias de Córdoba*, **6**, 1–1027.

Ameghino, F. 1891. Mamíferos y aves fósiles argentinas: especies nuevas, adiciones y correcciones. *Revista Argentina de Historia Natural*, **1**, 240–259.

Ameghino, F. 1894. Enumération synoptique des espèces de mammifères fossiles des formations éocènes de Patagonie. *Boletín de la Academia Nacional de Ciencias en Córdoba*, **13**, 259–445.

Ameghino, F. 1895. Sur les oiseaux fossiles de Patagonie et la faune mammalogique des couches à *Pyrotherium*: première contribution à la connaissance de la faune mammalogique des couches à *Pyrotherium*. *Boletín del Instituto Geográfico Argentino*, **15**, 603–660.

Ameghino, F. 1897. Mammifères crétacés de l'Argentine: deuxieme contribution à la connaissance de la fauna mammalogique des couches à *Pyrotherium*. *Boletín del Instituto Geográfico Argentino*, **18**, 406–429, 431–521.

Ameghino, F. 1901. Notices préliminaires sur des ongulés nouveaux des terrains crétacés de Patagonie. *Boletín de la Academia Nacional de Ciencias en Córdoba*, **16**, 349–426.

Ameghino, F. 1902. Cuadro sinóptico de las formaciones sedimentarias terciarias y cretáceas de la Argentina en relación con el desarrollo y descendencia de los mamíferos. *Anales del Museo Nacional de Buenos Aires*, **8**, 1–12.

Ameghino, F. 1904. Nuevas especies de mamíferos cretáceos y terciarios de la República Argentina. *Anales de la Sociedad Científica Argentina*, **56**, 162–175.

Ameghino, F. 1906. Les formations sédimentaires du Crétacé supérieur et du Tertiaire de Patagonie avec un parallèle entre leurs faunes et celles de l'ancien continent. *Anales del Museo Nacional de Buenos Aires*, **15**(8), 1–568.

Billet, G., B. Patterson, and C. De Muizon 2009. Craniodental anatomy of late Oligocene archaeohyracids (Notoungulata, Mammalia) from Bolivia and Argentina and new phylogenetic hypotheses. *Zoological Journal of the Linnean Society*, **155**, 458–509.

Bond, M., M. G. Vucetich, and R. Pascual 1984. Un nuevo Notoungulata de la Formación Lumbrera (Eoceno) de la provincia de Salta, Argentina. *I Jornadas Argentinas de Paleontología Vertebrados, Resúmenes*, 20.

Bremer, K. 1990. Combinable component consensus. *Cladistics*, **6**, 369–372.

Cerdeño, E. and M. Bond 1998. Taxonomic revision and phylogeny of *Paedotherium* and *Tremacyllus* (Pachyrukhinae, Hegetotheriidae, Notoungulata) from the late Miocene to Pleistocene of Argentina. *Journal of Vertebrate Paleontology*, **18**, 799–811.

Cerdeño, E. and C. I. Montalvo 2001. Los Mesotheriinae (Mesotheriidae, Notoungulata) del Mioceno Superior de La Pampa, Argentina. *Revista Española de Paleontología*, **16**, 63–75.

Cerdeño, E., G. M. López, and M. A. Reguero 2008. Biostratigraphic considerations of the Divisaderan faunal assemblage. *Journal of Vertebrate Paleontology*, **28**, 574–577.

Cifelli, R. 1985. South American ungulate evolution and extinction. In Stehli, F. and Webb, S. D. (eds.), *The Great American Biotic Interchange*. New York: Plenum, pp. 249–266.

Cifelli, R. L. 1993. The phylogeny of the native South American ungulates. In Szalay, F. S., Novacek, M. J., and McKenna, M. C. (eds.), *Mammal Phylogeny: Placentals*. New York: Springer-Verlag. pp. 195–216.

Coddington, J. A. and N. Scharff 1994. Problems with zero-length branches. *Cladistics*, **10**, 415–423.

Croft, D. A. 1999. Placentals: South American ungulates. In Singer, R. (ed.), *Encyclopedia of Paleontology*. Chicago, IL: Fitzroy-Dearborn, pp. 890–906.

Croft, D. A. 2000. Archaeohyracidae (Mammalia, Notoungulata) from the Tinguiririca Fauna, central Chile, and the evolution and paleoecology of South American mammalian herbivores. Ph.D. thesis, University of Chicago.

Croft, D. A. and F. Anaya 2004. A new hegetotheriid from the middle Miocene of Quebrada Honda, Bolivia, and a phylogeny of the Hegetotheriidae. *Journal of Vertebrate Paleontology*, **24**(3 Suppl.), 48–49A.

Croft, D. A. and F. Anaya 2006. A new middle Miocene hegetotheriid (Notoungulata: Typotheria) and a phylogeny of the Hegetotheriidae. *Journal of Vertebrate Paleontology*, **26**, 387–399.

Croft, D. A., M. Bond, J. J. Flynn, M. A. Reguero, and A. R. Wyss 2003. Large archaeohyracids (Typotheria, Notoungulata) from central Chile and Patagonia including a revision of *Archaeotypotherium*. *Fieldiana, Geology*, n.s., **49**, 1–38.

Croft, D. A., J. J. Flynn, and A. R. Wyss 2004. Notoungulata and Litopterna of the early Miocene Chucal Fauna, northern Chile. *Fieldiana, Geology*, n.s., **50**, 1–52.

Goloboff, P. A. 1993. Estimating character weights during tree search. *Cladistics*, **9**, 83–91.

Goloboff, P. 1995. Parsimony and weighting: a reply to Turner and Zandee. *Cladistics*, **11**, 91–104.

Goloboff, P. 1997. Self-weighted optimization: tree searches and character state reconstructions under implied transformation costs. *Cladistics*, **13**, 225–245.

Goloboff, P., J. S. Farris, and K. Nixon 2003a. *TNT Tree Analysis Using New Technology. Version 1.0*, program and documentation. www.zmuc.dk/public/phylogeny.

Goloboff, P., J. S. Farris, M. Källersjö, B. Oxelman, M. J. Ramírez, and C. A. Szumik 2003b. Improvements to resampling measures of group support. *Cladistics*, **19**, 324–332.

Hitz, R. 1997. Contributions to South American mammalian paleontology: new interatheres (Notoungulata) from Chile and Bolivia, typothere (Notoungulata) phylogeny, and paleosols from the late Oligocene Salla Beds. Ph.D. thesis, University of California, Santa Barbara.

Hitz, R., M. Reguero, A. R. Wyss, and J. J. Flynn 2000. New interatherines (Interatheriidae, Notoungulata) from the Paleogene of Central Chile and Southern Argentina. *Fieldiana, Geology*, n.s., **42**, 1–26.

Kraglievich, L. 1931. Cuatro notas paleontológicas (sobre "*Octomylodon aversus*" Amegh., "*Argyrolagus palmeri*" Amegh., "*Tetrastylus montanus*" Amegh., y "*Muñizia paranensis*" n.gen., n. sp.). *Physis*, **10**, 242–266.

Lee, D. G. and H. N. Bryant 1999. A reconsideration of the coding of inapplicable characters: assumptions and problems. *Cladistics*, **15**, 373–378.

Lopardo, L. 2005. Phylogenetic revision of the spider genus *Negayan* (Araneae, Anyphaenidae, Amaurobioidinae). *Zoologica Scripta*, **34**, 245–277.

Lydekker, R. 1886. *Catalogue of the Fossil Mammalia in the British Museum (Natural History)*, vol. 3. London: British Museum.

Lydekker, R. 1894. Contribuciones al conocimiento de los vertebrados fósiles de la Argentina. 1. Observaciones adicionales sobre los ungulados argentinos. *Anales del Museo de La Plata, Paleontología*, **2**, 1–91.

Norell, M. A. and W. C. Wheeler 2003. Missing entry replacement data analysis: a replacement approach to dealing with missing data in paleontological and total data sets. *Journal of Vertebrate Paleontology*, **23**, 275–283.

Pascual, R., M. G. Vucetich, and J. Fernández 1978. Los primeros mamíferos (Notoungulata, Henricosborniidae) de la Formación Mealla (Grupo Salta, Subgrupo Santa Bárbara): sus implicancias filogenéticas, taxonómicas y cronológicas. *Ameghiniana*, **15**, 366–390.

Patterson, B. 1934a. Upper premolar–molar structure in the Notoungulata with notes on taxonomy. *Field Museum of Natural History*, geol. ser., **6**, 91–111.

Patterson, B. 1934b. *Trachytherus*, a typotherid from the Deseado beds of Patagonia. *Field Museum of Natural History, geol. ser.*, **6**, 119–139.

Patterson, B. 1936. The internal structure of the ear in some notoungulates. *Geological Series of the Field Museum of Natural History*, **6**, 199–227.

Patterson, B. and R. Pascual 1972. The fossil mammal fauna of South America. In Keast, A., Erk, F. C., and Glass, B. (eds.), *Evolution, Mammals and Southern Continents*. Albany, NY: State University of New York Press, pp. 247–309.

de Queiroz, K. and J. Gauthier 1990. Phylogeny as a central principle in taxonomy: phylogenetic definitions of taxon names. *Systematic Zoology*, **39**, 307–322.

Ramírez, M. J. 2003. The spider subfamily Amaurobioidinae (Araneae, Anyphaenidae): a phylogenetic revision at generic level. *Bulletin of the American Museum of Natural History*, **277**, 1–262.

Reguero, M. A. 1994. Filogenia y clasificación de los Typotheria Zittel 1893 y Hegetotheria Simpson 1945 (Mammalia, Notoungulata). *Novenas Jornadas Argentinas de Mastozoología, Resúmenes*, **92**.

Reguero, M. A. 1999. El problema de las relaciones sistemáticas y filogenéticas de los Typotheria y Hegetotheria (Mammalia, †Notoungulata): análisis de los taxones de Patagonia de la Edad-mamífero Deseadense (Oligoceno). Ph.D. thesis, University of Buenos Aires.

Reguero, M. A. and P. V. Castro 2004. Un nuevo Trachytheriinae (Mammalia, †Notoungulata) del Deseadense (Oligoceno tardío) de Patagonia, Argentina: implicancias en la filogenia, biogeografía y bioestratigrafía de los Mesotheriidae. *Revista Geológica de Chile*, **31**, 45–64.

Reguero, M. and E. Cerdeño 2005. New late Oligocene Hegetotheriidae (Mammalia, Notoungulata) from Salla, Bolivia. *Journal of Vertebrate Paleontology*, **25**, 674–684.

Reguero, M. A. and R. L. Cifelli 1997. Deseadan Archaeohyracidae from Salla Bolivia. *Ameghiniana*, **34**, 539.

Reguero, M., M. Bond, and G. López 1996. *Campanorco inauguralis* (Typotheria, Notoungulata): an approach to the phylogeny of the Typotheria. *Journal of Vertebrate Paleontology*, **16**, 59A.

Reguero, M. A., M. Ubilla, and D. Perea 2002. A new species of *Eopachyrucos* (Mammalia, Notoungulata, Interatheriidae) from the late Oligocene of Uruguay. *Journal of Vertebrate Paleontology*, **23**, 445–457.

Reguero, M. A., D. A. Croft, J. J. Flynn, and A. R. Wyss 2003. Small archaeohyracids from Chubut, Argentina and Central Chile: implications for trans-Andean temporal correlation. *Fieldiana, Geology*, n.s., **48**, 1–17.

Reguero, M. A., M. T. Dozo, and E. Cerdeño 2007. A poorly known rodentlike mammal (Pachyrukhinae, Hegetotheriidae, Notoungulata) from the Deseadan (late Oligocene) of Argentina: paleoecology, biogeography, and radiation of the rodentlike ungulates in South America. *Journal of Paleontology*, **81**, 1298–1304.

Reig, O. A. 1981. Teoría del origen y desarrollo de la fauna de mamíferos de América del Sur. *Monographiae Naturae*, **1**, 1–162.

Roth, S. 1899. Aviso preliminar sobre mamíferos mesozoicos encontrados en Patagonia. *Revista del Museo de La Plata*, **9**, 381–388.

Roth, S. 1902. Notas sobre algunos nuevos mamiferos fósiles. *Revista del Museo de La Plata*, **10**, 251–56.

Roth, S. 1903. Los ungulados sudamericanos. *Anales del Museo La Plata, Paleontología Argentina*, **5**, 1–36.

Scott, W. B. 1913. *A History of Land Mammals in the Western Hemisphere*. New York: Macmillan.

Simpson, G. G. 1935. Occurrence and relationships of the Río Chico fauna of Patagonia. *American Museum Novitates*, **818**, 1–21.

Simpson, G. G. 1945. The principles of classification and a classification of mammals. *Bulletin of the American Museum of Natural History*, **85**, 1–350.

Simpson, G. G. 1948. The beginning of the Age of Mammals in South America. I. *Bulletin of the American Museum of Natural History*, **91**, 1–232.

Simpson, G. G. 1967. The beginning of the Age of Mammals in South America. II. *Bulletin of the American Museum of Natural History*, **137**, 1–259.

Simpson, G. G., J. L. Minoprio, and B. Patterson 1962. The mammalian fauna of the Divisadero Largo Formation, Mendoza, Argentina. *Bulletin of the Museum of Comparative Zoology*, **127**, 139–293.

Sinclair, W. J. 1909. Typotheria of the Santa Cruz beds. *Reports of the Princeton University Expedition to Patagonia*, **6**, 1–110.

Stirton, R. A. 1952. *Medistylus*, new name for *Phanophilus* Ameghino, not Sharp. *Journal of Paleontology*, **26**(3), 351.

Strong, E. E. and D. Lipscomb 1999. Character coding and inapplicable data. *Cladistics*, **11**, 363–371.

Zittel, K. A. von 1893. *Handbuch der Palaeontologie* Part I, *Palaeozoologie*, vol. IV, *Vertebrata (Mammalia)*. Munich: R. Oldenbourg.

11 The Leontiniidae (Mammalia, Notoungulata) from the Sarmiento Formation at Gran Barranca, Chubut Province, Argentina

Ana Maria Ribeiro, Guillermo M. López, and Mariano Bond

Abstract

At Gran Barranca, pyroclastic sediments comprising lithostratigraphic units of the Sarmiento Formation include several fossil-bearing levels containing Leontiniidae: three in the Upper Puesto Almendra Member (GBV-19 and GBV-34 in Unit 3, and GBV-35 in Unit 4) and five in the Lower Fossil Zone of the Colhue-Huapi Member (GBV-8, GBV-9, GBV-10, GBV-38, and GBV-43). Two new taxa occur at GBV-19; one of these, *Scarrittia barranquensis* n. sp., also occurs at GBV-34. This new taxon shares with *S. canquelensis* and *S. robusta* enlarged and caniniform incisors I1/i3, but is smaller and the premolar morphology is distinct. The second new taxon is *Henricofilholia vucetichia* n. sp. from GBV-19, much smaller than *S. barranquensis* and lacking labial and lingual cingulids on the premolars and molars. A right dentary fragment from GBV-35 with worn m1–3 is here assigned to cf. *S. canquelensis*. The genus *Scarrittia* is cited for the first time at Gran Barranca. The only leontiniid from the Lower Fossil Zone of the Colhue-Huapi Member (type Colhuehuapian) is *Colpodon distinctus*.

Resumen

La secuencia de sedimentos piroclásticos desarrollada en Gran Barranca se refiere a la Formación Sarmiento y en ella se reconocen tres niveles fosilíferos (GBV-19, GBV-34 y GBV-35) para el Miembro Puesto Almendra Superior y varios (e.g. GBV-8, GBV-9, GBV-10, GBV-38 and GBV-43) para el Miembro Colhue-Huapi. En este trabajo se presenta una revisión del material referido a Leontiniidae exhumado en estas localidades, las que se refieren a las edades Deseadense y Colhuehuapense respectivamente. En la localidad GBV-19 se reconocen dos nuevos taxones, uno de los cuales, *Scarrittia barranquensis* n. sp., también se registra en GBV-34. Este nuevo taxón presenta I1/i3 agrandados y caniniformes al igual que *S. canquelensis* y *S. robusta*, pero es más pequeño en tamaño y difiere en la morfología premolar. El segundo taxón de GBV-19, *Henricofilholia vucetichia* n. sp., es significativamente más pequeño y

The Paleontology of Gran Barranca: Evolution and Environmental Change through the Middle Cenozoic of Patagonia, eds. R. H. Madden, A. A. Carlini, M. G. Vucetich, and R. F. Kay. Published by Cambridge University Press.

carece de cingúlidos labial y lingual en los premolares y molares. En el nivel GBV-35 se registra un fragmento mandibular derecho con m1–3 muy gastados, y es aquí presentado como cf. *S. canquelensis*. También se discute el registro de *H. lustrata* para niveles deseadenses de Gran Barranca. El género *Scarrittia* es citado por primera vez para Gran Barranca. El único leontínido del Colhuehuapense es *Colpodon* y los materiales registrados en Gran Barranca (GBV-38 y GBV-43) son referidos a *C. distinctus*.

Introduction

Leontiniidae are endemic South American herbivores of medium to large size with distinctive brachydont to mesodont dentitions. In 1885, Burmeister described *Colpodon propinquus*, which later was included in the family that Ameghino (1895) proposed for notoungulates with enlarged and caniniform upper and lower third incisors (e.g. *Leontinia gaudryi* Ameghino 1895). With more complete material Ameghino (1897) observed that it was actually the second and not the third upper incisor that was enlarged. Over a ten-year period around the turn of the last century, Ameghino (1895, 1897, 1901, 1902, 1904a) described another 12 genera and 30 species for the family.

Martinmiguellia fernandezi Bond and Lopez 1995, a generalized leontiniid of medium size with a complete dental series and brachydont dentition, is the oldest known member of the family and comes from Eocene Casa Grande Formation of northwestern Argentina. Other important older material has been presented by Powell and Deraco (2003), Deraco and Powell (2004), and Deraco *et al.* (2008) from the upper levels of the Lumbrera Formation in Salta Povince, Argentina. The skull and mandible of *Coquenia bondi* present few characters distinct from *M. fernandezi*, and these two taxa are distinct from Leontiniidae elsewhere. Most leontiniids come from Deseadan age deposits in Argentina, Bolivia, Brazil and Uruguay. During the Colhuehuapian, leontiniids are represented only by *Colpodon*. Subsequently, leontiniids disappear from the Argentine fossil record and all younger records come from the Miocene of Colombia and Brazil at more equatorial latitudes.

170

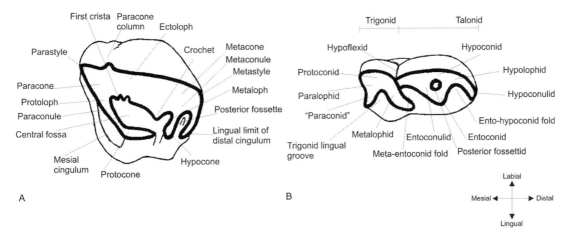

Fig. 11.1. Notoungulate tooth crown nomenclature (from Van Frank 1957; Van Valen 1966; Szalay 1969; Bond 1988). (A) left upper molar; (B) right lower molar.

Much new material of Leontiniidae has been collected at Gran Barranca over the last ten years, with stratigraphic, paleomagnetic, and geochronologic control (e.g. Kay *et al.* 1999; Ré *et al.* 2005; Ré *et al.* Chapter 3, this volume). In this work we describe new material of Leontiniidae from levels at Gran Barranca broadly understood as Colhuehuapian and Deseadan, and compare these with material from other Patagonian and extra-Patagonian localities.

Abbreviations

Institutional
AMNH, American Museum of Natural History, New York; **FMNH**, Field Museum of Natural History, Chicago, Illinois; **MACN**, Museo Argentino de Ciencias Naturales "Bernardino Rivadavia," Argentina; **MLP**, Museo de La Plata (Departamento Paleontología de Vertebrados), La Plata, Argentina; **MNHN**, Muséum National d'Histoire Naturelle, Paris, France; **MPEF-PV**, Museo Paleontológico "Egidio Feruglio," Trelew, Chubut, Argentina.

Anatomical
Dental terminology follows the convention of using uppercase letters for the upper incisors, canines, premolars, and molars and lower-case for the lower series. Deciduous teeth are prefixed with "d." Notoungulate tooth crown nomenclature used to describe Leontiniidae in this work is illustrated in Fig. 11.1.

Systematic paleontology

Order **NOTOUNGULATA** Roth 1903
Suborder **TOXODONTIA** Owen 1853
Family **LEONTINIIDAE** Ameghino 1895
Genus *Scarrittia* Simpson in Chaffee 1952

Scarrittia Simpson 1934, p. 2; Chaffee 1952, p. 517, plates 7, 8, 9 (Figs. 1, 2, 3, 4, 5), 10 (Figs. 1, 2), 11 (Figs. 1, 2), and 12 (Figs. 1,2); Simpson 1945, p. 127; Ubilla, Perea and Bond 1994, p. 94; McKenna and Bell 1997, p. 458.

Type species *Scarrittia canquelensis* Simpson in Chaffee 1952, p. 517

Referred species *Scarrittia canquelensis* Simpson in Chafee 1952; *S. robusta* Ubilla, Perea and Bond 1994; *S. barranquensis* n. sp.

Revised diagnosis Large Leontiniidae with a complete dental formula without diastema; I1 and i3 enlarged and compressed mesiodistally; small I2; continuous mesiolingual cingulum on the upper premolars; lacking a lingual groove on the P2–4 protocone; short mandible with narrow symphyseal region.

Geographic and chronologic distribution Argentina and Uruguay, Deseadan SALMA.

Scarrittia barranquensis n. sp.
Fig. 11.2.

Holotype MPEF-PV 7753, an incomplete skull with right I3–C, P2–M2 and alveoli of I1–2, C, and P1; left P2–4 and alveoli for I1–3, C, and P1.

Referred material MPEF-PV 6108, right M3; MPEF-PV 6109, right M1; MPEF-PV 6125, left maxilla fragment with P2–4; MPEF-PV 6126, left dentary fragment with p3–m3; MPEF-PV 6131, right P2; MPEF-PV 6132, right dentary fragment with p3–m1; MPEF-PV 6133, left fragment dentary with p4–m2 and m3 broken; MPEF-PV 6134, right P3; MPEF-PV 6137, left M1; MPEF-PV 6139, right P1; MPEF-PV 6141, left dentary fragment with m1–3; MPEF-PV 6144, right M3; MPEF-PV 6143, left P2–4; MPEF-PV 6149, left M1; MPEF-PV 6152, left p2 and P2; MPEF-PV 6227, right m3; MPEF-PV 6229, right

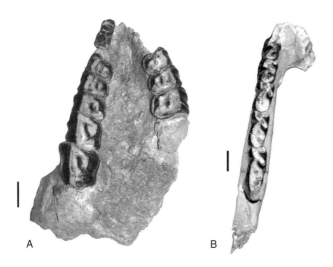

Fig. 11.2. *Scarrittia barranquensis* n. sp. (A) MPEF-PV 7753 (holotype), skull in palatal view with right I3–C, P2–M2 and I1–2, C, P1 alveoli; left P2–4 and I1–3, C, P1 alveoli; (B) MPEF-PV 6609, left mandible with p3–m3 in occlusal view. Scale bar 2 cm.

dentary fragment with p2–m2, MPEF-PV 6231, right dentary fragment with p4–m2; MPEF-PV 6232, left m3; MPEF-PV 6244, right p3; MPEF-PV 6245, right P4; MPEF-PV 6250, left P4; MPEF-PV 6252, right P2; MPEF-PV 6258, right I3; MPEF-PV 6259, right i1 and p3; MPEF-PV 6266, right P1; MPEF-PV 6267, fragmentary left maxilla with P3–4; MPEF-PV 6270, right M3; MPEF-PV 6271, isolated right and left I3 and C; MPEF-PV 6276, right I3 and left P4; MPEF-PV 6609, left mandibular ramus with p3–m3; MPEF-PV 6611, fragmentary right maxilla with P3–M2 and M3 broken; MPEF-PV 6620, right M1; MPEF-PV 6684, mandibular ramus with left i1,c, p1–m1 and right i1–3, p3–m1; MPEF-PV 6858, right p4; MPEF-PV 6953, left p4; MPEF-PV 6955, right P2–3; MPEF-PV 7082, right I1 and I3; MPEF-PV 7086, partial right mandibular ramus with m1–3, and left p3–m3; MPEF-PV 7346, left P3–4; MPEF-PV 7697, fragmentary right maxilla with P2–4; MPEF-PV 7701, right maxilla fragment with M1–3; MPEF-PV 7708, right maxilla fragment with M1–3; MPEF-PV 7718, left maxilla fragment with P2–3; MPEF-PV 7723, partial right mandibular ramus with p4–m3; MPEF-PV 7728, left dentary fragment with m3; MPEF-PV 7730, fragmentary right premaxilla with I2–3, left I3, right maxilla with P2–3; left maxilla with P2–4, M2–3, and right dentary fragment with m1–2; MPEF-PV 7736, right dentary fragment with p4–m3; MPEF-PV 7739, right dentary fragment with p3, m1–3; MPEF-PV 7752, right maxilla fragment with P2–4; MPEF-PV 7944, right maxilla fragment with P4-M3.

Diagnosis Smaller than *S. canquelensis* and *S. robusta*; the molars with mean occlusal areas

22% smaller than in *S. canquelensis*; dentition more brachydont than in *S. canquelensis*, hypocone absent on P1–3, and much reduced on P4, premolars with a distinct paracone column and an indistinct metacone column, labial face relatively flat.

Type locality Gran Barranca Vertebrate locality GBV-19 "La Cantera," Stratigraphic Profile A, and GBV-34, both in Unit 3 of the Upper Puesto Almendra Member, Sarmiento Formation at Gran Barranca, Chubut Province, Argentina.

Etymology In reference to geographic provenance.

Description *Skull* The holotype MPEF-PV 7753 (Fig. 11.2A) preserves the rostrum and from this it can be inferred that *Scarrittia barranquensis* n. sp. had a skull slightly smaller than that of *S. canquelensis* and *Leontinia gaudryi* and a facial region shorter that *L. gaudryi*. The anterior foramen of the infraorbital canal is located at the level of M1. The osseous palate is elongate, concave, and probably extended to the level of M3. The area of attachment of the masseter muscle is prominent (see also MPEF-PV 7730, MPEF-PV 7944) and situated ventrolaterally as in *S. canquelensis*, *L. gaudryi*, and *Taubatherium paulacoutoi*, but differs from *Ancylocoelus frequens* and *Colpodon propinquus*, wherein this attachment area has a more ventral position.

Mandible MPEF-PV 6229, 6684, 7730, and 6609 present a mandibular corpus about 54 mm deep at the level of m1, while *Scarrittia canquelensis* is about 65 mm and *Leontinia gaudryi* about 60–62 mm. The inferior margin of the corpus in *S. barranquensis* n. sp. is nearly horizontal and the symphyseal region is wider and relatively more elongate than in *S. canquelensis*, about 50 mm in length and extending to the level of p4. A mental foramen appears beneath p2. The posterior margin of the ascending ramus is vertical, similar to *S. canquelensis*, and differing from *L. gaudryi* and *Ancylocoelus frequens* where it is more rounded. As in *A. frequens* and *S. canquelensis* the coronoid process is vertical, differing from *L. gaudryi* where it projects anteriorly.

Upper dentition The upper teeth (Fig. 11.2A) are brachydont but become more mesodont in M1–3. The dentition is complete (3.1.4.3), without diastema. General crown morphology is similar to *Scarrittia canquelensis*. I1 (MPEF-PV 7082) is enlarged and slightly caniniform, triangular in cross-section, with a convex labial face much higher than the lingual, and a distolingual wear facet. Because the material is fragmentary, no labial or lingual cingula can be observed. I2 is not preserved, but its alveolus is smaller than that of I1. I3 (MPEF-PV 7753) is very much smaller than I1 and similar in shape to C in that both present a convex labiolingual cusp with strong labial and lingual cingula. Premolars are more brachydont than those of *S. canquelensis*. In MPEF-PV 6139, P1 is much smaller than P2–4 and has a single root. There is a strong protocone and labial and lingual cingula. The labial face is slightly convex in P1, flat in P2–4, differing from *S. canquelensis* wherein it is more convex. In MPEF-PV 7753, P2 is smaller than P3, of rectangular aspect, and narrower mesiodistally. The paracone column is well delineated by a median groove. A parastyle and a labial cingulum are both present. The strong protocone is united to the metaconule and in turn to the ectoloph, thereby forming a metaloph. The small paraconule also is united to the ectoloph. The central fossa is open and only closes in advanced wear. Mesial and lingual cingula are shallow, continuous, and separated from the distal cingulum by the protocone crest. P3 and P4 are larger than P2. Paraconule becomes united to the ectoloph and protocone by wear, thereby forming a protoloph that closes the central fossa mesially. The hypocone is absent on P2–3 and much reduced on P4; the parastyle and both labial and distal cingula are present, and the mesial and lingual cingula are continuous.

The molars of *Scarrittia barranquensis* n. sp. are similar to those of *S. canquelensis*, but more brachydont and the labial face is flatter. Both species present a variable lingual cingulum that is strong when present. M2 is larger than M1, with a proportionally larger protocone and smaller hypocone separated by the lingual aperture of an oblique central fossa that becomes obliterated with wear. Parastyle is also stronger on M2. At the lingual limit of the ectoloph the crests formed by enamel folds disappear quickly with wear. There is a persistent mesial cingulum. The distal cingulum becomes united to the hypocone very early in wear, forming a posterior fossette that is deeper on M2. The labial cingulum is absent. M3 (e.g. MPEF-PV 7944) has a reduced hypocone and the aperture of the central fossa is in a more distal position. M3 has a conspicuous parastyle and metastyle, a variable lingual cingulum mesially, and no labial cingulum.

Lower dentition In MPEF-PV 6684 the i1–2 are small and of similar size, with convex labial and lingual faces. Labial and lingual cingula are present. The i3 is larger than i1–2 and its root more compressed mesiodistally. The labial and lingual faces are convex, with a labial cingulum and very strong lingual cingulum. The i2 and i3 display a strong median ridge on the lingual face. In *Scarrittia canquelensis* the i3 is three times larger than the i1–2, whereas in *S. barranquensis* n. sp. it is only twice as large. In the available material the lower canine was not preserved.

The premolars and molars are morphologically similar to those of *Scarrittia canquelensis*, but much

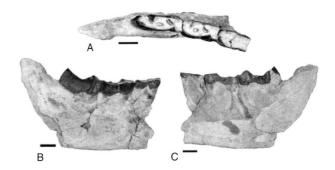

Fig. 11.3. *Scarrittia* cf. *S. canquelensis* Simpson in Chafee 1952. MPEF-PV 6207, right mandibular ramus fragment with m1–3 in occlusal (A), labial (B), and lingual (C) views. Scale bar 2 cm.

smaller in size. The p1 presents the premolar pattern, but the talonid is short and simple, and the labial and lingual cingulids are strong. The p2–4 have protoconids and an elongate distolingually directed metaconids. On each tooth, a small entoconid unites distally with the hypolophid to form a short entolophid and meta-entoconid fold mesially. There is a strong hypoflexid, and labial and lingual cingulids. The p3 and p4 display a small mesiolingual cusp that forms a delicate paralophid. With regard to the molars in MPEF-PV 6609, the m1–3 trigonids are much shorter than the talonids and the metaconid is elongate distolingually. The lingual groove of the trigonid and the meta-entoconid fold are deeper in m3, while the ento-hypoconid fold and hypoflexid are shallower. A posterior fossettid is present. The labial cingulid is only weakly developed in the m2–3. The lingual cingulid of m1–3 sometimes extends to the distal end of the crown.

Scarrittia cf. *S. canquelensis* Simpson in Chaffee 1952
Fig. 11.3.

Referred material MPEF-PV 6207, right mandibular ramus fragment with m1–3; MPEF-PV 6689, left M1.

Locality and stratigraphic horizon GBV-35, Unit 4 of the Upper Puesto Almendra Member, Profile J, Sarmiento Formation at Gran Barranca, Chubut Province, Argentina.

Comments Although MPEF-PV 6207 is very fragmentary, molars present the typical leontiniid posterior fossettid in the talonid, the mandibular corpus is 63 mm deep at the level of m1, similar to *Scarrittia canquelensis* and *Leontinia gaudryi*, but differs from the latter in being thinner or delicate, as in *S. canquelensis*.

Genus *Henricofilholia* Ameghino 1901
Henricofilholia Ameghino 1901, pp. 404–405; *Parastrapotherium* McKenna and Bell 1997, p. 467 (*partim*).

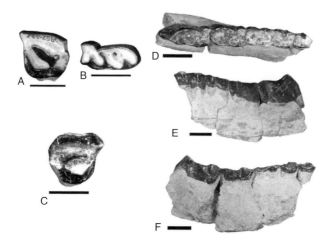

Fig. 11.4. *Henricofilholia lustrata* Ameghino 1901. (A) MACN A52–557 (lectotype), left upper molar, possibly M1, occlusal view; (B) MACN A52–538 (paralectotype), right lower molar, possibly m2, occlusal view. *H. vucetichia* n. sp. (C) MPEF-PV 6129 (holotype), right upper M1, occlusal view; (D–F) MPEF-PV 7717 (paratype) left mandibular fragment with p3–m3 in occlusal (D), labial (E), and lingual (F) views. Scale bar 2 cm.

Type species *Henricofilholia lustrata* Ameghino 1901, p. 405

Referred species *Henricofilholia lustrata* Ameghino 1901; *H. vucetichia* n. sp.

Revised diagnosis Smaller than *Ancylocoelus frequens* and *Taubatherium paulacoutoi*, lacking the first crista on the maxillary molars.

Geographic and chronologic distribution Patagonia, Argentina, Deseadan SALMA.

Henricofilholia lustrata Ameghino 1901
Fig. 11.4A, B.

Henricofilholia lustrata Ameghino 1901, p. 405.

Lectotype MACN A52–557: left upper molar, possibly M1.

Paralectotype MACN A52–538: right lower molar, possibly m2.

Revised diagnosis Leontiniidae having maxillary molars with a reduced parastyle, a mesial cingulum, and no labial or lingual cingula; and with the lower molar lingual cingulid extending to the entolophid.

Type locality *Pyrotherium* beds ("Piso Pyrotheriense"), Sarmiento Formation, Gran Barranca "Barranca Sur del Colhue-Huapi"), Chubut Province, Argentina.

Description The upper molar (MACN A52–557, possibly an M1) of *Henricofilholia lustrata* displays a protoloph larger than the metaloph separated by an oblique central fossa that is open lingually, parastyle is reduced, the mesial cingulum present but not a labial cingulum. The lower molar (MACN A52–

538, m1 or m2) presents a lingual cingulid that extends distally to the entolophid. The upper molar differs from those of *Ancylocoelus frequens* by being smaller, and lacking a first crista projecting into the central fossa from the ectoloph; it also lacks a strong mesiolingual cingulum.

Henricofilholia vucetichia n. sp.
Fig. 11.4C–F.

Holotype MPEF-PV 6129, right upper M1.

Paratype MPEF-PV 7717, left dentary fragment with p3–m3.

Referred material MPEF-PV 6135, right lower m2–3; MPEF-PV 6248, left m2; MPEF-PV 6269, right dentary fragment with m2–3; MPEF-PV 7700, associated right lower p2–m1.

Diagnosis Similar to *Henricofilholia lustrata* in size and in lacking a first crista, but differs by displaying a strong parastyle and mesial and lingual cingula in the upper molars; weakly developed or no labial and lingual cingulids on the lower premolars and no labial or lingual cingulids in the lower molars.

Type locality Gran Barranca Vertebrate locality GBV-19 "La Cantera," Profile A, Unit 3 of the Upper Puesto Almendra Member, Sarmiento Formation at Gran Barranca, Chubut Province, Argentina.

Etymology In honor of Dr. María Guiomar Vucetich (MLP).

Description The M1 (MPEF-PV 6129, Fig. 11.4C) is small, and the first crista and labial cingulum are absent as in *Henricofilholia lustrata*. This tooth differs from *H. lustrata* by the presence of a strong mesiolingual cingulum, whereas in *H. lustrata*, this cingulum is very weak and restricted to the mesial aspect, as in *Taubatherium paulacoutoi*. It also differs from *Leontinia gaudryi, Scarrittia barranquensis* n. sp., *S. canquelensis* and *Ancylocoelus frequens* by being very much smaller.

The lower premolars (e.g. MPEF-PV 7717, Fig. 11.4D) are rectangular in outline, with the trigonid and talonid having the same mesiodistal diameter. The entoconid on p2 is much reduced, but in p3–4 becomes united with the hypolophid labially and to the hypoconulid distolingually, forming a false fossettid that eventually disappears with wear. The lingual and labial cingulids are very weak (MPEF-PV 7717, Fig. 11.4E, F), and the latter is restricted to the base of the crown and does not have the same conformation observed in *Leontinia gaudryi, Scarrittia canquelensis* or *Ancylocoelus frequens*, where it is strong and extends across the labial face of the crown. The m1–3 trigonids are much shorter mesiodistally than the talonids; the metaconid is elongate distolingually and the meta-entoconid fold is deeper than the ento-hypoconid fold. The posterior fossettid is present and hypoflexid shallow. The lingual cingulid of the molars (MPEF-PV 6135, MPEF-PV 6269, and MPEF-PV 7717, Fig. 11.4F) is absent, and in this regard *Henricofilholia vucetichia* differs from the type of *H. lustrata* (MACN A52–538), where the lingual cingulid is evident, crenulated, and runs along the base of crown to the entolophid. In *H. vucetichia* the labial cingulid is absent as in *Taubatherium paulacoutoi* and *Huilatherium pluriplicatum*, but it differs from *H. lustrata*, *Leontinia gaudryi*, *Scarrittia barranquensis* n. sp., *S. canquelensis*, and *Ancylocoelus frequens* where the labial cingulid extends across the entire labial face of m1, although is reduced on the talonid of the m2–3.

Genus *Colpodon* Burmeister 1885

Colpodon Burmeister 1885, pp. 161–169, est. III, Fig. 46; Burmeister 1891, pp. 389–399, est.VII, Figs. 4–10; Ameghino 1902, pp. 108–109; Ameghino 1904a, pp. 232–233; Ameghino 1904b, p. 437; Ameghino 1904c, p. 239, Fig. 316; Simpson 1945, p. 127; Soria and Bond 1988, p. 36; McKenna and Bell 1997, p. 459. *Baenodon* Ameghino, 1892, p. 461. Syn. n.

Type species *Colpodon propinquus* Burmeister 1885, pp. 161–169.

Revised diagnosis Leontiniidae presenting a combination of characters including: concave area of origin of the masseter in a ventral position as in *Ancylocoelus frequens*; upper canine vestigial or absent, lower canine absent, but without diastema; in the premolars, the paraconule is close to protocone and labiodistal fossette; the maxillary molar fossettes by union of the first crista, second crista, and crochet.

Geographic and chronologic distribution Patagonia, Argentina, Colhuehuapian SALMA.

Colpodon distinctus Ameghino 1902
Fig. 11.5.

Colpodon distinctus Ameghino 1902, pp. 108–109; Ameghino 1904b, p. 437; Patterson, 1952, *in litteris*.
Colpodon divisus Ameghino 1904a, p. 232 (error)
Colpodon plicatus Ameghino 1904a, pp. 232–233; Ameghino 1904c, p. 239, Fig. 316; Patterson 1952, *in litteris* (sin.).

Lectotype MACN A52–574, premaxilla and maxilla fragments with right and left I1–3; right P1, right and left P2–M3, mandible with right and left i1–3, left p1, right and left p2–m3.

Referred material MACN A52–576, maxilla fragment with left DM2–4 (holotype of *Colpodon plicatus*); AMNH 29688, left dentary fragment with p2–m3; AMNH 29721, right dentary fragment with dm2–m2; FMNH-13304, anterior region of skull

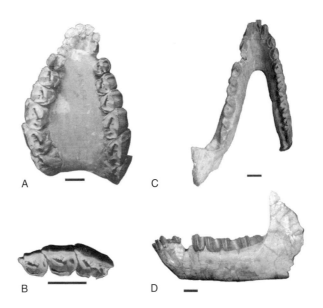

Fig. 11.5. *Colpodon distinctus* Ameghino 1902. (A, C, D) MACN A52–574, (lectotype); (B) MACN A52–576 (type of *C. plicatus*). (A) skull in palatal view with right I1–3, P1–M3 and left I1–3, P2–M3; (B) left DM2–4 in occlusal view; (C–D) mandible with left i1–3, p1–m3, and right i1–3, p2–m3 in occlusal (C) and lateral (D) views. Scale bar 2 cm.

fragment; FMNH-13310, skull fragment with left P2–M3; FMNH-13316, dentary fragment with right p2–m3; FMNH-13591, MACN A52–579, right and left maxilla fragments with P1–4; MACN A52–580, right maxilla fragment with M1–3; MACN A52–581, left maxilla fragment with M1–3; MPEF-PV 6346, left dentary fragment with m1–2; MPEF-PV 6448, left isolated dp4?; MPEF-PV 7014, left maxilla fragment with worn P3–4, M1; MPEF-PV 7080, left maxilla fragment with P4; MPEF-PV 7134, two associated teeth, MPEF-PV 7306, lower premolar; MPEF-PV 7465, maxilla; MPEF-PV 7837, upper molar fragment; MLP 82-V-2–40, right and left maxilla fragments with P4–M3 and temporal region; left P4–M3; MNHN 1900–18, three specimens: left P3–M2; right P1–M2; left M3.

Locality and stratigraphic horizon Gran Barranca, "*Colpodon* beds," GBV-8 (level A), GBV-9 (level B), GBV-10 (level C), GBV-38 (level Z), and GBV-43 (level 16), Lower Fossil Zone, Colhue-Huapi Member, Sarmiento Formation, Chubut Province, Argentina. The biostratigraphic range of *Colpodon distinctus* encompasses the full thickness of local magnetozone N2 of Profile A-1, interpreted as Chron C6An.1n (20.0 and 20.2 Ma) (Ré *et al.* Chapter 4, this book).

Revised diagnosis *Colpodon distinctus* is smaller than *C. propinquus*; with molar mean occlusal area

32% smaller than in C. *propinquus;* P2–4 protocone proportionally larger than in *C. propinquus*; P2 central fossa opens mesiodistally, and mesially in P3.

Description *Mandible* MACN A52–574 (Fig. 11.5C, D) displays a delicate horizontal corpus that is 42.4 mm deep at the level of p4, and that differs from the robust mandible of *Colpodon propinquus* (MPEF-PV 1104, Gaiman) that is 45 mm deep at the level of p4. The mandible is synostosed; and relatively narrow symphyseal region extends nearly to the level of p3.

Upper dentition All teeth are mesodont.

Deciduous upper dentition In MACN A52–576 (Fig. 11.5B), in size, dM2 < dM3 < dM4. The protocone is united to the paraconule and the ectoloph to form a protoloph. DM2 metaloph evident in advanced wear. Multiple crista are present along the lingual margin of the ectoloph. The parastyle is strong and separated from the paracone by a deep groove. The oblique central fossa is open lingually but closes with wear. The posterior fossette appears distally and there is a mesiolingual cingulum.

Permanent upper dentition The upper incisor is not markedly enlarged as in *Leontinia gaudryi*, but I2 is the largest. The incisors in MACN A52–574 (Fig. 11.5A) are somewhat fractured, but I1 has a triangular outline and a weak labial cingulum; I2 is larger than I1, with a subtriangular outline, a labial cingulum on the distal portion of labial face, and a lingual cingulum. I3 is smaller than I1–2, with a smaller labiolingual diameter, and strong labial and lingual cingula. There is no alveolus for the upper canine in MACN A52–574, nor in other material from Gran Barranca, but MLP 49-XI-21–15 (from Punta Magagna, Chubut, ?Trelew) displays small roots between I3 and P1, suggesting that the upper dental formula of *Colpodon* in this specimen was complete, including a vestigial C. P1 is small and in MACN A52–579 and presents little wear compared with the other premolars. P1 has a rudimentary protocone united to the mesial and distal cingula, a mesiodistally short ectoloph, and a central fossa, wide and open both mesially and distally, convex labial face, and labial and lingual cingula. P2–4 are rectangular in outline. P2 is smaller than P3 and slightly narrower mesiodistally. The P2 protocone is united lingually to the distolingual cingulum and distally to the metaconule: a "metaloph" forms in advanced wear. The paraconule is much reduced and united to the protocone. Mesiodistally the central fossa is open. There is a strong paracone column. The mesiolingual cingulum is continuous, a parastyle is present, as are distal and labial cingula. The P3–4 protocone is united to the paraconule to form an oblique protoloph, which when heavily worn in P3 becomes united to the ectoloph, closing the central fossa

mesially. The P3 hypocone is united to the metaconule and this in turn to the ectoloph thereby forming a metaloph. There is a strong paracone column. The labiodistal fossette is formed by the union of the crochet with the lingual end of the ectoloph; enamel folds in the ectoloph form isolated enamel fossettes with wear similar to those observed in *Huilatherium pluriplicatum* and *Purperia cribatidens*. There is a strong mesiolingual cingulum near the base of the crown. The deeper distal cingulum is separated from the mesiolingual cingulum by a crest that projects from the hypocone. A parastyle is present, as are lingual and distal cingula. The labial cingulum is present only on the mesial portion of the labial face. The premolar protocone of *Colpodon* differs from that of other leontiniids. This morphology may be a consequence of the closer approximation between paraconule and protocone. The mesiolingual cingulum is continuous, as in *Scarrittia*, but very different from *Leontinia* and *Ancylocoelus*. The labiodistal fossa is observed only in *Colpodon*. The premolars of *C. distinctus* are similar to the *C. propinquus*, but differ in more rectangular outline, and presenting a larger protocone more separated from the hypocone.

M1 and M2 are quadrangular in outline and with wear become mesiodistally shortened. The protocone is larger than the hypocone and separated by the lingual opening of the oblique central fossa. There is a strong parastyle. The first and second crests are observed on the ectoloph lingually; a strong crochet projects from the metaloph; crests form small enamel fossettes, the posterior fossette is conspicuous and the lingual cingulum is tenuous or absent; a labial cingulum is absent. M3 has a strong protoloph and much reduced hypocone and a central fossa that opens lingually; first and second crests are united to the crochet of the metaloph forming small fossettes with wear. Parastyle and metastyle are strong, the mesiolingual cingulum is tenuous or absent. There is no labial cingulum; a distal cingulum forms the posterior fossette. In the upper molars the crochet and second crest form conspicuous fossettes, differing in this regard to other Oligocene leontiniids and more nearly resembling *Huilatherium pluriplicatum* and *Purperia cribatidens*.

Permanent lower dentition The i1 and i2 are similar in size and much smaller than i3. The incisor crowns are narrow and high, with the labial face slightly flat and with a labial cingulid. A narrow longitudinal crest and cingulid appear on the lingual face. The i3 is enlarged and turned lingually; its labial and lingual faces are convex, the latter with a longitudinal crest medially. There are labial and lingual cingulids. The incisors are similar to those of

Colpodon propinquus, but more delicate and smaller, mesiodistally narrow, and mesodont. The c1 is absent, as in *Ancylocoelus frequens*. The p1 is very small, smaller than i1–2, and has a triangular and pointed crown with strong labial and lingual cingulids. The check teeth do not present significant morphological differences when compared with *C. propinquus* or the other leontiniids. The p2 trigonid is distinct from the talonid. A protoconid is present. There is a short groove between the trigonid and metaconid lingually and a strong hypoflexid labially. A small entoconid is present on the talonis and labial and lingual cingulids are present. The p3 is larger than p2. The p2 talonid slightly larger than trigonid; the entoconid is united with the hypolophid distally, the hypoflexid is deep, smaller meta-entoconid fold; labial and lingual cingulids present. The p4 is larger than p3. The p4 talonid slightly larger than trigonid; the entoconid is connected to the medial portion of the hypolophid; labial and lingual cingulids are present. The m1–3 trigonids are narrower mesiodistally than the talonid; metaconid slightly elongate distolingually, trigonid groove shallow; deep meta-entoconid fold; ephemeral ento-hypoconid fold on m1, persistent on m2, open and deep on m3; elongate hypoconulid and more mesodont crown; posterior fossettid present; shallow hypoflexid; no labial cingulid; lingual cingulid on m1–3 at the level of the entolophid.

Discussion

Systematics

Five taxa of Leontiniidae occur at Gran Barranca, two at GBV-19 (Unit 3, Upper Puesto Almendra Member) and GBV-34 (Unit 3, Upper Puesto Almendra Member), a third taxon at GBV-35 (Unit 4, Upper Puesto Almendra Member), and another purportedly from the "*Pyrotherium* beds" (Ameghino 1901, 1906), and the fifth in the Lower Fossil Zone of the Colhue-Huapi Member (GBV-8, 9, 10, 38, and 43).

Two new taxa occur at GBV-19 and GBV-34 (Unit 3) that are of different size and morphology. The larger taxon, *Scarrittia barranquensis* n. sp., presents many diagnostic characters that refer it unequivocally to *Scarrittia*. This Gran Barranca species is smaller of *S. canquelensis* from Chubut Province and *S. robusta* from Uruguay, with cheek teeth that are shorter mesiodistally (Fig. 11.6). Morphologically, it is very similar to the other species of *Scarrittia*, except in having more brachydont premolars and molars, and maxillary premolars with a more marked paracone column, a reduced hypocone, and a flatter labial face. By contrast, maxillary premolars in *S. canquelensis* display a well-marked metacone column, stronger hypocone (on P3–4), and a convex labial face.

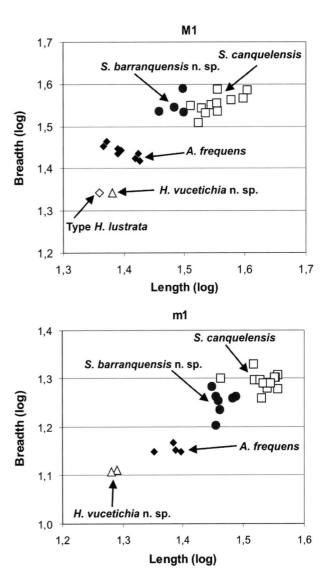

Fig. 11.6. Plot of upper and lower M1/m1 length and breadth measurements for specimens of *Scarrittia barranquensis* n. sp. from Gran Barranca (GBV-19 "La Cantera" and GBV-34), *S. canquelensis* from Scarritt Pocket, Los Búlgaros, Laguna Payahilé localities, *Henricofilholia vucetichia* n. sp. from GBV-19, *H. lustrata* from Gran Barranca, and *Ancylocoelus frequens* from La Flecha.

The fragmentary material MPEF-PV 6207 from locality GBV-35 (Unit 4, Upper Puesto Almendra Member) is tentatively refered to *Scarrittia* cf. *S. canquelensis* on the basis of its large size and dentary morphology. The discovery of *Scarrittia* at Gran Barranca is noteworthy, especially given that the Ameghino collection includes no material of this genus.

Three species of smaller Deseadan leontiniid were proposed by Ameghino (1895, 1901), *Leontinia garzoni* in 1895, and *Ancylocoelus minor* and *Henricofilholia lustrata* in 1901. *Leontinia garzoni* was described on the basis of two specimens from La Flecha (MACN A52–599

and MACN A52–600bis) analyzed by Patterson (1952) and
revised by Ribeiro (2003), who considered both specimens
very similar to *Ancylocoelus frequens*. *Ancylocoelus minor*
was based on a right maxillary fragment with M2–3
(MACN A52–551) and left M3, both from La Flecha.
According to the original diagnosis (Ameghino 1901) this
species is smaller than *A. frequens* with molars shorter
mesiodistally and a continuous lingual cingulum around
the protoloph. As the molars are only slightly smaller than
A. frequens and the occlusal morphology differs only in the
absence of the first crista in the central fossa of M3, Patterson (1952) and Ribeiro (2003) referred it to *A. frequens*.
Finally, *Henricofilholia lustrata* was proposed by Ameghino (1901). Five additional species were eventually
described (*H. circumdata*, *H. cingulatum*, *H. inaequilatera*,
H. intercincta, *H. lemoinei*) (Ameghino 1901, 1904a).
Patterson (1952) and Ribeiro (2003) synonymize *H. circumdata* from Cabeza Blanca with *Leontinia gaudryi*. They
consider *H. cingulatum* and *H. inaequilatera* from La
Flecha and *H. intercincta* from Monte Espejo as junior
synonyms of *A. frequens*. *Henricofilholia lemoinei* is an
Astrapotheriidae. *Henricofilholia lustrata* from Gran
Barranca differs in size and morphology and is therefore
considered a valid taxon. AMNH 29607 from Scarritt
Pocket with M1 morphology similar to the *H. lustrata*
(described by Chaffee 1952, p. 518, Fig. 3 as cf. Leontiniidae indet.) may also belong to *H. lustrata*.

    Henricofilholia lustrata was collected by Carlos
Ameghino at the "*Pyrotherium* beds" of Gran Barranca,
and it differs from the small leontiniid from GBV-19 in
important details: on M1 there is a strong mesiolingual
cingulum; the lingual cingulid of the lower molars is weak
or absent and there is no labial cingulid. On the basis of
these features, we conclude that the small leontiniid from
GBV-19 belongs to another new species, *Henricofilholia
vucetichia* n. sp.

    Leontiniids are solely represented in the Colhuehuapian
by *Colpodon*. Ameghino (1902) claimed that the
labial roots are bifurcate, well separated and divergent in
C. propinquus, whereas P1–2 in *C. distinctus* present single
convex labial roots and P3 a single labial and another
lingual root. Restudy of the original material reveals that
premolar roots are not preserved in MACN A-967 and
the premolar roots of the type of *C. distinctus* (MACN
A52–574) are poorly preserved. Nevertheless, while the
more proximal portion in *C. distinctus* is unique in structure, without benefit of further preparation or radiography,
Ameghino (1902) could not have excluded the possibility
that the roots were bifurcate within the alveolus (other
specimens in the Ameghino collection preserve bifurcate
roots on P2–3, e.g. MACN A52–579). One must conclude
then that this character (single labial root in the maxillary
premolars) of Ameghino (1902) is not observable in the

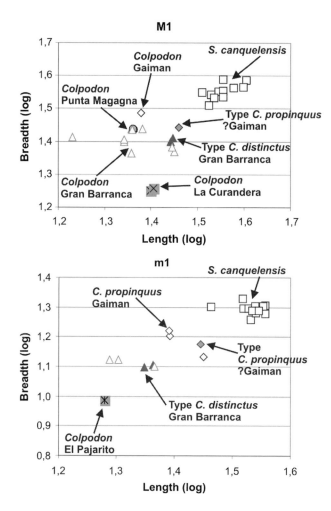

Fig. 11.7. Plot of upper and lower M1/m1 length and breadth
measurements for specimens of *Scarrittia canquelensis* from Scarritt
Pocket, Los Búlgaros, Laguna Payahilé localities, *Colpodon
propinquus* from Gaiman, *C. distinctus* from Gran Barranca,
Colpodon from Punta Magagna, and *Colpodon* sp. from La
Curandera and El Pajarito.

type specimen of *C. distinctus*, and thus, not a diagnostic
character of the species. Other than the fact that
Colpodon distinctus premolars are more rectangular in
outline (they have a larger protocone and somewhat more
separated hypocone), the cheek teeth of *C. distinctus* do
not present important morphological differences from
C. propinquus. So similar are they, in fact, that Soria
and Bond (1988) considered *C. distinctus* conspecific
with *C. propinquus*, but noted that the type material of
C. propinquus was somewhat larger. Dental measurements
of the types and all material subsequently assigned to
Colpodon from Gran Barranca, Gaiman, La Curandera,
and El Pajarito, confirm that the type of *C. propinquus* is
larger than the type of *C. distinctus*. However, this size
difference is not always observed in the upper molars, as
these teeth change size with occlusal wear (Fig. 11.7).

Table 11.1. *Localities of Deseadan age in Argentina (Patagonia and Mendoza province), Uruguay, Bolívia (Salla), and Brazil (Tremembé Formation) where the Leontiniidae are represented*

| Taxon | Gran Barranca | | | | Other localities | | | | | Uy | Bol | Bz |
| | UPA | | | | Argentina | | | | | | | |
	GBV-19	GBV-34	GBV-35	PB	SP	CB	EP	QF	LF			
Scarrittia canquelensis					x							
Scarrittia cf. *S. canquelensis*			x					x				
Scarrittia barranquensis n. sp.	x	x										
S. robusta										x		
Leontinia gaudryi						x	x		x			
Henricofilholia vucetichia n. sp.	x											
H. lustrata				x								
Ancylocoelus frequens						x	x		x			
Taubatherium paulacoutoi												x
Anayatherium fortis											x	
Anayatherium ekecoa											x	

Notes: LF, La Flecha; UPA, Upper Puesto Almendra; PB, "*Pyrotherium* beds"; SP, Scarritt Pocket; CB, Cabeza Blanca; EP, El Pajarito; QF, Quebrada Fiera; Uy, Uruguay, Fray Bentos Formation; Bol, Salla, Bolivia; Bz, Taubaté, Tremembé Formation, Brazil.

While most of the material of *C. distinctus* from Gran Barranca consists of lower molars, some have been found in association with maxillary teeth, and these associations lead us to conclude that the species at Gran Barranca is indeed *C. distinctus*.

Finally, in passing, we note that material of *Colpodon* from the Colhuehuapian localities at El Pajarito and La Curandera are demonstrably smaller than material from either Gran Barranca or Gaiman.

Biogeography, biostratigraphy, and paleoecology

Leontiniidae are represented in many localities of Deseadan age in Argentine Patagonia and in Mendoza Province, and elsewhere in South America at Salla in Bolívia and the Tremembé Formation of Brazil (Table 11.1). The largest number of Deseadan taxa occur in Patagonia and include *Leontinia gaudryi*, *Ancylocoelus frequens*, *Scarrittia canquelensis*, *Henricofilholia lustrata*, and the two new species described herein, *S. barranquensis* n. sp. and *H. vucetichia* n. sp.

The geographic distribution of these Deseadan species is remarkable. Whereas *Henricofilholia lustrata* and *H. vucetichia* n. sp. are leontiniids of relatively small size and occur only at Gran Barranca and possibly Scarritt Pocket (Chubut Province), *Leontinia gaudryi* and *Ancylocoelus frequens* (larger leontiniids) are restricted to the northern part of Santa Cruz Province (La Flecha, Cerro Alto, Piedra Negra, Monte Espejo,

Estancia 8 de Julio) and central Chubut Province (Cabeza Blanca, Rincón del Zampal, Las Cascadas) to as far north in Patagonia as El Pajarito. *Scarrittia* is the best-known leontiniid, and occurs in localities extending from the latitude of central Chubut (*S. canquelensis*) to as far north as Mendoza (*S.* cf. *S. canquelensis*) and Uruguay (*S. robusta*).

It is noteworthy that *Leontinia gaudryi* and *Ancylocoelus frequens* do not occur at any localities or levels where *Scarrittia* occurs, whether *S. barranquensis* n. sp., *S. canquelensis*, or *S. robusta*.

Scarrittia canquelensis comes from the Scarritt Pocket locality along with a faunal assemblage assigned to the Deseadan SALMA. *Scarrittia canquelensis* and the rodent *Platypittamys brachyodon* are the only taxa occurring in the Scarritt Pocket quarry (Locality I of Marshall *et al.* 1986) in stratigraphic relationship with basalts that were dated to between 23.4 Ma and about 21.0 Ma.

This contrast in composition may indicate differences in the age of these faunas, although most of these localities remain undated. What can be established is that *S. barranquensis* n. sp. occurs at two localities or levels at Gran Barranca, GBV-19 "La Cantera" and GBV-34 that are between 33.7 and 29.5 Ma in age, whereas GBV-34 occurs in Unit 4 of the Upper Puesto Almendra Member on a discontinuity surface that postdates basalts dated between about 29.2 and 26.3 Ma. The material assigned to *S.* cf. *S. canquelensis* is higher-crowned and somewhat larger and occurs at GBV-35 in a laterally extended lenticular fossil-rich homeoconglomerate bed (Ré *et al.* Chapter 3, this volume). Neither *L. gaudryi* nor *A. frequens* occur at these localities or levels at Gran Barranca. The only dated locality where *L. gaudryi* and *A. frequens* occur is at La Flecha where as yet unpublished dates are between 21.5 and 23.4 Ma (R. Madden, personal communication).

During the Colhuehuapian, leontiniids are represented only by *Colpodon* in a distribution restricted to central Patagonian localities in Chubut (Gran Barranca, Gaiman, Punta Magagna, and Sacanana) and Río Negro provinces (Paso Córdoba, Chichinales Formation).

The most outstanding features of leontiniids include their canine-like incisors and low-crowned, brachydont to meso-dont cheek teeth. The Deseadan leontiniids were herbivores of medium (greater than a tapir) to large (smaller than a rhinoceros) size with large, brachydont cheek teeth and for *S. canquelensis* Chaffee (1952) described broadly splayed digits with flat hooves, a combination of morphologies sug-gesting a browsing habitus on soft substrate. The Colhuehua-pian *Colpodon* is mesodont, having somewhat higher crowns than other known Leontiniidae. In general, while the diversity of Colhuehuapian ungulates is less than in the Deseadan, browsing forms like *Colpodon* are more common (Bond 1986).

ACKNOWLEDGEMENTS
The authors are grateful to the editors for the invitation to contribute with this work on the Leontiniidae of Gran Bar-ranca. They thank Marcelo Reguero, Alejandro Kramarz, and Eduardo Ruigómez for the permission and information supplied from the collections of the Museo de La Plata, the Museo Argentino de Ciencias Naturales "Bernardino Riva-davia," and the Museo Paleontológico "Egidio Feruglio" respectively, and also Richard Madden, Richard Kay, and Jorge Ferigolo for the discussion and revision of the text. Guiomar Vucetich and Cecilia Dechamps are thanked for their help with information on the material of Gran Bar-ranca. Field research was supported by US National Science Foundation grants EAR-0087636, BCS-0090255, and DEB-9907985 to Richard F. Kay and Richard H. Madden.

REFERENCES
Ameghino, F. 1892. Répliques aux critiques du Dr. Burmeister sur quelques genres de mammifères fossiles de la République Argentine. *Boletín de la Academia Nacional de Ciencias en Córdoba*, **12**, 437–470.
Ameghino, F. 1895. Sur les oiseaux fossiles de Patagonie et la faune mammalogique des couches à *Pyrotherium*. II. Première contribution la connaissance de la faune mammalogique de couches à *Pyrotherium*. *Boletín del Instituto Geográfico Argentino*, **15**, 603–660.
Ameghino, F. 1897. Mammifères crétacés de l'Argentine: deuxième contribution à la connaissance de la faune mammalogique de couches à *Pyrotherium*. *Boletín Instituto Geográfico Argentino*, **18**, 406–521.
Ameghino, F. 1901. Notices préliminaires sur des ongulés nouveaux des terrains crétacés de Patagonie. *Boletí n de la Academia Nacional de Ciencias en Córdoba*, **16**, 349–426.
Ameghino, F. 1902. Première contribution à la connaissance de la faune mammalogique des couches à *Colpodon*. *Boletín de la Academia Nacional de Ciencias en Córdoba*, **17**, 71–138.
Ameghino, F. 1904a. Nuevas especies de mamíferos, cretáceos y terciarios de la República Argentina. *Anales de la Sociedad Científica Argentina, Buenos Aires*, **58**, 225–240.
Ameghino, F. 1904b. La perforación astragaliana en los mamíferos no es un caráter originariamente primitivo. *Anales de Museo Nacional de Buenos Aires*, **4**(3), 349–460.
Ameghino, F. 1904c. Rechérchès de morphologie phylogénétique sur les molaires supérieures des ongulés. *Anales de Museo Nacional de Buenos Aires*, **9**(3a), 1–541.
Ameghino, F. 1906. Le formations sédimentaires du Crétacé supérieur et du Tertiaire de Patagonie avec un parallèle entre leurs faunes mammalogiques et celles de l'ancien continent. *Anales de Museo Nacional de Buenos Aires*, **3**, 1–568.
Bond, M. 1986. Los ungulados fósiles de Argentina: evolución y paleoambientes. *Actas IV Congreso Argentino de Paleontología y Bioestratigrafia (Mendoza)*, **2**, 173–185.
Bond, M. 1988. Consideraciones sobre la morfología de los molariformes inferiores de los Notoungulata. *Resúmenes V Jornadas Argentinas de Paleontología de Vertebrados, La Plata*, 76–77.
Bond, M. and G. M. López 1995. Los mamíferos de la Formación Casa Grande (Eoceno) de la Provincia de Jujuy, República Argentina. *Ameghiniana*, **32**, 301–309.
Burmeister, H. 1885. Exámen crítico de los mamíferos y reptiles fósiles denominados por D. Augusto Bravard y mencionados en su obra precedente. *Anales del Museo Nacional de Buenos Aires*, **3**(14), 95–174.
Burmeister, H. 1891. Adiciones al exámen crítico de los mamíferos fósiles tratados en el artículo IV anterior. *Anales del Museo Nacional de Buenos Aires*, **3**, 357–400.
Chaffee, R. G. 1952. The Deseadan vertebrate fauna of the Scarritt Pocket, Patagonia. *Bulletin of the American Museum of Natural History*, **98**, 509–562.
Deraco, M. V. and J. E. Powell 2004. Nuevas evidencias de leontínidos eocenos del Noroeste Argentino. *Ameghiniana*, **41**(4) (Supl.), 43R.
Deraco, M. V., J. E. Powell, and G. López 2008. Primer leontínido (Mammalia, Notoungulata) de la Formación Lumbrera (Subgrupo Santa Bárbara, Grupo Salta-Paleógeno) del noroeste argentino. *Ameghiniana*, **45**, 83–91.
Kay, R. F., R. Madden, M. G. Vucetich, A. Carlini, M. Mazzoni, and G. Ré 1999. Revised geochronology of the Casamayoran South American Land Mammal Age: climatic and biotic implications. *Proceedings of the National Academy of Sciences USA*, **96**, 13 235–13 240.

Marshall, L. G., R. L. Cifelli, R. E. Drake, and G. H. Curtis 1986. Vertebrate paleontology, geology, and geochronology of the Tapera de López and Scarritt Pocket, Chubut Province, Argentina. *Journal of Paleontology*, **60**, 920–951.

McKenna, M. C. and S. K. Bell 1997. *The Classification of Mammals above the Species Level.* New York: Columbia University Press.

Patterson, B. 1952. *Catálogo de Deseado y Colhue-Huapi: Paleozoología (Vertebrados).* Buenos Aires: Museo Argentino de Ciencias Naturales "Bernardino Rivadavia." (Unpublished.)

Powell, J. E. and M. V. Deraco 2003. Un nuevo leontínido (Mammalia, Notoungulata) del Miembro Superior de la Formación Lumbrera (Subgrupo Santa Bárbara) del Noroeste Argentino. *Ameghiniana*, **40**(4) (Supl.), 68R–69R.

Ré, G., R. Madden, M. Heizler, J. F. Vilas, M. E. Rodriguez, R. F. Kay, and A. Carlini 2005. Polaridad magnética preliminar de las sedimentitas de la Formación Sarmiento (Gran Barranca del Lago Colhue-Huapi, Chubut, Argentina). *Actas XVI Congreso Geológico Argentino (La Plata)*, **4**, 387–394.

Ribeiro, A. M. 2003. Contribuição ao conhecimento da Família Leontiniidae (Mammalia, Notoungulata, Toxodontia): aspectos anatômicos e filogenéticos. Ph.D. thesis, Universidade Federal do Rio Grande do Sul, Porto Alegre.

Simpson, G. G. 1934. A new notoungulate from the early Tertiary of Patagonia. *American Museum Novitates*, **735**, 1–3.

Simpson, G. G. 1945. The principles of classification and a new classification of mammals. *Bulletin of the American Museum of Natural History*, **85**, 1–350.

Soria, M. F. and M. Bond 1988. Asignación del género *Colpodon* Burmeister, 1885 a la familia Notohippidae Ameghino, 1894 (Notoungulata, Toxodonta). *Resúmenes V Jornadas de Paleontología de Vertebrados, La Plata*, **36**.

Szalay, F. S. 1969. Mixodectidae, Microsyopidae, and the insectivore–primate transition. *Bulletin of the American Museum of Natural History*, **140**(4), 193–330.

Ubilla, M., D. Perea, and M. Bond 1994. The Deseadan Land Mammal Age in Uruguay and the report of *Scarrittia robusta* nov. sp. (Leontiniidae, Notoungulata) in the Fray Bentos Formation (Oligocene–?Lower Miocene). *Geobios*, **27**, 95–102.

Van Frank, R. 1957. A fossil collection from northern Venezuela. I. Toxodontidae (Mammalia, Notoungulata). *American Museum Novitates*, **1850**, 1–38.

Van Valen, L. 1966. Deltatheridia, a new order of mammals. *Bulletin of the American Museum of Natural History*, **132**(1), 1–26.

12 Colhuehuapian Astrapotheriidae (Mammalia) from Gran Barranca south of Lake Colhue-Huapi

Alejandro G. Kramarz and Mariano Bond

Abstract

In this contribution the taxonomic status of the numerous astrapothere species described by Ameghino for his *Colpodon* beds (Colhuehuapian) are revised, and the diversity and distribution of the Colhuehuapian astrapotheres are examined in light of new and more complete material, and with more precise stratigraphic information. The new material discussed here comes from the Lower Fossil Zone of the Colhue-Huapi Member of the Sarmiento Formation; previously described material is known or presumed to have come from this same level and belongs to the Colhuehuapian SALMA, the *Colpodon* beds of Ameghino. The Lower Fossil Zone is assigned to the early Miocene, about 20 Ma from radiometric and magnetic polarity data (Ré *et al.* Chapter 4, this book).

Our revision reveals the presence of three species: *Astrapotherium? ruderarium* (Ameghino 1902), *Parastrapotherium symmetrum* (Ameghino 1902), and *Parastrapotherium martiale* Ameghino 1901 (the latter also known from the older Deseadan SALMA). *Parastrapotherium herculeum* may represent a fourth very large species but the type material of Ameghino (1889), supposed to come from Colhuehuapian beds, could not be found, and no available specimen can be certainly referred to it. Finally, a single upper canine seems to represent a fifth unnamed species, perhaps a uruguaytheriine astrapothere. Examination of old and new collections suggests that *Astrapothericulus* is not recorded at these levels, although the genus does occur at other Colhuehuapian localities.

The following names are brought into synonymy with *Astrapotherium? ruderarium*:

Parastrapotherium paucum Ameghino 1902

Parastrapotherium crassum Ameghino 1902 (*partim*)

Astrapothericulus minusculus Ameghino 1902

Astrapothericulus laevisculus Ameghino 1902

Astrapotherium triangulidens Ameghino 1902

Prochalicotherium patagonicum Ameghino 1902.

Astrapotherium? ruderarium is the only recognized species of *Astrapotherium* in the Colhuehuapian Age.

The Paleontology of Gran Barranca: Evolution and Environmental Change through the Middle Cenozoic of Patagonia, eds. R. H. Madden, A. A. Carlini, M. G. Vucetich, and R. F. Kay. Published by Cambridge University Press. © Cambridge University Press 2010.

Astrapotherium? ruderarium is also the most abundant astrapothere in Colhuehuapian levels of Gran Barranca.

Resumen

En esta contribución se revisa el status taxonómico de las numerosas especies de astrapoterios descriptos por Ameghino para sus capas con *Colpodon* (Colhuehuapense), y se examina la diversidad y la distribución de los astrapoterios colhuehuapenses a la luz de nuevos y más completos materiales y con información estratigráfica más precisa. El nuevo material discutido aquí proviene de la *Lower Fossil Zone* del Miembro Colhue-Huapi de la Formación Sarmiento; el material descripto previamente se presume como proveniente de este mismo nivel y corresponde a la Edad Colhuehuapense, las Capas con *Colpodon* de Ameghino. La *Lower Fossil Zone* está asignada al Mioceno temprano, aproximadamente 20 Ma. a partir de datos radimétricos y polaridad magnética. Nuestra revisión revela la presencia certera de tres especies: *Astrapotherium? ruderarium* (Ameghino 1902), *Parastrapotherium symmetrum* (Ameghino 1902), y *Parastrapotherium martiale* Ameghino, 1901 (el último también conocido para la inmediatamente más antigua Edad Deseadense). *Parastrapotherium herculeum* puede representar una cuarta especie muy grande, pero el material tipo de Ameghino, supuestamente proveniente de capas colhuehuapenses, no pudo ser localizado, y ningún ejemplar disponible pudo ser referido a ésta con certeza. Finalmente, un canino superior aislado parece representar una quinta especie innominada, tal vez un astrapoterio uruguayterino. El examen de viejas y nuevas colecciones sugiere que *Astrapothericulus* no se registra en estos niveles, aunque el género ocurre en otras localidades colhuehuapenses.

Los siguientes nombres son pasados a sinonimia con *Astrapotherium? ruderarium*:

Parastrapotherium paucum Ameghino 1902

Parastrapotherium crassum Ameghino 1902 (*partim*)

Astrapothericulus minusculus Ameghino 1902

Astrapothericulus laevisculus Ameghino 1902

Astrapotherium triangulidens Ameghino 1902

Prochalicotherium patagonicum Ameghino 1902.

Astrapotherium? ruderarium es la única especie reconocida de *Astrapotherium* en la Edad Colhuehuapense.

Astrapotherium? ruderarium también es el astrapoterio más abundante en los niveles colhuehuapenses de Gran Barranca.

Introduction

Astrapotheriidae (*sensu* Cifelli 1993) is the most derived family within the Order Astrapotheria, an extinct group of herbivorous South American land mammals (Scott 1937). The family is among the largest and most specialized mammals among the Tertiary native faunas. Their stratigraphic range is from the Mustersan (late Eocene) to the Laventan SALMA (middle Miocene) (Soria 1984; Cifelli 1993; Johnson and Madden 1997). Ameghino (1899, 1902) described eight species for his "couches à *Colpodon*" at the Gran Barranca south of Lake Colhue-Huapi (Colhuehuapian SALMA, early Miocene: Flynn and Swisher 1995; Ré *et al.* Chapter 3, this book), which range in from the size of a living Neotropical peccary to as large as an African rhinoceros. He grouped these species in the genera *Parastrapotherium*, *Astrapotherium*, and *Astrapothericulus*. On this basis, it is traditionally accepted that there are three Oligocene to Miocene Patagonian genera and that they co-occur during the Colhuehuapian Age (Scott 1937; Pascual and Odreman Rivas 1971; Marshall *et al.* 1983; Pascual *et al.* 1996; Johnson and Madden 1997). Unfortunately, most of the Colhuehuapian species described by Ameghino (1902) are based on fragmentary dental remains, which are not directly comparable.

Museo de La Plata – Duke University expeditions to Gran Barranca provided new specimens of Colhuehuapian astrapotheriids. These materials complement the abundant and more complete materials recovered by the Muséum National d'Histoire Naturelle (1899), Field Museum of Natural History (1923–24), and American Museum of Natural History (1930) expeditions to Gran Barranca, as well the specimens collected by A. Bordas (1940–41), and A. Castellanos (1944). In this contribution the astrapotheriids that come from Colhuehuapian levels at Gran Barranca are analyzed, the status of the species previously described for these beds is evaluated, and the diversity of the Colhuehuapian astrapotheres is re-examined.

Institutional abbreviations

AMNH, American Museum of Natural History; MACN, Museo Argentino de Ciencias Naturales "Bernardino Rivadavia"; MLP, Museo de La Plata (Argentina); MPEF, Museo Paleontológico "Egidio Feruglio" (Trelew, Chubut Province, Argentina); MUFYCA, Museo Universitario "Florentino y Carlos Ameghino" (Rosario, Argentina); MNHN, Muséum National d'Histoire Naturelle (Paris, France); FMNH, Field Museum of Natural History (Chicago, USA); YPM PU, Yale Peabody Museum, Princeton University (New Haven, USA).

List of specimens used for comparisons

(1) *Parastrapotherium holmbergi* Ameghino: MACN A 52–509, MACN A 52–504, MACN A 52–515, MACN A 52–518 (Syntypes), and additional materials from the Deseadan La Flecha locality, Santa Cruz Province, (MLP 95-III-10–74, 95-III-10–90, and 95-III-10–103; FMNH 13329, 13343, 13354, 13364, 13365, 13369, 13462,13473, 13491, 13492, and 13579).

(2) *Parastrapotherium martiale* Ameghino: MACN A 52–604 (Holotype) and additional materials from the Deseadan beds of the Upper Puesto Almendra Member (Sarmiento Formation) at Gran Barranca (MLP 93-XI-18–41, 93-XI-18–45, 93-XI-18–43, 93-XI-18–42, 93-XI-18–9, 93-XI-18–39, 93-XI-18–10, 93-XI-18–14, 93-XI-18–30, 93-XI-18–40, 93-XI-18–7, and 93-XI-18–5; MPEF PV 7129, 7133, 7135, 7128, and 7807; AMNH 29565; FMNH 13427, 13428, and 13529).

(3) *Astrapothericulus iheringi* Ameghino: MACN A 52–408 to 414, 52–417, 52–419, 52–421, 52–422, and 52–605 (Syntypes) and abundant additional materials from the Pinturas Formation, Santa Cruz Province, at the MACN.

(4) *Astrapotherium magnum* (Owen): MACN A 3207 (Ameghino 1894, Fig. 20), 3210, 3214, 3216–3220, 3296, 3279–3281, 3295–3298, 8580–8581, 8603, 11250 (Ameghino 1904, Fig. 226); MACN PV 14512; AMNH 9278 (Scott 1928, plates, 13–14); FMNH 13170, 13173, 14251, 14259; YPM PU 15142, 15332 (Scott 1928, plate 14). These materials were referred to this species following interpretations provided by Ameghino (1894).

Systematic paleontology

Order **ASTRAPOTHERIA** Lydekker 1894
Family **ASTRAPOTHERIIDAE** Ameghino 1887
Genus *Parastrapotherium* Ameghino 1895
Type species *Parastrapotherium holmbergi* Ameghino 1895
Distribution Argentina. Late Oligocene to early Miocene.

Parastrapotherium symmetrum (**Ameghino** 1902) **nov. comb.**
Fig. 12.1A, A′.

Astrapotherium? symmetrum; Ameghino 1902
Holotype MACN A 52–507a, an isolated lower incisor.
Referred material MPEF PV 7923, an isolated lower incisor.
Diagnosis Similar in size to *Parastrapotherium holmbergi*. Incisors with a prominent medial longitudinal crest on the lingual face.
Provenance According to Ameghino (1902), the holotype comes from the "*Colpodon* beds" at the

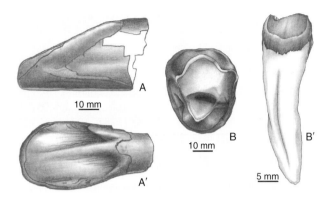

Fig. 12.1. *Parastrapotherium* spp. from Colhuehuapian beds at Gran Barranca. *P. symmetrum* (Ameghino, 1902), MACN A 52–507a, (holotype), incisor in (A) lateral view, (A′) lingual view. *?P. herculeum* (Ameghino, 1899), MACN A 52–516a, left P3 in (B) occlusal view, (B′) labial view.

Gran Barranca south of Lake Colhue-Huapi, Chubut Province, Argentina. The only referred specimen comes from the Lower Fossil Zone (Gran Barranca Colhuehuapian West locality, Level B), Colhue-Huapi Member of the Sarmiento Formation (Spalletti and Mazzoni 1979; Bellosi this book).

Comments The diagnostic character of this incisor (Fig. 12.1), already indicated by Ameghino (1902), is distinct from those of all the remaining astrapotheres and justifies its specific distinction. Moreover, the wear produces a regular curve around the tip and most of the sides of the crown, unlike the other species. As in *Parastrapotherium holmbergi* and *P. martiale*, the base of the crown of this incisor is bucco-lingually much broader than in the species of *Astrapotherium*, and therefore it is herein transferred to the genus *Parastrapotherium*. Ameghino also assigned to this species an isolated upper premolar (MACN A 52–507b) that he interpreted as a P3, pointing out that the crown have quadrangular contour and fused roots. However, this tooth is a P4, as evidenced by the presence of a wear facet on the anterior face for the P3. This tooth has very prominent labial fold and styles, as typically in the species of *Astrapotherium*, but lacks a cingulum at the base of the labial fold, as in the species of *Parastrapotherium*. Moreover, this tooth is not physically associated to the type material (Ameghino, 1902), and there is no other evidence to refer it to *P. symmetrum*.

Parastrapotherium herculeum Ameghino 1899
nov. comb.
Fig. 12.1B, B′.
Astrapotherium herculeum [Ameghino 1899]
Parastrapotherium herculeum (Ameghino 1901)

Astrapotherium herculeum Ameghino 1901, Ameghino 1902
Comments Ameghino (1899) based this species upon a mandibular fragment with p3–m3, a lower canine, and a P4. According to Ameghino (1899), the type materials come from "la Formación Patagónica del interior del Deseado y del Lago Musters" (Colhuehuapian or Astrapotericulan). In 1902 he described new materials as coming from his "couches à *Colpodon*" (but see below), suggesting that the type material comes from the same horizon. The type materials of *Astrapotherium herculeum* are not found in the paleontological collections of MACN, but according to the original description (Ameghino 1899) this species has two permanent lower premolars, thus the more proper generic assignation is to *Parastrapotherium*, a conclusion later arrived at by Ameghino (1901). In his original description, Ameghino (1899) did not indicate any specific diagnostic features of this taxon, other than to note that this is a very large species. Indeed, the measurements given by Ameghino (1899) for the m3 and for the complete p3–m3 length reveal that it is significantly larger than *P. martiale* and *P. superabile* (the largest Deseadan species of *Parastrapotherium*; see Table 12.1). Later, Ameghino (1902) described the P4 included in the type series, pointing out that it differs from other species by having fused tooth roots. However, this condition (also present in the premolars of the types of the Deseadan species *Parastrapotherium holmbergi* and *Traspoatherium convexidens*) could be merely individual variation, as Loomis (1914) contends, and the characters indicated of the crown do not differ from those observed in the P4 of *P. superabile*. Moreover, the size of this P4 is proportionally smaller than the lower teeth of the type series of *P. herculeum*, and they could not be conspecific. In 1902 Ameghino described several isolated incisors (MACN A 25–516b) and an isolated P3 (MACN A 25–516a). The incisors do not differ significantly from those of *P. martiale*; they are herein included in the list of specimens referred to the latter species (see below). We agree that the P3 should be referred to *P. herculeum*. The tooth has fused roots, as does the P4 of the type series (see Fig. 12.1B, B′). The ectoloph shows a conspicuous wear facet on the labial side from occlusion with the p3 (a tooth that is absent in *Astrapotherium*). MACN A 25–516a has a continuous labial cingulum at the base of the labial wall, as in the species of *Astrapotherium*, but the base of the fold is very broad and little prominent, as in *P. holmbergi* and *P. martiale*. This particular combination of characters, not noted by Ameghino (1899, 1902) for the

Table 12.1. *Dental measurements (in cm) for* Parastrapotherium herculeum *(after Ameghino, 1899), from the Gran Barranca south of Lake Colhue-Huapi, Colhuehuapian SALMA, and* Parastrapotherium martiale, *Deseadan and Colhuehuapian? SALMAs*

	Parastrapotherium herculeum	*Parastrapotherium martiale*			
		MACN 52–604 (Holotype)	MNHN COL 1 (Colhuehuapian?)	MLP 93-XI-18–14 (Deseadan)	MPEF 6693 (Deseadan)
m3 APL	10	8.1	8.2	8.2	–
p3–m3 length	28	23.8	–	–	–
Lower canine width	10	6.2	7.6	–	6.9

Note: APL, anteroposterior length.

P4 of the type series, is not observed in upper premolars of other known astrapotheriids. Following Ameghino's proposal, this P3 is tentatively referred to *P. herculeum.*

Because it is impossible to compare the type specimens with the other species, we provisionally accept the validity of *Parastrapotherium herculeum* as a species distinct from other *Parastrapotherium,* the only diagnostic feature being its significantly larger size than the remaining species of *Parastrapotherium.* The possibility that the type of *P. herculeum* is an extreme variant of *P. martiale* should continue to be considered.

Parastrapotherium martiale Ameghino 1901
Referred materials from Colhuehuapian levels at Gran Barranca MACN A 52–516b, seven isolated incisors; MACN Pv 12833, an incomplete P4; MPEF PV 7924, an incomplete incisor and fragment of lower molar (not associated); MPEF PV 7333, an isolated p4; MPEF PV 5500, a maxillary fragment with left P3, M1–3, and right P3; MNHN COL 1, both mandibular rami with canines, roots for p3, p4–m3; MNHN COL unnumbered, an isolated left M3.
Remarks *Parastrapotherium martiale* is a typical Deseadan species (Ameghino 1900–02), and it is also well represented at Deseadan levels at Gran Barranca. According to a recent revision of *Parastrapotherium* (Kramarz and Bond 2008) *P. martiale* is the largest Deseadan species of the genus, but neither the type (MACN A 52–604, jaw and palate with an almost complete dentition) nor the referred Deseadan specimens are as large as *P. herculeum.* We herein refer to *P. martiale* those specimens described by Ameghino (1902) as *Astrapotherium herculeum,* but not included in the type series, that match in size with *P. martiale.* Similarly, the MNHN specimens from Gran Barranca (collected by Tournouër) catalogued as *Astrapotherium herculeum,* match in size and morphology with the type of *P. martiale* (see

Table 12.1). The stratigraphic level of Ameghino's specimens from Gran Barranca is unknown. On the contrary, the MNHN specimens almost surely are derived from Colhuehuapian levels (Gaudry 1906). Similarly, the more recently collected MPEF specimens (PV 7924, 7333, and 5500) come positively from the Lower Fossil Zone, Colhue-Huapi Member of the Sarmiento Formation. The survival of this species into the Colhuehuapian is also documented in levels of equivalent age of the Cerro Bandera Formation at Neuquén Province (Kramarz *et al.* 2005).

Genus *Astrapotherium* Burmeister 1879
Type species *Astrapotherium magnum* (Owen 1853)
Distribution Argentina and Chile. Early to middle Miocene.
Astrapotherium? ruderarium (Ameghino 1902)
Fig. 12.2.
Parastrapotherium ruderarium Ameghino 1902
Parastrapotherium paucum Ameghino 1902
Parastrapotherium crassum Ameghino 1902 (*partim*)
Astrapothericulus minusculus Ameghino 1902
Astrapothericulus laevisculus Ameghino 1902
Astrapotherium triangulidens Ameghino 1902
Prochalicotherium patagonicum Ameghino 1902
Lectotype MACN A 52–524, a right mandibular fragment with dp3–dp4, the posterior portion of m1, a complete m2, and erupting m3 (Fig. 12.2).
Types in synonymy MACN A 52–525 (syntype of *Parastrapotherium paucum*), three incisors and two isolated cheek teeth; MACN A 52–512 (holotype of *Astrapothericulus laevisculus*), probably associated dp4, four deciduous incisors, a fragment of a lower canine, and a fragment of an upper canine; MACN A 52–511 (holotype of *Astrapothericulus minusculus*), probably associated DP3, two deciduous incisors, a fragment of a lower canine and a fragment of dp3; MACN A 52–521 (syntype of *Parastrapotherium crassum*), a right mandibular ramus with erupting

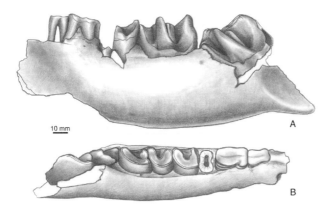

Fig. 12.2. *Astrapotherium? ruderarium* (Ameghino, 1902) MACN A 52–524, lectotype, right mandibular fragment with dp3–m3 in (A) lingual view, (B) occlusal view.

p4 and m1, an associated lower canine; an m2, two P4s, and one M2 of different individuals; MACN A 52–523 (type of *Astrapotherium triangulidens*), two probably associated fragments of upper canines; MACN A 52–337 (lectotype of *Prochalicotherium patagonicum*), an isolated right P4.

Remarks Although all the species here are referred to *Astrapotherium? ruderarium* were described in the same contribution (Ameghino 1902), we consider *Parastrapotherium ruderarium* the senior synonym because it is the species based on more complete materials and with verifiable diagnostic characters.

Referred material MACN A 52–513 (syntype of *P. ruderarium*), a P4 and two incisors; MACN A 52–522 (syntype of *P. ruderarium*), a fragment of an upper canine, p4, m1, m2, three incisors, a fragment of a lower juvenile canine, an M1, an M2, an M3, and an astragalus; MACN A 52–514d, an isolated m2; MACN A 52–520, a right mandibular ramus with dp4–m2 and alveoli for dp2?, dp3, and m3; MACN A 52–519a, two isolated p4s, two P4s, and a lower molar; MACN Pv 11244, one lower molar and three fragments of upper molars; MACN Pv 12830a, two fragments of upper canines and three fragments of lower canines; MACN Pv 12829b, five isolated incisors; MACN PV 12831a, an isolated m2; MACN Pv 14543, a right mandibular ramus with m2, three fragments of upper molars, three incisors, two fragments of lower canines, and one fragment of an upper canine; FMNH 13307, both mandibular rami with p4–m3; FMNH P13426, mandible with a complete dentition; FMNH P13429, a partial skull with C, P3, DP4, M1–2, and erupting M3; AMNH 29716, a juvenile mandible with c1, dp3–dp4, m1–2, and encrypted p4 and m3; AMNH 29717, an associated left mandibular ramus with m1–3 and right mandibular ramus with

m2–3; AMNH 29720, a left maxillary fragment with M1–3; AMNH 29723, a right maxillary with P4–M1, a left maxillary with P3–M2, and an upper canine; AMNH 29724, a palate with right P3–M3, left P3, M2–3 and a canine; MPEF PV 1134, an isolated lower canine; MPEF PV 1135, isolated upper canine; MPEF PV 1276, isolated M2; MPEF PV 5656, three isolated incisors; MPEF PV 7915, both maxillaries with a complete dentition and both mandibular rami with, c, p4–m3, and two incisors; MPEF PV 7918, an associated incisor, two lower molars, and one incomplete upper molar; MPEF PV 7919, an associated left M1 and M2; MPEF PV 7920, two fragments of upper canines, two fragments of lower canines, three incisors, a DP4, a P4, an incomplete upper molar, and an incomplete lower molar; MPEF PV 7921, a left maxillary fragment with DP4, M1, M2, P3, and an erupting P4; MPEF PV 7922, an associated right and left DP3, left DP4, two deciduous incisors, right dp3–dp4, left dp2–dp4, and several fragments of skull; MPEF PV 5635, an isolated ?m1; MLP 93-XI-18–4, a left lower molar; MLP 93-XI-18–23, an isolated left P4; MLP 93-XI-18–35, an incomplete left upper molar; MUFYCA 803, both mandibular rami with m1–3.

Diagnosis An astrapotheriid slightly larger than *Astrapothericulus iheringi*, nearly 25% smaller than *Astrapotherium magnum* and *Parastrapotherium holmbergi*. Permanent dental formula as in the species of *Astrapotherium* and *Astrapothericulus*. Cheek teeth comparatively lower crowned than in *Astrapotherium magnum*, but higher than in *Parastrapotherium holmbergi*. Basal cingula of all cheek teeth less prominent than in *Astrapothericulus iheringi*. Upper premolars with labial fold more prominent and with a narrower base than in *Parastrapotherium holmbergi*, and with cingulum present at the base of this fold, as in *Astrapotherium magnum* and *Astrapothericulus iheringi*. Upper molars with an anterolingual pocket less marked than in *Astrapothericulus iheringi*. Incisors broader and with more robust roots than in *Astrapothericulus iheringi*. The p4 with a labial flexid, thought not as penetrating as in *Astrapotherium magnum*. Lower molars with well-developed hypoflexids and lingual cingula present at the bases of the metaconids, as in *Astrapotherium magnum* and *Astrapothericulus iheringi*. The ever-growing upper canines have smooth enamel covering the lateral walls up to the base of the tooth. The lower canines are more robust than in *Astrapothericulus iheringi* and with more convex lingual wall than in other Astrapotheriidae.

Geographic and stratigraphic provenance According to Ameghino (1902) all the MACN

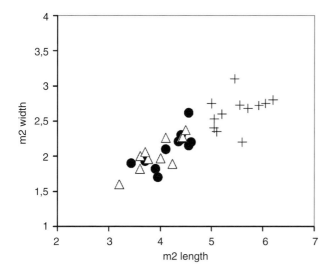

Fig. 12.3. Bivariate plot of m2 crown length and width (in cm) for *Astrapotherium? ruderarium* (circles), from the Gran Barranca south of Lake Colhue-Huapi (Colhuehuapian SALMA), *Astrapothericulus iheringi* (triangles), and *Astrapotherium magnum* (plus signs). Measured specimens of *Astrapotherium? ruderarium*: MACN A 52–514, 52–520, 52–521, 52–522, 54–524, MACN Pv 12831a, 14543, MUFYCA 803a, MPEF PV 7915, FMNH P13426, AMNH 29717. Measurements for *Astrapothericulus iheringi* taken from MACN specimens; for *Astrapotherium magnum* taken from FMNH, AMNH, YPM PU, and MACN specimens.

A specimens are from the "*Colpodon* beds" at the Gran Barranca south of Lake Colhue-Huapi. The MLP specimens, MPEF PV specimens, AMNH 29716, 29717, 29720, and FMNH 13426, 13429 come from the Colhue-Huapi Member, Lower Fossil Zone (Colhuehuapian) at Gran Barranca. MACN Pv 14543, 11244, 12830a, 12829b, 12831a, AMNH 29724, 29723, FMNH 13307, and MUFYCA 803 also come from Gran Barranca, but their exact position in the stratigraphic section is uncertain.

Description and comparisons The cheek teeth are slightly larger (nearly 6%) than those of *Astrapothericulus iheringi* (Ameghino 1899) from the "Astrapothericulan beds" (Ameghino 1900–02, 1906), assigned to the late early Miocene (Fleagle *et al.* 1995). However, the cheek teeth show a wide range of size, with variation of nearly 25% in m2 measurements. The measured specimens cluster in two groups (Fig. 12.3). The absence of other relevant morphological differences suggests that these two groups could represent males and females of a single species. A similar dispersion is also observed for the measurements of the m2 of *Astrapothericulus iheringi* (Fig. 12.3).

The height of the crowns nearly agrees with that of *Astrapothericulus iheringi*, which is intermediate between that of *Parastrapotherium holmbergi* and *Astrapotherium magnum* (Santacrucian). However, *Astrapotherium? ruderarium* clearly differs from *Astrapothericulus iheringi* by having all cheek teeth with much more delicate basal cingula, as in *Astrapotherium magnum*, whereas *Astrapothericulus iheringi* have much more prominent and crenulated basal cingula, which is the most characteristic feature of this species (Ameghino 1902). Moreover, the upper molars (Fig. 12.4A) differ from those of *Astrapothericulus iheringi* by having less penetrating fold of the anterior wall of the protocone; thus the anterolingual pocket is less conspicuous. As distinct from the species of *Parastrapotherium*, P3 and P4 (Fig. 4.B) have a continuous cingulum at the base of the labial fold, and the fold is more prominent and with narrower base. This is a resemblance to *Astrapothericulus iheringi* and *Astrapotherium magnum*. As in *Astrapotherium magnum*, the upper canines are evergrowing, and completely columnar in adults, with a subtriangular cross-section, the anterior face lacks enamel and it is flat in juvenile stages or it has a longitudinal furrow in adults (Fig. 12.4C), and the enamel extends up to near the base of the lateral walls. However, the enamel is comparatively smoother than in other astrapotheriids. Upper canines of *Astrapothericulus iheringi* are rooted and proportionally smaller.

The incisors (Fig. 12.4D, D′) are larger and comparatively broader and the base is more robust than in *Astrapothericulus iheringi*, resembling those of the species of *Parastrapotherium*. The lower canines (Fig. 12.4E, E′) are more robust than in *Astrapothericulus iheringi*, strongly extroverted, as in *Astrapotherium magnum* and the species of *Parastrapotherium*. In adult stages the canines are rooted, as in *Astrapothericulus iheringi* and the species of *Parastrapotherium* (but probably the same condition is present in *Astrapotherium magnum*). The lingual face is more convex than in those species. The p3 is absent. The p4 (Fig. 12.4F) is typically bicrescentic and has a conspicuous labial flexid as in *Astrapotherium magnum*, unlike *Astrapothericulus iheringi* and the species of *Parastrapotherium*, but the flexid is not as penetrating as in *Astrapotherium magnum*. Unlike the species of *Parastrapotherium* but as in *Astrapotherium magnum* and *Astrapothericulus iheringi*, the lower molars have a well-developed and deep labial flexid, and lingual cingulids are present at the bases of the metaconids (Figs. 12.2, 12.4G). All lower molars have a conspicuous column attached to posterior wall of the metalophid ("pillar" after Scott

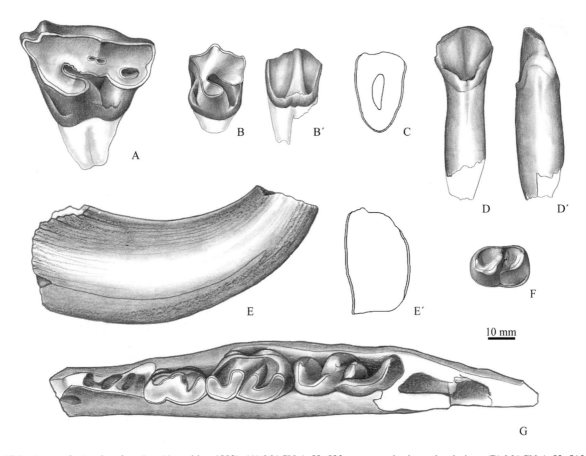

Fig. 12.4. *Astrapotherium? ruderarium* (Ameghino 1902). (A) MACN A 52–522, upper molar in occlusal view; (B) MACN A 52–513, P4 in occlusal view, (B′) in labial view; (C) MACN A 52–523 (holotype of *Astrapotherium triangulidens* Ameghino 1902), upper canine in schematic cross-section; (D) MACN A 52–522, incisor in lingual view, (D′) in lateral view; (E) MACN A 52–521 (syntype of *Parastrapotherium crassum* Ameghino 1902), lower canine in labial view, (E′) schematic cross-section; (F) MACN A 52–522, p4 in occlusal view; (G) MACN A 52–520, left mandibular fragment with dp4, m1, and m2 and alveoli for dp2, dp3, and m3.

1928, 1937), as in the species of *Parastrapotherium* and *Astrapothericulus iheringi*, but it has more bunoid appearance than in *Astrapotherium magnum*.

MPEF PV 7922 has associated DP3 and DP4. Both are molariform and the former lacks the anterolingual projection characteristic of an anteriormost deciduous premolar, suggesting that this specimen had at least another anterior deciduous premolar (DP2), as in *Parastrapotherium holmbergi*, *Astrapotherium magnum*, and *Astrapothericulus iheringi*. The juvenile mandibular fragment herein designed lectotype (MACN A 52–524) has five cheek teeth, the posteriormost is evidently the m3 in eruption (Fig. 12.2). The two anteriormost teeth are very worn dp3 and dp4. A diastema anterior to the dp3 shows no trace of an alveolus for another deciduous premolar. Deep inside the dentary there is an unerupted tooth below the dp4, but there is not a corresponding tooth for replacement below the dp3. Another juvenile mandibular fragment in the MACN Ameghino collection

(MACN A 52–520, Fig. 12.4G), also referable to *Astrapotherium? ruderarium*, is a slightly younger juvenile than the lectotype; it shows two small alveoli before the alveoli for the dp3, corresponding to a minute dp2. These two juvenile specimens reveal that *Astrapotherium? ruderarium* has only one lower permanent premolar, as in *Astrapotherium magnum* and *Astrapothericulus iheringi*, but the lower deciduous formula agrees with that of *Parastrapotherium*, though with a more ephemeral dp2. All deciduous premolars are similar in structure to that of *Astrapothericulus iheringi*, but differ in having more delicate basal cingula, as in the molars.

The mandible is partially preserved in the adult MPEF PV 7915; it is much more robust and the diastema is much longer than in *Astrapothericulus iheringi*, and proportionally slightly longer than in *A. magnum* and *P. martiale*. The strongly extroverted implantation of the canines and other features of the dentary bear no significant differences with *A. magnum*.

The juvenile skull FMNH P13429 (with erupting P3 and M3) was previously referred to *Astrapotheriiculus* by Johnson and Madden (1997). However, the molars differ from those of the species of *Astrapotheriiculus* and match with *Astrapotherium? ruderarium* in having delicate basal cingula and poorly developed anterolingual pocket. The canines have smooth enamel, as in other specimens referred to *Astrapotherium? ruderarium*. The preserved portions of the skull, which is partially restored, show that the frontals are proportionally narrower, the skull roof is somewhat more curved anteroposteriorly, and the zygomatic arches are less flared than in a skull of *Astrapotherium magnum* in similar ontogenetic stage (MLP 38-X-30–1). These differences are also observed in adult skulls of *Astrapotherium magnum*. Moreover, the occipital region (not preserved in the MLP 38-X-30–1) is lower than in adults of *Astrapotherium magnum*. Because of the ontogenetic variations of the skull in the Astrapotheria are very poorly known, the taxonomic significance of these cranial features remains uncertain.

Comments Ameghino's (1902) description of *Parastrapotherium ruderarium* is based upon a juvenile mandibular fragment with five teeth (MACN A 52–524), herein designed the lectotype (Fig. 12.2), and two lots of teeth (MACN A 52–522 and MACN A 52–513) corresponding to more than one individual. All the cheek teeth and incisors included in these lots are coherent in morphology and size with the lectotype. However, the upper canine described by Ameghino (1902) included in the lot MACN A 52–522 is significantly different from other canines that are positively associated with molars referable to this species. This canine is treated as a separate taxon in a section below.

Ameghino (1902) concluded that the juvenile mandibular fragment herein designed lectotype has four deciduous premolars, of which only two were replaced. As we mentioned above, this specimen has preserved three molars and only two deciduous premolars (dp3–dp4) but only the dp4 is replaced.

According to Ameghino (1902), the incisors of this species are characterized by being nearly similar in shape and size. Actually, the incisors he interpreted as i1 and i2 (MACN A 52–522) very probably are right and left to i2s.

Ameghino (1902) based *Parastrapotherium paucum* on two very worn cheek teeth and three isolated incisors (MACN A 52–525), and concluded that it was a separate species even smaller than *Parastrapotherium ruderarium*. However, the cheek teeth are in fact barely smaller than those described as

Parastrapotherium ruderarium, and equal in size to other specimens herein referred to this species, and they are not distinct morphologically. One of the incisors described as *Parastrapotherium paucum* is very similar in size and morphology to the i2 included in the original type series of *Parastrapotherium ruderarium*, but more worn. Another incisor which is smaller is probably the i1. The remaining incisor is much more slender and smaller and could be a deciduous tooth.

Astrapothericulus minusculus and *Astrapothericulus laevisculus* Ameghino 1902 are based on several deciduous teeth. The lower premolars of both types do not differ from those of the lectotype of *Astrapotherium? ruderarium*.

Ameghino (1902) based *Astrapotherium triangulidens* on an isolated fragment of an upper canine (MACN A 52–523), noting that it has smooth enamel, triangular cross-section and unlimited growth. These same features also occur in the specimen MACN Pv 14543, associated with molars and incisors clearly assigned to *Astrapotherium? ruderarium*.

According to Ameghino (1902), *Parastrapotherium? crassum* is a giant species, as large as *Astrapotherium giganteum* (Santacrucian SALMA), although he included in the original description a small lower canine (MACN A 52–521) that he interpreted as belonging to a juvenile individual. This canine (Fig. 12.4E) is associated with a mandibular fragment with p4–m1 and other cheek teeth referable to *Astrapotherium? ruderarium*.

Ameghino (1902) described *Prochalicotherium patagonicum* as a member of Homalodotheriidae (Notoungulata), and mentioned that this species is identifiable from features of the upper molars, but unfortunately he only described two upper premolars, a canine, and an incisor. Ameghino noted that the most characteristic element is the upper premolar MACN A 52–337 (Ameghino 1904, Fig. 391) that he understood to be a left P3. We find that this tooth, herein designed lectotype of *Prochalicotherium patagonicum*, has vertically banded enamel and corresponds to a right P4 of an astrapotheriid (Patterson 1952) very similar in shape and size to others here assigned to *Astrapotherium? ruderarium*, differing only in having a slightly less prominent labial fold and no lingual cingulum at the base of the protocone, features we consider to be merely variation at the population level. The remaining teeth described by Ameghino (1902) as *Prochalicotherium patagonicum* (MACN A 52–339, 52–534, and 52–550) have uncertain affinities, but they surely do not belong to an astrapothere.

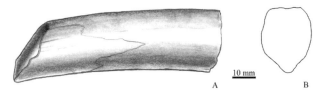

Fig. 12.5. Astrapotheriidae gen. et sp. nov, MACN A 52–522 (syntype of *Parastrapotherium ruderarium* Ameghino, 1902). (A) upper canine in lateral view, (B) schematic cross-section.

Affinities of *Astrapotherium? ruderarium*
Ameghino (1902) referred *Astrapotherium? ruderarium* to *Parastrapotherium* based on the erroneous assumption that the specimen herein designed lectotype has two lower permanent premolars. The dentition of this species does in fact share with *Parastrapotherium holmbergi* the presence of three lower deciduous premolars and the robust incisors with broader bases than in other astrapotheriids. On the other hand, *Astrapotherium? ruderarium* differs in significant ways from *Parastrapotherium holmbergi*, for example by having a cingulum at the base of the labial fold in all the upper premolars and at the lingual base of the metaconid in the lower molars, in lacking a p3 and in having more hypsodont teeth. *Astrapotherium magnum* and *Astrapothericulus iheringi* share with *Astrapotherium? ruderarium* the presence of a "pillar" in all the lower cheek teeth, the reduced lower premolar formula, the higher degree of hypsodonty of the cheek teeth; some or all these features are derived characters. Therefore, we conclude that the original assignation of *Astrapotherium? ruderarium* to *Parastrapotherium* is not justified. Indeed, in some ways *Astrapotherium? ruderarium* is more derived than *Astrapothericulus iheringi*: it shares with *Astrapotherium magnum* upper ever-growing canines and the derived presence of a labial flexid in p4, both features being absent in *Astrapothericulus iheringi*. Based on these characters we provisionally refer this species to the genus *Astrapotherium*.

Astrapotheriidae gen. et sp. nov.
Referred material MACN A 52–522b, an isolated upper canine (syntype of *Parastrapotherium ruderarium*).
Provenance According to Ameghino (1902), it comes from the "*Colpodon* beds" at the Gran Barranca south of Lake Colhue-Huapi, Colhuehuapian SALMA (early Miocene: Flynn and Swisher 1995; Ré *et al.* Chapter 4, this book), Chubut Province, Argentina.
Comments This canine (Fig. 12.5), described by Ameghino (1902) as belonging to *Parastrapotherium*

ruderarium, differs from all canines positively associated to molars of this species because it is much less curved and compressed, the cross-section is nearly subcircular, and the enamel does not extend up to the preserved base of the tooth. This tooth resembles an upper canine (associated with upper cheek teeth) more recently collected from Colhuehuapian sediments of the Cerro Bandera Formation, northwest Patagonia (Leanza and Hugo 1997; Kramarz *et al.* 2005). The associated cheek teeth have characters that suggest uruguaytheriine affinities (Kramarz and Bond 2005).

Parastrapotherium crassum (Ameghino 1902)
Parastrapotherium crassum was described by Ameghino (1902) as being from his "*Colpodon* beds" (Colhuehuapian). This species is based on several isolated specimens that clearly belong to different species. The type specimens have black, polished enamel and dark dentine. This kind of preservation is typical of the fossils teeth from the Deseadan levels at the Gran Barranca, and different from the characteristic brown and light-coloured teeth from Colhuehuapian levels at the Barranca. The remaining specimens referred by Ameghino to *P. crassum*, which preservation agrees with those more recently recovered at the Colhue-Huapi Member, are herein referred to *Astrapotherium? ruderarium*. Therefore, *P. crassum* should not be listed among the Colhuehuapian species.

Conclusions

The revision of astrapotheres from Colhuehuapian levels at Gran Barranca south of Lake Colhue-Huapi reveals the certain presence of three species: *Astrapotherium? ruderarium* (Ameghino 1902), *Parastrapotherium symmetrum* (Ameghino 1902), and *Parastrapotherium martiale* Ameghino 1901 (the latter also known from the Deseadan SALMA). *Parastrapotherium herculeum* may represent a fourth very large species but the type material of Ameghino (1889), supposed to come from Colhuehuapian beds, cannot be located, and no available specimen can be certainly referred to it. Finally, a single upper canine (part of the original syntype of *Parastrapotherium ruderarium*) seems to represent a fifth unnamed species, perhaps a uruguaytheriine. The number of species, then, is significantly less than the 12 originally proposed by Ameghino (1902). Six of Ameghino's species have been brought into synonymy with the remaining five. The seventh species, *Parastrapotherium crassum*, is a valid taxon from the Deseadan, but is not known in the Colhuehuapian.

The dentition of *Astrapotherium? ruderarium* shows a confusing combination of characters, in part transitional between *Parastrapotherium holmbergi* on one hand, and *Astrapotherium magnum* and *Astrapothericulus iheringi* on the other hand. We assign this species to *Astrapotherium* until further evidence is known.

Both Colhuehuapian species of *Astrapothericulus* described by Ameghino (1902) are here considered synonyms of *Astrapotherium? ruderarium.* Although some specimens (e.g. AMNH 29687, FMNH 15049) resemble *Astrapothericulus iheringi* in having lower molars with basal cingula somewhat more developed than in the lectotype of *Astrapotherium? ruderarium,* no examined specimen from Colhuehuapian levels at Gran Barranca is positively referable to *Astrapothericulus.* However, the record of *Astrapothericulus* is well documented in levels of equivalent age of the Cerro Bandera Formation in Neuquén Province (Kramarz and Bond 2005; Kramarz *et al.* 2005).

The record of *Astrapotherium* in Colhuehuapian beds is revised. In our revision *Astrapotherium triangulidens* is made a junior synonym of *Astrapotherium? ruderarium. Astrapotherium? symmetrum* and *Astrapotherium herculeum* are transferred to *Parastrapotherium.* This leaves only *Astrapotherium? ruderarium.*

Astrapotherium? ruderarium, the smallest astrapotheriid recorded at Colhuehuapian levels at Gran Barranca, is also the most abundantly represented. *Parastrapotherium symmetrum* is identified only through two isolated incisors; this is a middle-sized species equivalent in size to *Astrapotherium magnum.* The largest astrapotheres are *Parastrapotherium martiale* and *Parastrapotherium herculeum,* supposedly the largest astrapothere known so far.

ACKNOWLEDGEMENTS
The authors thank R. Madden (Duke University) and G. Vucetich (MLP) for allowing us to study material obtained during the MLP – Duke University expeditions to Gran Barranca. The authors are grateful to E. Ruigómez (MEF), M. Reguero (MLP), J. Flynn (AMNH), and P. Makovicky (FMNH) for access to materials under their care, and to C. de Muizon and G. Billet (MNHN) for providing casts of specimens in the Tournouër collection. The critical comments of P. O. Antoine (Université Paul Sabatier, Toulouse, France) and an anonymous reviewer improved this paper considerably. Drawings were made by the artist Jorge González. This work was supported by PICT 32344 awarded to Dr. Viviana Barreda (MACN) and Consejo Nacional de Investigaciones Científicas y Técnicas (CONICET). Field research was supported by US National Science Foundation grants EAR-0087636, BCS-0090255, and DEB-9907985 to Richard F. Kay and Richard H. Madden.

REFERENCES

Ameghino, F. 1894. Enumération synoptique des espèces de mammifères fossiles des formations éocènes de Patagonie. *Boletín de la Academia Nacional de Ciencias en Córdoba,* **13**, 259–455.

Ameghino, F. 1895. Première contribution à la connaissance de la faune mammalogique des couches à *Pyrotherium. Boletín Instituto Geográfico Argentino,* **15**, 306–660.

Ameghino, F. 1899. *Sinópsis geológico – paleontológica. Suplemento (Adiciones y Correcciones).* La Plata: Imprenta La Libertad.

Ameghino, F. 1900–02. L'âge des formations sédimentaires de Patagonie. *Anales de la Sociedad Científica Argentina,* **50**, 109–131, 145–165, 209–229; **51**, 20–39, 65–91; **52**, 198–197, 244–250; **54**, 161–180, 220–240, 283–342.

Ameghino, F. 1901. Notices préliminaires sur des ongulés des terrains Crétacés de Patagonie. *Boletín de la Academia de Ciencias en Córdoba,* **16**, 349–426.

Ameghino, F. 1902. Première contribution à la connaissance de la faune mammalogique des couches à *Colpodon. Boletín de la Academia Nacional de Ciencias en Córdoba,* **17**, 71–138.

Ameghino, F. 1904. Recherches de morphologie phylogénétique sur les molaires supérieures des ongulés. *Anales del Museo Nacional de Buenos Aires,* **3**, 1–541.

Ameghino, F. 1906. Les formations sédimentaires du Crétacé supérieur et du Tertiaire de Patagonie. *Anales del Museo Nacional de Buenos Aires,* **8**, 1–358.

Burmeister, G. 1879. *Description physique de la République Argentine d'après des observations personnelles et étrangères,* vol. 3, *Animaux vertebrés, Part 1, Mammifères vivants et éteints.,* Buenos Aires: P. E. Coni.

Cifelli, R. L. 1993. The phylogeny of the native South American ungulates. In Szalay, F. S., Novacek, M. J., and McKenna, M. C. (eds.), *Mammal Phylogeny.* New York: Springer-Verlag, pp. 195–216.

Fleagle, J. G., T. M. Bown, C. Swisher, and G. Buckley 1995. Age of the Pinturas and Santa Cruz formations. *Actas VI Congreso Argentino de Paleontología y Bioestratigrafía,* 129–135.

Flynn, J. J. and C. C. Swisher III 1995. Cenozoic South American Land Mammal Ages: correlation to global geochronologies. In Berggren, W. A., Kent, D. V., Aubry, M.-P., and Hardenbol, J. (eds.), *Geochronology Times Scales and Global Stratigraphic Correlation,* Special Publication no. 54. Tulsa, OK: Society for Sedimentary Geology, pp. 317–333.

Gaudry, A. 1906. Fossiles de Patagonie: étude sur une portion du monde Antartique. *Annales de Paleontologie,* **1**, 101–143.

Johnson, S. C. and R. H. Madden 1997. Uruguaytheriine Astrapotheres of Tropical South America. In Kay, R., Madden, R., Cifelli, R., and Flynn, J. (eds.) *Vertebrate Paleontology in the Neotropics: The Miocene Fauna of La Venta, Colombia.* Washington, DC: Smithsonian Institution Press, pp. 355–381.

Kramarz, A. and M. Bond 2005. Los Astrapotheriidae (Mammalia) de la Formacion Cerro Bandera, Mioceno temprano de Patagonia septentrional. *Ameghiniana,* **42**, 72R–73R.

Kramarz, A. and M. Bond 2008. Revision of *Parastrapotherium* (Mammalia, Astrapotheria) and other Deseadan astrapotheres of Patagonia. *Ameghiniana*, **45**, 537–551.

Kramarz, A., A. Garrido, A. Forasiepi, M. Bond, and C. Tambussi 2005. Estratigrafía y vertebrados (Mammalia – Aves) de la Formación Cerro Bandera, Mioceno Temprano de la provincia del Neuquén, Argentina. *Revista Geológica de Chile*, **32**, 273–291.

Leanza, H. A. and C. A. Hugo 1997. Hoja geológica 3969 – III Picun Leufú. SEGEMAR, *Boletín* N° **218**, pp. 1–135. Buenos Aires.

Loomis, F. B. 1914. *The Deseado Formation of Patagonia.* Concord, NH: Rumford Press.

Lydekker, R. 1894. Contribution to the knowledge of the fossil vertebrates of Argentina. III. A study of extinct Argentine ungulates. *Anales del Museo de La Plata, Paleontología Argentina*, **2**, 1–91.

Marshall, L. G., R. Hoffstetter, and R. Pascual 1983. Mammals and stratigraphy: geochronology of continental mammals-bearing Tertiary of South America. *Paleovertebrata, Mémoire Extraordinaire*, 1–93.

Owen, R. 1853. Description of some species of the extinct genus *Nesodon* with remarks of the primary group (Toxodontia) of the hoofed quadrupeds to which the genus is referable. *Philosophical Transactions of the Royal Society of London*, **143**, 291–309.

Pascual, R. and O. E. Odreman Rivas 1971. Evolución de las comunidades de vertebrados del Terciario argentino: los aspectos paleozoogeográficos y paleoclimáticos relacionados. *Ameghiniana*, **8**, 372–412.

Pascual, R., E. Ortíz Jaureguízar, and J. L. Prado 1996. Land mammals: paradigm for Cenozoic South American geobiotic evolution. *Abhandlungen Münchner Geowissenchaftliche, Reihe A, Geologie und Palaontologie*, **30**, 265–319.

Patterson, B. 1952. *Catálogo de los mamíferos del Deseadoano y del Colhueluapiano en la Colección Ameghino.* Buenos Aires: Argentino Museo de Ciencias Naturales "Bernardino Rivadavia." (Unpublished.)

Scott, W. B. 1928. Mammalia of the Santa Cruz beds. Volume VI, Paleontology. Part IV, Astrapotheria. In Scott, W. B. (ed.), *Reports of the Princeton Expedition to Patagonia 1896–99.* Stuttgart: E. Schweizerbart'sche Verlagshandlung (E. Nägele), pp. 301–351.

Scott, W. B. 1937. The Astrapotheria. *Proceedings of the American Philosophical Society*, **77**, 300–393.

Soria, M. F. 1984. Eoastrapostylopidae: diagnosis e implicaciones en la sistemática y evolución de los Astrapotheria preoligocénicos. *Actas II Congreso Argentino de Paleontología y Bioestratigrafía*, 175–182.

Spalletti, L. A. and M. M. Mazzoni 1979. Estratigrafía de la Formación Sarmiento en la barranca sur del lago Colhue-Huapi, provincia del Chubut. *Revista de la Asociación Geológica Argentina*, **34**, 271–281.

13 The rodents from La Cantera and the early evolution of caviomorphs in South America

María Guiomar Vucetich, Emma Carolina Vieytes,
María Encarnación Pérez, and Alfredo A. Carlini

Abstract

Here we describe the oldest rodents from Patagonia, found in the La Cantera level at Gran Barranca, intermediate in age between Tinguiririrican and Deseadan South American Land Mammal Age (SALMA). The represented taxa are the Octodontoidea *Draconomys verai* gen. et sp. nov., *Vallehermosomys mazzonii* gen. et sp. nov., and *Vallehermosomys? merlinae* sp. nov., the Cavioidea cf. *Eobranisamys*, Dasyproctidae gen. et sp. indet., and Chinchilloidea? gen. et sp. indet. Isolated teeth, P4 and Dp4? that cannot be referred to any superfamily, are also described. The rodents from La Cantera are essentially brachyodont; only one single fragment of an upper molar referred to Chinchilloidea? gen. et sp. indet. has conspicuous unilateral hypsodonty. The enamel microstructure of incisors agrees with that of other caviomorph taxa, only multiserial Hunter–Schreger bands with interprismatic matrix in acute angle and with an intermediate subtype between acute and rectangular has been found. The absence of rodents in fossiliferous levels immediately below (called the La Cancha fauna) referable to the Tinguiririrican SALMA suggests that the arrival of rodents to the latitude of Gran Barranca occurred during the early Oligocene. The arrival to the continent was not much before the early Oligocene. The geographic and temporal distribution of the oldest caviomorphs suggests a north–south dispersion route within South America. This route agrees with the hypothesis of the arrival of rodents via a trans-Atlantic crossing from Africa.

Resumen

Se describen los roedores más antiguos de Patagonia provenientes de un nivel de Gran Barranca, La Cantera, que tiene una antigüedad intermedia entre el Tinguiririquense y el Deseadense. Los taxa representados son los Octodontoidea *Draconomys* gen. et sp. nov., *Vallehermosomys mazzonii* gen. et sp. nov. y *Vallehermosomys? merlinae* sp. nov., los Cavioidea Cf. *Eobranisamys*, Dasyproctidae gen. et sp. indet. y Chinchilloidea? gen. et sp.
indet. También se describen dientes aislados referidos a P4 y Dp4? que no pueden referirse con seguridad a ninguna superfamilia. Los roedores de La Cantera son esencialmente braquiodontes; solo un fragmento de un molar superior referido a Chinchilloidea? gen. et sp. indet. presenta una conspicua hipsodoncia unilateral. La microestructura del esmalte de los incisivos concuerda con lo hallado para otras faunas de caviomorfos; solo se ha hallado HSB multiserial con la matriz interprismática en ángulo agudo y con un subtipo intermedio entre ángulo agudo y rectangular. La ausencia de roedores en los niveles fosilíferos inmediatamente inferiores (La Cancha), referidos a una edad Tinguiririquense sugiere que el arribo de los roedores a la latitud de Gran Barranca ocurrió durante el Oligoceno temprano. Se propone que el arribo al continente no debe haber sido muy anterior al Oligoceno temprano. La distribución geográfica y temporal de los más antiguos caviomorfos sugiere una ruta norte-sur de dispersión dentro de América del Sur. Esta ruta concuerda con la hipótesis de un ingreso de los roedores vía transoceánica desde África.

Introduction

One of the critical events in the history of South American land mammals was the immigration of caviomorph rodents (South American hystricognaths) and platyrrhine primates during the early Cenozoic. The geographic and phyletic origin of caviomorphs has been hotly disputed in the past (Wood and Patterson 1959; Lavocat 1976; Patterson and Wood 1982), but currently there is a relative consensus that caviomorphs are more closely related to Old World hystricognaths than to any other group of rodents, whether they are considered monophyletic (Huchon and Douzery 2001; Marivaux *et al.* 2004; Poux *et al.* 2006) or not (Bryant and McKenna 1995; Candela 1999). Their time of arrival in South America and whether more than one immigration event took place are still controversial. Likewise, the first stages of the evolution of caviomorphs in South America and the interaction of these newcomers with previously established mammal groups are not known.

During a large part of the past century, the oldest rodents of South America were represented by those of

The Paleontology of Gran Barranca: Evolution and Environmental Change through the Middle Cenozoic of Patagonia, eds. R. H. Madden, A. A. Carlini, M. G. Vucetich, and R. F. Kay. Published by Cambridge University Press. © Cambridge University Press 2010.

the Deseadan (late Oligocene) South American Land Mammal Age (SALMA), first described by Ameghino (1897, 1902) from La Flecha and Cabeza Blanca in Patagonia (Loomis 1914; Wood 1949; Wood and Patterson 1959; Vucetich 1989) (Fig. 13.1). Since 1970, Deseadan rodents have also been described from the Salla–Luribay basin in Bolivia (Hoffstetter and Lavocat 1970; Lavocat 1976; Patterson and Wood 1982), and also through a few specimens from Nueva Palmera in Uruguay (Kraglievich 1932; Mones and Castiglione 1979), Curuzú Cuatiá and Quebrada Fiera in central Argentina (Gorroño *et al.* 1979; Bond *et al.* 1998), Lacayani in Bolivia (Hoffstetter *et al.* 1971; Vucetich 1989), and the Taubaté Basin of Brazil (Vucetich *et al.* 1993; Vucetich and Ribeiro 2003) (see Fig. 13.1).

The degree of diversification already achieved by caviomorphs in the Deseadan (almost 20 genera referred to about six families) and the dental specializations acquired by some of them (e.g. *Eoviscaccia* and *Cephalomyopsis* attained a high degree of hypsodonty with a correlated highly simplified occlusal pattern) suggested that they must have entered the continent well before the late Oligocene (Patterson and Wood 1982, p. 471; Vucetich *et al.* 1999).

It was not until the 1990s that the first caviomorph rodent older than Deseadan was discovered in the Tinguirirican early Oligocene of Central Chile (Wyss *et al.* 1993) (Fig. 13.1). Recently, Frailey and Campbell (2004) described a rich and diverse rodent fauna from Santa Rosa in Peru (Fig. 13.1), which they considered late Eocene/early Oligocene, although other interpretations have been proposed for its age. In addition, Vucetich *et al.* (2005) briefly mentioned the record of rodents in a level of the Gran Barranca informally known as La Cantera, which, in view of its stratigraphic position and content of other mammals is regarded as older than the typical Patagonian Deseadan, but younger than Tinguirirican (Carlini *et al.* 2005; Vucetich *et al.* 2005). The La Cantera rodent fauna is highly informative from a faunal perspective compared to the rodent faunas of Tinguiririca and Santa Rosa because it is placed within a very complete and rich mammal sequence.

In this chapter we describe the rodents found in La Cantera, and discuss their significance in the understanding of the early evolution of caviomorphs.

Nomenclature and abbreviations

For tooth nomenclature we follow Candela (1999) and Marivaux *et al.* (2004) (Fig. 13.2). LACM, Los Angeles County Museum, USA; MPEF-PV, Museo Paleontológico "Egidio Feruglio," paleovertebrate collection, Trelew, Argentina. Lower teeth are designated by lower-case letters, the upper teeth by upper-case letters (e.g. m1 and M1).

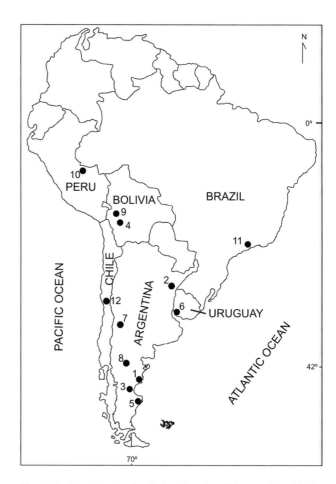

Fig. 13.1. Map showing South American localities mentioned in the text. 1, Cabeza Blanca (Chubut); 2, Curuzú Cuatiá (Corrientes); 3, Gran Barranca (Chubut); 4, Lacayani (Bolivia); 5, La Flecha (Santa Cruz); 6, Nueva Palmera (Uruguay); 7, Quebrada Fiera (Mendoza); 8, Rocas Bayas (Río Negro); 9, Salla–Luribay (Bolivia); 10, Santa Rosa (Peru); 11, Taubaté Basin (Brazil); 12, Tinguiririca (Chile).

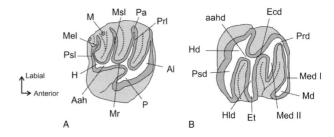

Fig. 13.2. Occlusal dental morphology (modified after Wood and Wilson 1936 and Marivaux *et al.* 2004). (A) Upper molar: Aah, anterior arm of the hypocone; Al, anteroloph; H, hypocone; M, metacone; Mel, metaloph; Mr, mure; Msl, mesoloph; P, protocone; Pa, paracone; Prl, protoloph; Psl, posteroloph. (B) Lower molar: aahd, anterior arm of the hypoconid; Ecd, ectolophid; Et, entoconid; Hld, hypolophid; Hd, hypoconid; Md, metaconid; Med I, metalophulid I; Med II, metalophulid II; Psd, posterolophid; Prd, protoconid.

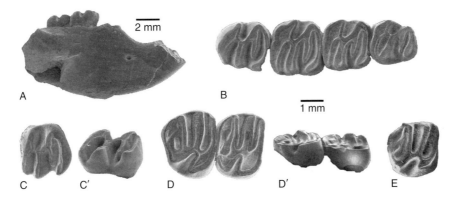

Fig. 13.3. *Draconomys verai* (gen. et sp. nov.). (A) MPEF-PV 7756, right mandible; (B) MPEF-PV 7506, right p4–m3 (holotype, reversed); (C–C′) MPEF-PV 7947, right m1 or m2 (reversed), (C) occlusal view, (C′) lingual view; (D–D′) MPEF-PV 7792, (D) labial view, (D′) occlusal view; (E) MPEF-PV 7946, occlusal view.

Stratigraphy, faunal association, and age of the La Cantera fauna

La Cantera is a rich fossil-bearing horizon at the west end of Gran Barranca, about 100 meters west of the line of Simpson's Profile A-2 (Simpson 1930). This fossil mammal locality was found during our field trips to Gran Barranca. La Cantera occurs at the level of Simpson's 6-meter thick unit of yellowish clay and tuff (Simpson's R10) which in turn crops out beneath a thin, hard, orange channel bed, and above a massive grey tuff (Simpson's R7). The rodents occur in association with a faunal assemblage containing diverse cingulates, marsupials, and ungulates (Carlini *et al.* 2005, this book; López *et al.* 2005; Goin *et al.* this book; Ribeiro *et al.* this book).

Ré *et al.* (Chapter 4, this book) estimated that La Cantera corresponds to either Chron C12n (30.627–31.116 Ma), C11n.2n (29.853–30.217 Ma), or C11n.1n (29.451–29.740 Ma), i.e. between 31.1 and 29.5 Ma, meaning that it is older than conventionally assumed for the Deseadan, which is in turn estimated to be about 25–29 Ma old (Flynn and Swisher 1995). Concordantly, mammals from this fauna provide rather strong evidence for a basal Deseadan or a pre-Deseadan age for the La Cantera fauna (Carlini *et al.* 2005, this book; López *et al.* 2005; Ribeiro *et al.* this book). The marsupial association in particular, suggests a pre-Deseadan age, intermediate in time between the Tinguiririican and Deseadan SALMAs (Goin *et al.* this book).

Systematic paleontology

Order **RODENTIA** Bowdich 1821
Infraorder **HYSTRICOGNATHI** Tulberg 1899
Superfamily **OCTODONTOIDEA** Waterhouse 1839
Family indet.
Genus *Draconomys* gen. nov.

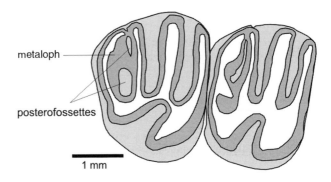

Fig. 13.4. *Draconomys verai* (MPEF-PV 7792), M2–3 in occlusal view. White, dentine; dark gray, worn enamel; light gray, unworn enamel (lateral walls, flexi, and fossettes).

Type species *Draconomys verai* gen. et sp. nov.
Diagnosis Small size, brachyodont with cusps submerged into high and sharp crests; p4 tetralophodont with well-developed hypolophid; m1–3 subrectangular in shape, tetralophodont with well-developed metalophulid II although not reaching the lingual margin of the tooth, metaconid anteroposteriorly elongate; M2–3 subquadrangular in shape, pentalophodont with the metaloph weakly developed, represented by a thickening of the labial end of the posteroloph, and a spur from the middle of the latter.
Distribution Early Oligocene of Patagonia.
Etymology From Cerro Dragón, an oil camp of PanEnergy in Chubut Province, in gratitude for the support during long years of work at Gran Barranca.

Draconomys verai gen. et sp. nov.
Figs. 13.3, 13.4.
Holotype MPEF-PV 7506, right mandibular fragment with part of the incisor and p4–m3.

Hypodigm The holotype and MPEF-PV 7756, right mandibular fragment with m1–2; MPEF-PV 7756a, isolated left m2; MPEF-PV 7757, nine isolated and fragmentary lower cheek teeth; MPEF-PV 7947, right m1 or m2; MPEF-PV 7948, left m1 or m2; MPEF-PV 7799, isolated right p4; MPEF-PV 7960, left m1 or m2; MPEF-PV 7961, fragmentary left m1 or m2; MPEF-PV 7792, fragment of right palate with M2–3; MPEF-PV 7946, isolated right M1 or M2, and several fragmentary tooth remains.

Diagnosis As for the genus by monotypy.

Etymology In honor of the Vera family for permission to work on their land.

Description The mandible (Fig. 13.3A) is robust with well-developed masseteric fossae; the upper side of the diastema is gently curved and the mental foramen is large and placed a little in front of the p4.

The p4 (Fig. 13.3B) is somewhat longer than wide, with the trigonid only slightly narrower than the talonid, and the anterior wall plane and continuous (Vucetich and Ribeiro 2003). The metalophulid II is transversely oriented occupying a little more than half the width of the occlusal surface. In the holotype it continues labially through a low and thin crestlet, almost up to the labial side. From the base of the metalophulid II a short spur forms the anterior portion of the ectolophid. From the entoconid, the hypolophid runs anterolabially to reach the spur; it connects with the hypoconid only in advanced stages of wear.

The m1–2 (Figs. 13.3B, C, C′) are longer than wide; the anterior and posterior walls are straight and parallel. The hypoflexid is posteriorly oriented; at early stages of wear it is continuous with the posteroflexid, but in more worn specimens, a transverse crest formed by the anterior arm of the hypolophid separates the two. The m3 is very similar to m1–2, but smaller and with a shorter and convex posterolophid (Fig. 13.3B).

Upper molars are more transverse than lower ones, and have five lophs (Fig. 13.3D, D′, E; Fig. 13.4; Table 13.1). The metaloph is formed by two short portions (Fig. 13.3D, E; Fig. 13.4), one shown by a thickening of the labial end of the posteroloph (see Vucetich and Verzi 1993) and the other by a short spur from the middle portion of the latter. When somewhat worn, the fusion of both these parts of the metaloph with the mesoloph produces a complex and shallow structure that includes one or two posterofossettes (Fig. 13.4).

Remarks The upper teeth vaguely resemble LACM 143267, one of the isolated teeth referred to *Eosallamys paulacoutoi* (Frailey and Campbell 2004, p. 120), in the degree of reduction and configuration of the metaloph, the long lophs, and short hypoflexus. But in the remaining material referred to *E. paulacoutoi* and

also *E. simpsoni* the hypoflexus is longer, lophs are shorter, and the metaloph is better developed.

Genus *Vallehermosomys* gen. nov.

Type species *Vallermosomys mazzonii*

Etymology From Valle Hermoso, another camp of PanEnergy at Chubut, in gratitude for their support during long years of work at Gran Barranca.

Diagnosis Small octodontoid with brachyodont and tetralophodont upper teeth, paracone and protocone are weakly or not connected to each other, and thus, anteroflexus and hypoflexus are conjoined; posterofossette closed in early stages of wear, at least in M3.

Vallehermosomys mazzonii gen. et sp. nov.
Fig. 13.5A–C.

Holotype MPEF-PV 7955, left M3.

Etymology In memory of our friend and colleague Dr. Mario M. Mazzoni.

Diagnosis The biggest species referred to the genus, with a well-differentiated metacone closing externally the posterofossette; protocone and paracone not in contact, and thus, anteroflexus and hypoflexus are interconnected. The posterofossette large, with a small knob at the bottom.

Description The holotype and only material known of this species has a relative low crown but clear unilateral hypsodonty. Cusps are submerged into thin crests, except the metacone which is partially isolated. The anteroloph is long and curved, forming an anterior–labial wall, whereas protoloph, mesoloph, and posterolophs are much shorter. This relationship gives the hypoflexid an almost anteroposterior orientation. Protocone and paracone do not contact to each other and thus the anteroloph is isolated from the other crest by a long hypoflexus plus paraflexus. Posteroloph, mesoloph, and metacone enclose a large and relatively shallow posterofossette, at the bottom of which there is a small knob.

Vallehermosomys? *merlinae* sp. nov.
Fig. 13.5D, E.

Holotype MPEF-PV 7954, right M3.

Etymology From the name of Miss E. Merlina Donato, the little daughter of one of the authors (ECV).

Diagnosis The smallest species referred to the genus, about 15% smaller than *V. mazzonii*; metacone not distinguishable; protocone and paracone weakly connected to each other, and thus, anteroflexus and hypoflexus only partially connected; posterofossette small.

Description The holotype and only known material of this species differs from the holotype of *V. mazzonii* in its smaller size, proportionally shorter anteroloph

Table 13.1. *Crown measurements of La Cantera rodent teeth*

| | *Draconomys verai* | | | | | | | *Vallehermosomys* | |
	MPEF 7506	MPEF 7756	MPEF 7799	MPEF 7947	MPEF 7948			*V. mazzoni* MPEF 7955	*V? merlinae* MPEF 7954
p4 AP	2.0		2.2						
AW	1.3		1.5				M3? AP	2.3	2.0
PW	1.8		1.9				AW	2.4	2.0
m1 AP	2.2	2.2		2.1	2.2		PW	1.6	1.4
AW	2.0	1.8		1.9	2.1				
PW	2.2	2.1		2.0	1.7			cf. *Eobranisamys*	
m2 AP	2.4	2.4						MPEF 7800	MPEF 7949
AW	2.3	2.2							
PW	2.4	2.3					m3 AP	2.7	3.1
m3 AP	2.3	2.3					AW	2.6	3.8
AW	2.1	2.0					PW	2.2	2.4
PW	2.0	2.2							
i AP	1.5	1.3						Gen. et sp. indet. 1	
W	2.0	2.1						MPEF 7950	
	MPEF 7792	MPEF 7946					M3 AP	2.7	
M1? AP		2.2					AW	2.7	
AW		2.0					PW	1.8	
PW		3.0							

			Caviomorpha indet.				
				MPEF 7952	MPEF 7953	MPEF 7956	MPEF 7957
M2 AP	2.3						
AW	2.5						
PW	2.3		DP3? AP	1.9	2.0		
M3 AP	2.3		AW	1.6	1.7		
AW	2.4		PW	1.5	1.6		
PW	1.9		P4 AP			1.7	1.5
			W			1.7	1.7

Notes: AP, anteroposterior length; AW, anterior width; PW, posterior width; W, incisor width.

and longer posteroloph making the hypoflexus obliquely rather than anteroposteriorly oriented. The mure is longer than in *V. mazzonii* and thus protoloph and mesoloph are more separate from each other. The metacone is not distinguishable, but the labial ends of posteroloph and mesoloph merge, closing a small posterofossette.

Superfamily **CAVIOIDEA**
Family **DASYPROCTIDAE**?
Genus Cf. ***Eobranisamys***
Fig. 13.5F, G.

Material MPEF-PV 7800, right m3, and MPEF-PV 7949, left deeply worn m3.

Description MPEF-PV 7800 is considered to be an m3 because the posterolophid is very curved and

short, and there is no sign of an interstitial wear facet on the posterior wall of the crown. MPEF-PV 7800 has a slight unilateral hypsodonty; it has four lophids, and cusps are not distinguishable. Metalophulid I is formed by two portions which, at this stage of wear, contact each other near the midpoint of the anterior wall: a short one from the protoconid, and a longer one from the metaconid. The metalophulid II is long, and reaches the labial wall, at least at this stage of wear; it joins the metaconid through an anteroposteriorly oriented crest, which encloses the anterofossettid. The hypolophid is long and widens at the lingual end indicating the presence of entoconid. At this stage of wear, the hypolophid is still incompletely fused to the anterior arm of the hypolophid. The posterolophid is short and has a sigmoid trajectory. The protoconid is

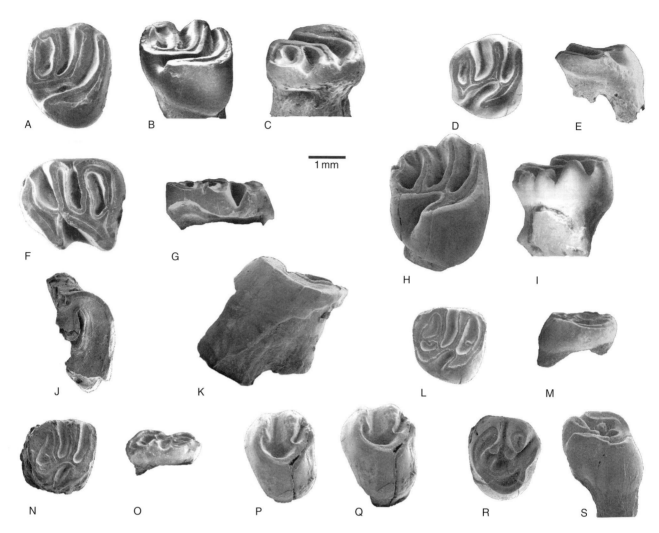

Fig. 13.5. (A–C) *Vallehermosomys mazzoni*, MPEF-PV 7955 (holotype), (A) occlusal view, (B) labial view, (C) lingual view (A and B reversed); (D, E) *Vallehermosomys? merlinae*, MPEF-PV 7954, (D) occlusal view, (E) posterior view; (F, G) cf. *Eobranisamys*, MPEF-PV 7800, (F) occlusal view, (G) labial view; (H, I) Dasyproctidae gen. et sp. indet. 1, MPEF-PV 7950, (H) occlusal view (reversed), (I) labial view; (J, K) Chinchilloidea gen. et sp. indet. 2, MPEF-PV 7951, (J) occlusal view, (K) posterior view; (L–O) Caviomorpha *incertae sedis* gen. et sp. indet. 3 (L, M) MPEF-PV 7952, (L) labial view, (M) occlusal view; (N, O) MPEF-PV 7953, (N) occlusal view (reversed), (O) labial view; (P–S) Caviomorpha *incertae sedis* gen. et sp. indet. 4, (P, Q) MPEF-PV 7956, (P) occlusal view, (Q) lingual view; (R, S) MPEF-PV 7957, (R) occlusal view, (S) lateral view.

labial with respect to the hypoconid. Wear extends onto the lingual wall of protoconid and hypoconid.

Remarks Several Oligocene genera (Dasyproctidae? from Tinguiririca, Chile, and *Branisamys*, *Eobranisamys*, *Paulacoutomys*) share the gross morphology of these teeth: four crests (or with a small fifth one) brachyodont, robust, with relative thick lophids and with an anteroposteriorly oriented crestlet closing a relative large anterofossettid. Cf. *Eobranisamys* is similar in size to *Eobranisamys* (Santa Rosa, ?early Oligocene, Peru: Frailey and Campbell 2004), but differs from the Santa Rosa

species of this genus in being proportionally wider, and more brachyodont. Cf. *Eobranisamys* differs from *Branisamys* in its smaller size and in the absence of a fifth lophid. It differs from *Paulacoutomys* Vucetich *et al.* 1993 because in the latter the protoconid is slightly more lingual than the hypoconid, instead of being more labial. Cf. *Eobranisamys* is much smaller and brachyodont than the Dasyproctidae? from Tinguiririca (Wyss *et al.* 1993), and in the latter hypoflexid and posteroflexid do not become separated by the anterior arm of the hypolophid until deeply worn.

Gen. et sp. indet. 1
Fig. 13.5

Material MPEF-PV 7950, left M3.

Description MPEF-PV 7950 is little worn allowing observation of some details of the gross morphology that disappear with wear. It is pentalophodont with a marked unilateral hypsodonty, with long, narrow, and deep flexi. It is subquadrangular in shape, with the anteroposterior diameter somewhat longer than the transverse one (Table 13.1). It has three well-differentiated labial cusps, the largest and more external being the paracone. The second cusp is much smaller than the paracone and because of its position, it is interpreted as the mesostyle; the posterior cusp is equal in size as the mesostyle, but somewhat more lingual, and is not connected to any crest; it is interpreted as the metacone. The anteroloph extends up to the labial side not reaching the paracone, and leaving the anteroflexus open. In this stage of wear the anterior portion of the mure already separates the antero-flexus and hypoflexus. The protoloph runs from the paracone slightly anterolabial–posterolingually, contacting the mure a little behind the protocone. The third crest is continuous, oblique, and connects with the mure somewhat in front of the hypocone. Except for the mesostyle, neither a cusp nor any interruption can be distinguished; hence there is no evidence that this loph is composed of adjoining crests as in erethizontids (Candela 1999); thus we consider it to be a mesoloph. The posteroloph has a high lingual portion that rises from the hypocone forming approximately 50% of the length, and a very low labial portion. At the transition point between both portions emerges a small crest, here considered a metaloph, which runs towards the labial wall, ending behind the metacone. Within this interpretation of homologies the metaloph is disconnected from the metacone.

Remarks This is the single upper tooth of the sample similar in size to the material referred to cf. *Eobranisamys*. However, it is not referred to this taxon because in *Eobranisamys* anteroflexus and hypoflexus are always joined, even in advanced stages of wear (Frailey and Campbell 2004). More and better materials will shed light on the relationships of the taxa represented by these specimens.

Superfamily **CHINCHILLOIDEA**?
Gen. et sp. indet. 2
Fig. 13.5J, K.

Material MPEF-PV 7951, fragment of upper molar.

Description The preserved fragment is a posterior lobe of a left upper molar with conspicuous unilateral hypsodonty. It has the highest crown of the whole sample. Its occlusal outline is vaguely heart-shaped and has a minute mesofossette on the labial side, and a long, narrow, and curved metafossette. There are no enamel discontinuities in the preserved fragment.

Remarks The assignment of MPEF-PV 7951 to Chinchilloidea? is based on the flat occlusal surface with a lobular appearance, and the presence of a curved mesofossette together with a relative high degree of unilateral hypsodonty.

Caviomorpha indet.
Gen. et sp. indet. 3
Fig. 13.5L–O.

Material MPEF-PV 7952, right DP4? MPEF-PV 7953, left DP4?

Description Both teeth are four crested and very brachyodont. In MPEF-PV 7952 (Fig. 13.5L, M) a backward directed transverse crestlet from the mesoloph almost divides the posterofossette into a large labial portion and a much smaller inner one; the labial tips of anteroloph and protoloph join closing the anterofossette; the anterofossette bears a small knob on the posterior wall of the anteroloph. MPEF-PV 7953 (Fig. 13.5N, O) is similar to the other teeth in size and degree of wear but it displays some differences in morphology; there is not a crestlet crossing the posterofossette, but a knob in a similar position; labial tips of protoloph and anteroloph do not join so the anteroflexus remains open.

Remarks The complicated occlusal surfaces of these teeth, with conules and anteroposteriorly oriented crestlets, and their different colour from the rest of the sample suggests they are examples of DP4. All are so similar in size and structure that they could belong to a single taxon.

Gen. et sp. indet. 4
Fig. 13.5P–S.

Material MEPF-PV 7956, right P4; MPEF-PV 7957, right P4.

Description None of these P4s has interdental facets to establish their orientation; we tentatively consider the biggest labial cusp as the paracone. The paracone of MPEF-PV 7956 (Fig. 13.5P, Q) is elongated as a short loph and united to the anteroloph through a very low transverse crestlet; the posteroloph bears a very short knob close to its labial end. MPEF-PV 7957 (Fig. 13.5R, S) is more complex as has also two central and isolated cuspules and short knobs from both antero- and posterolophs. These two P4 are very similar in size (MPEF-PV 7957 is slightly shorter) (Table 13.1), but their different occlusal morphologies, suggest they represent different species, one of which could be *Draconomys verai*.

Incisor enamel microstructure

Incisor enamel microstructure is considered a reliable tool for rodent phylogeny and systematics (Korvenkontio 1934; Boyde 1978; Martin 1992, 1997). The *Schmelzmuster* of rodent incisors is generally formed by two enamel layers, one inner portion (PI) with Hunter–Schreger bands (HSB) and an external portion (PE) formed by radial enamel (RE) (Boyde 1978; Koenigswald and Clemens 1992). Three types of HSB have been recognized: pauciserial (the most primitive), uniserial, and multiserial (Koverkontio 1934; Wahlert 1968; Martin 1992, 1993, 1997). Among hystricognath rodents only multiserial HSB have been found. Multiserial HSB are formed by several prisms (3 to 7), and the interprismatic matrix (IPM) may run parallel (the most primitive subtype), in acute angle (approximately 45°) or at right angles (rectangular) respect to the prisms, this latter considered the most derived (Martin 1992, 1993, 1994a). Among caviomorphs, the first two subtypes are present in the superfamilies Chinchilloidea, Cavioidea, and Erethizontoidea, whereas the most derived subtype is restricted to the Octodontoidea and considered a synapomorphy of this superfamily (Martin 1992). However, a transitional stage between the acute and rectangular subtypes was described for several Deseadan to Colloncuran (late Oligocene – middle Miocene) octodontoids (Martin 1994b, p. 126; Vieytes 2003; Vucetich and Vieytes 2006). In this subtype the angle between the IPM and the prisms is higher than 45° (up to 70°) in the upper incisors, and reaches 90° only in some sectors of the lower incisors (between 60° and 90°).

A few rodent incisor fragments were found in La Cantera. In order to analyse the enamel microstructure, we selected four of them displaying different gross morphologies (Fig. 13.6). In accordance with previous papers (Martin 2004, 2005) only multiserial HSB were found. Two of the specimens have the less derived subtypes of multiserial HSB: MPEF-PV 7962 parallel to acute IPM

(Fig. 13.7A), and MPEF-PV 7964 acute IPM. The other two (MPEF-PV 7963, MPEF-PV 7965: Fig 13.7B) have the transitional IPM subtype. Measurements and characteristics of the enamel of each specimen are reported in Table 13.2.

Discussion

Suprageneric systematics

Fossil caviomorphs have been assigned, with a few exceptions, to families with living representatives. However, these assignments have not been sufficiently tested with modern methods, and many of the classificatory schemes lacking explicit criteria have proved to be unstable. This is most dramatic for the Oligocene – lowest Miocene caviomorphs. For example, Wood and Patterson (1959, p. 301) supplied a list of dental characters, which they considered significant to separate octodontids from echimyids of the Deseadan–Colhuehuapian interval (late Oligocene – early Miocene), and referred all the Oligocene – early Miocene Octodontoidea to the families Octodontidae and Echimyidae. More recent results questioned the validity of those characters, as well as the assignment of some of the Deseadan–Colhuehuapian octodontoids to the families Octodontidae and Echimyidae (Vucetich and Kramarz

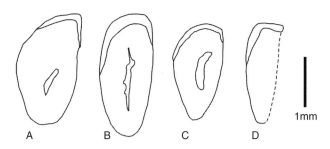

Fig. 13.6. Incisors in occlusal view. (A) MPEF-PV 7962; (B) MPEF-PV 7964; (C) MPEF-PV 7963; (D) MPEF-PV 7965.

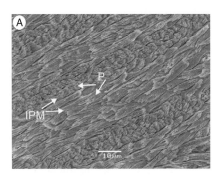
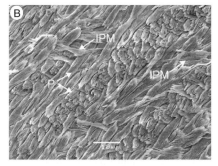

Fig. 13.7. Scanning electron micrographs of the enamel of two incisors from La Cantera in longitudinal section. (A) MPEF-PV 7962, detail of the Hunter–Schreger bands (HSB) with parallel to acute interprismatic matrix (IMP); (B) MPEF-PV 7963, detail of the HSB with transitional IPM. P, prism.

Table 13.2. *Incisor enamel features of specimens from La Cantera*

Specimen	Enamel thickness (μm)	HSB	Prisms per HSB	Inclination of HSB	PE thickness	PE prisms inclination	IPM in PI
MPEF 7962	150	multiserial	3–5	20°	20%	75°	parallel to acute (0°–20°)
MPEF 7963	190	multiserial	3–4	35°	20%	70°	transitional (40°–70°)
MPEF 7964	115	multiserial	3	25°	20%	65°	acute (15°–30°)
MPEF 7965	165	multiserial	4–5	30°	20%	80°	transitional (40°–65°)

Notes: HSB, Hunter–Schreger bands; PE, external portion; IPM, interprismatic index; PI, inner portion.

2003). Another enlightening example of the uncertainties of the relationships among fossil caviomorph genera is that of the Deseadan *Branisamys*. This genus has been referred alternatively to the Dasyproctidae (Lavocat 1976), Dinomyidae (Patterson and Wood 1982), and Erethizontidae (Candela 2002), families that in turn are referred to three different superfamilies. For this reason, in this chapter we decided to maintain an open classification – with a few precise suprageneric assignments – which reflects more clearly the actual state of the knowledge about the relationships among the most ancient caviomorphs.

Despite these systematic uncertainties, the La Cantera rodent fauna shows enough different dental morphological types as to shed some light on the early caviomorph evolution.

Comparisons with other old hystricognaths
The La Cantera rodents are mostly brachyodont, and display the primitive tooth pattern proposed by Marivaux *et al.* (2004) for the caviomorphs. They differ from the late Eocene – early Oligocene African phiomorphs (Wood 1968; Holroyd 1994) by having a small metacone and a relatively large mesostyle, as large as the metacone. The mesocone is absent or at least imperceptible, hypoconulid absent or unnoticeable, and the protospur is absent. These characters enable us to consider the La Cantera rodents as typical caviomorphs.

Among the oldest South American rodents, those from Tinguiririca are especially interesting, in this case. They are represented by only two specimens, one referred to Dasyproctidae? and the other to Chinchillidae (Wyss *et al.* 1993; Flynn *et al.* 2003). Both taxa display more derived dental characters than the La Cantera material; Dasyproctidae? has cheek teeth with straight walls indicative of a higher degree of hypsodonty than most of the La Cantera species, with the possible exception of Chinchilloidea? gen. et sp. indet. 2. The case of the Tinguiririrican Chinchillidae is particularly interesting as it presents a high degree of hypsodonty and a strong lamination comparable to that of the Deseadan *Eoviscaccia*; however, it has a more irregular outline of

the laminae (A. Wyss, personal communication to MGV, and personal observations; Vucetich 1989; Kramarz 2001) here interpreted as more primitive than those of the Deseadan genus, in accordance with its older age. On the other hand, this reinforces previous ideas about chinchilloids as the first caviomorphs to develop hypsodonty.

The age of the oldest rodent records in South America
Based upon geochronological data, Ré *et al.* (Chapter 4, this book) refer the La Cantera fossil-bearing levels (Unit 3 of the Upper Puesto Almendra Member: Bellosi this book) to either Chron C12n (30.627–31.116 Ma), C11n.2n (29.853–30.217 Ma), or C11n.1n (29.451–29.740 Ma), i.e. between 31.1 and 29.5 Ma. Consistent with this interpretation, the morphological and biostratigraphic evidence provided by marsupials, cingulates, and "ungulates" (Carlini *et al.* this book; Goin *et al.* this book; López *et al.* this book) indicates that the La Cantera mammals are more related to Deseadan than to Tinguirirican assemblages or are intermediate between them. Among the other putatively old caviomorphs (those from Tinguiririca in Chile and Santa Rosa in Peru), only those from Tinguiririca, with an estimated age not younger than 31.5 Ma (Flynn *et al.* 2003), are most probably older. The age of the Santa Rosa fauna in Peru, which supposedly included the oldest rodents in South America, is uncertain. Campbell (2004), Goin *et al.* (2004), and Frailey and Campbell (2004) claimed that the Santa Rosa fauna is middle to late Eocene or, less probably, early Oligocene in age, based upon the stage of evolution of the marsupials and rodents. However, the rodent *Eodelphomys almeidacamposi* – represented by a single tooth (Frailey and Campbell 2004) – suggests a younger age for this fauna, as it has degrees of molar lamination and hypsodonty comparable to those of Deseadan *Xylechimys* (Patterson and Pascual 1968). Moreover, the similarities between *Eobranisamys*, *Eoincamys*, and *Eosallamys* from Santa Rosa, with *Branisamys*, *Incamys*, and *Sallamys* from Salla, respectively, suggest that the Santa Rosa fauna is not as old as

proposed by Campbell and collaborators. Other mammals from Santa Rosa such as the notoungulates, also suggest that it may be as young as Deseadan or late Oligocene (Shockey *et al.* 2004; R. Madden personal communication to MGV 2007). Hence, the caviomorphs from La Cantera are not only the most ancient rodents of Argentina, but are also among the most ancient rodents of South America.

The La Cantera rodents and the time and route of the immigration event

A much-debated point about the origin of caviomorphs is the time of their arrival in South America, which has been estimated as middle to late Eocene (Wood and Patterson 1959; Patterson and Wood 1982; Houle 1999; Vucetich *et al.* 1999; Poux *et al.* 2006). However, and in spite of the lack of caviomorph remains in the rich mammal-bearing Eocene of Patagonia, a much older age for the rodents in the continent has also been proposed (Frailey and Campbell 2004 and literature therein). The La Cantera rodents help in clarifying this issue.

Flynn *et al.* (2003) suggest that the scarce record of rodents in Tinguiririca is due to their low abundance in the first stages of diversification, and that their absence in Tinguiriran localities of Argentina may be due to the paucity of specimens so far recovered. Currently, there is a large collection of mammals from a level at Gran Barranca, La Cancha, referred to the Tinguiriran SALMA (Goin *et al.* this book; Carlini *et al.* this book; Ré *et al.* Chapter 3, this book). In this level, there are lots of small mammal remains, including a large number of isolated rodentiform incisors of marsupials (Goin *et al.* this book). However, no remains from La Cancha can be referred to the Order Rodentia. We consider that the absence of rodent fossils at La Cancha reflects a case of real absence, not a sampling artifact. This fact indicates that rodents arrived at the latitude of Gran Barranca probably after the Tinguiriran at Gran Barranca, but prior to the deposition of the level of La Cantera, where they already display moderate diversification and abundance. When better collections are available, the possibility of finding rodents in Tinguiriran localities in northern Patagonia (e.g. Rocas Bayas, Fig. 13.1) must not be discounted.

Determining a lower temporal limit for the arrival of rodents in central Patagonia helps constrain the time of arrival in the continent, and suggests it was not an early Paleogene event. If rodents were in the continent during the early Paleogene, as proposed by some authors (e.g. Frailey and Campbell 2004), they should have been already present in central Patagonia during the late Eocene, since there were no climatic or environmental reasons or geographic barriers that would have impeded the arrival of rodents in Patagonia at that time (Bellosi 1995; Malumian 1999). This conclusion is in agreement with molecular data estimating the age of the first caviomorph radiation in late Eocene – early Oligocene (Opazo 2005; Poux *et al.* 2006).

The fact that rodents are not present in La Cancha (Tinguiriran), but are at La Cantera (post-Tinguiriran) and at Tinguiririca (about 12° latitude north of Gran Barranca) suggests that their initial migration within South America followed a southward route. Several routes of entry have been proposed for caviomorphs, including an Antarctic route (Huchon and Douzery 2001; Marivaux *et al.* 2002; Poux *et al.* 2006). The proposed southward migration route within South America supports the hypothesis of a trans-Atlantic migration from Africa, inasmuch as no hystricognath remains are known from North America to support a North American route (Martin 2006 and literature therein).

The study of the incisor enamel structure supplies additional evidence to interpret the first steps of caviomorphs evolution. Multiserial HSB characterizes the Hystricognathi, some Eocene and post-Eocene Ctenodactyloidea and *Pedetes*, and Phiomorpha (African hystricognaths) since the late Eocene (Martin 1992, 1993, 1994a). The three classical multiserial HSB subtypes (parallel, acute, and rectangular) are already present in some of the most ancient caviomorphs (pre-Deseadan? of Santa Rosa, and Deseadan: Martin 1992, 2004, 2005) – the incisor enamel of the Tinguiririca rodents has not been studied yet. On this basis, Martin (2004, 2005) postulated two different scenarios for the caviomorph origin. In the first, they entered South America substantially earlier than the most ancient record, and the most derived multiserial HSB subtypes would have developed within the continent. The other scenario would imply a more recent arrival, through several pulses of basal lineages, each one bearing different subtypes of multiserial HSB.

The improvement in the knowledge of the Octodontoidea enamel, which displays the most derived subtype among caviomorphs, allows the reassessment of the scenarios proposed by Martin. The incisor enamel of most Octodontoidea lineages – Octodontidae, Echimyidae (including *Myocastor* and *Capromys*), Abrocomidae, and Acaremyidae – has multiserial HSB with rectangular IPM (Martin 1992, 1993, 1994a, Vieytes 2003). However, some Deseadan to "Colloncuran" (middle Miocene) Octodontoidea genera (e.g. *Sallamys, Caviocricetus, Protadelphomys, Plesiacarechimys*: Ameghino 1902; Hoffstetter and Lavocat 1970; Vucetich *et al.* 1992; Martin 1994b; Vucetich and Verzi 1996; Vucetich and Vieytes 2006), representatives of different lineages, display a transitional subtype between acute and rectangular (Martin 1994b; Vieytes 2003; Vucetich and Vieytes 2006). The presence of transitional IPM in different octodontoid lineages, together with rectangular IPM in lineages not closely related to each other (e.g. Acaremyidae and Octodontidae: Vucetich and Kramarz 2003) suggests that the most derived subtype of HSB would have developed independently

(Saether 1979; Koenigswald 1997) in different lineages of this monophyletic group, after the caviomorphs arrived in South America. In this context, an alternative proposal to those postulated by Martin would be that, as newcomers, and without competition from other rodents, caviomorphs would have needed only a short time to develop the most derived subtype of multiserial HSB. We propose that the ancestors of caviomorphs could have arrived in South America shortly before the most ancient record of rectangular IPM.

Impact of early caviomorphs on the "ancient inhabitants"

The impact of caviomorphs and primates on established mammal communities in South America has been scarcely studied. Simpson (1980) considered that caviomorphs competed with and replaced different herbivore lineages, but more importantly, they succeeded in filling empty niches when they arrived.

Goin *et al.* (this book), describing marsupials from La Cancha and La Cantera, report a strong decline in diversity, and ascribe it to climate change, especially temperature decrease, more intense at higher latitudes. The competition with the newcomers – rodents arrived at Gran Barranca between the deposition of La Cancha and La Cantera – might have somehow reinforced the climatic influence on marsupial pauperization.

ACKNOWLEDGEMENTS
This study was supported by funds from Consejo Nacional de Investigaciones Científicas y Técnicas (CONICET) and Agencia Nacional de Promoción Científica y Tecnológica (ANPCyT) to M. G. Vucetich. Field research was supported by U.S. National Science Foundation grants EAR-0087636, BCS-0090255, and DEB-9907985 to Richard F. Kay and Richard H. Madden. We thank C. Deschamps and D. Verzi for contributions in reading and commenting on the manuscript.

REFERENCES
Ameghino, F. 1897. Mamíferos cretáceos de la Argentina: segunda contribución al conocimiento de la fauna mastológica de las capas con restos de Pyrotherium. *Boletín del Instituto Geográfico Argentino*, **18**, 406–521.
Ameghino, F. 1902. Première contribution à la connaissance de la faune mammalogique des couches à *Colpodon*. *Boletín de la Academia Nacional de Ciencias de Córdoba*, **17**, 71–138.
Bellosi, E. S. 1995. Paleogeografía y cambios ambientales de la Patagonia central durante el Terciario medio. *Boletín de Informaciones Petroleras (Yacimientos Petrolíferos Fiscales)*, **44**, 50–83.
Bond, M., G. M. López, M. A. Reguero, G. J. Scillato-Yané, and M. G. Vucetich 1998. Los mamíferos de la Fm. Fray Bentos (Oligoceno Superior?) de las provincias de Corrientes y Entre Ríos, Argentina. In *Paleógeno de América del Sur y de la Península Antártica*, Special Publication no. 5. Buenos Aires: Ameghiniana, pp. 41–50.
Boyde, A. 1978. Development of the structure of the enamel of the incisor teeth in three classical subordinal groups of the Rodentia. In Butler, P. M., and Josey, K. A. (eds.), *Development, Function and Evolution of Teeth*. London: Academic Press, pp. 43–58.
Bryant, J. D. and M. C. McKenna 1995. Cranial anatomy and phylogenetic position of *Tsaganomys altaicus* (Mammalia, Rodentia) from the Hsanda Gol Formation (Oligocene), Mongolia. *American Museum of Natural History Novitates*, **3156**, 1–42.
Campbell, K. E. Jr. 2004. The Santa Rosa local fauna: a summary. In Campbell, K. E. Jr. (ed.), *The Paleogene Mammalian Fauna of Santa Rosa, Amazonian Peru*. Los Angeles, CA: Natural History Museum of Los Angeles County, pp. 155–163.
Candela, A. M. 1999. The evolution of the molar pattern of the Erethizontidae (Rodentia, Hystricognathi) and the validity of *Parasteiromys* Ameghino, 1904. *Palaeovertebrata*, **28**, 53–73.
Candela, A. M. 2002. Lower deciduous tooth homologies in the Erethizontidae (Rodentia, Hystricognathi): evolutionary significance. *Acta Paleontologica Polonica*, **47**, 717–723.
Carlini, A. A., M. Ciancio, and G. J. Scillato-Yané 2005. Los Xenarthra de Gran Barranca, más de 20 Ma de historia. *Actas XVI Congreso Geológico Argentino*, **4**, 419–424.
Flynn, J. J. and C. C. Swisher III 1995. Chronology of the Cenozoic South American Land Mammal Ages. In Berggren, W. A., Kent, D. V., Aubry, M.-P., and Hardenbol, J. (eds.), *Geochronology, Time-Scales, and Global Stratigraphic Correlation*, Special Publication no. 54. Tulsa, OK: Society for Sedimentary Geology, pp. 317–333.
Flynn, J. J., A. R. Wyss, D. A. Croft, and R. Charrier 2003. The Tinguiririca Fauna, Chile: biochronology, paleoecology, biogeography, and a new earliest Oligocene South American Land Mammal "age." *Palaeogeography, Palaeoclimatology, Palaeoecology*, **195**, 229–259.
Frailey, C. D. and K. E. Campbell Jr. 2004. Paleogene rodents from Amazonian Peru: the Santa Rosa Local Fauna. In Campbell, K. E. Jr. (ed.) *The Paleogene Mammalian Fauna of Santa Rosa, Amazonian Peru*. Los Angeles, CA: Natural History Museum of Los Angeles County, pp. 71–130.
Goin, F. J. and A. M. Candela 2004. New Palaeogene mansupials from the Amazon basin of Eastern Peru. In Campbell, K. E. Jr. (ed.), *The Paleogene Mammalian Fauna of Santa Rosa, Amazonian Peru*. Los Angeles, CA: Natural History Museum of Los Angeles County, pp. 15–60.
Goin, F. J., E. C. Vieytes, M. G. Vucetich, A. A. Carlini, and M. Bond 2004. Enigmatic mammal from the Paleogene of Peru. In Campbell, K. E. Jr. (ed.), *The Paleogene Mammalian Fauna of Santa Rosa, Amazonian Peru*. Los Angeles, CA: Natural History Museum of Los Angeles County, pp. 145–153.

Gorroño, R., R. Pascual, and R. Pombo 1979. Hallazgo de mamíferos eógenos en el Sur de Mendoza: su implicancia en la datación de los "Rodados Lustrosos" y del primer episodio orogénico del Terciario de la región. *Actas VII Congreso Geológico Argentino*, **2**, 125–136.

Hoffstetter, R. and R. Lavocat 1970. Découverte dans la Deséadien de Bolivie des genres pentalophodontes appuyant les affinités africaines des rongeurs caviomorphes. *Comptes Rendus de l'Académie des Sciences, Paris*, **271**, 172–175.

Hoffstetter, R., C. Martinez, M. Mattauer, and P. Tomasi 1971. Lacayani, un nouveau gisement bolivien de mammifères déséadiens (Oligocène inférieur). *Comptes Rendus de l'Académie des Sciences, Paris*, **273**, 2215–2218.

Holroyd, P. A. 1994. An examination of dispersal origins for Fayum Mammalia. Ph.D. dissertation, Duke University, North Carolina.

Houle, A. 1999. The origin of platyrrhines: an evaluation of the Antarctic scenario and the floating island model. *American Journal of Physical Anthropology*, **109**, 541–559.

Huchon, D. and E.J.P. Douzery 2001. From the Old World to the New World: a molecular chronicle of the phylogeny and biogeography of hystricognath rodents. *Molecular Phylogenetics and Evolution*, **20**, 238–251.

Kramarz, A.G. 2001. Registro de Eoviscaccia (Rodentia, Chinchillidae) en estratos colhuehuapenses de Patagonia, Argentina. *Ameghiniana*, **38**, 237–242.

Koenigswald, W. von 1997. Evolutionary trends in the differentiation of mammalian enamel ultrastructure. In von Koenigswald, W. and Sander, P.M. (eds.), *Tooth Enamel Microstructure*. Rotterdam: Balkema, pp. 203–235.

Koenigswald, W. von and W.A. Clemens 1992. Levels of complexity in the microstructure of mammalian enamel and their application in studies of systematics. *Scanning Microscopy*, **6**, 195–218.

Korvenkontio, V.A. 1934. Mikroskopische Untersuchungen an Nagerinzisiven unter Hinweis auf die Schmelzsstruktur der Backenzahne. *Annales Zoologici Societatis Zoologicae–Botanicae Fennicae Vanamo*, **2**, 1–274.

Kraglievich, L. 1932. Nuevos apuntes para la geología y paleontología uruguayas. *Anales del Museo de Historia Natural de Montevideo*, **3**, 257–231.

Lavocat, R. 1976. Rongeurs caviomorphes de l'Oligocéne de Bolivie. II. Rongeurs du Bassin Déséadien de Salla–Luribay. *Paleovertebrata*, **7**(3), 15–90.

Loomis, F.B. 1914. *The Deseado Formation of Patagonia*. Concord, NH: Rumford Press.

López, G.M., M. Bond, M.A. Reguero, J.N. Gelfo, and A.G. Kramarz 2005. Los ungulados del Eoceno–Oligoceno de la Gran Barranca, Chubut. *Actas XVI Congreso Geoló gico Argentino*, **4**, 415–418.

Malumian, N. 1999. La sedimentación y el volcanismo terciarios en la Patagonia extraandina. 1. La sedimentación en la Patagonia extraandina. *Anales Instituto de Geología y Recursos Minerales*, **29**, 557–578.

Marivaux, L., M. Vianey-Liaud, J.-L. Welcomme, and J.-J. Jaeger 2002. The role of Asia in the origin and diversification of hystricougnathous rodents. *Zoologica Scripta*, **31**, 225–329.

Marivaux, L., M. Vianey-Liaud, and J.-J. Jaeger 2004. High-level phylogeny of early Tertiary rodents: dental evidence. *Zoological Journal of the Linnean Society*, **142**, 105–134.

Martin, T. 1992. Schmelzmikrostruktur in den inzisiven alt- und neuweltlicher Hystricognather nagetiere. *Palaeovertebrata, Mémoire Extraordinaire*, 1–168.

Martin, T. 1993. Early rodent incisor enamel evolution: phylogenetic implications. *Journal of Mammalian Evolution*, **1**, 227–254.

Martin, T. 1994a. African origin of caviomorph rodents is indicated by incisor enamel microstructure. *Paleobiology*, **20**, 5–13.

Martin, T. 1994b. On the systematic position of *Chaetomys subspinosus* (Rodentia: Caviomorpha) based on evidence from the incisor enamel microstructure. *Journal of Mammalian Evolution*, **2**, 117–131.

Martin, T. 1997. Incisor enamel microstructure and systematics in rodents. In von Koenigswald, W. and Sander, P.M. (eds.), *Tooth Enamel Microstructure*. Rotterdam: Balkema, pp. 163–175.

Martin, T. 2004. Incisor enamel microstructure of South America's earliest rodents: implications for Caviomorph origin and diversification. In Campbell, K.E. Jr. (ed.), *The Paleogene Mammalian Fauna of Santa Rosa, Amazonian Peru*. Los Angeles, CA: Natural History Museum of Los Angeles County, pp. 131–140.

Martin, T. 2005. Incisor schmelzmuster diversity in South America's oldest rodent fauna and early caviomorph history. *Journal of Mammalian Evolution*, **12**, 405–417.

Martin, T. 2006. Incisor enamel microstructure of Ischyromyoidea and the primitive rodent schmelzmuster. *Palaeontographica A*, **277**, 53–66.

Mones, A. and L.R. Castiglione 1979. Additions to the knowledge on fossil rodents of Uruguay (Mammalia: Rodentia). *Paläontologische Zeitschrift*, **53**, 77–87.

Opazo, J.C. 2005. A molecular timescale for caviomorph rodents (Mammalia, Hystricognathi). *Molecular Phylogenetics and Evolution*, **37**, 932–937.

Patterson, B. and R. Pascual 1968. New echimyid rodents from the Oligocene of Patagonia and a synopsis of the family. *Breviora*, **401**, 1–14.

Patterson, B. and A.E. Wood 1982. Rodents from the Deseadan Oligocene of Bolivia and the relationships of the Caviomorpha. *Bulletin of the Museum of Comparative Zoology*, **149**, 370–543.

Poux, C., P. Chevret, D. Huchon, W.W. de Jong, and E.J.P. Douzery 2006. Arrival and diversification of caviomorph rodents and platyrrhine primates in South America. *Systematic Biology*, **55**, 228–237.

Saether, O.A. 1979. Underlying synapomorphies and anagenetic analysis. *Zoologica Scripta*, **8**, 305–312.

Shockey, B.J., R. Hitz, and M. Bond 2004. Paleogene notoungulates from the Amazon Basin of Peru. In Campbell, K.E. Jr. (ed.), *The Paleogene Mammalian Fauna of Santa Rosa, Amazonian Peru*. Los Angeles, CA: Natural History Museum of Los Angeles County, pp. 61–69.

Simpson, G. G. 1930. *Scarritt-Patagonian Exped. FieldNotes.* New York: American Museum of Natural History. (Unpublished.) Available at http://paleo.amnh.org/notebooks/index.html

Simpson, G. G. 1980. *Splendid Isolation: The Curious History of South American Mammals.* New Haven, CT: Yale University Press.

Vieytes, E. C. 2003. Microestructura del esmalte de roedores Hystricognathi sudamericanos fósiles y vivientes: significado morfofuncional y filogenético. Ph.D. thesis, Universidad Nacional de La Plata.

Vucetich, M. G. 1989. Rodents (Mammalia) of the Lacayani fauna revisited (Deseadan, Bolivia): comparison with new Chinchillidae and Cephalomyidae from Argentina. *Bulletin du Muséum National d'Histoire Naturelle, Paris,* **11,** 233–247.

Vucetich, M. G. and A. G. Kramarz 2003. New Miocene rodents of Patagonia (Argentina) and their bearing in the early radiation of the octodontiform octodontoids. *Journal of Vertebrate Paleontology,* **23,** 435–444.

Vucetich, M. G. and A. M. Ribeiro 2003. A new and primitive rodent from the Tremembé Formation (late Oligocene) of Brazil, with comments on the morphology of the lower premolars of caviomorph rodents. *Revista Brasileira de Plaeontologia,* **5,** 73–82.

Vucetich, M. G. and D. H. Verzi 1993. Las Homologías en los diseños oclusales de los roedores Caviomorpha: un modelo alternativo. *Mastozoología Neotropical,* **1,** 61–72.

Vucetich, M. G. and D. H. Verzi 1996. A peculiar octodontoid (Rodentia, Caviomorpha) with terraced molars from the Lower Miocene of Patagonia (Argentina). *Journal of Vertebrate Paleontology,* **16,** 297–302.

Vucetich, M. G. and E. C. Vieytes 2006. A Middle Miocene primitive octodontoid rodent and its bearing on the early evolutionary history of the Octodontoidea. *Palaeontographica A,* **277,** 81–91.

Vucetich, M. G., D. H. Verzi, and M. T. Dozo 1992. El "status" sistemático de *Gaimanomys alwinea*

(Rodentia, Caviomorpha, Echimyidae). *Ameghiniana,* **29,** 85–86.

Vucetich, M. G., F. L. Souza Cunha, and H. M. F. de Alvarenga 1993. Un roedor Caviomorpha de la Formación Tremembé (Cuenca de Taubaté), Estado de Sao Paulo, Brasil. *Anales de la Academia Brasileira de Ciencias,* **65,** 247–251.

Vucetich, M. G., D. H. Verzi, and J.-L. Hartenberger 1999. Review and analysis of the radiation of the South American Hystricognathi (Mammalia, Rodentia). *Comptes Rendus de l'Académie des Sciences, Paris,* **329,** 763–769.

Vucetich, M. G., E. C. Vieytes, A. G. Kramarz, and A. A. Carlini 2005. Los roedores caviomorfos de Gran Barranca: aportes bioestratigráficos y paleoambientales. *Actas XVI Congreso Geológico Argentino,* **4,** 413–414.

Wahlert, J. H. 1968. Variability of rodent incisor enamel as viewed in thin section, and the microstructure of the enamel in fossil and recent rodent groups. *Breviora,* **309,** 1–18.

Wood, A. E. 1949. A new Oligocene rodent genus from Patagonia. *American Museum of Natural History Novitates,* **1435,** 1–54.

Wood, A. E. 1968. Early Cenozoic mammalian faunas, Fayum Province, Egypt. II. The African Oligocene Rodentia. *Bulletin of the Peabody Museum of Natural History,* **28,** 23–105.

Wood, A. E. and B. Patterson 1959. The rodents of the Deseadan Oligocene of Patagonia and the beginnings of South American rodents' evolution. *Bulletin of the Museum of Comparative Zoology,* **120,** 280–428.

Wood, A. E. and R. W. Wilson 1936. A suggested nomenclature for the cusps of the cheek teeth of rodents. *Journal of Paleontology,* **10,** 388–391.

Wyss, A. R., J. J. Flynn, M. A. Norell, C. C. Swisher III R. Charrier, M. J. Novacek, and M. C. McKenna 1993. South American's earliest rodent and the recognition of a new interval of mammalian evolution. *Nature,* **365,** 434–437.

14 Colhuehuapian rodents from Gran Barranca and other Patagonian localities: the state of the art

María Guiomar Vucetich, Alejandro G. Kramarz, and Adriana M. Candela

Abstract

Almost 20 years of fieldwork in the early Miocene of Patagonia reveals that the various faunas assigned to the Colhuehuapian SALMA share numerous rodent taxa, but they do not share most of them with faunas of SALMAs immediately younger or older. This supports assignment to a single biochronologic unit. Taxonomic differences especially between the classical localities of Bryn Gwyn and Gran Barranca are uncommon and may result largely from environmental variation. New finds result in a dramatic increase of the estimates of diversity for this time. Indeed, the Colhuehuapian caviomorphs currently constitute the most diverse Cenozoic rodent fauna of South America. This high taxonomic diversity could result from the coexistence of old Patagonian lineages together with others of probable northern origin that entered Patagonia in a post-Deseadan event or events. This diversity expresses itself in varying degrees of hypsodonty, a variety of occlusal designs, and a relatively great size range, showing that Colhuehuapian caviomorphs had already developed considerable dietary niche breadth. Colhuehuapian rodent faunas are characterized by the richness of small octodontoids with a mosaic of derived and primitive characters, and the great diversity of erethizontids that have their acme at this time. Most of the Colhuehuapian rodents belong to lineages not recorded in post middle Miocene faunas; only eocardiids and chinchillids are certainly closely related to extant representatives. After the Colhuehuapian there occurred a decline in taxonomic diversity, variety of occlusal designs, and degree of hypsodonty, suggesting a reduction of adaptive types.

Resumen

Casi 20 años de trabajos de campo en el Mioceno temprano de Patagonia revelaron que todas las faunas asignadas a la Edad Colhuehuapense comparten numerosos taxones de roedores no conocidos para las edades inmediatamente más jóvenes o más antiguas, corroborando su asignación a una única unidad biocronológica.

The Paleontology of Gran Barranca: Evolution and Environmental Change through the Middle Cenozoic of Patagonia, eds. R. H. Madden, A. A. Carlini, M. G. Vucetich, and R. F. Kay. Published by Cambridge University Press. © Cambridge University Press 2010.

Las diferencias taxonómicas, especialmente entre las localidades clásicas de Bryn Gwyn y Gran Barranca, son leves y podrían deberse mayormente a variaciones ambientales. Nuevos hallazgos produjeron un incremento drástico en las estimaciones de diversidad para este lapso y en consecuencia, los caviomorfos colhuehuapenses constituyen actualmente la fauna de roedores cenozoicos más diversa de América del Sur. Esta alta diversidad taxonómica podría deberse a la coexistencia de linajes patagónicos antiguos con otros de origen septentrional que habrían ingresado a Patagonia durante uno o más eventos post deseadenses. Asimismo, esta diversidad también se refleja en una amplia variedad de grados de hipsodoncia, una importante diversidad de diseños oclusales y un rango relativamente amplio de tamaño que muestra que los caviomorfos colhuehuapenses ya habían desarrollado una gran variedad de tipos adaptativos y de estrategias en el uso de los recursos alimenticios. La fauna de roedores colhuehuapenses se caracteriza por la riqueza de pequeños octodontoideos con un mosaico de caracteres primitivos y derivados, y por la gran diversidad de eretizóntidos que tienen su acmé en este momento. La mayoría de los roedores colhuehuapenses pertenece a linajes no registrados en faunas post Mioceno medio; sólo los eocárdidos y chinchíllidos están estrechamente relacionados con representantes actuales. Luego del Colhuehuapense disminuyen la diversidad taxonómica y la variedad de diseños oclusales y grados de hipsodoncia, sugiriendo una disminución de tipos adaptativos.

Introduction

The Colhuehuapian South American Land Mammal Age (SALMA; early Miocene) was defined by Simpson (1940) upon the fossil mammal assemblage of Ameghino's "couches à *Colpodon*" at central Chubut (Ameghino 1902). Colhuehuapian rodents, in particular, were first described by Ameghino (1902, 1904) based on specimens presumably from the southern cliffs of Lake Colhue-Huapi, known worldwide as Gran Barranca (Fig. 14.1). Ameghino recognized 18 species corresponding to nine genera, grouped in six families. Later, Patterson (1958), Patterson and Pascual (1968), and Vucetich and Kramarz (2003)

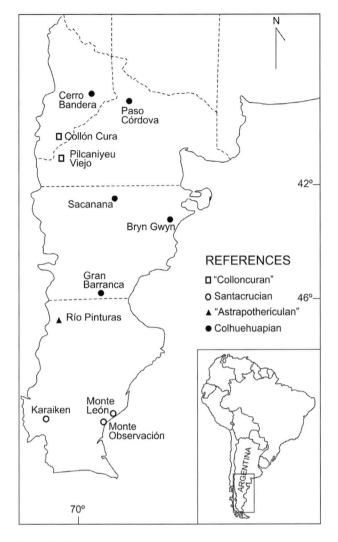

Fig. 14.1. Location map.

described a few other species from this stratigraphic level. Other localities assigned to the Colhuehuapian SALMA (i.e. Bryn Gwyn, Sacanana, Paso Córdova, and Cerro Bandera, Fig. 14.1) have also provided several rodent species (Bordas 1939; Vucetich and Bond 1984; Vucetich 1985; Vucetich and Verzi 1991, 1993, 1996; Kramarz 1998, 2001a, 2001b, 2005; Candela 2003; Vucetich and Kramarz 2003; Pérez *et al.* in press), but at present there are no comparative studies of Colhuehuapian rodents as a whole.

During the last 10 years, numerous field projects at Gran Barranca produced one of the most important collections of South American Miocene rodents from a single stratigraphic interval, and efforts to collect small mammals has yielded numerous rare species not previously known. Thus, these new collections significantly improved our knowledge of the diversity of Colhuehuapian rodents and allowed a more accurate comparison with those of the other

mentioned localities. The present knowledge of the rodents from Colhuehuapian levels at Gran Barranca and elsewhere in Patagonia, are here summarized in order to better understand the evolutionary and biogeographic history of the South American Hystricognathi.

Institutional abbreviations

MACN, Museo Argentino de Ciencias Naturales "Bernardino Rivadavia," Buenos Aires; MLP, Museo de La Plata, La Plata; MPEF, Museo Paleontológico "Egidio Feruglio," Trelew.

Colhuehuapian rodents at Gran Barranca

Superfamily Erethizontoidea

Family Erethizontidae

Erethizontidae includes the New World porcupines, medium-sized and heavy-set rodents, mostly with arboreal habits at least in extant species (Woods 1984). The cheek teeth are lophodont and low-crowned with a characteristic broad enamel covering. Erethizontids at Gran Barranca exhibit an unusual abundance and diversity, both taxonomic and adaptive.

Eosteiromys Ameghino 1902 has cheek teeth with wide and rather deep fossettes/ids; the uppers have five and the lowers four transverse crests; the incisors are narrow with convex anterior face. *Eosteiromys segregatus* (Ameghino 1902) originally referred to the Santacrucian genus *Steiromys* Ameghino 1887, differs from the type species *E. homogenidens* Ameghino 1902 (Fig. 14.2A) by being smaller and more gracile. A third unnamed species, also smaller than *E. homogenidens*, has the hypoflexus and anteroflexus connected (Candela 2000). This feature is also present in *E. annectens* (Ameghino 1901) from the "Astrapothericulan" (late early Miocene: Kramarz 2004), suggesting a close relationship between these species.

Parasteiromys Ameghino, 1904 has pentalophodont upper cheek teeth with a characteristic communication of the hypoflexus and mesoflexus and retains the DP3 during juvenile stages (Candela 1999). *Parasteiromys friantae* Candela 1999 differs from the type species *P. uniformis* Ameghino 1903 in its smaller size and shallower flexi, and in having the metaloph isolated from the posteroloph and oriented towards the hypocone. According to Candela (1999), *Parasteiromys* is close to the Deseadan *Protosteiromys* Wood and Patterson 1959 and its molar structure is considered to be the most generalized among the erethizontids.

Hypsosteiromys Patterson 1958 is the only erethizontid with a clear tendency to hypsodonty (Candela and Vucetich 2002). The upper and lower molars have four oblique crests separated by compressed, deep valleys; the lower molars

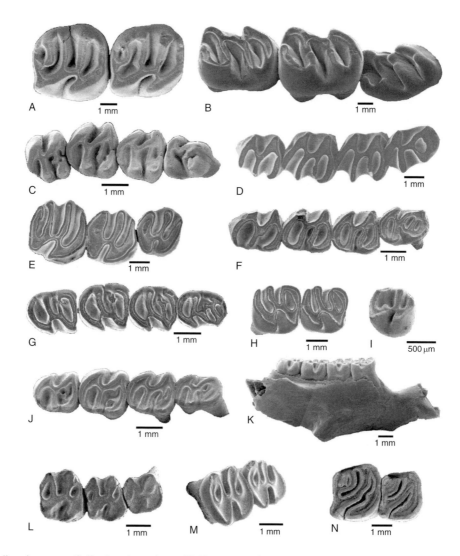

Fig. 14.2. Dental diversity among Colhuehuapian rodents. (A) *Eosteiromys homogenidens* MPEF 6097, right M1–2; (B) *Hypsosteiromys axiculus* MACN A 52–171, right m1–3 (holotype of *Steiromys axiculus*); (C) *Galileomys antelucanus* MPEF 5418, right p4–m3 (reversed); (D) *Prospaniomys priscus* MPEF 5627, left p4–m3; (E) Prospaniomys sp.1 MPEF 7574a, right P4–M2; (F) *Protacaremys prior* MPEF 5662, left dp4–m3; (G) *Protacaremys avunculus* MACN A 52–126 right p4–m3 (holotype); (H) *Protacaremys*? sp. nov. MEF 7557a, right m1–2; (I) *Acarechimys* sp. MLP 82-V-2–40, right DP4; (J,K) *Acarechimys pulchellus* comb. nov. MACN A 52–128 (holotype of *P. pulchellus*), (J) left p4–m3, (K) left mandible in labial view; (L) *Caviocricetus* MPEF 5419, right DP4–M2; (M) Octodontoidea gen. et sp. nov. MPEF 5420, left m2–3; (N) *Protadelphomys* morphotype A, MLP 82-V-2–30, left M1–2.

have a shallow anterior notch isolating the protoconid from the anterolophid; and the incisors are delicate (Patterson 1958). *Hypsosteiromys nectus* (Ameghino 1902) has lower molars with shallower anterior notches, a shorter p4, and narrower incisors than the type species *H. axiculus* (Ameghino 1902) (Fig. 14.2B; Candela and Vucetich 2002). This genus would have differentiated early from the remaining extinct porcupines (Candela 2000).

Branisamyopsis Candela 2003 has very low-crowned cheek teeth. The upper molars are pentalophodont, wider than long, with the hypoflexus and anteroflexus connected.

The lower molars are also pentalophodont and have a well-developed accessory lingual cusp (neoconid) posterior to the metaconid. The p4 has six transverse crests. The mandibular corpus is massive, the incisors are wide, and at least the lowers have a planar anterior face (Candela 2003). *Branisamyopsis* is related to *Eosteiromys* and to the Santacrucian *Steiromys* (Candela 2003). *Branisamyopsis* is represented at Gran Barranca only by the type species *B. australis* Candela 2003, but other species of the genus are recorded in early to middle Miocene "Astrapotericulan" and "Colloncuran" deposits (Candela 2003; Kramarz 2004).

Superfamily Octodontoidea

Family Acaremyidae

Acaremyids range from the Colhuehuapian to the "Colloncuran" (early to middle Miocene). They are small animals with mesodont figure-eight-shaped cheek tooth crowns, and non-molarized P4/4. Vucetich and Kramarz (2003) restricted the concept of Acaremyidae to the genera *Galileomys* Vucetich and Kramarz 2003, *Acaremys* Ameghino 1887, and *Sciamys* Ameghino 1887, excluding the Deseadan *Platypittamys* Wood 1949. Thus, the family is first recorded in the Colhuehuapian. Acaremyids are represented at Gran Barranca by *Galileomys antelucanus* Vucetich and Kramarz 2003 (Fig. 14.2C), with cheek teeth less hypsodont, more cuspidate and terraced than the remaining acaremyids, and by species of *Acaremys* Ameghino 1887 close to the Santacrucian *A. murinus* Ameghino 1887 (Vucetich and Kramarz 2003).

Family Echimyidae

The echimyids are the most diverse group among hystricognath rodents (Vucetich and Verzi 1991). Most of the extant species are arboreal or semi-arboreal and inhabit tropical and subtropical forested areas (Woods 1984; Emmons and Feer 1990), a few species are adapted to more xeric habitats, some are semifossorial.

Colhuehuapian Echimyidae at Gran Barranca are represented exclusively by the "Adelphomyinae" (pre-Deseadan to Laventan; early? Oligocene to middle Miocene). Species of the subfamily have mesodont cheek teeth, some with slight unilateral hypsodonty, crests tending to become oblique, dp4/4 is retained, shortened lower incisors (Vucetich *et al.* 1993), and a small mental foramen (Patterson and Pascual 1968; Vucetich *et al.* 1993).

Protacaremys Ameghino 1902 is smaller and higher crowned than *Prospaniomys*, with more oblique crests, and dp4 with five transverse crests. The type species *P. prior* Ameghino 1902 (Fig. 14.2F) differs from *P. avunculus* Ameghino 1902 (Fig. 14.2G) by being larger and having a less prominent mandibular masseteric crest. A fragmentary right mandible with m1–2 more quadrangular (Fig. 14.2H) represents a new species tentatively referred to this genus. Another species, "*P.*" *pulchellus* Ameghino 1902, is transferred in this chapter to the genus *Acarechimys* (see below).

The monotypic genus *Paradelphomys* Patterson and Pascual 1968, only known through the type specimen of *P. fissus* Patterson and Pascual 1968, differs by the absence of mesolophid, in having much more oblique crests, and a hypoconid isolated from the hypolophid. Currently this specimen is missing.

Octodontoidea *incertae sedis*

Here we note those basal octodontoids with uncertain affinities (Vucetich and Kramarz 2003; Vucetich and Vieytes 2006), previously referred to Octodontidae or Echimyidae.

Prospaniomys Ameghino 1902 has mesodont and tetralophodont cheek teeth; the uppers with the anteroloph separate from the paracone; the lowers with acuminated labial projections of the protocone and hypocone, and mesolophid and posterolophid transversely shorter than the anterolophid and hypolophid; dp4 with four transverse crests. Only one species, *P. priscus* Ameghino 1902 (Fig. 14.2D), has been named, but some specimens show upper molars with a more ephemeral metafossette and a more curved anterolingual angle, suggesting they may belong to a distinct species (*Prospaniomys* sp. 1, Fig. 14.2E).

Acarechimys Patterson 1965 (in Kraglievich 1965) includes one of the smallest hystricognath rodents so far known. The cheek teeth are brachyodont. Lower molars have three main transverse crests separated by broad and shallow lingual flexids which turn into fossettids early in wear. The upper molars have four transverse crests; the anterior and posterior labial flexi become fossettes early in wear. Deciduous premolars are retained throughout life, as in the Adelphomyinae and modern octodontoids. The dp4 has a wide and subcircular anterior basin, usually opened on the labial side. This genus has a wide temporal range, from Colhuehuapian (Vucetich *et al.* 1993) to Laventan (Walton 1997) SALMAs. The three recognized species come from the Santacrucian SALMA. The material referred to *Acarechimys* coming from SALMAs other Santacrucian has not been yet assigned to species level. *Acarechimys* is represented at Gran Barranca by three isolated cheek teeth (Fig. 14.2I) similar in size to *A. minutissimus* (Ameghino 1887).

Acarechimys pulchellus comb. nov. (Fig. 14.2J, K), is a rare species. It differs from the species of *Protacaremys* in having cheek teeth with a less columnar appearance, a much shorter trigonid, broader and less oblique crests separated by shallower flexids that turn into enamel lakes early in wear, and reduced mesolophids on m2–3. These dental characters are diagnostic features of *Acarechimys*, and it is herein transferred to this genus. It is slightly larger than *A. minutus* (Ameghino 1887).

The minute *Caviocricetus* Vucetich and Verzi 1996 (Fig. 14.2L) has very low-crowned and strongly terraced cheek teeth, an extreme condition unknown among the other Hystricognathi (Vucetich and Verzi 1996). The dp4 has a distinctive V-shaped anterior border. The lower molars have three transverse crests separated by very shallow and wide labial flexids, which do not become enamel lakes with wear. A low, independent knob is present in the anterolingual flexid. The upper molars are longer than wide, with four transverse crests; the anteroloph and mesoloph are shorter than the remaining crests.

A still unnamed genus and species nov. (Fig. 14.2M) is a small octodontoid with low-crowned and slightly terraced cheek teeth occurring both at Gran Barranca and Bryn Gwyn. The dp4 is structurally similar to *Acarechimys*, but the taxon differs from *Acarechimys* in having four transverse crests on m1–2. Also, the mesolophids and posterolophids are shorter than the remaining lophids, and the lingual portion of the mesolophid curves forward and reaches the posterolabial slope of the metaconid, resembling the condition observed in *Platypittamys* and *Galileomys* (Vucetich and Kramarz 2003). The m3 is trilophodont while all the upper cheek teeth are tetralophodont. Differences in size suggest that there are two species, the larger of which occurs at Gran Barranca.

Protadelphomys Ameghino 1902 has a peculiar systematic history (Ameghino 1902; Vucetich and Bond 1984; Vucetich *et al.* 1992; Vucetich and Verzi 1994). The type of *P. latus*, the sole species of this genus, when described, was a mandibular corpus with p4–m3 of uncertain provenance (Ameghino 1902); currently it has only the m3. *Protadelphomys* is very common in Bryn Gwyn, but very rare in Gran Barranca. We recognize a smaller morph from Gran Barranca as *P. latus* morphotype A (Fig. 14.2N), and a larger morph from Bryn Gwyn (and Sacanana) as *P. latus* morphotype B. The two probably represent different species. The single tooth (m3) preserved in the holotype of *P. latus* Ameghino 1902 provides too few clues to elucidate to which morphotype it belongs. *Protadelphomys* has mesodont cheek teeth, the uppers with conspicuous unilateral hypsodonty with broad and slightly oblique crests separated by a narrow valley. The upper molars have four crests, but the hypoflexus is connected to the anterolabial valleys, which in turn close labially early in wear. The lingual end of the mesoloph is curved backward to merge with the posteroloph, and it is isolated from the hypocone; the mesofossette is ephemeral. This arrangement produces a characteristic transitory S-shaped occlusal pattern. The lower molars are three-crested; the hypoflexid is continuous with the posterolingual flexid. The DP3 is present during juvenile stages. In contrast to geologically younger octodontoids and contemporaneous "Adelphomyines," this genus exhibits normal replacement of the deciduous dentition; both upper and lower permanent premolars are simpler than molars. The incisors are very broad and long, the lowers with a flat anterior face, the uppers are strongly curved and have a conspicuous longitudinal crest on the anterior face. *Protadelphomys* is closely related to the Colhuehuapian *Willidewu* Vucetich and Verzi 1991 and the Deseadan *Sallamys* Hoffstetter and Lavocat 1970 (Vucetich and Verzi 1991).

Superfamily Cavioidea

Family Eocardiidae

Eocardiids include several small to middle-sized hystricognath rodents with biprismatic, mesodont to euhypsodont cheek teeth. The family has traditionally been considered as the Oligocene–Miocene group out of which the modern caviids arose (Ameghino 1898; Scott 1905; Kraglievich 1934, 1940; Landry 1957). Eocardiids are represented in Gran Barranca only by *Luantus initialis* Ameghino 1901 (Fig. 14.3A), with lower-crowned cheek teeth and less molarized p4 than the type species *L. propheticus* Ameghino 1899 ("Astrapothericulan").

Superfamily Chinchilloidea

Family Neoepiblemidae

Neoepiblemids include large to giant late Miocene rodents characterized by euhypsodont multilaminar cheek teeth with thick interlaminar layers of cement. The only representative of this group in Gran Barranca is *Perimys* Ameghino 1887. Besides its smaller size, *Perimys* differs from *Neoepiblema* Ameghino 1889 and *Phoberomys* Kraglievich 1926 by its bilaminate cheek teeth (except the M3, which is trilaminate), and the morphology of the P4 which opens on the labial side, in contrast to the upper molars. This genus is also recorded in the Santacrucian SALMA, although represented by different species. *Perimys* is closely allied to the Deseadan *Scotamys* Loomis 1914.

Ameghino (1902) described four species for Gran Barranca. *Perimys incavatus* Ameghino 1902 (Fig. 14.3B) is slightly smaller than the living plains viscacha, and differs from the Santacrucian species by having cheek teeth with broader and more rounded laminae, and upper molars lacking labial enamel cover and labial flexus. *Perimys transversus* Ameghino 1902 is only known through the type specimen, similar in size to the type of *P. incavatus*, but the lower cheek teeth are more quadrangular, and the laminae are less oblique; these could be synonyms. *Perimys dissimilis* Ameghino 1902 (Fig. 14.3C) is much smaller than the other two; the lower cheek teeth have more delicate and strongly curved laminae, and very conspicuous lingual flexids. The upper molars have also thin laminae and the labial walls have a vertical enamel band between two labial flexi. This species differs from all the remaining species of *Perimys* by having an anterior projection of the labial end of the posterior lamina in the lower molars and a posterior projection of the lingual end of the anterior lamina in the uppers. "*Perimys incurvus*" Ameghino 1902 is only known through the type specimen, which at present is missing from the MACN collections. According to Ameghino (1902), this species is even smaller than *P. dissimilis*, the laminae are also much curved, and the p4 is reduced. These features suggest that "*P. incurvus*" was based on a juvenile of *P. dissimilis*. Some specimens from Gran Barranca have cheek teeth similar in morphology to those of *P. incavatus* but they are even smaller than juveniles of the latter; they could belong to a still unnamed species.

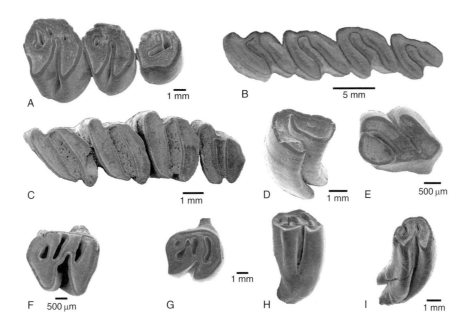

Fig. 14.3. Dental diversity among Colhuehuapian rodents (cont.). (A) *Luantus initialis* MPEF 6056, right P4–M2; (B) *Perimys incavatus* MPEF 6755, right p4–m3; (C) *Perimys dissimilis* MPEF 6754, right p4–m3; (D) *Eoviscacia australis* MPEF 5422a, left p4 (inverted); (E) *Eoviscacia australis* MPEF 5422b, left DP4; (F) *Soriamys* sp. MPEF 6934c, right DP4?; (G, H) Cephalomyidae indet. MACN A 52–163, right upper molar, (G) occlusal view, (H) lingual view; (I) Chinchilloidea *incertae sedis* MPEF 5421, upper molar.

Family Chinchillidae
Chinchillids include rather delicate rabbit-like rodents (Woods 1984) with protohypsodont to euhypsodont, bilaminar or trilaminar cheek teeth. There are only three extant genera grouped in two subfamilies: Lagostominae (*Lagostomus*) and Chinchillinae (*Chinchilla* and *Lagidium*). The fossil record is almost entirely restricted to the Lagostominae (but see Croft *et al.* 2004), characterized by bilaminar cheek teeth (except M3). The family is represented at Gran Barranca only by three isolated cheek teeth referable to *Eoviscaccia australis* Vucetich 1989 (Fig. 14.3D, E), originally described in the Deseadan. This species has more hypsodont cheek teeth than *E. boliviana* Vucetich 1989 (Deseadan of Bolivia); it differs from the species of *Prolagostomus* Ameghino 1887 and *Pliolagostomus* Ameghino 1887 (Santacrucian – "Colloncuran"), by its broader and less penetrating hypoflexus/-id, more persistent fossettes/tids, and a less developed posterior lobe in the M3.

Family Cephalomyidae
Cephalomyids include extinct rodents with different degrees of hypsodonty, characterized by an asymmetric dental pattern (i.e. the occlusal pattern of the lower cheek teeth does not mirror that of the uppers), probably of chinchilloid affinities (Vucetich 1985, 1989; Kramarz 2001b, 2005). Cephalomyids are diverse and very well represented in other Colhuehuapian local faunas, but at present they are known in Colhuehuapian levels of Gran

Barranca only through two isolated teeth. One of them is a mesodont, rooted, bilobed upper cheek tooth, similar to the DP4 of *Soriamys* Kramarz 2001b (Fig. 14.3F). The other is an isolated hypsodont and pentalophodont upper molar (MACN A 52–163) (Fig. 14.3G, H) interpreted by Wood and Patterson (1959) as a dasyproctid allied to *Neoreomys* Ameghino 1887. However, this molar differs from *Neoreomys* but resembles *Cephalomys* in its wider fossettes and acuminated lingual apices (Kramarz 1998).

Chinchilloidea incertae sedis
A genus and species of uncertain position within the superfamily is represented at Gran Barranca by an upper protohypsodont cheek tooth, with three small roots (Fig. 14.3I). An extended hypoflexus divides the occlusal surface into an anterior laminar lobe and a subtrapezoidal posterior one. This posterior lobe has a short and shallow flexus open to the labial wall.

Discussion

Comparison among Colhuehuapian local faunas and paleoenvironmental inferences
Strictly speaking, the fauna of Gran Barranca defines the Colhuehuapian. Other local faunas from central and northern Patagonia (i.e. Bryn Gwyn, Cerro Banderas, Sacanana, Paso Córdova; Fig. 14.1) were assigned to this SALMA based upon mammals other than rodents. Collections made

during the last 20 years in these localities permit the first analysis of the diversity of the rodents of this interval. Other faunas outside Argentina have also been assigned to the Colhuehuapian (e.g. Flynn *et al.* 1995), but rodents have not been reported from them, or they have not yet been adequately described.

Among Colhuehuapian faunas, the rodent assemblage at Gran Barranca is the best known; with at least 18 genera it is the most diverse rodent fauna for a single locality. The stratigraphic section bearing Colhuehuapian fauna (the Lower Fossil Zone: Ré *et al.* Chapter 3, this book) is about 40 meters thick and spans a single normal polarity interval of about 0.2 million years (Chron C6An.1n) between 20.0 and 20.2 Ma (Ré *et al.* Chapter 4, this book). Twelve of these genera (*Eosteiromys, Hypsosteiromys, Branisamyopsis, Galileomys, Acarechimys, Prospaniomys, Protacaremys, Acarechimys, Caviocricetus*, Octodontoidea gen. et sp. nov., *Luantus*, and *Perimys*) are represented together in one of the levels more intensively worked with different collecting techniques. Until now the richest caviomorph fossil assemblage was that of La Venta (middle Miocene, Colombia) with 13 genera, 11 of which are represented in the richest level (Walton 1997).

At Gran Barranca the rodent assemblage is dominated by brachydont or mesodont octodontoids, comprising nine genera, which represent 50% of the total generic diversity. The erethizontids, with four genera (including eight species) achieved their greatest known diversity, representing 22% of the total diversity. Thus, the rodent assemblage at Gran Barranca is dominated by low-crowned species, suggesting varied forested environments. Morphofunctional studies show that at least some Miocene erethizontids of Patagonia had climbing and grasping abilities, in agreement with the presence of forested habitats (Candela and Picasso 2008). Neither evolutionary trends nor the prevalence of a special adaptive type in any level can be seen along the stratigraphic sequence, suggesting relative environmental stability during its deposition and also in agreement with the short temporal interval sampled. This stability is also suggested by the record of several genera (*Hypsosteiromys, Protacaremys, Prospaniomys,* and *Perimys*) throughout the Lower Fossil Zone.

Among euhypsodont genera, *Perimys* is the most abundant and diverse. Although euhypsodont species are often considered to have inhabited open environments, *Perimys* could have lived in proximity to water bodies, as did the giant neoepiblemids of the late Miocene. These latter, sharing with *Perimys* the basic gross dental morphology (laminar cheek teeth with a thick cement interlaminar layer and a thin enamel surrounding layer), are always recorded in fluvial sediments. Sedimentological evidence (Bellosi this book) also suggests a fluvial depositional setting for the Colhuehuapian levels at Gran Barranca. The remaining

high-crowned taxa (*Eoviscacia australis, Luantus initialis*, Chinchilloidea *i.s.*, Cephalomyidae? indet., and *Soriamys* sp.) could also indicate of open environments, but these taxa are comparatively rare and might indicate that open environments were present distally.

The best-known Colhuehuapian fauna outside Gran Barranca is that of Bryn Gwyn, but there are fewer taxa than at Gran Barranca (Table 14.1). Although most genera are found in both localities, *Willidewu* (Octodontoidea), *Cephalomyopsis* (Cephalomyidae), and *Australoprocta* (Dasyproctidae) are recorded in Bryn Gwyn but not in Gran Barranca. Given the large number of rodents found in Gran Barranca (within this project alone, almost 400 specimens were collected) and the variety of collecting techniques used, we think that the lack of records reflects actual absence. Among the taxa not recorded in Bryn Gwyn, *Protacaremys* is important because it is so abundant in Gran Barranca. The frequencies of shared taxa are quite different: *Eoviscaccia, Protadelphomys, Caviocricetus*, and Octodontoidea gen. et sp. nov. are abundant in Bryn Gwyn, but more rare at Gran Barranca. Another important difference is the much greater diversity achieved by erethizontids in Gran Barranca, whereas cephalomyids are much better represented in Bryn Gwyn. Consequently, the proportion of hypsodont rodents is higher in Bryn Gwyn than in Gran Barranca. These differences in composition show that, unlike in Gran Barranca, more open environments might have prevailed in Bryn Gwyn. Sedimentological and paleopedological data agree with this interpretation (Bellosi this book; Bellosi and González this book).

Some authors (Simpson 1940) suggested these two faunas are slightly diachronic, that of Bryn Gwyn being somewhat younger, and recommended to distinguish it as "Trelewense" (Kraglievich 1930). However, our comparison of both rodent assemblages shows they are very similar overall, sharing several taxa even at the species level. Moreover, most of the shared genera are not known for other SALMAs. Consequently, the rodents as a whole do not suggest a temporal difference sufficient to justify separating them into different biochronological units. On the contrary the compositional differences above mentioned seem to be mostly the result of environmental differences. Temporal differences, if any, would be limited.

The local faunas of Paso Córdova, Sacanana, and Cerro Banderas, as well as their respective environmental conditions, are still too poorly known to permit an accurate analysis. As far as the present evidence goes, their composition is more similar to that of Bryn Gwyn than Gran Barranca. With the exception of *Banderomys*, which is exclusive from Cerro Bandera, all the recorded genera occur also in Bryn Gwyn whereas two taxa are not at Gran Barranca (Table 14.1).

Table 14.1. *Colhuehuapian rodent record*

Taxa			Taxa collected by Ameghino	Distribution according to new collections				
Family	Genus	Species		Gran Barranca	Bryn Gwyn	Sacanana	Paso Córdoba	Cerro Bandera
Erethizontidae	*Eosteiromys*	*E. homogenidens*	X	X	X			
		E. segregatus	X	X	X			
		E.? sp. nov		X	X			X
	Hypsosteiromys	*H. axiculus*	X	X	X			
		H. nectus	X	X				
	Parasteiromys	*P. uniformis*	X	X				
		P. friantae		X				
	Branisamyopsis	*B. australis*		X				
Acaremyidae	*Galileomys*	*G. antelucanus*		X	X			
	Acaremys	*A.* cf. *A. murinus*		X				
Echimyidae	*Protacaremys*	*P. prior*	X	X				
		P. avunculus	X	X				
	Protacaremys?	*P.? sp. nov.*		X				
	Paradelphomys	*P. fissus*		X				
Octodontoidea incertae sedis	*Acarechimys*	*A. pulchellus*	X	X				
		A. cf. *A. minutissimus*		X	X			
	Caviocricetus	*C. lucasi*		X	X		X	X
	Gen. nov.	sp. large		?	X			
		sp. small			X			
	Willidewu	*W. esteparius*			X		X	
	Protadelphomys	*P. latus* morph. A	?	X				
		P. latus morph. B	?		X	X		
	Prospaniomys	*P. priscus*	X	X	?	X		
		P. sp. 1		X				
		P. sp. 2			X			
		P. sp. 3				X		
Dasyproctidae	*Australoprocta*	*A. fleaglei*			X	X	X	
Eocardiidae	*Luantus*	*L. initialis*	X	X				
		L. sp. nov?			X			

Table 14.1. (*cont.*)

Taxa			Taxa collected by Ameghino	Distribution according to new collections				
Family	Genus	Species		Gran Barranca	Bryn Gwyn	Sacanana	Paso Córdoba	Cerro Bandera
Neoepiblemidae	*Perimys*	*P. dissimilis*	X	X				?
		P. transversus	X	X	?			?
		P. incavatus	X	X	?			X
Chinchillidae	*Eoviscaccia*	*E. australis*	X	X	X		X	X
Cephalomyidae	*Cephalomyopsis*	*C. hipselodontus*			X			
	Soriamys	*S. gaimanensis*			X			
		S. ganganensis				X		
		Soriamys sp.		X				
	Banderomys	*B. leanzai*	X	?				X
	Cephalomyidae indet.							
Chinchilloidea *incertae sedis*	Gen. indet.	gen. et sp. indet.		X	?			

214

Among the exclusive Colhuehuapian rodent taxa, *Caviocricetus lucasi* has the widest geographic distribution, as it is recorded in four of the five local faunas; eventually it might serve as a guide taxon of this SALMA.

Evolutionary significance of the Colhuehuapian rodent assemblage

The family composition of the entire Colhuehuapian rodent assemblage is similar to that of the other late Oligocene to middle Miocene Patagonian faunas, as it includes families that are currently restricted to the Brazilian Subregion (*sensu* Hershkovitz 1958: Erethizontidae, Echimyidae, and Dasyproctidae), others constrained to the Patagonian Subregion (Chinchillidae), the primitive caviids traditionally grouped in the Eocardiidae, and several extinct ones, such as the Acaremyidae (not yet recorded in the late Oligocene Deseadan), Cephalomyidae, and Neoepiblemidae. In fact, taking into account mammal family composition, the Colhuehuapian SALMA has been long considered as part of a major faunistic unit (the Pansantacrucian Cycle of Pascual *et al.* 1996 and references therein).

On the contrary, the generic composition (Table 14.1) is quite distinctive, especially because of the great diversity and abundance of small octodontoids, and the diversity of erethizontids and cephalomyids. Thirteen of the 23 rodent genera are found exclusively in the Colhuehuapian (Table 14.1), and notably three of them (*Hypsosteiromys*, *Caviocricetus*, and *Soriamys*) represent three lineages so far only recorded during this period.

The octodontoids, with ten genera belonging to several lineages, are the most enigmatic and interesting Colhuehuapian group. Oligocene to middle Miocene octodontoids have been referred traditionally to Octodontidae or Echimyidae (see Patterson and Wood 1982). Modern Echimyidae and Octodontidae, as well as the remaining extant octodontoids, share the retention of Dp4/4 and a distinctive incisor enamel microstructure with rectangular interprismatic matrix in multiserial Hunter–Schreger bands (Martin 2005). However, half of the Colhuehuapian octodontoid genera (*Caviocricetus*, the clade *Protadelphomys–Willidewu*, and the acaremyids *Galileomys* and *Acaremys*) display different combinations of "primitive" and "derived" states of these characters; therefore they cannot be easily allocated to the modern groups. It seems likely that the early evolution of the octodontoids was more complex than the simple dichotomy Octodontidae–Echimyidae implies, and that these taxa represent different branches of the early "stem" radiation of the Octodontoidea (Vucetich and Kramarz 2003; Vucetich and Vieytes 2006).

Acarechimys has alternately been classified within the Echimyidae or Octodontidae (Patterson and Wood 1982; Vucetich and Verzi 1991; Vucetich *et al.* 1993; Walton 1997; Verzi 2002; Carvalho and Salles 2004). At present

there is no strong evidence supporting either one or the other hypothesis. *Acarechimys* still retains a very primitive dental morphology, but it has the derived characters of modern octodontoid families mentioned above, suggesting it could be close to the base of the Echimyidae–Octodontidae group.

Among Colhuehuapian octodontoids, the adelphomyines are the closest to the modern Echimyidae. At least three different lineages can be identified among genera traditionally included in this subfamily: (1) *Spaniomys* (Santacrucian)–*Maruchito* ("Colloncuran": Ameghino 1902; Vucetich *et al.* 1993); the extant genus *Callistomys*, usually considered an Echimyinae, is proposed as a survivor of this lineage (Emmons and Vucetich 1998); (2) *Protacaremys–Prostichomys* ("Astrapothericulan")–*Stichomys* and *Adelphomys* (Santacrucian: Kramarz 2001c); and (3) *Xylechimys* (Deseadan)–*Paradelphomys–Ricardomys* (Laventan)–*Olallamys* (extant, Dactylomyinae: Patterson and Pascual 1968; Walton 1997; Carvalho and Salles 2004); *Eodelphomys* (pre-Deseadan, Santa Rosa, Peru: Frailey and Campbell 2004) could be related to this third lineage. We provisionally accept the subfamily "Adelphomyinae" as a monophyletic group (Vucetich and Verzi 1991; Carvalho and Salles 2004) with the proviso that these proposals are yet to be tested within a broader systematic context.

On the contrary, it is clear that Colhuehuapian erethizontids are not closely related to the extant species, most of which are confined to the Brazilian Subregion. They belong to a Patagonian clade differentiated from modern representatives at least by late Oligocene (Candela 2003; Candela and Morrone 2003). It is only through the great taxonomic and adaptive diversity achieved by Colhuehuapian erethizontids (especially recorded in Gran Barranca) that the magnitude of this southern radiation is manifested.

Among hypsodont rodents, cephalomyids have their acme and are the most diversified group during the Colhuehuapian (Kramarz 2001b, 2005). Two well-differentiated lineages are recognized; one is represented by the euhypsodont *Cephalomyopsis* Vucetich 1985 with simplified figure-eight-shaped lower molar pattern (perhaps related to the Deseadan *Cephalomys* Ameghino 1897 and *Litodontomys* Loomis 1914) and the other by the protohypsodont *Soriamys* with trilaminar lower molar pattern. Both lines would have radiated in pre-Deseadan times from an ancestor with a dentition very close to that of *Banderomys* (Kramarz 2005). Except for the record of a cephalomyid close to the *Cephalomys* from the middle Miocene Quebrada Honda Group of Bolivia (Frailey 1980), this family is not represented in faunas younger than Colhuehuapian.

Colhuehuapian eocardiids and chinchillids are less diversified than cephalomyids. The eocardiid *Luantus initialis* is structurally antecedent to euhypsodont Santacrucian eocardiids (Kramarz 2006), which in turn are the ancestral stock of the modern caviids (Ameghino 1898; Scott 1905; Kraglievich 1934, 1940; Landry 1957). The chinchillid *Eoviscaccia australis* is structurally the ancestor of *Prolagostomus* and *Pliolagostomus* (Santacrucian – "Colloncuran"), which are closely related to the extant lagostomines (Chinchillinae probably have an extra-Patagonian origin: Croft *et al.* 2004).

Several Colhuehuapian rodent genera belong to lineages already represented in Deseadan faunas of Patagonia, although mostly by different genera, i.e. Eocardiidae (*Asteromys–Luantus*), Chinchillidae, Lagostominae (*Eoviscaccia*), Cephalomyidae (*Cephalomys–Cephalomyopsis*), Neoepiblemidae (*Scotamys–Perimys*), Echimyidae, Adelphomyinae (*Xylechimys–Paradelphomys*), and Erethizontidae (*Protosteiromys–Eosteiromys*). Others have no known ancestors in Patagonia: *Caviocricetus, Protadelphomys–Willidewu,* Acaremyidae, Octodontoidea gen. et sp. nov., and *Hypsosteiromys*. At least one of these lineages, the clade *Protadelphomys–Willidewu* (Vucetich and Verzi 1991) is already represented in the Deseadan closer to the equator (*Sallamys*). Overall then, it appears that the high diversity of Colhuehuapian rodents result from the coexistence of surviving Patagonian lineages with others that entered in post-Deseadan times. Based on palynological data, Barreda and Palazzesi (2007, this book) propose that early Miocene climate amelioration allowed the southward dispersal of Neotropical elements. However, the validity of this hypothesis has to be tested after more intensive prospecting of Deseadan levels of Patagonia, which have not been worked as intensively as those with Colhuehuapian faunas.

After the Colhuehuapian a decrease of taxonomic diversity, in the variety of occlusal designs and in the range of hypsodonty, suggests an impoverishment of niche breadth. This is especially notable for the Santacrucian, known mainly from the southern end of Patagonia, during which the prevailing taxa were proto- and euhypsodont. This could be the result of a progressive climate deterioration in Patagonia during the early to middle Miocene as proposed by Barreda and Palazzesi (2007, this book). But it is also possible that these differences are magnified by the latitudinal differences between Colhuehuapian and Santacrucian localities (Fig. 14.1). In fact, "Colloncuran" faunas from northwestern Patagonia, somewhat younger than Santacrucian, show slightly higher diversity of brachyodont forms than the Santacrucian ones (Vucetich *et al.* 1993, and unpublished data).

Conclusions

The efforts of fieldwork in the early Miocene of Patagonia during the last 20 and especially the last ten years in Gran Barranca provide a more integrative knowledge of the caviomorph evolution and diversity during this interval. The new findings reveal that all the faunas assigned to the Colhuehuapian SALMA share numerous rodent taxa, not known from faunas assigned to immediately younger or older SALMAs, This situation provides support for the assignment of all faunas to a single biochronologic unit. Despite the faunas from northern Patagonia being more similar to that of Bryn Gwyn than to that of Gran Barranca, the taxonomic differences are minor and could result mainly from environmental variation.

The new finds have dramatically increased the known rodent diversity for the Colhuehuapian. Numerous new taxa (not named yet) are here described; several others are first recorded at Gran Barranca. Colhuehuapian caviomorphs are currently the most diverse Cenozoic rodent fauna of South America. Some old Patagonian lineages survive from the Deseadan; others entered Patagonia from the north in one or more post-Deseadan events. This diversity is also reflected in a wide range of morphological diversity in the observed range of hypsodonty, a variety of occlusal designs (cuspidate, terraced, laminar, sigmoid, octodontiform, precordiform, and variants within each of these types), as well as a relatively wide range of body size, comparable to that of the late early Miocene Santacrucian SALMA. This morphologic diversity shows that for the Colhuehuapian caviomorphs had already developed a large variety of adaptive types and strategies to use of food resources.

The Colhuehuapian rodent fauna is notable for the large number of small octodontoids possessing a confusing mosaic of "derived" and "primitive" characters, and by the great diversity of erethizontids that reach their peak of diversity in the fossil record. Hypsodont taxa are represented by lineages and taxa unrelated to those that prevailed in younger faunas. Among hypsodont Colhuehuapian taxa, hypsodont eocardiid and chinchillid clades persisted and diversified into the succeeding Pinturan and Santacrucian.

Several questions concerning Colhuehuapian rodents are still poorly explored: (1) which factors promoted one of the highest peaks in diversity of fossil rodents in South America at this particular time, and what particular conditions in Patagonia allowed the existence of such a wide diversity of octodontoids and erethizontids, (2) what selective pressures may have conditioned the disappearance of some lineages like cephalomyids, and the survival of others until today (chinchillids and caviids). Only paleobiological studies can identify the ecomorphological features that would have favored the selective survival of some hypsodont taxa but the extinction of others.

These and other questions will be answered as new data enlarge the knowledge of rodents from previous and subsequent faunas, and when more inclusive phylogenetic studies, especially of the enigmatic octodontoids, are available.

ACKNOWLEDGEMENTS
Cecilia Deschamps kindly helped in the preparation of this chapter. This study was supported in different ways by CONICET, Universidad Nacional de La Plata (grants to MGV) and National Science Foundation (EAR-0087636, BCS-0090255, and DEB-9907985 to Richard F. Kay and Richard H. Madden, Duke University). L. Flynn, D. Verzi, and R. Kay helped to improve this manuscript.

REFERENCES

Ameghino, F. 1887. Enumeración sistemática de las especies de mamíferos fósiles coleccionados por Carlos Ameghino en los terrenos eocenos de Patagonia austral y depositados en el Museo de La Plata. *Boletín del Museo de La Plata*, **1**, 1–26.

Ameghino, F. 1889. Contribución al conocimiento de los mamíferos fósiles de la República Argentina. *Actas Academia Nacional de Ciencias en Córdoba*, **6**, 1–1027.

Ameghino, F. 1897. Mammifères crétacés de l'Argentine: deuxième contribution à la connaissance de la faune mammalogique des couches à *Pyrotherium. Boletín del Instituto Geográfico Argentino*, **18**, 406–429, 431–521.

Ameghino, F. 1898. *Sinopsis geológico – paleontológica de la Argentina*. La Plata: Imprenta La Libertad.

Ameghino, F. 1899. *Sinopsis geológico – paleontológica. Suplemento (Adiciones y correcciones)*. La Plata: Imprenta La Libertad.

Ameghino, F. 1901. L'âge des formations sédimentaires de Patagonie. *Anales de la Sociedad Científica Argentina*, **52**, 189–197, 244–250.

Ameghino, F. 1902. Première contribution à la connaissance de la faune mammalogique des couches à *Colpodon. Boletín de la Academia Nacional de Ciencias en Córdoba*, **17**, 71–138.

Ameghino, F. 1903. Los diprotodontes del orden de los Plagiaulacoideos y el origen de los roedores y de los polimastodontes. *Anales Museo Nacional Buenos Aires*, **9**, 81–192.

Ameghino, F. 1904. Nuevas especies de mamíferos cretáceos y terciarios de la República Argentina. *Anales de la Sociedad Científica Argentina*, **58**, 35–41, 56–71, 182–192, 225–291.

Barreda, V. and L. Palazzesi 2007. Patagonian vegetation turnovers during the Paloegene – early Neogene: origin of arid-adapted floras. *Botanical Review*, **73**, 31–50.

Bordas, A. F. 1939. Diagnosis sobre algunos mamíferos de las capas con *Colpodon* del Valle del Río Chubut (República Argentina). *Physis*, **14**, 413–433.

Candela, A. M. 1999. The evolution of the molar pattern of the Erethizontidae (Rodentia, Hystricognathi) and the validity of *Parasteiromys* Ameghino, 1904. *Palaeovertebrata*, **28**, 53–73.

Candela, A. M. 2000. Los Erethizontidae (Rodentia, Hystricognathi) fósiles de Argentina. Sistemática e historia evolutiva evolutiva y biogeográfica. Ph.D. thesis, Universidad Nacional de La Plata, Argentina.

Candela, A. M. 2003. A new porcupine (Rodentia, Hystricognathi, Erethizontidae) from the Early – Middle Miocene of Patagonia. *Ameghiniana*, **40**, 483–494.

Candela, A. M. and J. J. Morrone 2003. Biogeografía de puercoespines neotropicales (Rodentia, Hystricognathi): integrando datos fósiles y actuales a través de un enfoque panbiogeográfico. *Ameghiniana*, **40**, 361–378.

Candela, A. M. and M. G. Vucetich 2002. *Hypsosteiromys* (Rodentia, Hystricognathi) from the Early Miocene of Patagonia (Argentina), the only Erethizontidae with tendency to hypsodonty. *Geobios*, **35**, 153–161.

Candela A. M. and M. Picasso 2008. Inferring locomotor behavior in the Miocene porcupines (Rodentia, Erethizontidae). *Journal of Morphology*, **269**, 552–593.

Carvalho, G. A. and L. O. Salles 2004. Relationships among extant and fossil echimyids (Rodentia, Hystricognathi). *Zoogical Journal of the Linnean Society*, **142**, 445–477.

Croft, D., J. Flynn, and A. Wyss 2004. Notoungulata and Litopterna of the Early Miocene Chucal Fauna, northern Chile. *Fieldiana, Geology*, n.s., **50**, 1–52.

Emmons, L. H. and F. Feer 1990. *Neotropical Rainforest Mammals: A Field Guide*. Chicago, IL: University of Chicago Press.

Emmons, L. H. and M. G. Vucetich 1998. The identity of Winge's *Lasiuromys villosus* and the description of a new genus of echimyid rodent (Rodentia: Echimyidae). *American Museum of Natural History Novitates*, **3223**, 1–12.

Flynn, J. J., A. R. Wyss, R. Charrier, and C. C. Swisher III 1995. An Early Miocene anthropoid skull from the Chilean Andes. *Nature*, **373**, 603–607.

Frailey, C. D. 1980. Studies on the Cenozoic vertebrata of Bolivia and Peru. Ph.D. thesis, University of Kansas, Lawrence.

Frailey, C. D. and K. E. Campbell Jr. 2004. Paleogene rodents from Amazonian Peru: the Santa Rosa Local Fauna. In Campbell, K. E. Jr. (ed.), *The Paleogene Mammalian Fauna of Santa Rosa, Amazonian Peru*. Los Angeles, CA: Natural History Museum of Los Angeles County pp. 71–130.

Hershkovitz, P. 1958. A geographic classification of Neotropical mammals. *Fieldiana, Zoology*, **36**, 581–620.

Hoffstetter, R. and R. Lavocat 1970. Découverte dans la Déséadien de Bolivie de genres pentalophodontes appuyant les affinités africaines des rongeurs caviomorphes. *Comptes Rendus de l'Académie des Sciences, Paris*, **271**, 172–175.

Kraglievich, J. L. 1965. Spéciation phylétique dans les rongeurs fossiles du genre *Eumysops* Amegh. (Echimyidae, Heteropsomyidae). *Mammalia*, **29**, 258–267.

Kraglievich, L. 1926. Los grandes roedores terciarios de la Argentina y sus relaciones con ciertos géneros pleistocenos de las Antillas. *Anales del Museo Nacional de Historia Natural*, **34**, 121–135.

Kraglievich, L. 1930. La formación Friaseana del río Frías, río Fénix, Laguna Blanca, etc. y su fauna de mamíferos. *Physis*, **10**, 127–161.

Kraglievich, L. 1934. *La antigüedad pliocena de las faunas de Monte Hermoso y Chapadmalal, deducidas de su comparación con las que precedieron y sucedieron.* Montevideo: Imprenta "El Siglo Ilustrado."

Kraglievich, L. 1940. Descripción detallada de diversos roedores argentinos terciarios clasificados por el autor. In Torcelli, A.J. (ed.), *Obras de Geología y Paleontología*, vol. 1. La Plata: Ministerio de Obras Públicas de la Provincia de Buenos Aires, pp. 297–330.

Kramarz, A.G. 1998. Un nuevo dasyproctidae (Rodentia, Caviomorpha) del Mioceno inferior de Patagonia. *Ameghiniana*, **35**, 181–192.

Kramarz, A.G. 2001a. Registro de *Eoviscaccia* (Rodentia, Chinchillidae) en estratos colhuehuapenses de Patagonia, Argentina. *Ameghiniana*, **38**, 237–242.

Kramarz, A.G. 2001b. Revision of the family Cephalomyidae (Rodentia, Caviomorpha) and new cephalomyids from the Early Miocene of Patagonia. *Palaeovertebrata*, **30**, 51–88.

Kramarz, A.G. 2001c. *Prostichomys bowni*, un nuevo roedor Adelphomyinae (Hystricognathi, Echimyidae) del Mioceno medio – inferior de Patagonia, Argentina. *Ameghiniana*, **38**, 163–168.

Kramarz, A.G. 2004. Octodontoids and erethizontoids (Rodentia, Hystricognathi) from the Pinturas Formation, early – middle Miocene of Patagonia, Argentina. *Ameghiniana*, **41**, 199–216.

Kramarz, A.G. 2005. A primitive cephalomyid hystricognath rodent from the Early Miocene of northern Patagonia, Argentina. *Acta Paleontologica Polonica*, **50**, 249–258.

Kramarz, A.G. 2006. Eocardiids (Rodentia, Hystricognathi) from the Pinturas Formation, late Early Miocene of Patagonia, Argentina, *Journal of Vertebrate Paleontology*, **26**, 770–778.

Landry, S.O. 1957. The interrelationships of the New and Old World hystricomorph Rodents. *University of California Publications in Zoology*, **56**, 1–118.

Loomis, F.B. 1914. *The Deseado Formation of Patagonia.* Concord, NH: Rumford Press.

Martin, T. 2005. Incisor Schmelzmuster diversity in South America's oldest rodent fauna and early caviomorph history. *Journal of Mammalian Evolution*, **12**, 405–417.

Pascual, R., E. Ortiz Jaureguizar, and J.L. Prado 1996. Land mammals: paradigm for Cenozoic South American geobiotic evolution. *Münchner Geowissenschaftlich Abhandlungen A*, **30**, 265–319.

Patterson, B. 1958. A new genus of erethizontid rodents from the Colhuehuapian of Patagonia. *Breviora Museum of Comparative Zoology*, **92**, 1–4.

Patterson, B. and R. Pascual 1968. New echimyid rodents from the Oligocene of Patagonia and a Synopsis of the family. *Breviora Museum of Comparative Zoology*, **401**, 1–14.

Patterson, B. and A.E. Wood 1982. Rodents from the Deseadan Oligocene of Bolivia and the relationships of the Caviomorpha. *Bulletin of the Museum of Comparative Zoology*, **149**, 371–543.

Pérez, M.E., M.G. Vucetich, and A. Kramarz in press. The first Eocardiidae (Rodentia) in the Colhuehuapian (early Miocene) of Bryn Gwyn (northern Chubut, Argentina) and the early evolution of the peculiar cavioid rodents. *Journal of Vertebrate Paleontology*.

Scott, W.B. 1905. Mammalia of the Santa Cruz Beds. Volume V, Paleontology. Part III, Glires. In Scott, W.B. (ed.), *Reports of the Princeton University Expeditions to Patagonia, 1896–1899*. Stuttgart: E. Schweizerbart'sche Verlag (E. Nägele), pp. 384–490.

Simpson, G.G. 1940. Review of the mammal-bearing Tertiary of South America. *Proceedings of the American Philosophical Society*, **83**, 649–709.

Verzi, D.H. 2002. Patrones de evolución morfológica en Ctenomyinae (Rodentia, Octodontidae). *Mastozoología Neotropical*, **9**, 309–328.

Vucetich, M.G. 1985. *Cephalomyopsis hipselodontus* gen. et sp. nov. (Rodentia, Caviomorpha, Cephalomyidae) de la Edad Colhuehuapense (Oligoceno tardío) de Chubut, Argentina. *Ameghiniana*, **22**, 243–245.

Vucetich, M.G. 1989. Rodents (Mammalia) of the Lacayani fauna revisited (Deseadan, Bolivia): comparison with new Chinchillidae and Cephalomyidae from Argentina. *Bulletin du Muséum National d'Histoire Naturelle, Paris*, **4**, 233–247.

Vucetich, M.G. and M. Bond 1984. Un nuevo Octodontoidea (Rodentia, Caviomorpha) del Oligoceno tardío de la provincia de Chubut (Argentina). *Ameghiniana*, **21**, 105–114.

Vucetich, M.G. and A.G. Kramarz 2003. New Miocene rodents of Patagonia (Argentina) and their bearing in the early radiation of the octodontiform octodontoids. *Journal of Vertebrate Paleontology*, **23**, 435–444.

Vucetich, M.G. and D.H. Verzi 1991. Un nuevo Echimyidae (Rodentia, Hystricognathi) de la Edad Colhuehuapense de Patagonia y consideraciones sobre la sistemática de la familia. *Ameghiniana*, **28**, 67–74.

Vucetich, M.G. and D.H. Verzi 1993. Un nuevo Chinchillidae del Colhuehuapense (Mioceno Inferior?) de Gaiman (Chubut): su aporte a la comprensión de la dicotomía vizcachas – chinchillas. *Ameghiniana*, **30**, 115.

Vucetich, M.G. and D.H. Verzi 1994. The presence of *Protadelphomys* (Rodentia, Echimyidae) in the Colhuehuapian of South Barrancas of Lake Colhue-Huapi (Chubut). *Ameghiniana*, **31**, 93–94.

Vucetich, M.G. and D.H. Verzi 1996. A peculiar octodontoid (Rodentia, Caviomorpha) with terraced molars from the lower Miocene of Patagonia (Argentina). *Journal of Vertebrate Paleontology*, **16**, 297–302.

Vucetich, M.G. and E.C. Vieytes 2006. A Middle Miocene primitive octodontoid rodent and its bearing on the early evolutionary history of the Octodontoidea. *Palaeontographica Abteilung A*, **27**, 79–89.

Vucetich, M.G., D.H. Verzi, and M.T. Dozo 1992. El "status" sistemático de *Gaimanomys alwinea* (Rodentia, Caviomorpha, Echimyidae). *Ameghiniana*, **29**, 85–86.

Vucetich, M.G., M.M. Mazzoni, and U.F.J. Pardiñas 1993. Los roedores de la Formación Collón Cura (Mioceno Medio), y la Ignimbrita Pilcaniyeu. Cañadón del Tordillo, Neuquén. *Ameghiniana*, **30**, 361–381.

Walton, A. H. 1997. Rodents. In Kay, R. F., Madden, R. H., Cifelli, R. L., and Flynn, J. J. (eds.), *Vertebrate Paleontology in the Neotropics: The Miocene Fauna of La Venta, Colombia.* Washington, DC: Smithsonian Institution Press: pp. 392–409.

Wood, A. E. 1949. A new Oligocene rodent genus from Patagonia. *American Museum of Natural History Novitates,* **1435**, 1–54.

Wood, A. E. and B. Patterson 1959. Rodents of the Deseadan Oligocene of Patagonia and the beginnings of South American rodent evolution. *Bulletin of the Museum of Comparative Zoology,* **120**, 281–428.

Woods, C. A. 1984. Hystricognath rodents. In Anderson, S. and Jones, J. K. (eds.), *Orders and Families of Recent Mammals of the World.* New York: John Wiley, pp. 384–446.

15 A new primate from the early Miocene of Gran Barranca, Chubut Province, Argentina: paleoecological implications

Richard F. Kay

Abstract

New specimens from the Gran Barranca, south of Lake Colhue-Huapi, in Patagonian Argentina document the presence of a new genus and species of primate, *Mazzonicebus almendrae*. *Mazzonicebus* specimens comprise mandibular and maxillary fragments and teeth of a medium-sized platyrrhine monkey. The new taxon comes from the Colhue-Huapi Member (type section of the Colhuehuapian land mammal age; early Miocene). Data presented by Ré *et al.* (Chapter 4, this book) establish a chronological age for this taxon, and for the Colhuehuapian fauna at Gran Barranca, of about 20 Ma. Thus, *Mazzonicebus* joins *Dolichocebus* and *Tremacebus* as the oldest record of primates from Argentina.

Mazzonicebus is closely related to *Soriacebus* from the younger Miocene rocks of the Pinturas Formation in Santa Cruz Province. For the most part it is more primitive than *Soriacebus* although several morphological aspects do not support a direct ancestor–descendant relationship. Nevertheless, the two offer the first well-documented sister taxon relationship between a Colhuehuapian and a (early) Santacrucian primate, spanning ~3 million years.

Mazzonicebus and *Soriacebus* show specializations for fruit-husking and seed predation, a dietary niche similar to that seen in the more derived members of the living platyrrhine subfamily Pitheciinae (sakis and uakaries), a fact that has lead some researchers to conclude that *Soriacebus* is an early representative of that clade. However, the more primitive characters of the dentition of *Mazzonicebus* compared with *Soriacebus* further document and reinforce the hypothesis that these two Argentine early Miocene taxa are an independently evolved clade that is an ecological "vicar" of the extant platyrrhine seed predation dietary niche.

Resumen

Nuevos especímenes provenientes de Gran Barranca, al sur del lago Colhue-Huapi (Patagonia Argentina) documentan la presencia de un nuevo género y especie de primate, *Mazzonicebus almendrae*. *Mazzonicebus* es un mono platirrino de tamaño mediano, que está representado por dientes y fragmentos de maxila y mandíbula. El nuevo taxón proviene del Miembro Colhue-Huapi (sección tipo de la Edad Colhuehuapense, Mioceno temprano). Los datos presentados por Ré *et al.* (Capítulo 4), establecen una edad de alrededor de 20,0 a 20,2 Ma. para este taxón, y para la fauna Colhuehuapense en Gran Barranca. Por lo tanto, *Mazzonicebus* representa, junto con *Dolichocebus* y *Tremacebus*, el registro más antiguo de primates de Argentina.

Mazzonicebus está estrechamente relacionado con *Soriacebus,* proveniente del Mioceno más joven de la Formación Pinturas (provincia de Santa Cruz). *Mazzonicebus* es en general más primitivo que *Soriacebus,* aunque varios aspectos morfológicos no apoyan una relación directa ancestro-descendiente. Sin embargo, ellos ofrecen la primera evidencia bien documentada de una relación de grupo-hermano entre un primates Colhuehuapense y uno del Santacrucense (temprano), que abarca unos 3 millones de años.

Mazzonicebus y *Soriacebus* muestran especializaciones para pelar frutos de cáscara dura y consumir semillas, un nicho dietario similar al observado en los miembros más derivados de platirrinos vivientes de la subfamilia Pitheciinae (sakis y uakaries). Esto llevó a algunos investigadores a concluir que *Soriacebus* es un representante temprano de dicho clado. Sin embargo, los caracteres más primitivos de la dentición de *Mazzonicebus* comparados con los de *Soriacebus* documentan y refuerzan la hipótesis de que estos dos taxones del Mioceno temprano de Argentina son un clado evolucionado independientemente, que representa un "vicariante" ecológico de los platirrinos vivientes consumidores de semillas.

Introduction

On the cliffs south of Lake Colhue-Huapi at 45° South latitude, in Chubut Province, Argentina, are exposed a layered sequence of fossil-bearing volcanic ashes that have figured prominently in the history of South American paleontology and continue to do so today (Fig. 15.1). The fossil potential was first recognized and exploited by Carlos Ameghino and his more famous brother Florentino more than 100 years ago. Locally, the cliffs are called a gran barranca (high escarpment) but to paleontologists they have won worldwide acclaim as *The* Gran Barranca. The foremost twentieth-century North American paleontologist George Gaylord

The Paleontology of Gran Barranca: Evolution and Environmental Change through the Middle Cenozoic of Patagonia, eds. R. H. Madden, A. A. Carlini, M. G. Vucetich, and R. F. Kay. Published by Cambridge University Press. © Cambridge University Press 2010.

Fig. 15.1. Members of the Elmer Riggs expedition at their camp, Puesto Almendra, south of Lake Colhue-Huapi, about 1923. (Photograph courtesy of: Field Museum of Natural History.)

Simpson made an important collection of fossil mammals there in the early 1930s and recognized the scientific value of these fossil beds.

Early collectors did not find any primates in spite of the richly fossiliferous nature of the beds. Perhaps this was a consequence of collecting bias towards the lower levels of the geologic section. Simpson's party, for example, concentrated their efforts on collecting fossils from the lower fossiliferous levels of the Barranca, especially the Casamayoran levels, then thought to represent early Eocene. It was not until December 1967 that Robert G. Goelet collected a mandible from the Colhuehuapian levels of the Barranca. A mandibular fragment with P_4 and root sockets for the anterior teeth entered into the paleontology collections of the Museo de La Plata in 1969 as MLP 69-III-12–1, "*Homunculus* sp." This specimen was figured and described by Hershkovitz, who allocated it to cf. *H. patagonicus* Ameghino (Hershkovitz 1981). Fleagle (1990) noted the similarities of this specimen to *Soriacebus* from the Pinturas Formation and provisionally allocated it to *Soriacebus* cf. *ameghinorum*. In 1977 an upper canine (MLP 77-VI-13–6) from Gran Barranca was identified in the La Plata collections as a primate. Beginning in 1995, a joint field party from Museo de la Plata and Duke University began intensive collecting efforts at the Barranca. To date, we have recovered numerous additional specimens of jaws and teeth, all belonging to the same primate species from the Colhuahuapian levels.

This collection of primates represents the oldest known occurrence of primates from Argentina, co-equal in age (by faunal correlation with the Colhuehuapian South American Land Mammal Age) with finds further north at Gaiman (*Dolichocebus*), Sacanana (*Tremacebus*), and (by radiometric age determination of ~19–20 Ma) with *Chilecebus* from Chile. Only the fossil primate *Branisella* from the Deseadan fauna at Salla, Bolivia (at about 26 Ma: Kay *et al.* 1998b) is older on that continent. Deseadan primates have not yet been recovered in Argentina and further sampling is needed to test this apparent regional absence.

A notable aspect of the dental structure of the Barranca species described below is the unmistakable imprint of its adaptation for fruit husking (sclerocarpy: Kinzey *et al.* 1990) and for eating (chewing up) the seeds of ripe and unripe fruit – *Mazzonicebus* was apparently a seed predator. Seed predation, that is chewing up seeds as distinct from swallowing them whole, is a feeding niche in South America occupied today by several mammalian groups (various rodents including *Sciurus*; peccaries of the genus *Tayassu*: Emmons *et al.* 1990) and among primates by the Pitheciinae (sakis and uakaries: Kinzey *et al.* 1990; Kinzey 1992). A second notable feature of the cheek teeth is the very high molar crowns. These may well be signs of scansorial habits – a niche that is notably absent among living platyrrhines but one that apparently also existed in the Deseadan *Branisella* (Kay *et al.* 2001) and is common among Old World monkeys, albeit at much larger body size.

Debate has arisen as to the phyletic position of the Argentine primate *Soriacebus* (early Miocene, Pinturas Formation, Santa Cruz Province, Argentina). This issue will be pertinent to our discussion because *Soriacebus* is closely related to our new species. To set the stage for the discussions that follow, I review the phyletic arrangement of living platyrrhine taxa, especially the Pitheciidae, based on molecular studies and the points of agreement and disagreement about fossil pitheciids.

The monophyly of the extant Pitheciinae, *Pithecia* (the saki), *Chiropotes* (the bearded saki), and *Cacajao* (the uakari) is well established from molecular and morphologic data (Kay 1990; Barroso *et al.* 1997; Schneider *et al.* 1993, 1996, 2001). Rosenberger advocates inclusion of *Callicebus* (the titi) and *Aotus* (the owl monkey) in the Pitheciidae (Rosenberger 1981, 2002; Setoguchi *et al.* 1987), a view resisted by Kay (1990). Rosenberger's view that *Callicebus* is a basal pitheciid has now received strong corroboration from phylogenetic analyses of DNA nucleotide sequences. However, the molecular evidence equally strongly rejects his hypothesis that *Aotus* is a pitheciid (Schneider *et al.* 1993, 1996). In recent studies of short interspersed elements (SINEs), which have a unidirectional mode of evolution (Hillis 1999; Ray *et al.* 2006), Ray *et al.* (2005) identifies three SINEs supporting the linkage of *Callicebus* to sakis and uakaries, none of which is found in *Aotus*. Furthermore, *Aotus* shares seven SINEs with the cebid *Saimiri*, none of which is found in pitheciines. Thus, *Callicebus* certainly is a pitheciid and *Aotus* certainly is not. The classification of Schneider *et al.* (1993) reflects these cladogenetic events and is followed here. The Pitheciidae, then, represents a clade including the monophyletic Pitheciinae with the extant genera *Pithecia*, *Chiropotes*, and *Cacajao*, and the Callicebinae for *Callicebus* alone.

A number of fossil taxa are agreed by all to be pitheciids. Takai *et al.* (2001) describe *Miocallicebus*, a middle Miocene callicebine pitheciid from Colombia. *Cebupithecia* and *Nuciruptor* (both middle Miocene, Colombia) are pitheciines (Meldrum *et al.* 1997). Another possible middle Miocene pitheciine is *Proteropithecia* from the middle Miocene of Argentina (Kay *et al.* 1998a, 1999a). Allocation of this taxon to pitheciines is less certain because it is known from just a few isolated teeth and a talus. Finally, mention also should be made of Quaternary primates of Cuba and Dominica *Xenothrix* (Rosenberger 1975; MacPhee *et al.* 2004), *Antillothrix* (MacPhee *et al.* 1995), and *Paralouatta* (Horovitz *et al.* 1999) which are possible pitheciids (but see Hershkovitz 1977; Rivero *et al.* 1991).

A second group of fossil primates has been assigned to Pitheciidae on the basis of the claim that *Aotus* is a pitheciid. For example, *Tremacebus* (early Miocene, Argentina) might be related to *Aotus* and for that reason is included in pitheciids (Rosenberger 1979, 2002; Rosenberger *et al.* 1990). But with *Aotus* removed from pitheciids, so should be *Tremacebus*. Early Miocene *Homunculus* from southern Patagonia also has been accorded pitheciid status based upon cranial and dental similarities to *Callicebus* (Rosenberger 1979; Rosenberger *et al.* 1990; Tauber 1991; Fleagle *et al.* 2002; Tejedor *et al.* 2006).

Two other taxa, *Soriacebus* and *Mohanamico*, are of uncertain affinities but have been proposed as pitheciines. *Mohanamico* from the middle Miocene of Colombia was suggested to have pitheciid affinities (Luchterhand *et al.*

1986). Rosenberger *et al.* (1990) presented evidence that this taxon was related to the Callitrichidae. Meldrum and Kay (1997) argue that *Mohanamico* presents significant 'crossing specializations' making it an unlikely candidate for membership in Pitheciidae (see also Horovitz 1999).

Some have maintained that *Soriacebus* (an early Miocene platyrrhine from the Pinturas Formation in Santa Cruz Province) is a pitheciine (Rosenberger 2000, 2002; Fleagle *et al.* 2002; Tejedor 2005) whilst others maintain that this genus is a primitive platyrrhine possibly a stem taxon sister to all living platyrrhines, or at least not specially related to pitheciids (Fleagle *et al.* 1987; Fleagle 1990; Kay 1990; Meldrum *et al.* 1997; Kay *et al.* 1998a, 2008). It is this taxon that is of special interest here since it appears that our new monkey is an older more primitive version of *Soriacebus*.

In summary, it is agreed by all that pitheciid fossil history goes back at least to the middle Miocene in Colombia (13.7 Ma or younger: Flynn *et al.* 1997; *Miocallicebus*, *Cebupithecia*, and *Nuciruptor*). *Proteropithecia* from Argentina would extend this record back to the earliest part of the middle Miocene, about 15.7 Ma (Kay *et al.* 1998a). If *Soriacebus* is a pitheciid, the record goes even further back to the later early Miocene (18.7 and 19.7 Ma; Ré *et al.* Chapter 4). Finally, If Fleagle has correctly allocated primate remains from the Gran Barranca to *Soriacebus*, and if *Soriacebus* is a pitheciine, the age of first appearance of Pitheciinae would be pushed back to ~20 Ma (Kay *et al.* 1999b). The purpose of this chapter is to diagnose and describe this new primate species to interpret its morphology in adaptive terms, and to reopen the question of the antiquity of pitheciine clade: is the oldest known pitheciine 15.7 Ma, ~17 Ma, or as old as ~20 Ma?

Abbreviations

AMNH, American Museum of Natural History, mammalogy collection; AR, field number of specimen belonging to Museo de La Plata, La Plata, Argentina; BMNH, Natural History Museum, London; IGM, Geological Museum of INGEOMINAS, Bogota, Colombia; MACN, Museo National de Ciencias Naturales "Bernardino Rivadavia," Buenos Aires, Argentina – SC, MACN specimens from Santa Cruz Province, Argentina, CH, MACN specimens from Chubut Province, Argentina; MLP, Museo de La Plata, La Plata, Argentina; MNHN, Museo Nacional de Historia Natural, La Paz, Bolivia; MPEF, Museo Paleontológico "Egidio Feruglio" de Trelew, Chubut Province, Argentina; MSP, Museu de Zoologia da Universidade de São Paulo, Brazil; and USNM, Smithsonian Institution, National Museum of Natural History, vertebrate zoology collections.

Systematic paleontology

Order **PRIMATES** Linnaeus 1758
Semiorder **HAPLORHINI** Pocock 1918

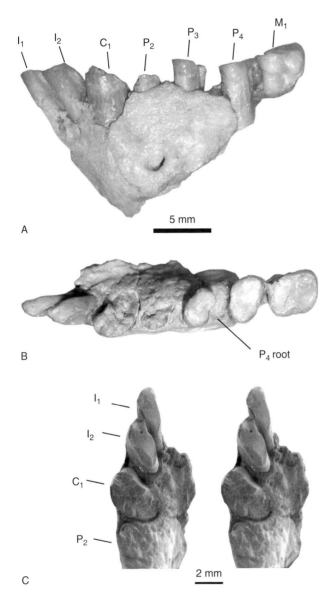

Fig. 15.2. Type specimen of *Mazzonicebus almendrae*, MPEF-PV 6752. (A) Lateral view of mandible; (B) occlusal view of mandible; note that the root socket of P$_4$ was filled with matrix postmortem and P$_4$ is shifted distally; (C) Occlusal view of I$_1$–P$_2$ (stereopair).

Suborder **ANTHROPOIDEA** Mivart 1864
Infraorder **PLATYRRHINI** Geoffroy 1812
Family **HOMUNCULIDAE** Ameghino 1894
Subfamily **SORIACEBINAE** subfamily nov.
Type genus *Soriacebus* Fleagle *et al.* 1987
Included genera *Soriacebus* Fleagle *et al.* 1987; *Mazzonicebus*, gen. nov.
Distribution Early Miocene (Colhue-Huapi Member, Sarmiento Formation, Gran Barranca, southern Chubut Province (Re *et al.* Chapter 3, this book) and

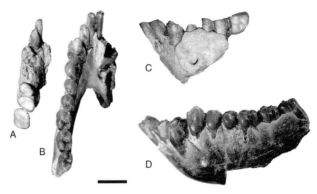

Fig. 15.3. Occlusal and lateral views of mandibles of *Mazzonicebus almendrae* (MPEF-PV 6752) (A, C) and *Soriacebus ameghinorum* (MACN SC 2) (B, D). MACN SC 2 is photographically reversed for conparison. Scale bar 5 mm.

Lower and Middle Sequence, Pinturas Formation, northern Santa Cruz Province, Argentina (Kramarz *et al.* 2005).
Diagnosis Homunculids with: lower incisors elongate, procumbent and mesiodistally compressed; lower molars with paraconids, low occlusal relief, reduced entoconids, and shallow hypoflexids.

Mazzonicebus gen. nov

cf. *Homunculus patagonius* Ameghino: Hershkovitz 1981, p. 182, figs. 3, 5.
Soriacebus cf. *ameghinorum*: Fleagle 1990, p. 79, fig. 15.
Type species *Mazzonicebus almendrae* sp. nov.
Included species The type species only.
Distribution As for the type and only species.
Diagnosis As for the type and only species.
Etymology named in honor of Professor Mario Mazzoni, our late esteemed colleague.

Mazzonicebus almendrae sp. nov
Figs. 15.2–15.10.

Type specimen MPEF-PV 6752, a mandible with the symphysis and left I$_1$–M$_1$ (Figs. 15.2, 15.3; dental dimensions in Tables 15.1 and 15.2).
Hypodigm See list of specimens in Appendix 15.1.
Horizon and locality Colhue-Huapi West locality, Gran Barranca, Chubut Province, Argentina, Age: Lower Fossil Zone, Colhue-Huapi Member, Sarmiento Formation; type specimen collected by Damien Glaz. All specimens in the hypodigm come from closely spaced stratigraphic levels in the Lower Fossil Zone (LFZ). The LFZ is an interval of normal polarity corresponding to Chron C6An.1n of the Global Magnetic Polarity Time Scale, an interval that is between 20.0 and 20.2 Ma (Ré *et al.* Chapter 4, this book).

Table 15.1. *Measurements of the lower teeth of* Mazzonicebus almendrae

Specimen number	I_1		I_2			C_1			P_2		P_3		P_4		M_1			M_2		
	m–d root	b–l	m–d	m–d root	b–l	m–d	b–l	height	m–d	b–l	m–d	b–l	m–d	b–l	m–d	trigonid b–l	talonid b–l	m–d	trigonid b–l	talonid b–l
MPEF-PV 5466	.	.	.	.	.	.	.	.	3.18	2.43	.	.	.	.	.	.	.	.	.	.
MPEF-PV 6752	1.36	2.91	.	.	.	4.33	2.78	.	.	.	.	.	3.23	3.88	4.20	3.36	3.88	.	.	.
MPEF-PV 5345	.	.	.	.	.	.	.	.	4.13	2.91	.	.	.	.	.	.	.	.	.	.
MPEF-PV 5346	.	.	2.13	1.61	2.58	.	.	.	.	.	.	.	.	.	.	.	.	.	.	.
MPEF-PV 5348	.	.	.	.	.	4.65	3.62	7.81	.	.	.	.	.	.	.	.	.	3.75	3.16	3.62
MPEF-PV 5349	.	.	.	.	.	4.46	3.23	.	.	.	.	.	.	.	.	.	.	.	.	.
MPEF-PV 5341	.	.	.	.	.	.	.	.	.	.	.	.	.	.	.	.	.	.	.	.
MPEF-PV 6518	.	.	.	.	.	.	.	.	3.75	2.78	.	.	.	.	.	.	.	.	.	.
MLP 69-III-12-1	.	.	.	.	.	.	.	.	.	.	.	.	2.78	3.04	.	.	.	.	.	.
MLP 82-V-8-2	.	.	1.88	1.42	2.78	.	.	.	.	.	.	.	.	.	.	.	.	.	.	.
MLP 82-V-8-3	.	.	.	.	.	.	.	.	4.00	2.91	.	.	.	.	.	.	.	.	.	.
MPEF-PV 6484	.	.	.	.	.	.	.	.	.	.	.	.	.	.	4.21	3.53	3.78	.	.	.

Notes: All measurements are in millimeters.

m–d, mesiodistal; b–l, buccolingual.

MPEF-PV 5466, a right dP$_4$: m–d, 4.04; trigonid b–l, 2.89; talonid b–l, 2.92.

Table 15.2. *Measurements of the upper teeth of* Mazzonicebus almendrae

Specimen number	I¹ m–d	I¹ m–d root	I¹ b–l	I² m–d	I² m–d root	C¹ m–d	C¹ b–l	C¹ height	P² m–d	P² b–l	P³ m–d	P³ b–l	P⁴ m–d	P⁴ b–l	M¹ m–d	M¹ b–l	M² m–d	M² b–l
MPEF-PV 5468	.	.	.	.	.	.	.	.	.	.	.	.	.	.	.	.	3.52	5.11
MPEF-PV 5462	.	.	.	.	.	.	3.41	8.79	.	.	.	.	.	.	.	.	.	.
MPEF-PV 7021	.	.	.	.	.	3.91	.	.	3.29	.	.	.	.	.	.	.	.	.
MPEF-PV 5372	.	.	.	.	.	.	.	.	.	4.53	.	.	.	.	.	.	.	.
MPEF-PV 5342	.	.	.	.	.	.	.	.	.	.	.	.	.	.	.	.	3.86	4.68
MPEF-PV 5343	.	.	.	.	.	.	.	.	.	.	2.65	4.20	.	.	.	.	.	.
MPEF-PV 5344	.	.	.	.	.	.	.	.	.	.	2.52	4.26	2.65	4.33	.	.	.	.
MPEF-PV 5347	.	.	.	.	.	4.84	3.62	.	3.10	3.62	.	.	.	.	4.18	5.60	4.02	4.89
MPEF-PV 5350	.	.	.	.	.	3.88	3.49	8.40	.	.	.	.	.	.	.	.	.	.
MPEF-PV 5352	.	.	.	.	.	4.26	4.20	7.88	.	.	.	.	.	.	.	.	.	.
MPEF-PV 5391	.	.	.	.	.	.	.	.	.	.	.	.	.	.	.	.	4.11	4.96
MPEF-PV 5699	.	.	.	.	.	.	.	.	.	.	2.58	4.59	2.58	4.59	.	.	.	.
MPEF-PV 6282	.	.	.	.	.	3.86	3.53	.	.	.	.	.	2.79	5.54	.	.	.	.
MPEF-PV 7112	.	.	.	.	.	.	.	.	.	.	.	.	.	.	.	.	.	.
MPEF-PV 7195	.	.	.	.	.	.	.	.	.	.	.	.	.	.	4.03	5.45	.	.
MPEF-PV 7063	.	.	.	.	.	.	.	.	.	.	.	.	.	.	3.87	4.88	.	.
MLP 82-V-8-1	3.36	2.39	2.33	.	.	.	.	.	.	.	.	.	.	.	.	.	.	.
MLP 77-VI-13-6	.	.	.	.	.	5.17	4.46	8.28	.	.	.	.	.	.	.	.	.	.

Notes: All measurements are in millimeters.
m–d, mesiodistal; b–l, buccolingual.

Diagnosis *Mazzonicebus* resembles *Soriacebus* and is derived relative to other Oligocene and early Miocene platyrrhines (*Branisella*, *Dolichocebus*, *Carlocebus*, *Homunculus*) in having procumbent, styliform and high-crowned (elongate) lower incisors with crowns and roots extremely compressed mesiodistally. The P_4 metaconid is close to the protoconid and set distolingually to it. M^{1-2} have strong preprotocristas but lack a paraconule (also absent in *Soriacebus*, but present in *Dolichocebus*, *Homunculus*, and *Carlocebus*). An upper molar postprotocrista spur is present as in *Soriacebus*, but absent in the above-mentioned taxa, and also absent in *Tremacebus* and *Chilecebus*. As in *Soriacebus*, together with the postprotocrista spur, the prehypocrista walls off the talon lingually. In other Oligocene – early Miocene taxa, a prehypocrista is present but a sulcus separates hypocone from protocone.

Mazzonicebus differs from *Soriacebus* in a number of ways, some primitive and some synapomorphous (state of the trait in *Soriacebus* given in parentheses): The incisors canine and P_2 of are much smaller relative to the first molar (vs. very large). The projective height of P_2 is similar to that of P_3 and P_4 (vs. P_2 projecting above the P_3 and P_4). The P_4 trigonid is mesiodistally short, lingually open, and has a paraconid (vs. a more elongate, lingually closed trigonid lacking a paraconid). The lower molars are "squared" (vs. more elongate), the trigonid is short (vs. elongate), the trigonid breadth is much less than that of the talonid (vs. similar breadths), and the first molar has a paraconid (vs. absent). The entoconid is moderate in size and there is no hypoconulid (vs. reduced entoconid and hypoconulid present). Also, the molars are strikingly high crowned (vs. low crowned). *Mazzonicebus* has single rooted P^{2-4} (vs. a root count of 2–3–3).

Etymology Named for the now abandoned homeplace, Puesto Almendra, south of the type locality (Fig. 15.1).

Description and comparisons

Where possible, comparative information is provided for Oligocene and early Miocene platyrrhines *Branisella, Dolichocebus, Chilecebus, Soriacebus, Carlocebus,* and *Homunculus* based on publications and personal study of specimens (Hershkovitz 1981, 1984; Fleagle *et al.* 1987, 2002, 2006; Fleagle 1990; Tauber 1991; Flynn *et al.* 1995; Takai *et al.* 1996, 2000; Kay *et al.* 2001, 2005, 2006a, 2008; Tejedor 2005; Tejedor *et al.* 2006).

Mandibular morphology
MLP 69-III-12–1, MPEF-PV 5351, and MPEF-PV 6752 (Figs. 15.2, 15.3, 15.5).

MLP 69-III-12–1 is a left mandible with a complete symphysis showing an elongate and procumbent *planum*

Fig. 15.4. Lower canine and P_2. (A) MPEF-PV 5349 lower canine, lingual view; (B) MPEF-PV 5345, P_2, stereopair.

alveolare with a rounded superior transverse torus and a weak inferior transverse torus separated by a shallow genial fossa. The inferior border of the symphysis reaches to the posterior edge of the root socket for P_3. All the above descriptive features are resemblances to *Soriacebus, Carlocebus,* and *Homunculus*. The mandible of *Dolichocebus* and *Chilecebus* are unknown. That of *Tremacebus* is likewise unknown unless the specimen MACN CH-354 from Sacanana, Chubut Province, belongs to *Tremacebus* (see below).

MLP 69-III-12–1 and MPEF-PV 5351 preserve sufficient parts of the incisor sockets to show that the breadth between the canines was very narrow. No diastema separates the lateral incisor root from the canine root. The absence of a diastema between the I_2 and lower canine is very similar to the condition seen in *Soriacebus* and *Homunculus*.

In MLP 69-III-12–1, the cheek tooth rows diverge substantially showing that the lower dental arcade was V-shaped, as in *Branisella, Soriacebus,* and *Homunculus* but unlike middle Miocene to Recent pitheciids. The inferior border of the mandible of MPEF-PV 6752 is incomplete, so the total depth of the mandible cannot be observed. Nor can we judge whether the jaw depth increased distally, as in *Soriacebus*, middle Miocene to Recent pitheciines, and some atelids, or was of approximately uniform depth, as in *Homunculus* and *Aotus*.

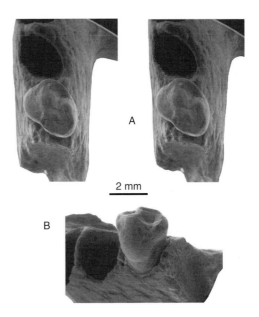

Fig. 15.5. MPEF-PV 5351 fragment of mandibular symphysis with right canine (mesial and apical enamel broken away). (A) Occlusal view; (B) view from anterior perspective showing the mesiodistally compressed incisor sockets.

Fig. 15.6. MLP 69-III-12–1. (A) Occlusal stereopair showing P_4 and the root socket for P_3; (B) occlusobuccal view.

Lower incisors

MPEF-PV 6752, left I_{1-2}; MPEF-PV 5346, right I_2; MLP 82-V-8–2, left I_2. MPEF-PV 5351 and MLP 69-III-12–1 preserve parts of the incisor root sockets (Figs. 15.2, 15.5).

The lower incisors are procumbent, styliform, and high crowned (elongate). In MPEF-PV 6752 I_1 is smaller than I_2. MPEF-PV 5346, an isolated unworn I_2, has well-developed lingual enamel. There is a weak distolingual marginal cingulum but a lingual heel is absent. The crown and roots of the incisors are extremely compressed mesiodistally: the ratio of mesiodistal to buccolingual lateral incisor root diameters ranges from 0.51 to 0.62 in three specimens. This is also the case in *Soriacebus* (0.49; $n = 1$). Incisors of *Dolichocebus* are not as high crowned nor mesiodistally compressed. Compared with *Soriacebus*, the two incisors are much smaller relative to the first molar (Fig. 15.3) (see Table 1 in Fleagle *et al.* 1987). Mesiodistal incisal compression and a reduced lingual heel are characteristic of middle Miocene to Recent pitheciines.

Lower canine

MPEF-PV 6752, C_1 root and buccal base of crown; MPEF-PV 5349, complete isolated left C_1; MPEF-PV 5351, complete C_1 embedded in a symphyseal fragment. MPEF-PV 7021, left C_1 (Figs. 15.2, 15.4, 15.5).

C_1 has a rounded oval cross-section. MPEF-PV 5349 and MPEF-PV 5351 have a strong lingual cingulum but MPEF-PV 7021 does not. MPEF-PV 5351 also has a weak but complete buccal cingulum, a resemblance to *Soriacebus*

ameghinorum, but MPEF-PV 5349 and MPEF-PV 7021 do not. All specimens have a strong mesial groove that bisects the cingulum and reaches the base of the crown. Also all have a rounded concave entocristid ending in a weak distostylid. *Mazzonicebus* has a much smaller canine proportionally than *Soriacebus*. The ratio of canine cross-sectional area to molar cross-sectional area in *Mazzonicebus* is 0.74, whereas a similar ratio for *Soriacebus ameghinorum* is 1.69. This likely is a species-specific difference, not one to be accounted for by sexual dimorphism because it continues the same pattern seen in the incisor size where *Mazzonicebus* has proportionately smaller incisors than *Soriacebus*.

Lower premolars

MPEF-PV 6752, left P_2 (root only), P_3 and P_4 crowns, very heavily worn; MPEF-PV 5345, right P_2; MPEF-PV 6518, left P_2; MLP 82-V-8–3, right P_2; MPEF-PV 5466, left P_2; MLP 69-III-12–1, left P_4 (Figs. 15.4, 15.6).

All P_2 specimens are similar in appearance to one another and very similar in shape to the P_2 of *Soriacebus*. Each tooth has a single root. The trigonids dominate these teeth. Crests running mesially and distally from the protoconid are only deflected very slightly lingually. The buccal side of the crown bulges over the root. The talonids are very small without a discrete hypoconid. Trigonids are single cusped and lack any evidence of a metaconid. P_2 of *Mazzonicebus* differs is several notable ways compared with that of *Soriacebus*. In *Soriacebus*, the pattern of disproportionate size of the anterior dentition is continued in the P_2. The ratio of the cross-sectional area of P_2 to the area of M_1 is 1.26 whereas in *Mazzonicebus* it is 0.69. Moreover, P_2 of *Soriacebus* is massive and projects above the P_3 and P_4 whereas in *Mazzonicebus* the projective height of P_2 is similar to that of P_3 and P_4. As in all known platyrrhines, the talonid is small and lacks a hypoconid and entoconid.

Resemblance to *Dolichocebus* is noted in the single root, the lack of a buccal cingulum, and oval occlusal outline with a buccal flare at the base of the protoconid (less in *Dolichocebus* than in *Mazzonicebus* or *Soriacebus*).

P_3 is preserved only in an extremely worn state in MPEF-PV 6752. P_4 is represented in its unworn state in MLP 69-III-12–1. The trigonid is large with a weakly developed mesiolingually placed paraconid at the terminus of a mesially oriented paracristid. Metaconid is slightly smaller than, and positioned distolingually from, the protoconid. The trigonid basin is well developed and is open lingually. The talonid has a small hypoconid, an indistinct entoconid and is enclosed lingually (i.e. there is a continuous lingual wall mesial to the entoconid). The crown has sloping sides especially the mesiobuccal margin is greatly expanded. Thus, the occlusal surfaces are narrow relative to the size of the tooth. There is no buccal cingulum. This tooth differs from P_4s of *Soriacebus* in having a paraconid, a lingually open trigonid, and in the trigonid being mesiodistally short (*Soriacebus* lacks a paraconid, and has a lingually closed trigonid and a mesiodistally more elongate trigonid). Also, P_4s of *Soriacebus*, are much narrower buccolingually, with the metaconid closer to the protoconid.

Many resemblances are also noted between the P_4 of *Mazzonicebus* and that of *Dolichocebus*. Like *Dolichocebus*, the tooth is single-rooted. The trigonid is widely open lingually. Unlike *Mazzonicebus*, the P_4 crown of *Dolichocebus* is mesiodistally short and buccolingually broad, protoconid and metaconid are more widely spaced, and there is no paraconid.

Lower molars

MPEF-PV 6484, right M_1; MPEF-PV 6752 mandible including left M_1, MPEF-PV 5348, left M_2 (Fig. 15.7).

MPEF-PV 5348 and MPEF-PV 5350 are complete and almost unworn specimens; MPEF-PV 6752 is heavily worn and provides no information about the crown morphology. The M_1 crown is "squared," having a ratio of mesiodistal length to buccolingual breadth of 1.11. This ratio is similar to *Dolichocebus* (1.13) and much lower than *Soriacebus* species (1.27 for *S. ameghinorum*, and 1.22 for *S. adriane*). The trigonid is mesiodistally elongate. The ratio of mesiodistal trigonid length to total mesiodistal length is 38% and 40%, more elongate than *Dolichocebus* (36%) but shorter than in *Soriacebus ameghinorum* (42%). The trigonid breadth is much less than that of the talonid; breadths are more nearly equal in *Soriacebus* and *Dolichocebus*. The

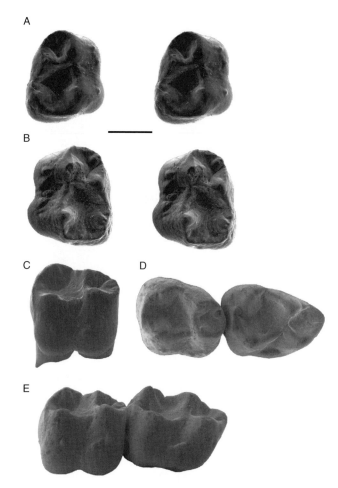

Fig. 15.7. Lower molars. (A) MPEF-PV 5348, right M_1, occlusal stereopair; (B) MPEF-PV 6484, left M_2 occlusal stereopair; (C) MPEF-PV 5348, right M_1, occlusobuccal view; (D) MPEF-PV 5348 and MPEF-PV 5469 M_1 and dP_4 occlusal view; (E) MPEF-PV 5348 and MPEF-PV 5469, M_1 and dP_4, occlusobuccal view. Scale bar 2 mm.

occlusal surface, especially in the area of the trigonid, is quite compressed and the sides of the marginal cusps bulge, a feature *Mazzonicebus* shares with *Dolichocebus* and *Soriacebus*. Low cusp relief and smooth enamel characterize the crowns of all species examined.

The M_{1-2} trigonids are three-cusped, with a small cuspate paraconid on the mesiolingual edge of the trigonid. Cuspate lower molar paraconids are unknown in living platyrrhines and unusual among Miocene platyrrhines. One does not occur in *Soriacebus* but is present in *Dolichocebus*. The metaconid is positioned distolingual to the protoconid as in *Soriacebus* and *Dolichocebus*. The paracristid is short and mesiodistally oriented. On M_1 the premetacristid is separated from the paraconid by a distinct notch so that the trigonid is "open" lingualy. On M_2 the premetacristid encloses the trigonid lingually. Lingual and buccal protocristids form a complete distal trigonid wall. Wear surface "X" is absent.

Fig. 15.8. MPEF-PV 6282, left maxillary fragment with P$^{3-4}$ showing the position of the infraorbital foramen and the ventral edge of the orbit. The zygomatic root, present on the specimen, is not shown.

The trigonid and talonid are similar in height. The molars are strikingly high crowned compared to any other early Miocene taxon. The entoconid is moderate in size and positioned distolingual to the hypoconid. *Dolichocebus* has a much larger entoconid but *Soriacebus* has a more reduced one. The distal edge of the talonid supports a raised ridge confluent with the hypocristid, but lacks a hypoconulid. *Dolichocebus* and *Soriacebus* have small lingually positioned hypoconulids.

The M$_{1-2}$ cristid obliquas reach the trigonid wall distolingual to the protoconid and extends part way up the wall but do not reach to the protocristid whereas the cristid obliqua does not reach to the protoconid on M$_1$, but it does on M$_2$s of *Soriacebus* and *Dolichocebus*. The lingual margin of the talonid basin is notched where the postmetacristid abuts the pre-entocristid (as in *Dolichocebus*, which has a distinct notch, but not in *Soriacebus*). A weak and incomplete buccal cingulum wraps around the protoconid and spans the hypoflexid, but disappears lingual to the hypoconid. Similar cingulum development is found *Dolichocebus* but not in other early Miocene platyrrhines.

Maxilla

MPEF-PV 6282 is a maxillary fragment with P^3–P^4, preserving a substantial part of the inferior orbital region and part of maxillary root of the zygomatic arch (Fig. 15.8). The infraorbital foramen is situated dorsally between P^3 and P^4. A small part of the ventral margin of the orbit also is preserved. The infraorbital region is very deep dorsoventrally. The distance between the alveolar margin between P^3 and P^4 and the ventral orbital rim is 12.83 mm. A ratio

between this dimension and P^3–P^4 mesiodistal length is 2.55. This is a deeper face than in *Cebus* or *Aotus*: the comparable ratio in a specimen of *Cebus* is 2.39 and in *Aotus* 2.01. The orbit must have been small: species with large orbits (*Tremacebus* and *Aotus*) have much a reduced distance between the alveolar margin and the ventral orbital margin and this clearly was not the case in *Mazzonicebus*.

Upper incisors

MLP 82-V-8–1, an isolated right I^1; MPEF-PV 6495, isolated left I^2 (Fig. 15.9).

The upper central incisor is broadly spatulate. The lingual surface is delineated by a weak and discontinuous cingulum. No central lingual pillar is present. The upper central incisor is buccolingually compressed (md:bl ratio is 1.44), as is that of *Dolichocebus* (1.36). The tooth is very small compared with upper first molar area—the ratio is 0.36 comparing isolated specimens; the ratio is even smaller in *Dolichocebus* (0.30). I^1 is slightly larger than I^2.

Upper canine

MPEF-PV 5350 (right), MPEF-PV 5352 (right), AR 95–338 (right), MPEF-PV 7112 (left), MLP 77-VI-13-6 (left) (Fig. 15.9).

All specimens have a rounded to oval cross-section. As in *Soriacebus*, there is a deep mesial groove and a weak lingual cingulum. Also as in *Soriacebus* there is no crest leading lingually from the cusp apex. Compared with *Soriacebus*, the canine of *Mazzonicebus* is more rounded in cross-section.

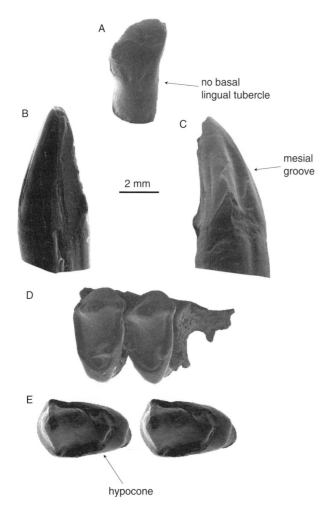

Fig. 15.9. Upper incisors, canines, and premolars. (A) MLP 82-V-8–1, lingual view of I¹; (B) MLP 77-VI-13–6, upper canine, lingual view; (C) MPEF-PV 5391, upper canine, mesial view; (D) MPEF-PV 5344, right P³⁻⁴, occlusal view; (E) MPEF-PV 5343, P⁴, stereopair.

With a sample size of five specimens, considerable variation is noted in the size of the canine. The coefficient of variation for maximum occlusal diameter is 13% and the ratio of the largest to the smallest specimen for that measurement is 1.28.

Upper premolars

P²: MPEF-PV 5372 (right); MPEF-PV 5347 (right, also preserves M¹⁻²); P³⁻⁴: MPEF-PV 5344; MPEF-PV 6282; P⁴ alone: MPEF-PV 5343 (Figs. 15.8, 15.9).

P² has a single root and the roots of P³⁻⁴ are likewise singular but form a figure-eight in cross-section, suggesting that the primitive condition was two roots, a supposition that is supported by the 1–2–2 root counts of *Branisella*, *Dolichocebus*, and *Tremacebus*. In *Soriacebus* the root count is 2–3–3. The area of P² is more than 15% smaller than P³ in *Branisella* and *Dolichocebus* whereas it is only slightly smaller in *Mazzonicebus* and *Soriacebus*. P² has a distinct protocone lingually, as in *Soriacebus*; *Dolichocebus* has only a lingual swelling. As in *Soriacebus* and *Dolichocebus*, P³⁻⁴ have well-developed cingula that wrap lingually around the protocone. The cingula support hypocones. The postprotocrista is absent (P³) or very short (P⁴), as in *Dolichocebus* and *Soriacebus*.

MPEF-PV 5344 and MPEF-PV 6282 are similar in morphology overall but the former is buccolingually less transverse than the latter. Also MPEF-PV 6282 has better developed hypocones, especially on P⁴. This offers the tantalizing prospect that more than one species of monkey is represented at Gran Barranca. For the moment, given the absence of two morphs at the other tooth positions, I include MPEF-PV 6282 within the hypodigm of *Mazzonicebus*.

Upper molars

M¹: MPEF-PV 5342 (left); MPEF-PV 5699 (right); MPEF-PV 7063 (right); MPEF-PV 7195 (right); M¹⁻²: MPEF-PV 5347 (right) (Fig. 15.10). No example of M³ is preserved.

Although we have no specimens identified as M³, this tooth certainly was present in life because there is an interstitial wear facet on the posterior aspect of M² in MPEF-PV 5347. The molars have three roots but the two buccal roots of all specimens are closely approximated (partially fused in MPEF-PV 5699). The roots are more widely separate in *Soriacebus* and *Dolichocebus*. The area of the first molar is ~20% greater than that of the second molar in both *Mazzonicebus* and *Soriacebus*. The first molar is much broader than the second in MPEF-PV 5347, the single specimen for which we have the two teeth together. *Soriacebus* has more nearly identical M¹⁻² shape.

M¹⁻² have strong preprotocristas but lack paraconules (also absent in *Soriacebus*, but present in *Dolichocebus*, *Homunculus*, and *Carlocebus*). As in the above-mentioned taxa, metaconule and pericone are absent. The M¹ hypocone is large, is connected to a very strong lingual cingulum, and is distal and slightly lingual to the protocone on both teeth. This pattern is also found in *Soriacebus*, *Dolichocebus*, and *Tremacebus* except as note below. On M², the hypocone is smaller. *Soriacebus* has smaller hypocones on both teeth but demonstrates a similar pattern with the M¹ hypocone larger than that of M².

The postprotocrista runs distally and connects with a spur that is directed towards the hypocone; postprotocrista then continues lingually to the base of the metacone where it joins the lateral posterior transverse crista (= hypometacrista). A postprotocrista spur is also present and more strongly developed in *Soriacebus*, but is absent in *Dolichocebus*. As in *Soriacebus*, together with the postprotocrista spur, the prehypocrista walls off the talon lingually. In *Dolichocebus* a prehypocrista is present but a sulcus separates hypocone from protocone.

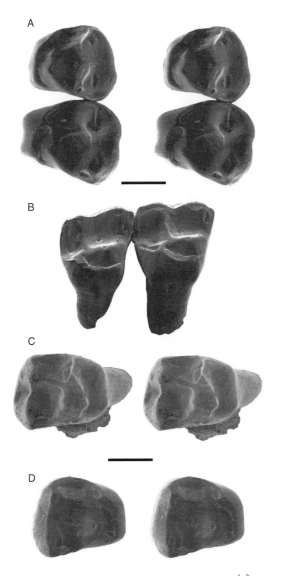

Fig. 15.10. Upper molars. (A) MPEF-PV 5347, right M^{1-2} occlusal stereopair; (B) MPEF-PV 5347 occlusolingual view; (C) MPEF-PV 7063, left M^1, occlusal stereopair; (D) MPEF-PV 5699, left M^2, occlusal stereopair. Scale bars 2 mm.

Mazzonicebus, *Soriacebus*, and *Dolichocebus* have weak and discontinuous buccal cingula. M^{1-2} mesostyles are absent (as in *Soriacebus*; present in *Dolichocebus*). In *Mazzonicebus*, M^2 is ~20% smaller than M^1, virtually the same size difference as is present in *Soriacebus*.

Deciduous molars

dP_4: MPEF-PV 5469 (left) (Fig. 15.7D, E).

An isolated right dP_4 has two roots. The tooth is low crowned (i.e. the distance from the fork of the two roots and the deepest part of the talonid basin is short) but exhibits high cuspal relief. The trigonid is slightly narrower than the

talonid. The trigonid is triangular in occlusal view. It has a strong centrally placed cuspate paraconid separated by a sulcus from the metaconid leaving the trigonid lingually open. The metaconid is placed far distal to the protoconid. The hypoconid supports a strong hypocristid leading to a medially positioned small to moderate-sized, cuspate hypoconulid. Leading mesiolingually from the hypoconid, the cristid obliqua is strong and reaches up the distal trigonid wall to the protocristid midway between the metaconid and protoconid. The entoconid is large and positioned lingually opposite the hypoconid across a broad talonid basin. A distinct postentoconid sulcus is present separating the entoconid from the medially positioned hypoconulid. The *Mazzonicebus* dP_4 resembles that of *Homunculus* in having a well-developed paraconid and lingually open trigonid. It is readily distinguished from that of *Homunculus* because the latter has an indistinct hypoconulid and lacks a post entoconid sulcus. The tooth differs strikingly from all extant platyrrhines, especially in the in the construction of the trigonid: all extant platyrrhines have an indistinct or absent paraconid and the trigonid basin is always closed lingually by a bordering crest running from the metaconid to the mesial margin of the tooth. It is further distinct from all extant pitheciids in the orientation of the cristid obliqua, which in pitheciids runs mesially to the protoconid (mesiolingual in *Mazzonicebus* and *Homunculus*) leaving a shallow (vs. deep) ectoflexid.

Status of Sacanana mandible

MACN CH 354, figured by Fleagle and Bown (1983), is a mandibular fragment with a broken P_4 and complete but worn M_1 from Sacanana, the locality of *Tremacebus*. Allocation has been suggested to either *Soriacebus* or (by default) to *Tremacebus* because the type skull of the latter comes from Sacanana as well. MACN CH 354 should not be allocated to *Mazzonicebus*, *Dolichocebus*, or *Soriacebus* because it lacks several derived features that characterize each. The lower molar of MACN CH 354 lacks the derived features of *Mazzonicebus*: it has low crowns (vs. high crowns) and a small lingually placed hypoconulid (vs. loss of hypoconulid). Likewise, MACN CH 354 lacks the derived features of *Soriacebus* including the greatly reduced entoconid (vs. well-developed entoconid) and the distally placed metaconid (vs. transversely placed). Distinctness from *Dolichocebus* is less obvious although it appears that MACN CH 354 has a derivedly mesiodistally elongate trigonid lacking a paraconid (vs. a short trigonid with a small paraconid). We do not have an associated mandibular specimen of *Tremacebus*. However, this specimen is most plausibly assigned to *Tremacebus* based on the rarity of occurrence of primates and their overall low species richness in Patagonian localities where they occur.

Soriacebines: stem platyrrhines or Pitheciinae?

Soriacebine evolution has been placed in the context of two dramatically different evolutionary scenarios – what have been called the "deep-time" versus the "layered-evolution" hypotheses of platyrrhine evolution (Kay *et al.* 2008). Rosenberger and colleagues (Rosenberger 2002; Tejedor *et al.* 2006) read the fossil record of platyrrhine evolution through the lens of deep time and assert that broad adaptive niches of platyrrhines evolved just once and had been set by the early Miocene or earlier. From this theoretical perspective, soriacebines could only be part of a basal radiation of Pitheciinae because of their similar adaptations. In contrast, Kay and colleagues view the same morphological evidence as indicative of a "layered-evolution" hypothesis.

The resemblances between soriacebines and pitheciines, especially in the anterior dentition, are undeniable and profound: both groups have adopted a distinctive and unusual adaptive syndrome for prying open fruits with tough protective rinds (sclerocarps) to access the nutritious seeds within. They are seed predators. The difference in interpretation is over whether these sclerocarp-peeling seed predators are close relatives of one another or whether the adaptation evolved independently in unrelated groups of platyrrhines. Two lines of evidence, one molecular and the other morphological, suggest that the two groups were separately evolved.

With respect to the molecular evidence, it is established in this chapter that soriacebines extended back in the fossil record at least to the early Miocene, about 20 Ma, far earlier than the supposed branch times between callicebines and pitheciines. That latter event has been suggested to occur between 13.5 Ma and 16.7 Ma (Barroso *et al.* 1997). The molecular clock in this case would have to be off by between 20% and 50%.

The second line of evidence is more direct: the phylogentetic interpretation of morphology gained through a balanced consideration of all characters collectively. If soriacebines are close relatives of pitheciines then they ought to share: (1) the derived characters diagnostic of the last common ancestor (LCA) of crown platyrrhines; (2) the derived characters of the LCA of extant Pitheciidae (*Callicebus*, *Pithecia*, *Chiropotes*, and *Cacajao*); and (3) the derived characters special to the pitheciine clade (just *Pithecia*, *Chiropotes*, and *Cacajao*). Contrariwise, if the resemblances between soriacebines and pitheciines are restricted mainly to pitheciines alone but include few if any resemblances at deeper nodes of the crown platyrrhine clade, then soriacebine–pitheciine similarities are better viewed as homoplasies.

Many examples have occurred in the literature of anthropoid evolution where paleontologists have inferred phylogenetic links, which, with increased fossil evidence, have proved incorrect on account of homoplasy being interpreted

as symplesiomorphy. The early Oligocene parapithecids were incorrectly interpreted to be cercopithecoid monkeys on account of convergence on a bilophodont molar pattern. *Pliopithecus* from the early Miocene of Europe was interpreted as a gibbon (Hylobatidae) on account of its convergent adaptations for suspensory locomotion. Species of what was then called *Proconsul* from the early Miocene of Africa were interpreted as lineal ancestors of *Gorilla* and *Pan* owing to adaptations in the cheek teeth and face for folivory and frugivory. *Sivapithecus* from the middle Miocene of Asia was interpreted as an early hominid because of its thickly enameled teeth and massive jaws. In each case, as a fuller appreciation was gained of the overall morphology, the resemblances upon which the phylogenetic interpretation hinged were shown to represent homoplasy as a result of adaptive convergence rather than symplesiomorphy. Parapithecids are stem anthropoids. *Pliopithecus* is a stem catarrhine. *Proconsul* is a stem hominoid. *Sivapithecus* is related to *Pongo*, the orangutan. In each case the shift in views came about because the shared resemblances were mainly restricted to a few characters related to an adaptive specialization – bilophodonty, suspensory locomotion, folivory versus frugivory and hard-object feeding – embedded within an overall more primitive anatomical organization. It is the same with soriacebines and pitheciines.

In a review of the phylogeny of Miocene Patagonian platyrrhines, Kay *et al.* (2008) provide a character–taxon matrix from which can be reconstructed, using the parsimony criterion, the hypothetical LCAs of crown platyrrhines, pitheciids, and pitheciins. Using this data as a guide, and adding some new observations about the anatomy of the deciduous P_4 (see above) it is possible to evaluate proposed soriacebine–pitheciin clade. For soriacebines to be confirmed as sister to pitheciins, the following character distributions should be expected (the numbering below corresponds to that presented above):

(1) *Soriacebus* and *Mazzonicebus* should possess not only the synapomorphous features of pitheciins but also those of the LCA of crown platyrrhines. However, soriacebines do not possess many of these crown platyrrhine synapomorphies. Kay *et al.* (2008) reconstruct the crown platyrrhine LCA as having parallel-sided tooth rows. The infraorbital foramen was above P^2. P_3 had a large metaconid widely spaced from the protoconid. The P_4 metaconid was large and transverse to the protoconid. The M_1 paraconid was lost and the premetactristid enclosed the trigonid lingually. The M_3 was single-rooted and did not have a third lobe or heel. The dP_4 trigonid was enclosed lingually and lacked a paraconid. In contrast, *Soriacebus* and *Mazzonicebus* have the stem platyrrhine condition of these features – one shared by most or all early Miocene platyrrhines. The lower tooth rows diverge (are V-shaped). The infraorbital foramen is further back,

Table 15.3. *Female body mass estimates from lower first molars*

Species	Age	Sample size	Mean female body mass (g)
Callicebus moloch	Modern	19	956
Callicebus personatus	Modern	7	1380
Callicebus torquatus	Modern	21	1210
Pithecia pithecia	Modern	4	1580
Pithecia monachus	Modern	10	2110
Cacajao melanocephalus	Modern	6	2710
Chiropotes satanas	Modern	19	2580
Nuciruptor rubricae	Middle Miocene	1	2044
Cebupithecia sarmientoi	Middle Miocene	1	1792
Mazzonicebus ameghinorum	Early Miocene	2	1602
Soriacebus ameghinorum	Early Miocene	1	1483
Soriacebus adrianae	Early Miocene	2	852

Note: Data on mean female body mass for extant taxa from Smith and Jungers (1997). Estimated body mass of extinct taxa based on formula in text.

between P^3 and P^4. P_{3-4} metaconids are small and positioned distolingual to the protoconids. We do not have an M_3 of *Mazzonicebus* but in *Soriacebus*, the M_3 has a pronounced heel and the root is divided at its base. Notably, *Mazzonicebus* has a very strongly developed dP_4 paraconid separated from the metaconid by a deep groove. The discovery of *Mazzonicebus* shows that some features of resemblance between *Soriacebus* and crown platyrrhines are convergent: the M_1 paraconid is absent in *Soriacebus*, an apparent synapomorphy with crown platyrrhines but it is present in the more primitive *Mazzonicebus* demonstrating that paraconid loss was achieved in parallel in soriacebines and crown platyrrhines. The trigonid of *Soriacebus* is closed lingually, a similarity to crown platyrrhines that must have evolved independently because the trigonid is open in *Mazzonicebus*.

(2) Soriacebines possess some but not all features the LCA of the pithecid clade (i.e. *Callicebus*, *Pithecia*, *Chiropotes*, and *Cacajao*). The pithecid LCA had posteriorly deepening mandible. It had an M_1 cristid obliqua that reached the protoconid. Its upper central incisor has a central lingual pillar and the upper canine had a shallow mesial groove that is separated from the root by an intervening cingulum. The upper molars of the pithecid LCA had a strong prehypocrista. *Mazzonicebus* and *Soriacebus* agree in the shape of the jaw and the completeness of the M_1 protocristid but lack all the other features – the mesial groove of the upper canine is strong and reaches the root, and the hypocone is reduced and lacks a prehypocrista.

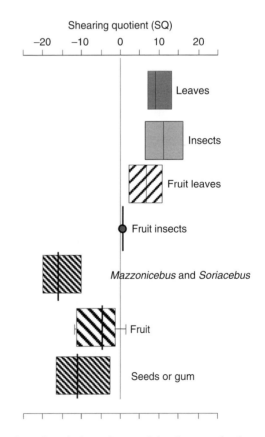

Fig. 15.11. Quantitative estimates of shearing-crest development, expressed as "shearing quotients" (SQ). *Mazzonicebus* resembles *Soriacebus* species in having low SQs, from which we may infer that fruit, gum, or seeds made up a significant component of its diet.

Table 15.4. *Shearing quotients and inferred diets of selected Miocene platyrrhines*

Species	n	M$_1$ length	Sum, M$_1$ shear	Expected shear	Shear quotient	Major dietary feature
Callimico goeldii	3	2.60	5.48	4.72	16.14	Insects
Brachyteles arachnoides	9	7.22	15.19	3.10	15.93	Leaves
Alouatta palliata	10	6.92	13.91	12.56	10.76	Leaves
Alouatta caraya	6	6.72	13.09	12.20	7.33	Leaves
Alouatta fusca	6	6.70	12.94	12.16	6.42	Leaves
Aotus trivirgatus	10	3.06	6.16	5.55	10.92	Fruit/leaves
Saimiri sciureus	5	2.87	5.54	5.21	6.36	Insects/fruit
Lagothrix lagotricha	8	5.47	10.12	9.93	1.94	Fruit/leaves
Leontopithecus rosalia	5	3.09	5.62	5.61	0.22	Fruit/insects
Ateles geoffroyi	10	5.26	9.31	9.55	−2.47	Fruit
Callicebus moloch	10	3.18	5.50	5.77	−4.70	Fruit
Saguinas mystax	5	2.52	4.03	4.57	−11.88	Fruit/insects
Callithrix argentata	4	2.22	4.08	4.03	1.27	Fruit/gum
Cebuella pygmaea	4	1.78	3.26	3.23	0.92	Gum/fruit
Pithecia monachus	4	4.00	6.78	7.26	−6.60	Fruit/seeds
Cebus apella	5	4.79	7.71	8.69	−11.31	Fruit/seeds
Chiropotes satanas	5	3.64	5.50	6.61	−15.53	Seeds/fruit
Cacajao melanocephalus	2	3.97	5.90	7.20	−17.70	Seeds/fruit
Mazzonicebus almendrae	1	4.19	6.49	7.60	−14.61	Seeds/fruit
Soriacebus ameghinorum	1	3.60	5.87	6.53	−10.15	Seeds/fruit
Soriacebus adriane	1	3.23	4.67	5.86	−20.33	Seeds/fruit
Aotus dindensis	1	3.23	6.14	5.86	4.74	Insects/fruit
Mohanamico hershkovitzi	1	3.20	4.96	5.81	−14.59	Seeds/fruit
Cebupithecia sarmientoi	1	3.52	5.15	6.39	−19.38	Seeds/fruit

Notes: The estimate of shearing development is based on measurements of six lower molar crests (see Kay 1975 for further details). A line was assigned to a bivariate cluster of the natural log of M$_1$ length (lnML) versus the natural log of the sum of the measured shearing crests (lnSH). The line was assigned a slope of 1.0 (slope of isometry) and passed through the mean lnML and mean lnSH for extant taxa. The equation expressing this line is: lnSH = 1.0(lnML) + 0.596. For each taxon, the expected lnSH was calculated from this equation. The observed (measured) lnSH for each species was compared with the expected and expressed as a residual (shearing quotient, or SQ): SQ = 100 * (observed − expected)/(expected). Extant and extinct taxa are listed separately according to dietary categories (Fleagle *et al.* 1997). Diet is inferred for extinct taxa by comparion to a modern analog with a similar SQ. All measurements are in millimeters.

(3) Finally, soriacebines must have derived features of the LCA of the pitheciin clade. They do. In support of this they share: elongate, procumbent mesiodistally compressed lower incisors, and low-crowned molars with shallow hypoflexids and molars in which the protocristid does not reach the metaconid.

In summary, the patterning of the apparent symplesiomorphies shared by soriacebines and pitheciines points to homoplasy: soriacebines lack a number of shared-derived features of crown platyrrhines more generally and also many of the symplesiomorphies of Pitheciidae. They possess, for the most part only characters of the most specialized pitheciines. This is the classic pattern expected for parallel evolution.

Adaptations

As noted above, one specimen of *Mazzonicebus*, MPEF-PV 6282, preserves a substantial part of the inferior orbital region including a bit of the ventral orbital rim. Although the orbit cannot be measured, it must have been small: as noted above, taxa with large orbits (*Aotus*) have a much reduced distance between the alveolar margin and the ventral orbital margin and this clearly was not the case in *Mazzonicebus*. The presence of small orbits allows us to infer that *Mazzonicebus* was diurnal.

An estimate of body mass for *Mazzonicebus* may be based on its molar occlusal area. From body masses of females of 15 platyrrhine species found in the literature

Table 15.5. *Crown height of lower first molars*

Species	Specimen number	M$_1$ length (mm)	Buccal talonid height (mm)	M$_1$ length/ buccal talonid ht
Alouatta palliata	USNM 258308	6.97	3.09	0.44
Alouatta palliata	USNM 257308	6.65	2.61	0.39
Brachyteles arachnoides	MSP 11102	7.28	2.77	0.38
Brachyteles arachnoides	MSP 1159	6.97	3.09	0.44
Lagothrix lagotricha	USNM 269839	5.62	2.38	0.42
Lagothrix lagotricha	USNM 538105	5.54	2.85	0.51
Ateles geoffroyi	USNM 292207	5.38	3.17	0.59
Ateles geoffroyi	USNM 292240	5.46	3.17	0.58
Aotus trivirgatus	USNM 335442	3.33	1.50	0.45
Aotus trivirgatus	USNM 335441	3.40	1.66	0.49
Saimiri oerstedii	USNM 363148	2.61	1.19	0.45
Saimiri oerstedii	USNM 291047	2.69	1.27	0.47
Cebus apella	USNM 518266	5.15	2.53	0.49
Cebus apella	USNM 397959	4.67	1.90	0.41
Callicebus moloch	BMNH 20.7.14.10	3.33	1.58	0.48
Callicebus moloch	BMNH 26.5.5.16	3.33	1.50	0.45
Pithecia monachus	USNM 518223	4.12	2.14	0.52
Chiropotes satanas	USNM 388168	3.64	1.58	0.43
Chiropotes satanas	USNM 406594	3.80	1.98	0.52
Cacajao rubicundus	USNM 395027	4.35	2.06	0.47
Cacajao melanocephalus	USNM 406427	3.96	2.06	0.52
Mazzonicebus almendrae	MPEF-PV 6484	4.12	2.69	0.65
Soriacebus ameghinorum	MACN SC 2	3.40	1.98	0.58
Soriacebus adriane	MACN SC 344	3.88	2.06	0.53
Dolichocebus gaimanensis	MPEF 5146	4.04	1.98	0.49
Mohanamico hershkovitzi	IGM 181500	3.25	1.66	0.51
Cebupithecia sarmientoi	UCMP 38762	3.64	1.66	0.46
Nuciruptor rubricae	IGM 251074	3.96	1.90	0.48
Branisella boliviana	MNHN 3468	2.93	4.24	0.70

Notes: measurements are: M$_1$ length, mesiodistal crown length; buccal talonid height, distance from cemento-enamel junction to the deepest point of the hypoflexid in the parasaggital plane. Measurements made with reticle on binicular microscope at 12× magnification.

and molar areas of the same species, Kay and Meldrum (1997) derived the following least-squares regression with an r^2 of 0.935:

ln female body mass = ln M$_1$ area (1.565) + 3.272

From this equation, based on our sample of two lower first molars, the mean body mass of *Mazzonicebus* was 1602 grams. This places *Mazzonicebus* within the size range of living pitheciins (Table 15.3).

The molar crowns have poorly developed shearing crests. Quantitative estimates of shearing-crest development (expressed as "shearing quotients": SQ) allow some dietary groups of living platyrrhines to be distinguished from one another (Fig. 15.11, Table 15.4). Insectivores and folivores have the highest SQs; part of their adaptation for processing of chitin and cellulose, respectively. Fruit-eaters, gum-eaters, and seed-eaters have intermediate and low SQs. *Mazzonicebus* resembles *Soriacebus* species in having low SQs, from which we may infer that fruit, gum, or seeds made up a significant component of its diet.

A more refined interpretation of the diet of *Mazzonicebus* may be gained from examination of the jaw and anterior teeth. The procumbent and mesiodistally compressed incisors and robust canine set in a stoutly buttressed mandibular symphysis greatly resembles living pitheciins like *Pithecia* and *Cacajao*. Pitheciins use their front teeth for piercing and gnawing the outer rinds (scler-ocarps) of tough fruits (Kinzey 1992) and it appears that *Mazzonicebus*, like *Soriacebus* (Kay 1990; Kay *et al.* 2006b), did the same.

The molars of *Mazzonicebus* differ in one respect from those of any living platyrrhine: the crown heights are extraordinarily large. Table 15.5 summarizes provides a summary of lower molar crown height in small samples of 11 genera of living platyrrhines and seven species of Miocene platyrrhines. Mean crown height of the extant species is 0.47 with a standard deviation of 0.055 and the range is 0.38 to 0.59. Six taxa of Miocene platyrrhines fall within the range of the extant forms with a mean of 0.51 and a range of 0.46 to 0.58. *Mazzonicebus* has exceptionally high crowns with a ratio of 0.65, outside that of all measured living and extinct platyrrhines. Only one other platyrrhine, the late Oligocene species *Branisella boliviana*, has higher molar crowns. It has been proposed that *Branisella* was terrestrial or scansorial based on its high molar crowns and heavy wear rate (Kay *et al.* 2001). A similar hypothesis might be entertained for *Mazzonicebus*.

Conclusions

The following general conclusions are made in this chapter:

- *Mazzonicebus* is the sister taxon to the geologically younger taxon *Soriacebus*. Together they comprise the Soriacebinae, a monophyletic clade of stem platyrrhines.
- *Mazzonicebus* had a body mass of about 1600 grams. Its cheek teeth suggest that it was specialized for fruit- or gum-eating. It had specialized incisors and canines as an adaptation for sclerocarpic foraging, as do living sakis and uakaries. These many similarities to living pitheciines (sakis and uakaries) evolved in parallel in the two groups. More speculatively, the high-crowned cheek teeth of *Mazzonicebus* suggest that it may have been scansorial.

ACKNOWLEDGEMENTS
This study was supported by funds from CONICET (PID3–005900) to Dr. Maria Guiomar Vucetich, and grants from the US National Science Foundation (EAR-0087636, BCS-0090255, and DEB-9907985) to Richard F. Kay and Richard H. Madden. I thank R. H. Madden, U. Pardiñas, M. G. Vucetich, A. A. Carlini, S. Vizcaíno, B. A. Williams, and M. R. L. Anthony for contributions in the field and/or reading and commenting on the manuscript.

Appendix 15.1 List of specimens

MLP 69-III-12–1, left mandible with alveoli for C_1, $P_{2–3}$; P_4 complete. Locality: Colhuehuapi level, Gran Barranca; collected by Robert Goulet.

MLP 77-VI-13–6, left C^1. Locality: Gran Barranca.

MLP 82-V-8–1, right I^1. Locality: Screen-wash picking from Colhuehuapi West, 1995 season; Gran Barranca.

MLP 82-V-8–2, left I_2. Locality: Screen-wash picking from Colhuehuapi West, 1995 season; Gran Barranca.

MLP 82-V-8–3, right P_2. Locality: Screen-wash picking from Colhuehuapi West, 1995 season; Gran Barranca.

MPEF-PV 5342 (field number: AR 95-195), right M^1 or M^2. Locality: Colhuehuapi East, east and stratigraphically 1.5 meters above astrapothere quarry, Gran Barranca; collected by Jorge Noriega.

MPEF-PV 5343 (field number: AR 95-196), right upper premolar (cf. P^3). Locality: Colhuehuapi East, on point east of astrapothere quarry, Gran Barranca; collected by Georgina Erra.

MPEF-PV 5344 (field number: AR 95-197), maxilla with right P^3–P^4. Locality: Colhuehuapi West, Gran Barranca; collected by Richard Madden.

MPEF-PV 5345 (field number: AR 95-198), right P_2. Locality: Colhuehuapi West, Gran Barranca; collected by Anna Maria Ribeiro.

MPEF-PV 5346 (field number: AR 95-199), right I^2. Locality: Colhuehuapi West, Gran Barranca; collected by Richard Kay.

MPEF-PV 5347 (field number: AR 95-200), associated right P^2, M^1, M^2. Locality: Colhuehuapi West, Gran Barranca; collected by Valeria Bertoia.

MPEF-PV 5348 (field number: AR 95-201), right M_2. Locality: Colhuehuapi East, near astrapothere quarry, Gran Barranca.

MPEF-PV 5349 (field number: AR 95-202), left C_1. Locality: Colhuehuapi West, Gran Barranca.

MPEF-PV 5350 (field number: AR 95-203), right C^1. Locality: Colhuehuapi West, Gran Barranca.

MPEF-PV 5351 (field number: AR 95-204), mandible with sockets for left I_1–C_1; right $I_{1–2}$; right C_1 complete. Locality: Colhuehuapi West, Gran Barranca.

MPEF-PV 5352 (field number: AR 95-205), right C^1. Locality: Colhuehuapi West, Gran Barranca.

MPEF-PV 5372 (field number: AR 95-312), right P^2. Locality: Colhuehuapi East, between Mazzoni levels 20 and 19 (1/2 m below 20), Gran Barranca; collected 3.12.1995.

MPEF-PV 5391 (field number: AR 95-338), right C^1. Locality: Gran Barranca.

MPEF-PV 5466 (no field number; screen-wash specimen 1), left P_2. Locality: Colhuehuapi West, Gran Barranca.

MPEF-PV 5467 (no field number; screen-wash specimen 2), left M^2. Locality: Colhuehuapi West, Gran Barranca.

MPEF-PV 5468 (no field number; screen-wash specimen 3), right P^4. Locality: Colhuehuapi West, Gran Barranca.

MPEF-PV 5469 (no field number; screen-wash specimen 4), left dP_4. Locality: Colhuehuapi West, Gran Barranca.

MPEF-PV 5699 (field number: AR 97-152), left M^1 (other material with this field number is not primate). Locality: Gran Barranca.

MPEF-PV 6282 (field number: AR 99-03), maxilla with right $P^{3–4}$. Locality: Colhuehuapi West, along road, level certainly below the "monkey hill," perhaps equivalent to level A, Gran Barranca, collected by G. Vucetich.

MPEF-PV 6484 (field number: AR 99-203), left M_1. Locality: Colhuehuapi West, level C, Gran Barranca.

MPEF-PV 6484 (field number: AR 99-240), lower molar fragment. Locality: Surface collected from level B or above, Gran Barranca.

MPEF-PV 6495 (field number: AR 99-250), left I^2. Locality: Surface collected from level B or above, Gran Barranca.

MPEF-PV 6518 (field number: AR 99-274), left P_2. Locality: Surface collected from level C or above, Gran Barranca.

MPEF-PV 6752 (field number: AR 2000-70), mandible with left I_1–M_1. Locality: Colhuehuapi West, Gran Barranca; collected by Damien Glaz.

MPEF-PV 7021 (field number: AR 2000-363), left C^1. Locality: Colhuehuapi levels, Gran Barranca.

MPEF-PV 7063 (field number: AR 2001-36), left M^1. Locality: Colhuehuapi West, about level A, Gran Barranca.

MPEF-PV 7112 (field number: AR 2001-117), left C^1. Locality: Colhuehuapi West, level 19, Gran Barranca.

MPEF-PV 7195 (field number: AR 2001-217), left M^1. Locality: Colhuehuapi West, surface below level C, Gran Barranca.

REFERENCES

Barroso, C. M. L., H. Schneider, M. P. C. Schneider, I. Sampaio, M. L. Harada, J. Czelusniak, and M. Goodman 1997. Update on the phylogenetic systematics of New World monkeys: further DNA evidence for placing pygmy marmoset (*Cebuella*) within the genus *Callithrix*. *International Journal of Primatology*, **18**, 651–674.

Emmons, L. H. and F. Feer 1990. *Neotropical Rainforest Mammals: A Field Guide*. Chicago, IL: University of Chicago Press.

Fleagle, J. G. 1990. New fossil platyrrhines from the Pinturas Formation, southern Argentina. *Journal of Human Evolution*, **19**, 61–85.

Fleagle, J. G. and T. M. Bown 1983. New primate fossils from late Oligocene (Colhuehuapian) localities of Chubut Province, Argentina. *Folia Primatologica*, **41**, 240–266.

Fleagle, J. G. and R. F. Kay 2006. A new humerus of *Homunculus* from the Santa Cruz Formation (early–middle Miocene, Patagonia). *Journal of Vertebrate Paleontology*, **26**, 62A.

Fleagle, J. G. and M. F. Tejedor 2002. Early platyrrhines of southern South America. In Hartwig, W. C. (ed.), *The Primate Fossil Record*. Cambridge, UK: Cambridge University Press, pp. 161–173.

Fleagle, J. G., D. W. Powers, G. C. Conroy, and J. P. Watters 1987. New fossil platyrrhines from Santa Cruz Province, Argentina. *Folia Primatologica*, **48**, 65–77.

Fleagle, J. G., T. M. Bown, C. C. Swisher III, and G. A. Buckley 1995. Age of the Pinturas and Santa Cruz formations. *Actas VI Congreso Argentino de Paleontologia y Bioestratigrafia*, 129–135.

Fleagle, J. G., R. F. Kay, and M. R. L. Anthony 1997. Fossil New World monkeys. In Kay, R. F., Madden, R. H., Cifelli, R. L., and Flynn, J. J. (eds.), *Mammalian Evolution in the Neotropics*. Washington, DC: Smithsonian Institution Press, pp. 473–495.

Flynn, J. J., A. R. Wyss, R. Charrier, and C. C. Swisher III 1995. An early Miocene anthropoid skull from the Chilean Andes. *Nature*, **373**, 603–607.

Flynn, J. J., J. Guerrero, and C. C. Swisher III 1997. Geochronology of the Honda Group. In Kay, R. F., Madden, R. H., Cifelli, R. L., and Flynn, J. J. (eds.), *Vertebrate Paleontology in the Neotropics*. Washington, DC: Smithsonian Institution Press, pp. 44–60.

Hershkovitz, P. 1977. *Living New World Monkeys (Platyrrhini) with an Introduction to Primates*, vol. 1. Chicago, IL: University of Chicago Press.

Hershkovitz, P. 1981. Comparative anatomy of platyrrhine mandibular cheek teeth dpm4, pm4, m1, with particular reference to those of *Homunculus* (Cebidae), and comments on platyrrhine origins. *Folia Primatologica*, **35**, 179–217.

Hershkovitz, P. 1984. More on the *Homunculus* dpm4 and m1 and comparisons with *Alouatta* and *Stirtonia* (Primates, Platyrrhini, Cebidae). *American Journal of Primatology*, **7**, 261–283.

Hillis, D. M. 1999. SINEs of the perfect character. *Proceedings of the National Academy of Sciences USA*, **96**, 9979–9981.

Horovitz, I. 1999. A phylogenetic study of living and fossil platyrrhines. *American Museum Novitates*, **3269**, 1–40.

Horovitz, I. and R. D. E. MacPhee 1999. The Quaternary Cuban platyrrhine *Paralouatta varonai* and the origin of Antillean monkeys. *Journal of Human Evolution*, **36**, 33–68.

Kay, R. F. 1990. The phyletic relationships of extant and fossil Pitheciinae (Platyrrhini, Anthropoidea). *Journal of Human Evolution*, **19**, 175–208.

Kay, R. F. and D. J. Meldrum 1997. A new small platyrrhine from the Miocene of Colombia and the phyletic position of Callitrichinae. In Kay, R. F., Madden, R. H., Cifelli, R. L., and Flynn, J. J. (eds.), *Vertebrate Paleontology in the Neotropics*. Washington, DC: Smithsonian Institution Press, pp. 435–458.

Kay, R. F., D. J. Johnson, and D. J. Meldrum 1998a. A new pitheciin primate from the middle Miocene of Argentina. *American Journal of Primatology*, **45**, 317–336.

Kay, R. F., B. J. MacFadden, R. H. Madden, H. Sandeman, and F. Anaya 1998b. Revised age of the Salla beds, Bolivia, and its bearing on the age of the Deseadan South American Land Mammal 'Age'. *Journal of Vertebrate Paleontology*, **18**, 189–199.

Kay, R. F., D. J. Johnson, and D. J. Meldrum 1999a. *Proteropithecia*, new name for *Propithecia* Kay, Johnson and Meldrum, 1998 non Vojnits 1985. *American Journal of Primatology*, **47**, 347.

Kay, R. F., R. H. Madden, M. Mazzoni, M. G. Vucetich, G. Ré, M. Heizler, and H. Sandeman 1999b. The oldest Argentine primates: first age determinations for the Colhuehuapian South American Land Mammal 'Age'. *American Journal of Physical Anthropology*, Suppl. **28**, 166.

Kay, R. F., B. A. Williams, and F. Anaya 2001. The adaptations of *Branisella boliviana*, the earliest South American monkey. In Plavcan, J. M., van Schaik, C., Kay, R. F., and Jungers, W. L. (eds.), *Reconstructing Behavior in the Primate Fossil Record*. New York: Kluwer Academic, pp. 339–370.

Kay, R. F., S. F. Vizcaino, A. Tauber, M. S. Bargo, B. A. Williams, C. Luna, and M. W. Colbert 2005. Three newly discovered skulls of *Homunculus patagonicus* support its position as a stem platyrrhine and establish its diurnal arboreal folivorous habits. *American Journal of Physical Anthropology*, Suppl. **40**, 127.

Kay, R. F., E. C. Kirk, M. Malinzak, and M. W. Colbert 2006a. Brain size, activity pattern, and visual acuity in *Homunculus patagonicus*, an early Miocene stem platyrrhine: the mosaic evolution of brain size and visual acuity in Anthropoidea. *Journal of Vertebrate Paleontology*, **26**, 83A–84A.

Kay, R. F., M. Takai, and D. J. Meldrum 2006b. Pitheciidae and other platyrrhine seed predators: the dual occupation of the seed predator niche during platyrrhine evolution. *XXI International Congress of Primatology*, Entebbe, Uganda.

Kay, R. F., J. G. Fleagle, T. R. Mitchell, M. Colbert, T. Bown, and D. W. Powers 2008. The anatomy of *Dolichocebus gaimanensis*, a stem platyrrhine monkey from Argentina. *Journal of Human Evolution*, **54**, 323–382.

Kinzey, W. G. 1992. Dietary and dental adaptations in the Pitheciinae. *American Journal Physical Anthropology*, **88**, 499–514.

Kinzey, W. G. and M. A. Norconk 1990. Hardness as a basis of fruit choice in two sympatric primates. *American Journal of Physical Anthropology*, **81**, 5–15.

Kramarz, A. G. and E. S. Bellosi 2005. Hystricognath rodents from the Pinturas Formation, early–middle Miocene of Patagonia: biostratigraphic and paleoenvironmental implications. *Journal of South American Earth Sciences*, **18**, 199–212.

Luchterhand, K., R. F. Kay, and R. H. Madden 1986. *Mohanamico hershkovitzi*, gen. et sp. nov., un primate du Miocène moyen d'Amérique du Sud. *Comptes Rendus de l'Académie des Sciences, Paris, Ser. II*, **303**, 1753–1758.

MacPhee, R. D. E. and I. Horovitz 2004. New craniodental remains of the Quaternary Jamaican monkey *Xenothrix mcgregori* (Xenotrichini, Callicebinae, Pitheciidae), with a reconsideration of the *Aotus* hypothesis. *American Museum Novitates*, **3434**, 1–51.

MacPhee, R. D. E., I. Horovitz, O. Arredondo, and O. Jiménez Vázquez 1995. A new genus for the extinct Hispaniolan monkey *Saimiri bernensis* Rímoli, 1977, with notes on its systematic position. *American Museum Novitates*, **3134**, 1–21.

Meldrum, D. J. and R. F. Kay 1997. *Nuciruptor rubricae*, a new pitheciin seed predator from the Miocene of Colombia. *American Journal of Physical Anthropology*, **102**, 407–427.

Ray, D. A., J. Xing, D. J. Hedges, M. A. Hall, M. E. Laborde, B. A. Anders, B. R. White, N. Stoilova, J. D. Fowlkes, K. E. Landry, L. G. Chemnik, O. A. Ryder, and M. A. Batzer 2005. Alu insertion loci and platyrrhine primate phylogeny. *Molecular Phylogenetics and Evolution*, **35**, 117–126.

Ray, D. A., J. Xing, S. Abdel-Halim, and M. A. Batzer 2006. SINEs of a nearly perfect character. *Systematic Biology*, **55**, 928–935.

Rivero, M. and O. Arredondo 1991. *Paralouatta varonai*, a new Quaternary platyrrhine from Cuba. *Journal of Human Evolution*, **21**, 1–12.

Rosenberger, A. L. 1975. On the distinctiveness of *Xenothrix*. *American Journal of Physical Anthropology*, **42**, 326.

Rosenberger, A. L. 1979. Phylogeny, evolution and classification of New World monkeys (Platyrrhini, Primates). Ph.D. thesis, City University of New York.

Rosenberger, A. L. 1981. Systematics: the higher taxa. In Coimbra-Filho, A. F. and Mittermeier, R. A. (eds.), *Ecology and Behavior of Neotropical Primates*, vol. 1. Rio de Janeiro: Academia Brasileira de Ciências, pp. 9–27.

Rosenberger, A. L. 2000. Pitheciinae. In Delson, E., Tattersal, I., and Van Couvering, J. (eds.), *Encyclopedia of Human Evolution and Prehistory*, 2nd edn. New York: Garland, pp. 562–3.

Rosenberger, A. L. 2002. Platyrrhine paleontology and systematics: the paradigm shifts. In Hartwig, W. C. (ed.), *The Primate Fossil Record*. Cambridge, UK: Cambridge University Press, pp. 151–159.

Rosenberger, A. L., T. Setoguchi, and N. Shigehara 1990. The fossil record of callitrichine primates. *Journal of Human Evolution*, **19**, 209–236.

Schneider, H., F. C. Canavez, I. Sampaio, M. A. M. Moreira, C. H. Tagliaro, and H. N. Seuánez 2001. Can molecular data place each neotropical monkey in its own branch? *Chromosoma*, **109**, 515–523.

Schneider, H., M. P. C. Schneider, I. Sampaio, M. L. Harada, M. Stanhope, J. Czelusniak, and M. Goodman 1993. Molecular Phylogeny of the New World monkeys (Platyrrhini, Primates). *Molecular Phylogenetics and Evolution*, **2**, 225–242.

Schneider, H., I. Sampaio, M. L. Harada, C. M. L. Barroso, M. P. C. Schneider, J. Czelusniak, and M. Goodman 1996. Molecular phylogeny of the New World monkeys (Platyrrhini, Primates) based on two unlinked nuclear genes: IRBP intron 1 and I-globin sequences. *American Journal of Physical Anthropology*, **100**, 153–179.

Setoguchi, T. and A. L. Rosenberger 1987. A fossil owl monkey from La Venta, Colombia. *Nature*, **326**, 692–694.

Smith, R. J. and W. L. Jungers 1997. Body mass in comparative primatology. *Journal of Human Evolution*, **32**, 523–559.

Takai, M. and F. Anaya 1996. New specimens of the oldest fossil platyrrhine, *Branisella boliviana*, from Salla, Bolivia. *American Journal of Physical Anthropology*, **99**, 301–318.

Takai, M., F. Anaya, N. Shigehara, and T. Setoguchi 2000. New fossil materials of the earliest New World monkey, *Branisella boliviana*, and the problem of platyrrhine origins. *American Journal of Physical Anthropology*, **111**, 263–281.

Takai, M., F. Anaya, H. Suzuki, N. Shigehara, and T. Setoguchi 2001. A new platyrrhine from the middle Miocene of La Venta, Colombia, and the phyletic position of Callicebinae. *Anthropological Science (Japan)*, **109**, 289–307.

Tauber, A. 1991. *Homunculus patagonicus* Ameghino, 1891 (Primates, Ceboidea), Mioceno Temprano, de la costa Atlantica Austral, Prov. de Santa Cruz, Republica Argentina. *Academia Nacional de Ciencias en Córdoba, Argentina*, **82**, 1–32.

Tejedor, M. F. 2005. New specimens of *Soriacebus adrianae* Fleagle, 1990, with comments on pitheciin primates from the Miocene of Patagonia. *Ameghiniana*, **41**, 249–251.

Tejedor, M. F., A. A. Tauber, A. L. Rosenberger, C. C. Swisher III, and M. E. Palacios 2006. New primate genus from the Miocene of Argentina. *Proceedings of the National Academy of Sciences USA*, **103**, 5437–5441.

16 Bats (Mammalia: Chiroptera) from Gran Barranca (early Miocene, Colhuehuapian), Chubut Province, Argentina

Nicholas J. Czaplewski

Abstract

Bats are extremely rare as fossils in Argentina, where they are known only from the early Eocene and Pleistocene–Holocene. Bat specimens recently were found in the middle of the Colhue-Huapi Member of the Sarmiento Formation (type section of the Colhuehuapian South American Land Mammal Age [SALMA]; early Miocene) at the Gran Barranca, south of Lake Colhue-Huapi, Argentina. An isolated m3 represents an unidentified large leaf-nosed bat (Phyllostomidae, Phyllostominae) and is the oldest record of this family in South America. A partial dentary with p3–m2 and an isolated m2 are most similar to those in the genera *Mormopterus* and *Tomopeas*, and represent a new species of Molossidae, *Mormopterus barrancae* sp. nov. The new species has myotodont molars, an advanced p4 and a moderately shortened anterior portion of the jaw, with lower premolars more crowded than in the Deseadan species *Mormopterus faustoi* and less crowded than in the early Miocene European species *M. stehlini*. An isolated m2 probably represents a second, unidentified species of *Mormopterus*, 20%–30% smaller than *M. barrancae*. The presence of the phyllostomine suggests a warm moist climate and the availability of lowland forest habitat in the Gran Barranca area about 20 million years ago, much different from the cool dry temperate climate of this area today and consistent in part with paleohabitats inferred from early Miocene rodents and primates.

Resumen

Los murciélagos son extremadamente raros como fósiles en la Argentina, en donde solo se conocen en niveles del Eoceno temprano y del Pleistoceno-Holoceno. En este trabajo se describen los especímenes encontrados recientemente en niveles medios del Miembro Colhue-Huapi en la parte superior de la Formación Sarmiento (sección tipo de la Edad-mamífero Colhuehuapense; Mioceno temprano) en Gran Barranca, al sur del Lago Colhué-Huapí, Argentina. Un m3 aislado representa un filostomino (Phyllostomidae, Phyllostominae) indeterminado; el mismo representa el más antiguo registro de un miembro de esta familia en América del Sur. Un fragmento de dentario,

The Paleontology of Gran Barranca: Evolution and Environmental Change through the Middle Cenozoic of Patagonia, eds. R. H. Madden, A. A. Carlini, M. G. Vucetich, and R. F. Kay. Published by Cambridge University Press. © Cambridge University Press 2010.

incluyendo los p3-m2, así como también un m2 aislado, representan una nueva especie de Molossidae (*Mormopterus barrancae* sp. nov.), similar a las de las especies de *Mormopterus* y *Tomopeas*. La nueva especie tiene molares de aspecto myotodontoide, un p4 avanzado y un moderado acortamiento anterior de las mandíbulas, con los premolares más bajos y comprimidos que aquel de la especie deseadense *Mormopterus faustoi*, así como también menos apretadamente dispuestos que los de *M. stehlini*, del Mioceno temprano de Europa. Un m2 aislado representa probablemente una segunda especie no identificada de *Mormopterus*, la cual es un 20% a 30% más pequeña que *M. barrancae*. La presencia del filostomino sugiere la disponibilidad de un habitat forestado, húmedo y cálido para el área de Gran Barranca hace unos 20 Ma, muy distinto del clima templado-seco y fresco de esta área en la actualidad y consistente en parte con los paleoambientales deducidos de roedores y de primates del Mioceno temprano.

Introduction

Despite the remarkably rich record of fossil vertebrates in Argentina, bats have an extremely poor fossil record there, consisting of two Eocene specimens and a few Pleistocene – early Holocene occurrences, with a huge blank in between. The Eocene specimens are of indeterminate family and were found at the Laguna Fría locality, Chubut, in rocks of the volcaniclastic complex of the middle Río Chubut (early Eocene, Casamayoran: Tejedor *et al.* 2004, 2005). The Pleistocene–Holocene specimens comprise (1) an upper canine tooth of *Noctilio* sp. from a middle Pleistocene locality Las Grutas, Necochea, in coastal Buenos Aires province (Merino *et al.* 2007), (2) a possible *Histiotus* jaw fragment with m3 in the late Pleistocene at an archeological site Cueva Epullán Grande, Collón Curá Department, Neuquén Province (Iudica *et al.* 2003), (3) an isolated tooth of the extinct vampire *Desmodus draculae* from the middle to late Holocene of Centinela del Mar, Buenos Aires Province (Pardiñas and Tonni 1996), and (4) unidentified Chiroptera in a Holocene archeological site known as Cueva Tixi, Buenos Aires Province (Tonni *et al.* 1988; Quintana and Mazzanti 1996). Ameghino (1880) also mentioned Pleistocene–Holocene bats in Argentinean sediments, but gave no details (*fide* Pardiñas and Tonni 2000). Recently, a few specimens of bats were found in the type section of the Colhuehuapian SALMA (early Miocene) in the Gran Barranca south of Lake Colhue-Huapi. They constitute the first fossil bats from the Neogene of Patagonia. The Gran Barranca bat fossils are the southernmost pre-Pleistocene records of bats in the western hemisphere. One is described below as a new species.

The specimens were discovered at Gran Barranca, in the Colhue-Huapi Member of the Sarmiento Formation, which is mainly fluvial in origin (see Ré *et al.* Chapter 3, this book; Bellosi this book). New ^{40}Ar/^{39}Ar dates within and immediately below the fossil-bearing horizon of this unit, as well as other dates above it and a magnetostratigraphic column (Kay *et al.* 1999; Carlini *et al.* 2004; Madden *et al.* 2004; Ré *et al.* Chapter 4, this book) indicate that the Colhuehuapian fauna in Gran Barranca correlates with paleomagnetic Chron C6An.1n (20.0 to 20.2 Ma).

Systematic paleontology

Order **Chiroptera** Blumenbach 1779
Family **Phyllostomidae** Gray 1825
Subfamily **Phyllostominae** Gray 1838
Genus indeterminate

Referred specimen Museo de Ciencias Naturales de La Plata, MLP 93-XI-18–16, right m3 (Fig. 16.1A).

Description The m3 represents a large bat (Table 16.1); the tooth is about the size of the m3 in the extant greater spear-nosed bat *Phyllostomus hastatus*. It is identical in its morphology to m3s of *Tonatia* and *Lophostoma*, but is far larger than any known species in those genera. It is slightly less similar to the m3 in *Chrotopterus*, *Macrotus*, *Phylloderma*, *Phyllostomus*, *Trachops*, and *Trinycteris*. Although no m3 is known for species of *Notonycteris*, the trigonid portion in the m3 from Gran Barranca is the same size and morphology as the m2 trigonid in *Notonycteris sucharadeus* from the La Venta fauna, Colombia (Czaplewski *et al.* 2003); it is much smaller than would be expected in an m3 of *Notonycteris magdalenensis*.

Remarks This specimen is the oldest known member of the Phyllostomidae in South America. With an age of about 20 Ma, it exceeds the records of other phyllostomids *Notonycteris, Tonatia*, and *Palynephyllum* in the La Venta fauna, Colombia (about 12 Ma; Laventan SALMA) by 7 or 8 Ma (Savage 1951; Czaplewski 1997; Czaplewski *et al.* 2003).

Family **Molossidae** Gervais 1856
Genus ***Mormopterus*** Peters 1865
Mormopterus barrancae sp. nov.

Holotype Museo de Ciencias Naturales de La Plata MLP 93-XI-18–15, left dentary fragment including

Table 16.1. *Dental measurements (mm) of bat fossils from Gran Barranca*

Measurement	*Mormopterus barrancae* sp. nov.		*Mormopterus* sp.	*Phyllostominae* sp.
	MLP 93-XI-18–15	MLP 93-XI-18–18	MLP 93-XI-18–19	MLP 93-XI-18–16
p2 apl	0.87			
p2 tw	0.80			
p4 apl	1.07			
p4 tw	1.03			
m1 apl	1.87	1.87		
m1 trigw	1.17	1.05		
m1 talw	1.35	1.25		
m2 apl	1.72		1.31	
m2 trigw	1.16		0.92	
m2 talw	1.27		0.77	
m3 apl				2.72
m3 trigw				2.02
m3 talw				1.25

Abbreviations: apl, anteroposterior length; talw, talonid width; trigw, trigonid width; tw, transverse width; MLP, Museo Ciencias Naturales de La Plata.

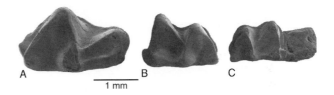

Fig. 16.1. Scanning electron micrographs of Colhuehuapian bats from Gran Barranca. (A) Phyllostominae indet., right m3 (MLP 93-XI-18–16); (B) *Mormopterus barrancae*, left m1 (MLP 93-XI-18–18); (C) *Mormopterus* sp. indet., right m2 in small fragment of dentary (MLP 93-XI-18–19).

mandibular symphysis with p2, p4, m1, and m2 *in situ*, and with vacant alveoli for two incisors, canine, and m3 (Figs. 16.2, 16.3).

Hypodigm The holotype plus MLP 93-XI-18–18, isolated left m1 (Fig. 16.1B).

Type locality and age In the cliffs ("Gran Barranca") south of Lake Colhue-Huapi, Chubut Province, Argentina. Sarmiento Formation; Colhue-Huapi Member (type section of the Colhuehuapian SALMA), early Miocene. MLP 93-XI-18–18 is from "nivel 'C' (gris)" (level "C" [gray]).

Etymology In reference to the derivation of these fossils from Gran Barranca.

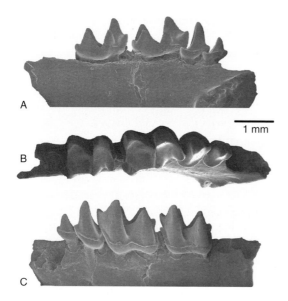

Fig. 16.2. *Mormopterus barrancae* sp. nov. left dentary with p3–m2 (MLP 93-XI-18–15) holotype, scanning electron micrographs in (A) lingual; (B) occlusal; and (C) labial views.

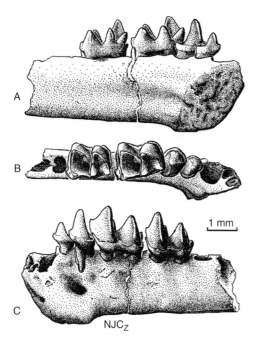

Fig. 16.3. *Mormopterus barrancae* sp. nov., MLP 93-XI-18–15, drawing of holotype dentary in (A) lingual; (B) occlusal; and (C) labial views.

Diagnosis Differs from *Tomopeas ravus* in much larger size; in having a weak lingual cingulum at the base of the trigonid valley of m1; in having less obliquely staggered roots of p4; in having a weak

posterolingual crest on p4 with no metaconid; m2 shorter than m1; cristid obliqua of m1 and m2 connects with the posterior wall of the trigonid instead of dropping down just before reaching the trigonid wall; and in having the lateral mental foramen positioned between the roots of p2 and p4 rather than between c1 and p2. Differs from *Mormopterus faustoi* in having premolar portion of lower toothrow much shorter; in having m1 slightly larger than m2; in having a weak lingual cingulum at the foot of the trigonid valley in m1 and m2; in having p2 with roots coalesced into one rather than separated; and in having a less gracile horizontal ramus of the dentary. Differs from *Mormopterus nonghenensis* in lacking a rudimentary metaconid on p4. Differs from *M. stehlini* and *M. helveticus* in having lower premolars much less anteroposteriorly compressed. Differs from the extant species *Mormopterus phrudus* and *M. kalinowskii* in larger size; in having m1 slightly larger than m2 rather than equal-sized; and in having nearly straight rather than curved entocristids on m1 and m2. Further differs from *M. phrudus* in lacking a vestigial metaconid on p4. Differs from all other known genera of Molossidae in having the following unique combination of characters, at least some of which may be primitive characters shared with other bats: lower incisors two in number; lower premolars two in number; p2 single-rooted; p4 lacks anterolingual cingular cusp and heel; p4 bears two roots obliquely oriented to the axis of the toothrow; lower molars myotodont (Menu and Sigé 1971); lower molars with tiny lingual cingulum at base of trigonid valley; m2 shorter than m1; m1 entocristid straight in functional view.

Description Measurements (in mm) of the type specimen are: depth of dentary beneath lower edge of anterior alveolus of m1, 1.77; length of alveolar row, 7.3; additional dental measurements are given in Table 16.1. The dental formula of the lower tooth row is 2, 1, 2, 3. The horizontal ramus of the dentary is moderately robust. A large lateral mental foramen is present below p4; a single, smaller, anterior mental foramen below the incisors is situated at a level slightly lower than the lateral foramen. Along the ventral edge of the ramus a faint groove, possibly for insertion of the digastricus muscle, extends forward to a point beneath the anterior root of m3.

The lower incisors are not preserved, although a fragment of the root of i1 is retained within its alveolus; alveoli for i1 and i2 are subequal in size. The canine also is unknown; its alveolus is relatively large.

The anterior lower premolar (p2) is single-rooted although in anterior view the visible portion of the root is deeply grooved, possibly indicating that it is the product of two coalesced roots. The p2 is about one-half the size of p4 in its crown height or basal cross-sectional area. Its crown height is about equal to its length, and its anteroposterior length is greater than its width. The tooth bears a strong cingulum that almost completely encircles the base except for a tiny lingual segment. No cingular cusps are present. The main cusp of p2 is simple and subconical, without strong ridges or crests.

The posterior premolar (p4) has two roots that are arranged oblique to the long axis of the dentary. The height of the crown is about equal to its length. Crown height is less than that of the protoconid of m1. In occlusal view the crown is wider than long. There is one main cusp (protoconid) whose apex is recurved slightly posterolingually. In horizontal section this cusp has a rounded triangular cross-section. The anterolabial face is convex, the lingual face is slightly convex, and the posterior face is flattened. Rounded vertical crests occur on the posterolabial and posterolingual portions of the main cusp. The cingulum is strong and completely encircles the base of the crown, although it is narrowed along a short anterolingual segment. A low cingular swelling is present on the anterior margin of the tooth. No posterolingual cingular cusp is present. In occlusal view, the anterior portion of the cingulum strongly overlaps the posterolabial portion of the p2. In turn, the posterolabial portion of p4 is strongly overlapped by the paraconid of m1. The crown of the p4 has no ventrolabial constriction or emargination along its base (that would make the tooth in labial view appear to be incipiently separated into anterior and posterior moieties supported by different roots, as in many other bats) and the posterolabial corner of the tooth is neither squarish nor expanded posteroventrally.

The isolated m1 (MLP 93-XI-18–18) is virtually identical to the m1 in the holotype jaw. The m1 and m2 are similar to one another and also generally similar to the same teeth in other molossids. Each has two roots. The trigonids are anteroposteriorly compressed and are transversely narrower than the talonids. The m2 is slightly shorter than m1. Both molars exhibit myotodont morphology in their talonids (the hypoconid is connected via the postcristid to the entoconid and not to the hypoconulid). In each molar, the paraconid is about three-fourths the height of the metaconid. Cingula are well developed anteriorly, labially, and posteriorly, but lingually only a tiny, weak remnant of a cingulum is present at the foot of the trigonid valley. The anterior cingulum in m1 lacks an anterior projection; that of m2 has a moderate anterior projection. The angles formed

between the paracristid and protocristid in both m1 and m2 are acute. In occlusal view, the cristid obliqua is straight and it meets the posterior wall of the trigonid in the middle, below the notch in the protocristid. The entocristid (pre-entocristid) is straight and is ridged or moderately crested. Entoconids are equal in height to metaconids. Carnassial-like notches are not developed in the cristid obliqua and postcristid. The hypoconulid is low but not as low as the distal cingulum. Although the hypoconulid is displaced lingually so that it and the entoconid are "twinned," it is nevertheless positioned slightly labial from the lingual wall of the m1 and m2 relative to the hypoconulid in other *Mormopterus* (in which it is flush with the lingual wall of the tooth). The m3 is not preserved in MLP 93-XI-18–15, but is represented by two large alveoli.

Comparisons and discussion In its possession of myotodont first and second lower molars, *Mormopterus barrancae* is derived with respect to most other molossids. In the earliest known (late Eocene) North American molossid *Wallia*, the lower molars are unknown. In lower molars of the early Paleogene European molossid *Cuvierimops* the postcristid connects the hypoconid with the hypoconulid (nyctalodont condition). Only three Paleogene genera of bats are known to have myotodont lower molars. These are the philisid *Philisis* in the early Oligocene of Africa–Arabia (Sigé 1985; Sigé *et al.* 1994), the vespertilionid *Myotis* (*Leuconoe*) in the early Oligocene of Europe (Sigé and Menu 1995), and the molossid *Mormopterus faustoi* in the late Oligocene of South America (Paula Couto and Mezzalira 1971; Paula Couto 1983; Legendre 1985). In myotodonty and many other characters, *M. barrancae* is very similar to other species of *Mormopterus*. In characters other than the configuration of the postcristid, *Philisis* and *Myotis* (*Leuconoe*) differ from *Mormopterus barrancae* in many other ways.

Among Neogene Chiroptera, few known taxa have myotodont first and second lower molars. Those that do belong to the families Mystacinidae, Vespertilionidae, and Molossidae. The middle Miocene (Barstovian) North American genus *Ancenycteris* (Vespertilionidae) had myotodont molars but differs from *M. barrancae* in having three lower incisors and in having p4 with a well-developed heel. North American Pliocene *Anzanycteris* and Miocene–Recent *Antrozous* differ in having a p4 with its posterolabial corner expanded posteroventrally. *Mormopterus barrancae* shares the myotodont condition of its molars with the Miocene Australian mystacinid *Icarops* but differs in lacking a fused mandibular symphysis, in lacking a ventral mandibular shelf, and in other characters (Hand *et al.* 1998). Molossid genera known in the Neogene are *Petramops, Mormopterus, Potamops, Eumops,* and *Tadarida* (Legendre 1984a, 1984b, 1985; Hand 1990; Hand *et al.* 1999; Czaplewski 1993, 1997; Czaplewski *et al.* 2003). Of these, the lower molars of *Potamops* are unknown, those of *Petramops* and *Tadarida* are nyctalodont, and those of other Neogene *Mormopterus* are variable in this character, depending on the species. The lower teeth of *Mormopterus colombiensis* are unknown. (Czaplewski 1997). Most fossil forms of *Mormopterus* (*M. faustoi, M. stehlini,* and *M. helveticus*) are myotodont, but *M. nonghenensis* is myotodont to submyotodont with an occasional individual showing nyctalodonty (Legendre 1984b, 1985; Legendre *et al.* 1988). Modern species of *Mormopterus* also are nyctalodont to myotodont (see below). *Eumops* and *Mormopterus (Micronomus)* are presently recognized in the Neogene only by isolated or fragmentary upper teeth (Czaplewski 1993, 1997; Hand *et al.* 1999), but modern members of these two taxa are nyctalodont.

Among extant Chiroptera, few genera possess myotodont lower first and second molars. *Mormopterus barrancae* differs from the Emballonuridae, Rhinopomatidae, Craseonycteridae, Rhinolophidae, Hipposideridae, Megadermatidae, Nycteridae, Phyllostomidae (although some non-specialized forms are submyotodont), Mormoopidae, Natalidae, Furipteridae, and most Molossidae (i.e. *Eumops, Nyctinomops, Tadarida, Chaerephon, Mops, Myopterus, Otomops, Promops, Molossus,* and *Molossops*) in having myotodont rather than nyctalodont lower molars. Modern bats that do have myotodont (or submyotodont) molars are Mystacinidae (*Mystacina*), Thyropteridae (*Thyroptera*), Myzopodidae (*Myzopoda*), Noctilionidae (*Noctilio*), many Vespertilionidae (*Kerivoula, Histiotus, Myotis, Eptesicus,* some *Lasiurus, Rhogeessa, Antrozous,* and *Bauerus*), and some Molossidae (*Cheiromeles,* some *Mormopterus,* and *Tomopeas*). The Gran Barranca bat shows persistent other differences from all these except for the molossids *Tomopeas* and *Mormopterus* (subgenera *Platymops, Sauromys,* and *Mormopterus*).

The Colhuehuapian bat differs from *Tomopeas ravus*, the extant and only known species of the molossid subfamily Tomopeatinae (Sudman *et al.* 1994), in much larger size, in less obliquely staggered roots of p4, and in having a weak posterolingual crest on p4 that has no hint of a metaconid (*sensu* Legendre 1984a, p. 410 and Fig. 76). *Mormopterus barrancae* further differs from *Tomopeas* in the following characters: m2 shorter than m1; cristid obliqua of m1 and m2 connects with the posterior wall of the

trigonid instead of dropping down just before reaching the trigonid wall; lateral mental foramen opens more posteriorly (between roots of p2 and p4 rather than between roots of c1 and p2).

With respect to various species of *Mormopterus*, it is difficult to make comparisons between *M. barrancae* and *M. faustoi* because of damage to the lower teeth in the type and only specimen of *M. faustoi* (Paula Couto 1956; Legendre 1985). Nevertheless, it is possible to observe small portions of two separate, obliquely oriented alveoli for the p2, indicating that the p2 has two separate roots in *M. faustoi* instead of roots coalesced into one as in *M. barrancae*. Moreover, despite the fact that the lower molars of these two species are similar in size, the proportional contribution of the teeth in the anterior portion of the toothrow and jaw is different in *M. barrancae*. The length of the lower premolars and canine alveolus in *M. barrancae* is approximately the same as the length of the lower premolars alone in *M. faustoi*. Thus, the Barranca bat had a shorter muzzle.

Mormopterus barrancae differs from other extant and extinct *Mormopterus* species in which lower teeth are known (Legendre 1984a, 1985; Legendre *et al.* 1988) in having a straighter pre-entocristid and in having m2 shorter than m1. In early to middle Miocene *Mormopterus nonghenensis* of Thailand the p4 metaconid is rudimentary (Legendre *et al.* 1988), but in early Miocene *Mormopterus stehlini* and middle Miocene *M. helveticus* of Europe the p4 metaconid is absent (Engesser 1972; Legendre 1984b; Hand 1990). In the lack of a p4 metaconid *M. barrancae* is dentally more derived than its contemporary *M. nonghenensis* in southeast Asia, but is equivalent to its contemporary *M. stehlini* in Europe. In *M. barrancae* the lower premolars are not so anteroposteriorly compressed (transversely oriented) as in *M. stehlini* and *M. helveticus* (Engesser 1972; Legendre 1984a, 1985). *Mormopterus barrancae* is much larger than the extant South American species *Mormopterus phrudus, M. kalinowskii,* and *Tomopeas ravus*; it is similar in size to the extinct *M. stehlini, M. helveticus,* and *M. nonghenensis* (Engesser 1972; Legendre *et al.* 1988).

In summary, the Colhuehuapian new species *Mormopterus barrancae* is most similar in dental morphology to other species of *Mormopterus* and to *Tomopeas*. When better specimens are known, *M. barrancae* and *M. faustoi* may bear evidence relevant to the phylogeny of the Tomopeatinae, which has no fossil record, or of other Molossidae. In its foreshortened lower premolar row, the Patagonian Miocene fossil is slightly more derived than (and possibly

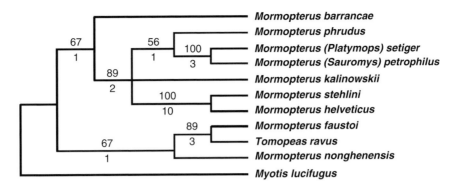

Fig. 16.4. Hypothesis of the phylogenetic relationship of *Mormopterus barrancae* to *Tomopeas* and other *Mormopterus* species, based on unweighted parsimony analysis of 31 dental and osteological characters (see Appendix 16.1). Tree topology represents 50% majority-rule consensus of nine most parsimonious trees using bootstrap method with branch-and-bound search. Tree length = 62 steps; consistency index = 0.645; retention index = 0.542. Numbers above the branch lines are bootstrap support values; numbers below the branches are decay (Bremer support) values.

descended from) the Deseadan *M. faustoi*, but less derived than European Aquitanian *M. stehlini*, which has even more anteroposteriorly compressed lower premolars. In its loss of a metaconid on p4, *M. barrancae* is more derived than, and hence unlikely to be ancestral to, the living South American *M. phrudus* and *M. kalinowskii*.

Phylogenetic analysis of *M. barrancae* A simple parsimony analysis of morphological characters in the fossil and several extinct and extant species of *Mormopterus* was done using the tomopeatine molossid *Tomopeas ravus* and the vespertilionid *Myotis lucifugus* as outgroups. A branch-and-bound search of 31 characters (Appendix 16.1; see caption of Fig. 16.4 for details) yielded nine equally parsimonious trees. In the 50% majority-rule consensus of these nine trees, *Mormopterus barrancae* clusters within a clade composed of Miocene European and extant African and South American species of *Mormopterus* (Fig. 16.4). Interestingly, it does not cluster with the Deseadan *M. faustoi*. Except for the two European Miocene species *M. stehlini* and *M. helveticus*, which are strongly linked, there is generally weak support for this tree. The Barranca bat is sister to a clade comprised of all the extant species of *Mormopterus* plus the two European Miocene species. This clade, in turn, has as its sister group a small clade consisting of the southern Asian Miocene species *M. nonghenensis*, the Brazilian Oligocene species *M. faustoi*, and extant *Tomopeas ravus*.

Mormopterus **sp. indet.**
Referred specimen MLP 93-XI-18–19, small fragment of right dentary with m2 (Fig. 16.1C).
Remarks Morphology of the specimen is identical to that of the m2 in *M. barrancae*, but MLP 93-XI-18–19

represents a much smaller bat. Depth of the dentary below the anterior alveolus of m2 is 1.20 mm. In this dimension and in measurements of the tooth, the specimen represents a species of *Mormopterus* that is about 20%–30% smaller than *M. barrancae* (Table 16.1).

Implications for paleobiogeography and paleoecology

Gran Barranca is presently located at nearly 46° South latitude. In the early Miocene it was probably only a few degrees different than today, but even farther south. The fossil phyllostomine and *Mormopterus* have some implications for the paleoenvironment of the area.

The Phyllostomidae are well known as the most trophically diverse family of bats. The subfamily Phyllostominae are basal within Phyllostomidae and have relatively primitive morphology compared with other phyllostomid clades. Basal phyllostomines were insectivorous or mixed insectivorous/frugivorous; some large-bodied, dentally derived members of the clade (tribe Vampyrini) are largely carnivorous. By comparison, other more derived subfamilies have specializations for nectar- and pollen-feeding (Glossophaginae), frugivory (Sturnirinae and Stenodermatinae), and blood-feeding (Desmodontinae) (Gardner 1977; Ferrarezzi and Gimenez 1996; Freeman 2000; Simmons and Conway 2003; Patterson *et al.* 2003; Bonato *et al.* 2004). In terms of roosting and habitat selection, phyllostomines roost mainly in hollow trees and caves, and sometimes also in decayed tree stumps, fallen tree trunks, or arboreal termite and ant nests (Nowak 1994; Kunz and Lumsden 2003). They occupy primarily humid lowland forested habitats often near streams, pools, swamps, and estuaries. Many occupy multistratal evergreen forests, but several species also utilize dry tropical deciduous forest, forest clearings and edges, and even cloud forest. In terms of thermal biology and biogeography, like most phyllostomids, the phyllostomines are virtually restricted to the tropics and low-latitude subtropics; they are incapable of tolerating the cooler temperatures at higher latitudes. Some such as Linnaeus' false vampire *Vampyrum spectrum* reach only to the southern limits of the Amazon Basin at about 19° South latitude, but several other phyllostomines occur southward along the Atlantic Coast (where the ocean may mediate the climate) to about 29° South, just outside the tropic zone, although the wooly false vampire *Chrotopterus auritus* reaches about 33° South (Koopman 1982). In the northern hemisphere subtropics, the California leaf-nosed bat *Macrotus californicus* penetrates the southwestern North American deserts northward to about latitude 37° North, but only into areas in which geothermally heated caves are available for year-round roosting at about 29 °C, because *Macrotus* is unable to use torpor (Bell *et al.* 1986). No caves are known in central Patagonia in which phyllostomines could roost.

Using a phylogenetic approach to interpret the fossil phyllostomine's ecology (the approach of Owen *et al.* 2000), based on homologous behavior and ecology of its presumed relatives (*Chrotopterus, Glyphonycteris, Lampronycteris, Lonchorhina, Lophostoma, Macrophyllum, Macrotus, Micronycteris, Mimon, Neonycteris, Phylloderma, Phyllostomus, Tonatia, Trachops, Trinycteris,* and *Vampyrum*), the Gran Barranca large phyllostomine was likely insectivorous/frugivorous, roosted and foraged in lowland rainforest, and tolerated only tropical climate. No paleokarst is known in central Patagonia, and few carbonate rocks in which paleocaves could have formed; thus cave roosting was very unlikely for the Colhuehuapian bats. This part of Patagonia is characterized by strong and constant winds, and during the late twentieth century (1961–90) had an annual mean temperature of 14.5 °C and an annual frost frequency of 69 days (Hulme and Sheard 1999). Tectonic and recent geochemical evidence suggests that the Drake Passage separating South America and Antarctica opened and circum-Antarctic ocean circulation began by the late Eocene. These events were accompanied by abrupt deterioration of the early Cenozoic greenhouse climate, resulting in widespread and permanent glaciation of Antarctica (Zachos *et al.* 2001, 2008; Scher and Martin 2006). The presence of a cold-intolerant fossil phyllostomine in the Colhuehuapian at Gran Barranca thus suggests a very different early Miocene paleoenvironment in Patagonia than today, with a much warmer, wetter early Miocene climate and rainforest vegetation.

Extant molossid bats are exclusively insectivorous. The family Molossidae is one of only four modern bat families that broadly occupy the temperate zones. The genus *Mormopterus* has a relatively extensive fossil record compared to other molossids and seems to be a basal member of the

family (Freeman 1981; Legendre 1984a, 1984b, 1985). Unfortunately, the biology of modern *Mormopterus* species is poorly known. The two extant *Mormopterus* species in South America have very restricted (perhaps relict) distributions and may not be representative of the genus. The Incan little mastiff bat *M. phrudus* occurs in the Andes in southeastern Peru where it may occur only at middle elevations, at about 1850 m (holotype from Machu Picchu, Cuzco: Handley 1956; Willig *et al.* 2003). Kalinowski's mastiff bat *M. kalinowskii* occupies the Atacama Desert of northern Chile, plus coastal desert and interandean valleys in Peru. Most other extant *Mormopterus* species scattered in Cuba, Africa, Madagascar, Indonesia, New Guinea, and Australia are primarily tropical, but several species in Australia occur in the temperate zone (Churchill 1998; Simmons 2005; Goodman *et al.* 2008). Their natural (i.e. other than human-made) roosting sites are mainly in hollows in trees and sometimes mangrove branches, occasionally under dead palm fronds or tree bark, and rarely in caves. They usually forage in open air above the canopy but occasionally also over waterholes, and sometimes they attack prey on the ground (Nowak 1994). Habitat selection is poorly documented in most areas, but in Australia they utilize a wide variety of habitats ranging from rainforest to savanna woodland, arid shrublands, mangroves, floodplain forest, and grasslands.

Again using extant phylogenetic bracketing to reconstruct their behavior and ecology, the Colhuehuapian *Mormopterus* from Gran Barranca likely foraged for insects in open air above the canopy and roosted in trees or caves. Because they probably tolerated a wide variety of habitats and tropical and temperate climate zones, they are not otherwise helpful in interpreting the Colhuehuapian paleoenvironment.

Interpretation of the Colhuehuapian paleoenvironment of Gran Barranca as warm, humid, and forested is consistent in part with similar prior conclusions for parts of Patagonia based on fossils of arboreal echimyid and erethizontid rodents and primates (Vucetich 1985, 1986; Kay *et al.* 1999; Kay this book), although patagoniid marsupials and potentially fossorial rodents were taken to indicate that at least some portions of the habitat were subtropical savanna with open or xeric patches, sand dunes, or a parkland–savanna (Pascual *et al.* 1985; Pascual 1986; Pascual and Carlini 1987; Vucetich and Verzi 1991, 1994).

ACKNOWLEDGEMENTS
The specimens in this report were collected by the Museo de La Plata/Duke University expeditions; I extend my sincere appreciation to Alfredo A. Carlini and others at the Universidad Nacional de La Plata, and Richard F. Kay and Richard H. Madden of Duke University for collecting and

screen-washing fossiliferous matrix that produced the specimens described herein, providing catalog numbers for the specimens, and supporting my research through loans of specimens and the invitation to participate in this volume. Their field work was supported by the National Science Foundation. Thanks to Alexander W. A. Kellner and Deise D. R. Henriques (Museu Nacional, Universidade Federal do Rio de Janeiro), James L. Patton and B. R. Stein (University of California Museum of Vertebrate Zoology), Linda K. Gordon (National Museum of Natural History, Smithsonian Institution), and Nancy B. Simmons (American Museum of Natural History) for loans and access to comparative specimens in their care; to Bernard Sigé and Christian de Muizon for providing a cast of *Mormopterus faustoi*; to Amanda McCoy and Christina Stewart for molding and casting specimens; and to Brian Davis for scanning electron microscopy. Funding from the National Science Foundation DEB-9981512 to NJC and Gary S. Morgan partly supported this project. Scanning electron microscopy and image processing were aided in part by funding from NSF DBI-0100317 to R. Lupia, R. Cifelli, NJC, and S. Westrop. I thank Francisco J. Goin and G. S. Morgan for helpful critical reviews that improved the manuscript.

Appendix 16.1 Morphological characters and data matrix

Character states for extant species of bats are recorded from museum specimens; those for extinct bats were recorded from the literature or from casts. Modern species included in the analysis are *Mormopterus kalinowskii*, *M. phrudus*, *M. (Platymops) setiger*, and *M. (Sauromys) petrophilus*, plus *Tomopeas ravus* (Molossidae: Tomopeatinae) and *Myotis lucifugus* (Vespertilionidae) as outgroups. Extinct species comprised *M. barrancae*, *M. faustoi*, *M. stehlini*, *M. helveticus*, and *M. nonghenensis*. Literature sources included Legendre *et al.* (1988) for *Mormopterus nonghenensis*; Revilliod (1920), Engesser (1972), and Legendre (1982) for *M. helveticus*; Revilliod (1920) for *M. stehlini*; and Paula Couto (1956) and a cast for *M. faustoi*. I originally scored the specimens for 43 characters, of which 12 were constant and thus excluded. The remaining 31 characters used in the final analysis are given below.

1. Hypocristid pattern on m1 and m2 nyctalodont (0); submyotodont (1); or myotodont (2).
2. Talonid of m3 bears three cusps, hypoconid, entoconid, and hypoconulid (0); two cusps, hypoconid and entoconid (1).
3. Lower molars relative size m1=m2=m3 (0); m1=m2 >m3 (1); m1>m2>m3 (2).
4. Lingual cuspids of lower molars aligned (0); metaconid lingual (1).
5. m1 with paraconid tall (0); short (1) relative to metaconid.
6. Lingual cingulum at foot of trigonid valley of m1 strong (0); weak (1); absent (2).
7. Lingual cingulum at foot of trigonid valley of m2 strong (0); weak (1); absent (2).
8. Cristid obliqua on m1 meets posterior wall of trigonid centrally, below the notch in postcristid (0); labial to notch (1).
9. Shape of entocristid of m1/m2 in occlusal view nearly straight (0); curved, concave lingually (1).
10. Notches in cristid obliqua and postcristid of m1 absent (0); weakly developed (1).
11. Labial cingulum of lower molars narrow (0); wide and regular in height (1).
12. Number of lower premolars three (0); two (1).
13. p2 with two roots (0); one root (1).
14. p4 with two roots (0); one root (1).
15. Lower premolars in occlusal view laterally compressed (0); not compressed (1); anteroposteriorly compressed (2).
16. Lower premolars aligned with axis of toothrow (0); oblique to axis (1).
17. p4 molariform, with talonid-like heel (0); talonid reduced or absent, vestigial metaconid present (1); talonid and metaconid absent (2).
18. Number of lower incisors three (0); two (1).
19. Number of upper premolars three (0); two (1); one (2).
20. P4 with talon weakly developed or absent (0); moderately developed (1); elongated posterolingually (2).
21. Hypocone on M1 and M2 absent (0); cuspate (1); cristate (2).
22. M1 and M2 talon expanded (0); talon weakly developed (1).
23. Upper molar hypoconal crest not confluent with cingulum of talon or absent (0); confluent (1).
24. M1 and M2 conule(s) present (0); conules absent (1).
25. M1 and M2 with paraloph and metaloph (0); paraloph only (1); metaloph only (2); neither paraloph nor metaloph (3).
26. Postprotocrista does not reach metacone and lacks hypoconal swelling (0); extends to base of metacone (1); extends to metaconule (2); extends to hypocone (3); extends to metastyle (confluent with postcingulum), bypassing hypocone (4).
27. M1 and M2 lingual cingulum absent (0); weak (1); strong (2).
28. Lateral mental foramen positioned between roots of c1 and anteriormost lower premolar (0); between roots of p2 and next posterior premolar (1).
29. Height of coronoid process tall (0); moderate (1); short (2).
30. Horizontal ramus of dentary moderate in thickness (about as deep as teeth are tall) (0); gracile (shallower than the teeth are tall) (1); robust (deeper than the teeth are tall) (2).
31. Deltopectoral ridge length short (0); long (1).

Data matrix

Characters 1–31 are listed sequentially. Where character states are polymorphic, the different states are listed between parentheses, beneath the character number. For examples, in *Mormopterus nonghenensis*, character 1 occurs as states 0, 1, and 2; and in *Mormopterus stehlini*, character 30 occurs as states 0 and 2.

Taxon/Character	1	2	3	4	5	6	7	8	9	10	11	12	13	14	15	16	17	18	19	20	21	22	23	24	25	26	27	28	29	30	31
Mormopterus barrancae	2	?	2	0	0	1	1	0	0	0	1	1	1	0	1	1	2	1	?	?	?	?	0	?	?	?	?	1	?	0	?
Mormopterus faustoi	2	1	1	0	1	2	2	0	0	0	1	1	0	0	?	1	2	?	?	1	1	1	0	2	?	4	0	1	0	1	?
Mormopterus phrudus	2	1	1	0	1	2	2	0	0	0	1	1	0	0	1	1	1	1	1	2	1	0	1	1	1	4	1	1	?	0	?
Mormopterus kalinowskii	2	0	1	0	0	1	2	0	1	0	1	1	1	0	1	1	2	1	2	2	1	0	1	1	3	1	1	1	?	0	?
Mormopterus (P.) setiger	2	0	1	1	1	2	?	1	1	0	0	0	?	1	2	1	1	1	2	2	(12)	1	0	1	2	4	1	0	0	0	?
Mormopterus (S.) petrophilus	2	1	1	1	1	1	2	0	1	0	1	1	?	?	1	1	1	1	1	2	1	0	0	1	0	4	0	1	?	0	?
Mormopterus nonghenensis	(012)	0	1	0	?	1	1	0	1	0	1	1	1	0	2	1	?	1	2	1	1	1	0	1	0	3	1	1	?	?	?
Mormopterus stehlini	2	?	1	0	0	?	?	0	0	?	1	1	(01)	0	2	1	2	1	2	1	2	0	1	1	1	3	2	1	0	(02)	0
Mormopterus helveticus	2	0	1	0	0	1	2	0	1	0	1	1	1	1	2	1	2	?	2	1	2	0	1	1	1	3	2	1	?	2	0
Tomopeas ravus	2	0	1	0	0	2	1	0	0	0	1	1	1	0	2	1	2	1	2	1	1	1	0	1	0	3	0	0	1	1	1
Myotis lucifugus	2	0	1	0	0	1	1	0	1	1	0	0	1	0	1	0	2	0	0	2	2	1	0	0	0	3	1	0	0	1	?

REFERENCES

Ameghino, F. 1880. *La antiguedad del Hombre en el Plata*, vol. 2. Paris: published by the author.

Bell, G. P., G. A. Bartholomew, and K. A. Nagy 1986. The roles of energetics, water economy, foraging behavior, and geothermal refugia in the distribution of the bat, *Macrotus californicus*. *Journal of Comparative Physiology B*, **156**, 441–450.

Blumenbach, J. F. 1779. *Handbuch der Naturgeschichte*. Göttingen: Johann Christian Dieterich.

Bonato, V., K. G. Facure, and W. Uieda 2004. Food habits of bats of subfamily Vampyrinae in Brazil. *Journal of Mammalogy*, **85**, 708–713.

Carlini, A., M. Bond, G. López, M. Reguero, A. Scarano, and R. Madden 2004. Mammalian biostratigraphy and biochronology at Gran Barranca: the standard reference section for the continental middle Cenozoic of South America. *Journal of Vertebrate Paleontology*, **24**(Suppl. 3), 43A–44A.

Churchill, S. 1998. *Australian Bats*. Sydney: Reed New Holland.

Czaplewski, N. J. 1993. Late Tertiary bats (Mammalia: Chiroptera) from the southwestern United States. *Southwestern Naturalist*, **38**, 111–118.

Czaplewski, N. J. 1997. Chiroptera. In Kay, R. F., Madden, R. H., Cifelli, R. L., and Flynn, J. J. (eds.), *Vertebrate Paleontology in the Neotropics: The Miocene Fauna of La Venta, Colombia*. Washington, DC: Smithsonian Institution Press, pp. 410–431.

Czaplewski, N. J., G. S. Morgan, and T. M. Naeher 2003. Molossid bats from the late Tertiary of Florida with a review of the Tertiary Molossidae of North America. *Acta Chiropterologica*, **5**, 61–74.

Czaplewski, N. J., M. Takai, T. M. Naeher, N. Shigehara, and T. Setoguchi 2003. Additional bats from the middle Miocene La Venta fauna of Colombia. *Revista de la Academia Colombiana de Ciencias Exactas, Fisicas y Naturales*, **27**, 263–282.

Engesser, B. 1972. Die obermiozäne Säugetierfauna von Anwil (Baselland). *Tatigkeitsberichte der Naturforschenden Gesellschaft Baselland*, **28**, 370–363.

Ferrarezzi, H. and E. do A. Gimenez 1996. Systematic patterns and the evolution of feeding habits in Chiroptera (Archonta: Mammalia). *Journal of Comparative Biology*, **1**, 75–94.

Freeman, P. W. 1981. A multivariate study of the family Molossidae (Mammalia, Chiroptera): morphology, ecology, evolution. *Fieldiana, Zoology*, **7**, 1–173.

Freeman, P. W. 2000. Macroevolution in Microchiroptera: recoupling morphology and ecology with phylogeny. *Evolutionary Ecology Research*, **2**, 317–335.

Gardner, A. L. 1977. Feeding habits. In Buker, R. J., Jones, J. K. Jr, and Carter, D. C. (eds.), *Biology of Bats of the New World Family Phyllostomatidae, Part II*. Lubbock, TX: Museum of Texas Tech University, pp. 293–350.

Gervais, P. 1856. Deuxième mémoire. Documents zoologiques pour servir à la monographie des chéiroptères Sud-Américains. In Gervais, F. L. P. (ed.), *Mammifères*, pp. 25–88, In *Animaux nouveaux ou rares recueillis pendant l'expédition dans les parties centrales de l'Amérique du Sud, de Rio de Janeiro à Lima, et de Lima au Parà; exécutée par ordre du gouvernement Français pendant les années 1843 à 1847, sous la direction du Comte Francis de Castelnau*, ed. F. de Castelnau. Paris: P. Bertrand.

Goodman, S. M., B. J. van Vuuren, F. Ratrimomanarivo, J.-M. Probst, and R. C. K. Bowie 2008. Specific status of populations in the Mascarene Islands referred to *Mormopterus acetabulosus* (Chiroptera: Molossidae), with description of a new species. *Journal of Mammalogy*, **89**, 1316–1327.

Gradstein, F. M., J. G. Ogg, and A. Smith (eds.) 2004. *A Geological Time Scale 2004*. Cambridge, UK: Cambridge University Press.

Gray, J. E. 1825. An attempt at a division of the family Vespertilionidae into groups. *Zoological Journal*, **2**(6), 242–243.

Gray, J. E. 1838. A revision of the genera of bats (Vespertilionidae), and the description of some new genera and species. *Magazine of Zoology and Botany*, **2**, 486–505.

Hand, S. 1990. First Tertiary molossid (Microchiroptera: Molossidae) from Australia: its phylogenetic and biogeographic implications. *Memoirs of the Queensland Museum*, **28**, 175–192.

Hand, S. J., P. Murray, D. Megirian, M. Archer, and H. Godthelp 1998. Mystacinid bats (Microchiroptera) from the Australian Tertiary. *Journal of Paleontology*, **72**, 538–545.

Hand, S. J., B. S. Mackness, C. E. Wilkinson, and D. M. Wilkinson 1999. First Australian Pliocene molossid bat: *Mormopterus* (*Micronomus*) sp. from the Chinchilla Local Fauna, southeastern Queensland. *CAVEPS Symposium, Records of the Western Australian Museum*, **57**(Suppl.), 291–298.

Handley, C. O., Jr. 1956. A new species of free-tailed bat (genus *Mormopterus*) from Peru. *Proceedings of the Biological Society of Washington*, **69**, 197–202.

Hulme, M. and N. Sheard 1999. *Climate Change Scenarios for Argentina*. Norwich, UK: Climatic Research Unit, University of East Anglia. www.cru.uea.ac.uk/~mikeh/research/wwf.argent.pdf

Iudica, C. A., J. Arroyo-Cabrales, T. J. McCarthy, and U. F. J. Pardiñas 2003. An insect-eating bat (Mammalia: Chiroptera) from the Pleistocene of Argentina. *Current Research in the Pleistocene*, **20**, 101–103.

Kay, R. F., R. H. Madden, M. Mazzoni, M. G. Vucetich, G. Ré, M. Heizler, and H. Sandeman 1999. The oldest Argentine primates: first age determinations for the Colhuehuapian South American Land Mammal 'Age'. *American Journal of Physical Anthropology*, **108** (Suppl. 28), 166.

Koopman, K. F. 1982. Biogeography of the bats of South America. In Mares, M. A. and Genoways, H. H. (eds.), *Mammalian Biology in South America*. Pymatuning Laboratory of Ecology, University of Pittsburgh, Pittsburgh, PA: pp. 273–302.

Kunz, T. H. and L. F. Lumsden 2003. Ecology of cavity and foliage roosting bats. In Kunz, T. H. and Fenton, M. B. (eds.), *Bat Ecology*. Chicago, IL: University of Chicago Press, pp. 3–89.

Legendre, S. 1982. Étude anatomique de *Tadarida helvetica* (Chiroptera, Molossidae) du gisement burdigalien de Port-la-Nouvelle (Aude): denture et squelette appendiculaire. *Zoologische Jahrbücher, Anatomie*, **108**, 263–292.

Legendre, S. 1984a. Étude odontologique des représentants actuels du groupe *Tadarida* (Chiroptera, Molossidae): implications phylogénique, systématiques et zoogéographiques. *Revue Suisse de Zoologie*, **91**, 399–442.

Legendre, S. 1984b. Identification de deux sous-genres et compréhension phylogénique du genre *Mormopterus* (Molossidae, Chiroptera). *Comptes Rendus de l'Académie de Sciences, Paris*, **298**, 715–720.

Legendre, S. 1985. Molossidés (Mammalia, Chiroptera) cénozoïques de l'Ancien et du Nouveau Monde: statut systématique, intégration phylogénique des données. *Neues Jahrbuch für Geologie und Paläontologie, Abhandlungen*, **170**, 205–227.

Legendre, S., T. H. V. Rich, P. V. Rich, G. J. Knox, P. Punyaprasiddhi, D. M. Trumpy, J. Wahlert, and P. Napawongse Newman 1988. Miocene fossil vertebrates from the Nong Hen-I(A) exploration well of Thai Shell Exploration and Production Company Limited, Phitsanulok Basin, Thailand. *Journal of Vertebrate Paleontology*, **8**, 278–289.

Madden, R., R. Kay, M. Heizler, J. Vilas, and G. Ré 2004. Geochronology of the Sarmiento Formation at Gran Barranca and elsewhere in Patagonia: calibrating middle Cenozoic mammal evolution in South America. *Journal of Vertebrate Paleontology*, **24**(Suppl. 3), 87A.

Menu, H. and B. Sigé 1971. Nyctalodontie et myotodontie, importants caractères de grades évolutifs chez les chiroptères entomophages. *Comptes Rendus de l'Académie des Sciences, Paris*, **272**, 1735–1738.

Merino, M. L., M. A. Lutz, D. H. Verzi, and E. P. Tonni 2007. The fishing bat *Noctilio* (Mammalia, Chiroptera) in the middle Pleistocene of central Argentina. *Acta Chiropterologica*, **9**, 401–407.

Nowak, R. M. 1994. *Walker's Bats of the World*. Baltimore, MD: Johns Hopkins University Press.

Owen, P. R., C. J. Bell, and E. M. Mead 2000. Fossils, diet, and conservation of black-footed ferrets (*Mustela nigripes*). *Journal of Mammalogy*, **8**, 422–433.

Pardiñas, U. F. L. and E. P. Tonni 1996. El primer vampiro fosil de la Argentina (Mammalia, Chiroptera), significación paleoambiental. *Ameghiniana*, **33**, 468.

Pardiñas, U. F. L. and E. P. Tonni 2000. A giant vampire (Mammalia, Chiroptera) in the late Holocene from the Argentinean pampas: paleoenvironmental significance. *Palaeogeography, Palaeoclimatology, Palaeoecology*, **160**, 213–221.

Pascual, R. 1986. Evolución de los vertebrados cenozoicos: sumario de los principales hitos. *Actas IV Congreso Argentino de Paleontología y Bioestratigrafía*, **2**, 209–218.

Pascual, R. and A. A. Carlini 1987. A new superfamily in the extensive radiation of South American Paleogene marsupials. *Fieldiana, Zoology*, **39**, 99–110.

Pascual, R., M. G. Vucetich, G. J. Scillato-Yané, and M. Bond 1985. Main pathways of mammalian diversification in South America. In Stehli, F. G. and Webb, S. D. (eds.), *The Great American Biotic Interchange*. New York: Plenum Press, pp. 219–247.

Patterson, B. D., M. R. Willig, and R. D. Stevens 2003. Trophic strategies, niche partitioning, and patterns of ecological organization. In Kunz, T. H. and Fenton, M. B. (eds.), *Bat Ecology*. Chicago, IL: University of Chicago Press, pp. 536–579.

Paula Couto, C. de 1956. Une chauve-souris fossile des argiles feuilletées Pléistocènes de Tremembé, État de São Paulo (Brésil). *Actes IV Congrès International du Quaternaire, Roma–Pisa 1953*, **1**, 343–347.

Paula Couto, C. de 1983. Geochronology and paleontology of the basin of Tremembé–Taubaté, state of São Paulo. *Iheringia, séries Geologia, Porto Alegre*, **8**, 5–31.

Paula Couto, C. de and S. Mezzalira 1971. Nova conceituação geocronológica de Tremembé, Estado de São Paulo, Brasil. *Anais da Academia Brasileira de Ciências*, **43**(Supl.), 473–488.

Peters, W. C. H. 1865. Über die brasilianischen, von Spix beschreibenen Flederthiere. *Monatsberichte der Königlich Preussischen Akademie der Wissenschaften zu Berlin*, **1865**, 505–588.

Quintana, C. A. and D. L. Mazzanti 1996. Secuencia faunistica del sitio arqueológico Cueva Tixi (Pleistoceno tardio–Holoceno) provincia de Buenos Aires. *Actas VI Jornadas Pampeanas de Ciencias Naturales*, **187–194**.

Revilliod, P. 1920. Contribution a l'étude des chiroptères des terrains tertiaires. II. *Mémoires de la Société Palé ontologique Suisse, Genève*, **44**, 62–128.

Savage, D. E. 1951. A Miocene phyllostomatid bat from Colombia, South America. *University of California Publications in Geological Sciences*, **28**, 357–366.

Scher, H. D. and E. E. Martin 2006. Timing and climatic consequences of the opening of Drake Passage. *Science*, **312**, 428–430.

Sigé, B. 1985. Les chiroptères oligocènes du Fayum, Egypte. *Geologica et Palaeontologica*, **19**, 161–189.

Sigé, B. and H. Menu 1995. Le Garouillas et les sites contemporains (Oligocene, MP 25) des phosphorites du Quercy (Lot, Tarn-et-Garonne, France) et leurs faunes de vertébrés. V. Chiroptera. *Palaeontographica Abteilung A*, **236**, 77–124.

Sigé, B., H. Thomas, S. Sen, E. Gheerbrant, J. Roger, and Z. Al-Sulaimani 1994. Les chiroptères de Taqah (Oligocène inférieur, Sultanat d'Oman): premier inventaire systématique. *Münchner Geowissenschaftliche Abhandlungen A*, **26**, 35–48.

Simmons, N. B. 2005. Chiroptera. In Wilson, D. E. and Reeder, D. M. (eds.), *Mammal Species of the World A Taxonomic and Geographic Reference*, 3rd edn. Baltimore, MD: Johns Hopkins University Press, pp. 312–529.

Simmons, N. B. and T. M. Conway 2003. Evolution of ecological diversity in bats. In Kunz, T. H. and Fenton, M. B. (eds.), *Bat Ecology*. Chicago, IL: University of Chicago Press, pp. 493–535.

Sudman, P. D., L. J. Barkley, and M. S. Hafner 1994. Familial affinity of *Tomopeas ravus* (Chiroptera) based on protein electrophoretic and cytochrome B sequence data. *Journal of Mammalogy*, **75**, 365–377.

Tejedor, M. F., N. J. Czaplewski, F. J. Goin, and E. Aragón 2004. Los quirópteros más antiguos de Sudamérica. *Resumenes XX Jornadas Argentinas de Paleontología de Vertebrados, La Plata.*

Tejedor, M. F., N. J. Czaplewski, F. J. Goin, and E. Aragón 2005. The oldest record of South American bats. *Journal of Vertebrate Paleontology*, **25**, 990–993.

Tonni, E. P., M. S. Bargo, and J. L. Prado 1988. Los cambios ambientales en el Pleistoceno tardio y Holoceno del sudeste de la provincia de Buenos Aires a través de una secuencia de mamíferos. *Ameghiniana*, **25**, 99–110.

Vucetich, M. G. 1985. *Cephalomyopsis hypselodontus* gen. et sp. nov. (Rodentia, Caviomorpha, Cephalomyidae) de la edad Colhuehuapense (Oligoceno tardio) del Chubut, Argentina. *Ameghiniana*, **22**, 243–245.

Vucetich, M. G. 1986. Historia de los roedores y primates en Argentina: su aporte al conocimiento de los cambios ambientales durante el cenozoico. *Actas VI Congreso Argentino de Paleontología y Bioestratigrafía*, **2**, 157–164.

Vucetich, M. G. and D. H. Verzi 1991. Un nuevo Echimyidae (Rodentia, Hystricognathi) de la edad Colhuehuapense de Patagonia y consideraciones sobre la sistematica de la familia. *Ameghiniana*, **28**, 67–74.

Vucetich, M. G. and D. H. Verzi 1994. The presence of *Protadelphomys* (Rodentia, Echimyidae) in the Colhuehuapian of the South Barrancas of Lake Colhue-Huapi (Chubut). *Ameghiniana*, **31**, 93–94.

Willig, M. R., B. D. Patterson, and R. D. Stevens 2003. Patterns of range size, richness, and body size in the Chiroptera. In Kunz, T. H. and Fenton, M. B. (eds.), *Bat Ecology*. Chicago, IL: University of Chicago Press, pp. 580–621.

Zachos, J., M. Pagani, L. Sloan, E. Thomas, and K. Billups 2001. Trends, rhythms, and aberrations in global climate 65 Ma to present. *Science*, **292**, 686–693.

Zachos, J. C., G. R. Dickens, and R. E. Zeebe 2008. An early Cenozoic perspective on greenhouse warming and carbon-cycle dynamics. *Nature*, **451**, 279–283.

PART III PATTERNS OF EVOLUTION AND ENVIRONMENTAL CHANGE

17 The Mustersan age at Gran Barranca: a review

Mariano Bond and Cecilia M. Deschamps

Abstract

The Mustersan South American Land Mammal Age (SALMA), based on the "Astraponotéen" of Ameghino (late Eocene), represents the first moment when native ungulate mammals showed an increase in the number of high-crowned taxa. Ameghino initially distinguished two levels in his "Astraponotéen": the "Astraponotéen" and an "Astraponotéen le plus supérieur," but later abandoned this subdivision. After Ameghino, other faunas came to be included in the "Astraponotéen," such as those from Cerro del Humo and La Gran Hondonada. New collections made at Gran Barranca with precise stratigraphic control, isotopic dates, and magnetostratigraphy allow resuscitation of the "Astraponotéen le plus supérieur" and demonstrate its equivalence to the Tinguiririran of Chile, in both content and age. New studies show that many Mustersan ungulates with high-crowned cheek teeth come from the "Astraponotéen le plus supérieur." This explains the mixture of faunas of notoungulates with different degrees of hypsodonty. The Mustersan SALMA is represented at Gran Barranca by assemblages from both the Rosado and Lower Puesto Almendra members, dated between 39.0 and 36.5 Ma. These members are separated by unconformities from the underlying Gran Barranca Member and the overlying Vera Member, containing faunas ascribed to the Casamayoran and "Astraponotéen le plus supérieur," respectively. The classical Mustersan site at "Colhue-Huapi Norte" where several Mustersan taxa were found (among them the type of *Astraponotus assymmetrus*) is the same site as Cerro del Humo. The Rosado Member at Profile J (Gran Barranca) is here proposed as the source of the assemblage of taxa that serve as the type fauna of the Mustersan SALMA, because it has the characteristic taxa, precise upper and lower boundaries, and is placed within a well-dated and very complete mammal sequence.

Resumen

La Edad Mamífero Mustersense, basada en el "Astraponotéen" de Ameghino (Eoceno tardío), representa el primer momento en que los ungulados nativos muestran un incremento en el número de taxa de corona alta. Ameghino primero mencionó dos niveles el "Astraponotéen" y el "Astraponotéen le plus supérieur", pero más tarde abandonó esta subdivisión. Después de Ameghino se incluyeron en estas faunas, aquellas provenientes de otros sitios como Cerro del Humo o La Gran Hondonada. Nuevas colecciones realizadas en Gran Barranca con adecuado control estratigráfico, dataciones isotópicas y magnetoestratigrafía, permitieron resucitar el "Astraponotéen le plus supérieur" y comprobar su equivalencia con el Tinguiririquense (Eoceno tardío-Oligoceno temprano) de Chile. Los nuevos estudios muestran que muchos de los ungulados Mustersenses con coronas más altas provienen del "Astraponotéen le plus supérieur". Esto explica la supuesta mezcla de faunas de notoungulados con diferentes grados de hipsodoncia, considerada típica del Mustersense. La Edad Mustersense está representada en la Gran Barranca por el Miembro Rosado y el Miembro Puesto Almendra Inferior, datados entre 39,0 y 36,5 Ma. Estas unidades están separadas por dos discordancias del infrayacente Miembro Gran Barranca y el suprayacente Miembro Vera (que contienen faunas asignadas respectivamente al Casamayorense y al "Astraponotéen plus supérieur" = Tinguiririquense). También se comprobó que el sitio Mustersense clásico "Colhue-Huapi Norte", de donde provienen varios de los taxones Mustersenses, entre ellos el tipo de *Astraponotus assymmetrus*, que da nombre al "Astraponotense" (=Mustersense), es sin duda el mismo que Cerro del Humo. Se propone considerar al Miembro Rosado, como nivel tipo de la Edad Mustersense debido a que tiene los fósiles característicos, cuenta con límites inferior y superior precisos y está incluido en una secuencia estratigráfica datada y muy completa.

Introduction

Since the work of Florentino Ameghino more than a century ago, many of the localities and the biostratigraphic and geochronologic units defined by him have received different denominations. Sometimes Ameghino's units were neglected or joined together, and even the original concept was altered. As Simpson (1933) noted, the "Astraponotéen" of Ameghino, currently the Mustersan South American Land Mammals Age (SALMA), has been "happy" to receive little attention from researchers after Ameghino, as there was no multiplication of names that could confuse its original concept. However, the problem with the Mustersan arose at the very beginning by its original definition. In Ameghino's original concept, this fauna apparently involved a "mixture"

The Paleontology of Gran Barranca: Evolution and Environmental Change through the Middle Cenozoic of Patagonia, eds. R. H. Madden, A. A. Carlini, M. G. Vucetich, and R. F. Kay. Published by Cambridge University Press. © Cambridge University Press 2010.

of related taxa with very different degrees of hypsodonty, and he changed his mind about the existence of more than one level within the "Astraponotéen." Even the original locality and assemblage used in its definition was not made clear. Gran Barranca was assumed to be the type locality, but as years passed, the concept of the Mustersan SALMA came to be based mainly on mammals from localities other than Gran Barranca. Thus, until very recently the faunal assemblage of the Mustersan SALMA, although defined or exemplified in new richer localities, continued being characterized by a mixture of taxa of different age.

The rediscovery that the Mustersan SALMA includes more than one faunal assemblage prompts us to review the Mustersan faunas at Gran Barranca. New collections with precise biostratigraphic context allow a revision of its faunal composition. Here we review this history in order to understand and interpret accurately the original concept in light of the new relationships among elements of this fauna, and propose a new type locality for the Mustersan.

The beginnings

Between 1887 and 1906, knowledge of the geology of the lower and middle Tertiary of Patagonia changed radically from what was known by D'Orbigny (1842), Darwin (1846), or Doering (1882). This change was almost solely due to the effort of the Ameghino brothers, Carlos in the field and Florentino in the cabinet (a museum's cabinet or an adapted store room!). We owe an enduring debt to them for much of our understanding of the Cenozoic faunal succession in Patagonia. The succession they established is not only still in use in Argentina but has become the reference standard for SALMAs (Pascual *et al.* 1965, 1966; Marshall *et al.* 1983; Cione and Tonni 1995).

Between 1889 and 1895 Carlos Ameghino discovered first in Chubut and later in Santa Cruz Province fossil mammals that became Florentino's "*Pyrotherium* fauna" and the basis for his "Étage Pyrothéreén" (presently the Deseadan SALMA). This "Étage" was the first pre-Santacrucian biochronologic unit recognized in Patagonia. According to Simpson (1967a), the "*Pyrotherium* fauna" was found by Carlos Ameghino in his 1895–96 expedition (but perhaps even as early as 1893: Ameghino 1897), when he explored south of Lake Colhue-Huapi (Fig. 17.1). Only later, during his 1898–99 expedition, did Carlos realize that the original collection of fossils described as the *Pyrotherium* fauna actually included material of a distinctly older fauna from beds which eventually were named the "Couches à *Notostylops*." Among the pre-Deseadan mammals described in 1897 from the "couches à *Pyrotherium*," some may have been collected at Gran Barranca. In two letters to Florentino (15 February 1899 and 9 June 1899, in Ameghino 1900; Simpson 1967a) Carlos reported a good

exposure in "Colhue-Huapi" where the "*Notostylops*" and "*Pyrotherium*" beds were perfectly concordant stratigraphically but not "paleontologically" as there were sediments devoid of fossils between them (Fig. 17.2). But in his expedition of 1899–1900, he found mammals in beds he formerly supposed to be sterile. Florentino Ameghino (1901) named these new mammal-bearing beds and its fauna the "faune des couches à *Astraponotus*" and the corresponding stage as the "Étage Astraponotéen" (currently the Mustersan SALMA). However, he provided no precise geographic location. Consequently, we only know through specimen labels that the type of the taxon characterizing this stage, *Astraponotus assymmetrus* (Fig. 17.3A–C), was not collected at Gran Barranca but at "Colhue-Huapi Norte," a locality north of Lake Colhue-Huapi, sometimes written as "Colhuapi" or "Coluapi Norte." Thus, Carlos collected "Astraponotéen" fossils not only at Gran Barranca but also at other sites (Fig. 17.1).

All these mammal-bearing levels were included in Ameghino's "Formation Guaranienne." He still used the term "Formation" to mean a coeval group of beds or "couches," or a chronological unit, but not as was beginning to be used worldwide following recommendations of geological congresses, a lithological unit of lower hierarchy. He also used the term "couches" as a unit defined by fossils (nearly equivalent to biostratigraphic units), as did many authors at that time. Moreover, sometimes his "couches" were equivalent to a Substage or a Stage, representing an "Étage" with its corresponding "Age."

Descriptions of the new taxa from the "Faune Astraponotéenne" and a faunal list were published in several papers (Ameghino 1901, 1902, 1904, 1905), but unfortunately without detailed geographic location. Nearly all of them were found in the "Couches à *Astraponotus*," but three, *Pseudopachyrucos foliiformis*, *Interhippus deflexus*, and *Proplatyarthrus longipes* (Ameghino 1901, 1902, 1905) came from the "partie la plus supérieure des couches à *Astraponotus*." Thus, between 1901 and 1906, F. Ameghino distinguished two fossil mammal-bearing levels within the "couches à *Astraponotus*."

In his geology of Patagonia, Ameghino (1906) considered the "Astraponotéen" fauna transitional between the "*Notostylops*" and "*Pyrotherium*" faunas, and with "stratification parfaitement concordante." "Astraponotéen" exposures around Lakes Musters and Colhue-Huapi were as extensive as those of the Notostylopéen, but because they were less fossiliferous it was difficult to determine the upper and lower boundaries of the "Astraponotéen" beds. The location of the outcrops in maps (Ameghino 1906) is far from exact. One of the localities, with molds of land snails (*Strophocheilus*), was figured on a profile on the left margin of the "río Chico del Chubut" on the upper third of its valley (Ameghino 1906) (Fig. 17.1). It was considered

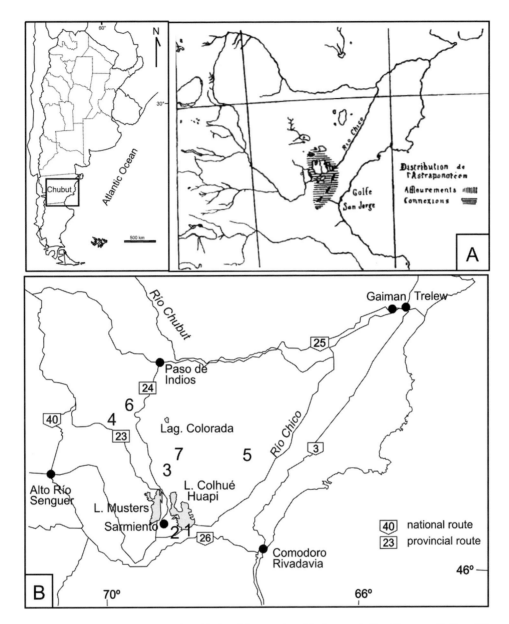

Fig. 17.1. Location map. (A) Original map of Ameghino; (B) localities mentioned in the text: 1, Gran Barranca; 2, Cerro Blanco (west flank of Gran Barranca); 3, Colhue-Huapi Norte = Cerro del Humo = Cretáceo superior del Lago Musters = El Pajarito; 4, Laguna del Mate; 5, Sierra Cuadrada; 6, Gran Hondonada; 7, Cerro Talquino.

important because carnivorous dinosaur remains (*Genyodectes serus*) were mentioned in the uppermost Notostylopéen or Astraponotéen beds. Ameghino (1906, Figs. 31, 58) included a profile of Gran Barranca (Fig. 17.2), based on field observations by Carlos, showing the "Astraponotéen" composed of greenish and whitish clays between the "notostylopéen supérieur" and the "pyrothéréen," and below a mantle of eruptive basalt. In that paper, Florentino stated that the evolutionary transition between the *Astraponotus* and *Pyrotherium* faunas seemed to be very gradual and the temporal hiatus probably not very long. Hence,

Ameghino abandoned the intermediate level ("partie la plus supérieure des couches à *Astraponotus*"), but ambiguously said that a transitional fauna might be found. Instead, he found stronger evidence for a paleontological "hiatus" between the "Astraponotéen" and the underlying "Faunes Notostylopéennes."

Ameghino (1906) characterized the "Faune Astraponotéenne" pointing out the independent appearance in many lineages of ungulates of a strong tendency toward molar hypselodonty ("vers l'hypselodontie"). Among other faunal distinctions, he reported the decrease in the diversity of

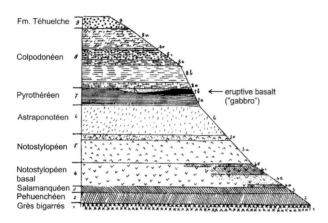

Fig. 17.2. Profile from the Gran Barranca, south of Lake
Colhue-Huapi, according to C. Ameghino (from F. Ameghino 1906,
figure 81, page 112).

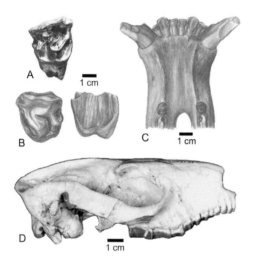

Fig. 17.3. Examples of typical Mustersan mammals. (A–C)
Astraponotus assymmetrus and (D) *Puelia coarctatus*. (A) Type of
Astraponotus dicksoni, MLP 12–2217, right M1 or M2; (B) MLP
67-II-27–46, left M1 or M2; (C) MLP 82-V-7–2, mandibular
symphysis with right and left I1–3, C1 and P2–3; (D) MLP 67-II-
27–27, skull in lateral view.

"condylarths" and notostylopids and the appearance of the
earliest pyrotheriids. Hence, the "Faune Astraponotéenne"
appeared as an intermediate stage between the more brachyodont ungulates of the "Faunes Notostylopéennes" and
the more hypsodont or even hypselodont (or euhypsodont)
taxa of the "Faune Pyrotheréenne." Thus, as early as 1906
the stratigraphic position and general features of the "Faune
Astraponotéenne" were established, but neither the precise
location of its main localities nor the provenance of its
original materials was ever reported.

How many of Ameghino's fossil specimens of Mustersan
age really came from Gran Barranca? At the time Carlos

Ameghino was collecting in the "Astraponotéen," his rival
Santiago Roth of the Museo de La Plata collected Mustersan mammals at three sites, especially one north of Lake
Colhue-Huapi (called Musters by Roth). Like Carlos, Roth
reported no precise locations for his Mustersan localities
(Roth 1899, 1901, 1903), but later he indicated them on an
unpublished map (Simpson 1936a). The fossils found in
the most productive of these sites were labeled "C. S. M.,"
"Cretáceo Superior Lago Musters," as Roth thought these
beds were much older (Simpson 1936a). Roth also collected
at another locality, Cañadón Blanco (probably in central
Chubut), that he supposed to be lower Tertiary in age,
probably "Pyrotheréen." Marshall (1982) confused Roth's
Cañadón Blanco with Simpson's Mustersan locality of
Cerro Blanco west of Colhue-Huapi (Fig. 17.1B).

After Ameghino: between Simpson and the SALMAs

For 20 years after 1906 nobody studied the "Faune Astraponotéenne." Feruglio (1929) and Frenguelli (1930) used
the term "Capas con *Astraponotus*," but Kraglievich (1930)
used "Mustersense," and all later authors eventually
accepted this name.

After the Ameghinos, George Gaylord Simpson is one
of the most influential figures in the study of Patagonian
Paleogene mammals. No other author has ever made such
an exhaustive review of the early mammalian faunas of
Argentina and South America. From 1930 to 1934 Simpson
led the Scarritt Expeditions to Patagonia, making new collections from new localities and also from the classic ones
of Ameghino and Roth, with accurate field notes and
sketches of the geology and stratigraphy. (Curiously the
first references of Simpson to the Mustersan were related
to diseases or dinosaurs. Studying a pathological specimen
of the Roth Collection in the Museo de La Plata from the
"C. S. M.," Simpson (1932a) said that it probably came from
the *Astraponotus* beds near "El Pajarito," west of Cerro del
Humo (Fig. 17.1B). On the other hand, Simpson (1932b)
disproved the occurrence of dinosaurs in the *Notostylops*
and *Astraponotus* beds, which he considered as Tertiary in
age.) Simpson (1933, 1940) accepted that the *Astraponotus*
beds contained a distinctive fauna different from those of
the "Notostylopéen" and "Pyrotheréen," and "officialized"
the use of Mustersan (also spelled Mustersian) for this Age
and Stage, and suggested that it could be subdivided in
minor "faunules or facies." He proposed Musters Formation
for the mammal-bearing sediments, as part of the Sarmiento
Group. For Simpson (1941) the Mustersan fauna was a
distinctive biostratigraphic zone, and as he supposed the
Casamayoran as early Eocene, he assigned the Mustersan
to the late Eocene. Some of the Mustersan exposures appear
as channels cut into the underlying Casamayoran, and in

some localities it is absent between the Casamayoran and Deseadan; because of this, Simpson (1941, 1948) inferred a temporal hiatus separating the Mustersan from the Deseadan fauna. Simpson (1940, 1941) considered that the Mustersan fauna was represented at Cerro del Humo, Colhue-Huapi (Gran Barranca), Cañadón Colorado and perhaps other sites. In his study of the Roth collections from the "C. S. M." and "Cañadón Colorado" Simpson (1936a, 1967a) found that many of the taxa were synonyms of those of Ameghino, and that other different taxa were represented in collections made by Carlos Ameghino at Colhue-Huapi Norte. Simpson (1936a) concluded that the "C. S. M." was the same as his Cerro del Humo and also probably "Colhue-Huapi Norte," although Carlos emphatically denied this (Simpson 1967a).

Later, Simpson (1967b) did not recognize distinct levels within the Mustersan, despite the fact that these had been mentioned by F. Ameghino, and Simpson's own statements of 1933 and 1940. He included some taxa from "Cañadón Blanco" in the Mustersan fauna, even though they were more evolved than those from typical Mustersan levels of Cerro del Humo and Colhue-Huapi Norte. The bulk of Simpson's Mustersan specimens come from Cerro del Humo, but some from Gran Barranca, Cerro Blanco (on the westward flank), Profile A, and the upper levels of Profile M (Simpson 1930).

Shortly after Simpson, Bordas (1943, 1945) made a good collection of Mustersan fossils at Cerro del Humo, Gran Barranca, Sierra Cuadrada, and especially at Cerro Talquino (Fig. 17.1B). Feruglio's (1949) geological description of Patagonia did not change what was previously known for the Mustersan.

Sometime before 1946, Tomás Suero, a geologist of the national petroleum company (YPF) discovered a locality in central Chubut (El Pozón or La Gran Hondonada) (Fig. 17.1B), that proved to be extremely rich in fossil mammals. During the 1960s and early 1970s, this locality, and the nearby Laguna del Mate (Fig. 17.1B), were intensively prospected and rich collections of Mustersan mammals were made (Odreman Rivas 1978; Cladera *et al.* 2004). In their classic work about Tertiary vertebrate evolution of Argentina, Pascual and Odreman Rivas (1971) (see also Pascual 1970; Pascual *et al.* 1965) considered the Mustersan to be middle Eocene, as they regarded the "Divisaderan" as late Eocene and the Deseadan as early Oligocene (see López this book, concerning the Divisaderan fauna). Their faunal list was based almost solely on fossil mammals from Cerro del Humo, Colhue-Huapi Norte, and La Gran Hondonada. This faunal list is nearly the same as that of Simpson (1967b), and also includes a mixture of related taxa with high-crowned to brachyodont cheek teeth. Pascual and Odreman Rivas (1971, 1973) interpreted the absence of reptiles in the Mustersan as due to more temperate climates than those of the Casamayoran. They interpreted the presence

of hypsodont notoungulates as an indication of a herbaceous steppe paleoenvironment. The discontinuous distribution of Mustersan sediments and the presence of channel-beds cutting the underlying Casamayoran suggested that the low flat plains of the Casamayoran had been modified by an uplift-induced erosion event (Pascual and Odreman Rivas 1973).

The new Mustersan fossil localities of La Gran Hondonada and Laguna del Mate gradually enlarged the mammal list of this SALMA, which became more and more based on remains from these localities and Cerro del Humo. For example, in the *Stratigraphical Lexicon of the Argentine Republic* (Servicio Geológico Nacional 1976), Gran Barranca is only secondarily mentioned as an exposure for the Mustersan.

Spalletti and Mazzoni (1979) studied the stratigraphy of the Sarmiento Formation at Gran Barranca; they mention Casamayoran and Deseadan fossils, but they do not mention Mustersan ones.

Then, Marshall *et al.* (1983), in a classic synthesis of the geochronology of the mammal-bearing Tertiary of South America, tentatively assigned the Mustersan to the middle Eocene. They repeated the misunderstanding that the "*Astraponotus* fauna" was first found and recognized by specimens collected by Carlos Ameghino in his 1895–96 expedition, and suggested Gran Barranca as the type locality of the Mustersan "land-mammal age," also because a "majority" of Ameghino's specimens came from there.

In a biostratigraphic study of the Casamayoran, Richard Cifelli (1985) had access to Simpson's unpublished field notes of Gran Barranca. He reported two sites with Mustersan mammals, the first in his Profile I (Simpson's Profile A as in this book) occurs in the base of Simpson's Channel Series, and the second in Profile V (Simpson's Profile M as in this book) with the notohippid ?*Eomorphippus pascuali* collected above and below an unconformity 20 m higher than the Channel Series of Profile V (= Profile M). The unconformity corresponds to the erosional base of Simpson's Upper Channel Beds (Simpson 1967b) currently assigned to the Tinguirirican SALMA.

Legarreta and Uliana (1994), clearly influenced by the observations of earlier authors on the channel series of Gran Barranca, stated again that the Mustersan beds were filling channels incised in the Casamayoran, with hiatuses between the Upper Casamayoran (Barrancan) and the Deseadan. These authors considered the Mustersan to be late middle Eocene in age and referred it to the Bartonian.

Flynn and Swisher (1995) assigned the Mustersan to the interval 45–50 Ma. These authors commented on the new fauna from Tinguiririca in central Chile which eventually led to the recognition of a new interval of time, the Tinguirirican SALMA, not younger than 31.5 Ma (Wyss *et al.* 1994; Flynn *et al.* 2003). This new SALMA filled the gap between the Mustersan and the Deseadan, traditionally filled in part by the controversial "Divisaderan" SALMA (López this book). At about the same time, restudy of the Ameghino and Roth collections led Bond *et al.* (1996, 1997) to resuscitate Ameghino's original subdivision of the "Astraponotéen" and to recognize that many of the mammals with a higher degree of hypsodonty probably came from levels higher than those yielding the original "Astraponotéen" fauna. These higher levels could be equivalent to the upper levels of "Cañadón Blanco" and to the Tinguiririran SALMA. The revival of the "Astraponotéen le plus supérieur" and recognition of its equivalence to the Tinguiririran make clear that the supposed mixture of notoungulates with very different degrees of hypsodonty is the result of joining two different faunas from different stratigraphic levels and ages.

Gran Barranca revisited and the Mustersan site

Recently, a series of joint expeditions from Duke University and the Museo de La Plata, together with the Museo Argentino de Ciencias Naturales and Universidad de Buenos Aires, have made new collections of fossils with stratigraphic and geochronologic control. The completeness of the fossil record of Gran Barranca, together with the new geochronology, clarify the timing of the faunal succession between the Barrancan and Tinguiririran SALMAs.

In 1999, a revised geochronology at Gran Barranca using new isotopic age determinations, indicated an age for the Barrancan between 35.3 and 37.6 Ma (see Kay *et al.* 1999), that is late Eocene. This would imply that the Mustersan was late Eocene and the "Astraponotéen plus supérieur" early Oligocene. Recently, additional isotopic determinations indicate an older age for the Barrancan of 42.1–38.3 Ma, or middle Eocene (Ré *et al.* Chapter 4, this book). The Tinguiririran SALMA or "Astraponotéen plus supérieur" at Gran Barranca is present in the Vera Member (Bellosi and Madden 2005; Bellosi this book), estimated as not older than 33.3 Ma (Ré *et al.* Chapter 4, this book). Many of the taxa described by Roth from "Cañadón Blanco" that were placed in synonymy with taxa described by Ameghino from "Colhue-Huapi" (Gran Barranca?) are now in fact recorded from the Vera Member. Among them, *Eomorphippus* is very common. The Mustersan SALMA is represented in the Rosado and Lower Puesto Almendra members (Ré *et al.* Chapter 3, this book).

Unraveling systematics and provenance

The systematic history of some notoungulates illustrates the causes of the confusion concerning the Mustersan concept.

Ameghino (1901) described *Eomorphippus obscurus*, as coming from his "Couches à *Astraponotus*" without further

details as to provenance, but it is noteworthy that the original material (MACN 10917) comes from a level very rich in manganese. This taxon shows molar crowns higher than any other Mustersan toxodontid. Simpson (1967b) considered *Eomorphippus obscurus* as senior synonym of *Pseudostylops subquadratus* Ameghino 1901, *Eurystomus stehlini* Roth 1901, and *Lonkus rugei* Roth 1901. The last two taxa come from the Roth's "Formación terciaria inferior" of Cañadón Blanco, which he said it was clearly different and younger than his "C. S. M.," the designation for fossils that later became the most representative example of the Mustersan fauna. Because of this synonymy, Simpson (1967b) listed *Eomorphippus obscurus* in the Mustersan, disregarding the stratigraphic provenance which Roth gave for the junior synonyms *Eurystomus* and *Lonkus*. *Eomorphippus* was also found in upper levels of the putative Mustersan at Gran Barranca, in a section where the Italian geologist Egidio Feruglio (1927) had also found it (Simpson 1936b). The notohippid *?Eomorphippus pascuali*, relatively lower-crowned than *E. obscurus* but more advanced than the reputedly ancestral notohippid *Puelia* (see below), was also found stratigraphically higher than the levels with the "typical" Mustersan fauna (Simpson 1967b; Cifelli 1985).

Ameghino (1902) described the notohippid *Interhippus deflexus* as coming from the "partie la plus supérieure des couches à *Astraponotus*." Simpson (1967b) said that it could have been collected from the overlaying Deseadan beds, and thus he listed it with doubt in the Mustersan fauna.

Ameghino (1902) reported the Archaeohyracidae "*Archaeohyrax sulcidens*" ("*Bryanpattersonia sulcidens*," Simpson 1967b) as coming from the "Couches à *Astraponotus*." This species is a synonym of "*Archaeohyrax gracilis*" Roth 1903 (= *Protarchaeohyrax gracilis*: Reguero *et al.* 2003a) from his "Formación terciaria inferior" at Cañadón Blanco. As Roth considered the fossils from Cañadón Blanco to be Deseadan in age, Simpson (1967b) concluded that this fauna probably included a mixture of Casamayoran, Mustersan, and Deseadan fossils based also on Bryan Patterson's personal communication (Simpson, 1967b).

In this way, the Mustersan became a mixed assemblage of taxa from different stratigraphic levels, notably, the higher-crowned taxa came from "Cañadón Blanco" or from geographically imprecise and stratigrafically poorly documented "Astraponotéen" levels.

Another interesting case is that of *Astraponotus*. This genus is represented by a few remains from the Mustersan beds at Gran Barranca and from other places. By contrast, *Astraponotus* is very common in the Mustersan beds at Laguna del Mate, La Gran Hondonada (Cladera *et al.* 2004), and Cerro del Humo. Whilst revising the old collections, I have observed that the type of *Astraponotus assymmetrus* from Ameghino's "Colhue-Huapi Norte" is the *same individual* as a specimen collected by Roth in the "C. S. M." (Cerro del Humo of Simpson). This demonstrates that Cerro del Humo and "Colhue-Huapi Norte" are certainly a single locality, and confirms Simpson's suspicion (1967a).

The Mustersan fauna at Gran Barranca

Based on new collections made in the Rosado Member, Carlini *et al.* (2005a) recorded Dasypodidae: Dasypodinae: Stegotheriini, Astegotheriini (with a new species of *Stegosimpsonia*), Euphractinae: Euphractini, Eutatini (*Meteutatus* cf. *attonsus*), and also larger xenarthrans as the glyptodontoid *Machlydotherium*.

The Astrapotheria are represented by fragmentary isolated teeth of *Astraponotus assymmetrus* and by the Trigonostylopidae *Trigonostylops gegembauri* (Roth 1899), very similar to and comparable in size to the larger specimens of *T. wortmanni*. The type of *T. gegembauri* comes from the Mustersan of "C. S. M."

The Litopterna are represented by fragmentary remains, mainly isolated teeth of a medium to large brachyodont and bunodont taxon similar to *Xesmodon* and *Decaconus* (Proterotheriidae). There are also remains similar to *Lambdaconus*, a medium- to large-sized didolodontid "condylarth." The Mustersan is the last moment in which these bunodont forms are recorded (see Gelfo this book).

The Notoungulata are represented by several groups. The Notostylopidae are represented by *Otronia muhlbergi* (described by Roth (1901) from "C. S. M."), and the Isotemnidae (basal toxodontids) by fragmentary remains of *Periphragnis* and *Rhyphodon*, two common taxa in beds of Mustersan age.

Another Mustersan notoungulate present at Gran Barranca is *Puelia* originally described by the type species *P. plicata* Roth 1899 from "C. S. M." New material from the Mustersan beds at La Gran Hondonada, allowed us to refer another species to this genus, *P. coarctatus* (Fig. 17.3D) (Bond and López 1993). Both species are present in the Rosado Member. They have been considered either as notoungulates *incertae sedis*, isotemnids similar to or somehow related to the Notohippidae (Simpson 1967b), either as basal Notohippidae (Bond and López 1993) or basal Toxodontia (Shockey 1997). Whatever their phylogenetic position, these species have been recorded so far only in Mustersan beds. Their morphology is what might be expected for a structural ancestor of generalized Notohippidae such as the Tinguirirican *Eomorphippus obscurus*.

Other notoungulates recorded in the Rosado Member are Oldfieldthomasiidae, Archaeohyracidae, and notopithecine Interatheriidae. The Oldfieldthomasiidae probably pertain to the *Ultrapithecus–Tsamnichoria* group. The archaeohyracids are represented by *Pseudhyrax* sp. and *P. strangulatus*, a mesodont form, very frequent and typical of the Mustersan SALMA, also recorded in the Tinguirirican SALMA

(Croft *et al.* 2003). Notopithecine Interatheriidae is represented by *Guilielmoscottia,* typical of the Mustersan SALMA (Simpson 1967b).

The only incongruence in the Rosado fauna is the presence of an interatheriid Interatheriinae, cf. *Eopachyrucos*, which is a hypsodont taxon characteristic of the Tinguirirican SALMA (Hitz *et al.* 2003; Reguero *et al.* 2003b). If its provenance is confirmed, it will represent the oldest record of an advanced interatheriid.

Conclusions

Although Florentino Ameghino stated that the "Couches à *Astraponotus*" and the "Astraponotéen" were recognized in a section of the Gran Barranca found by Carlos Ameghino, from this review it is clear that the bulk of fossil mammals that were used to define the Mustersan SALMA came from the localities of Cerro del Humo (or Colhue-Huapi Norte), and later from Cerro Blanco, Cerro Talquino, La Gran Hondonada, and Laguna del Mate (Fig. 17.1B). The case of the holotype of *Astraponotus assymmetrus* mentioned above proves that the concept of "Astraponotéen" was constructed mainly on the basis of materials coming from north of Lake Colhue-Huapi and not from Gran Barranca. As we have also seen, some of the taxa reported by Ameghino for the "Astraponotéen" which very probably came from Gran Barranca have been recognized as components of the Tinguirirican SALMA or "Astraponotéen le plus supérieure." The recognition of the "Astraponotéen plus supérieur" or Tinguirirican fauna as distinct from those of the Mustersan, and their stratigraphic and temporal separation, allows discrimination of the more advanced higher-crowned taxa of the Tinguirirican from the lower-crowned taxa from the Mustersan and the Barrancan, and documents the transition of these faunas to more open and grass-dominated habitats.

The type locality of the Mustersan Stage/Age should be defined at Gran Barranca – the locality with the best and continuous mammal record for the middle Eocene – early Miocene – in Profile J (Bellosi this book) where the Rosado Member has both the characteristic fossils and well-defined upper and lower limits.

ACKNOWLEDGEMENTS

The authors thank the editors for the invitation to contribute with this chapter, M. Reguero and A. Kramarz for access to the collections of the Museo de La Plata and the Museo Argentino de Ciencias Naturales "Bernardino Rivadavia" respectively and information supplied, R. H. Madden for information on some aspects of the material, and M. G. Vucetich and R. F. Kay for valuable comments on the manuscript. Field research was supported by US National Science Foundation grants EAR-0087636, BCS-0090255, and DEB-9907985 to Richard F. Kay and Richard H. Madden.

REFERENCES

Ameghino, F. 1897. Mammifères crétacés de l'Argentine: deuxième contribution à la connaissance de la faune mammalogique des couches à *Pyrotherium. Boletín del Instituto Geográfico Argentino*, **18**, 406–521.

Ameghino, F. 1900–02. l'Age des formations sédimentaires de Patagonia. *Anales del la Sociedad Científica Argentina*, **50**, 109–130, 145–165, 209–229; **51**, 20–39, 65–91; **52**, 189–197, 244–250; **54**, 161–180, 220–249, 283–342.

Ameghino, F. 1901. Notices préliminaires sur de ongulés nouveaux de terrains crétacés de Patagonie. *Boletín de la Academia Nacional de Ciencias en Córdoba*, **16**, 349–426.

Ameghino, F. 1902. Notice préliminaire sur des mammifères nouveaux des terrains crétacés de Patagonie. *Boletín de la Academia Nacional de Ciencias en Córdoba*, **17**, 5–70.

Ameghino, F. 1904. Nuevas especies de mamíferos cretáceos y terciarios de la República Argentina. *Anales de la Sociedad Científica Argentina*, **56**, 193–208; **57**, 162–175, 327–341; **58**, 35–71, 182–192, 225–291.

Ameghino, F. 1905. La faceta articular inferior única del astrágalo de algunos mamíferos no es un carácter primitivo. *Anales del Museo Nacional de Buenos Aires*, **3**, 1–64.

Ameghino, F. 1906. Les formations sédimentaires du Crétacé supérieur et du Tertiaire de Patagonie avec un paralléle entre leurs faunes mammalogiques et celles de l'ancien continent. *Anales del Museo Nacional de Buenos Aires*, **15**, 1–568.

Bellosi, E. S. and R. H. Madden 2005. Estratigrafia física preliminar de las secuencias piroclásticas terrestres de la Formación Sarmiento (Eoceno-Mioceno) en la Gran Barranca, Chubut. *Actas XVI Congreso Geológico Argentino*, **4**, 427–432.

Bond, M. and G. M. López 1993. El primer Notohippidae (Mammalia, Notoungulata) de la Formación Lumbrera (Grupo Salta) del noroeste argentino: consideraciones sobre la sistemática de la familia Notohippidae. *Ameghiniana*, **30**, 59–68.

Bond, M., G. López, and M. A. Reguero 1996. "Astraponotéen plus supérieure" of Ameghino: another interval in the Paleogene record of South America. *Journal of Vertebrate Paleontology*, **16**(Suppl. 3), 23A.

Bond, M., M. A. Reguero, G. M. López, A. A. Carlini, F. J. Goin, R. H. Madden, M. G. Vucetich, and R. F. Kay 1997. The"Astraponotéen plus supérieur" (Paleogene) in Patagonia. *Ameghiniana*, **34**, 533.

Bordas, A. F. 1943. Contribución al conocimiento de las bentonitas argentinas. *Revista Minera, Geología y Mineralogía*, **14**, 1–60.

Bordas, A. F. 1945. Geología estratigráfica de algunas zonas de Patagonia. *Anales del Museo de la Patagonia*, **1**, 139–184.

Carlini, A. A., M. Ciancio, and G. J. Scillato-Yané 2005a. Los Xenarthra de Gran Barranca: más 20 Ma de Historia. *Actas XVI Congreso Geológico Argentino*, **4**, 419–424.

Cifelli, R. L. 1985. Biostratigraphy of the Casamayoran, Early Eocene, of Patagonia. *American Museum Novitates*, **2820**, 1–26.

Cione, A. L. and E. P. Tonni 1995. Chronostratigraphy and "Land-Mammal Ages" in the Cenozoic of southern South America: principles, practices and the "Uquian" problem. *Journal of Paleontology*, **69**, 135–159.

Cladera, G., E. Ruigomez, E. Ortiz Jauireguizar, M. Bond, and G. M. López 2004. Tafonomía de La Gran Hondonada (Formación Sarmiento, edad-mamífero Mustersense, Eoceno Medio) Chubut, Argentina. *Ameghiniana*, **41**, 315–330.

Croft, D. A., M. Bond, J. J. Flynn, M. A. Reguero, and A. R. Wyss 2003. Large archaeohyracids (Typotheria, Notoungulata) from Central Chile and Patagonia, including a revision of *Archaeotypotherium. Fieldiana, Geology*, n.s., **49**, 1–38.

Darwin, C. R. 1846. *Geological Observations on South America, being the Third Part of the Geology of the voyage of* Beagle, *under the Command of Captain Fitzroy, R. N. during the Years 1832 to 1836*. London: Smith Elder and Co.

D'Orbigny, A. 1842. *Voyage dans l'Amérique Méridionale* vol. 3, Geologie. Paris: P. Bertrand.

Doering, A. 1882. Geología. *Informe Oficial de la Comisión Científica Agregada al Estado Mayor General de la Expedición al Río Negro*, **3**, 299–530.

Feruglio, E. 1929. Apuntes sobre la constitución geológica del la region del Golfo de San Jorge. *Anales de la Sociedad Argentina de Estudios Geográficos "Gaea,"* **3**, 395–486.

Flynn, J. J. and C. C. Swisher III 1995. Chronology of the Cenozoic South American Land Mammal Ages. In Berggren, W. A., Kent, D. V., Aubry, M. P., and Hardenbol, J. (eds.), *Geochronology, Time-Scales, and Global Stratigraphic Correlation*, Special Publication no. 54. Tulsa, OK: Society for Sedimentary Geology, pp. 317–333.

Flynn, J. J., A. R. Wyss, D. A. Croft, and R. Charrier 2003. The Tinguirírica Fauna, Chile: biochronology, paleoecology, biogeography, and a new earliest Oligocene South American Land Mammal 'Age'. *Palaeogeography, Palaeoclimatology, Palaeoecology*, **195**, 229–259.

Frenguelli, J. 1930. Nomenclatura estratigráfica patagónica. *Anales de la Sociedad Científica de Santa Fé*, **3**, 1–117.

Hitz, R. B., M. A. Reguero, A. R. Wyss, and J. J. Flynn 2000. New Interatheriines (Interatheriidae, Notoungulata) from the Paleogene of Central Chile and Southern Argentina. *Fieldiana, Geology*, n.s., **42**, 1–26.

Kay, R. F., R. H. Madden, M. G. Vucetich, A. A. Carlini, M. M. Mazzoni, G. H. Ré, M. Heizler, and H. Sandeman 1999. Revised geochronology of the Casamayoran South Land-Mammal Age: climatic and biotic implications. *Proceedings of the National Academy of Sciences USA*, **96**, 13 235–13 240.

Kraglievich, L. 1930. La Formación Friaseana del río Frías, río Fénix, Laguna Blanca, etc. y su fauna de mamíferos. *Physis*, **10**, 127–161.

Legarreta, L. and M. A. Uliana 1994. Asociaciones de fósiles y hiatos en el Supracretácico–Neógeno de Patagonia: una perspectiva estratigráfico–secuencial. *Ameghiniana*, **31**, 257–281.

Marshall, L. G., R. Hoffstetter, and R. Pascual 1983. Mammals and stratigraphy: geochronology of the continental mammal-bearing tertiary of South America. *Palaeovertebrata, Mémoire Extraordinaire*, 1–93.

Odreman Rivas, O. 1978. Sobre la presencia de un Polydolopidae (Mammalia, Marsupialia) en capas de Edad Mustersense (Eoceno Medio) de Patagonia. *Obra del Centenario del Museo de La Plata*, **5**, 29–38.

Pascual, R. 1970. Evolución de las comunidades, cambios faunísticos e integraciones biocenóticas de los vertebrados cenozoicos de Argentina. *Actas IV Congreso Latinoamericano de Zoología*, **2**, 991–1088.

Pascual, R. and O. Odreman Rivas 1971. Evolución de las comunidades de los vertebrados del Terciario argentino: los aspectos paleozoogeográficos y paleoclimáticos relacionados. *Ameghiniana*, **8**, 372–421.

Pascual, R. and O. Odreman Rivas 1973. Las unidades estratigráficas del Terciario portadoras de mamíferos, su distribución y sus relaciones con los acontecimientos diastróficos. *Actas V Congreso Geológico Argentino*, **3**, 293–338.

Pascual, R., E. J. Ortega Hinojosa, D. Gondar, and E. P. Tonni 1966. Las Edades del Cenozoico mamalífero de la Provincia de Buenos Aires. In *Paleontografía Bonaerense Vertebrata*. Buenos Aires: Comisión de Investigaciones Científicas, vol. 4, pp. 1–27.

Reguero, M. A., D. A. Croft, J. J. Flynn, and A. R. Wyss 2003a. Small archaeohyracids (Typotheria, Notoungulata) from Chubut Province, Argentina, and Central Chile: implications for Trans-Andean temporal correlation. *Fieldiana, Geology*, n.s., **48**, 1–17.

Reguero, M. A., M. Ubilla, and D. Perea 2003b. A new species of *Eopachyrucos* (Mammalia, Notoungulata, Interatheriidae) from the Late Oligocene of Uruguay. *Journal of Vertebrate Paleontology*, **23**, 445–457.

Roth, S. 1899. Aviso preliminar sobre Mamíferos Mesozoicos encontrados en Patagonia. *Revista del Museo de La Plata*, **9**, 381–388.

Roth, S. 1901. Notas sobre algunos nuevos mamíferos fósiles. *Revista del Museo de La Plata*, **10**, 251–256.

Roth, S. 1903. Noticia preliminar sobre nuevos mamíferos fósiles del Cretáceo Superior y Terciario Inferior de la Patagonia. *Revista del Museo de La Plata*, **11**, 133–158.

Servicio Geológico Nacional 1976. *Léxico Estratigráfico de la República Argentina*. Buenos Aires: Secretaría de Estado de Minería, Servicio Geológico Nacional.

Shockey, B. J. 1997. Two new notoungulates (Family Notohippidae) from the Salla Beds of Bolivia (Deseadan: Late Oligocene): systematics and functional morphology. *Journal of Vertebrate Paleontology*, **17**, 584–599.

Simpson, G. G. 1930. *Scarritt-Patagonian Exped. FieldNotes*. New York: American Museum of Natural History. (Unpublished.) Available at http://paleo.amnh.org/ notebooks/index.html.

Simpson, G. G. 1932a. The most ancient evidence of diseases among South American mammals. *American Museum Novitates*, **543**, 1–4.

Simpson, G. G. 1932b. The supposed association of dinosaurs with mammals of Tertiary type in Patagonia. *American Museum Novitates*, **566**, 1–21.

Simpson, G. G. 1933. Stratigraphic nomenclature of the early Tertiary of Patagonia. *American Museum Novitates*, **644**, 1–13.

Simpson, G. G. 1936a. Notas sobre los mamíferos más antiguos de la Colección Roth. *Instituto del Museo de la Universidad Nacional de La Plata, Obra del Cincuentenario*, **2**, 63–94.

Simpson, G. G. 1936b. A specimen of *Pseudostylops subquadratus* Ameghino. *Memoria dell'Instituto Geologico Reale Universita di Padova*, **11**, 1–12.

Simpson, G. G. 1940. Review of the mammal-bearing Tertiary of South America. *Proceedings of the American Philosophical Society*, **83**, 649–709.

Simpson, G. G. 1941. The Eogene of Patagonia. *American Museum Novitates*, **1120**, 1–15.

Simpson, G. G. 1948. The beginning of the Age of Mammals in South America. I. *Bulletin of the American Museum of Natural History*, **91**, 1–232.

Simpson, G. G. 1967a. The Ameghinos' localities for early Cenozoic mammals in Patagonia. *Bulletin of the Museum of Comparative Zoology*, **136**(4), 63–76.

Simpson, G. G. 1967b. The beginning of the Age of Mammals in South America. II. *Bulletin of the American Museum of Natural History*, **137**, 1–259.

Spalletti, L. A. and M. M. Mazzoni 1979. Estratigrafía de la Formación Sarmiento en la barranca sur del Lago Colhué Huapí, provincia del Chubut. *Revista de la Asociación Geológica Argentina*, **34**, 271–281.

Wyss, A., J. J. Flynn, M. A. Norell, C. C. Swisher III, M. J. Novacek, M. C. McKenna, and R. Charrier 1994. Paleogene mammals from the Andes of Central Chile: a preliminary taxonomic, biostratigraphic, and geochronologic assessment. *American Museum Novitates*, **3098**, 1–31.

18 A new mammal fauna at the top of the Gran Barranca sequence and its biochronological significance

Alejandro G. Kramarz, María Guiomar Vucetich, Alfredo A. Carlini, Martín R. Ciancio, María Alejandra Abello, Cecilia M. Deschamps, and Javier N. Gelfo

Abstract

The top of the stratigraphic sequence of the fossil-bearing Sarmiento Formation at Gran Barranca has yielded a distinctive assemblage of mammalian taxa. Fossils occur in strata located 25 m above the uppermost level bearing typical Colhuehuapian mammals. This association is represented by fragmentary remains of small to middle-sized mammals belonging to paucituberculatan marsupials, dasypodids, glyptodonts, tardigrades, hegetotheres, interatheres, proterotheriids, and hystricognath rodents. Although this fauna contains at least two new rodent species not recorded in other faunas, it is more similar to that recorded at the lower and middle sections of the Pinturas Formation in northwest Santa Cruz Province (the Ameghinos' Astrapotericulan fauna) than to any other known fossil mammal assemblage. The strata yielding this fauna (Upper Fossil Zone, UFZ) is estimated to be in Chron C6n (19.7 to 18.7 Ma), but the taxic similarity with the younger Astrapotericulan fauna (dated between 17.5 and 16.5 Ma) suggests that the UFZ is placed at the youngest extreme of the Chron C6n. The Astrapotericulan and the UFZ assemblages would correspond to a single post-Colhuehuapian – pre-Santacrucian biochronological unit (the "Pinturan") which would span from ∼18.75 to ∼16.5 Ma. This is the youngest faunal unit within the standard sequence of the middle Cenozoic SALMAs at Gran Barranca. The assignment to this unit of other early Miocene faunas elsewhere in South America (at present poorly or partially known) depends on further descriptions and future findings.

Resumen

El tope de la secuencia estratigráfica de la Formación Sarmiento en Gran Barranca ha brindado una singular asociación de taxones de mamíferos. Los fósiles se registran en estratos ubicados 25 metros por encima de los niveles más altos con fauna colhuehuapense. Esta asociación está representada por restos fragmentarios de mamíferos pequeños a medianos, correspondientes a marsupiales paucituberculados, dasipódidos, gliptodontes, tardígrados, hegetoterios, interaterios, proterotéridos y roedores histricognatos. Aunque esta asociación contiene al menos dos especies nuevas de roedores no registradas en otras faunas, es más similar a la registrada en las secuencias inferior y media de la Formación Pinturas en el NW de la provincia de Santa Cruz (fauna Astrapotericulense de Ameghino) que a cualquier otra asociación conocida de mamíferos fósiles. Se estima que los estratos portadores de esta fauna (Upper Fossil Zone —UFZ—) están en el Cron C6n (19.7 a 18.7 Ma), pero las afinidades sistemáticas con la fauna más joven Astrapotericulense (datada entre 17.5 y 16.5 Ma) sugiere que la UFZ se encuentra en el extremo más joven del Cron C6n. Las asociaciones del Astrapotericulense y de la UFZ corresponderían a una única unidad biocronológica post-colhuehuapense-pre-santacrucense (el "Pinturense") que se extendería desde ca. 18.75 a 16.5 Ma. Esta es la unidad faunística más joven dentro de la secuencia estándar de SALMAs del Cenozoico medio en Gran Barranca. La asignación de otras faunas del Mioceno temprano en otras partes de Sudamérica (hasta ahora poco o parcialmente conocidas) a esta unidad depende de nuevas descripciones y futuros hallazgos.

The Paleontology of Gran Barranca: Evolution and Environmental Change through the Middle Cenozoic of Patagonia, eds. R. H. Madden, A. A. Carlini, M. G. Vucetich, and R. F. Kay. Published by Cambridge University Press. © Cambridge University Press 2010.

Introduction

The Sarmiento Formation at Gran Barranca is probably the most complete Cenozoic mammal-bearing sequence in South America (Simpson 1940; Madden *et al.* 2005b). As recognized by Ameghino (1906), this unit documents the occurrence of four superposed faunal zones (Casamayoran, Mustersan, Deseadan, and Colhuehuapian). Field work over the last ten years has revealed other mammal-bearing horizons documenting the existence of more faunal assemblages than previously thought (Carlini *et al.* 2005, this book; López *et al.* 2005; Vucetich *et al.* 2005; Goin *et al.* this book). Among these mammal-bearing levels, herein is examined a new one located at the top of the Colhue-Huapi Member section at this locality (Upper Fossil Zone, UFZ: Ré *et al.* Chapter 3, this book; GBV-54 and 55: Bellosi 2005, this book), 25 meters above the uppermost horizon bearing typical Colhuehuapian mammals (Lower Fossil Zone, LFZ: Ré *et al.* Chapter 3, this book). Based on radioisotopic and paleomagnetic data, the UFZ is referred to the early Miocene (Ré *et al.* Chapter 4, this book). According to Bellosi and González (this book) this level "is constituted by thick tephric loessites and scarce, weakly developed calcic paleosols, representing a semiarid eolian environment." The recovered materials include very fragmentary specimens with partial dentitions, isolated teeth, and cingulate osteoderms. After a preliminary examination of the rodents, Vucetich *et al.* (2005) concluded that this assemblage is similar to that described by Ameghino (1900–02, 1906) for his "Couches à *Astrapotericulus*", in the Pinturas valley, northwest Santa Cruz Province, Argentina (Fig. 18.1), and assigned a "Pinturan" – early Santacrucian age to these fossil-bearing deposits (see also Carlini *et al.* 2005; Madden *et al.* 2005a).

In this contribution we analyse in detail the taxa recorded in the UFZ (Appendix 18.1), and we compare them with those recorded in other early–middle Miocene South American faunas. Likewise, we discuss the biochronological implications of these faunas in the refinement of the chronological calibration of the South American early Miocene.

Abbreviations

MACN, Museo Argentino de Ciencias Naturales; MPEF-PV, Museo Paleontológico "Egidio Feruglio" Paleontología Vertebrados (Trelew, Chubut Province, Argentina); SALMA, South American Land Mammal Age.

Systematic paleontology

Class **MAMMALIA** Linnaeus 1758
Infraclass **MARSUPIALIA** Illiger 1811
Order **PAUCITUBERCULATA** Ameghino 1894
Family **PALAEOTHENTIDAE** Sinclair 1906
Genus *Titanothentes* Rae, Bown and Fleagle 1996
Titanothentes sp. nov.
Fig. 18.2A.

Referred materials MPEF-PV 4541, isolated left M1 and MPEF-PV 4544, left maxillary fragment with partial M1 and complete M2.

Comments *Titanothentes* was originally considered as a member of the Palaeothentinae paleothentids (Rae *et al.* 1996). However, as a result of a recent revision of the Paucituberculata (Abello 2007) this genus is currently included in the Decastinae (= Acdestinae Bown and Fleagle 1993; see Abello 2007). Likewise, a new species was described for *Titanothentes*, which includes a group of specimens previously referred to

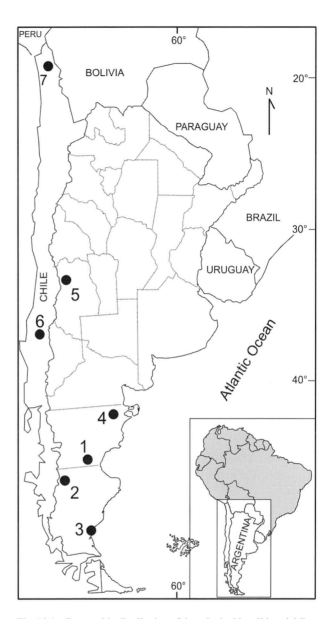

Fig. 18.1. Geographic distribution of the principal localities yielding early Miocene fossil mammals. 1, Gran Barranca; 2, Pinturas Valley; 3, Monte León – Monte Observación; 4, Gaiman; 5, Divisadero Largo; 6, Laguna del Laja (after Wertheim *et al.* 2006); 7, Chucal.

Acdestis oweni (Bown and Fleagle 1993) found in the lower and middle levels of the Pinturas Formation.

The material from UFZ shows a large development of the metaconule and of the cusp of the premetaconular crest of the M1, interpreted by Abello (2007) as diagnostic characters of *Titanothentes*. In addition, these specimens are referred to *Titanothentes* sp. nov. because they agree in size with the specimens from the Pinturas Formation, which are 20% smaller than *Titanothentes simpsoni*, the type species of the genus. *Titanothentes* sp. nov. is recorded exclusively

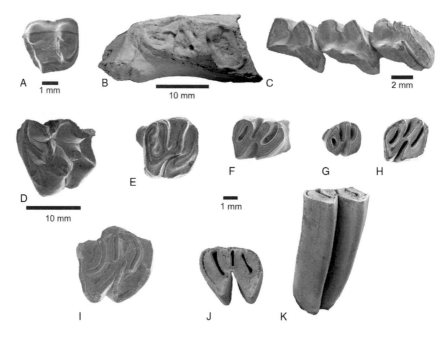

Fig. 18.2. Fossil mammals from top of the Gran Barranca section. (A) *Titanothentes* sp. nov., MPEF-PV 4541, left M1; (B) *Nematherium* sp., MPEF-PV 7289, right mandibular fragment with m3–4; (C) *Interatherium* sp., MPEF-PV 7296, left maxillary fragment with P4–M2; (D) *Tetramerorhinus* sp., MPEF-PV 8128, fragmentary left M1 or M2; (E) *Eosteiromys*? sp., MPEF-PV 7981, right M1 or M2; (F) *Prostichomys bowni*, MPEF-PV 7980, left m1 or m2; (G, H) *Prostichomys* cf. *P. bowni*: (G) MPEF-PV 6066, right M1 or M2; (H) MPEF-PV 7318d, isolated right or left M3; (I) *Neoreomys pinturensis*, MPEF-PV 7986, fragmentary right M1 or M2; (J, K) *Luantus propheticus*, MPEF-PV 6062a, right m1 or m2; (J) occlusal view, (K) lingual view. Scale bar 1 mm for A, C, and E–K.

in the lower and middle levels of the Pinturan Formation, and in the UFZ of Gran Barranca.

Genus *Palaeothentes* Ameghino 1887a
Palaeothentes minutus?

Referred materials MPEF-PV 4543, isolated right m2.

Comments This specimen shows two diagnostic characters of the genus *Palaeothentes*: distal end of the oblique cristid contacts the notch of the protocristid and the entocristid is curved towards the talonid basin. The crown is gracile and the cusps are sharp, resembling *Palaeothentes minutus* (recorded in both Astrapothericulan and Santacrucian beds), but the available material is too fragmentary to allow an accurate specific determination.

Family **ABDERITIDAE** Ameghino 1889
Genus *Abderites* Ameghino 1887a
Abderites meridionalis Ameghino 1887a

Referred materials MPEF-PV 4104, isolated left m2; MPEF-PV 4103, isolated right M2; MPEF-PV 4105, isolated right M2 (incomplete).

Comments These specimens are referred to *A. meridionalis* because they lack the paraconid on m2 and metacone on M2, as distinct from the Colhuehuapian species *A. crispus*. Moreover, the m2 has

longer trigonid and bigger protoconid, and the metaconid is less anteriorly displaced than in *A. crispus*. The size of m2/M2 agrees with *A. meridionalis* (nearly 30% larger than in *A. crispus*). This species is the best-represented marsupial in the UFZ. *Abderites meridionalis* is recorded in the Astrapothericulan and Santacrucian (including Notohippidean) beds.

Infraclass **EUTHERIA** Gill 1872
Superorder **XENARTHRA** Cope 1889
Order **CINGULATA** Illiger 1811
Family **DASYPODIDAE** Gray 1821
Genus *Stegotherium* Ameghino 1887a
Stegotherium tessellatum Ameghino 1887a

Referred materials MPEF-PV 7315D, 8118B, and 8122A; lot of fixed and movable isolated osteoderms.

Comments As reported by Ameghino in the original description of *S. tessellatum*, these osteoderms are small and thick (9.5 mm long, 6.5 mm wide, 3.8 mm thick); the external surface is smooth with punctuations and many peripheral pits around the exposed area, with anterior and posterior beveled margins. The movable osteoderms are proportionally longer and narrower (21 mm long, 7.3 mm wide) than the fixed ones; some have a hardly insinuated medial keel. This species was so far reported for the Santacrucian SALMA (Ameghino 1887a; Scott 1903–05).

Stegotherium variegatum Ameghino 1902
Referred materials MPEF-PV 8118A and MPEF-PV 8122F; lot of fixed and movable isolated osteoderms.
Comments These osteoderms differ from those of *S. tessellatum* and match with the syntype of *S. variegatum* (MACN A-12680) by having rough external surface, a well-developed medial keel, and larger number of peripheral pits. Ameghino (1902) pointed out that the osteoderms of the syntype of *S. variegatum* exhibit different morphologies (besides the presence of a well-developed medial keel) and suggested they could belong to more than one species; thus, an accurate diagnosis awaits more detailed taxonomic revision. *Stegotherium variegatum* was described for the Colhuehuapian SALMA (Ameghino 1902).

Genus *Prozaedyus* Ameghino 1891a
Prozaedyus sp.
Referred materials MPEF-PV 7315C, 8119, and 8122B; lot of fixed and movable isolated osteoderms.
Comments These osteoderms are referred to *Prozaedyus* by their small size (fixed osteoderms 8 mm long, 5 mm wide; movables 12 mm long, 4.5 mm wide) among the euphractines, the elongated and narrow central figure, surrounded by peripheral figures forming two lateral rows, and very few and small piliferous pits on the posterior border. The fixed osteoderms agree with those of the Colhuehuapian species *P. humilis* in having peripheral figures with flat surfaces, posterior figures 40% longer than in the Santacrucian *P. proximus* and *P. exilis*, and only a single piliferous pit on the center of the posterior margin of the osteoderm. However, peripheral figures are better defined than in *P. humilis*, and the central figure reaches the posterior margin, resembling the Santacrucian species. The osteoderms bear six or seven peripheral figures, whereas *P. humilis* has only six, and the Santacrucian species have seven to nine. These combined features, partially intermediate between the Colhuehuapian and Santacrucian species, suggest that these osteoderms would belong to a distinct and probably new species, more derived than *P. humilis*.

Genus *Stenotatus* Ameghino 1891b
Stenotatus cf. *S. patagonicus* (Ameghino 1887a)
Referred materials MPEF-PV 7315B, 8121, and 8122C; lot of isolated osteoderms.
Comments The fixed osteoderm match with those of *Stenotatus patagonicus* (Santacrucian – "Friasian" – this latter encompasses the typical Friasian of Chile, and Colloncuran and Mayoan of Argentina) in having convex and well-defined figures, an elliptical central figure surrounded by four to six peripheral figures, and small pits on the groove delimiting the anterior

margin of the central figure. On the posterior border of the osteoderm there are three or four large piliferous pits, intercalated dorsally with small piliferous ones; all of these pits are placed in a transverse furrow. However, all the osteoderms are significantly smaller (8–11 mm long and 5–7 mm wide, that is 20% smaller), than the smallest Santacrucian specimens of this species.

Genus *Proeutatus* Ameghino 1891a
Proeutatus sp.
Referred materials MPEF-PV 7314A, 7315A, and 8122D; lot of fixed and movable isolated osteoderms.
Comments The fixed osteoderms are thick and massive, with a bottle-shaped central figure and four anterior peripheral figures, and large piliferous pits on the posterior margin, which are characteristic of the genus. They differ from the Colhuehuapian species *P. postpuntum* in lacking the two characteristic holes on the posterior portion of the osteoderm, from the Santacrucian *P. lagena* and *P. distans* in being smaller, and from *P. carinatus* in lacking a conspicuous keel. However, the available material is not distinctly different from the Santacrucian *P. oenophorus* or *P. robustus*.

Family **PELTEPHILIDAE** Ameghino 1894
Genus *Peltephilus* Ameghino 1887a
Peltephilus pumilus Ameghino 1887a
Referred material MPEF-PV 8120C, a fragmentary fixed osteoderm.
Comments This osteoderm shows a very rough external surface with conspicuous tubercles and a prominent medial keel with two large foramina on the anterior portion, characteristic of the genus *Peltephilus*. Size (10 mm wide) and tubercles morphology agree with the Santacrucian *P. pumilus* (species based on isolated osteoderms), which is typically 50% smaller than the smallest remaining peltephilids (with the only exception of *P. nanus*).

Family **GLYPTODONTIDAE** Gray 1869
Subfamily **PROPALAEHOPLOPHORINAE** Ameghino 1891b
Propalaehoplophorinae indet.
Referred material MPEF-PV 7281, 7314B, 7350, 7521, and 8120B; lot of isolated osteoderms.
Comments The osteoderms are thin, flat, and externally punctuated, with an oval central figure. Surrounding the central figure there are one or two rows of peripheral, generally pentagonal figures. A single pit is located in the intersection of the sulcus delimiting the central figure and the radial grooves that separate peripheral figures. These features reveal an evolutionary

stage typical of the Oligocene – middle Miocene glyptodonts, traditionally grouped within the Propalaehoplophorinae. However, the fragmentary condition of the available materials prevents a more accurate taxonomic identification.

Order **PHYLLOPHAGA** Owen 1842
Family **MYLODONTIDAE** Gill 1872
Subfamily **NEMATHERIINAE** Mercerat 1891
Genus *Nematherium* Ameghino 1887a
Nematherium sp.
Fig. 18.2B.

Referred materials MPEF-PV 7289, right mandibular fragment with m3–4.

Comments The recovered specimen matches with *Nematherium* (Santacrucian) in size and in the mandibular and dental morphology. The position of the posterior foramen of the dental channel and the shape of the preserved teeth resemble *N. angulatum*. However, the m4 is proportionally shorter, and the labial sulcus is more superficial than in the Santacrucian specimens (e.g. PU 15530, see Scott 1903–05, pl. LXIII, fig. 4), although probably this feature is variable in this species (Scott 1903–05). All the remaining nematherines are known only from the typical coastal Santacrucian localities of the Santa Cruz Formation, except for both *N. birdi* Simpson 1941 (from Palomares – or Santacrucian – Formation in southern Chile), and specimens referred to this genus from levels older than 18 Ma at central Chile (Flynn *et al.* 2008).

Order **NOTOUNGULATA** Roth 1903
Family **INTERATHERIIDAE** Ameghino 1887b
Genus *Interatherium* Ameghino 1887b
(ex Moreno 1882)
Interatherium sp.
Fig. 18.2C.

Referred materials MPEF-PV 7291, left mandibular fragment with m3; MPEF-PV 7294, left maxillary fragment with two cheek teeth; MPEF-PV 7296, left maxillary fragment with P4–M2; MPEF-PV 8126, two upper and four lower cheek teeth; MPEF-PV 8124, one upper and nine lower cheek teeth.

Comments The available cheek teeth have columnar, ever-growing crowns. They differ from those of *Protypotherium* and *Cochilius* by being smaller and showing characteristic features of the genus (Sinclair 1909): upper molars with a deep lingual sulcus and conspicuous anterolabial crests, posterior upper premolars molariform, lower molars bilobated, nearly figure-eight-shaped, m3 with an insinuated posterior lobe. These materials do not allow precise assignment to any of the numerous described species, which, in

turn, are in need of a deep taxonomic revision. Sinclair (1909) interpreted *Icochilus* Ameghino as synonym of *Interatherium*, and transferred all the Santacrucian species. However, no authority has discussed the validity and the generic assignment of *Icochilus ulter*, described by Ameghino (1899) for his Astrapothericulan beds. Preliminary examination of the type specimen of *Icochilus ulter* (MACN A-11601) suggests that it is congeneric with the Santacrucian species of *Icochilus*, independently of its validity. *Interatherium* is also recorded at Colloncuran beds (Roth 1899; also described as *Icochilus*).

Genus *Protypotherium* Ameghino 1885
Protypotherium sp.

Referred materials MPEF-PV 7282, a maxillary fragment with partial P3–M3.

Comments It is assigned to *Protypotherium* because of the characteristic reduction of the upper premolar anterior prism (Sinclair 1909). Compared with the three Santacrucian species interpreted as valid by Tauber (1996), the preserved cheek teeth are similar in size to *P. australe* (Moreno 1882), but their imbrication is not as marked, resembling *P. praerutilum* Ameghino. This genus is recorded from the Colhuehuapian (Bordas 1939; de Barrio *et al.* 1989; Kramarz *et al.* 2005) to the late Miocene "Piso Mesopotamiense", from where it was originally described (Ameghino 1885).

Family **HEGETOTHERIIDAE** Ameghino 1894
Genus *Pachyrukhos* Ameghino 1885
Pachyrukhos sp.

Referred materials MPEF-PV 7299, left mandibular fragment with two cheek teeth; MPEF-PV 7287, left maxillary fragment with P3–M1; MPEF-PV 8123, five upper and four lower isolated cheek teeth; MPEF-PV 8125, one upper and three lower isolated cheek teeth.

Comments The available teeth differ from *Hegetotherium* by being much smaller (almost 50%) and more delicate, the lowers with nearly subequal prisms, the m3 with well-developed third prism. These are characteristic features of *Pachyrukhos* (Ameghino 1889). The genus is recorded from the Colhuehuapian (Ameghino 1902), up to the lower levels of the Arroyo Chasicó Formation (late Miocene: Cerdeño and Bond 1998).

Order **LITOPTERNA** Ameghino 1889
Family **PROTEROTHERIIDAE** Ameghino 1887a
Genus *Tetramerorhinus* Ameghino 1894
Tetramerorhinus sp.
Fig. 18.2D.

Referred materials MPEF-PV 8128, an incomplete, isolated left M1 or M2.

Comments The labial side of this tooth is broken, hence only the lingual enamel edge of the paracone and metacone are preserved. The relative cusp size and distribution match with those of the species of *Tetramerorhinus*. The size of the preserved part of the crown is similar to the corresponding in the Santacrucian species *T. lucarius* (Ameghino 1894) and *T. prosistens*, described for the Astrapothericulan beds (Ameghino 1899, 1900–02). However, we cannot verify the presence of the labial folding of the paracone and metacone, diagnostic characters of *T. prosistens* (Soria 2001), precluding species assignment.

Order **RODENTIA** Bowdich 1821
Suborder **HYSTRICOGNATHI** Tullberg 1899
Family **ERETHIZONTIDAE** Thomas 1897
Genus *Eosteiromys*?
Eosteiromys? sp.
Fig. 18.2E.

Referred materials MPEF-PV 7981, isolated right M1 or M2; MPEF-PV 6064a, incomplete left p4; MPEF-PV 6064d, incomplete lower right molar; MPEF-PV 6064e, incomplete right m3.

Comments The upper molar MPEF-PV 7981 is low-crowned, with five transverse crests separated by rather deep and wide fossettes. It differs from the named species of *Eosteiromys* by being smaller (probably the smallest early Miocene erethizontid from Patagonia) and by having the hypoflexus in communication with the paraflexus, even in heavily worn stages. These features are shared with a species previously identified and referred tentatively to *Eosteiromys* by Candela (2000), and reported as *Eosteiromys*? sp. nov. from several Colhuehuapian localities (Vucetich *et al.* this book). However, it differs from those by having the protoloph isolated from the paracone. The remaining recovered materials (MPEF-PV 6064a, d, and e) are incomplete lower cheek teeth, which match in size with the MPEF-PV 7981.

Family **ACAREMYIDAE** Ameghino 1902
Genus *Acaremys* Ameghino 1887a
Acaremys sp.

Referred materials MPEF-PV 7989, isolated left p4; MPEF-PV 7982a isolated left m1 or m2

Comments These cheek teeth are figure-eight-shaped, higher crowned than in *Galileomys*, but not as much as in *Sciamys*. The lower molar (MPEF-PV 7982a) differs from those of *Galileomys* because it has no conspicuous cusps and the fossetids are more ephemeral. The p4 (MPEF-PV 7989) has a flexid on the anterior wall,

which is much shallower than in the species of *Galileomys*. This p4 lacks an anterior accessory flexid, present in *A. tricarinatus* Ameghino 1894, but the available material does not allow definite assignment to some of the remaining species proposed by Scott (1905). *Acaremys* was originally described for the Santacrucian (Ameghino 1887a) and also in Colhuehuapian levels of the Sarmiento Formation at Gran Barranca (Vucetich and Kramarz 2003).

Family **ECHIMYIDAE** Miller and Gidley 1918
Genus *Acarechimys* Patterson (in Kraglievich 1965)
Acarechimys sp.

Referred materials MPEF-PV 6063, isolated left upper molar.

Comments The available molar is similar in size and morphology to specimens identified as *A. minutissimus* from the Santa Cruz Formation, but the validity and extension of the three recognized species of the genus requires further study. The genus *Acarechimys* has a long biochron, from the Colhuehuapian (Vucetich *et al.* this book) to Laventan (Walton 1997).

Genus *Prostichomys* Kramarz 2001
Prostichomys bowni Kramarz 2001
Fig. 18.2F.

Referred materials MPEF-PV 7990, incomplete right dp4; MPEF-PV 7980, isolated left m1 or m2; MPEF-PV 7976, isolated left m1 or m2; MPEF-PV 7318a, isolated left DP4.

Comments These cheek teeth match in size, hypsodonty, and occlusal morphology with those described for *P. bowni*, recorded in the Astrapothericulan levels of the Pinturas Formation (Kramarz 2001, 2004; Kramarz and Bellosi 2005), and it was considered as characteristic of these levels. Wertheim *et al.* (2006) reported the presence of a new (still undescribed) species of *Prostichomys* at the lowest levels of the Miocene Cura Mallín Formation (Chile), but did not provide biochronological considerations.

Prostichomys cf. *P. bowni*
Fig. 18.2G, H.

Referred materials MPEF-PV 6066, isolated right M1 or M2; MPEF-PV 7318c isolated right M1 or M2; MPEF-PV 7318d, isolated right or left M3.

Comments These cheek teeth are similar in occlusal morphology but are slightly larger than those of *Prostichomys bowni*. Particularly, the crown of the little-worn upper molar MPEF-PV 6066 has more columnar appearance, but not as much as in the Santacrucian genus *Stichomys*, suggesting that it could belong to a different species.

Family **DASYPROCTIDAE** Smith 1842
Genus *Neoreomys* Ameghino 1887a
Neoreomys pinturensis Kramarz 2006b
Fig. 18.2I

Referred materials MPEF-PV 7986, incomplete right M1 or M2.

Comments This molar is referred to *N. pinturensis* because the crown is lower than that of *N. australis*. Specimens MPEF-PV 6062b and c, 6065a and b, also belong to *Neoreomys*, but they are too incomplete to make a definite specific assignment. Both species occur in Astrapothericulan levels of the Pinturas Formation (Kramarz 2006b), but only the latter is recorded at the Santacrucian levels. *Neoreomys? huilensis* Fields 1957, from the La Victoria Formation, Laventan SALMA, late middle Miocene, in Colombia, is significantly smaller than the Patagonian species.

Family **EOCARDIIDAE** Ameghino 1891c
Genus *Luantus* Ameghino 1899
Luantus propheticus Ameghino 1899
Figs. 18.2J, K.

Referred materials MPEF-PV 6062a, isolated right m1 or m2.

Comments This cheek tooth is clearly referable to *L. propheticus* because the crown is higher than in *L. initalis* (Colhuehuapian) but lower than in *L. toldensis* (Notohippidan?) (Kramarz 2006a). The lingual basal areas of exposed dentine have moderate height, as in specimens from Astrapothericulan levels of the Pinturas Formation. Other teeth from UFZ (MPEF-PV 7979, 7978, 7985) are very incomplete and/or too worn to make a definite specific determination, but they have cement in senile stages indicating that they cannot be allocated to *L. initialis*. *Luantus propheticus* is a characteristic species from the Astrapothericulan levels of the Pinturas Formation (Kramarz and Bellosi 2005; Kramarz 2006a). Wertheim *et al.* (2006) reported the presence of a new species (still undescribed) of *Luantus* from the lowest levels of the Miocene Cura Mallín Formation (Chile).

Family **DINOMYIDAE** Alston 1876
Gen. et sp. nov.

Referred materials MPEF-PV 6068a, isolated left m3; MPEF-PV 6068b, isolated left M1 or M2; MPEF-PV 7977, isolated right M3.

Comments These remains belong to a genus not recorded in any other fauna. It is much smaller and lower-crowned than the species of *Scleromys* (Astrapothericulan–Santacrucian). The m3 has four oblique crests separated by deep, compressed valleys. The hypoflexid merges with the metaflexid even in a senile stage. The hypolophid is continuous with the ectolophid. The labial end of the mesolophid contacts the anterolophid, but not the ectolophid or protoconid. This occlusal pattern is similar to that of *Scleromys*, but the anterofossettid is comparatively larger and persists until more advanced stages of wear, even more than in the Astrapothericulan *S. quadrangulatus* (Kramarz 2006b), resembling species of "*Scleromys*" from La Venta. The upper molars are pentalophodont (the metalophid of M3 is rudimentary). As in the lower molar, the crests are oblique and separated by deep, compressed valleys. The hypoflexus is continuous with the paraflexus, the paraloph is continuous with the anterior arm of the hypocone (there is no distinct mure), and the mesoloph contacts the posteroloph. This pattern also resembles the Laventan "*Scleromys*" *colombianus* (Fields, 1957, fig. 14, 5 for the M3) rather than the Patagonian species of *Scleromys*.

The age and correlations of the Upper Fossil Zone fauna

The early Miocene is one of the best-known periods (in terms of temporal calibrations, faunal diversity, etc.) within the continental Tertiary sequence of southern South America. Simpson (1940) recognized two faunal units within this interval in the hierarchy of Land Mammal Ages – Colhuehuapian (essentially equivalent to the Colpodonéen of Ameghino 1902) and Santacrucian (including the Notohippidéen of Ameghino) – which are still valid today. A post-Colhuehuapian – pre-Santacrucian faunal zone, the Astrapothericulan, was recognized by Ameghino (1900–02) based on the mammal assemblage from the continental Miocene deposits at the upper valley of the Pinturas River (northwest Santa Cruz Province, Fig. 18.1), currently known as the Pinturas Formation (Bown *et al.* 1988). Castellanos (1937) coined the term "Pinturense" for the faunal interval represented by the Ameghino's Astrapothericulan fauna. But Simpson (1940) concluded that there was not enough geological or paleontological evidence to determine whether the Pinturense was younger than Colhuehuapian or older than Santacrucian. However, Wood and Patterson (1959, pp. 366–367) considered the Astrapothericulan not "to be anything but a Santacrucian local fauna." This interpretation was accepted by later authors (Pascual *et al.* 1965; Pascual and Odreman Rivas 1971; Marshall 1976; Marshall *et al.* 1983). Consequently, most authors do not recognize the "Pinturense" as a distinct unit (e.g. Pascual and Ortiz Jaureguizar 1990; Flynn and Swisher 1995; Croft *et al.* 2004; Ortiz Jaureguizar and Cladera 2006), whereas a few recognized it as a subage within the Santacrucian SALMA (Soria 2001; Pascual *et al.* 2002). Recent systematic studies based upon marsupials, rodents, and litopterns (Bown and Fleagle 1993; Kramarz and Bellosi 2005; Abello 2007) conclude that the Astrapothericulan fauna is composed of distinct taxa, generally more primitive than those of the typical localities of the Santa Cruz Formation. In spite of this, the recognition of the Pinturan as a distinct biochronologic unit is still controversial.

Previous discussions about the age and validity of these biochronologic units were based mainly on the evolutionary stage of the faunas, and when possible, on stratigraphic relationships. Currently, geochronological data and new faunal and biostratigraphic information is now available for many early Miocene faunas and these provide new evidence helpful for understanding their relationships. The fauna of the UFZ studied in this chapter again brings this matter under discussion, and motivates a revision, taking into account all this new faunal and geochronological evidence.

Geochronological context

The UFZ occurs at the top of the Gran Barranca sequence, in the Colhue-Huapi Member, 25 m above the highest level bearing typical Colhuehuapian mammals (e.g. *Palaepanorthus* [see Abello 2007], *Pseudostegotherium, Cramauchenia, Colpodon, Prospaniomys, Caviocricetus, Protadelphomys*). The sterile sediments between both zones are fine-grained and form thick, massively bedded deposits without identifiable discontinuities. Based on radioisotopic and paleomagnetic data, the UFZ is estimated to be in Chron C6n (19.7–18.7 Ma) by Ré *et al.* (Chapter 4, this book),

The Colhuehuapian fossil zone in the Colhue-Huapi Member at Gran Barranca (LFZ) would span most of the C6An.1n (20.0 and 20.2 Ma) following Ré *et al.* (Chapter 4, this book). However, the chronological limits of the Colhuehuapian SALMA have not been defined, and other Colhuehuapian localities in central and northern Patagonia (i.e. Gaiman, Sacanana, Chichinales, and Cerro Bandera) have been interpreted to be slightly younger (Simpson 1940; Vucetich *et al.* this book).

The geochronology of the Santacrucian is provided by a tuff near the base of the Santa Cruz Formation at Monte León and Monte Observación (Fig. 18.1), which yields an age of 16.5 Ma. (Fleagle *et al.* 1995). Typical Santacrucian mammal taxa occur below the dated tuff, indicating that the Santacrucian extends to somewhat older than 16.5 Ma. Based upon some faunal elements, Croft *et al.* (2004) referred a mammal fauna from the Chucal Formation in northern Chile (Fig. 18.1), bracketed between 19 and 17.4 Ma (Croft *et al.* 2004, 2007), to the Santacrucian SALMA and concluded that the Santacrucian extended to a slightly older age than previously believed according to the Patagonian record.

The Astrapothericulan fauna is restricted to the lower and middle sections of the Pinturas Formation (Bown and Larriestra 1990; Kramarz and Bellosi 2005; Kramarz and Bond 2005). Fleagle *et al.* (1995) reported a radioisotopic date of

17.5 Ma from rocks from the lower sequence of the Pinturas Formation, below all localities with Astrapothericulan mammals except one (Lower Carmen: Bown and Larriestra 1990), and a date of 16.5 Ma for a tuff near the top of the middle sequence of the Pinturas Formation, above all localities bearing Astrapothericulan mammals. Consequently, the Astrapothericulan section of the Pinturas Formation is positively older than most of the Santacrucian levels at Monte León, but it is still uncertain whether the lower part the Santacrucian at Monte León and Monte Observación partially overlaps the Pinturan section. It is also uncertain how old is the lower chronological limit of the Astrapothericulan.

These data suggest that the UFZ at Gran Barranca represents an interval of time absent in other fossil localities, at least in Patagonia. The understanding of its relationships to other faunas must then be made on faunal evidence, in addition to stratigraphic position and geochronology.

Faunal context

The fossils from the UFZ correspond mostly to small- to medium-sized mammals. This is evidently a partial representation of the fauna, probably biased by taphonomic factors in addition to the small extent of the exposures of the fossil-bearing sediments. Most of the 20 identified genera have long biochrons and are not useful for subtle chronological calibration. However, *Titanothentes, Nematherium, Interatherium, Tetramerorhinus*, and *Neoreomys* indicate a post-Colhuehuapian age, whereas *Eosteiromys* and *Luantus* indicate a pre-Santacrucian age (Table 18.1). A post-Colhuehuapian – pre-Santacrucian age is also suggested by the occurrence of *Prostichomys*, only shared with the Astrapothericulan fauna (Table 18.1). Accordingly, a calculation of the Simpson's coefficient of faunal resemblance (Simpson 1960) at the generic level (Table 18.2) indicates that the UFZ fauna is more similar to the Astrapothericulan than to the remaining analyzed assemblages.

In turn, interpretations based on analyses of the UFZ fauna at the species level have a different degree of certainty, depending on the depth of systematic knowledge about each taxonomic group and the available material. Although proterotheriids were exhaustively studied by Soria (2001), the only available specimen in the UFZ is too fragmentary for a specific determination. On the other hand, taxonomic revisions of interatheriids and hegetotheriids from the early Miocene are still incomplete. Current knowledge of the early Miocene dasypodids only permits us to infer a post-Colhuehuapian age for the UFZ. Marsupials and rodents have been more intensively studied, and consequently allow a more precise chronological calibration. Among rodents, only *Eosteiromys*? sp. is shared with the Colhuehuapian (Vucetich *et al.* this book). The remaining identified species are shared neither with the Colhuehuapian nor with the

Table 18.1. *Chronological distributions within the early Miocene of the genera recorded at the top of the Gran Barranca.*

Taxon	Older levels	C	UFZ	APF	S	Younger levels
Pre-Santacrucian genera						
Eosteiromys		x	x	x		
Luantus						
Long-biochron genera						
Palaeothentes		x	x	x	x	
Abderites		x	x	x	x	
Stegotherium		x	x	x	x	
Prozaedyus		x	x		x	
Stenotatus	x	x	x	x	x	x
Proeutatus		x	x	x	x	x
Peltephilus		x	x	x	x	x
Protypotherium		x	x	x	x	x
Pachyrukhos		x	x		x	x
Acaremys		x	x		x	
Acarechimys		x	x	x	x	x
Post-Colhuehuapian genera						
Titanothentes			x	x		
Nematherium			x		x	
Interatherium			x	x	x	x
Tetramerorhinus			x	x	x	
Neoreomys			x	x	x	x
"Pinturan" genera						
Prostichomys			x	x		
Exclusive genera						
Dinomyidae gen. nov.			x			

Note: APF, "Astrapothericulan" at Pinturas Formation; C, Colhuehuapian; S, Santacrucian; UFZ, Upper Fossil Zone, top of the Gran Barranca sequence.

typical Santacrucian faunas. Notably, three of them – *P. bowni*, *L. propheticus*, and *N. pinturensis* – are exclusively shared with the Astrapothericulan fauna, whereas the Dinomyidae gen. et sp. nov. and *Prostichomys* cf. *P. bowni* are only known for the UFZ. Marsupial species are shared both with the Santacrucian and Astrapothericulan faunas, except for *Titanothentes* sp. nov. which is only known from the latter (Abello 2007).

Table 18.2. *Faunal resemblance among early Miocene faunas based on the Simpson coefficient of faunal resemblance (= C × 100/n, where C is the number of shared genera and n is the number of genera in the smallest fauna) for the Upper Fossil Zone (UFZ) fauna, the "Astrapothericulan" assemblage at Pinturas Formation (APF), Colhuehuapian (C) (including all the referred Patagonian local faunas), and Santacrucian (S) (including all the referred local faunas at Santa Cruz Province)*

Simpson index	C	UFZ	APF
UFZ	63.16	–	–
APF	58.62	94.74	–
S	38.03	84.21	75.86

Notes: Faunal list for Colhuehuapian taken from Marshall *et al.* (1983), Bown and Fleagle (1993), Goin *et al.* (2007), Soria (2001), and Vucetich *et al.* (this book). For the Santacrucian, taken from Marshall *et al.* (1983) (excluding *Luantus, Scotaeumys, Olenopsis,* and *Epipatriarchus*), modified after Bown and Fleagle (1993), and Soria (2001). Faunal list for the "Astrapothericulan" at Pinturas Formation taken from Ameghino (1990–02), complemented from Bown and Fleagle (1993), Bown and Larriestra (1990), de Barrio *et al.* (1984), González *et al.* (2006), Kramarz and Bellosi (2005), Soria (2001), Kramarz and Bond (2005), and preliminary identifications of undescribed materials at MACN. Genera recorded at the Colhuehuapian and the Santacrucian but still not recorded at the intermediate "Astrapothericulan" were scored as positive occurrences.

Biochronologic implications

The evidence discussed here suggests that the mammal fauna from the UFZ is more similar to the Astrapothericulan fauna than to that of the LFZ (Colhuehuapian) or any other fossil mammal fauna. These results confirm the occurrence of a faunal unit distinct from and younger than the Colhuehuapian within the sequence of Gran Barranca, as suggested by Vucetich *et al.* (2005), Carlini *et al.* (2005), and Madden *et al.* (2005a).

The current geochronological interpretation for the UFZ suggests this faunal unit is placed between 19.7 and 18.7 Ma (Ré *et al.* Chapter 4, this book). The systematic affinities with the Astrapothericulan fauna suggest that the UFZ is placed at the youngest extreme of the above-mentioned interval.

Based upon the close taxonomic affinities between the Astrapothericulan fauna and that of the UFZ, these assemblages should be assigned to a single post-Colhuehuapian biochronological unit, for which the name Pinturan (Castellanos 1937) is available. Therefore the Pinturan, as here understood, would span from about 18.75 to somewhat older than 16.5 Ma. Among the best-studied mammals of these faunas, we propose the echimyid rodent *Prostichomys* as the guide fossil for the "Pinturan," instead of *Astrapotericulus* because this latter is also recorded in the Colhuehuapian and the Santacrucian. The occurrence of *Prostichomys* in the Laguna del Laja fauna (Cura Mallín Formation, south central Chile; Fig. 18.1), although represented by species different from those recorded in southern Patagonia (Wertheim *et al.* 2006), suggests that this fauna could be also assigned to the Pinturan. Cerdeño and Vucetich (2007) assigned the still poorly known mammal assemblage from the middle member of the Mariño Formation (Divisadero Largo, Mendoza Province; Fig. 18.1) to the Pinturan because of the occurrence of a species of *Scleromys* (Rodentia, Dinomyidae) closer to the Pinturan *S. quadrangulatus* than to the Santacrucian species.

Certainly, there are arguments against the recognition of the Pinturan as a distinct unit. On the one hand, there are close taxonomic affinities – particularly among middle- and large-sized mammals– between the Pinturan (Astrapotericulan + the UFZ faunas) and the typical Santacrucian assemblages (Table 18.2), supporting the notion that they all should be grouped in a single faunal unit of long duration. However, taxonomic differences, particularly in small-sized mammals, consistent with geochronological differences should not be ignored, as they can be effective elements to determine low-rank faunal and chronological differences. Even if the Pinturan fauna is not different enough to be the basis for a distinct SALMA, it should be considered as a distinct early phase within the Santacrucian SALMA – the Santacrucian in a broad sense – whereas the typical Santacrucian (or Santacrucian in a strict sense) would correspond to a younger one.

The putative Santacrucian Chucal fauna is dated as partially synchronic with the Astrapotericulan interval of the Pinturas Formation and with the UFZ. The assignment of the Chucal fauna to the Santacrucian SALMA was based on the record of *Neoreomys*, *Hegetotherium* cf. *H. mirabile*, and particularly *Nesodon* (Croft *et al.* 2004). But these three taxa also occur at the lower and middle levels of the Pinturas Formation associated to typical Astrapotericulan mammals (Ameghino 1900–02 and AGK observations) and in the middle Miocene Collón Cura Formation associated to typical Colloncuran taxa (Vucetich *et al.* 1993 and AGK observations). Thus, *Neoreomys*, *Hegetotherium mirabile*, and *Nesodon* have a long biochron, instead of being typical Santacrucian taxa; they probably indicate a post-Colhuehuapian age, but they do not necessarily indicate that the typical Santacrucian overlaps the Pinturan. The few micro-mammals recorded at Chucal are still either undescribed or endemic (e.g. chinchilline rodents) and can not be compared with those of high-latitude localities. This fact precludes determining if the Chucal fauna is closer to the Pinturan than to the typical Santacrucian fauna.

In sum, the formal recognition of the Pinturan as an Age intermediate between the Colhuehuapian and Santacrucian, or as a Subage of the Santacrucian SALMA, depends on further systematic, biostratigraphical and geochronological studies. Meanwhile, in accordance to Ameghino and Castellanos's position, we propose to retain the Pinturan at least as an informal biochronological unit, to recognise the subtle faunal and chronological differences from the Colhuehuapian and typical Santacrucian faunas.

ACKNOWLEDGEMENTS
This study was supported in different ways by CONICET, Universidad Nacional de La Plata (grants to MGV and to AAC), ANPCyT (to AAC) and by US National Science Foundation grants EAR-0087636, BCS-0090255, and DEB-9907985 to Richard F. Kay and Richard H. Madden. Revisions by D. Croft and R. Kay improved the manuscript.

Appendix 18.1 Systematic list of taxa recorded in the Upper Fossil Zone

Class MAMMALIA Linnaeus 1758
  Infraclass MARSUPIALIA Illiger 1811
    Order PAUCITUBERCULATA Ameghino 1894
      Family PALAEOTHENTIDAE Sinclair 1906
      Genus *Titanothentes* Rae, Bown and Fleagle 1996
        Titanothentes sp. nov.
      Genus *Palaeothentes* Ameghino 1887a
        Palaeothentes minutus?
      Family ABDERITIDAE Ameghino 1889
      Genus *Abderites* Ameghino 1887a
        Abderites meridionalis Ameghino 1887a
  Infraclass EUTHERIA Gill 1872
    Superorder XENARTHRA Cope 1889
    Order CINGULATA Illiger 1811
      Family DASYPODIDAE Gray 1821
      Genus *Stegotherium* Ameghino 1887a
        Stegotherium tessellatum Ameghino 1887a
        Stegotherium variegatum Ameghino 1902
      Genus *Prozaedyus* Ameghino 1891a
        Prozaedyus sp.
      Genus *Stenotatus* Ameghino 1891b
        Stenotatus cf. *S. patagonicus* (Ameghino 1887a)
      Genus *Proeutatus* Ameghino 1891a
        Proeutatus sp.
      Family PELTEPHILIDAE Ameghino 1894
      Genus *Peltephilus* Ameghino 1887a
        Peltephilus pumilus Ameghino 1887a
      Family GLYPTODONTIDAE Gray 1869
      Subfamily PROPALAEHOPLOPHORINAE Ameghino 1891b
        Propalaehoplophorinae indet.
    Order PHYLLOPHAGA Owen 1842
      Family MYLODONTIDAE Gill 1872
      Subfamily NEMATHERIINAE Mercerat 1891
      Genus *Nematherium* Ameghino 1887a
        Nematherium sp.
    Order NOTOUNGULATA Roth 1903
      Family INTERATHERIIDAE Ameghino 1887b
      Genus *Interatherium* Ameghino 1887b (ex Moreno 1882)
        Interatherium sp.
      Genus *Protypotherium* Ameghino 1885
        Protypotherium sp.
      Family HEGETOTHERIIDAE Ameghino 1894
      Genus *Pachyrukhos* Ameghino 1885
        Pachyrukhos sp.
    Order LITOPTERNA Ameghino 1889
      Family PROTEROTHERIIDAE, Ameghino 1887a
      Genus *Tetramerorhinus* Ameghino 1894
        Tetramerorhinus sp.
    Order RODENTIA Bowdich 1821
    Suborder HYSTRICOGNATHI Tullberg 1899
      Family ERETHIZONTIDAE Thomas 1897
      Genus *Eosteiromys*?
        Eosteiromys? sp.
      Family ACAREMYIDAE Ameghino 1902
      Genus *Acaremys* Ameghino 1887a
        Acaremys sp.
      Family ECHIMYIDAE Miller and Gidley 1918
      Genus *Acarechimys* Patterson (in Kraglievich 1965)
        Acarechimys sp.
      Genus *Prostichomys* Kramarz 2001
        Prostichomys bowni Kramarz 2001
        Prostichomys cf. *P. bowni*
      Family DASYPROCTIDAE Smith 1842
      Genus *Neoreomys* Ameghino 1887a
        Neoreomys pinturensis Kramarz 2006b
      Family EOCARDIIDAE Ameghino 1891c

Genus *Luantus* Ameghino 1899
Luantus propheticus Ameghino
1899
Family DINOMYIDAE Alston 1876
Gen. et sp. nov.

REFERENCES

Abello, M. A. 2007. Sistemática y bioestratigrafía de los Paucituberculata (Mammalia, Marsupialia) del Cenozoico de América del Sur. Ph.D. thesis, Universidad Nacional de La Plata, Argentina.

Alston, E. R. 1876. On the classification of the order Glires. *Proceedings of the Zoological Society of London*, 61–98.

Ameghino, F. 1885. Nuevos restos de mamíferos oligocenos recogidos por el Profesor Pedro Scalabrini y pertenecientes al Museo Provincial de la ciudad de Paraná. *Boletín de la Academia Nacional de Ciencias en Córdoba*, **8**, 5–207.

Ameghino, F. 1887a. Enumeración sistemática de las especies de mamíferos fósiles coleccionados por Carlos Ameghino en los terrenos eocenos de Patagonia austral y depositados en el museo de La Plata. *Boletín del Museo de La Plata*, **1**, 1–26.

Ameghino, F. 1887b. Observaciones generales sobre el orden de mamíferos extinguidos Sud-Americanos llamados toxodontes (Toxodontia) y sinopsis de los géneros y especies hasta ahora conocidos. *Anales del Museo de La Plata*, 1–66.

Ameghino, F. 1889. Contribución al conocimiento de los mamíferos fósiles de la Repábica Argentina. *Actas Academia Nacional de Ciencias en Córdoba*, **6**, 1–32 + 1–1027.

Ameghino, F. 1891a. Nuevos restos de mamíferos fósiles descubiertos por Carlos Ameghino en el Eoceno inferior de la Patagonia austral: especies nuevas, adiciones y correcciones. *Revista Argentina de Historia Natural*, **1**, 289–328.

Ameghino, F. 1891b. Mamíferos y Aves fósiles argentinas: especies nuevas, adiciones y correcciones. *Revista Argentina de Historia Natural*, **1**, 240–259.

Ameghino, F. 1891c. Caracteres diagnósticos de cincuenta especies nuevas de mamíferos fósiles argentinos. *Revista Argentina de Historia Natural*, **1**, 129–167.

Ameghino, F. 1894. Enumération synoptique des espèces de mammifères fossiles des formations éocènes de Patagonie. *Boletín de la Academia de Ciencias de Córdoba*, **13**, 259–452.

Ameghino, F. 1899. *Sinopsis geológico–paleontológica: Suplemento (adiciones y correcciones)*. La Plata: Imprenta La Libertad.

Ameghino, F. 1900–02. L'âge des formations sédimentaires de Patagonie. *Anales de la Sociedad Científica Argentina*, **50**, 109–131, 145–165, 209–229; **51**, 20–39, 65–91; **52**, 198–197, 244–250; **54**, 161–180, 220–240, 283–342.

Ameghino, F. 1902. Première contribution à la connaissance de la faune mammalogique des couches à *Colpodon*. *Boletín de la Academia Nacional de Ciencias en Córdoba*, **17**, 71–138.

Ameghino, F. 1906. Les formations sédimentaires du Crétacé supérieur et du Tertiaire de Patagonie. *Anales del Museo Nacional de Buenos Aires*, **8**, 1–358.

Barrio, C., A. A. Carlini, and F. J. Goin 1989. Litogénesis y antigüedad de la Formación Chichinales de Paso Córdoba (Río Negro, Argentina). *Actas IV Congreso Argentino de Paleontología y Bioestratigrafía*, **4**, 149–156.

Bellosi, E. S. 2005. Sedimentology and depositional setting of the Sarmiento Formation (Eocene-Miocene) in central Patagonia. *Actas XVI Congreso Geológico Argentino*, **4**, 439–440.

Bonaparte, C. L. J. L. 1838. Synopsis vertebratorum systematis. *Nuovi Annales Scientiae Naturae*, **2**, 105–133.

Bordas, A. F. 1939. Diagnosis sobre algunos mamíferos de las capas con *Colpodon* del Valle del Río Chubut (Repábica Argentina). *Physis*, **14**, 413–433.

Bowdich, T. E. 1821. *An Analysis of the Natural Classifications of Mammalia for the Use of Students and Travellers*. Paris: J. Smith.

Bown, T. M. and J. G. Fleagle 1993. Systematics, biostratigraphy and dental evolution of the Palaeothentidae, later-Oligocene to early–middle Miocene (Deseadan–Santacrucian) caenolestoid marsupials of South America. *Paleontological Society Memoir*, **29**, 1–76.

Bown, T. M. and C. N. Larriestra 1990. Sedimentary paleoenvironments of fossil platyrrhine localities, Miocene Pinturas Formation, Santa Cruz province, Argentina. *Journal of Human Evolution*, **19**, 87–119.

Candela, A. M. 2000. Los Erethizontidae (Rodentia, Hystricoganthi) fósiles de Argentina: sistemática e historia evolutiva y biogeográfica. Ph.D. thesis, Universidad Nacional de la Plata. Argentina.

Carlini, A. A., M. Ciancio, and G. J. Scillato-Yané 2005. Los Xenarthra de Gran Barranca: más de 20 Ma de historia. *Actas XVI Congreso Geológico Argentino*, **4**, 419–424.

Castellanos, A. 1937. Ameghino y la antigüedad del hombre sudamericano. *Asociación Cultural de Conferencias de Rosario, Ciclo de Carácter General*, **2**, 47–192.

Cerdeño, E. and M. Bond 1998. Taxonomic revision and phylogeny of *Paedotherium* and *Tremacyllus* (Pachyrukhinae, Hegetotheriidae, Notoungulata) from the Late Miocene to the Pleistocene of Argentina. *Journal of Vertebrate Paleontology*, **18**, 799–811.

Cerdeño, E. and M. G. Vucetich 2007. The first rodent from the Mariño Formation (Miocene) at Divisadero Largo (Mendoza, Argentina) and its biochronological implications. *Revista Geológica de Chile*, **34**, 199–207.

Cope, E. D. 1889. The Edentata of North America. *American Naturalist*, **23**, 657–664.

Croft, D. A., J. Flynn, and A. Wyss 2004. Notoungulata and Litopterna of the Early Miocene Chucal Fauna, northern Chile. *Fieldiana, Geology*, n.s., **50**, 1–52.

Croft, D. A., J. J. Flynn, and A. R. Wyss 2007. A new basal glyptodontid and other Xenarthra of the early Miocene Chucal Fauna, northern Chile. *Journal of Vertebrate Paleontology*, **27**, 781–797.

de Barrio, R. E., G. J. Scillato-Yané, and M. Bond 1984. La Formación Santa Cruz en el borde occidental del macizo del

Deseado (provincia de Santa Cruz) y su contenido paleontológico. *Actas IX Congreso Geológico Argentino*, 539–556.

Fields, R. W. 1957. Hystricomorph rodents from the Late Miocene of Colombia, South America. *University of California Publications in Geological Sciences*, **32**, 273–404.

Fleagle, J. G., T. M. Bown, C. Swisher, and G. Buckley 1995. Ages of the Pinturas and Santa Cruz Formations. *Actas VI Congreso Argentino de Paleontología y Bioestratigrafía*, 129–135.

Flynn, J. J. and C. C. Swisher III 1995. Cenozoic South American Land Mammal Ages: correlation to global geochronologies. In Berggren, W. A., Kent, D. V., Aubry, M.-P., and Hardentol, J. (eds.), *Geochronology Times Scales and Global Stratigraphic Correlation, Special Publication* no. 54. Tulsa, OK: Society for Sedimentary Geology, pp. 317–333.

Flynn, J. J., R. Charrier, D. A. Croft, P. B. Gans, T. M. Heriott, J. A. Wertheim, and A. R. Wyss 2008. Chronological implications of New Miocene mammals from the Cura-Mallin and Trapa Trapa formations, Laguna del Laja region, south central Chile. *Journal of South American Earth Sciences*, **26**, 412–423.

Gill, T. 1872. Arrangement of the families of mammals with analytical tables. *Smithsonian Miscellaneous Collections*, **11**, 1–98.

Goin, F. J., M. A. Abello, E. S. Bellosi, R. F. Kay, R. H. Madden, and A. A. Carlini 2007. Los Metatheria sudamericanos del comienzo del Neógeno (Mioceno Temprano, Edad-mamífero Colhuehuapense). I. Introducción, Didelphimrphia y Sparassodonta. *Ameghiniana*, **44**, 29–71.

González, L., M. Tejedor, and G. J. Scillato-Yané 2006. Los Dasypodidae de la Formación Pinturas (Mioceno Inferior), Provincia de Santa Cruz. *Ameghiniana*, **43**, 40R–41R.

Gray, J. E. 1821. On the natural arrangement of vertebrose animals. *The London Medical Repository Monthly Journal and Review*, **15**, 296–310.

Gray, J. E. 1869. *Catalogue of Carnivorous, Pachydermatous, and Edentate Mammalia in The British Museum*. London: British Museum.

Illiger, C. 1811. *Prodromus systematis mammalium et avium additis terminis zoographicis utriudque classis*. Berlin: C. Salfeld.

Kraglievich, J. L. 1965. Spéciation phylétique dans les rongeurs fossiles du genre *Eumysosps* Amegh. (Echimyidae, Heteropsomyinae). *Mammalia*, **29**, 258–267.

Kramarz, A. G. 2001. *Prostichomys bowni*, un nuevo roedor Adelphomyinae (Hystricognathi, Echimyidae) del Mioceno medio – inferior de Patagonia, Argentina. *Ameghiniana*, **38**, 163–168.

Kramarz, A. G. 2004. Octodontoids and Erethizontoids (Rodentia, Hystricognathi) from the Pinturas Formation, early – middle Miocene of Patagonia, Argentina. *Ameghiniana*, **41**, 199–216.

Kramarz, A. G. 2006a. Eocardiids (Rodentia, Hystricognathi) from the Pinturas Formation, late Early Miocene of Patagonia, Argentina. *Journal of Vertebrate Paleontology*, **26**, 770–778.

Kramarz, A. G. 2006b. *Neoreomys* and *Scleromys* (Rodentia, Hystricognathi) from the Pinturas Formation, late Early

Miocene of Patagonia, Argentina. *Revista del Museo Argentino de Ciencias Naturales*, n.s., **8**, 53–62.

Kramarz, A. G. and E. S. Bellosi 2005. Hystricognath rodents from the Pinturas Formation, early – middle Miocene of Patagonia: biostratigraphic and paleoenvironmental implications. *Journal of South American Earth Sciences*, **18**, 199–212.

Kramarz, A. G. and M. Bond 2005. Los Litopterna (Mammalia) de la Formación Pinturas, Mioceno Temprano – Medio de Patagonia. *Ameghiniana*, **42**, 611–625.

Kramarz, A. G., A. Garrido, A. Forasiepi, M. Bond, and C. Tambussi 2005. Estratigrafía y vertebrados (Mammalia – Aves) de la Formación Cerro Bandera, Mioceno Temprano de la provincia del Neuquén, Argentina. *Revista Geológica de Chile*, **32**, 273–291.

Linnaeus, C. 1758. *Systema naturae per regna tria naturae, secundum classes, ordines, genera, species, cum characteribus, differentiis, synonymis, locis*, 10th edn. Stockholm: Laurentii Salvii.

López, G. M., M. Bond, M. A. Reguero, J. Gelfo, and A. G. Kramarz 2005. Los ungulados del Eoceno–Oligoceno de la Gran Barranca, Chubut. *Actas XVI Congreso Geológico Argentino*, 415–418.

Madden, R. H., E. S. Bellosi, A. A. Carlini, M. Heizler, J. J. Vilas, G. H. Ré, R. F. Kay, and M. G. Vucetich 2005a. Geochronology of the Sarmiento Formation at Gran Barranca and elsewhere in Patagonia: calibrating middle Cenozoic mammal evolution in South America. *Actas XVI Congreso Geológico Argentino*, 411–412.

Madden, R. H., A. A. Carlini, M. G. Vucetich, and R. F. Kay 2005b. The paleontology of Gran Barranca: evolution and environmental change through the Middle Cenozoic of Patagonia. *Actas XVI Congreso Geológico Argentino*, 409–410.

Marshall, L. G. 1976. Fossil localities for Santacrucian (Early Miocene) mammals, Santa Cruz Province, Southern Patagonia, Argentina. *Journal of Paleontology*, **50**, 1129–1142.

Marshall, L. G., R. Hoffstetter, and R. Pascual 1983. Mammals and stratigraphy: geochronology of continental mammals-bearing Tertiary of South America. *Paleovertebrata, Mémoire Extraordinaire*, 1–93.

Mercerat, A. 1891. Datos sobre restos de mamíferos pertenecientes a los Bruta conservados en el Museo de La Plata y procedentes de los terrenos eocenos de Patagonia. *Revista del Museo de La Plata*, **2**, 5–46.

Miller, G. S. and J. W. Gidley 1918. Synopsis of the suprageneric groups of rodents. *Journal of Washington Academy of Sciences*, **8**, 431–438.

Moreno, F. P. 1882. Patagonia, resto de un antiguo continente hoy sumergido. *Anales de la Sociedad Científica Argentina*, **14**, 97–131.

Ortiz Jaureguizar, E. and G. Cladera 2006. Paleoenvironmental evolution of southern South America during the Cenozoic. *Journal of Arid Environments*, **66**, 489–532.

Owen, R. 1842. *Description of the Skeleton of an Extinct Gigantic Sloth*, Mylodon robustus, *Owen, with Observations on the*

Osteology, Natural Affinities, and Probable Habits of the Megatherioid Quadrupeds in General. London: Royal College of Surgeons of England.

Pascual, R. and O. E. Odreman Rivas 1971. Evolución de las comunidades de vertebrados del Terciario argentino: los aspectos paleozoogeográficos y paleoclimáticos relacionados. *Ameghiniana*, **8**, 372–412.

Pascual, R. and E. Ortiz Jaureguizar 1990. Evolving climates and mammal faunas in Cenozoic South America. *Journal of Human Evolution*, **19**, 23–60.

Pascual, R., E. Ortega Hinojosa, E. Gondar, and E. P. Tonni 1965. Las edades del Cenozoico mamalífero de la Argentina, con especial atención a aquellas del territorio bonaerense. *Anales Comisión Investigaciones Científicas, Buenos Aires*, **6**, 165–193.

Pascual, R., A. A. Carlini, M. Bond, and F. J. Goin 2002. Mamíferos Cenozoicos. *Relatorio del XV Congreso Geológico Argentino*, **2**, 533–544.

Rae, T. C., T. M. Bown, and J. G. Fleagle 1996. New palaeothentid marsupials (Caenolestoidea) from the early Miocene of Patagonian Argentina. *American Museum Novitates*, **3165**, 1–10.

Roth, S. 1899. Apuntes sobre la geología y paleontología de los territorios de Río Negro y Neuquén (diciembre de 1895 a junio de 1896). *Revista del Museo de La Plata*, **9**, 141–197.

Roth, S. 1903. Los ungulados sudamericanos. *Anales del Museo de La Plata, Sección Paleontología*, **5**, 1–36.

Scott, W. B. 1903–05. Mammalia of the Santa Cruz Beds. Volume V, Paleontology. Part I, Edentata. In Scott, W. B. (ed.), *Reports of the Princeton Expedition to Patagonia 1896–99.* Stuttgart: E. Schweizerbart'sche Verlagshandlung, pp. 1–364.

Scott, W. B. 1905. Mammalia of the Santa Cruz Beds. Volume V, Paleontology. Part III, Glires. In Scott, W. B. (ed.), *Reports of the Princeton Expedition to Patagonia 1896–99.* Stuttgart: E. Schweizerbart'sche Verlagshandlung, pp. 348–489.

Simpson, G. G. 1940. Review of the mammal-bearing Tertiary of South America. *Proceedings of the American Philosophy Society*, **83**, 649–709.

Simpson, G. G. 1941. A Miocene sloth from Southern Chile. *American Museum Novitates*, **1156**, 1–6.

Simpson, G. G. 1960. Notes on the measurement of faunal resemblance. *American Journal of Science*, **258A**, 300–311.

Sinclair, W. J. 1906. Mammalia of the Santa Cruz beds. Volume VI, Paleontology. Part III, Marsupialia of the Santa Cruz beds. In Scott, W. B. (ed.), *Reports of the Princeton Expedition to Patagonia 1896–99.* Stuttgart: E. Schweizerbart'sche Verlagshandlung, pp. 333–459.

Sinclair, W. J. 1909. Mammalia of the Santa Cruz beds. Volume VI, Paleontology. Part I, Typotheria of the Santa Cruz beds. In Scott, W. B. (ed.), *Reports of the Princeton Expedition to Patagonia 1896–99.* Stuttgart: E. Schweizerbart'sche Verlagshandlung, pp. 1–110.

Smith, C. H. 1842. *Mammalia: Introduction to Mammals.* London: Chatto & Windus.

Soria, M. F. 2001. Los Proterotheriidae (Litopterna, Mammalia), sistemática, origen y filogenia. *Monografías del Museo Argentino de Ciencias Naturales*, **1**, 1–167.

Tauber, A. A. 1996. Los representantes del género *Protypotherium* (Mammalia, Notoungulata, Interatheriidae) del Mioceno Temprano del sudeste de la provincia de Santa Cruz, República Argentina. *Academia Nacional de Ciencias (Córdoba, Argentina), miscelánea*, **95**, 3–29.

Tedford, R. H., L. B. Albright III, A. D. Barnosky, I. Ferrusquia-Villafranca, R. M. Hunt Jr., J. E. Storer, C. C. Swisher III, M. R. Voorhies, S. D. Webb, and D. P. Whistler 2004. Mammalian biochronology of the Arikareean through Hemphillian interval (Late Oligocene through early Pliocene epochs). In Woodburne, M. O. (ed.), *Late Cretaceous and Cenozoic Mammals of North America: Biostratigraphy and Geochronology*, New York: Columbia University Press, pp. 169–230.

Thomas, O. 1897. On the genera of rodents: an attempt to bring up to date the current arrangement of the order. *Proceedings of the Zoological Society of London*, **1897**, 1012–1028.

Tullberg, T. 1899. Über das System der Nagethiere: eine phylogenetische Studie. *Nova Acta Regiae Societatis Scientiarum Upsaliensis*, **3**, 1–514.

Vucetich, M. G. and A. G. Kramarz 2003. New Miocene rodents of Patagonia (Argentina) and their bearing in the early radiation of the octodontiform octodontoids. *Journal of Vertebrate Paleontology*, **23**, 435–444.

Vucetich, M. G., E. M. Vieytes, A. G. Kramarz, and A. A. Carlini 2005. Los roedores caviomorfos de la Gran Barranca: aportes bioestratigráficos y paleoambientales. *Actas XVI Congreso Geológico Argentino*, 413–414.

Walton, A. H. 1997. Rodents. In Kay, R., Madden, R., Cifelli, R., and Flynn, J. (eds.), *Vertebrate Paleontology in the Neotropics: The Miocene Fauna of La Venta, Colombia.* Washington, DC: Smithsonian Institution Press, pp. 392–409.

Wertheim, J., T. Herriott, D. A. Croft, J. J. Flynn, and P. Gans 2006. Unusual fossil rodent faunas from south central Chile. *Journal of Vertebrate Paleontology*, **26**, 137A.

Wood, A. E. and B. Patterson 1959. Rodents of the Deseadan Oligocene of Patagonia and the beginnings of South American rodent evolution. *Bulletin of the Museum of Comparative Zoology*, **120**, 281–428.

19 Loessic and fluvial sedimentation in Sarmiento Formation pyroclastics, middle Cenozoic of central Patagonia

Eduardo S. Bellosi

Abstract

The pyroclastic Sarmiento Formation (middle Eocene to early Miocene) at its type locality (Gran Barranca) is divided into six members, delimited by erosive and non-depositional unconformities, and subordinate discontinuity surfaces. The sedimentary history during Sarmiento time was characterized by subaerial distal ash falls (tephric loessites), eolian and fluvial reworking and deposition, soil formation during landscape stability, and events of deep fluvial erosion. These processes were controlled by changes in climate, and subordinately by volcanism, subsidence, and sea-level variation. The Gran Barranca Member (middle Eocene) is mostly composed of tephric loessites deposited in subhumid to semi-arid rolling plains, with ephemeral ponds in areas of poor drainage. Similar depositional processes are also found for the Rosado Member (late middle Eocene), a strongly developed paleosol which records the driest and more stable landscape period during Sarmiento time. Coarse intraformational deposits and paleosols of the Lower Puesto Almendra Member (late Eocene) represent the installation of a braided, probably ephemeral fluvial system in a subhumid environment. A significant modification in sedimentation style is interpreted for the overlying Vera Member (late Eocene to earliest Oligocene). During this interval, laterally continuous and very thick loessic sediments accumulated rapidly, with few interruptions in a steady semi-arid environment. Subsequently, the development of an incised valley, related to a climate change (increase in rainfall) and probably to extensional faulting, eroded up to 100 m of sediments. The infill deposits constitute the Upper Puesto Almendra Member (Oligocene). In a first stage, the valley was filled with fluvial sediments, accumulated in braided channels, with some eolian deposits. Basalt flows occupied some of these channels during a relatively short volcanic episode. The sedimentation of the second stage (late Oligocene) occurred in a meandering fluvial system under wetter conditions. A new period of valley incision and infilling is recorded in the Colhue-Huapi Member (early Miocene). During Colhue-Huapi time, depositional settings evolved from braided fluvial and loessic in a subhumid seasonal climate, to exclusively loessic in a semi-arid environment.

Resumen

En su localidad tipo (Gran Barranca), la Formación Sarmiento se divide en seis miembros delimitados por discordancias erosivas o no-depositacionales y discontinuidades menores. La historia sedimentaria estuvo regida por caídas distales de cenizas (loessitas téfricas), retrabajo y acumulación eólica y fluvial, pedogénesis en lapsos de estabilidad ambiental y eventos de profunda erosión fluvial. Tales procesos habrían sido controlados por cambios climáticos, y secundariamente por el vulcanismo, subsidencia tectónica y nivel del mar. El Miembro Gran Barranca (Eoceno medio) incluye principalmente loessitas originadas en planicies onduladas con lagunas efímeras, en condiciones subhúmedo-semiáridas. Una sedimentación similar es interpretada para el Miembro Rosado (Eoceno medio tardío), el cual registra el período más seco y de mayor estabilidad ambiental. Los depósitos intraformacionales gruesos y paleosuelos del Miembro Puesto Almendra inferior (Eoceno superior) representan la instalación de un sistema fluvial entrelazado, probablemente efímero, en un ambiente subhúmedo-estacional. A continuación se produjo un notable cambio del paisaje y estilo de sedimentación. El Miembro Vera (Eoceno superior-Oligoceno basal) registra la rápida acumulación de potentes mantos de loess téfrico en un ambiente semi-árido. La subsiguiente incisión de un valle fluvial debido al cambio del clima (aumento de lluvias) y probablemente al fallamiento extensional, produjo una profunda erosión (100 m verticales). Los rellenos de valle constituyen el Miembro Puesto Almendra superior (Oligoceno). En la primera etapa se acumularon sedimentos fluviales en canales entrelazados, juntamente con cenizas volcánicas por acción eólica. Un simultáneo y relativamente corto episodio volcánico local generó derrames basálticos, los cuales ocuparon algunos de los canales fluviales. En la segunda etapa (Oligoceno tardío) el sistema fluvial se volvió meandrante bajo condiciones húmedo-subhúmedas. El Miembro Colhue-Huapi (Mioceno inferior) registra un

The Paleontology of Gran Barranca: Evolution and Environmental Change through the Middle Cenozoic of Patagonia, eds. R. H. Madden, A. A. Carlini, M. G. Vucetich, and R. F. Kay. Published by Cambridge University Press. © Cambridge University Press 2010.

nuevo período de incisión y relleno de valles, vinculado a cambios del nivel del mar. La sedimentación se inició en un sistema fluvial entrelazado con participación subordinada de acumulaciones loéssicas, en un clima subhúmedo estacional. El cambio hacia condiciones semiáridas transformó finalmente el escenario sedimentario en una llanura loéssica.

Introduction

The Sarmiento Formation or Sarmiento Tuffs (*Tobas de Sarmiento*) is an emblematic continental unit of the Cenozoic of southern Argentina. It is widely exposed in central and northern Patagonia and has been investigated by paleontologists and geologists since the 1880s. Mainly, these studies have concentrated on the abundant fossil mammals and the biostratigraphic relationships and geologic age of the mammal-bearing sequences (Marshall *et al.* 1983). Without doubt the most impressive and continuous outcrop of the Formation is at Gran Barranca, in Chubut Province. The paleontological importance of this site was highlighted by Cifelli (1985), who considered it to be one of the most important fossil localities in the world.

Sedimentologic and lithostratigraphic analyses of this unit are few. Early studies described this unit as a poorly stratified, light-colored and homogeneous succession of unconsolidated tuffs and bentonites, with a few basalt flows (Feruglio 1949). More recently, a more detailed sedimentologic description was provided by Andreis (1972, 1977), Andreis *et al.* (1975), Spalletti and Mazzoni (1977, 1979), and Mazzoni (1979, 1985, 1994).

This contribution describes and interprets the facies of the Sarmiento Formation at its type locality Gran Barranca. This facies characterization, together with the analysis of the physical stratigraphy (Bellosi this book) and paleosols (Bellosi and González this book) serve as the basis for a reconstruction of sedimentary environments of central Patagonia from the middle Eocene to the early Miocene. This succession provides a unique record of the terrestrial accumulation of fine-grained eolian sediments, in this case tephric dust. Examples of pre-late Cenozoic loessites are uncommon in the literature, particularly those unrelated to glaciations (Parrish 1998). The rich and diverse fossil content of the Sarmiento Formation is typical of loess–paleosol sequences elsewhere (Behrensmeyer and Hook 1992), and includes frequent vertebrates (mammals), plant microfossils (phytoliths), ichnofossils (insects), terrestrial invertebrates (gastropods), and even plant macrofossils.

Geological setting

In extra-Andean Patagonia today, the Sarmiento Formation and correlated pyroclastic successions crop out from Neuquén Province in the north to Santa Cruz Province in the south, and from the Atlantic coast westward up to 400 km (Mazzoni 1985; Franchi and Nullo 1986; Panza *et al.* 1994; Lizuaín *et al.* 1995; Ardolino *et al.* 1999). The area of distribution of these deposits is estimated at 350 000 km^2 (Fig. 19.1). This region comprises several geological provinces, from north to south: the Somun Cura Massif, the Patagonian Precordillera, Bernardides, the San Jorge Basin, and the Deseado Massif.

The thickest development of the Sarmiento Formation is in the San Jorge Basin. Gran Barranca is the most complete exposure, attaining nearly 170 m of exposed thickness and 319 m of cumulative thickness. A compressional tectonic regime has been proposed for the western extreme of Central Patagonia from the middle Eocene through the Oligocene (Suárez and de la Cruz 2002; Morata *et al.* 2005), while the eastern region supported extensional conditions. The total "Cenozoic" infill of the San Jorge Basin is 1200 m thick, and includes marine and continental units. The stratigraphic characteristics of this succession have been summarized by Legarreta and Uliana (1994), Bellosi (1995), and Malumián (1999).

Bellosi (this book) presents a detailed study of the physical stratigraphy of the Sarmiento Formation based on a hierarchy of different types of unconformities and discontinuity surfaces. At Gran Barranca, the Sarmiento Formation conformably overlays the continental early to middle Eocene Koluel-Kaike Formation (Feruglio 1949) and is covered by the Plio-Pleistocene "Rodados Patagónicos." At Valle Hermoso ("Km 147" locality) this upper contact with the early Miocene marine Chenque Formation is transitional (Bellosi 1987). Bedding is horizontal or dips slightly 2° to the south.

Along Gran Barranca, extending 7 km from west to east, Simpson (1930) selected several profiles identified with letters A to N, from west to east. These profiles, plus a few incorporated in this work, are used for this study (Fig. 19.2). Spalletti and Mazzoni (1979) selected a type section (Profile MMZ) near the west end of Gran Barranca, where they defined three members, Gran Barranca, Puesto Almendra, and Colhue-Huapi. New observations have led to a reorganization of the stratigraphy of the Sarmiento Formation (Bellosi this book), adding new members, and subdividing the Puesto Almendra Member into two parts and five units (Fig. 19.3).

Rock types and interpretation

Figure 19.4 illustrates the facies logs of each member of the Sarmiento Formation at Gran Barranca. As Mazzoni (1979) noted, the lithologic homogeneity of Sarmiento Formation often mentioned by early authors is an illusion. Facies differentiation is rendered difficult by the uniform coloration and manner of rock weathering. According to Spalletti and Mazzoni (1979) and Mazzoni (1979), most of the

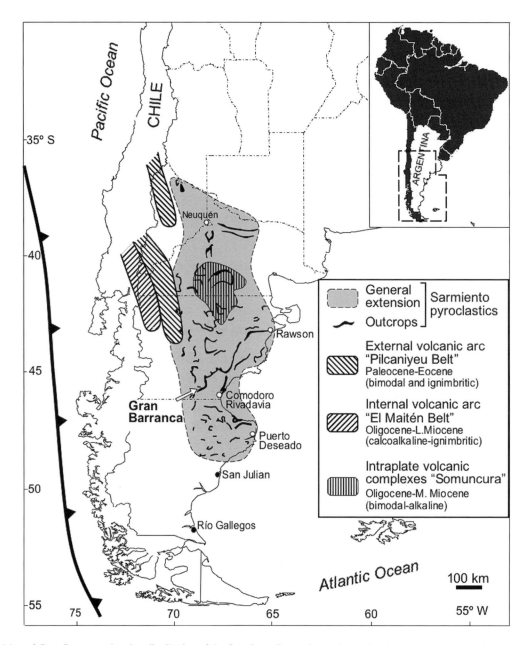

Fig. 19.1. Map of Gran Barranca showing distribution of the Sarmiento Formation and correlated pyroclastic deposits in extra-Andean north and central Patagonia. Eocene–lower Miocene acidic–intermediate volcanic rocks are also mapped. Data from Mazzoni (1985), Franchi and Nullo (1986), Rapela *et al.* (1988), and Ardolino *et al.* (1999).

Sarmiento sediments in Gran Barranca correspond to volcanic dust or very fine ash. Noting that there is no unique denomination for the lithified equivalent of very fine ash (synonymous with "volcanic mudrock," "fine tuff," "pyroclastic mudstone," etc.), Teruggi *et al.* (1978) proposed the term "chonite" (Gk) to distinguish the lithified equivalent of fine to very fine ash from "tuff" (a term used to describe a coarser sand-sized fraction). As the term "chonite" has not found wide application internationally, in this contribution these fine-grained deposits are called pyroclastic mudstones.

Massive pyroclastic mudstones and tuffs

Most pyroclastic mudstones have a silt grain-size, and include the "chonitas cineríticas" and "chonitas *sensu stricto*" distinguished by Spalletti and Mazzoni (1979). In the Gran Barranca, Vera and upper section of the Colhue-Huapi members, pyroclastic mudstones are the dominant facies, while fine tuffs are much less frequent (Fig. 19.4). Pyroclastic mudstones are massive or weakly stratified and poorly consolidated (Fig. 19.5: 1–2, 6), except where they have been modified and cemented by soil processes. They

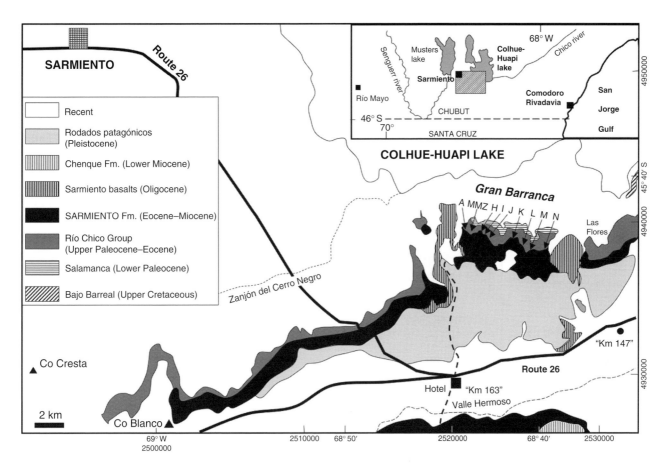

Fig. 19.2. Geological map of Gran Barranca locality and neighboring outcrops of the Sarmiento Formation (from Bellosi *et al.* 2002a). Simpson's profiles A to N along the escarpment are shown.

show very light chroma, ranging from white, very light gray, to yellowish, pinkish, or greenish white. Bed thickness varies from a few decimeters to several meters but always less than 10 m. Beds are laterally continuous, showing tabular geometry with little variation in thickness along several hundreds of meters to kilometers of exposure. Pedoturbation and associated soil bioturbation are common, as evidenced by nodules, soil structures, duripans (Fig. 19.5: 2), color changes, and trace fossils. In this chapter, such facies are described as paleosols (see also Bellosi and González this book). In the Gran Barranca Member, pyroclastic mudstones include diagenetic concretions of silica and gypsum.

A detailed description of the mineralogy and texture of the pyroclastic mudstone facies is provided by Mazzoni (1979). The coarse fraction (coarse silt to very fine sand) comprises uncolored glass shards and pumice fragments (61–99%), fresh andesine plagioclase (10–38%), and felsic volcanic rock fragments (1%). Sanidine and quartz are rare (<1%). Weathering of vitric fragments is variable. In many samples fresh and altered glass clasts coexist, indicating sediment mixing. Fresh glass is dominant in the Gran

Barranca and Vera Members. In the Lower and Upper Puesto Almendra and Colhue-Huapi Members, weathered vitroclasts generally show a peripheral brown stained band, or may be colored and wholly birefringent. In several analyzed samples, the volcanic glass was partially transformed to smectite, which is the dominant clay mineral.

Grass and palm silica phytoliths, and subordinately diatoms or sponge spicules, are important organic components of pyroclastic mudstones. According to Mazzoni (1979), the mean concentration of phytoliths is about 2% up to 30% in the Gran Barranca Member (with mostly palm morphotypes), and 7% in Upper Puesto Almendra Member (with mostly grass morphotypes).

The massive pyroclastic mudstones represent subaerial distal ash fall deposits or tephric loessites (Feruglio 1949; Pascual and Odreman 1971; Andreis 1972; Andreis *et al.* 1975). This interpretation is supported by: (a) high angular shape of vitric fragments indicating low abrasion, (b) the dominantly well-sorted silt grain-size resulting from eolian transport, (c) mantle bedding, (d) massive stratification resulting from slow suspension or fallout, and (e) frequent

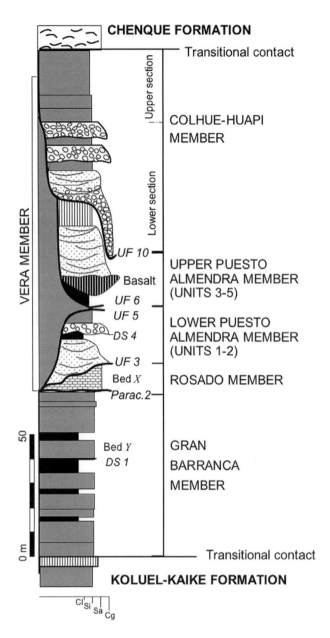

Fig. 19.3. Reconstructed profile of the Sarmiento Formation at Gran Barranca. Numbers of unconformities (UF) and discontinuity surfaces (DS) are shown. See Fig. 19.4 for symbols.

pedogenesis represent quiescent lapses between eruption events or periods (Fisher and Schmincke 1984). Most of the pyroclastic mudstones and tuffs do not represent pure ash falls, but rather show some, albeit weak, evidence of mixing resulting from eolian and fluvial transport or bioturbation.

Spalletti and Mazzoni (1977, 1979) compared this facies to that of the late Cenozoic Pampean Loess that accumulated in settings peripheral to deserts (Spalletti 1992). Vicars and Breyer (1981), Speed *et al.* (2002), and Rose *et al.* (2003) also described similar well-sorted, distal fine ash beds as loessites. Such homogeneous deposits are common in distant fallout tephra blankets (Fisher and Schmincke 1984) and resemble wind-blown dust or loess (Fisher 1966). The absence of graded bedding is the consequence of homogeneity in grain-size and density and results in uniform settling velocities. The observed changes in mineral composition between some beds (Mazzoni 1979) would have originated by eolian fractionation or differentiation within the traveling ash cloud or by reworking.

Bentonites

Bentonites or "chonitas bentoníticas" (Spalletti and Mazzoni, 1979), are clay-rich layers resulting from the alteration of vitric fine ash, regardless of mineral composition (Fischer and Shmincke 1984). This facies is thicker, more common, and laterally more continuous (over some kilometers) in the Gran Barranca Member. It is less frequent in the Lower and Upper Puesto Almendra Members, and even less frequent and thinner in the Vera and Colhue-Huapi Members (Fig. 19.4).

Bentonites are generally massive and olive brown (5Y5/4) to dusky yellow (5Y6/4) in color. They are the darker rocks in the succession and are interbedded with unaltered vitric pyroclastic mudstones (Fig. 19.5: 2). Some beds show a mixed composition of bentonite and volcanic ash (McCartney 1933). Bed thickness varies from few decimeters up to 6.5 m. The lower contact is flat and transitional to sharp, and the top contact is sharp or erosive when it underlies conglomerates or tufoarenites. Lateral changes to unaltered tuffaceous beds were not observed. Slickensides are frequent in some beds. Fossil vertebrates and bioturbation are uncommon. Swelling bentonites are formed by montmorillonite, and subordinately beidellite, glass shards, plagioclase, and quartz (McCartney 1933; Mazzoni 1979).

Weaver (1978) and Fischer and Shmincke (1984) indicated that bentonites form by early diagenesis or chemical alteration ("devitrification") of glassy felsic material (high silica content). Glass shard alteration occurred most probably in freshwater lacustrine environments under low pH conditions (Hay 1978). Thus, the thicker and laterally continuous bentonite beds that occurred in the Gran Barranca and Puesto Almendra Members probably originated in small shallow lakes.

intercalation of paleosols. Generally, bed contacts resulted from pedogenic modification and bioturbation. Variation in the eolian sedimentation rate can be inferred from changes in the frequency and degree of development of the intercalated paleosols. Pedogenic modification of these eolian sediments occurred in a stable landscape without deposition or when the rate of sedimentation was lower than the rate of pedogenesis (Bown and Larriestra 1990; Kraus 1999). At other times, fine volcaniclastics modified by

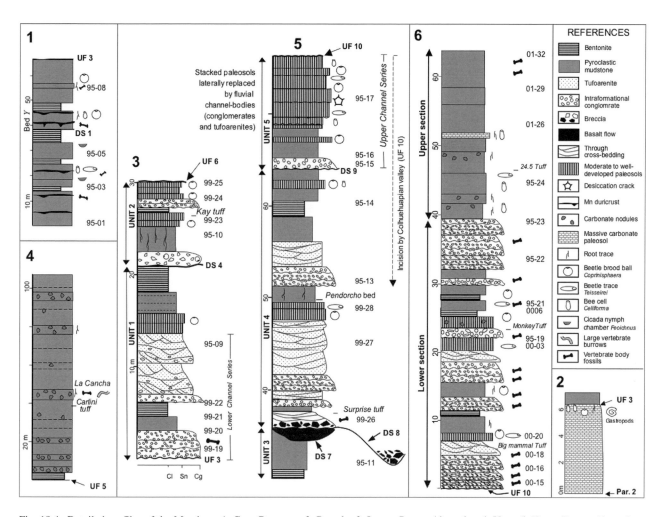

Fig. 19.4. Detailed profiles of the Members. 1, Gran Barranca; 2, Rosado; 3, Lower Puesto Almendra; 4, Vera; 5, Upper Puesto Almendra; 6, Colhue-Huapi. All columns correspond to the Profile MMZ, with the exception of the Members Rosado (Profile J) and Vera (Profile K). Number of stakes are shown.

Cross-bedded tufoarenites

Tufoarenites are secondary sand-size pyroclastic rocks, with scarce vitric clasts (<25%), not mixed with epiclastic sediments (Teruggi *et al.* 1978). Reworking is evident because they are richer in crystals and rock fragments, appearing as lithic or feldspathic sandstones. They are common (22%) in Lower and Upper Puesto Almendra Members, and scarce in the lower section of the Colhue-Huapi Member.

Tufoarenites are fine- to medium-grained, showing dominant trough cross-stratification (Fig. 19.5: 3–4). They occur with lenticular bed geometry (like the conglomerates), some tens to a hundred meters wide. Sets are 0.5–1.5 m thick and fining-upward. Angular intraclasts and scattered blocks (1–20 cm in diameter) can be present, generally overlying erosive surfaces as lag deposits (Fig. 19.5: 4). This facies represents channel bodies (Fig. 19.5: 5–6), classified as multi-story narrow sheets (100 > width/thickness > 15). In Unit 1 of the Lower Puesto Almendra Member, channel

bodies are 3–5 m thick and 100–200 m wide, and show subhorizontal undulating bounding surfaces. A change in architectural arrangement is observed eastward between Profiles A and J, from laterally amalgamated to isolated channel bodies embedded in loessites. In the Upper Puesto Almendra Member, channel bodies of Units 3 and 4 are oriented NNW, while cross-bedding measurements indicate a NNE paleoflow. At Profiles L–M, some exhibit rooted surfaces internally. Channel bodies of Unit 5 are amalgamated, and show large-scale inclined surfaces and clearly preserved margins (Fig. 19.5: 6).

The tufoarenite facies originated as three-dimensional dunes or megaripples in low-sinuosity, sand-bedded rivers. The association with intraformational conglomerates suggests that discharge fluctuated frequently, probably according to seasonal rainfall. Ephemeral currents are inferred by rooted pause-planes observed in Units 3 and 4. Typical crevasse or overbank deposits are not recognized.

284	E. S. Bellosi

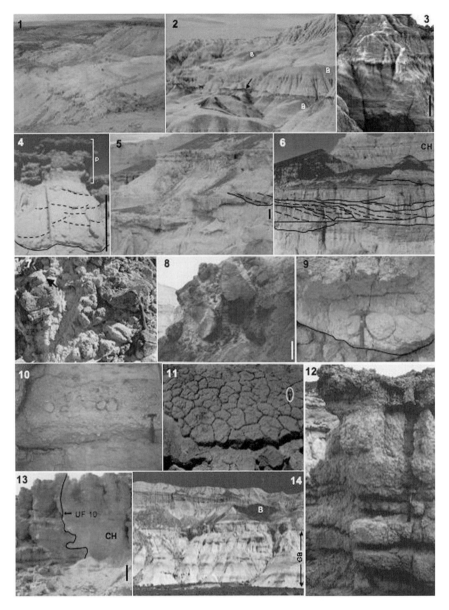

Fig. 19.5. 1, Thick and laterally continuous massive pyroclastic mudstones (loessites) of the GB; 2, pyroclastic mudstones interbedded with bentonites (B), arrow indicates a pedogenic manganese duricrust in Bed Y; 3, cross-bedded tufoarenites (Unit 1, LPA); 4, finning-upward channel fill, strongly pedified (P) in the upper part (lower Colhue-Huapi Mbr); 5, fluvial channel bodies of Unit 4 (UPA); 6, amalgamated fluvial channel bodies (Unit 5, UPA), at background, pyroclastic mudstones (loessites) of lower Colhue-Huapi Mbr (CH); 7, breccia of weathered basalt clasts, arrow shows a bone fragment (Unit 4, UPA); 8, large vertical blocks of basalt in a white tuffaceous matrix, adjacent to a basalt flow (idem 7); 9, intraformational conglomerate with indurated paleosol fragments (UPA); 10, Matrix-supported intraformational conglomerate (Unit 2, LPA); 11, polygonal desiccation cracks in a surface horizon of a reddish paleosol (Alfisol–Ultisol) (Unit 5, UPA); 12, stacked and indurated, strongly developed paleosols, with granular ped structure and intense bioturbation, elaborated in conglomerates (Unit 5, UPA); 13, irregular vertical wall (Unconformity 10) of the Colhuehuapian incised valley (CH) filled with conglomerates; 14, lenticular basalt flow (B) in UPA, outcrops of the GB in the lower part. GB, Gran Barranca Mbr; LPA, Lower Puesto Almendra Mbr; UPA, Upper Puesto Almendra Mbr. Bar 1 m, hammer 30 cm.

The development of paleosols indicates channel abandonment. The association of facies suggests that sedimentation on interfluves was governed mostly by eolian processes. The inclined surfaces observed in channel bodies of Unit 5 are interpreted as lateral accretion surfaces or point bars.

This feature and the fining-upward trend are consistent with deposition in channels with higher sinuosity, at least in some parts of the river. The different characteristics of Unit 1 suggest that its fluvial deposits formed under conditions of lower discharge.

Intraformational conglomerates and breccias

Clast-supported intraformational conglomerates are frequent in the Lower and Upper Puesto Almendra Members and in the lower section of the Colhue-Huapi Member, overlying unconformities in each case (Fig. 19.5: 13). They are massive, normal graded or with horizontal stratification. Basal contacts are erosive and bed thickness varies between 0.4 and 0.8 m (Fig. 19.5: 9). They occur stacked in intervals 2–4 m thick, commonly interbedded with tufoarenites. In the lower Colhue-Huapi Member they can reach up to 8 m in thickness. Modification by pedogenesis is frequent, particularly in Unit 5 where some conglomerates exhibit desiccation cracks on top (Fig. 19.5: 11). They are moderately to well sorted (mean: 3 cm). Maximum clast size varies from 1 to 30 cm. Intraclasts mostly correspond to soil fragments or peds. Polished *Coprinisphaera* balls, a beetle trace fossil, also appear. The distinction between some intraformational conglomerates and paleosols is subtle when the paleosols show coarse granular peds similar to intraformational clasts. This problem was solved by assessing bed geometry, contacts, color changes, and other pedogenic features. The origin of intraformational clast-supported conglomerates is similar to that of the fluvial tufoarenites, i.e. both constitute channel fill. Soil aggregates, insect ichnofossils, and bank fragments were entrained and transported as bedload.

A distinct matrix-supported intraformational conglomerate is present in Unit 2 (Fig. 19.5: 10). This is lenticular (few hundred meters), 1.4 m thick, and poorly stratified. It is formed by subangular and poorly sorted soil fragments (5–35 cm in diameter). The maximum clast size is observed in the middle part of the bed. This deposit may represent an extraordinary debris flow, possibly related to high increase in rainfall. These processes are frequent in regions that received large amounts of volcanic ash in short lapses of time (Smith 1991).

Lenticular to wedge-shaped coarse breccias are present in Unit 4. They are composed of weathered or fresh basalt clasts or blocks, 0.1–2.0 m in diameter, in a white fine tuffaceous matrix (Fig. 19.5: 7–8). Basalt breccias occur in three situations: (1) lateral and adjacent to basalt flows, (2) overlying irregular erosive surfaces on top of basalts (DS 8, Fig. 19.4: 5), and (3) as much as a few meters below and lateral to the basalts where they constitute the infill of narrow channels (30 m wide, 5 m thick), which also yield large bones (Deseadan SALMA). Laterally, this facies grades into cross-bedded tufoarenites. Basalt breccias are interpreted to be residual (with fresh blocks) or slightly transported fluvial accumulations.

Paleosols

Owing to their significance to stratigraphy, paleoenvironments, and paleoclimate, paleosols are assessed in a separate chapter (Bellosi and González this book). Paleosols

occur in all members and generally are moderately to highly indurated. They have been characterized in the field by soil horizons, texture, structure, contacts, color changes, and bioturbation, and in the laboratory by thin sections. X-ray diffraction and geochemical analysis were also performed on some samples. The classification of paleosols was made according to the U.S. soil taxonomy system as adapted to fossil soils by Retallack (2001). Parent material (vitric ash) is relatively uniform throughout the studied succession. Trace fossils, particularly those produced by insects, are an outstanding feature of Sarmiento paleosols (Bellosi *et al.* 2001; Bellosi *et al.* this book). Most of the trace fossil associations belong to the *Coprinisphaera* ichnofacies (Bellosi and Genise 2004; Genise *et al.* 2004; Cosarinsky *et al.* 2005), with several ichnospecies of *Coprinisphaera* (dung beetle brood balls: Laza 2006), *Feoichnus* (cicada nymph chambers: Krause *et al.* 2008), and *Celliforma* (bee cells).

According to their diagnostic features, Sarmiento Formation paleosols include Andisols, Alfisols, Aridisols, calcic Entisols, and Vertisols. The stratigraphic and lateral distribution of paleosol types is a consequence of change in the time of formation and environmental conditions of climate, vegetation, and topography. Moderately developed Alfisols are associated with fluvial facies of the Lower and Upper Puesto Almendra members, and lower Colhue-Huapi Member (Fig. 19.5: 12), suggesting discontinuous sedimentation and a wetter climate. Calcic Entisols and Aridisols, intercalated in loessites of the Vera and upper Colhue-Huapi Members, are related to a dryer environment, and a faster and more continuous (eolian) accumulation.

Basalt flows

Alkaline basalt flows are present in the Sarmiento Formation at many localities in extra-Andean central Patagonia. These volcanic rocks are exposed at Gran Barranca, and elsewhere at Cerro Blanco, Cerro Negro, Valle Hermoso, Los Leones, Gran Bajo Oriental, Pico Truncado, south of Deseado river (at Basalto Alma Gaucha), and in the San Bernardo Range (Feruglio 1949; Bellosi 1995; Panza and Franchi 2002). Associated with these lavas and distributed throughout the San Jorge Basin, there are numerous hypabyssal intrusions (sills and dykes) in the Rio Chico Group, Salamanca Formation, and Chubut Group, as observed in both outcrops and subsurface (Bellosi 1995).

At Gran Barranca, five lenticular basalts occur in Unit 3 of the Upper Puesto Almendra Member, intercalated in fluvial channel deposits (Fig. 19.5: 14). The two largest define the east (Profile A) and west (Profile N) ends of the cliff (Simpson 1930). They are 1100–1600 m wide, 6–18 m thick, and extend parallel to channel bodies (approximately N–S). Internally they are zoned (0.3–0.7 m thick) by changes in size and proportion of vesicles. Internal weathered surfaces were not observed. The smaller basalts are 8–35 m wide and 4–8 m thick, with a concave-up lower contact. Subaerial exposure of the basalts is evident from weathered upper surfaces and the presence of associated breccias. Several basalt lenses also outcrop nearby in Valle Hermoso, showing similar orientation, geometry, and stratigraphic position. They have been related to basaltic necks intruding through the Gran Barranca Member and Koluel-Kaike Formation.

The lava flows at Gran Barranca and Valle Hermoso could be correlated with the brown volcaniclastic deposits (30 m thick) cropping out at Route 26 (near Cerro Dragón), 30 km east of Gran Barranca. These basaltic deposits show high-angle bedding, and include matrix-supported debris flows with large blocks (0.5–1.0 m), cross-bedded conglomerates with lapilli and bombs, and thick clast-supported tuffaceous breccias. They represent volcanic edifices constructed by pyroclastic mass flows with autoclastic fragments. This association of lava flows and explosive activity is similar to the cinder cones described by Mazzoni (1994) from Scarritt Pocket (Deseadean SALMA), 170 km north of Gran Barranca.

Isotopic age and stratigraphy of Gran Barranca basalts has been discussed by Marshall *et al.* (1986). New Ar/Ar dates indicate an isotopic age range of 29.2–26.3 Ma (Ré *et al.* Chapter 4, this book). The basalt flows at Gran Barranca were interpreted as tongues that filled shallow "valleys" (McCartney 1933). Field observations at Gran Barranca and Valle Hermoso suggest that lava flowed along N–S-oriented fluvial channels. Subsequently, intensive and differential erosion produced inversions of local relief, as demonstrated by the presence of basalt breccias topographically below basalts.

The late Oligocene volcanic episode recorded by the lava flows and the construction of cinder cones seems to have been related to normal faulting and formation of half-grabens (Flores 1954; Bitschene *et al.* 1991; Bellosi 1995).

Tephra ejection and dispersion

Felsic ash sequences of the Sarmiento Formation are the distal or basinal facies of tephric loess of the northwest Patagonia Volcanic Province (Fig. 19.1), represented in the source area by Pilcaniyeu, Ventana, and Huitrera Formations (Paleocene–Miocene). Two sub-parallel Andean volcanic arcs producing silicic facies have been recognized by Rapela *et al.* (1988). The more eastern Pilcaniyeu Belt was more active from the Paleocene to the middle Eocene (60–42 Ma), while the western El Maiten Belt developed during the Oligocene to early Miocene (33–21 Ma). In extra-Andean central Patagonia, Eocene to Miocene silicic volcanic rocks are nearly absent. Small outcrops of middle Eocene (40.8 Ma) dacitic lava flows were reported by Guido *et al.* (2004) in northern Santa Cruz Province.

The Oligocene to early Miocene bimodal effusive complexes of the extra-Andean Somun Cura Massif in northern Patagonia had been proposed as a secondary source of felsic ash in Sarmiento Formation (Franchi and Nullo 1986). However, the alkaline signature of this particular volcanism (Ardolino *et al.* 1999) differs in composition from the dacitic composition of Sarmiento ash (Mazzoni 1985). Unfortunately, regional field mapping of grain-size, sorting, mineral and trace element composition or thickness is lacking, which would provide further information about ancient wind patterns, eruptive centers, dynamics of explosions, eruptive energy, and others volcanologic characteristics (Fisher and Schmincke 1984). Geochemical studies indicate that Sarmiento distal tephras resulted from high-energy explosive Plinian to sub-Plinian volcanism (Mazzoni 1985), although phreatoplinian eruptions cannot be excluded because of the dominantly very fine grain-size even in the most proximal deposits near the centers (Self and Sparks 1978; Zimanowsky *et al.* 2003). Dispersion of volcaniclastics was probably controlled by wind direction and speed, and eruption style (e.g. ejecta fragmentation, volume and height of the eruption column).

Judging from their ages, pyroclastic deposits of the Gran Barranca, Rosado, and Lower Puesto Almendra Members resulted from the activity of the external Pilcaniyeu Belt (Fig. 19.1) with bimodal calc–alkaline composition and with abundant ignimbritic facies in proximal areas (Ardolino *et al.* 1999). On the other hand, the Oligocene to early Miocene pyroclastics of the Vera, Upper Puesto Almendra, and Colhue-Huapi Members would have been ejected from the internal El Maitén Belt (Fig. 19.1). The activity of this volcanic arc was simultaneous with intraplate basic alkaline magmatism in central and eastern Patagonia.

The thickness of ash deposits in northern and central Patagonia varies due to relief, erosional unconformities, and different tectonic settings. Irregular topographies can be recognized locally by deep erosion surfaces as at Gran Barranca (Bellosi *et al.* 2002b; Bellosi this book) or Scarritt Pocket (Marshall *et al.* 1986; Mazzoni 1994). Fallout tephra drapes hillslopes and valleys and can be deposited on very steep surfaces (>25°). The absence of deposition on deep scouring surfaces at Gran Barranca indicates that ash accumulation, if it occurred on slopes, was rapidly transported onto low and flat areas by wind, water, or gravity. The notorious scarcity of felsic debris flow deposits suggests that such sediment transport was more probably eolian.

Temporal changes in depositional settings

From the early Eocene (~52 Ma) to the early Miocene (~17 Ma), extra-Andean central and north Patagonia were subjected to subaerial ash falls, sediment accumulation and/ or reworking through eolian and fluvial processes, erosion events, and periods of landscape stability and pedogenesis. These processes were interrelated, for example, thick ash falls could destroy vegetation and thereby promote increased rates of surface runoff and erosion during rainstorms.

Different types of evidences suggest that these middle Cenozoic terrestrial scenarios evolved largely in response to allocyclic factors such as climate, volcanism, subsidence, and sea level, which controlled sediment input, accommodation space, and erosion rate (Bellosi 1995). The greater thickness of the Sarmiento Formation at Gran Barranca indicates that this part of the San Jorge Basin underwent a higher subsidence by comparison to other sectors of central Patagonia. Climate played a cardinal role during Sarmiento time, controlling sediment supply, transport, and deposition. Important global climatic variation occurred in the late Eocene, at the Eocene–Oligocene boundary, and in the late Oligocene and early to middle Miocene (Zachos *et al.* 2001). Of these, one of the more significant instances of climate shifts is claimed to be related to be at the Eocene–Oligocene boundary – often referred to as a change from a "greenhouse" to an "icehouse" world. Enhanced explosive volcanism with the attendant rapid, thick, and extensive accumulation of ash must also have had a great impact on landscapes, biota (Axelrod 1981; Fisher and Smith 1991; Rose *et al.* 2003) and climate (Rampino 1991). The consequences of local tectonism during Sarmiento time are poorly known and probably were more evident during extensional periods in the Oligocene. Legarreta and Uliana (1994) claimed that accommodation changes linked to global eustatism had a critical impact on the stratigraphy of the Sarmiento Formation. However, sea-level fluctuations most likely affected continental settings closer to the marine paleocoast (Ethridge *et al.* 1998). Thus, eustatic fluctuations are more evident in the San Jorge Basin during the early Miocene, in association with the Leonian and Superpatagonian transgressions, which are correlated temporally with the Colhue-Huapi Member and its basal unconformity, formed by a sea-level fall (Bellosi *et al.* 2002b).

Changes in the depositional settings of the Sarmiento Formation are reflected by variation of the sedimentary processes seen in successive members. During the time of the Gran Barranca Member (middle Eocene), a subaerial loessic sedimentation prevailed, represented by massive pyroclastic mudstones. Tephric loessites with their associated paleosols formed in a subhumid to semi-arid, seasonal climate. Eolian transport of the sediments by suspension is supported by the abundant unbroken glass shards (Spalletti and Mazzoni 1977, 1979). Parrish (1998) considers eolian dust or loess deposits as a general indicator of dry environments or desert peripheries. Similarly, an arid to semi-arid fluctuating climate was interpreted for an analogous Permian loessite–paleosol sequence (Kessler *et al.* 2001). The absence of fluvial reworking in this member indicates

a water-balance deficit. Bottomlands were occupied by ephemeral small lakes, as suggested by bentonite beds. The limited occurrence and weak developement of Andisols (Bellosi and González this book) indicate that eolian accumulation was relatively continuous. Paleosols represent moist and/or warmer conditions between dryer periods of loess accumulation (Kemp 2001). The presence of Andisols with thick placic horizons (Mn crusts) and fossil crocodiles (Simpson 1940) in Bed Y indicate waterlogged conditions. The environment can be envisaged as broad rolling plains with ponded sectors (Simpson 1940). Fossil reptiles and mammals of the Barrancan Subage (Pascual and Ortíz-Jaureguizar 1990) together with abundant palm remains (Zucol *et al.* this book) suggest a warm–temperate climate. Smectite-bearing paleosols indicate a mean annual precipitation (MAP) <1000 mm (Evans 1992; Macias and Chesworth 1992); however, the absence of carbonate nodules indicates conditions above 500 mm MAP. Similar paleoenvironments developed in other places of central Patagonia (Feruglio 1949; Andreis *et al.* 1975; Spalletti and Mazzoni 1977; Legarreta and Uliana 1994). In Cañadón Hondo, Andreis (1977) interpreted a lacustrine–palustrine depocenter fed by ephemeral streams, although the higher proportion of paleosols points to a more discontinuous accumulation.

This period of uniform loessic sedimentation stopped towards the end of the middle Eocene, and was followed by a stage of marked landscape stability and dryness, represented by the Rosado Member (late middle Eocene), a thick calcareous paleosol (Aridisol) formed in pyroclastic mudstones. Mustersan mammals, diverse and abundant land snails, and insect trace fossils suggest a heterogeneous environment (Bellosi *et al.* 2002a, Bellosi *et al.* this book; Miquel and Bellosi 2007). The contrasting characteristics between this and the earlier period were the consequence of change from subhumid–humid to extreme xeric conditions. Terrestrial gastropods of diverse environmental requirements indicate a high frequency of climate variation during the time-span of the Rosado Member (Bellosi *et al.* 2002a). Similarly diverse terrestrial gastropod associations accumulate in loess–paleosol successions by eolian deflation and time-averaging (Behrensmeyer and Hook 1992). Arid conditions are coincident with the disappearance of crocodilians, the increase of hypsodonty in some mammal herbivores (Pascual and Ortíz-Jaureguizar 1990), and the variation in phytolith composition (Mazzoni 1979; Zucol *et al.* 2006), suggesting a shift from a palm-dominated Barrancan subage to grass-dominated vegetation in Mustersan time.

A significant modification in the sedimentation style is again recorded in the late Eocene Lower Puesto Almendra Member, after of an erosive event (Discontinuity 3). Facies association of the Units 1 and 2 of the Lower Puesto Almendra Member indicate a fluvial setting, suggesting an increase in precipitation. Conglomerates and sandstones accumulated in braided probably ephemeral channels. Subordinated eolian accumulation of fine volcaniclastics contributed to the aggradation on adjacent plains, which included some shallow and small lakes because the presence of bentonites. Alfisols intercalated in this member indicate a subhumid environment (Bellosi and González this book). The debris flow of Unit 2 would represent a catastrophic event, probably related to a sudden increase in rainfall. Correlated sections at Gran Hondonada and Laguna del Mate, north of Gran Barranca, also suggest an ephemeral fluvial system with small channels, and subordinated eolian accumulation on plains (Cladera *et al.* 2004), where argillic paleosols developed under a seasonal semi-arid climate (Andreis 1972).

An important erosive event occurred subsequently in the late Eocene (Discontinuity 5), eliminating nearly 20 m of previous deposits, mostly in the western sector of Gran Barranca. The change in the sedimentary scenario and accumulation rate is represented by the late Eocene to early Oligocene Vera Member, a thick and homogeneous succession of tephric loessites, with a few thin bentonite beds and very weakly developed calcic paleosols. Discontinuity surfaces or alluvial deposits were not observed in this member. This uniform sequence records a continuous subaerial sedimentation on plains, interrupted by occasional and short lapses of pedogenesis. Thus, a semi-arid climate is suggested. Thickness, lithologic homogeneity and absence of discontinuities indicate an accelerated sediment accumulation rate caused by an increase in the volcanic arc activity and/or eolian sediment supply. The steady and rapid subaerial aggradation and scarcity of paleosols could be the causes of the virtual absence of fossil vertebrates, owing to increased stratigraphic dispersion of vertebrate remains and minimal chemical modification of deposits by pedogenesis. Accordingly, the only fossiliferous bed in Vera Member (La Cancha Bed, Tinguirirican SALMA) corresponds to the more developed paleosol.

The most intense episode of intraformational erosion and landscape destabilization occurred in the early Oligocene. This erosion event is represented by Discontinuity 6 (Bellosi this book). The removal of a 100-m vertical column from Vera and Lower Puesto Almendra Members resulted in the formation of a deep valley, which was probably controlled by extensional faulting (Barreda and Bellosi 2003). Fine tephras emplaced rapidly in dry environments as thick mantling layers are susceptible to deep erosion by fluvial processes, because destruction of sediment-stabilizing vegetation (Smith 1991). The infill deposits of the valley constitute the Oligocene Upper Puesto Almendra Member, representing a new significant change in the sedimentary style with progressive increase in pedogenesis. The dominant facies association of Units 3 and 4 (Upper Puesto Almendra Member) originated in a NNE-oriented multi-

channel fluvial system, probably braided, with major (1–2 km wide) and narrow (15–40 m wide) rivers. Adjacent plains were covered sporadically by loess deposits. Moderately developed paleosols (Alfisols) included in these Deseadan age units formed in a subhumid seasonal climate (Bellosi and González this book). In addition, slightly pedified surfaces observed in some channel fills indicate subaerial exposure of fluvial bars, and an ephemeral regime. An episode of alkaline basalt flows and construction of autoclastic volcanic cones (~27.5 Ma), related to a NNW-oriented rifting, developed simultaneously (Bellosi 1995). The mid-Oligocene extensional regime and associated basic magmatism were related to readjustment of Nazca–South America plate convergence, from slower oblique to faster near orthogonal subduction (Cande and Leslie 1986; Morata *et al.* 2005). Unit 5 of the Upper Puesto Almendra Member (uppermost Oligocene) consists of amalgamated fluvial channel bodies accumulated in a meandering, probably perennial fluvial system. Fluvial deposits were moderately to strongly pedified, constituting stacked orangish Alfisols–Ultisols. These paleosols indicate a landscape stabilized by vegetation with a relatively low sediment accumulation rate and a subhumid–humid seasonal climate (Bellosi and González this book). Late Oligocene vegetation was dominated by forests of Podocarpaceae, Nothofagaceae, and Araucariaceae, abundant ferns, diverse angiosperms, and relatively rare shrubby elements (Barreda 1997; Barreda and Palamarczuk 2000), suggesting warm–temperate and humid conditions. This climate amelioration would be coincident with the late Oligocene global event mentioned by Zachos *et al.* (2001).

Since the early Miocene, the San Jorge Basin was partially flooded by the Leonian and Superpatagonian marine transgressions, favoring the development of coastal plain environments (Bellosi 1995). The early Miocene Colhue-Huapi Member records the last stage of Sarmiento sedimentation between both transgressions, established after an episode of deep fluvial incision (Discontinuity 10), caused by a sea-level fall (Bellosi and Barreda 1993; Legarreta and Uliana 1994). At Gran Barranca, the lower section of this member is mainly constituted by coarse facies originated in a bedload fluvial system, and subordinated loessites. Accordingly to geographic distribution and stratigraphic relationships, these deposits concentrated in valley fills (Bellosi *et al.* 2002b; Goin *et al.* 2007). The small size of conglomerate and sandstone bodies suggest shallow and narrow channels, which preserve very abundant fossil vertebrates, corresponding to Colhuehuapian SALMA. Andic Alfisols are frequent and developed on both fluvial and eolian sediments. These characteristics are compatible with a seasonal subhumid climate and an increased sedimentation rate. Early Miocene vegetation experienced the expansion of shrubby components, and the restriction of forests and megathermal plants, in response to less humid conditions (Barreda and Bellosi 2003). However, closed riparian forests remained in this region (Barreda and Palazzesi 2007, this book; Bellosi and González this book), which were probably occupied by the Colhuehuapian platyrrhines (Kay this book). The upper section of the Colhue-Huapi Member is constituted by thick tephric loessites and less abundant, weakly developed calcic paleosols, representing a semi-arid eolian environment. The Pinturan rodents recovered from this section show adaptations to dryer and/or cooler conditions (Vucetich *et al.* 2005). Early Miocene palynological data obtained from coeval marine deposits of the Chenque Formation also suggest a dryer vegetation in the coastal plain, with the spreading of xerophytic herbs and shrubs and sclerophyll trees (Barreda and Palazzesi 2007, this book; Palazzesi and Barreda 2007). The San Jorge Basin was subsequently invaded by the Superpatagonian Sea (Bellosi 1995) which marked the end of the continental Sarmiento sedimentation.

Conclusions

The Sarmiento Formation records the protracted loessic and fluvial accumulation of pyroclastic sediments in Central Patagonia from the middle Eocene into the early Miocene. Both types of paleoenvironments alternated several times, mainly in response to climate change. Sediment supply was a significant secondary factor, at a time governed by Plinian activity in the northwestern Patagonia volcanic arc (e.g. dynamics of explosions, eruptive energy) and by the eolian transport of fine tephras. Likewise, topography and landscape modifications were also related to sea-level fluctuation and extensional tectonism, particularly from the late Oligocene into the early Miocene. In general terms, the development of the Sarmiento Formation involved eolian and fluvial reworking and accumulation of distal ash fall deposits. Pedogenesis occurred during intervals of landscape stability. At times there were events of deep fluvial erosion. The end product of these processes is a set of sedimentary sequences bounded by erosive and non-depositional unconformities of different hierarchy.

Dryer periods (arid–semi-arid to subhumid climate), with relatively rapid eolian deposition, are mainly represented by pyroclastic mudstones (loessites) and very weakly to weakly developed calcic and non-calcic paleosols. Such intervals are recorded in the Gran Barranca, Rosado, Vera, and upper Colhue-Huapi Members. Wetter periods (subhumid to humid climate) dominated by fluvial deposition are recognized by channel facies (intraformational conglomerates and sandstones) and weakly to moderately developed non-calcic paleosols. These intervals of more discontinuous sedimentation predominate in the Lower and Upper Puesto Almendra, and lower Colhue-Huapi Members.

ACKNOWLEDGEMENTS
Field research was supported by US National Science Foundation grants EAR-0087636, BCS-0090255, and DEB-9907985 to Richard F. Kay and Richard H. Madden. I am grateful to T. Bown, R. F. Kay, and R. H. Madden for the review of the manuscript, and to Pan American Energy for the assistance during field work.

REFERENCES
Ameghino, F. 1906. Les formations sédimentaires du Crétacé súperieur et du Tertiaire de Patagonie. *Anales del Museo Nacional de Historia Natural*, **15**, 1–568.

Andreis, R. 1972. Paleosuelos de la Formación Musters (Eoceno medio), Laguna del Mate, prov. de Chubut, Rep. Argentina. *Revista de la Asociación Argentina de Mineralogía, Petrografía y Sedimentología*, **3**, 91–97.

Andreis, R. 1977. Geología del área de Cañadón Hondo, Escalante, prov. de Chubut, República Argentina. *Museo La Plata, Obra del Centenario*, **4**, 77–102.

Andreis, R., M. Mazzoni, and L. Spalletti 1975. Estudio estratigráfico y paleoambiental de las sedimentitas terciarias entre Pico Salamanca y Bahía Bustamante, provincia de Chubut, República Argentina. *Revista de la Asociación Geológica Argentina*, **30**, 85–103.

Ardolino, A., M. Franchi, M. Remesal, and F. Salani 1999. El volcanismo en la Patagonia extraandina. *Anales de Instituto de Geología y Recursos Minerales*, **29**, 579–612.

Axelrod, D. 1981. *Role of Volcanism in Climate and Evolution*, Special Paper no. 185. Bourder, CO: Geological Society of America.

Barreda, V. 1997. Palinoestratigrafía de la Formación San Julián en el área de Playa La Mina (Provincia de Santa Cruz), Oligoceno de la Cuenca Austral. *Ameghiniana*, **34**, 283–294.

Barreda, V. and E. Bellosi 2003. Ecosistemas terrestres del Mioceno temprano de la Patagonia central: primeros avances. *Revista del Museo Argentino de Ciencias Naturales*, n.s., **5**, 125–134.

Barreda, V. and S. Palamarczuk 2000. Palinoestratigrafía de depósitos del Oligoceno tardío–Mioceno, en el área sur del Golfo San Jorge, provincia de Santa Cruz, Argentina. *Ameghiniana*, **37**, 103–117.

Barreda, V. and L. Palazzesi 2007. Patagonian vegetation turnovers during the Paleogene–Early Neogene: origin of arid-adapted floras. *Botanical Review*, **73**, 31–50.

Behrensmeyer, A. and R. Hook 1992. Paleoenvironmental contexts and taphonomic modes. In Behrensmeyer, A., Damuth, J. D., and DiMichele, W. A. (eds.), *Terrestrial Ecosystems through Time*. Chicago, IL: University of Chicago Press, pp. 15–136.

Bellosi, E. 1987. Litoestratigrafía y sedimentación del "Patagoniano" en la Cuenca San Jorge, Terciario de Chubut y Santa Cruz. Ph.D. thesis, University of Buenos Aires.

Bellosi, E. 1995. Paleogeografía y cambios ambientales de la Patagonia central durante el Terciario medio. *Boletín de Informaciones Petroleras*, **44**, 50–83.

Bellosi, E. and V. Barreda 1993. Secuencias y palinología del Terciario medio en la Cuenca San Jorge, registro de oscilaciones eustáticas en Patagonia. *Actas XII Congreso Geológico Argentino y II Congreso de Exploración de Hidrocarburos*, **1**, 78–86.

Bellosi, E. and J. Genise 2004. Insect trace fossils from paleosols of the Sarmiento Formation (Middle Eocene–Lower Miocene) at Gran Barranca (Chubut Province). In Bellosi, E. and Melchor, R. (eds.), *Fieldtrip Guidebook from the First International Congress on Ichnology*, pp. 15–29.

Bellosi, E., J. Laza, and M. González 2001. Icnofaunas en paleosuelos de la Formación Sarmiento (Eoceno–Mioceno), Patagonia central. *Resumenes IV Reunión Argentina de Icnología y II Reunión de Icnología del Mercosur*, 31.

Bellosi, E., S. Miquel, R. Kay, and R. Madden 2002a. Un paleosuelo mustersense con microgastrópodos terrestres (Charopidae) de la Formación Sarmiento, Eoceno de Patagonia central: significado paleoclimático. *Ameghiniana*, **39**, 465–477.

Bellosi, E., M. González, R. Kay, and R. Madden 2002b. El valle inciso colhuehuapense de Patagonia central (Mioceno inferior). *Resumenes Reunión Argentina de Sedimentología*, 49.

Bitschene, P., R. Giacosa, and M. Márquez 1991. Geologic and mineralogic aspects of the Sarmiento alkaline province in Central eastern Patagonia, Argentina. *Actas VI Congreso Geológico Chileno*, 328–331.

Bown, J. and C. Larriestra 1990. Sedimentary paleoenvironments of fossil platyrrhine localities, Miocene Pinturas Formation, Santa Cruz province, Argentina. *Journal of Human Evolution*, **19**, 87–119.

Cande, S. C. and R. B. Leslie 1986. Late Cenozoic tectonics of southern Chile trench. *Journal of Geophysical Research*, **91B**, 471–496.

Cifelli, R. L. 1985. Biostratigraphy of the Casamayoran, Early Eocene, of Patagonia. *American Museum Novitates*, **2820**, 1–26.

Cladera, G., E. Ruigómez, J. E. Ortíz-Jaureguizar, M. Bond, and G. López 2004. Tafonomía de la Gran Hondonada (Formación Sarmiento, Edad-mamífero Mustersense, Eoceno medio) Chubut, Argentina. *Ameghiniana*, **41**, 315–330.

Cosarinsky, M., E. Bellosi, and J. Genise 2005. Micromorphology of modern epigean termite nests and possible termite ichnofossils: a comparative analysis. *Sociobiology*, **45**, 1–34.

Ethridge, F. G., L. J. Wood, and S. A. Schumm 1998. Cyclic variables controlling fluvial sequence development: problems and perspectives. *Society of Economic Paleontologists and Mineralogists Special Publications*, **59**, 17–29.

Evans, L. J. 1992. Alteration products at the earth's surface: the clay minerals. In Martini, I. and Chesworth, W. (eds.), *Weathering, Soils and Paleosols*. Amsterdam: Elsevier, pp. 107–125.

Feruglio, E. 1949. *Descripción Geológica de la Patagonia*, vol. 2. Buenos Aires: Dirección General de Yacimientos Petrolíferos Fiscales.

Fisher, R. 1966. Textural comparison of John Day volcanic siltstone with loess and volcanic ash. *Journal of Sedimentary Petrology*, **36**, 706–718.

Fisher, R. and H.-U. Schmincke 1984. *Pyroclastic Rocks*. Berlin: Springer-Verlag.

Flores, M. 1954. *Levantamiento Geológico–Estructural entre Cerro Dragón y Cañadón Grande, Chubut*. Buenos Aires: Dirección General de Yacimientos Petrolíferos Fiscales. (Unpublished.)

Franchi, M. and F. Nullo 1986. Las Tobas de Sarmiento en el Macizo de Somuncura. *Revista de la Asociación Geológica Argentina*, **41**, 219–222.

Genise, J., E. S. Bellosi, and M. González 2004. An approach to the description and interpretation of ichnofabrics in palaeosols. In McIlroy, D. (ed.), *The Application of Ichnology to Paleonvironmental and Stratigraphic Analysis*, Special Publication no. 228. London: Geological Society of London, pp. 355–382.

Genise, J., M. G. Mángano, L. Buatois, J. Laza, and M. Verde 2000. Insect trace fossils associations in paleosols: the *Coprinisphaera* ichnofacies. *Palaios*, **15**, 49–64.

Goin, F., A. Abello, E. Bellosi, R. Kay, R. Madden, and A. Carlini 2007. Los Metatheria sudamericanos de comienzos del Neógeno (Mioceno Temprano, Edad-mamífero Colhuehuapense). I. Introducción, Didelphimorphia y Sparassodonta. *Ameghiniana*, **44**, 29–71.

Guido, D., M. Escayola, R. de Barrio, I. Schalamuk, and K. Wemmer 2004. Formación Laguna Tordillo: volcanismo dacítico eoceno en el Macizo del Deseado, provincia de Santa Cruz. *Revista de la Asociación Geológica Argentina*, **59**, 763–770.

Hay, R. L. 1978. Volcanic ash diagenesis. In Fairbridge, R. and Bourgeois, J. (eds.), *The Encyclopedia of Sedimentology, Encyclopedia of Earth Sciences*, vol. 6. Stroudsburg, PA: Dowden, Hutchinson, & Ross, pp. 850–1.

Kemp, R. A. 2001. Pedogenic modification of loess: significance for paleoclimatic reconstructions. *Earth Sciences Reviews*, **54**, 145–156.

Kessler, J. L., G. S. Soreghan, and H. J. Wacker 2001. Equatorial aridity in Western Pangea: Lower Permian loessite and dolomitic paleosols in Northeastern New Mexico, U.S.A. *Journal of Sedimentary Research*, **71**, 817–832.

Kraus, M. 1999. Paleosols in clastic sedimentary rocks. *Earth Sciences Reviews*, **47**, 41–70.

Krause, J. M., T. Bown, E. Bellosi, and J. Genise 2008. Trace fossils of cicadas in the Cenozoic of Central Patagonia, Argentina. *Palaeontology*, **51**, 405–418.

Laza, J. 2006. Dung-beetle fossil brood balls: the ichnogenera *Coprinisphaera* Sauer and *Quirogaichnus* (Coprinisphaeridae). *Ichnos*, **13**, 217–235.

Legarreta, L. and M. Uliana 1994. Asociaciones de fósiles y hiatos en el Supracretácico–Neógeno de Patagonia: una perspectiva estratigráfico–secuencial. *Ameghiniana*, **31**, 257–281.

Lizuaín, A., D. Ragona, and A. Folguera 1995. *Mapa geológico de la provincia del Chubut*. Buenos Aires: Dirección Nacional del Servicio Geológico.

Macias, F. and W. Chesworth 1992. Weathering in humid regions, with emphasis on igneous rocks and their metamorphic equivalents. In Martini, I. and Chesworth, W. (eds.), *Weathering, soils and paleosols*. Amsterdam: Elsevier, pp. 283–306.

McCartney, G. 1933. The bentonites and closely related rocks of Patagonia. *American Museum Novitates*, **630**, 1–16.

Malumián, N. 1999. La sedimentación en la Patagonia extraandina. *Anales de Instituto de Geología y Recursos Minerales*, **29**, 557–578.

Marshall, L., R. Hoffstetter, and R. Pascual 1983. Mammals and stratigraphy: geochronology of the continental mammal-bearing Tertiary of South America. *Paleovertebrata, Mémoire Extraordinaire*, 1–93.

Marshall, L., R. Cifelli, R. Drake, and G. Curtis 1986. Vertebrate paleontology, geology and geochronology of the Tapera de López and Scarritt Pocket. *Journal of Paleontology*, **60**, 920–951.

Mazzoni, M. 1979. Contribución al conocimiento petrográfico de la Formación Sarmiento, barranca sur del lago Colhue-Huapi, provincia de Chubut. *Revista de la Asociación Argentina de Mineralogía, Petrografía y Sedimentología*, **10**, 33–54.

Mazzoni, M. 1985. La Formación Sarmiento y el vulcanismo paleógeno. *Revista de la Asociación Geológica Argentina*, **40**, 60–68.

Mazzoni, M. 1994. Conos de cinder y facies volcaniclásticas miocenas en la Meseta del Canquel (Scarritt Pocket), provincia de Chubut, Argentina. *Revista de la Asociación Argentina de Sedimentología*, **1**, 15–31.

Miquel, S. and E. Bellosi 2007. Microgasterópodos terrestres (Charopidae) del Eoceno medio de Gran Barranca (Patagonia Central, Argentina). *Ameghiniana*, **41**, 121–131.

Morata D., C. Oliva, R. de la Cruz, and M. Suárez 2005. The Bandurrias gabbro: late Oligocene alkaline magmatism in the Patagonian Cordillera. *Journal of South American Earth Sciences*, **18**, 147–162.

Palazzesi, L. and V. Barreda 2007. Major vegetation trends in the Tertiary of Patagonia (Argentina): a qualitative paleoclimatic approach based on palynological evidence. *Flora*, **202**, 328–337.

Panza, J. and M. Franchi 2002. Magmatismo basáltico cenozoico extrandino. *Resúmenes XV Congreso Geológico Argentino*, 201–236.

Panza, J., J. Cobos, and D. Ragona 1994. *Mapa geológico de la provincia de Santa Cruz*. Buenos Aires: Dirección Nacional del Servicio Geológico.

Parrish, J. T. 1998. *Interpreting Pre-Quaternary Climate from the Geologic Record*. New York: Columbia University Press.

Pascual, R. and O. Odreman Rivas 1971. Evolución de las comunidades de los vertebrados del Terciario argentino, los aspectos paleozoogeográficos y paleoclimáticos relacionados. *Ameghiniana*, **8**, 372–412.

Pascual, R. and J. E. Ortíz-Jaureguizar 1990. Evolving climates and mammal faunas in Cenozoic South America. *Journal of Human Evolution*, **19**, 23–60.

Rampino, M. 1991. Volcanism, climate change, and the geologic record. *Society of Economic Paleontologists and Mineralogists Special Publications*, **45**, 9–18.

Rapela, C., L. Spalletti, J. Merodio, and E. Aragón 1988. Temporal evolution and spatial variation of Early Tertiary volcanism in the Patagonian Andes (40° S–42° 30′ S). *Journal of South American Earth Sciences*, **1**, 75–88.

Retallack, G. 2001. *Soils of the Past*, 2nd edn. Oxford, UK: Blackwell.

Rose, W., C. Riley, and S. Dartevelle 2003. Sizes and shapes of 10-Ma distal fall pyroclasts in the Ogallala Group, Nebraska. *Journal of Geology*, **111**, 115–124.

Self, S. and R. Sparks 1978. Characteristics of wide-spread pyroclastic deposits formed by the interaction of silicic magma and water. *Bulletin of Volcanology*, **41**, 1–17.

Simpson, G. G. 1930. *Scarritt-Patagonian Exped. Field Notes*. New York: American Museum of Natural History. (Unpublished.) Available at http://paleo.amnh.org/ notebooks/index.html

Simpson, G. G. 1940. Review of the mammal-bearing Tertiary of South America. *Proceedings of the American Philosophical Society*, **83**, 649–709.

Smith, G. 1991. Facies sequences and geometries in continental volcaniclastic sequences. *Society of Economic Paleontologists and Mineralogists Special Publications*, **45**, 109–137.

Spalletti, L. 1992. El loess y el problema de la identificación de las loessitas. *Revista del Museo La Plata*, n.s. 2, *Geología*, **102**, 45–53.

Spalletti, L. and M. Mazzoni 1977. Sedimentología del Grupo Sarmiento en un perfil ubicado al sudeste del lago Colhue-Huapi, provincia de Chubut. *Museo La Plata, Obra del Centenario*, **4**, 261–283.

Spalletti, L. and M. Mazzoni 1979. Estratigrafia de la Formación Sarmiento en la barranca sur del lago Colhue-Huapi, provincia del Chubut. *Revista de la Asociación Geológica Argentina*, **34**, 271–281.

Speed, J., P. Shane, and I. Nairn 2002. Volcanic stratigraphy and phase chemistry of the 11 900 yr BP Waiohau eruptive episode, Tarawera Volcanic Complex, New Zealand. *New Zealand Journal of Geology and Geophysics*, **45**, 395–410.

Suárez, M. and R. de la Cruz 2002. Stratigraphic discontinuities in the Meso–Cenozoic of eastern Aysén, Chile (44°–47° S). *Actas XV Congreso Geológico Argentino*, **1**, 706–710.

Teruggi, M., M. Mazzoni, L. Spalletti, and R. Andreis 1978. Rocas piroclásticas, interpretación y sistemática. *Asociación Geológica Argentina, Publicaciones Especiales*, **5**.

Teruggi, M. and R. Andreis 1971. Microestructuras pedológicas: características, distribución en sedimentitas argentinas y posible aplicación a la sedimentología. *Revista de la Asociación Geológica Argentina*, **24**, 491–502.

Vicars, R. and J. Breyer 1981. Sedimentary facies in air-fall pyroclastic debris, Arikaree Group (Miocene), Northwest Nebraska, USA. *Journal of Sedimentary Petrology*, **51**, 909–921.

Vucetich, M., E. Vieytes, A. Kramarz, and A. Carlini 2005. Caviomorph rodents from Gran Barranca: biostratigraphic and paleoenvironmental contribution. *Resúmenes XVI Congreso Geológico Argentino*, 303.

Weaver, C. 1978. Bentonite. In Fairbridge, R. and Bourgeois, J. (eds.), *The Encyclopedia of Sedimentology, Encyclopedia of Earth Sciences*, vol. 6. Stroudsburg, PA: Dowden, Hutchinson, & Ross, pp. 59–60.

Zachos, J., M. Pagani, L. Sloan, E. Thomas, and K. Billups 2001. Trends, rhythms, and aberrations in global climate 65 Ma to present. *Science*, **292**, 686–692.

Zimanowski, B., K. Wohletz, P. Dellino, and R. Büttner 2003. The volcanic ash problem. *Journal of Volcanology and Geothermal Research*, **122**, 1–5.

Zucol, A., M. Brea, R. Madden, E. Bellosi, A. Carlini, and M. G. Vucetich 2006. Preliminary phytolith analysis of Sarmiento Formation in the Gran Barranca (Central Patagonia, Argentina). In Madella, M., and Zurro, D. (eds.), *Places, People and Plants: Using Phytoliths in Archaeology and Palaeoecology*. Oxford, UK: Oxbow Books, pp. 197–203.

20 Paleosols of the middle Cenozoic Sarmiento Formation, central Patagonia

Eduardo S. Bellosi and Mirta G. González

Abstract

In the loessic and fluvial pyroclastic deposits of the mammal-bearing Sarmiento Formation (middle Eocene – early Miocene) of the Gran Barranca (central Patagonia) several types of alkaline and oxidized paleosols are recognized. These are characterized by relative uniform parent material, diverse insect trace fossils (*Coprinisphaera* and *Celliforma* ichnofacies), and abundant opal phytoliths of grass and palms. The paleosol succession preserves changes in soil moisture, biogenic activity, chemical weathering, and sediment influx. Two maturation trends of the paleosols are identified on the basis of field, micromorphologic, and geochemical analyses. Both trends alternated several times during Sarmiento time. Trend A developed exclusively from tephric loessites, and includes calcic Entisols, calcic Andisols, and Aridisols. It corresponds to periods of a semi-arid–arid climate represented in the Gran Barranca (middle Eocene), Vera (late Eocene – early Oligocene), and upper Colhue-Huapi (early Miocene) Members. Most paleosols are very weakly to weakly developed, indicating continuous or fast eolian sediment influx. They probably formed in grasslands, shrubby grasslands, and palm savannas. A moderately to strongly developed Aridisol, identified in the Rosado Member (middle Eocene), represents the driest environmental conditions. Trend B is associated with fluvial facies (intraformational conglomerates and tufoarenites) and subordinately with loessic deposits, and includes non-calcic Andisols, Alfisols, and intermediate Alfisols–Ultisols. It records periods of seasonal, subhumid to humid climate, and woodland or wooded-grassland vegetation, corresponding to the Gran Barranca (partially), Lower Puesto Almendra (late Eocene), Upper Puesto Almendra (Oligocene), and lower Colhue-Huapi (lower Miocene) Members. The stronger development of these paleosols indicates more discontinuous or slower fluvial–eolian accumulation.

Resumen

En los depósitos piroclásticos loéssicos y fluviales de la Formación Sarmiento (Gran Barranca, Patagonia central) se intercalan diferentes tipos de paleosuelos alcalinos y oxidados. Estos se caracterizan por su material parental relativamente uniforme, diversas trazas fósiles de insectos (ichnofacies de *Coprinisphaera* y *Celliforma*) y abundantes fitolitos de gramíneas y palmeras. La sucesión de paleosuelos preserva los cambios en la humedad del suelo, actividad biogénica, meteorización química y tasa de sedimentación. Tales factores determinaron dos tendencias de maduración de suelos, identificadas sobre la base de estudios de campo, micromorfológicos y análisis geoquímicos. Durante la sedimentación de la Formación Sarmiento ambas tendencias se alternaron en varias oportunidades. La tendencia A, que incluye Entisoles y Andisoles cálcicos y Aridisoles, desarrollados exclusivamente en loessitas piroclásticas, corresponde a períodos de clima semiárido-árido. Estos están representados en los miembros Gran Barranca (Eoceno medio) y Vera (Eoceno cuspidal-Oligoceno basal) y en la sección superior del Miembro Colhue-Huapi (Mioceno inferior). Los paleosuelos son, en su mayoría, de bajo a muy bajo desarrollo, indicando una acumulación continua y/o rápida. Ellos se habrían formado en pastizales, pastizales arbustivos y sabanas con palmeras. El Aridisol del Miembro Rosado (Eoceno medio alto) registra las condiciones más secas de toda la sucesión. La tendencia B se conforma por Andisoles no-cálcicos, Alfisoles y Alfisoles-Ultisoles, asociados a facies fluviales (conglomerados intraformacionales y tufoarenitas) y subordinadamente loessitas. Esta tendencia registra lapsos de clima subhúmedo a húmedo y estacional, con vegetación boscosa abierta o de pastizal arbóreo. Ha sido identificada en los miembros Gran Barranca (parcialmente), Puesto Almendra Inferior (Eoceno superior), Puesto Almendra superior (Oligoceno) y en la sección inferior del Miembro Colhue-Huapi (Mioceno inferior). El más avanzado desarrollo de estos paleosuelos evidencia una tasa de acumulación más discontinua o lenta.

The Paleontology of Gran Barranca: Evolution and Environmental Change through the Middle Cenozoic of Patagonia, eds. R. H. Madden, A. A. Carlini, M. G. Vucetich, and R. F. Kay. Published by Cambridge University Press. © Cambridge University Press 2010.

Introduction

Paleosols of the Sarmiento Formation are a significant but a poorly known component of the middle Cenozoic terrestrial ecosystems of Patagonia. Very few authors have investigated the paleopedological characteristics of this vastly exposed and fossiliferous pyroclastic succession. The first study was made by Andreis (1972) from Laguna del Mate – Gran Hondonada. Later, Spalletti and Mazzoni (1977, 1979)

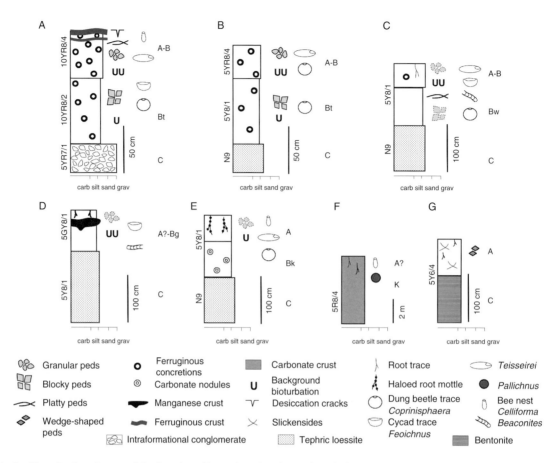

Fig. 20.1. Profiles of paleosol types of the Sarmiento Formation, showing pedogenic features and trace fossils. (A) Orangish Alfisol–Ultisol; (B) Alfisol; (C) non-calcic Andisol; (D) placic Andisol; (E) calcic Andisol; (F) thick, moderate-strongly developed Aridisol; (G) Vertisol.

mentioned some features of paleosols and subaerial paleo-surfaces from Valle Hermoso and Gran Barranca outcrops.

This contribution is dedicated to the characterization of fossil soils included in the Sarmiento Formation at Gran Barranca (Chubut). Paleopedological information is used to determine soil forming or environmental factors (e.g. rainfall and seasonality, organisms, geomorphological processes), and sedimentation rate.

According to Bellosi (Chapter 2, this book), several intraformational unconformities and discontinuity surfaces define six members in the Sarmiento Formation: Gran Barranca (middle Eocene), Rosado (middle Eocene), Lower Puesto Almendra (late Eocene), Vera (late Eocene – early Oligocene), Upper Puesto Almendra (Oligocene), and Colhue-Huapi (early Miocene). The Sarmiento Formation is mostly composed by tephric loessites, intraformational conglomerates, tufoarenites, bentonites, and paleosols (Spalletti and Mazzoni 1979; Bellosi Chapter 19, this book). Major unconformities correspond to high-angle or vertical erosive surfaces, a common feature in loess successions (Van Loon 2006). The relevance of Gran Barranca is enhanced by the occurrence of several mammal faunas

(Ameghino 1906). The high number of fossil vertebrates is closely related to the type of pedogenic modification of the pyroclastic deposits. Alkaline paleosols of dry climates, as those of the Sarmiento Formation, generally preserve the most complete record of fossil mammals (Retallack 2001).

General paleosol features

A particular characteristic of the paleosols of the Sarmiento Formation is the uniform parent material (Fig. 20.1). Homogeneity in composition, structure, and grain-size of the original sediment simplify the assessment of the degree of development of paleosols, and the paleowater-table depth. Dominant lithologies are pyroclastic (vitric) mudstones, tufoarenites, and intraformational conglomerates, the last formed by the same pyroclastic clasts or paleosol fragments (Spalletti and Mazzoni 1979; Bellosi Chapter 19, this book). Mineralogy and consolidation of these deposits are relatively uniform. The original soil sediment was white or very light gray, well-sorted, highly porous volcanic dust (silt > clay). Mineral composition is dominated by dacitic–rhyolitic glass shards (61% to 99%) and andesine plagioclase

(Mazzoni 1979). Exceptions to these features are bentonites and basalts, but few paleosols developed on bentonites, and non-pedified surfaces were observed on basalts. Chemical weathering, estimated for the different types of paleosols in 30 samples, ranges from very slight to moderately intense according to mineral composition, degree of alteration (etching), and geochemical analysis (weight percent of oxides of major elements). The color change to yellowish gray, reddish, or orangish reflects a stronger pedogenic modification of the original tephric loess due to iron oxide concentration, a property also observed in central China loess–paleosol sequences (Chen *et al.* 2002).

Biotic features are common in these paleosols (Bellosi *et al.* this book). The more relevant are trace fossils, opal phytoliths, small land snails, and vertebrates. Discrete ichnofossils, along with a background bioturbation, are abundant in many examples (Bellosi *et al.* 2001; Genise *et al.* 2004). Trace fossil associations correspond to *Coprinisphaera* and *Celliforma* ichnofacies. Most of ichnofossils were produced by coleopterans (*Coprinisphaera* ispp., *Teisseirei*, *Pallichnus*), bees (*Celliforma* ispp.), and cycadas (*Feoichnus challa*) (Krause *et al.* 2008). Thick-walled burrows correspond to crayfishes (*Lolloichnus baqueroensis*) (Bedatou *et al.* 2008). Other trace fossils were assigned to insects (e.g. cocoons) and invertebrates (*Lazaichnus*, *Beaconites*, and septate burrows). Root traces and large burrows attributed to subterranean mammals are also recognized (Bellosi *et al.* this book). *Coprinisphaera* ichnofacies have been related to open herbaceous ecosystems, from subtropical savannas to steppes (Genise *et al.* 2000). Some paleosols in the Lower and Upper Puesto Almendra and lower Colhue-Huapi Members exhibit an extraordinary density and diversity of *Coprinisphaera*, a dung beetle brood ball, suggesting very favorable conditions for living and reproduction (food resource, climate, soil characteristics, etc.). Species richness of tunneler and roller dung beetles is at a maximum in tropical grasslands (Hanski and Cambefort 1991). According to Halffter (1991), abundance of such beetles increases in warm temperate to warm (mean annual temperature [MAT] > 15 °C), and semi-arid to humid regions (mean annual precipitation [MAP] > 250 mm). Changes in trace fossil assemblages in the Sarmiento Formation parallel variations in paleosols types in response to climate and vegetation (Bellosi *et al.* this book).

The record of silica phytoliths in Sarmiento paleosols and in other facies is surprisingly high and diverse (Mazzoni 1979). Most of them correspond to grasses and palms, and subordinately to dicotyledons and sedges (Zucol *et al.* 2008, this book). Finally, preservation of shells of terrestrial gastropods, and bones and teeth of mammals, was favored by alkaline and well-drained conditions during pedogenesis (Bellosi *et al.* 2002; Miquel and Bellosi 2007).

After burial modification of the studied paleosols is considered to be slight. The presence of smectite and absence of illite suggests minimal diagenetic alteration (Bethke and Altaner 1986; Evans 1992; Robinson and Wright 1987). Burial reddening probably occurred in the Upper Puesto Almendra Member. The more significant process was the decomposition of organic matter. Considering the abundance of trace fossils and grass phytoliths, the expected original content of organic matter would have been relatively high. However, maximum weight percent of total organic carbon, measured in ten selected samples, is 0.03–0.09. Fast loss of organic matter was produced by oxidizing conditions and microbial activity during pedogenesis or soon after burial. There is no clear evidence of post-burial diagenetic calcite cementation. Cement of iron oxides and hydroxides is common in the studied samples, but their micromorphologic characteristics are associated to soil processes. Change in thickness by overburden compaction was low to moderate. Deformation of bones, shell snails, or ichnofossils is not observed. The reconstructed burial history for this sector of the basin can be estimated in a maximum depth of less than 400 m.

Types of paleosols

Paleosols of the Gran Barranca are described and sampled in fresh and continuous exposures of the escarpment (Fig. 20.2: 1). Macroscopic features observed in the field included the definition of horizons and contacts, grain-size, ped structure, glaebules (nodules, concretions), color changes, content of calcium carbonate, and bioturbation and/or ichnofossils. Root traces are not abundant or well preserved, but invertebrate ichnofossils are very common. Micromorphologic analysis of the fine-grained part (i.e. distinctive plasmic microfabric) (Bullock *et al.* 1985) served to recognize: grain-size, porosity, mineral composition, and small-scale pedofeatures (cutans, nodules, etc.). Clayey fabric is best expressed in paleosols with a relative high degree of development, but abundant opaque sesquioxides can masked this characteristic. Clay coating on peds and grains is considered to be the result of physical translocation (illuviation). Other processes (e.g. hydrothermal alteration, precipitation from solutions) may be discounted: allophane crystallization show transitions between amorphous and crystalline material (Jongmans *et al.* 1994), a feature not present in the studied samples.

Geochemical and X-ray diffraction analysis of selected samples were also performed. Molecular weathering ratios were used to approximate soil-forming chemical reactions such as acidification (alumina/bases), oxidation (total Fe+Mn/alumina), hydratation (silica/sesquioxides), calcification (Ca+Mg/alumina), and salinization (alkalis/alumina).

E. S. Bellosi and M. G. González

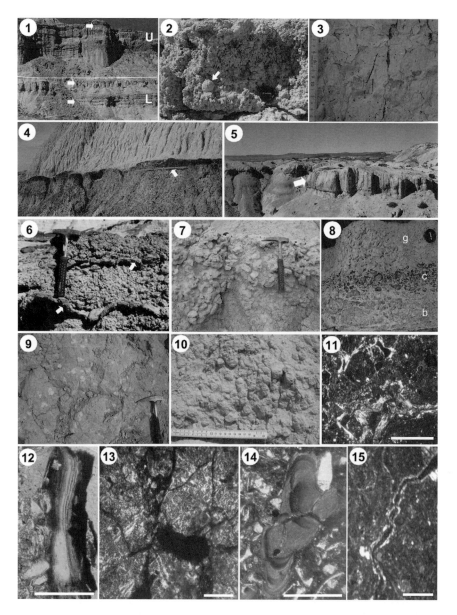

Fig. 20.2. Features of the Sarmiento Formation paleosols. 1, Stacked paleosols (arrows) in the Lower (L) and Upper (U) Puesto Almendra members, total thickness 60 m; 2, upper horizon of an intensely bioturbated (complex ichnofabric) Alfisol with granular peds and a dung beetle ichnofossil (arrow); 3, thin manganese root traces, Andisol, Vera Member; 4, Mn duricrust, placic Andisol, Gran Barranca Member, scale bar equals 1 m (arrow); 5, massive Aridisol (6 m thick), Rosado Member (arrow); 6, orangish Alfisol–Ultisol with folded ferruginous duricrust (arrow), Upper Puesto Almendra Member, hammer 30 cm; 7, nodular calcrete horizon, Vera Member, hammer 30 cm; 8, Alfisol–Ultisol showing granular (g) and blocky (b) peds, and Fe concretions (c), Upper Puesto Almendra Member, lens cap 4.5 cm.; 9, dispersed carbonate nodules in a calcic Andisol, Vera Member, hammer 30 cm; 10, well-formed granular peds in an Alfisol, lower Colhue-Huapi Member, scale in centimeters. Photomicrographs (scale bar 0.5 mm): 11, bright illuvial clay, Alfisol Bt horizon, Upper Puesto Almendra Member, crossed nicols; 12, clay and Fe–Mn cutans, Alfisol, lower Colhue-Huapi Member; 13, sepic microfabric formed by birefringent streaks in two directions at a high angle, Vertisol, lower Colhue-Huapi Member, crossed nicols; 14, thick laminated clay skin, Alfisol–Ultisol, Upper Puesto Almendra Member, crossed nicols; 15, grain-striated to undifferentiated–punctuated ground mass, blocky microstructure, Alfisol, Upper Puesto Almendra Member, crossed nicols.

Five types of paleosols are recognized in the Sarmiento Formation on the basis of their vertical profiles (Figs. 20.1, 20.2). These are classified using definitions of the Soil Survey Staff (1975), and adaptations to the stratigraphic record elaborated by Retallack (2001). Identification of A horizons, in cases where there is no evidence of erosion, is often difficult. Their properties can be modified by postburial oxidation, compaction, or rejuvenation by bases leached from overlying soil (Kemp 2001). The degree of development of the paleosols is not just assessed by destruction of the sedimentary fabric, because most of original beds were massive (e.g. tephric loessites). A combination of other features are important, such as the stage of carbonate accumulation, clayeyness and thickness of the B horizon, ped structure, presence of clay films, mineral weathering, color (e.g. rubification), and nodules (Retallack 1988; Bestland 1997). Physical and biological properties reflect variable drainage conditions and environmental/climatic contexts. In the following paragraphs, a description and interpretation of each paleosol type of the Sarmiento Formation is presented (Fig. 20.3).

Alfisols

Alfisols occur in the Lower and Upper Puesto Almendra, and lower section of Colhue-Huapi Members, frequently stacked and related to fluvial facies (Fig. 20.2: 1, 2). They are distinguished by reddish colors, high clayeyness and induration reflecting a high content in Fe oxides. Alfisols show two horizons separated by a transitional wavy contact (Fig. 20.1). The upper (A–B?) horizon is 50 cm thick, with coarse granular and coarse platy ped structure (Fig. 20.2: 10), and frequent to abundant ferruginous concretions (Fig. 20.2: 8). Colors vary from light brownish gray (5YR7/1) to orange gray (5Y7/2) and grayish orange (10YR7/4). Root traces are rarely preserved. In some cases this horizon includes folded, dark orange (10YR6/6) ferruginized duricrusts, up to 4 cm thick (Fig. 20.2: 6). Small manganese nodules are also present. Some of the more reddened levels exhibit polygonal desiccation cracks. The lower (Bt) horizon is 70 cm thick, more argillic, less indurated, and shows blocky ped structure, Fe concretions and similar but lighter colors (10YR8/2, 5Y8/1). Background bioturbation is low to intense in both horizons (Fig. 20.2: 2).

Microstructure is dominantly spongy and subordinately blocky. The ratio of coarse/fine material is 0.3–0.7. The coarse fraction is composed by silt to very fine-sand particles of weathered to fresh glass shards (>40%), plagioclase, and quartz. The fine fraction is smectite clay, brownish gray in color. The ground mass shows a stipple-speckled to grain-striated or pore-striated b-fabric (Fig. 20.2: 15), frequently obscured by iron–manganese oxides and hydroxides (Fig. 20.2: 11). Thick, laminated argilans and mangans formed by illuviation are very common (Fig. 20.2: 12, 14). Spherical biogenic clay microgranules are observed in horizons with granular peds (Cosarinsky *et al.* 2005). Chemical analyses show molecular weathering ratios of alumina/base around 2.0, with maximum values of 3.9. The total iron/alumina are between 0.30 and 0.45, with maximum value of 1.4. A third lowermost (C) horizon is occasionally preserved as massive

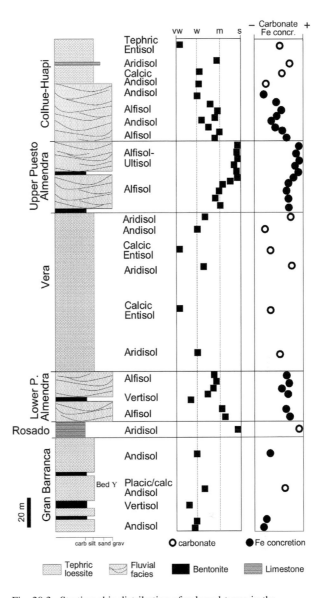

Fig. 20.3. Stratigraphic distribution of paleosol types in the Sarmiento Formation (reconstructed profile). Variation in degree of development (vw, very weak; w, weak; m, moderate; s, strong) and carbonate accumulations or ferric concretions are shown.

intraformational conglomerates, cross-bedded tufoarenites, or thin tephric loessites.

Ped structure and well-developed argillans in subsurface horizons are indicative of argillic soils. The above-mentioned features and the opaque microfabric form in well-drained soils. According to chemical and mineral composition and clayey horizons (Bt), these moderately developed paleosols are interpreted as Alfisols, and less probably as Ultisols. Spalletti and Mazzoni (1977) considered some pedogenic levels cropping out at

"Km 163" locality as non-calcic Mollisols. Although they have some quasi-mollic characteristics (granular peds), the absence of organic horizon, crumb peds, and thin root traces precludes such an interpretation. Original surface horizons were probably modified after burial, or eroded. Most of the studied cases have a higher proportion of glass shards, and very scarce feldspars, thinner cutans, and lower alumina/base ratio (<2) corresponding to Alfisols. The remaining cases with a lower percentage of weatherable minerals, thicker and more frequent ferro-argillans, presence of ferruginous duricrusts, more developed bright clay and opaque microfabric, higher alumina/base (>2), and total iron/alumina (>1.2) ratios are considered intermediate examples of Alfisols–Ultisols (moderately to strongly developed). In the Upper Puesto Almendra Member they show orange color.

Alfisols develop in forests and woodlands and are common on loess deposits, because prevailing decalcification and lessivage. The thickness of the described horizons (<1 m) better suggests a woodland vegetation (Retallack 2001). However, considering the abundant grass phytoliths (Mazzoni 1979; Zucol *et al.* 2006) and the *Coprinisphaera* ichnofacies (Bellosi *et al.* this book), grasses would have been dominant. Thus, wooded grassland would be the most probable vegetation structure. Granular structure in grassland soils has been attributed to earthworm activity (Pawluk and Bal 1985; Retallack 2001), which probably produced the background bioturbation.

Alfisols lacking carbonate concentrations are present in moist woodland mid-latitude subtropical settings, with MAP 750–1000 mm (Mack 1992) (Fig. 20.4). In subtropical Chaco plains of northern Argentina, Alfisols develop in warm–temperate, subhumid to humid (MAP 750–1400 mm) wooded grasslands. The exclusive presence of smectite and absence of kaolinite in moderately developed soils would not suggest humid weathering conditions and MAP <1000–1200 mm (Barshad 1966; Evans 1992). Seasonality of rainfall is indicated by thick, microlaminated argillans, ferruginous concretions, and desiccation cracks. Water-table fluctuations produced redoximorphic features as Fe segregation in the matrix and coatings on grains and pores when soil drainage in impeded (periods of higher water-table) with clay illuviation with improved soil drainage (periods of lower water-table) (Ashley and Driese 2000). The abundant ichnofossils (Bellosi *et al.* this book), fossil mammals, and palm phytoliths (Zucol *et al.* this book) indicate warm–temperate conditions. These paleosols are compared to non-calcareous, reddish, smectitic Alfisols from Eocene–Oligocene of USA, for which a MAP of 700–800 mm was inferred (Retallack *et al.* 2000). Consequently, a subhumid to humid, seasonal climate is interpreted. Probably slightly wetter conditions (humid to subhumid) correspond to intermediate Alfisol–Ultisols. The number of wet months

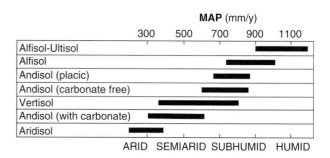

Fig. 20.4. Interpreted mean annual precipitation (MAP) for the Sarmiento paleosol types.

for developing Alfisols and Ultisols is 2.5 to 6 and >7, respectively (Cecil and Dulong 2003).

Andisols

Andisols are the most common type of paleosol in the Sarmiento Formation. It generally shows two incomplete argillic (cambic) and slightly colored horizons (Figs. 20.1, 20.3). The upper (A–B?) horizon is 0.5 m thick, moderately to intensely bioturbated, and light yellowish gray (5Y8/1) or orange gray (10YR8/2) in color. Poorly defined coarse granular to subangular blocky peds, scarce root traces, lower sharp contact, and few Fe–Mn oxides concretions also characterize this horizon. The middle (B) horizon is 0.7 m thick and lighter in color (N9, 5YR9/1). It is less bioturbated and massive, or shows indistinct ped structure, with root traces and a transitional lower contact (Fig. 20.2: 3). In some intervals, where peds are less developed, this horizon can include carbonate nodules (Bk horizon) (Fig. 20.2: 9). Occasionally, black manganese-cemented layers, or hardpans formed by agglutinated nodules, are preserved as subsurface levels (placic horizon, Bg) (Figs. 20.1D, 20.2: 4). They are 0.2–0.6 m thick, laterally continuous for hundreds of meters and show wavy upper and lower contacts. The lower (C) horizon is generally a massive pyroclastic mudstone, which ranges from 0.3 to 1.5 m in thickness. Trace fossils of insects and other invertebrates are frequent (Bellosi *et al.* this book). Surface horizons (A, O, or E) are probably modified by burial diagenesis. The microstructure of B horizons is spongy or slightly blocky. Ground mass shows a stipple-speckled b-fabric with patches or short streaks of highly birefringent plasma (mosaic-speckled), and are weakly obscured by Fe–Mn oxides, or with micrite. Ped and pore clay or mixed cutans are a common feature of Andisols, but thinner than in Alfisols. The mean coarse-to-fine ratio is similar to Alfisols. The coarse fraction (coarse silt > very fine sand) is composed of fresh to poorly weathered shards (>60%), plagioclase, and quartz. The fine fraction is composed of gray to yellowish gray smectite clay. Subordinated small Fe–Mn oxide nodules and mottles, and clay biogenic microgranules, are also present. Alumina/bases ratios range from 1.5 to 1.8.

Several features of these paleosols are shared with Alfisols. However, the low proportion of illuviation clay and the less developed ped structure impede distinguishing argillic horizons. The high percentage of fresh volcanic shards and the presence of slightly illuviated horizons (Bw) are diagnostic of Andisols. These poorly developed soils needed less time than Alfisols, or dryer and/or colder conditions. The low degree of development restricts paleoclimatic interpretation. By comparison with other paleosol types of the Sarmiento Formation, a semi-arid–subhumid, slightly seasonal climate ($600 < \mathrm{MAP} < 800\,\mathrm{mm}$) is suggested for Andisols without carbonate concentrations (Fig. 20.4). On the other hand, the presence of a Bk horizon allows defining less developed calcic Andisols, formed in semi-arid conditions ($\mathrm{MAP} < 600\,\mathrm{mm}$) (Machette 1985; Birkeland 1999; Royer 1999). Thick placic horizons (Bg) are an indicator of extended waterlogging. Manganese concentrates in poorly drained soils with common redox fluctuations (Dixon and Skinner 1992; Lynn and Austin 1998; Bestland 1997). Paleosols with abundant non-oxidized Mn nodules would record toxic bottomlands with stagnant ponds (Retallack *et al.* 2000), and have been related to particular manganese-accumulating plants. This environment would be related to the phytoliths recognized in the upper section of Gran Barranca Member, which suggest a swamp-dominated vegetation (Zucol *et al.* this book). Laterally, this Mn-rich paleosol transitions to a calcic Andisol. The transition reflects paleotopographic and soil drainage variation.

Entisols

Entisols are very poorly developed paleosols, frequently intercalated in tephric loessites of the Vera Member (Fig. 20.3). They generally show two light gray horizons (N9, 5YR8/1). The upper horizon (A) is massive, 0.5 to 2.0 m thick, with scarce fine root traces, and dispersed small carbonate nodules in the upper part. Animal bioturbation is absent to sparse. Smaller manganese nodules can be also present. The lower (C) horizon is 1 to 4 m thick, and shows an upper transitional contact. It is a massive or poorly bedded pyroclastic mudstone. In thin sections, glass shards from both horizons are very abundant and fresh. Plasmic microfabric is mostly asepic (argillasepic), lacking highly birefringent streaks, or calciasepic when micrite patches are present in the ground mass. The profile of these paleosols is generally complete because they occur in intervals without discontinuity surfaces.

Taking into account the single profile, mineral composition, low weathering, and the presence of carbonate nodules, these levels are interpreted as calcic vitric Entisols. They could be also considered Aridisols; but this soil

type commonly includes more abundant carbonate nodules or carbonate-cemented layers. The thickness would be related to continuous eolian supply of fine volcanic ashes (Bellosi Chapter 19, this book), thus they could be considered cumulic paleosols. Entisols are not appropriated for paleoclimatic interpretations, because of their poor development. However, their sedimentary paleoenvironmental context and the presence of shallow carbonate nodules suggest semi-arid conditions ($\mathrm{MAP} < 600\,\mathrm{mm}$) (Fig. 20.4).

Aridisols

Aridisols are characterized by a surface calcic horizon, less than 1 m deep (Fig. 20.1). The most significant Aridisol corresponds to the Rosado Member, with a pink (5R8/4) firmly carbonate-cemented bed (hardpan), and sharp upper and lower contacts (Figs. 20.2: 5, 20.3). Thickness (6 m) is uniform along several kilometres of exposure, in the central and eastern part of Gran Barranca (Bellosi *et al.* 2002). Wherever it occurs, it is massive and shows scarce irregular black mottles of manganese oxide. Bioturbation also is sparse. At or near the upper surface, a distinctive assemblage of insect trace fossils is recognized (*Celliforma* ichnofacies). These ichnofossils correspond to nests of solitary bees (*Celliforma rosellii*), dispersed or constituting local aggregations, scarab pupation chambers (*Pallichnus dakotensis*), and undetermined burrows (Bellosi *et al.* this book). This paleosol also preserves fossil mammals (Mustersan SALMA), and a diverse association of small land snails (Miquel and Bellosi 2007). Under the microscope, this level has a dense and homogeneous carbonate–argillic matrix (crystallitic b-fabric), with low porosity. It shows typical features of alpha calcretes (Wright 1990): crystalarias, floating grains, displacive growths, and diffuse Fe–Mn spots. The carbonate is micrite in high proportion ($>70\%$). Spar fills cracks and pores. The matrix contains subordinated calcified orthotubules and channels. Floating grains are plagioclase, quartz, tuffaceous lithoclasts, glass shards, and soil fragments. Evidence of recrystallization and replacement by calcite are observed. Together, these characteristics indicate a moderate to strongly development of the calcic horizon.

Another massive carbonate bed is located near the top of the Colhue-Huapi Member (Fig. 20.3). It is an indurated, thin (0.30 m thick) and white level, bearing fine manganese root traces. The only invertebrate ichnofossils corresponds to bee cells (*Celliforma*). It shows similar microscopic features to the Rosado bed, but is slightly richer in soil fragments and grains, and lacks Fe–Mn spots and cracks. In both cases, no evidence is present of karstification or brecciation.

The remaining examples occur in Vera Member, and in the upper section of the Colhue-Huapi Member (Fig. 20.3), intercalated with thick tephric loessites (Bellosi Chapter 19, this book). A single type of profile unifies them. The upper (Bk) horizon is a massive whitish pyroclastic mudstone, 0.6 to 1.8 m thick, with abundant subspherical to ovoid carbonate nodules (3–9 cm), generally grouped (nodular horizon), fine rhizoliths, and scarce Mn mottles (Fig. 20.2: 7). Ground mass shows a specked b-fabric. Illuviation is evident in some samples as thin clay skins around grains. Occasionally, this horizon passes laterally to a discontinuous carbonate crust. The lower (C) horizon is similar in composition to the upper one, but does not include nodules. It is 0.8 to 2.0 m thick, slightly darker, and has a transitional upper contact. Trace fossils are very rare. Only large-size deep burrows filled with carbonate are present in the Vera Member and are attributed to mammals (Bellosi *et al.* this book).

Carbonate concentrations (Bk) forming cemented crusts (petrocalcic horizon) or nodules (calcic horizon) are found in soils of arid or semi-arid regions with sparse vegetation. Thus, the described paleosols are comparable to Aridisols (Soil Survey Staff 1975). Alkaline solutions and a rate of evapotranspiration that greatly exceeds MAP are required for carbonate retention in a soil profile. High calcite concentration in soils without calcareous parent material, as in Sarmiento paleosols, tends to form in arid to semi-arid climates (Borchardt and Lienkaemper 1999). Necessary MAP to form pedocals is <500 mm in cool climate, or <600 mm in warm climate (Birkeland 1992, 1999), and less than 2.5 wet months (Crithfield 1974). Thus, it is inferred a MAP <400 mm (Fig. 20.4). Petrocalcic horizons with alpha fabrics observed in the Rosado Member and top of Colhue-Huapi Member, would represent more developed Aridisols, and probably originated in more xeric conditions with low biological activity (Wright and Tucker 1991).

Vertisols

Only a few Vertisols were recognized in the Gran Barranca, Lower and Upper Puesto Almendra and Colhue-Huapi Members (Fig. 20.1), intercalated in tephric loessites or fluvial deposits. They are uniform and occur in the upper part of some olive gray (5Y5/2) to grayish yellow (5Y6/6) bentonites beds (C horizon) (Fig. 20.3). Prominent features of these cracking clayey paleosols are rare fine root traces, and abundant slickensides that define wedge-shaped peds. The A horizon is 0.5 to 1.2 m thick, and non-calcareous. The ground mass is highly birefringent, and reticulate to cross-striated (Fig. 20.2: 13). X-ray diffractometer analyses reveal that smectite is the only clay mineral. These Vertisols developed in flat or undulating landscapes, where dominant vegetation ranged from grassland to open woodland. Climate was subhumid to semi-arid, with a marked seasonality in rainfall (Fig. 20.4).

Stratigraphic variations of Sarmiento paleosols

Pedogenic modification of the Sarmiento Formation pyroclastics took place in eolian–loessic and fluvial depositional scenarios during pauses in accumulation. In drier environments, pedogenesis, along with sediment supply, erosion, and vegetation, is strongly governed by climate regime (Cecil and Dulong 2003). Thus, the Gran Barranca succession of paleosols preserves a substantial part of the history of continental climate, soil biota, and sedimentation during a protracted period in the middle Cenozoic. Figure 20.5 shows the stratigraphic arrangement of paleosol types and the inferred rainfall regime of the Sarmiento Formation.

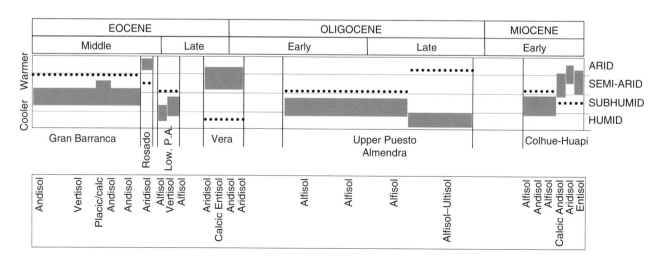

Fig. 20.5. Variations in rainfall regime (gray rectangles) and relative temperature (dot line) inferred from paleosols types, along the Sarmiento Formation.

Gran Barranca Member (middle Eocene)

This unit contains scattered weakly developed Andisols, and few calcic Andisols and Vertisols, intercalated in thick tephric loessites (Fig. 20.3). It records a relatively continuous and steady eolian sedimentation rate, in sub-humid to semi-arid, slightly seasonal conditions (MAP 600–800 mm). Paleosols with more advanced soil development (e.g. having higher clay illuviation and carbonate concentration) are generally more fossiliferous as a consequence of more alkaline conditions during pedogenesis. Early-successional paleosols as Andisols are not reliable indicators of plant formations. According to Retallack (2001), they usually support herbaceous vegetation (e.g. grasses). Gran Barranca Member vegetation can be partially reconstructed from the silica phytoliths content. Mazzoni (1979) and Zucol *et al.* (this book) recognized representatives of palms, along with grasses and dicots, few sedges, and Podostemaceae indicating a warm–temperate not-humid paleoflora, such as palm savanna, open savanna dominated by megathermal tall grasses (prairie). High temperatures are also corroborated by fossil land snails (Miquel and Bellosi this book) and reptiles. The intense bioturbation is compatible with a subtropical to tropical ecosystem (Retallack *et al.* 2000).

The fossiliferous Bed Y, a marker layer located in the upper section, shows lateral variations related to original topography and water-table depth (paleocatena). Along 5 km, this paleosol changes from a poorly clayey, highly bioturbated Andisol, to a calcic Andisol with a thick Mn placic (Bg) horizon, and finally to a more clayey and ferruginous Andisol. The Mn placic horizon indicates gleying processes in lowland waterlogged areas, with swamp vegetation (Zucol *et al.* this book). Better-drained soils with a poorly developed argillic horizon and sesquioxides would have occupied topographic highs. The remaining Andisol type developed in intermediate zones.

Rosado Member (middle Eocene)

This unit corresponds to a thick strongly developed Aridisol associated with the Unconformity 3 (Bellosi Chapter 2, this book) (Fig. 20.3). This calcic paleosol records a lengthy period of landscape stability and the driest conditions of the Sarmiento times. Preservation of the abundant land snails of Mustersan age (Bellosi *et al.* 2002) was favored by elevated pH soil conditions. The low bioturbation would suggest a less productive ecosystem probably related to cooler conditions, as also interpreted from the gastropod assemblage (Miquel and Bellosi this book). Middle–late Eocene vegetation from southern and northwest Patagonia is characterized by the presence of *Nothofagus*, along with other micro-mesothermal elements, and subordinated megathermal plants (Melendi *et al.* 2003; Barreda and Palazzesi 2007). Morphological studies on fossil leaves also indicate a strong decrease in rainfall and temperature (Hinojosa 2005). Consequently, semi-arid to arid (MAP 500–300 mm), probably temperate, conditions prevailed during pedogenesis (Fig. 20.5). The presence of *Celliforma* ichnofacies in this paleosol suggests a reduced herbaceous vegetation cover (Bellosi *et al.* this book).

Lower Puesto Almendra Member (late Eocene)

Paleosols from this unit denote a marked change in environmental conditions, also supported by sedimentary (Bellosi Chapter 19, this book) and trace fossil data (Bellosi *et al.* this book). However, fossil mammals belong to the same faunal interval (Mustersan SALMA). This section includes stacked, non-calcic grayish Alfisols intercalated in fluvial deposits (Fig. 20.3). Prevailing warm–temperate, subhumid to humid, seasonal conditions (MAP 650–950 mm) are inferred for this interval (Fig. 20.5). In general, these bioturbated paleosols are moderately developed. Gradual reduction in degree of development up-section suggests a progressive increase in sedimentation rate. Nodules and skins (mangans) of Fe–Mn are indicators of seasonal waterlogging. Typical plant formations associated to Alfisols are grassy woodlands or open forests (Retallack 2001). The abundant and diverse trace fossils of dung beetles indicate a temperate–warm climate (Halffter 1991) and a grass-dominated vegetation (Bellosi *et al.* this book). The high proportion of meso-microthermal grass phytoliths and the low amount of palm phytoliths support this interpretation (Mazzoni 1979; Zucol *et al.* 2006). In sum, the vegetation was composed of a mixture of trees that do not produce phytoliths and grasses that do, i.e. a wooded savanna with a forest canopy covering less than 20% (Pratt *et al.* 1966).

Vera Member (early Oligocene)

A significant change in type, degree of development, and frequency of paleosols is noted in this unit, which bears vertebrates of the Tinguirirican SALMA. The scattered Aridisols and calcic Entisols and Andisols identified in this interval are intercalated in thick tephric loessite sheets (Fig. 20.3). Paleosols are weakly or very weakly developed, and bioturbation is low to absent. This association records a lapse of steady, rapid, and pulsed eolian accumulation. A slight decrease in aggradation rate and more frequent interruptions in sedimentation can be estimated for the upper section due to the stronger development degree of paleosols. Palynological data from southern Patagonia localities of this age indicate a vegetation dominated by micro- and mesothermal plants, without megathermal components (Barreda and Palazzesi 2007). Consequently, the interpreted climate according to paleosol types and paleobotanical information fluctuated between non-seasonal semi-arid and arid (MAP 650–300 mm) (Fig. 20.5), and was probably cold–temperate.

Upper Puesto Almendra Member (Oligocene)

A succession of Alfisols that pass upwards to stacked orange ferruginous Alfisols–Ultisols intercalate in the alluvial deposits of this member (Fig. 20.3). Several of these well-drained oxidized paleosols show composite profiles indicating pulse-like depositional conditions. Upwards in the section, paleosols vary from moderately to moderate–strongly developed, in response to decrease in sedimentation rate and/or more intense chemical weathering related to climate fluctuation. Common ferruginous concretions provide evidence of seasonality and oxidizing conditions. The reddish hue and intense bioturbation is compatible with modern soils from high-productivity subtropical–tropical ecosystems (Birkeland 1999; Retallack *et al.* 2000). This increase in temperature and rainfall is confirmed by what is known about late Oligocene vegetation. Palynomorphs indicate the expansion of forests, ferns, palms, and other megathermal angiosperms (Barreda and Palamarczuk 2000); and phytolith assemblages record the dominance of megathermal grasses and reappearance of palms (Zucol *et al.* in press). Thus, Oligocene pedogenesis took place under a warm–temperate, seasonal, subhumid to humid (MAP 750–1100 mm) climate (Fig. 20.5). The plant formation would have been a wooded savanna, probably with a higher canopy cover than in the Lower Puesto Almendra Member.

Colhue-Huapi Member (early Miocene)

This member includes two contrasting types of paleosols (Fig. 20.3). Gradual increase in sedimentation rate, associated with the change in depositional mode along the unit, is reflected in paleosols that vary from moderate to weak. The lower section includes frequent, non-calcareous Alfisols and Andisols, associated to fluvial deposits. They are very rich in mammal remains (Colhuehuapian SALMA). The upper section contains scattered calcic Andisols and Aridisols, intercalated in tephric loessites. The content in fossil vertebrates is strongly reduced. Silica phytoliths are also common in the Colhue-Huapi Member, but in less proportion than in older units (Mazzoni 1979). The assemblages are largely dominated by grass phytoliths, with a mixture of macrothermal and micro-mesothermal components (Zucol *et al.* in press). Regional correlation of marine and continental successions and unconformities (Bellosi Chapter 19, this book) allows to link early Miocene pedological and floristic variations. Abundant palynomorphs collected from coeval marine deposits of the San Jorge Basin (Chenque Formation) indicate a floral modification in the central Patagonia coastal plain, in response to increased aridity (Barreda and Bellosi 2003). During the Aquitanian, there occurred a retreat of arboreal vegetation and concomitant expansion of shrubs and gallery forests (with rainforest trees). This trend was intensified in the late Aquitanian leading into the Burdigalian, with the predominance of xerophytic and halophytic shrubs and herbs (Barreda and Bellosi 2003; Barreda and Palazzesi 2007). Consequently, the lower section of the Colhue-Huapi Member records an interval of subhumid seasonal conditions (MAP 650–900 mm), and a fluvial landscape dominated by wooded grasslands with subordinated shrubs, and riparian forests (Fig. 20.5). The interpreted paleoenvironment of the upper section would be loessic shrubby grassland, with a semi-arid to arid (MAP 650–400 mm) and probably cooler climate (Fig. 20.5). This section can be correlated to the Colhuehuapian tephric loessites with calcic Entisols and calcrete horizons from Gaiman region (northeast Chubut) (Scasso and Bellosi 2004). The environmental variation through the Colhue-Huapi Member agrees with the ichnofossil analysis, which indicate a change from *Coprinisphaera* to *Celliforma* ichnofacies in accordance to reduction in plant cover (Bellosi *et al.* this book).

Discussion and conclusions

Frequency and types of the Sarmiento paleosols indicate that environmental factors, particularly climate, and sedimentation rate fluctuated several times during middle Eocene – early Miocene in central Patagonia. Both factors determined two contrasting trends of soil maturation or weathering (Fig. 20.6A). Paleosols of the Trend A developed exclusively on massive tephric (vitric) loessites, evolving from calcic Entisols (very weakly developed), to calcic Andisols (weakly developed), and finally to nodular or massive Aridisols (weakly to strongly developed). Trend B developed on intraformational conglomerates or tephric loessites, and evolved from non-calcic Andisols (weakly developed), to Alfisols (moderately developed), and intermediate Alfisol–Ultisols (moderately to strongly developed). Both maturation trends would have been governed by climate over time. Trend A occurred in a dryer regime (semi-arid–arid) where grasslands, shrubby grasslands, and palm savannas developed (Fig. 20.7A). The resulting soils have high concentration of calcium carbonate and moderate chemical weathering. The stratigraphic intervals that record this maturation trend are: Gran Barranca Member (partially), Rosado Member, Vera Member, and the upper section of the Colhue-Huapi Member. Trend B represents wetter (subhumid to humid), and more oxidizing environments with woodland or wooded grassland vegetation (Fig. 20.7B). The paleosols have a higher proportion of ferric concretions and clay, as a product of glass shard and plagioclase alteration. In addition, accumulation of alumina increases at the expense of silica and bases. This maturation trend is

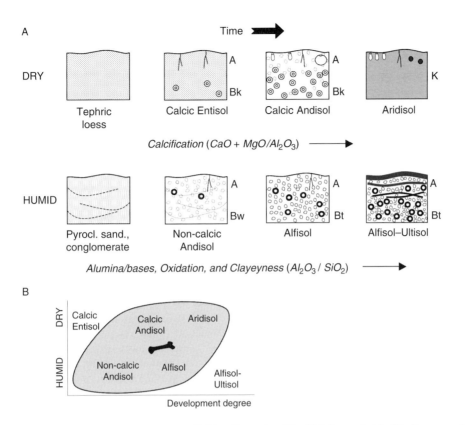

Fig. 20.6. (A) Schematic diagram of pedogenic trends controlled by climate (see Fig. 20.1 for symbols), (B) climate vs. development degree and relation to preservation of vertebrate remains in Sarmiento paleosols.

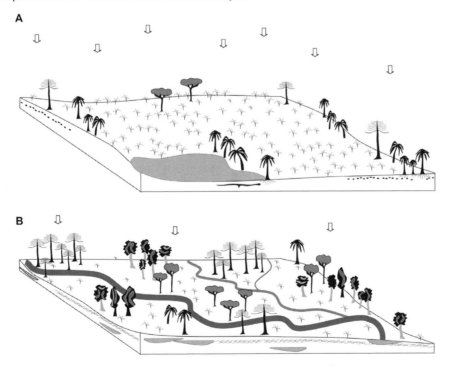

Fig. 20.7. Reconstructed paleoenvironments of the Sarmiento Formation. (A) Rolling plains formed in dryer periods (arid–semi-arid) with relatively rapid and uniform eolian deposition, represented by tephric loessites, and very weakly to weakly developed calcic and non-calcic paleosols. (B) Fluvial landscape favored in wetter periods (subhumid to humid) with slower and more discontinuous sedimentation. These intervals are recognized by channel facies (intraformational conglomerates and sandstones) and weakly to moderately developed non-calcic paleosols. Number of arrows indicates intensity of eolian ash supply.

recorded in the Gran Barranca, Upper and Lower Puesto Almendra, and in the lower section of Colhue-Huapi Member. With a few exceptions, the paleosols of Trend A reach a lesser degree of development than those of Trend B, suggesting that sediment accumulation during eolian–dryer periods was more continuous and/or faster than fluvial–wetter ones.

Pedogenic conditions that favoured the preservation of bones and teeth occurred in both maturation trends (Fig. 20.6B), because they depended on soil chemistry (alkaline conditions) and the amount of time over which paleosol formed. Aridisols, calcic and non-calcic Andisols, and Alfisols are the richest in vertebrate fossils of the Sarmiento Formation.

ACKNOWLEDGEMENTS
This research has been supported by US National Science Foundation grants EAR-0087636, BCS-0090255, and DEB-9907985 to Richard F. Kay and Richard H. Madden. We are grateful to Gregory Retallack and Richard Kay for revision of the manuscript, and to Pan American Energy for assistance during the field work.

REFERENCES

Ameghino, F. 1906. Les formations sédimentaires du Crétacé supérieur et du Tertiaire de Patagonie avec un parallèle entre leurs faunes mammalogiques et celles de l'ancien continent. *Anales del Museo Nacional de Historia Natural*, **15**, 1–568.

Andreis, R. 1972. Paleosuelos de la Formación Musters (Eoceno Medio), Laguna del Mate, Provincia de Chubut, R. Argentina. *Revista de la Sociedad Argentina de Mineralogía, Petrografía y Sedimentología*, **3**, 91–97.

Ashley, G. and S. Driese 2000. Paleopedology and paleohydrology of a volcaniclastic paleosol interval: implications for Early Pleistocene stratigraphy and paleoclimate record, Olduvai Gorge, Tanzania. *Journal of Sedimentary Research*, **70**, 1065–1080.

Barreda, V. and E. Bellosi 2003. Ecosistemas terrestres del Mioceno temprano de la Patagonia central: primeros avances. *Revista del Museo Argentino de Ciencias Naturales*, n.s., **5**, 125–134.

Barreda, V. and S. Palamarczuk 2000. Palinoestratigrafía de depósitos del Oligoceno tardío-Mioceno, en el área sur del Golfo San Jorge, provincia de Santa Cruz, Argentina. *Ameghiniana*, **37**, 103–117.

Barreda, V. and L. Palazzesi 2007. Patagonian vegetation turnovers during the Paleogene–early Neogene: origin of arid-adapted floras. *Botanical Review*, **73**, 31–50.

Barshad, I. 1966. The effects of variation in precipitation on the nature of clay mineral formation in soils from acidic and basic igneous rocks. *Proceedings, International Clay Conference*, **1**, 167–173.

Bedatou, E., R. Melchor, E. Bellosi, and J. Genise 2008. Crayfish burrows from Late Jurassic–Late Cretaceous continental deposits of Patagonia, Argentina: their palaeoecological, palaeoclimatic and palaeobiogeographical significance. *Palaeogeography, Palaeoclimatology, Palaeoecology*, **257**, 169–184.

Bellosi, E., J. Laza, and M. González 2001. Icnofaunas en paleosuelos de la Formación Sarmiento (Eoceno-Mioceno), Patagonia central. *Actas IV Reunión Argentina de Icnología and II Reunión Icnología del Mercosur*, 31.

Bellosi, E., S. Miquel, R. Kay, and R. Madden 2002. Un paleosuelo mustersense con microgastrópodos terrestres (Charopidae) de la Formación Sarmiento, Eoceno de Patagonia central: significado paleoclimático. *Ameghiniana*, **39**, 465–477.

Bestland, E. 1997. Alluvial terraces and paleosols as indicators of Early Oligocene climate change (John Day Formation, Oregon). *Journal of Sedimentary Research*, **67**, 840–855.

Bethke, C. and S. Altaner 1986. Layer-by-layer mechanism of smectite illitization and application to a new rate law. *Clay and Clay Minerals*, **34**, 136–142.

Birkeland, P. 1992. Quaternary soil chronosequences in various environments: extremely arid to humid tropical. In: Martini, I. and Chesworth, W. (eds.), *Weathering, Soils and Paleosols*. Amsterdam: Elsevier, pp. 261–281.

Birkeland, P. 1999. *Soils and Geomorphology*, 3rd edn. Oxford, UK: Oxford University Press.

Borchardt, G. and J. Lienkaemper 1999. Pedogenic calcite as evidence for an early Holocene dry period in the San Francisco Bay area, California. *Geological Society of America Bulletin*, **111**, 906–918.

Bullock, P., N. Fedoroff, A. Jongerius, G. Stoops, and T. Tursina 1985. *Handbook for Soil Thin-Section Description*. Albrighton, UK: Waine Research Publications.

Cecil, C. and F. Dulong 2003. *Precipitation Models for Sediment Supply in Warm Climates: Climate Controls on Stratigraphy*, Special Publication no. 77. Tulsa, OK: Society for Sedimentary Geology.

Chen, J., J. Ji, W. Balsam, Y. Chen, L. Liu, and Z. An 2002. Characterization of the Chinese loess–paleosol stratigraphy by whiteness measurement. *Palaeogeography, Palaeoclimatology, Palaeoecology*, **183**, 287–297.

Cosarinsky, M., E. Bellosi, and J. Genise 2005. Micromorphology of modern epigean termite nests and possible termite ichnofossils: a comparative analysis. *Sociobiology*, **45**, 1–34.

Crithfield, H. 1974. *General Climatology*. Englewood Cliffs, NJ: Prentice-Hall.

Dixon, J. and H. Skinner 1992. Manganese minerals in surface environments. *Catena* (Suppl.), **21**, 31–50.

Evans, L. 1992. Alteration products at the earth's surface: the clay minerals. In Martini, I. and Chesworth, W. (eds.), *Weathering, Soils and Paleosols*. Amsterdam: Elsevier, pp. 107–125.

Genise, J., M. Mángano, L. Buatois, J. Laza, and M. Verde 2000. Insect trace fossil associations in palaeosols: the *Coprinisphaera* ichnofacies. *Palaios*, **15**, 49–64.

Genise, J., E. Bellosi, and M. González 2004. An approach to the description and interpretation of ichnofabrics in palaeosols. In McIlroy, D. (ed.), *The Application of Ichnology to Paleonvironmental and Stratigraphic analysis*, Special Publication no. 228. London: Geological Society of London, pp. 355–382.

Halffter, G. 1991. Historical and ecological factors determining the geographical distribution of beetles (Coleoptera: Scarabaeidae: Scarabaeinae). *Folia Entomoló-gica Mexicana*, **82**, 195–238.

Hanski, I. and Y. Cambefort 1991. Species richness. In Hanski, I. and Cambefort, Y. (eds.), *Dung Beetle Ecology*, Princeton, NJ: Princeton University Press, pp. 350–365.

Hinojosa, L. 2005. Cambios climáticos y vegetacionales inferidos a partir de paleofloras cenozoicas del sur de Sudamérica. *Revista Geológica de Chile*, **32**, 95–115.

Jongmans, A., F. van Ort, P. Buurman, and A. Jaunet 1994. Micromorphology and submicroscopy of isotropic and anisotropic Al/Si coatings in Quaternary Allier terrace, France. In Ringrose, A. and Humpreys, S. (eds.), *Soil Micromorphology: Studies in Management and Genesis*. Amsterdam: Elsevier, pp. 285–291.

Kemp, R. 2001. Pedogenic modification of loess: significance for palaeoclimatic reconstructions. *Earth Sciences Reviews*, **54**, 145–156.

Krause, J., T. Bown, E. Bellosi, and J. Genise 2008. Trace fossils of cicadas from the Cenozoic of Central Patagonia, Argentina. *Paleontology*, **51**, 405–418.

Lynn, W. and W. Austin 1998. Oxymorphic manganese (iron) segregations in a wet soil catena in the Willamette Valley, Oregon. *Soil Science Society of America Special Publications*, **54**, 209–226.

Machette, M. 1985. Calcic soils of the southwestern United States. In *Soils and Quaternary Geology of the Southwestern United States*, Special Paper no. 203. Bouldev, CO: Geological Society of America, pp. 1–21.

Mack, G. 1992. Paleosols as an indicator of climate change at the Early–Late Cretaceous boundary, southwestern New Mexico. *Journal Sedimentary Petrology*, **62**, 483–494.

Mazzoni, M. 1979. Contribución al conocimiento petrográfico de la Formación Sarmiento, barranca S del lago Colhue-Huapi, provincia del Chubut. *Revista de la Sociedad Argentina de Mineralogía, Petrografía y Sedimentología*, **10**, 33–53.

Melendi, D., L. Scafati, and W. Volkheimer 2003. Palynostratigraphy of the Paleogene Huitrera Formation in NW Patagonia, Argentina. *Neues Jahrbuch für Geologie und Paläontologie, Abhandlungen*, **228**, 205–273.

Miquel, S. and E. Bellosi 2007. Microgasterópodos terrestres (Charopidae) del Eoceno Medio de Gran Barranca (Patagonia Central, Argentina). *Ameghiniana*, **44**, 121–131.

Pawluk, S. and L. Bal 1985. Micromorphology of selected mollic epipedons. *Soil Science Society of America Special Publications*, **15**, 63–83.

Pratt, D., P. Greenway, and M. Gwynne 1966. A classification of East Africa rangeland with an appendix on terminology. *Journal of Applied Ecology*, **3**, 369–382.

Retallack, G. 1988. Field recognition of paleosols. In Reinhardt, J. and Sigleo, W. (eds.), *Paleosols and Weathering through Geological Time: Techniques and Applications*, Special Paper no. 216. Boulder, CO: Geological Society of America, pp. 1–20.

Retallack, G. 2001. *Soils of the Past*, 2nd edn. London: Blackwell Science.

Retallack, G., E. Bestland, and T. Fremd 2000. *Eocene and Oligocene Paleosols of Central Oregon*, Special Paper no. 344. Boulder, CO: Geological Society of America.

Robinson, D. and V. Wright 1987. Ordered illite–smectite and kaolinite–smectite: pedogenic minerals in a Lower Carboniferous paleosol sequence, South Wales. *Clay Minerals*, **22**, 109–118.

Royer, D. 1999. Depth to pedogenic carbonate horizon as a paleoprecipitation indicator? *Geology*, **27**, 1123–1126.

Scasso, R. and E. Bellosi 2004. Cenozoic continental and marine trace fossils at the Bryn Gwyn Paleontological Park, Chubut: Bryn Gwyn guidebook. *Proceedings I International Congress on Ichnology*, 1–18.

Soil Survey Staff 1975. *Soil Taxonomy: A Basic System for Making and Interpreting Soil Surveys*. Washington, DC: U.S. Department of Agriculture.

Spalletti, L. and M. Mazzoni 1977. Sedimentología del Grupo Sarmiento en un perfil ubicado al sudeste del lago Colhue-Huapi, provincia de Chubut. *Museo de La Plata, Obra del Centenario*, **4**, 261–283.

Spalletti, L. and M. Mazzoni 1979. Estratigrafía de la Formación Sarmiento en la barranca sur del Lago Colhue-Huapi, provincia del Chubut. *Revista de la Asociación Geológica Argentina*, **34**, 271–281.

Van Loon, A. 2006. Lost loesses. *Earth Sciences Reviews*, **74**, 309–316.

Wright, V. 1990. A micromorphological classification of fossil and recent calcic and petrocalcic microstructures. In Douglas, L. (ed.), *Soil Micromorphology: A Basic and Applied Science*. Amsterdam: Elsevier, pp. 401–407.

Wright, V. and M. Tucker 1991. Calcretes: an introduction. In Wright, V. and Tucker, M. (eds.), *Calcretes*. Oxford, UK: Blackwell, pp. 1–22.

Zucol, A., M. Brea, R. Madden, E. Bellosi, A. Carlini, and M.G. Vucetich 2006. Preliminary phytolith analysis of Sarmiento Formation in the Gran Barranca (Central Patagonia, Argentina). In Madella, M. and Zurro, D. (eds.), *Places, People and Plants: Using Phytoliths in Archaeology and Palaeoecology*. Oxford, UK: Oxbow Books, pp. 197–203.

Zucol, A., M. Brea, R. Madden, E. Bellosi, A. Carlini, and G. Vucetich 2008. Análisis fitolítico del Miembro Colhue-Huapi (Formación Sarmiento), Chubut, Argentina. In Zucol, A., Osterrieth, M., and Brea, M. (eds.), *Fitolitos, estado actual de sus conocimientos en América del Sur*. Mar del Plata: Universidad Nacional Mar del Plata, EUDEM, pp. 157–165.

21 Ichnofacies analysis of the Sarmiento Formation (middle Eocene – early Miocene) at Gran Barranca, central Patagonia

Eduardo S. Bellosi, José H. Laza, M. Victoria Sánchez, and Jorge F. Genise

Abstract

The ichnofacies analysis of the Sarmiento Formation at Gran Barranca, along with older and younger adjacent Patagonian units, permits an assessment of the early Eocene to middle Miocene evolution of terrestrial ecosystems. Most ichnofossils found in Sarmiento Formation paleosols were made by insects: *Coprinisphaera* ispp. (dung beetles), *Celliforma* ispp. (solitary bees), *Teisseirei barattinia* and *Pallichnus dakotensis* (beetles), *Feoichnus challa* (cycads), and cocoons. Thick-walled vertical tubes (*Loloichnus baqueroensis*) were produced by crayfishes (Parastacidae). The nature of other traces is uncertain: *Lazaichnus fistulosus*, septate burrows, *Beaconites coronus*, and large burrows. Different types of root traces are also present. The lower middle Eocene Koluel-Kaike Formation contains *Feoichnus, Skolithos, Taenidium*, and root traces. In the middle Eocene Gran Barranca Member, *Feoichnus* is abundant, *Coprinisphaera, Teisseirei, Beaconites, Celliforma*, and cocoons are less common. The late middle Eocene Rosado Member, a pedogenic calcrete, only contains *Celliforma* and *Pallichnus*. The Lower Puesto Almendra Member (late Eocene) shows the greatest density and diversity of *Coprinisphaera*, and includes *Lazaichnus* and *Teisseirei*. Large burrows attributed to mammals and very few *Coprinisphaera* are found in the late Eocene to early Oligocene Vera Member. The Oligocene Upper Puesto Almendra Member records the greatest ichnodiversity in the whole succession, with relative high number of nearly all ichnofossils. The early Miocene Colhue-Huapi Member has two distinctive associations. Ichnodiversity of the lower and middle sections of the member is similar to that of the underlying member; but diversity and abundance decline in the upper section, where a calcrete contains only *Celliforma*. Elsewhere in west–central Patagonia, the early Miocene Pinturas Formation includes *Coprinisphaera, Syntermesichnus*, and *Palmiraichnus*. In the early to middle Miocene Santa Cruz Formation only *Celliforma* is recognized.

The Sarmiento Formation at Gran Barranca represents the oldest and one of the best examples worldwide of the *Coprinisphaera* ichnofacies. This ichnofacies is present in Gran Barranca, Puesto Almendra, and Colhue-Huapi Members, and elsewhere in the Pinturas Formation. The record of the *Coprinisphaera* ichnofacies, and particularly the high diversity and abundance of *Coprinisphaera* since the late Eocene, is interpreted as being the result of the appearance of grass-dominated warm–temperate environments with abundant dung produced by mammalian herbivores. The Rosado Member and the upper section of the Colhue-Huapi Member record the tentative *Celliforma* ichnofacies, probably indicating drier and/or colder environments with poor vegetation coverage. The Koluel-Kaike and Santa Cruz Formations and the Vera Member would have unfavorable conditions for the development of either *Coprinisphaera* or *Celliforma* ichnofacies.

Resumen

El análisis de icnofacies de la Formación Sarmiento en Gran Barranca, junto con otras unidades cenozoicas patagónicas, permite evaluar la evolución de los ecosistemas terrestres del Eoceno temprano-Mioceno medio. La mayoría de los icnofósiles se encuentran en paleosuelos y corresponden a insectos: *Coprinisphaera* ispp. (escarabajos estercoleros), *Celliforma* ispp. (abejas solitarias), *Teisseirei barattinia* y *Pallichnus dakotensis* (escarabajos), *Feoichnus challa* (cigarras) y capullos. Tubos verticales de pared gruesa (*Loloichnus baqueroensis*) son asignados a decápodos o langostas de río (Parastácidos). Otras trazas reconocidas son de afinidad incierta: *Lazaichnus fistulosus*, túneles septados, *Beaconites coronus* y túneles grandes. También aparecen diferentes trazas de raíces. La Formación Koluel-Kaike (Eoceno inferior-medio) contiene *Feoichnus, Skolithos, Taenidium* y trazas de raíces. El Miembro Gran Barranca (Eoceno medio) incluye abundantes *Feoichnus*, escasas *Coprinisphaera, Teisseirei, Beaconites, Celliforma* y capullos. El calcrete del Miembro Rosado (Eoceno medio superior) solo posee *Celliforma* y

The Paleontology of Gran Barranca: Evolution and Environmental Change through the Middle Cenozoic of Patagonia, eds. R. H. Madden, A. A. Carlini, M. G. Vucetich, and R. F. Kay. Published by Cambridge University Press. © Cambridge University Press 2010.

Pallichnus. El Miembro Puesto Almendra Inferior (Eoceno superior) exhibe la mayor diversidad y densidad de *Coprinisphaera*, acompañada por *Lazaichnus* y *Teisseirei*. El Miembro Vera (Eoceno superior-Oligoceno basal) presenta grandes túneles atribuidos a mamíferos y escasas *Coprinisphaera*. El Miembro Puesto Almendra Superior (Oligoceno) registra la mayor icnodiversidad de toda la sucesión y una alta proporción de ejemplares. El Miembro Colhue-Huapi (Mioceno inferior) muestra dos asociaciones. Las secciones inferior y media poseen similares características al miembro anterior. En la sección superior decrecen la diversidad y abundancia y se preserva un calcrete con solo *Celliforma*. La Formación Pinturas (Mioceno inferior), expuesta en el oeste de Patagonia central, incluye *Coprinisphaera*, *Syntermesichnus* y *Palmiraichnus*. La Formación Santa Cruz (Mioceno medio) sólo contiene *Celliforma*.

La Formación Sarmiento en Gran Barranca constituye el mejor y más antiguo ejemplo de icnofacies de *Coprinisphaera*. Su registro en los miembros Gran Barranca, Puesto Almendra Inferior y Superior, y Colhue-Huapi, así como en la Formación Pinturas, es un claro indicio de paleocomunidades herbáceas. Particularmente, la elevada diversidad y abundancia del icnogénero *Coprinisphaera* refleja, a partir del Eoceno tardío, la aparición de pastizales templado-cálidos con abundante estiércol de mamíferos herbívoros. El Miembro Rosado y la sección superior del Miembro Colhue-Huapi incluyen la tentativa icnofacies de *Celliforma*, probablemente originada en ambientes más secos y/o fríos con escasa vegetación. Las condiciones para el desarrollo de las icnofacies de *Coprinisphaera* y *Celliforma* serían desfavorables para las Formaciones Koluel-Kaike y Santa Cruz, y el Miembro Vera.

Introduction

The magnificent exposures of the middle Eocene to early Miocene Sarmiento Formation at Gran Barranca are full of insect trace fossils, representing an opportunity to use one of the most powerful tools of ichnology, the ichnofacies analysis, for studying regional and global changes of terrestrial ecosystems. In particular, this sequence is critical for understanding the evolution of South American Land Mammal Ages (SALMA), and the origin and expansion of grass-dominated ecosystems. Here it is also possible to integrate the information in a complete ichnostratigraphic scenario, taking into account the thickness of the exposures, a well-known regional context, and a detailed lithostratigraphy (Bellosi Chapter 2, this book), controlled by labeled stakes and isotopic ages for numerous beds (Ré *et al.* Chapter 4, this book).

Often different approaches to ichnology, such as the ichnofacies and ichnofabric or ichnoassemblage analyses, have been presented as mutually exclusive approaches, when in fact they are complementary (see McIlroy 2004 for a recent revision on this, and also Hasiotis 2004, Bromley *et al.* 2007). However, McIlroy (2004) notes that the approaches

chosen should depend on the sorts of questions being asked and the resolution at which data can be recorded and analyzed. Confusion on this point is, in part, responsible for the controversy (see Bromley *et al.* 2007). The purpose of this contribution is to provide new information on the ichnology of the continental Sarmiento Formation in order to outline the changing assemblages of trace fossils through time (middle Eocene–early Miocene), and to link these changes to proposed variation in Patagonian terrestrial ecosystems. For this study, a low-resolution, large-scale analysis was accomplished using the ichnofacies approach. A higher-resolution, smaller-scale approach is presently in progress (M. V. Sánchez *et al.* unpubl. data).

The ichnofacies approach is based on the recurrence in time and space of particular associations of trace fossils that are useful for recognizing and distinguishing paleoenvironments, which are termed archetypal or Seilacherian ichnofacies. The model was created by Seilacher (1967) mostly for marine rocks, and since then improved and discussed by many ichnologists (McIlroy 2004). The application of the model to continental deposits has been addressed only recently (Buatois and Mángano 1995; Genise *et al.* 2000, 2004a). Particularly important for this study is the *Coprinisphaera* ichnofacies, which is an association dominated by insect trace fossils, basically representative of Coprinisphaeridae and Celliformidae. This ichnofacies is indicative of terrestrial herbaceous communities in accordance with the ecological preference of the supposed producers. Based on a global-scale database, archetypal or Seilacherian ichnofacies are particularly useful for studying large-scale, regional, or global paleoenvironmental change. The expression of the *Coprinisphaera* ichnofacies (Genise *et al.* 2000) in the Sarmiento Formation is one of the best examples worldwide (Bellosi and Genise 2004).

Taxonomic allocation and the affinities of trace fossils are key to the analyses. The ichnotaxonomy and affinities of ichnofossils recorded for the Sarmiento Formation at Gran Barranca are reviewed in the first section of this work. Most of the trace fossils are reliably allocated to the animals that produced them, but a few remain in doubt, such as septate burrows or cocoons. Chambered trace fossils in paleosols have been reviewed recently by Genise (2000, 2004) where an ichnofamily arrangement was proposed.

In the next section, the ichnoassociations for stratigraphic subunits of the Sarmiento Formation at Gran Barranca are analyzed at a low-resolution scale. A preliminary version of this analysis was presented by Bellosi *et al.* (2001). Also, the ichnoassociations of adjacent older and younger formations of central Patagonia are examined, to record change through the Paleogene – early Neogene and to stress the regional context for the evolution of the terrestrial ecosystems in southern South America. Such purpose is particularly interesting because these formations comprise a protracted period (about 40 m.y.) characterized by significant global climate and environmental change.

The Sarmiento Formation is a volcaniclastic continental succession, composed of pyroclastic mudstones, andic paleosols, bentonites, intraformational conglomerates, pyroclastic sandstones, and basalt flows (Spalleti and Mazzoni 1979; Bellosi Chapter 19, this book). Grass and palm phytoliths are abundant throughout the unit (Mazzoni 1979; Zucol *et al.* 2006). At Gran Barranca, the Sarmiento Formation is divided into seven members: Gran Barranca, Rosado, Lower Puesto Almendra, Vera, Upper Puesto Almendra, and Colhue-Huapi (Bellosi Chapter 2, this book). Depositional settings include subaerial loessic plains (eolian deposits), ephemeral small lakes, and incised fluvial valleys, developed in subhumid to arid paleoclimatic conditions (Bellosi Chapter 19, this book). According to diagnostic features, recognized soil types are Andisols, Alfisols, Entisols, Aridisols, and a few Vertisols (Bellosi and González this book). A characteristic feature of these paleosols is well-preserved ichnofossils, particularly those of insects. At Gran Barranca, most of them occur in the type section (Profile MMZ), or westward at Profiles J and M (Bellosi Chapter 2, this book).

Ichnotaxonomy

Coprinisphaera

This ichnogenus is represented by spherical, subspherical, and pear-shaped chambers having a constructed wall, which may be completely pierced by a medium-sized hole or may show a secondary, smaller chamber connected with the main one by a narrow passage (Fig. 21.1: 1–4). The secondary chamber may be also connected to the exterior by a medium-sized hole. The internal cavities mostly contain passive fillings or in some cases are empty. Chambers are found isolated in contact with the paleosol matrix or as in an exceptionally preserved specimen surrounded by the original cavity in which they were constructed (Genise 2004, Laza 2006). The similarities between *Coprinisphaera* and modern dung beetle brood balls are so undisputable that since the former authors dealing with them there were no doubts in attributing these trace fossils to Scarabaeinae beetles (Frenguelli 1938; Roselli 1939; Sauer 1955; Halffter 1959; Laza 2006).

A brief review of the behavior of dung beetles taken from Halffter and Matthews (1966) and Halffter and Edmonds (1982) is useful to explain the morphology of these trace fossils and their importance to paleoecology. Mostly, dung beetles use excrement of herbivorous mammals as food for their larvae. Known ichnospecies of *Coprinisphaera*, namely *C. ecuadoriensis* Sauer, *C. murguiai* Roselli (Fig. 21.1: 1–3), *C. kheprii* Laza, *C. tonnii* Laza, and *C. kraglievichi* Roselli, are present in the Sarmiento Formation (Laza 2006). These traces are comparable with those constructed by dung beetles representative of the tribe Coprini (subtribes Dichotomiina and Phanaeina) and Scarabaeini (subtribe Canthonina). Scarabaeines are abundant in warm temperate (mean annual temperature [MAT] > 15 °C), and semi-arid to humid (mean annual precipitation [MAP] > 250 mm/yr) environments (Halffter and Edmonds 1982).

Celliforma

Celliforma comprises isolated oval chambers, having rounded rears and flat or conical tops in many cases with a spiral design (Fig. 21.1: 5). They may be preserved as empty chambers with polished linings or as ovoid casts (Genise 2000). In Gran Barranca these trace fossils are found as vertical empty chambers, or as casts included or detached from the rock matrix. This ichnogenus is attributed to the fossil cells of solitary bees, which in turn are considered as indicative of a poor plant cover because bees prefer to nest in dry soils (Genise and Bown 1994; Genise *et al.* 2004b). Adult females of most species prepare soil nests, digging up to a preferred depth to locate a cell. This cell is lined with waterproof secretions, provisioned with pollen and/or nectar, in which a single egg is laid, and then closed with a cap of soil material. In most bee taxa, this cap displays a spiral design, whereas in Colletinae, the cap is made with secretions. One or more cells may be connected with a single tunnel (Batra 1969). The absence of spiral caps in the material from Gran Barranca, and the vertical and isolated condition of cells, suggest that Colletinae may be potential producers.

Most material found can be tentatively attributed to *Celliforma rosellii* Genise and Bown (1994); however, the ichnotaxonomy of this ichnogenus is still under study (Genise 2000). A few possible specimens of *Celliforma germanica* Brown (1935) were also found at Gran Barranca Member.

Teisseirei barattinia

This trace fossil is, along with *Coprinisphaera*, one of the most common in Sarmiento Formation. It was originally described for the Asencio Formation (Roselli 1939), and also recorded from the Gran Salitral Formation (Melchor *et al.* 2002), and from the Sarmiento Formation of northern Chubut (Genise 2004; Scasso and Bellosi 2004).

This ichnospecies comprises depressed chambers slightly arched downwards and having constructed walls (Fig. 21.1: 4, 6). Some specimens may show a small, rounded antechamber. In Gran Barranca the most common preservation is that of complete specimens having the constructed walls, whereas, internal casts are present but few in numbers. This ichnotaxon is interpreted as a beetle pupation chamber. Representatives of Tenebrionidae, Curculionidae, Scarabaeidae among others, have free-living larvae, which feed on roots or organic matter in soils, and before pupation, construct a chamber to protect themselves during the quiescent period up to the time of emergence as adults.

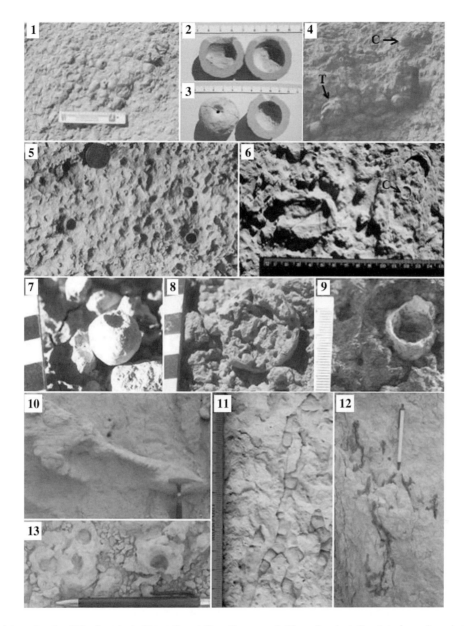

Fig. 21.1. Paleosol trace fossils of the Sarmiento Formation at Gran Barranca. 1, Very abundant *Coprinisphaera* ispp. in the Lower Puesto Almendra Member, scale 20 cm; 2, *C. ecuadoriensis*, scale in cm; 3, *C. murguiai*, scale in cm; 4, *Coprinisphaera* isp. (C) and *Teisseirei barattinia* (T), scale 15 cm; 5, *Celliforma rosellii*, coin 2.3 cm; 6, *Tesseirei barattinia* in longitudinal section, with *Celliforma* (C) inside, scale 15 cm; 7, *Pallichnus dakotensis*, scale in cm; 8, *Lazaichnus fistulosus* in a *Coprinisphaera* specimen, scale in cm; 9, *Feoichnus challa*, scale in cm; 10, Large-size burrows attributed to mammals, hammer 30 cm; 11, septate burrows, scale in cm; 12, manganese root traces, pen 13 cm; 13, *Loloichnus baqueroensis*, pen 13 cm.

Pallichnus dakotensis

A single specimen found in the Rosado Member (Fig. 21.1: 7), constitutes the third record for this ichnospecies worldwide. It was described from the Brule Formation (Retallack 1984), and from a Cretaceous Patagonian locality (Genise *et al.* 2007). This ichnospecies consists of lined, rounded chambers showing a circular scar about half the diameter of the chamber (Retallack 1984). Retallack attributed this trace fossil

tentatively to pupation chambers of Geotrupinae (Coleoptera, Scarabaeoidea).

Lazaichnus fistulosus

Some specimens of *Coprinisphaera* show holes piercing their wall, which continue into the infillings as an internal boxwork (Fig. 21.1: 8). Mikulá and Genise (2003) described them as *Lazaichnus fistulosus*, interpreting the internal boxwork structure as the work of cleptoparasites feeding on the provisions of chambers or their remains. A boxwork of tunnels is also developed in the surrounding paleosol, meaning that the producer was not an obligate cleptoparasite of beetle brood balls. Probably it was an organism feeding in the soil organic matter, which enters the balls due to their higher organic content. The boxwork in the paleosol cannot be considered as *Lazaichnus fistulosus* because this ichnotaxon is reserved for a trace within a trace. It was considered as "possible termite nest" (Bellosi *et al.* 2001; Bellosi and Genise 2004; Genise *et al.* 2004b); however, micromorphological examination (Cosarinsky *et al.* 2005) shows no clear evidence of termite activity.

The connection of *Lazaichnus fistulosus* with meniscate burrows and chambers in the surrounding paleosol attributable to estivation chambers of earthworms (Verde *et al.* 2007) revealed that these organisms would have been active cleptoparasites or detritivores in dung beetle fossil brood balls (Sánchez and Genise 2009).

Beaconites coronus

Beaconites is a rare trace fossil in the Sarmiento Formation whereas in some other Patagonian units, such as the Cretaceous Laguna Palacios Formation, it is quite common (Genise and Bellosi 2004). This ichnospecies comprises meniscate burrows showing a discrete wall, and menisci gently to moderately arcuate (Keighley and Pickerill 1994). It has not been attributed to any particular organism, and its presence in marine, lacustrine, and fluvial deposits points to no particular paleoenvironmental significance.

Feoichnus challa

This ichnofossil, previously known as "pan-shaped trace," is common in paleosols of the Sarmiento and other Cenozoic Patagonian formations, (Bellosi *et al.* 2001; Bellosi and Genise 2004; Genise *et al.* 2004b). It was recently characterized and their affinities studied by Krause *et al.* (2008), who also documented it in the Koluel-Kaike and Pinturas Formations.

The morphology of *Feoichnus challa* is simple although specimens range from showing defined shapes to irregular ones. It has a hemispherical to conical shape, a mostly subvertical orientation, smoothed internal lining with knobby texture and a rough and irregular external surface (Fig. 21.1: 9). The diameter and height of the hemi-chambers of the Sarmiento specimens are about 20–40 mm and 20–34 mm, respectively. The wall is always thicker at the base of the structure (2.2 to 3.5 mm). A thin and short groove with longitudinal striations is frequent on the wall surface.

This trace fossil generally occurs ungrouped and associated with *Coprinisphaera* and *Celliforma* in the same paleosol horizon. By comparisons with extant material, it has been attributed to feeding chambers of cicada nymphs (Hemiptera: Sternorrhyncha: Cicadidae). Diagnostic characters for this assignation are based on wall morphology, interior lining, and the presence of grooves which represent root traces, originally related to feeding activity of nymphs (Krause *et al.* 2008).

Loloichnus baqueroensis

Thick-walled tubes are relatively uncommon trace fossil in the Sarmiento Formation. These burrows are subcircular in cross-section, inclined or vertical, ranging from 1 to 2 cm in diameter. Their distinctive character is the presence of a thick wall (Fig. 21.1: 13). The fillings are passive or they may be empty. The length is up to 20 cm, and the extremes seem to be broken, suggesting a taphonomical bias or reworking that precludes the appreciation of the actual length and complete morphology. At one locality, short tubes occur in vertical clusters suggesting a position *in situ* probably related with the life habits of the producers. According to morphological characters and the paleoenvironment, Bedatou *et al.* (2008) interpreted *L. baqueroensis* as produced by crayfishes (Parastacidae) that inhabited soils.

Septate burrows

A paleosol presents vertical, relatively long (up to 40 cm) and thin (about 6 mm in diameter) slightly sinuous burrows, which may be branched or intersecting one another (Fig. 21.1: 11). The filling is divided in packets, whose length is about or little more than the burrow diameter. The packets have no clear meniscate shape. The composition of the infilling is homogeneous and similar to the rock matrix. The wall shows annulation that corresponds to the packets. This trace fossil resembles *Taenidium serpentinum*; however, the packets are not clearly meniscus-shaped, and they are slightly larger than the burrow width. Moreover, *T. serpentinum* is up to now only known from marine deposits (Keighley and Pickerill 1994).

Medium-sized to large burrows

Two distinctive types of medium-sized to large burrows are recognized. At Profile M occur sinuous, vertical, inclined, to almost horizontal burrows, about 7 cm in diameter passively filled. These traces of uncertain origin are included in fluvial channel sandstones of the Upper Puesto Almendra Member. The other type is larger and is present in "La Cancha" bed of the Vera Member (Profile K) (Bellosi Chapter 2, this book), a weakly developed calcic paleosol. These structures are sinuous, generally inclined, 20–50 cm in diameter, and up to 2.5 m in length (Fig. 21.1: 10). The fill is passive and calcareous in composition. Based on the size, morphology, and distribution, they can be interpreted as mammal dens (Bellosi *et al.* 2005).

Root traces

Rooting structures are not frequently observed in the Sarmiento Formation, and are best preserved in weakly developed paleosols (Fig. 21.1: 12). They are fine, branching, downward tapering, and discrete. Very occasionally, they form root mats or dense patterns. Original carbonaceous material is not preserved. Root traces differ according to paleosol types. Long (up to 2 m), very fine vertical root traces with a clayey infill (Genise *et al.* 2004b) are associated with well-drained soils. In the Gran Barranca Member, fine root and rootlet traces with a manganese core and a bluish grey halo suggest reducing waterlogging conditions. In a few cases, mineralized overgrowths and fillings produced siliceous rhizocretions.

Low-scale ichnostratigraphic analysis: ichnofacies through time

A low-scale ichnostratigraphic analysis is performed for the Paleogene trough early Neogene based upon representative continental units of central Patagonia (Table 21.1) including: (a) the underlying Koluel-Kaike Formation, (b–g) the members of the Sarmiento Formation at Gran Barranca, and (h–i), elsewhere in the Pinturas and Santa Cruz Formations.

(a) Koluel-Kaike Formation

The early to middle Eocene Koluel-Kaike Formation outcrops in central Patagonia. At Gran Barranca, this unit consists of silicified pyroclastic mudstones and bentonites. Orange, yellow, and red paleosols, with abundant kaolinite, root mottling, and Fe–Mn nodules, developed in warm, humid, probably seasonal, forested environments (Krause and Bellosi 2006; Krause *et al.* in press).

This unit is ichnologically almost unstudied, but the first field research on it shows a very different environmental scenario from that of the Sarmiento Formation. The recorded ichnofossils are *Feoichnus challa, Skolithos, Taenidium*, and root traces. A previously mentioned occurrence of dung beetle nests (Laza 1986) requires further documentation because they were found near the transitional contact with the Sarmiento Formation. Paleoenvironments probably changed at the end of Koluel-Kaike time, where *Feoichnus* appears.

(b) Gran Barranca Member

The middle Eocene Gran Barranca Member contains fossil mammals of the Barrancan subage, and is composed by pyroclastic mudstones and bentonites. These facies are interpreted as tephric loessites deposited on subaerial rolling plains having ephemeral ponds (Spalletti and Mazzoni 1979, Bellosi Chapter 19, this book). Trace fossils occur in weakly developed Andisols.

This unit marks the first appearance of several of the trace fossils that occur throughout the Sarmiento Formation, but in scarce numbers. The ichnofossils present are: *Coprinisphaera ecuadoriensis, Coprinisphaera murguiai, Teisseirei barattinia, Beaconites coronus, Celliforma rosellii, Celliforma germanica*, gray-haloed rhizoliths, and manganese rhizoconcretions. *Feoichnus challa* is probably the most abundant ichnofossil. With the exception of *Coprinisphaera*, the above mentioned trace fossils are present in the lower section, along with possible insect cocoons. *Teisseirei, Celliforma*, rhizocretions, and cocoons are preserved as gypsum or siliceous casts. In the upper section of the member, some *Coprinisphaera ecuadoriensis* and *C. murguiai* also occur.

(c) Rosado Member

The late middle Eocene Rosado Member is a strongly developed calcic paleosol, which includes fossil mammals (Mustersan SALMA) and land snails. This Aridisol records the driest and the more stable landscape period during Sarmiento time (Bellosi Chapter 2, this book).

The Rosado paleosol lacks most of the insect trace fossils of the underlying and overlying ones. It includes a few thin rhizoliths and discrete burrows. Only *Celliforma rosellii* is abundant, composing dense aggregations of vertical cells in some sectors of the upper paleosol surface. The estimated maximum density is about 120–150 cells/m^2 (Bellosi *et al.* 2002). A single specimen of *Pallichnus dakotensis* was found at 25 cm below the upper surface, along exposures totalling some thousand square meters.

(d) Lower Puesto Almendra Member

The late Eocene Lower Puesto Almendra Member is composed of conglomerates and sandstones, frequently modified as moderately to weakly developed paleosols, bearing Mustersan mammals. These deposits accumulated in braided, probably ephemeral, fluvial channels (Bellosi Chapter 2, this book).

This member shows the greatest density and diversity of *Coprinisphaera* in the Sarmiento Formation. The ichnospecies present are *C. ecuadoriensis, C. murguiai, C. kheprii, C. tonnii*, and *C. kraglievichi* (*C. kraglievichi* only occurs in this member). The density of *Coprinisphaera* measured in vertical sections may reach up to 100 specimens/m^2. Laterally in the same paleosol, density decrease and many *Coprinisphaera* balls are perforated by *Lazaichnus fistulosus*, the boxwork trace fossil which continues in the surrounding paleosol (Genise *et al.* 2004b). A few *Teisseirei barattinia* are also present in this member.

Table 21.1. *Stratigraphic distribution of trace fossils and ichnofacies in Sarmiento Formation and related Cenozoic continental units from Patagonia*

	Coprinisphaera	*Teisseirei*	*Celliforma*	*Palmiraichnus*	Cocoons	*Feoichnus*	*Pallichnus*	*Beaconites*	*Lazaichnus*	*Syntermesichnus*	Septate burrows	*Loloichnus*	Medium–large burrows	Root traces	*Coprinisphaera* ichnofacies	*Celliforma* ichnofacies
Santa Cruz Formation (Middle Miocene)	?		X											x		
Pinturas Formation (Early Miocene)	X			X		X				X			X	X	X	
Colhue-Huapi Mbr (Sarmiento Fm; Early Miocene)																
Upper section			X													x
Low–middle section	X	X	X			X						X		x		
Upper Puesto Almendra Mbr (Sarmiento Fm; Late Oligocene)	XX	X	X			X		X	X		X	X	x	x	X	
Vera Mbr (Sarmiento Fm; Early Oligocene)	x												x	x	X	
Lower Puesto Almendra Mbr (Sarmiento Fm; Late Eocene)	XX	x	X			X			X					X	X	
Rosado Mbr (Sarmiento Fm; Middle Eocene)			X				x							x		X
Gran Barranca Mbr (Sarmiento Fm; Middle Eocene)	X	X	X		X	X		X						X	X	
Koluel-Kaike Formation (Early–Middle Eocene)						x								X		

Notes: **XX**, very abundant; **X**, present; x, few; ?, probable.

(e) Vera Member

The late Eocene to early Oligocene Vera Member is a very thick and homogeneous sequence of tephric loessites with a few very weakly developed calcareous paleosols. Infrequent fossil vertebrates correspond to Tinguirirican SALMA. This unit records a continuous eolian accumulation of fine volcanic ash on plains (Bellosi Chapter 2, this book).

The Vera Member is almost devoid of invertebrate trace fossils, but contains some thin rhizoliths, and a few specimens of *Coprinisphaera* at the contact with the Upper Puesto Almendra Member (Profile M). The only abundant ichnofossils are large and inclined structures in La Cancha Bed (Profile K), interpreted as mammal burrows.

(f) Upper Puesto Almendra Member

Conglomerates, sandstones, pyroclastic mudstones, and reddish paleosols form the Upper Puesto Almendra Member of Oligocene age. These facies records the sedimentation of a fluvial system in a deeply incised valley (Bellosi Chapter 2, this book).

This member shows the highest ichnodiversity, and also the largest paleoenvironmental heterogeneity. The ichnofossils present are: *Coprinisphaera ecuadoriensis, C. murguiai, C. kheprii,* and *C. tonnii, Teisseirei barattinia, Beaconites coronus, Feoichnus challa, Celliforma rosellii, Lazaichnus fistulosus, Loloichnus baqueroensis,* septate burrows, and medium-sized burrows. *Teisseirei barattinia* becomes abundant, occurring in the same paleosol horizons containing *Coprinisphaera.* Vertically oriented specimens of *Celliforma rosellii* appear in monoichnospecific and dense aggregations, but also associated with other ichnofossils as scattered specimens or in dense aggregations in other levels. The paleosols at top of Unit 4 show probably the highest ichnodiversity of the Sarmiento Formation bearing: *Coprinisphaera ecuadoriensis, C. kheprii, Teisseirei barattinia, Feoichnus challa, Celliforma rosellii, Loloichnus baqueroensis,* and *Beaconites coronus.*

Loloichnus baqueroensis is present in different levels, but is abundant at Profile M, where vertical clusters of short tubes indicate their position *in situ.* In the same locality occur *Coprinisphaera ecuadoriensis, C. murguiai, C. kheprii,* and a single specimen of *C. tonnii,* unusually preserved with the surrounding cavity. Medium-sized tubes and *Teisseirei barattinia* are also present. Septate burrows only appear in a paleosol at Profile H, as the single component of a dense ichnofabric.

(g) Colhue-Huapi Member

The early Miocene Colhue-Huapi Member records a new episode of fluvial incision and aggradation. The lower and middle sections include fluvial channel deposits, tephric loessites, and paleosols, with abundant mammals of the Colhuehuapian SALMA. Similar eolian deposits and calcareous weakly developed paleosols, bearing "Pinturan" mammals, constitute the upper section (Bellosi Chapter 2, this book).

Ichnofossils of the lower and middle sections are similar in diversity to those in the Upper Puesto Almendra Member. Abundant trace fossils are: *Coprinisphaera ecuadoriensis, C. murguiai, C. kheprii, C. tonnii, Teisseirei barattinia, Feoichnus challa, Loloichnus baqueroensis, Celliforma rosellii,* and rhizocretions. These ichnofossils, with the only exception of *C. rosellii,* show different relative abundance laterally without a clear pattern. *Loloichnus* is lacking in the upper beds. The density of trace fossils is similar throughout this interval – about one specimen per lineal meter of vertical exposure of the level. Diversity and abundance of ichnofossils in the upper section decrease considerably. In a calcic paleosol close to the top of the member, *C. rosellii* is the only insect ichnofossils present, and it is found at a high density – 15/horizontal m^2. This paleosol shows the same characteristics as the Rosado Member.

(h) Pinturas Formation

The early to middle Miocene Pinturas Formation is an eolian succession accumulated in an upland setting of west–central Patagonia (Bown and Larriestra 1990). Fossil mammals of the lower and middle sequences correspond to the "Pinturan" association (as the upper section of Colhue-Huapi Member), and those of the upper sequence to the Santacrucian SALMA (Kramarz and Bellosi 2005). Tuffaceous mollic paleosols, with diverse trace fossils, are frequent. The lower sequence includes *Syntermesichnus fontanae,* attributed to termite nests (Bown and Laza 1990), *Coprinisphaera ecuadoriensis* (Genise and Bown 1994; Laza 2006), *Feoichnus challa* (Krause *et al.* 2008), and rhizoliths. Fossil bee cells (*Palmiraichnus pinturensis*) are recognized in the upper sequence (Genise and Bown 1994; Genise and Hazeldine 1998), along with fossil roots and stumps. Large burrows of rodents occur in the three sequences (Bown and Larriestra 1990).

(i) Santa Cruz Formation

The early to middle Miocene Santa Cruz Formation outcrops in central and southern Patagonia. It is composed by epiclastic sandstones and tuffs accumulated in a broad alluvial coastal plain (Bellosi 1995). Its abundant fossils mammals constitute the Santacrucian SALMA.

The ichnofossil record is very scarce. The only traces are from Monte Observation (500 km south of Gran Barranca). They are rhizoliths, a few specimens of *Celliforma rosellii,* and some dung beetle balls (Genise and Bown 1994). Although it is mostly unstudied ichnologically, past field research by some of us yielded no other ichnofossils.

Discussion

The listing of trace fossils (Table 21.1), in combination with lithological and paleopedological data, is a low-scale ichnological approach that allows the recognition of the ichnofacies present, their changes along the studied Cenozoic continental succession, and with that, the record of regional evolution of terrestrial ecosystems.

The Sarmiento Formation at Gran Barranca represents one of the best examples worldwide of the *Coprinisphaera* ichnofacies. The association of *Coprinisphaera* ichnospecies, *Teisseirei barattinia, Celliforma rosellii, Feoichnus challa*, and root traces in the Gran Barranca, Puesto Almendra and Colhue-Huapi Members and in the lower sequence of the Pinturas Formation are diagnostic of this ichnofacies (Table 21.1). Its development along these units, with exceptions in some intervals mentioned below, suggests regional environmental stability for this large-scale analysis.

The *Coprinisphaera* ichnofacies indicates environments dominated by herbaceous vegetation from steppes to savannas (Genise *et al.* 2000). These herbaceous communities may range from dry and cold steppes to humid and warm savannas, and it is possible to obtain additional paleoecological precision by considering the relative abundance of different traces. For example, the presence of termite nests in some assemblages would indicate wetter conditions, whereas the presence of fossil bee cells may indicate drier ones. Sarmiento Formation lacks any fossil termite nest, whereas *Celliforma* is a rather common trace fossil, suggesting drier environments. The abundance and diversity of insect ichnofossils in most levels indicates more likely warm than cold general conditions. The ichnogenus *Coprinisphaera* is particularly important for understanding the evolution of terrestrial ecosystems because it is produced by dung beetles of groups that are generally associated with grass-dominated environments, where herbivore dung is abundant (Halftter and Edmonds 1982).

The Sarmiento Formation, and particularly the Gran Barranca Member, marks the first appearance of the *Coprinisphaera* ichnofacies in the fossil record and could mark the increasing presence of grasses in the ecosystem. It is in the Lower Puesto Almendra Member where *Coprinisphaera* reaches its greatest abundance and diversity.

The Rosado Member represents one of the exceptions to the *Coprinisphaera* ichnofacies, and may be considered as an example of a tentative *Celliforma* ichnofacies (Table 21.1). The recurrence of an association of some insect trace fossils in calcareous paleosols, along with land snails and fruit endocarps, is interpreted as an incipient ichnofacies (Genise *et al.* 2000; Bellosi *et al.* 2002). Genise *et al.* (2004a) name this association the *Celliforma* ichnofacies. The cases recorded worldwide of the *Celliforma* ichnofacies are few in comparison with that of the *Coprinisphaera*

ichnofacies, lacking a broad data set to support the required recurrence in time and space to define it formally. The inferred plant associations for *Celliforma* ichnofacies range from scrubs to woodlands, with poor plant cover. Other potential continental ichnofacies are related similarly with the global classes of plant assemblages (Mueller-Dombois and Ellenberg 1980).

The calcareous paleosol of the upper section of the Colhue-Huapi Member shows the same particular sedimentologic and ichnofacies conditions, suggesting an environmental change, similar to that of the Rosado Member (Bellosi Chapter 2, this book). The Vera Member, which has calcareous paleosols, mostly lacks invertebrate trace fossils to be included in this analysis. The scarcity of insect ichnofossils in the Vera Member may be related with a rapid sedimentation rate and unfavorable environmental factors. A few specimens of *Coprinisphaera* were recorded in the strongest-developed paleosol at its top in Profile M.

In sum, the Rosado Member and the upper paleosol of the Colhue-Huapi Member show the tentative *Celliforma* ichnofacies suggestive of drier periods in which the herbaceous vegetation did not attain sufficient mass to support a critical number of herbivores, and consequently dung beetles, as inferred by the absence of *Coprinisphaera*. In these environments, with widespread bare soil, bees found appropriate sites for nesting. The Rosado Member represents a period of shift to drier conditions, and probably a dramatic decrease of available grasses, which recovered later. Whereas the upper paleosol of the Colhue-Huapi Member represents, with the present data, a more profound environmental change including the disappearance of most dung beetles from this region. Appropriate conditions for the development of the *Coprinisphaera* ichnofacies during these lapses would have shifted to wetter environments in west–central Patagonia, as for those interpreted for the lower sequence of the Pinturas Formation (Bown and Larriestra 1990).

Accordingly, the Sarmiento Formation at Gran Barranca records the appearance, evolution, and near disappearance due to a drier and/or colder climate of the ecosystems in which dung beetles dominated. The whole sequence denotes the dominance of the *Coprinisphaera* ichnofacies during the deposition of the Gran Barranca, Lower and Upper Puesto Almendra, and lower–middle sections of Colhue-Huapi Members, probably representing the abundance of grasses covering soils. This evolution would have been interrupted by two drier periods with low soil coverage. The first recorded by the Rosado Member and the second by the upper section of Colhue-Huapi Member, dominated by bee cells (potential *Celliforma* ichnofacies). Finally, during the deposition of the Vera Member, unfavorable conditions for the development of the *Coprinisphaera* ichnofacies were more probably linked to a local faster loess sedimentation, and major global cooling, related to the latest Eocene–early Oligocene growth of Antarctic ice-sheets (Miller *et al.* 1987; Zachos *et al.* 2001).

ACKNOWLEDGEMENTS
This work was supported by grants from the National Science Foundation USA (BCS-9318942, DEB-9907985, EAR-0087636, BCS-0090255) to R. F. Kay and R. H. Madden, and ANPCyT Argentina (PICT 13286) to J. Genise. We wish to thank the constructive comments of the reviewer G. Retallack, and field assistance of R. Madden, M. Krause, and M. González.

REFERENCES

Batra, S. 1969. Solitary bees. *Scientific American*, **250**, 120–187.

Bedatou, E., R. Melchor, E. Bellosi, and J. Genise 2008. Crayfish burrows from Late Jurassic–Late Cretaceous continental deposits of Patagonia, Argentina: their palaeoecological, palaeoclimatic and palaeobiogeographical significance. *Palaeogeography, Palaeoclimatology, Palaeoecology*, **257**, 169–184.

Bellosi, E. 1995. Paleogeografía y cambios ambientales de la Patagonia central durante el Terciario medio. *Boletín de Informaciones Petroleras*, **44**, 50–83.

Bellosi, E. and J. Genise 2004. Insect trace fossils from paleosols of the Sarmiento Formation (Middle Eocene–Lower Miocene) at Gran Barranca (Chubut Province). In Bellosi, E. and Melchor, R. (eds.), *Fieldtrip Guidebook from the First International Congress on Ichnology*, p. 15–29.

Bellosi, E., J. Laza, and M. González 2001. Icnofaunas en paleosuelos de la Formación Sarmiento (Eoceno-Mioceno), Patagonia central. *Resúmenes IV Reunión Argentina de Icnología y II Reunión de Icnología del Mercosur*, p. 31.

Bellosi, E., S. Miquel, R. Kay, and R. Madden 2002. Un paleosuelo mustersense con microgastrópodos terrestres (Charopidae) de la Formación Sarmiento, Eoceno de Patagonia central: significado paleoclimático. *Ameghiniana*, **39**, 465–477.

Bellosi, E., J. Genise, J. Laza, and M. Sánchez 2005. Terrestrial trace fossils and ichnostratigraphy of the Sarmiento Formation: implications for the oldest grass-dominated ecosystem. *Resumenes XVI Congreso Geológico Argentino*, **306**.

Bown, J. and C. Larriestra 1990. Sedimentary paleoenvironments of fossil platyrrhine localities, Miocene Pinturas Formation, Santa Cruz province, Argentina. *Journal of Human Evolution*, **19**, 87–119.

Bown, T. and J. Laza 1990. A fossil nest of a Miocene termite from southern Patagonia, Argentina, and the oldest record of the termites in South America. *Ichnos*, **1**, 73–79.

Bromley, R., L. Buatois, J. Genise, C. Labandeira, M. Mangano, R. Melchor, M. Schlirf, and A. Uchman 2007. Comments on "Reconnaissance of Upper Jurassic Morrison Formation ichnofossils, Rocky Mountain Region, USA:

paleoenvironmental, stratigraphic, and paleoclimatic significance of terrestrial and freshwater ichnocoenoses" by S. Hasiotis. *Sedimentary Geology*, **200**, 141–150.

Brown, R. 1935. Further notes on fossil larval chambers of mining bees. *Journal of the Washington Academy of Sciences*, **25**, 526–528.

Buatois, L. and M. Mángano 1995. The paleoenvironmental and paleoecological significance of the lacustrine *Mermia* ichnofacies: an archetypical subaqueous nonmarine trace fossil assemblage. *Ichnos*, **4**, 151–161.

Cosarinsky, M., E. Bellosi, and J. Genise 2005. Micromorphology of modern epigean termite nests and possible termite ichnofossils: a comparative analysis (Isoptera). *Sociobiology*, **45**, 1–34.

Frenguelli, J. 1938. Bolas de escarabeidos y nidos de véspidos. *Physis*, **12**, 348–352.

Genise, J. 2000. The ichnofamily Celliformidae for *Celliforma* and allied genera. *Ichnos*, **7**, 267–284.

Genise, J. 2004. Ichnotaxonomy and ichnostratigraphy of chambered trace fossils in palaeosols attributed to coleopterans, ants and termites. In McIlroy, D. (ed.), *The Application of Ichnology to Paleoenvironmental and Stratigraphic Analysis*, Special Publication no. 228. London: Geological Society of London, pp. 419–453.

Genise, J. and E. Bellosi 2004. Continental trace fossils of the Laguna Palacios Formation (Upper Cretaceous) from the San Bernardo Range (Chubut Province). In Bellosi, E. and Melchor, R. (eds.), *Fieldtrip Guidebook of the First International Congress on Ichnology*, pp. 33–43.

Genise, J. and T. Bown 1994. New Miocene scarabeid and hymenopterous nest and Early Miocene (Santacrucian) paleoenvironments, Patagonian Argentina. *Ichnos*, **3**, 107–117.

Genise, J. and P. Hazeldine 1998. The ichnogenus *Palmiraichnus roselli*; for fossil bee cells. *Ichnos* **6**, 151–166.

Genise, J., M. Mángano, L. Buatois, J. Laza, and M. Verde 2000. Insect trace fossils associations in paleosols: the *Coprinisphaera* ichnofacies. *Palaios*, **15**, 49–64.

Genise, J., G. Mángano, and L. Buatois 2004a. Ichnology moving out of the water: a model for terrestrial ichnofacies. *Abstracts I International Congress on Ichnology*, **38**.

Genise, J., E. S. Bellosi, and M. González 2004b. An approach to the description and interpretation of ichnofabrics in palaeosols. In McIlroy, D. (ed.), *The Application of Ichnology to Paleoenvironmental and Stratigraphic Analysis*, Special Publication no. 228. London: Geological Society of London, pp. 355–382.

Genise, J., R. Melchor, E. Bellosi, M. González, and M. Krause 2007. New insect pupation chambers (*Pupichnia*) from the Late Cretaceous of Patagonia (Argentina). *Cretaceous Research*, **28**, 545–559.

Halffter, G. 1959. Etología y paleontología de Scarabaeinae. *Ciencia*, **19**, 165–178.

Halffter, G. and W. Edmonds 1982. The nesting behavior of dung-beetles: an ecological and evolutive approach. *Instituto de Ecología de México Publicaciones*, **10**, 1–176.

Halffter, G. and G. Matthews 1966. The natural history of dung-beetles of the subfamily Scarabeinae. *Folia Entomológica Mexicana*, 12–14, 1–312.

Hasiotis, S. 2004. Reconnaissance of Upper Jurassic Morrison Formation ichnofossils, Rocky Mountain Region, USA: paleoenvironmental, stratigraphic, and paleoclimatic significance of terrestrial and freshwater ichnocoenoses. *Sedimentary Geology*, **167**, 177–268.

Keighley, D. and R. Pickerill 1994. The ichnogenus *Beaconites* and its distinction from *Ancorichnus* and *Taenidium*. *Paleontology*, **37**, 305–337.

Kramarz, A. and E. Bellosi 2005. Hystricognath rodents from the Pinturas Formation, Early–Middle Miocene of Patagonia: biostratigraphic and paleoenvironmental implications. *Journal of South American Earth Sciences*, **18**, 199–212.

Krause, J. M. and E. Bellosi 2006. Paleosols from the Koluel-Kaike Formation (Lower–Middle Eocene) in south-central Chubut: a preliminary analysis. *Resúmenes IV Congreso Latinoamericano de Sedimentología*, **125**.

Krause, J. M., T. Bown, E. Bellosi, and J. Genise, 2008. Trace fossils of cicadas from the Cenozoic of Central Patagonia, Argentina. *Palaeontology*, **51**, 405–418.

Krause, J. M., E. S. Bellosi, and M. S. Raigemborn In press. Lateritized tephric palaeosols from Central Patagonia Eocene, Argentina: a southern high-latitude archive of Palaeogene global greenhouse conditions. *Sedimentology*.

Laza, J. 1986. Icnofósiles de paleosuelos del Cenozoico mamalífero de Argentina. II. Paleógeno. *Boletín de la Asociación Paleontológica Argentina*, **15**, 19.

Laza, J. 2006. Dung-beetle fossil brood balls: the ichnogenera *Coprinisphaera* Sauer and *Quirogaichnus* n. igen. (Coprinisphaeridae). *Ichnos*, **13**, 217–235.

Mazzoni, M. 1979. Contribución al conocimiento petrográfico de la Formación Sarmiento, barranca sur del lago Colhue Huapi, provincia de Chubut. *Revista de la Asociación Argentina de Mineralogía, Petrografía y Sedimentología*, **10**, 33–54.

McIlroy, D. 2004. Some ichnological concepts, methodologies, applications and frontiers. In McIlroy, D. (ed.), *The Application of Ichnology to Paleonvironmental and Stratigraphic Analysis*, Special Publication no. 228. London: Geological Society of London, pp. 3–27.

Melchor, R., J. Genise and S. Miquel 2002. Ichnology, sedimentology and paleontology of Eocene calcareous paleosols from a palustrine sequence, south west La Pampa, central Argentina. *Palaios*, **17**, 16–35.

Miller, K., R. Fairbanks, and G. Mountain 1987. Tertiary oxygen isotope synthesis, sea level history, and continental margin erosion. *Paleoocenography*, **2**, 1–19.

Mikuláš, R. and J. Genise 2003. Traces within traces: holes, pits and galleries in walls and fillings of insect trace fossils in paleosols. *Geologica Acta*, **1**, 339–348.

Mueller-Dombois, D. and H. Ellenberg 1980. *Aims and Methods of Vegetation Ecology*. New York: John Wiley.

Retallack, G. 1984. Trace fossils of burrowing beetles and bees in an Oligocene paleosol, Badlands National Park, South Dakota. *Journal of Paleontology*, **58**, 571–592.

Roselli, F. 1939. Apuntes de geología y paleontología uruguaya: sobre insectos del Cretácico del Uruguay o descubrimiento de admirables instintos constructivos de esa época. *Boletín de la Sociedad de Amigos de Ciencias Naturales "Kraglievich-Fontana,"* **1**, 72–102.

Sánchez, M. V. and J. F. Genise 2009. Cleptoparasitism and detritivory in dung beetle fossil brood balls from Patagonia (Argentina). *Palaeontology*, **52**, 837–848.

Sauer, W. 1955. *Coprinisphaera ecuadorensis*, un fósil singular del Pleistoceno. *Boletín del Instituto de Ciencias Naturales del Ecuador*, **1**, 123–132.

Scasso, R. and E. Bellosi 2004. Cenozoic continental and marine trace fossils at the Bryn Gwyn Paleontological Park, Chubut. *Bryn Gwyn Guidebook, First International Congress on Ichnology*.

Seilacher, A. 1967. Bathymetry of trace fossils. *Marine Geology*, **5**, 413–428.

Spalletti, L. and M. Mazzoni 1979. Estratigrafia de la Formación Sarmiento en la barranca sur del lago Colhue-Huapi, provincia del Chubut. *Revista de la Asociación Geológica Argentina*, **34**, 271–281.

Verde, M., J. J. Jimenez, M. Ubilla, and J. F. Genise 2007. A new earthworm trace fossil from paleosols: aestivation chambers from the late Pleistocene Sopas Formation of Uruguay. *Palaeogeography, Palaeoclimatology, Palaeoecology*, **243**, 339–347.

Zucol, A., M. Brea, R. Madden, E. Bellosi, A. Carlini, and M. G. Vucetich 2006. Preliminary phytolith analysis of Sarmiento Formation in the Gran Barranca (Central Patagonia, Argentina). In Madella, M. and Zurro, D. (eds.), *Places, People and Plants: Using Phytoliths in Archaeology and Palaeoecology*, Oxford, UK: Oxbow Books, pp. 197–203.

Zachos, J., M. Pagani, L. Sloan, E. Thomas, and K. Billups 2001. Trends, rhythms, and aberrations in global climate 65 Ma to present. *Science*, **292**, 686–692.

22 Phytolith studies in Gran Barranca (central Patagonia, Argentina): the middle–late Eocene

*Alejandro F. Zucol, Mariana Brea, and
Eduardo S. Bellosi*

In memoriam to Mario M. Mazzoni

Abstract

The sedimentary sequence of the southern escarpment of Lake Colhue-Huapi (Gran Barranca, Chubut Province, Argentina) is composed of the Sarmiento Formation and Río Chico Group. At the type profile of the Sarmiento Formation at Gran Barranca (Profile MMZ), the sediments of the Gran Barranca Member form the lowest part of the Sarmiento Formation and are of middle Eocene age. Sediments of this member are analyzed for their phytolith content and these results are compared with phytolith assemblages from the underlying Koluel-Kaike Formation.

A detailed phytolith analysis of the Gran Barranca Member reveals an abundance of phytoliths that allowed us to establish a phytolith zonation of the member. A basal zone (GBI) clearly similar to the Koluel-Kaike Formation assemblages indicates a savanna community; the middle zone (GBII) with different compositional characteristics described by the basal predominance of palms associated with grasses and dicots, that gradually decrease (principally palms phytoliths) with more abundant woody dicotyledons elements in the middle levels. The upper section of GBII is characterized by the first presence of open megathermic grasslands with the rising abundance of panicoid elements. The upper zone (GBIII) shows some changes in the floristic composition including an increase in aquatic herbs.

Our results indicate the presence of C_4 elements during the middle Eocene and explore the significance of this early record in light of its implications for the origins of grass-dominated ecosystems.

Resumen

La secuencia sedimentaria de la Barranca sur de Lago Colhue-Huapi (Gran Barranca, Provincia del Chubut, Argentina) se encuentra formada por la Formación Sarmiento y el Grupo Río Chico. En el perfil tipo de la Formación Sarmiento (Perfil MMZ) en Gran Barranca, el Miembro Gran Barranca, de edad eocena media, conforma el miembro inferior de la Formación Sarmiento. En esta contribución el contenido fitolítico de los sedimento de este miembro ha sido analizados y estos resultados fueron comparados con las asociaciones fitolíticas observadas en los términos superiores del Grupo Río Chico (en particular Formación Koluel-Kaike).

El análisis fitolítico detallado del Miembro Gran Barranca revela una abundancia de fitolitos que permitió establecer una zonación de la secuencia sedimentaria. Una zona basal (GBI) claramente similar a las asociaciones de la Formación Koluel-Kaike, y cuyos elementos indican la presencia de una comunidad de sabana; la zona media (GBII) con algunas variaciones en su sección que indican el predominio basal de palmeras asociadas con dicotiledóneas y hierbas, las cuales gradualmente disminuyen y en los niveles medios se observa un incremento de indicadores de dictotiledóneas arbóreas. La sección superior de la zona GBII es caracterizada por la primera evidencia de pastizales megatérmicos abiertos con una abundancia creciente de elementos panicoideos. La zona superior (GBIII) muestra algunos cambios de la composición florística con un marcado aumento de hierbas acuáticas.

Nuestros resultados indican la presencia de elementos C_4 durante el Eoceno medio y analizamos el significado de este temprano registro a la luz de sus implicancias en los orígenes de los ecosistemas dominados por gramíneas.

Introduction

The escarpment south of Lake Colhue-Huapi (Chubut Province, Argentina) called Gran Barranca is renowned for the diversity of its fossil vertebrates. In 1979, Spalletti and Mazzoni made a detailed sedimentary study of the Sarmiento Formation at Gran Barranca. An accompanying petrographic analysis of the intraformational silts (Mazzoni 1979) noted for the first time that phytoliths were present and remarked upon their potential importance as a tool in the study of biodiversity through this sedimentary sequence. Zucol *et al.* (1999) show that phytoliths are present throughout the Gran Barranca sequence. More recent analysis documented the presence of grass-dominated ecosystems in the

The Paleontology of Gran Barranca: Evolution and Environmental Change through the Middle Cenozoic of Patagonia, eds. R. H. Madden, A. A. Carlini, M. G. Vucetich, and R. F. Kay. Published by Cambridge University Press. © Cambridge University Press 2010.

Sarmiento Formation at Gran Barranca in the Oligocene and Miocene (Zucol *et al.* 2007). In this chapter, we present our findings about the phytolith record near the base of the section, in an effort to shed light on when grasslands first appeared. Specifically, we describe the phytolith content of the Gran Barranca Member (middle Eocene), the lowest part of the Sarmiento Formation at Gran Barranca, and compare these results with previous studies of the Sarmiento Formation and the underlying Río Chico Group (Brea *et al.* 2008; Raigemborn *et al.* 2009). The Gran Barranca Member is a whitish, poorly consolidated, 67-meter thick succession of pyroclastic mudstones, bentonites, and poorly developed paleosols interpreted as Andisols (Spalletti and Mazzoni, 1979; Bellosi Chapter 2, this book). The base is transitional on the Koluel-Kaike Formation; and the top is marked by an erosive unconformity. Fossil mammals are abundant and correspond to the Barrancan Subage (Cifelli, 1985). Geochronologic studies elaborated by Ré *et al.* (Chapter 4, this book) on the basis of radiometric dates and magnetic polarity stratigraphy indicates a middle Eocene age (~41.6–39.0 Ma) (late Lutetian to early Bartonian). The Member is interpreted as a period of tephric loessite sedimentation where bottomlands were occupied by ephemeral small lakes and the weakly developed Andisols (Bellosi and González this book) formed in a subhumid to semi-arid, seasonal climate.

Materials and methods

In total, 14 samples were collected (Fig. 22.1) and analyzed at Profile MMZ (Ré *et al.* Chapter 3, this book). The two lowest samples (MMZ9501.0 and MMZ9501.5) were collected in Río Chico Group sediments (Koluel-Kaike Formation). The remaining 12 samples belong to the Gran Barranca Member, and represent a stratigraphic and temporal sequence (MMZ9502.0, MMZ9502.5, MMZ9503.0, MMZ9503.3, MMZ9503.5, MMZ9504.0, MMZ9505.0, MMZ9506.0, MMZ9506.5, MMZ9507.0, MMZ9507.5, and MMZ9508.0). The temporal interval represented by the sediments of the Gran Barranca Member extends from about 41.6 to 39.0 Ma, or about 2.6 m.y. (Ré *et al.* Chapter 4, this book). The portion of the section between levels MMZ 95–2.0 and 95–8.0 extends from about 40.5 to 39 Ma, or about 1.5 m.y.

Phytoliths were extracted using standard wet oxidation and heavy flotation techniques (Bonomo *et al.* 2009). Laboratory procedures can be summarized as follows: 20 g of sediments were washed and soluble salts were dissolved with distilled water; carbonate material was removed with dilute HCl; organic material was removed by adding H_2O_2 (30%) followed by deflocculation in $Na_4P_2O_7$ solution. The silt fraction was extracted by gravity sedimentation methods (<5-μm fraction) and then sieving to separate the >250-μm

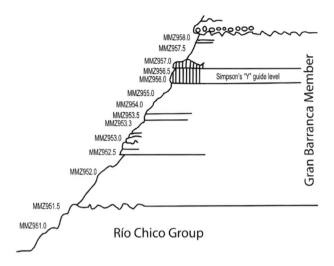

Fig. 22.1. Sedimentary sequence samples of Gran Barranca Member in the MMZ profile, made over Mazzoni's original drawn profile (modified from Kay *et al.* 1999). For more details see Bellosi (Chapter 2, this book).

and 5–250-μm fractions. Heavy liquid flotation was carried out on the 5–250-μm fraction using sodium polytungstate heavy liquid at $2.3 \, g/cm^3$. Phytoliths were mounted on microscope slides in liquid (oil immersion) and solid (Canada balsam) media and analyzed under a Nikon Eclipse E200 light microscope. Photographs were taken with a Nikon Coolpix 990 digital camera.

The different phytolith morphotypes present in each sample were noted, and counts were made of the number of phytoliths of each morphotype. In all 450 elements were counted, a number determined to be representative of the samples as whole. MMZ9507.0 had fewer than 450 elements and was treated separately (see below). Phytolith morphotypes were determined and classified according to previous morphological classifications (Twiss *et al.* 1969; Bertoldi de Pomar 1971; Piperno 1988, 1989; Bozarth 1992; Rapp and Mulholland 1992; Twiss 1992; Kondo *et al.* 1994; Piperno and Becker 1996; Zucol 1996, 1999; Runge 1999; Wallis 2003; Zucol and Brea 2005; Pearsall 2006) and described according to morphotype descriptors proposed by the IPCNWG (2005). A summary classification of all the morphotypes used for this analysis is presented in Appendix 22.1. For the description of stomatocysts of Chrysostomataceae, terminology and treatment follows Rull and Vegas-Vilarrúbia (2000) and Coradeghini and Vigna (2001). Data analysis and the phytolith diagrams were prepared using POLPAL Numerical Analysis Program (Walanus and Nalepka 1999a, 1999b, 2002; Nalepka and Walanus 2003). The zonification schemes were obtained with the constrained single link cluster analysis on square root (SQRT) transformed data. Margalef's richness index *sensu* Moreno (2001) was used to compare the diversity of elements in each sample.

Correspondence analyses (CA) were made with PAST (PAlaeontological STatistics, version 1.75) software (Hammer *et al.* 2007), using a database obtained by absolute morphotypes counting from Zucol (2005), Raigemborn *et al.* (2006, 2009), and this chapter, for comparing phytolith assemblages containing the counts of phytolith morphotaxa from Las Flores Formation, Koluel-Kaike Formation, and Gran Barranca Member of Sarmiento Formation. The CA routine finds the eigenvalues and eigenvectors for a matrix containing the chi-squared distances between all data points. The eigenvalue, giving a measure of the similarity accounted for by the corresponding eigenvector, is given for each eigenvector.

Processed sediment samples and microscope slides are deposited in the Laboratorio de Paleobotánica of the CICYTTP, Diamante, under the acronym CIDPalbo-mic.

Phytolith analysis of the Gran Barranca Member

Most of the phytoliths are well preserved, although in samples MMZ9501.0 and MMZ9501.5 a small percentage of elements has partial internal dissolution or superficial weathering. Nevertheless in many cases it is possible to recognize phytolith form and morphotype. Throughout the sequence the highest phytolith concentrations were found in the fine fraction. All samples in the analyzed sequence show elevated phytolith variability and abundance, with the exception of sample MMZ9507.0 (Fig. 22.1). Not enough elements were present in MMZ9507.0 to establish a minimum sample (Fig. 22.5, below). The small fraction of assemblage MMZ9507.0 is composed of rare spherical and ellipsoidal echinulate phytoliths (Gl), small prismatic phytoliths (Be and Me), and rondel phytoliths (Rn). Phytoliths originating from vascular elements (Ce), prismatic and polyhedral phytoliths (El and Po), and spherical smooth phytoliths (Mg01) were also present.

Phytolith morphotype abundances through the Gran Barranca Member are presented in Fig. 22.2, and graphically presented in Fig. 22.3. One group of phytolith elements comprised of Ps09, Be02, Ce04, Dm01, Gl02, Gl03, Gl04, Lb01, Me01, El08, El09, Ct01, Po01, Po03, and Po05 are present in great abundance throughout the sedimentary sequence. Constrained cluster analysis of the phytolith assemblages reveals three phytolith zones.

Zone I (GbZI – MMZ9501.0 and MMZ9501.5)
Many phytoliths in Zone I show an internal dissolution of opal matrix or various degrees of surface weathering. The basal Zone I phytolith assemblages present marked differences in composition compared with higher zones by the criterion of presence and abundance of different morphotypes. Zone I is characterized by the presence but relatively low abundance of phytolith types Bc03, Bc05, Cr02, Sa03, El07, Eb03, Co06, and Ct07. High percentages of small prismatic phytoliths (Be02 and Me01) are present, as are the larger prismatic types such as El09 (Fig. 22.4K), El03 and Po01 jigsaw (Lb01), vascular element (Ce04) and aspheric echinate (Gl04) phytoliths are observed. Several unidentified phytoliths are found in Zone I that seem to be derived from vascular elements with helical thickening and morphotypes originating from vessel elements with alternate intervessel pits.

Zone II (GbZII – MMZ9502.0 – MMZ9506.5)
The middle section presented a complex composition in their phytolith assemblages. Some morphotypes, Dm02, Ct04, Tr01, and Pp01, are found throughout this zone. Also, this section of the profile has the highest presence and variability of arecoid phytoliths (Gl02, Gl03, Gl04, and Gl05). The latter dominant elements are associated in all subzones with prismatic, faceted, point-shaped, fan-shaped, saddle, and dumbbell phytoliths. Other elements are more variable permitting subdivision into three subzones (Fig. 22.3).

Subzone IIA (GbZIIA – MMZ9502.0, MMZ9502.5, MMZ9503.0, MMZ9503.3, and MMZ9503.5)
The marker morphotypes of Subzone IIA (Fig. 22.2) are Ss01, Be01, Ce03, Fu01, Fu03, Ct02, Ta01, and Ca01. This subzone also has the highest abundance of spherical to ellipsoidal echinate or globular types (Gl02 and Gl04). These are associated with jigsaw phytoliths (Lb01). Prismatic and laminar papillae phytoliths (Lp01 and Pp01) and smooth prismatic phytoliths (Be02, El09, Po01, and Po03) are present in relatively high abundance. Some phytoliths with ciperoid affinities (such Ss01, El01, and El06), jointly with prismatic (Be01 and El04) and fan-shaped (Fs07) phytoliths, also characterize Subzone IIA. The abundance of saddle (Sa01 and Sa02) and dumbbell phytoliths (Dm01, Dm07, Dm09, and Dm10) increases in the highest stratigraphic sample. Truncated conical phytoliths with a predominance of Ct01, Ct03, and Ct05 increase notably from the stratigraphically basal samples upwards through Subzone IIA.

Subzone IIB (GbZIIB – MMZ9504.0 and MMZ9505.0)
Dumbbell morphotypes Dm06 and Pl01 are indicators for this subzone (Fig. 22.2). Although globular phytoliths are abundant, in contrast to ZIIA, echinates (Gl02 and Gl04) are less well represented, together with an increased percentage of those with smooth surfaces (Gl03 and Gl05). Faceted phytoliths Mc02, Mc03, and Mc04 also occur frequently as do truncated conical phytoliths Ct01, Ct03, and Ct04. Another diagnostic feature of Subzone IIB is the remarkable

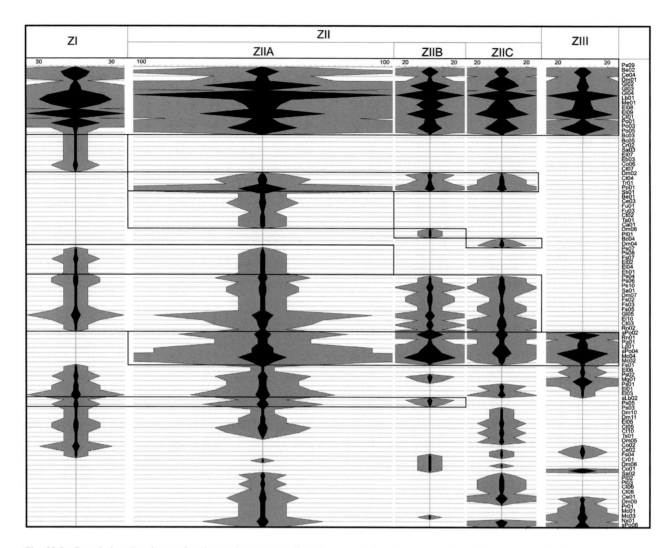

Fig. 22.2. Cumulative abundance of each morphotype according phytolith zonification (in number of elements counting for each morphotype). Acronym descriptions are explained in Appendix 22.1. In black, percentages; in gray, exaggeration (factor 5) of percentage (left).

increase in prismatic phytoliths Be02, Me01, and El09, and the gradual reduction in the presence of the polyhedrical morphotypes (Po01 and Po03). On the other hand, the smaller polyhedrical phytoliths that appear in the sedimentary sequence in the upper section of the GbZIIA become abundant (sPo02 and sPo04). It is also notable that Subzone IIB contains morphotypes derived from vessel elements with scaleriform perforation plates, vessel elements with alternate intervessel pits, vascular elements with helical thickening, and phytoliths derived from vessel elements with simple perforation plates.

Subzone IIC (GbZIIC – MMZ9506.0 and MMZ9506.5)

Morphotypes only occurring in this subzone are Bc04 and Dm04. This subzone shows the highest variability (Fig. 22.2), especially sample MMZ9506.0 from Simpson's

Y Tuff, which shows the highest morphotype richness of the samples from the Gran Barranca Member (Fig. 22.5). Dumbbell phytoliths show the highest percentage (Fig. 22.3) and variability in the sequence with the presence of Dm01, Dm02, Dm04, Dm05, Dm07, Dm08, Dm09, Dm10, and Dm11 (Fig. 22.2). Other panicoid morphotypes such as polylobates Pl02 and Pl03 are also abundant. This prevalence is observed in the lower of the two samples, jointly by the abundance of Ce01, El05, El08, Nx01, Ct03, Ts01, Ps02, Ps03, and Ps06, many of which are not present in the upper sample.

Some groups such as globular and prismatic phytoliths that are very frequent in samples from elsewhere in the Gran Barranca Member are present with relatively lower abundances. For example, saddle and truncated conical phytoliths are present in this subzone, but with their lowest percentages in comparison with the other subzones. Saddles

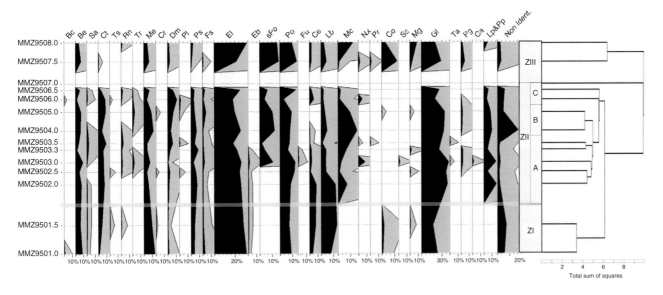

Fig. 22.3. Percentage data for major phytolith types groups present in Profile MMZ of the Gran Barranca, showing the cluster analysis dendrogram obtained with constrained single link with square root (SQRT) method transformation. In black, percentages; in gray, exaggeration of percentage (factor 5) drawn to observe more clearly the low percentages.

are not observed in GbZIII and truncated conical phytoliths are present only with very low frequencies.

Zone III (GbZIII – MMZ9507.5 and MMZ9508.0)

Short and elongate prismatic phytoliths (Be02, Me01, El08, and El09) are dominant in this zone, as they are in every sample analyzed. Polyhedric phytoliths of different sizes (Po01, Po03, sPo02, and sPo04) and echinate or smooth spherical phytoliths (Gl04 and Mg01) are most abundant. Morphotype Mg01, characterized by its large size, is most abundant. Jigsaw phytoliths (Lb01) are present, but are smaller in size than those observed in GbZI and the lowest samples in GbZII. Conical and hexagonal phytoliths (Co01 and Pg01, respectively) and prismatic phytoliths (El01 and El06) occur in this zone. Faceted phytoliths are abundant. Dumbbell phytoliths are scarce, and only two kinds (Dm01 and Dm09) have been observed.

Phytolith distribution and affinities

The phytolith composition across the analyzed profile (Fig. 22.2) shows no marked compositional discontinuities, but some notable variation contributes to understanding floristic change through the sequence.

First, regarding the major groups present throughout the sequence, small prismatic phytoliths (Me01 and Be02) are ubiquitous elements with the highest abundance in GbZI, GbZIIB, and GbZIII, while small prismatic phytoliths with undulate margin contour (Be01) are only present in GbIIA. Among the larger prismatic phytoliths, principally represented by the El group, the elongated element with smooth

contour (El09) is the most frequent in GbZI, GbZIIB, and GbZIII. The elongated element with undulate contour (El08) is present throughout the sequence but has its highest abundances in GbZIIC and GbZIII; whereas the prismatic elements with serrate and denticulate contour (El05 and El03) are principally present in GbZI and GbZII, although the last one is poorly represented in GbZIII and both are absent in GbZIIB.

Among the less abundant elongate types (El01 and El06), two prismatic morphotypes with ciperoid affinity are present in all three zones with similar low abundance, while El07 is characteristic of GbZI.

Polyhedrical phytoliths show a homogeneous distribution through the profile (especially Po01). Point-shaped (Ps) and fan-shaped (Fs) phytolith types possess heterogeneous distributions among their different subtypes. The point-shaped elements are the best represented group (with the most abundant subtypes Ps02, Ps04, Ps05, Ps06, Ps09, and Ps10 being most widely represented). Fan-shaped phytoliths are principally represented in GbZIIB and GbZIIC (with the dominance of Fs05, Fs03, and Fs02 types). Among phytoliths of graminoid affinity, bacillary phytoliths are poorly represented (Bc03 and Bc05 occur only in GbZI, and Bc04 only in GbZIIC). Small triangular phytoliths (Tr01) are observed in GbZII. Small polyhedral phytoliths are present in all Gran Barranca Member samples, including zones GbZIIA, GbZIIB, GbZIIC, and GbZIII.

With regard to grass types with high diagnostic significance, the festucoid boat or sinuate trapezoid type (Ts01) is present in GbZI, GbZIIA, and GbZIIC. Among pooid or

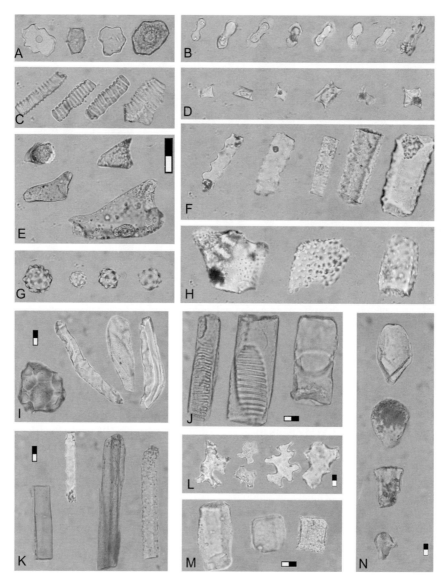

Fig. 22.4. Different phytoliths. (A) Conical and polygonal plate (Co and Pg); (B) dumbbell (Dm); (C) cylindrical vascular elements (Ce); (D) truncated cones (Ct); (E) point-shaped (Ps); (F) short elongate (Be and Me); (G) globular echinate (Gl); (H) papillose (Lp01 and Pp01); (I) multicavate or multifaceted elements (Mc); (J) other phytoliths originated in vascular elements; (K) long elongate (El); (L) lobular–jigsaw (LB); (M) polyhedral bulliform (Po); (N) fan-shaped (Fs). Scale bar 20 μm (in E valid for A–H).

rondel phytoliths, Rn01 is present in all phytolith zones of the Gran Barranca Member but is most abundant in GbZIIA, while Rn02 is present in GbZI and GbZII but not GbZIII. Saddle chloridoid phytoliths (Sa) are found in GbZI and GbZII, with Sa03 restricted to GbZI, Sa01 in GbZI and GbZII and Sa02 in GbZIIA and GbZIIC. Truncated conical phytoliths of danthonoid affinity are present in GbZI and GbZII, with higher representation in GbZII (Ct01, Ct02, Ct03, Ct04, and Ct05) but their abundance falls towards the end of GbZIIC and into GbZIII.

Among panicoid morphotypes, the principal cross phytolith morphotype (Cr01) is present with low frequency in

only three samples (GbZI, GbZIIA, and GbZIIB). Typical dumbbell phytoliths (Dm) are distributed throughout the profile, but their abundance gradually increases from the base to the top of GbZII, becoming most abundant in GbZIIC, with only Dm01 and Dm09 present in GbZIII. The complex dumbbell or polylobate phytoliths (Pl02 and Pl03) are also present in GbZII, especially in GbZIIC, in samples with the highest abundance of typical dumbbells.

Arecoid phytolith types (principally represented by Gl02 and Gl04) are the most abundant phytoliths in all samples, and their maximum relative percentage is in GbZIIA. Other globular subtypes (Gl03 and Gl05) show similar

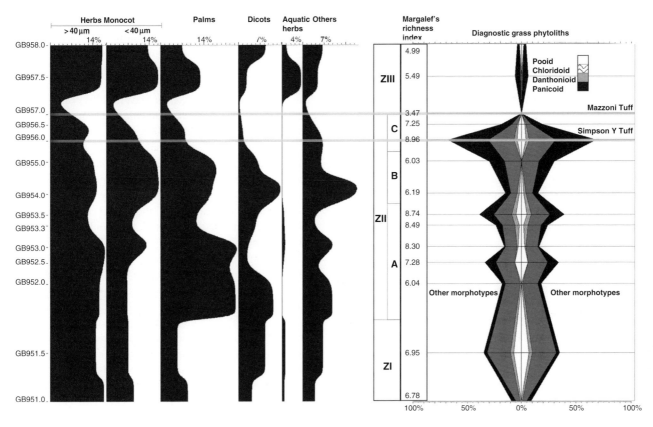

Fig. 22.5. Abundance diagram showing relative frequencies (percentage) of different life forms along the sedimentary sequence drawn by Gaussian smoothing; each class formed by the abundance of the following morphotypes. Monocot herbs greater than 40 μm: elongate (El), fan-shaped (Fs), point-shaped (Ps), and polyhedral bulliforms (Po). Monocot herbs less than 40 μm: micro-elongate (Me), brief elongate (Be), bacillar (Bc), polyhedral bulliforms (sPo), sinuate trapezoid (Ts), rondel (Rn), saddle (Sa), truncated cone (Ct), cross (Cr), dumbbell (Dm), and polylobate (Pl). Palms: globular (Gl). Dicots: lobular (Lb) and multicavate (Mc). Aquatic herbs: star-shaped (Ss), conical (Co), proteiform (Pr), fusiform (Fu), and some elongate morphotypes (suhc El01 and El06). Margalef's richness index value for each sample assemblage (middle). Cumulative abundance of diagnostic grasses phytoliths (right); each class formed by the abundance of the following morphotypes: Pooid, sinuate trapezoid (Ts), and rondel (Rn); Chloridoid, saddle (Sa); Danthonioid, truncated cone (Ct); Panicoid, cross (Cr), dumbbell (Dm), and polylobate (Pl).

nearly ubiquitous distribution but are less abundant and Gl05 does not occur in GbZIII. These last two morphotypes have dubious botanical affinity, as they have been assigned variously either to Chrysobalanaceae, Restionaceae, or certain woody dicotyledons (Piperno 1988; Kondo *et al.* 1994; Strömberg 2004; Zucol and Brea 2005; Pearsall 2006).

Fusiform (Fu01and Fu03) phytoliths related to Podostemaceae (Bertoldi de Pomar 1971) are only present in GbZIIA. In this same subzone occur Nx01 and Pr01, also related to the Podostemaceae (Bertoldi de Pomar 1971). Nx01 phytoliths are present in sample MMZ9506.0, and both Nx01 and Pr01 phytolith types are present in GbZIII.

Phytoliths with ciperoid affinity (Co02, Co06, El01, and El06) are present in GbZI. El01 and El06 phytoliths occur together in GbZIIA and GbZIII, Ss01 in GbZIIA, and Pg01 in all zones in the Gran Barranca Member. GbZIII shows the highest abundance of ciperoid phytoliths (Fig. 22.5) (Co01, Co02, El01, El06, and Pg01).

Jigsaw phytoliths are mostly represented by Lb01, a morphotype with close similarity to dicotyledonous plants, and present in all zones throughout the section. This morphotype presents variations of form and size that make us suspect multiple plant species origins. For example, a larger variant of this morphotype with a well-lobated outline is abundant in GbZI and the lower samples of GbZIIA, while a smaller variant is present in the upper part of the Gran Barranca Member.

Elongated branched phytoliths were observed in GbZI (Eb01 and Eb03) and GbZIIA (Eb01). Ta01, a phytolith morphotype that some authors have assigned mimosoid affinities (Kealhofer and Piperno 1998; Wallis 2003), is present in GbZIIA.

Faceted phytoliths (Mc) partially related to Annonaceae and Magnoliaceae (Kondo and Pearson 1981; Bozarth 1987; Piperno 1988; Runge 1999) are found in GbZII and GbZIII with a marked abundance in GbZIIB and in sample

MMZ9508.0 of GbZIII. Morphotypes Mc02 and Mc04 have a similar distribution in zones GbZII and GbZIII, while Mc01 and Mc03 are present in GbZIIA and in GbZIII.

Prismatic and laminar papillae phytoliths (Lp01 and Pp01) are almost entirely restricted to GbZII, with highest percentages in GbZIIA, but Lp01 also occurs in the highest level of GbZIII. These morphotypes are assigned to Burseraceae (Piperno 1988, 1989; Pearsall 1993).

Two morphotypes with uncertain affinities are the great spherical smooth phytoliths and phytoliths originated from vascular elements (Ce). Among these, only one type (Ce03) is restricted to a single zone (GbZIIA), while the others (Ce01, Ce02, and Ce04) appear intermittently throughout the sampled profile.

Discussion

Our data suggest three phytolith zones based on composition (Figs. 22.3, 22.6B). The basal zone (GbZI, comprising of samples MMZ9501.0 and MMZ9501.5), was considered by Mazzoni to belong to the Sarmiento Formation. Comparison with bona fide samples from the Gran Barranca Member reveals a marked contrast between these basal samples and the remaining Gran Barranca Member assemblages and a closer similarity in composition to phytolith assemblages from the Koluel-Kaike and Las Flores Formations (Zucol 2005; Raigemborn *et al.* 2006, 2009) (Fig. 22.6A). When correspondence analysis was made with the phytolith assemblages of the Gran Barranca Member without the phytolith data of the Río Chico Group samples, their variability shows the samples arranged in three groups (Fig. 22.6B), GbZI, GbZII, and GbZIII (Fig. 22.3).

Correspondence analysis (Fig. 22.6A) reveals GbZII and GbZIII to be distinct from GbZI, and GbZI associated with the Koluel-Kaike Formation assemblages. Morphotypes (with their weights or eigenvalues) Tr (-0.897), Rn (0.861), Ce (0.828), and El03 (-0.781) contributed to the separation along Axis 2. Gran Barranca Member samples (Fig. 22.6B) are separated into zones along Axis 1 by the following morphotypes (with their eigenvalues): Lb (0.853), Bc (0.841), Ss (-0.798), Ce (-0.742), Fu (-0.738), Rn (-0.710), Tr (-0.673), and Nx (-0.627). On the other hand, GbZIII is distinct from lower Gran Barranca Member phytolith zones GbZI and GbZII along Axis 2 by morphotypes with ciperoid affinities Co (-1.803), Pr (-1.532), and Pg (-0.917) and morphotype Mg (-1.110) with unknown affinity.

The phytolith assemblages of the GbZI reveal the presence of mixed vegetation with abundant palms associated with grasses, sedges, and dicotyledons, including a considerable number of woody elements. The grass morphotypes (Fig. 22.5) are represented mainly by chloridoid, danthonioid, festucoid, and pooid phytoliths; panicoid types

are more rare. The dominant grass groups have C_3 photosynthesis, with the exception of chloridoids, which have C_4 (NAD malic enzyme). The components of the BEP clade Pooideae (= Festucoideae *sensu* Grass Phylogeny Working Group 2001) as well as the PACCMAD clade (Chloridoideae and Danthonioideae *sensu* Stevens 2001) early differentiated into the principal C_4-dominated clades (Christin *et al.* 2008; Vicentini *et al.* 2008).

GbZIIA marks a vegetational transition from a mixed palm–grass assemblage in lower GbZIIA into an open community in upper GbZIIA where palms predominate. The dicotyledonous component is not clearly associated with woody plants, but rather by the presence of elements associated with dicot shrubs. Chloridoid, danthonioid, festucoid, and pooid phytoliths persist with some variation; for example, we note the scarcity of chloridoids complemented by an increased percentage of pooid morphotypes (Fig. 22.5). The presence of podostem and sedge phytoliths in upper GbZIIA may indicate the presence of aquatic biotopes and with perennial streams.

In GbZIIB, the compositional characteristics of phytolith assemblages suggest a more balanced co-occurrence of woody dicotyledonous plants, palms, and grasses. Danthonioid and panicoid phytoliths increase through GbZIIB, while chloridoid and pooid phytoliths decrease (Figs. 22.2, 22.5). Fan-shaped and polyhedral bulliform phytoliths, increase associated with conditions when plants are under stress. This feature, together with absence of aquatic herbs and elements associated with humid conditions, suggests the advent of seasonal dry periods. The basal sector of this subzone shows a high abundance of dicot phytoliths of which the most informative are vascular elements of arboreal dicots. In GbZIIC, for the first time, the herbaceous component is dominant over arboreal elements. GbZIIC shows rare aquatic herbs and lower percentages of palm and dicot indicators, whereas grass types become abundant with an increase in all different diagnostic grass morphotypes but with marked predominance of the panicoid types.

Arecoid morphotypes return to their former abundance in GbZIII, but now they are related to scarce diagnostic grasses, and abundant sedges and podostems. Dicotyledonous elements are also present in relatively high proportions in GbZIII, and lobular types have greater variation in shape and size compared with the same morphotype in lower zones.

With respect to grass phytoliths, the principal distinction we can make is between non-diagnostic (elongate, point-shaped, fan-shaped, and polyhedral types) and diagnostic types (short cell phytoliths). Among the diagnostic types, the BEP clade is represented by festucoid and pooid morphotypes of the Pooideae (sinuate trapezoid, rondel, and some elongate and bacillar types) while the PACCMAD clade is represented by morphotypes of the Danthoniodeae (mostly truncated cones), Chloridoideae (saddle chloridoid

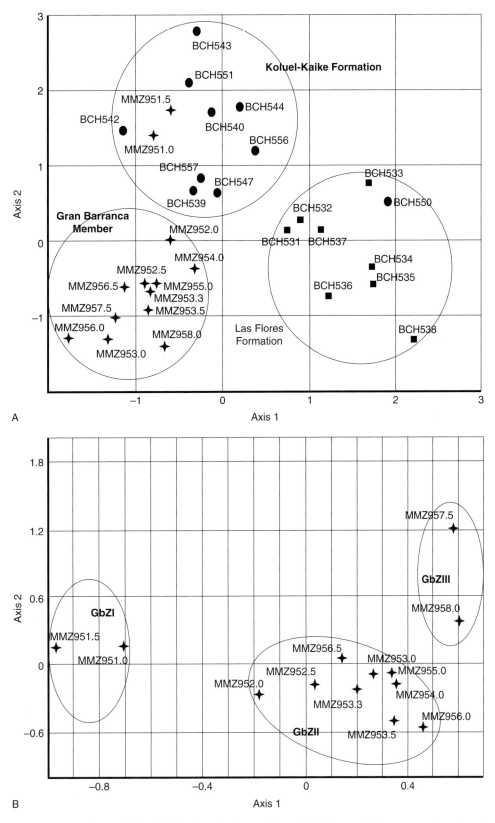

Fig. 22.6. (A) Correspondence analyses of phytolith abundance in three sedimentological terms of the sequence: Sarmiento Formation – Gran Barranca Member (crosses MMZ951–8), and Río Chico Group – Koluel-Kaike Formation (circles BCH539–557) and Las Flores Formation (squares BCH531–538). Eigenvalue in Axis 1: 0.24578 (23.06%), Axis 2: 0.11858 (11.13%). (B) Correspondence analyses of phytolith abundance in Gran Barranca Member. Eigenvalue in Axis 1: 0.17090 (29.18%), Axis 2: 0.11570 (19.76%).

types), and Panicoideae (cross and some dumbbell types) (see Appendix 22.1).

Among aquatic or humid-climate herbs stellate cells, some elongated types (such as El01 and El06), and some conical and polygonal plate phytoliths are here considered cyperoid morphotypes. Podostems are represented by fusiform, proteiform, and scrobicular phytoliths.

While palms (Arecaceae) are represented mainly by globular morphotypes, certain small conical types are also produced in some palms. Many dicot morphotypes do not have clearly established diagnostic value as a consequence of the lack of reference study. Some morphotypes such as lobular, globular with smooth or lightly rough surface, short tracheid, some cylindrical or irregular vascular elements, and some multifaceted phytoliths are heterogeneous types which lack information not only about their shape and size but also their presence/absence in different plant groups (Piperno 2006).

The botanical origin of small globular phytoliths with smooth or lightly roughened surface is another controversial topic that raises the uncertainty of our interpretation. Different authors assign them to Chrysobalanaceae (Piperno 1988; Runge 1999; Pearsall 2006), or Marantaceae when the surface is rough to crushed (Piperno 1988; Runge 1999; Pearsall 2006), or Cannaceae when the surface is rugulose (Piperno 1988), or to other woody dicotyledons (Kondo *et al.* 1994; Alexandre *et al.* 1997; Barboni *et al.* 1999; Runge 1999; Bremond *et al.* 2004, 2005; Strömberg 2004), or to rushes (Restionaceae *sensu* Kondo *et al.* 1994). According to Christin *et al.* (2008) rushes are a group that differentiated much earlier than the age of the sediments analyzed here, and this morphotype may indicate their presence. Interestingly, rushes exist in the Patagonian record since the Paleocene (Archangelsky 1973). Rushes are a plant group of Poales more closely related to grasses than sedges (Stevens 2001).

The significance of the Gran Barranca Member analyses in the Sarmiento Formation

Although the presence of graminoid elements has been reported previously for the middle Eocene Gran Barranca Member, published descriptions suggest they were a secondary component of plant communities in southern South America (Barreda and Palazzesi 2007; Palazzesi and Barreda 2007) as they were in records of the same age elsewhere in the world (Utescher and Mosbrugger 2007). The results of the present analyses not only affirm the presence of graminoid phytoliths in the Gran Barranca Member, but also identify graminoid elements as components of savanna and/or forest communities in some levels of the Gran

Barranca Member. On the other hand, the early coexistence of groups that at present are environmental antagonists (the pooids and chloridoids) and their continued co-occurrence in zone GbZII (A and B) but later dissociation in GbZIII, along with the presence of indicators of plant stress conditions, provides evidence for aridification. The dissociation of pooid and chloridoid grasses could be considered as the oldest evidence for the adaptation of chloridoid grasses to arid environmental conditions.

Implications for the origin of grass-dominated ecosystems

Clayton (1981) proposed that grasses originated from a Flagellariaceae/Commelinaceae ancestry linked to forest communities in Gondwanaland. In his scenario, grasses inhabited the basal stratum of these communities, and originated as herbaceous bambusoids from which the modern herbaceous and woody bamboos were derived. In the next stage of their evolutionary history, some grass groups adapted to drier, more direct-light environments in mixed arboreal/herbaceous communities. From this ancestral group of arundinoid grasses, all current components of Gondwanan savannas and grasslands (such as Paniceae, Andropogoneae, Aristideae, and Chloridoideae) and Euroamerican steppes (Pooideae) would have been derived.

Jacobs *et al.* (1999) compiled a list of indicators of grass-dominated ecosystems among botanical, faunal, and stable carbon isotope records. They state that while the oldest graminoid pollen comes from the "Paleocene" of South America, mammalian dentitions adapted to grazing appear later in Patagonia towards the end of the Eocene, and the oldest floral evidence for grass-dominated ecosystems along with a C_4 isotopic signal date from the Miocene. Since Andreis (1972) and Mazzoni (1979) first described grass phytoliths from Patagonia, much additional work has been done (Zucol *et al.* 1999, 2007; Raigemborn *et al.* 2009). The new work ratifies the presence of grass-dominated ecosystems during the middle Eocene, but also the presence of grass lineages from which C_4 grasses evolved.

Grasses had a Paleocene origin in several different parts of the world (Jacobs *et al.* 1999) corresponding to the first stage of grass evolution (Clayton 1981) and the differentiation of the principal graminoid clades (Anomochloaceae + Pharoideae + Puelioideae + PACCMAD + BEP clades [according Christin *et al.* 2008; Vicentini *et al.* 2008]). Some of these lineages began open-habitat colonization. Different studies (Retallack 2001, 2004; Strömberg 2004, 2005; Strömberg *et al.* 2007a; Christin *et al.* 2008; Vicentini *et al.* 2008) agree in showing that the PACCMAD + BEP clade diverged, and early appearance of C_4 lineages (C_4–NAD) appeared during the Eocene–Oligocene interval. Finally, during the Miocene these different lineages would begin to colonize different habitat types leading eventually to grassland expansion (Barreda and Palazzesi 2007; Palazzesi and Barreda 2007; Strömberg *et al.* 2007b).

Phylogenetic analysis of grasses (Christin *et al.* 2008; Vicentini *et al.* 2008) with several fossil calibration points make the common ancestor of the grass crown clade somewhat more ancient (late Cretaceous), in congruity with the pollen record (Muller 1981, 1984). Vicentini *et al.* (2008) used different kinds of fossil information for their calibration, including grass pollen (Linder 1986), Eocene grass spikelet fossils (Crepet and Feldman 1991), phytolith data from the North American Great Plains (Strömberg 2005), fossil grass epidermal remains from a dinosaur coprolite from India (Prasad *et al.* 2005) for the "Gondwana calibration," and the presence of chloridoid phytoliths (Strömberg 2005), *Dichanthelium* (Thomasson 1978), and *Setaria* (Elias 1942) in Cenozoic sediments of North America. Regardless of the calibration issue, both studies (Christin *et al.* 2008; Vicentini *et al.* 2008) agree in the first appearance of C_4 indicators at about 32 Ma. Given the coincidence of this age with events and perturbations in the Southern Ocean record, their results have engendered a debate about whether atmospheric carbon dioxide concentration was the major factor in their appearance.

Among arboreal elements, palms prevail from the basal Las Flores Formation (Brea *et al.* 2008; Raigemborn *et al.* 2009), through a large part of the Koluel-Kaike Formation (Zucol *et al.* 2007) and into the Gran Barranca Member. Along with the predominance of palms, herbaceous components are also present through this interval. Grass elements are scarce in the Koluel-Kaike Formation although at its top arboreal indicators decrease. This abundance of palms and herbaceous phytoliths (for example the truncated cones) continues in the Gran Barranca Member (GbZI).

Among diagnostic grass phytoliths, danthoniod elements increase from the top of the Koluel-Kaike Formation through GbZIIC. Pooid and chloridoid components occur together consistently through GbZI until GbZIIC. Panicoid components increase in abundance from GbZIIA through GbZIIC and predominate in sample MMZGB956.0, after which aquatic herbs become abundant in GbZIII. Indicator phytoliths suggest the presence of a grassland with the predominance of panicoid grass in GbZIIC times. In the other words, the oldest record of a megathermic grassland is found in GbZIIC of the Gran Barranca Member at Gran Barranca.

In the upper portion of GbZII grasses are the most abundant components of the phytolith assemblages, and chloridoid, danthonoid, and pooid phytoliths are gradually less abundant, together with an increase in the panicoid types.

The top of the sequence (GbZIII) shows similar conditions, with abundance of herbs and associated palms, but the botanical composition indicates more swampy vegetation.

These results agree with those obtained by Vicentini *et al.* (2008) for the appearance of the node G between 35 and 40 Ma with the first C_4–NAD grass lineage (the chloridoid clade). But our results also raise doubts with regard to the appearance of the most typical C_4 grass groups or Panicoideae, since the presence of megathermic grassland with abundant panicoid phytoliths does not necessarily indicate the first presence of C_4 grasslands, as panicoid phytolith morphotypes are present in both C_3 and C_4 (NAD, NADP, and PCK types) species. Nevertheless, our analysis indicates the first or oldest record of the presence of this clade (clade J *sensu* Vicentini *et al.* 2008) at around 39 Ma.

Another topic that should be mentioned in light of these new records is the assumption that C_4 grass origins are principally promoted by low atmospheric CO_2 concentrations (Christin *et al.* 2008; Vicentini *et al.* 2008). During the late Eocene, atmospheric CO_2 concentrations show very high amplitudes (Pagani *et al.* 2005). What might have been the best stage for the appearance of C_4 grass groups, whether conditions of gradually decreasing CO_2 concentration or alternating pulses of high and low concentrations? It would be risky to attempt to answer this question by means of these results, but progress in interdisciplinary study (Christin *et al.* 2008; Vicentini *et al.* 2008) may offer improvement on this complex subject.

Finally, the variation in vegetation described here includes the first record of an ecological niche that herbs began to occupy. The presence of this community type in the Middle Eocene suggests the precocious appearance of an ecological niche that herbivorous mammals would later exploit.

ACKNOWLEDGEMENTS

I want to express my sincere gratitude to M. M. Mazzoni, who invited me (AFZ) to participate in these studies and provided me with the necessary material and information to begin them. We would also like to express our thanks to Luis A. Spalletti (Centro de Investigaciones Geological CONICET–UNLP) for his disinterested collaboration in this work and for facilitating my access to M. M. Mazzoni's field notes, material, and information. The authors are grateful to two anonymous reviewers and to Mirta Arriaga, for the suggestions made in the early reading of this manuscript. We would like to express our thanks to editors for correcting the English text and for critically appraising the manuscript. This study was financed by National Science Foundation (grant EAR 00-87636 to R. F. Kay and R. H. Madden and Agencia Nacional de Promoción Científica y Tecnológica (ANPCyT – PICT 07–13864).

Appendix 22.1 Informal identification acronym of phytolith morphotypes found in Gran Barranca Member samples: brief description, equivalent names, and schematic drawings (Fig. 22.7) used for counting

Point-shaped (= Point-shaped, unciform hair) Ps

Equivalent names

Lithodontium p.p. (Ehrenberg 1841, 1854 emend. Deflandre 1963; Dumitrica 1973; Locker and Martini 1986, 1989); Hook-shaped phytoliths (Baker 1960); Silification of hairs and hair bases (Parry and Smithson 1964); Silicified hairs, p.p. (Parry and Smithson 1966); *Aculeolita* (Bertoldi de Pomar 1971; Parra and Flórez 2001; Zucol and Bonomo 2008: Bonomo *et al.* 2009); *Aculeolithus* Bert. (Taugourdeau-Lantz *et al.* 1976); Point-shape from prickle hair (Kondo 1977); Type II Trichomes (Brown 1984, 1986); Point-shaped class (Kondo and Sase 1986; Kondo *et al.* 1987, 1994); Trichome – hairs and papillae p.p. (Mulholland and Rapp 1992); Point-shaped phytolith (Twiss, 1992); Category 40 – Dermal appendages p.p. (Pearsall and Dinan 1992; Pearsall 2000); Point-shaped (Bardoni *et al.* 1999); G3 – Different types of trichomes, including prickles, hairs, and prickles bases (Runge 1999); Point class p.p. (Hart *et al.* 2000); Point-shaped class (Gallego and Distel 2004; Gallego *et al.* 2004); Class H2 Simple trachoma p.p. (Strömberg 2004). Unciform (ICPNWG 2005); *Aculeolithum* p.p. (Zucol and Brea 2005; Brea *et al.* 2008); Point-shaped (Bremond *et al.* 2004, 2005); Points (Marx *et al.* 2004; Lu *et al.* 2006); NDG p.p. (Strömberg *et al.* 2007b).

ID	Acronym	Brief description
1	Ps01	Point-shaped element originated in an acicular prickle-hair, with pointed and thin basal area and acicular barb.
2	Ps02	Point-shaped element characterized by oblong basal area with a short and recurved barb.
3	Ps03	Point-shaped element with small and circular basal area, big barb in relation to small body.

ID	Acronym	Brief description
4	Ps04	Point-shaped element with lengthened base with short antrorse barb.
5	Ps05	Element originated by the inner silicification of prickle or in an unbarbed prickle, with tear-shaped contour.
6	Ps06	Point-shaped element with prominent central barb, like a shark tooth.
7	Ps07	Point-shaped element with big and elongated basal area, and a small, short, and retrorse barb not standing out of the perimeter of the base.
8	Ps08	Point-shaped element with a cylindrical body ending in a sharp top; hollow or solid body.
9	Ps09	Point-shaped element with the body of prickle fused with a prominent barb giving to the contour aspect of parrot beak or rostrate.
10	Ps10	Point-shaped element with small and isodiametrical basal area and short barb, originating in hook (or crochet cell).

Star-shaped (= Stellate cell phytoliths) Ss

Equivalent names

Stellate trabeculate parenchyma cells (Gordon Gray *et al.* 1978); Stellate shape (ICPNWG 2005); *Asteriolita* sp1. (Bonomo *et al.* 2009; Zucol and Bonomo 2008).

ID	Acronym	Brief description
11	Ss01	Stellate element formed by a body and variable number of projections with flat or circular transverse section.

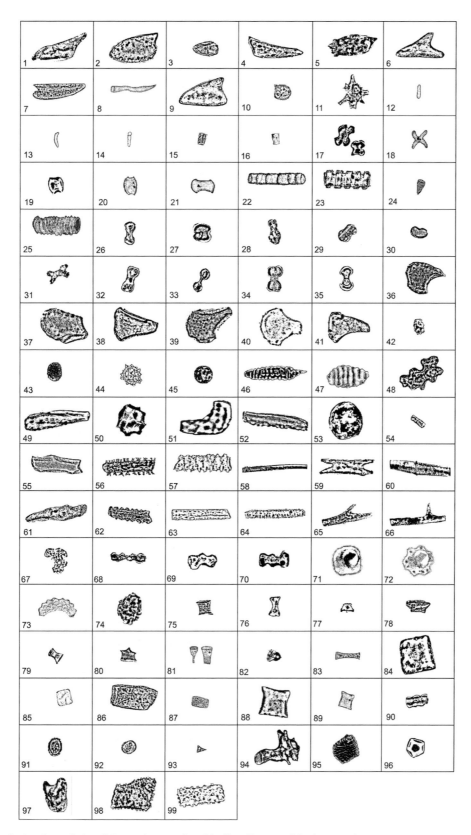

Fig. 22.7. Schematic drawings of phytolith morphotypes found in Gran Barranca Member samples.

Bacillar (= Bacillar or clavate) Bc

Equivalent names

Bacilolita (Bertoldi de Pomar 1971); *Capilusita* (Parra and Flórez 2001); Clavate p.p. (ICPNWG 2005).

ID	Acronym	Brief description
12	Bc03	Bacillary element, with cylindrical straight body and rounded top, less than 40 μm in length; oblong contour shape.
13	Bc04	Bacillary element, with cylindrical recurved body and rounded top; crescent contour shape.
14	Bc05	Bacillary element, with cylindrical body and rounded ends, one of them enlarged; virgate contour shape.

Brief elongate (= Brief prismatic) Be

Equivalent names

Braquiolita (Bertoldi de Pomar 1971; Parra and Flórez, 2001; Zucol and Bonomo 2008; Bonomo *et al.* 2009).

ID	Acronym	Brief description
15	Be01	Brief elongate element, less than 40 μm in length, with undulate margin contour.
16	Be02	Brief elongate element, less than 40 μm in length, with smooth margin contour.

Cross (= Cross-shaped) Cr

Equivalent names

Cross (Twiss *et al.* 1969; Twiss 1992); *Euhalteriolita cruciformata* (Bertoldi de Pomar 1971); Type VIII Crosses (Brown 1984, 1986); Cross (Mulholland and Rapp 1989, 1992; Barboni *et al.* 1999; Bremond *et al.* 2004, 2005); G1 – Short cell phytoliths: 2, Cross-shaped (Runge 1999; Pearsall 2000); Panicoide class p.p. (Gallego and Distel 2004; Gallego *et al.* 2004); Class F5 Cross (Strömberg 2004); Panicoideae p.p. (Lu *et al.* 2006; Piperno 2006); *Halteriolita cruciformata* (Zucol and Bonomo 2008; Bonomo *et al.* 2009).

ID	Acronym	Brief description
17	Cr01	Cross-shaped element with large shank and lobules as wide as long.
18	Cr02	Cross-shaped element with small shank and lobules thin and long.

Saddle (= Battle axe) Sa

Equivalent names

Lithodontium p.p. (Ehrenberg 1841, 1854, emend. Deflandre 1963; Dumitrica 1973; Locker and Martini 1986, 1989); Battle axes with a double edge (Prat 1936, 1948); Chloridoid class (Twiss *et al.* 1969; Kondo and Sase 1986; Kondo *et al.* 1987, 1994); *Doliolita* (Bertoldi de Pomar 1971; Parra and Flórez 2001; Zucol and Bonomo 2008; Bonomo *et al.* 2009); Type IV Saddles p.p. (Brown 1984, 1986); Saddle (Mulholland and Rapp 1989, 1992; Fredlund and Tieszen 1994; Bremond *et al.* 2004, 2005); Chloridoid phytolith (Twiss 1992); Chloridoid: battle axe saddle-shapes (Shulmeister *et al.* 1999; Carter 2000); Chloridoid type: saddle (Bardoni *et al.* 1999); G1 – Short cell phytoliths: 4, Saddle-shaped (Runge 1999); Chloridoide class (Gallego and Distel 2004; Gallego *et al.* 2004); Saddles (Chloridoideae) (Strömberg 2005); Chloridoid (Marx *et al.* 2004; Lu *et al.* 2006); Chloridoideae p.p. and Bambusoideae p.p. (Piperno 2006); PACCAD TOT p.p. (Strömberg *et al.* 2007b).

ID	Acronym	Brief description
19	Sa01	Saddle-shaped element, with width: length ratio 1:1.
20	Sa02	Saddle shaped element, with width: length ratio 1: +1.5.
21	Sa03	Saddle shaped element, with width: length ratio +1.5:1.

Cylindrical vascular element (= Cylindrical sulcate) Ce

Equivalent names

A2 – Elongated phytolith with rings, spiral or point-shaped ornamentation (Runge 1999); *Cylindrita* (Parra and Flórez 2001); Class E5 – Tracheary element p.p. and Class H5 – Monocot tracheid p.p. (Strömberg 2004); Cylindrical sulcate p.p. (ICPNWG 2005); NDG p.p. (Strömberg *et al.* 2007b); MT15-MT18 vascular element phytoliths (Brea *et al.* 2008).

ID	Acronym	Brief description
22	Ce01	Cylindrical element with thin annular thickenings.
23	Ce02	Cylindrical element with thick annular thickenings.
24	Ce03	Conical element with thin annular thickenings.
25	Ce04	Cylindrical element with thin helical thickenings.

Dumbbell (= Bilobate ?) Dm

Equivalent names

Lithomesites (Ehrenberg 1841, 1854, emend. Deflandre 1963; Dumitrica 1973; Locker and Martini 1986, 1989); Panicoid class p.p. (Twiss *et al.* 1969; Kondo and Sase 1986; Kondo *et al.* 1987, 1994); Panicoid form p.p. (Kondo 1977); *Euhalteriolita* p.p. (Bertoldi de Pomar 1971; Zucol and Bonomo 2008; Bonomo *et al.* 2009); *Halteriolithus* Bert. (Taugourdeau-Lantz *et al.* 1976); types VI – Bilobates (Brown 1984, 1986); Sinuous–bilobate/polylobate and Dumbbell – regular/complex (Mulholland and Rapp 1989, 1992); Panicoid phytolith p.p. (Twiss 1992; Lu *et al.* 2006); Convex long dumbbells, Convex short dumbbells, Flat long dumbbells, and Flat short dumbbells (Powers-Jones and Padmore 1993); Simply lobate and Panicoid-type (Fredlund and Tieszen 1994); Dumbbell (Alexandre *et al.* 1997; Bremond *et al.* 2004, 2005); Bilobate (Piperno and Pearsall 1998; ICPNWG 2005; Barboni *et al.* 2007); G1 – Short cell phytoliths: 1, Dumbbell (Runge 1999); Panicoide type, Dumbbell (Barboni *et al.* 1999); Panicoid dumbbell forms p.p. (Shulmeister *et al.* 1999; Carter 2000); Category 30 – Short cells p.p. (Pearsall and Dinan 1992; Pearsall 2000); *Halteriolita* and *Bilobulita* (Parra and Flórez 2001); Panicoid and Dumbbell (Marx *et al.* 2004); Panicoide class p.p. (Gallego and Distel 2004; Gallego *et al.* 2004); Class F4 Bilobate (Strömberg 2004, 2005); Panicoideae p.p., Aristoideae p.p., and Bambusoideae p.p. (Piperno 2006); PACCAD TOT p.p. (Strömberg *et al.* 2007b); *Halterios* (Brea *et al.* 2008).

ID	Acronym	Brief description
26	Dm01	Dumbbell element with two enlarged convex ends or heads, and poorly delimited shank insertion.
27	Dm02	Dumbbell element with botuliform body, and poorly delimited shank insertion.
28	Dm04	Dumbbell element with two enlarged convex ends or heads differently developed like a guitar body, and poorly delimited shank insertion.
29	Dm05	Dumbbell element with two enlarged convex ends or heads, and poorly developed shank constriction, giving an almost ovoid aspect to the outline.
30	Dm06	Dumbbell element with asymmetrical development of the shank side's concavity and two enlarged convex ends or heads.
31	Dm07	Dumbbell element with deeply excavated ends, bilobate and well-delimited shank insertion.
32	Dm08	Dumbbell element with two enlarged convex or straight ends or heads, with two appendages at the end of each of them.
33	Dm09	Dumbbell element with two globular heads, well-delimited shank insertion, and parallel shank sides.
34	Dm10	Dumbbell element with two heads closely cleft, bilobate, well-delimited shank insertion, and parallel shank sides.
35	Dm11	Dumbbell element with two enlarged convex ends or heads, well-delimited shank insertion, and symmetrically concave shank sides.

Fan-shaped (= Cuneiform ?) Fs

Equivalent names

Lithodontium p.p (Ehrenberg 1841, 1854, emend. Deflandre 1963; Dumitrica 1973; Locker and Martini 1986, 1989); *Flabelolita* (Bertoldi de Pomar 1971; Zucol and Bonomo 2008; Bonomo *et al.* 2009); *Flabelolithus* Bert. (Taugourdeau-Lantz *et al.* 1976); Fan-shaped from bulliform cell (Kondo 1977); Silicified bulliform cells (Parry and Smithson 1958, 1964); Fan-shaped class (Kondo and Sase 1986; Kondo *et al.* 1987, 1994); Fan-shaped phytolith (Twiss 1992); Bulliform cells – Enlarged thin-walled epidermal cells: Keystone shaped (Mulholland and Rapp 1992); Category 50 – Bulliform cells p.p. (Pearsall and Dinan 1992; Pearsall 2000); Motor cells silica bodies (Fujiwara 1993); Bulliform: fan-shaped (Shulmeister *et al.* 1999; Carter 2000); Fan-shaped (Bardoni *et al.* 1999); G2 – Bulliform phytoliths: 1, Fan-shaped (Runge 1999); *Flabellulita* (Parra and Flórez 2001); Fans (Marx *et al.* 2004); Fan-shaped and Polyhedrical class p.p. (Gallego and Distel 2004; Gallego *et al.* 2004); Class G Bulliform p.p. (Strömberg 2004); *Flabelolithum* p.p. (Zucol and Brea 2005; Brea *et al.* 2008); Fan-shaped (Alexandre *et al.* 1997; Barboni *et al.* 1999; Bremond *et al.* 2004, 2005; Lu *et al.* 2006); Cuneiform bulliform cell (Barboni *et al.* 2007); NDG p.p. (Strömberg *et al.* 2007b).

ID	Acronym	Brief description
36	Fs01	Fan-shaped element with asymmetrical body as consequence of the presence of concave and convex lateral sides of the basal sector of the fan.
37	Fs02	Fan-shaped element with campanulate or bell-shaped fan and basal sector less 1/3 of the fan length.

ID	Acronym	Brief description
38	Fs03	Fan-shaped element with cuneiform contour and straight lateral sides of the basal sector.
39	Fs04	Fan-shaped element with convex, semicircular fan end and a basal sector thin and three times longer than the fan length.
40	Fs05	Fan-shaped element with width: length ratio 1:1, semicircular fan end, and convex lateral sides of the basal sector.
41	Fs07	Fan-shaped element with asymmetrical body as consequence of the presence of straight and convex lateral sides of the basal sector of the fan.

Globular (= Spherical) Gl

Equivalent names

Lithodontium p.p. (Ehrenberg 1841, 1854, emend. Deflandre 1963; Dumitrica 1973; Locker and Martini 1986, 1989); Spherical (Tomlinson 1961); *Globulolita* (Bertoldi de Pomar 1971; Parra and Flórez 2001; Zucol and Bonomo 2008; Bonomo *et al.* 2009); *Globulolithus* Bert. (Taugourdeau-Lantz *et al.* 1976); Irregular sphere or oval shaped opals (Kondo and Peason 1981); Spherical smooth class p.p., Spherical verrucose class p.p., Spherical nodulose class p.p., and Spherical spinulose class p.p. (Kondo *et al.* 1994); Class V.B – Spherical to aspherical (Piperno 1988); B – Spherical phytoliths p.p. (Runge 1999; Bremond *et al.* 2004, 2005); Circular crenate and circular rugose (Barboni *et al.* 1999); Category 80 – Stegmata and other spherical or spheroidal bodies (Pearsall and Dinan 1992; Pearsall 2000); Spherical shaped (Carter 2000); *Globulolithum* (Zucol and Brea 2005); Class C Palm-type p.p., Class D1 Spherical bodies (Strömberg 2004); FI TOT p.p. (Strömberg *et al.* 2007b); MT1–MT6 globular phytoliths (Brea *et al.* 2008).

ID	Acronym	Brief description
42	Gl02	Globular element, prolate to globose with echinate surface, less than 12 μm in diameter.
43	Gl03	Globular element, prolate to globose with smooth or lightly rough surface, less than 12 μm in diameter.
44	Gl04	Globular element, globose with echinate surface, less than 12 μm in diameter.
45	Gl05	Globular element, globose with smooth or lightly rough surface, less than 12 μm in diameter.

Fusiform Fu

Equivalent names

Lithostylidium p.p. (Ehrenberg 1841, 1854, emend. Deflandre 1963; Dumitrica 1973; Locker and Martini 1986, 1989); *Longolita* p.p. (Bertoldi de Pomar 1971; Parra and Flórez 2001); *Longolithus* Bert. (Taugourdeau-Lantz *et al.* 1976); MT28 Fusiformes (Brea *et al.* 2008).

ID	Acronym	Brief description
46	Fu01	Element with clavate body with narrow thin striate ornamentation.
47	Fu03	Element with navicular body with thick striate ornamentation, corrugate.

Lobular (= Jigzaw) Lb

Equivalent names

Jigsaw shaped (Kondo 1977; Carter 2000); Puzzle piece-shaped (Runge 1999) [as silicified skeletons] Class E1 Dicotyledons epidermis (Strömberg 2004); *Lobulado* sp1 (Zucol and Bonomo 2008; Bonomo *et al.* 2009); MT9 Lobulados (Brea *et al.* 2008).

ID	Acronym	Brief description
48	Lb01	Jigsaw-puzzle element with variable lobules.

Multicavate (= Faceted) Mc

Equivalent names

VI.A – Multifaceted bodies (Piperno 1988); Category 110 – Sclereids p.p. (Pearsall and Dinan 1992; Pearsall 2000); Multi-faced and Sclereid class p.p. (Kondo *et al.* 1994); B4 – Spherical faceted phytoliths p.p., A5 – Slightly faceted phytoliths, and A4 – Elongate faceted phytoliths p.p. (Runge 1999); *Clavaetlita* p.p. (Parra and Flórez 2001); Facetate (ICPNWG 2005); NDG p.p. (Strömberg *et al.* 2007b); MT7–MT8 Faceted phytoliths (Brea *et al.* 2008).

ID	Acronym	Brief description
49	Mc01	Clavate element, more than 40 μm in length and faceted surface.
50	Mc02	Spherical to subspherical element, more or less isodiametric, with faceted surface and more than 35 μm in diameter in the spherical view.
51	Mc03	Element with elongate body, double angle in their middle section and faceted surface.

ID	Acronym	Brief description
52	Mc04	Element with elongate body, more than 40 μm in length, with faceted surface.

Macro-globular (= Macro-spherical) Mg

Equivalent names

V.B.4.b – Spherical to aspherical, with smooth surface, more than 12 μm in diameter (Piperno 1988); Spherical smooth class p.p. (Kondo *et al.* 1994); B1 – Spherical phytoliths with smooth or slightly rough surface p.p. (Runge 1999); Globular smooth p.p. (ICPNWG 2005).

ID	Acronym	Brief description
53	Mg01	Globular element, globose with smooth or lightly rough surface, more than 12 μm in diameter.

Micro-elongate (= Micro-elongate) Me

Equivalent names

Lithodontium p.p. (Ehrenberg 1841, 1854, emend. Deflandre 1963; Dumitrica 1973; Locker and Martini 1986, 1989); *Braquiolita* p.p. (Bertoldi de Pomar 1971; Zucol and Bonomo 2008); *Braquiolithus* Bert. (Taugourdeau-Lantz *et al.* 1976); Rectangle p.p. (Mulholland and Rapp 1992); Elongate smooth p.p. (ICPNWG 2005); Micro-prismatolita (Bonomo *et al.* 2009).

ID	Acronym	Brief description
54	Me01	Prismatic elongate element, with smooth edges, less than 20 μm in length.

Elongate (= Parallelepipedal or Prismatic) El

Equivalent names

Lithostylidium p.p. (Ehrenberg 1841, 1854, emend. Deflandre 1963; Dumitrica 1973; Locker and Martini 1986, 1989); Elongate class (Twiss *et al.* 1969; Kondo and Sase 1986; Kondo *et al.* 1987, 1994); *Prismatolita* (Bertoldi de Pomar 1971; Parra and Flórez 2001; Zucol and Bonomo 2008; Bonomo *et al.* 2009); *Prismatolithus* Bert. (Taugourdeau-Lantz *et al.* 1976); Elongate (Kondo 1977); Type I – Plates (Brown 1984, 1986); Elongate phytolith (Twiss 1992); Category 10 – Epidermal quadrilaterals (Pearsall and Dinan 1992; Pearsall 2000); A – Elongate phytoliths p.p. and G4 – Long cell phytoliths (Runge 1999); Elongate: rectangular (Shulmeister *et al.* 1999; Carter 2000); Elongate (Bardoni

et al. 1999; Marx *et al.* 2004; Lu *et al.* 2006); Elongate class (Gallego and Distel 2004; Gallego *et al.* 2004); Class H1 Elongate p.p. (Strömberg 2004); Elongate long cell (ICPNWG 2005); Elongate types (Bremond *et al.* 2004, 2005); *Macroprismatolithum* (Zucol and Brea 2005; Brea *et al.* 2008); NDG p.p. and NDO p.p. (Strömberg *et al.* 2007b).

ID	Acronym	Brief description
55	El01	Elongate element, more than 30 μm in length, generally smooth margin contour and a variation of their transversal section length generating margins in angles with asymmetrical development.
56	El02	Elongate element, more than 30 μm in length, dendriform margin contour and parallel lateral sides.
57	El03	Elongate element, more than 30 μm in length, denticulate margin contour and parallel lateral sides.
58	El04	Elongate element, with smooth or slightly undulating margin contour and parallel lateral sides, more than 30 μm in length, very long, with width: length ratio 1: +5.
59	El05	Elongate element, with smooth or slightly undulating margin contour, parallel lateral sides, and one or two concave ends.
60	El06	Elongate element, more than 30 μm in length, generally smooth margin contour and a variation of their transversal section length generating margins in angles with symmetrical development.
61	El07	Elongate element, more than 30 μm in length, generally smooth margin contour and a gradual bulge in the middle transversal section of their body.
62	El08	Elongate element, more than 30 μm in length, with undulating margin contour and parallel lateral sides.
63	El09	Elongate element, more than 30 μm in length, with smooth margin contour and parallel lateral sides.
64	El10	Elongate element, more than 30 μm in length, with serrate or serrulate margin contour and parallel lateral sides.

Elongated branched (= Y-shaped) Eb

Equivalent names

Lithostylidium p.p. (Ehrenberg 1841, 1854, emend. Deflandre 1963; Dumitrica 1973; Locker and Martini 1986, 1989); *Nasoprismatolita* (Bertoldi de Pomar 1971); *Prismatolithus*

Bert. p.p. (Taugourdeau-Lantz *et al.* 1976); Branched or Y-shaped phytolith (Kondo 1977); *Prismatolita* p.p. (Zucol and Bonomo 2008; Bonomo *et al.*, 2009).

ID	Acronym	Brief description
65	Eb01	Prismatic elongate element, branched at an acute angle to the principal axis.
66	Eb03	Prismatic elongate element, branched at right angles to the principal axis.

Scrobiluar (= Nuxolita) Nx

Equivalent names

Nuxolita (Bertoldi de Pomar 1971; Zucol and Bonomo 2008; Bonomo *et al.* 2009); *Nuxolithus* Bert. p.p. (Taugourdeau-Lantz *et al.* 1976).

ID	Acronym	Brief description
67	Nx01	Scrobicular element with bowl or "cuenco" shape.

Polylobate (= Dumbbell complex) Pl

Equivalent names

Lithomesites (Ehrenberg 1841, 1854, emend. Deflandre 1963; Dumitrica 1973; Locker and Martini 1986, 1989); *Plurihalteriolita* (Bertoldi de Pomar 1971); Panicoid class p.p. (Twiss *et al.* 1969; Kondo and Sase 1986; Kondo *et al.* 1987, 1994); Panicoid form p.p. (Kondo 1977); Sinuous – bilobate/polylobate and Dumbbell – regular/complex (Mulholland and Rapp 1989, 1992); Panicoid phytolith p.p. (Twiss 1992); G1 – Short cell phytoliths: 3, Polylobate (Runge 1999); Complex dumbbell shaped (Carter 2000); Panicoide class p.p. (Gallego and Distel 2004; Gallego *et al.* 2004); Polylobate (Bremond *et al.* 2004, 2005; ICPNWG 2005); PACCAD TOT p.p. (Strömberg *et al.* 2007b); *Halteriolita* p.p. (Zucol and Bonomo 2008; Bonomo *et al.* 2009).

ID	Acronym	Brief description
68	Pl01	Polylobate element with more than 2 shanks and 4 or more lobules arranged symmetrically.
69	Pl02	Polylobate element with more than 2 shanks and 4 or more lobules arranged asymmetrically.
70	Pl03	Polylobate element with 2 shanks and 3 lobules arranged symmetrically in each side.

Conical (= Papillae conical) Co

Equivalent names

Cyperaceous type (Mehra and Sharma 1965); Cones (Metcalfe 1971; Ollendorf 1992; Wallis 2003); *Pileolita* (Bertoldi de Pomar 1971); *Pileolithus* Bert. (Taugourdeau-Lantz *et al.* 1976); Conical-shaped (Piperno 1985); Hat-shaped (Piperno 1988); D5 – Conical to hat shaped, G7 – Conical phytolith, and G8 – Cyperaceae achene phytoliths (Runge 1999); Class A Sedge-type (Strömberg 2004); AQ p.p. (Strömberg *et al.* 2007b).

ID	Acronym	Brief description
71	Co01	Conical element with single apex and smooth surface, basal shape more or less circular with smooth basal margin.
72	Co02	Conical element with single apex, basal shape scutiform with undulating basal margin and peripheral satellite spines.
73	Co06	Conical element without single apex, panto or stephanomicroechinate and irregular basal shape.

Proteiform (= Irregular ovoid) Pr

Equivalent names

Proteolita (Bertoldi de Pomar 1971; Zucol and Bonomo 2008; Bonomo *et al.* 2009); *Proteolithus* Bert. (Taugourdeau-Lantz *et al.* 1976).

ID	Acronym	Brief description
74	Pr01	Proteiform element, with amoeboid to ovoid contour of up to 100 μm, with surface projections, in many cases with bifid ends.

Truncated cone (= Spool-shaped) Ct

Equivalent names

Lithodontium p.p. (Ehrenberg 1841, 1854, emend. Deflandre 1963; Dumitrica 1973; Locker and Martini 1986, 1989); *Estrobilolita* (Bertoldi de Pomar 1971; Parra and Flórez 2001; Zucol and Bonomo 2008; Bonomo *et al.* 2009); *Estro-bilolithus* Bert. (Taugourdeau-Lantz *et al.* 1976); Truncated cone shaped class (Kondo and Sase 1986); Chionochloid class (Kondo *et al.* 1994; Carter 2000); Chionochloid, spool shaped (Shulmeister *et al.* 1999); Rondel p.p. (Barboni *et al.* 1999, Strömberg 2004; ICPNWG 2005); Chionochloid and Truncated cones (Marx *et al.* 2004); Conical (Bremond *et al.* 2004, 2005); Rondel short cell (Barboni *et al.* 2007); Pooideae p.p. and Bambusoideae p.p. (Piperno 2006); POOID-ND p.p. (Strömberg *et al.* 2007b); Cónicos (Brea *et al.* 2008).

ID	Acronym	Brief description
75	Ct01	Truncated cone element, without demarcated shank, height: basal diameter ratio is 1:1 and major basal diameter/minor basal diameter ratio is 1.5:1.
76	Ct02	Truncated cone element, with demarcated shank, height: basal diameter ratio is more 3–2:1 and major basal diameter/minor basal diameter ratio is 1.5:1.
77	Ct03	Truncated cone element, without demarcated shank, flattened, height: basal diameter ratio is 1:2 and major basal diameter/minor basal diameter ratio is 1.5:1.
78	Ct04	Truncated cone element, without demarcated shank, very flattened, height: basal diameter ratio is 1:3 and major basal diameter/minor basal diameter ratio is 2:1.
79	Ct05	Truncated cone element, with lightly demarcated shank, height: basal diameter ratio is more 3–2:1 and major basal diameter/minor basal diameter ratio is 3:2–1.
80	Ct06	Truncated cone element, without demarcated shank, height: basal diameter ratio is 1:1 and major basal diameter/minor basal diameter ratio is 1.5:1; one basal side plane or concave and the other one carinate.
81	Ct07	Truncated cone element, with one basal side flattened.
82	Ct08	Truncated cone element, with one basal side acuminate in funnel shape.
83	Ct10	Truncated cone element, with demarcated shank, height: basal diameter ratio is more than 3:1, with plane or concave basal sides.

Polyhedral bulliforms (= Parallepipedal) Po or sPo

Equivalent names

G2 – Bulliform phytoliths: 2, Cubical, elongated (Runge 1999); Fan-shaped and Polyhedral class p.p. (Gallego and Distel 2004; Gallego *et al.* 2004); Class G Bulliform p.p. (Strömberg 2004); Parallepipedal (ICPNWG 2005); *Polié drico* sp1 (Zucol and Bonomo 2008); MT29 Poliédricos (Brea *et al.* 2008).

ID	Acronym	Brief description
84	Po01	Bulliform polyhedrical element with straight lateral edge of the major side, with width: length ratio 1:1 and more than 30 µm in length.
85	sPo02	Bulliform polyhedrical element with straight lateral edge of the major side, with width: length ratio 1:1 and less than 30 µm in length.
86	Po03	Bulliform polyhedrical element with straight lateral edge of the major side, with width: length ratio 1:2 and more than 30 µm in length.
87	sPo04	Bulliform polyhedrical element with straight lateral edge of the major side, with width: length ratio 1:2 and less than 30 µm in length.
88	Po05	Bulliform polyhedrical element with concave lateral edge of the major side, with width: length ratio 1:1–2 and more than 30 µm in length.
89	sPo06	Bulliform polyhedrical element with concave lateral edge of the major side, with width: length ratio 1:1–2 and less than 30 µm in length.

Sinuate trapezoid (= Trapezoid sinuate) Ts

Equivalent names

Boat shaped (Parry and Smithson 1964); Festucoid boat (Kondo *et al.* 1994; Zucol and Bonomo 2008; Bonomo *et al.* 2009); Elongate boat shaped with saw-tooth edges (Carter 2000); Class F3 Crenate (Strömberg 2004, 2005); Trapezoid sinuate (ICPNWG 2005); Festucoid p.p (Marx *et al.* 2004; Lu *et al.* 2006); Pooideae p.p. (Piperno 2006); POOID-D (Strömberg *et al.* 2007b).

ID	Acronym	Brief description
90	Ts01	Trapezoidal short element with crenate contour in one surface which finishes off in an elevated edge like a boat.

Rondel (= Rondel) Rn

Equivalent names

Festucoid class p.p. (Twiss *et al.* 1969; Kondo and Sase 1986; Kondo *et al.* 1987, 1994); Pooid (Festucoid) class p.p. (Twiss 1992); Rondel (Mulholland and Rapp 1992); Crenate (Fredlund and Tieszen 1994); G1 – Short cell phytoliths: 5,

Circular (Runge 1999); Rondel p.p. (Barboni *et al.* 1999); *Elipsoidita* (Parra and Flórez 2001); Class F2 Rondel (Strömberg 2004); Pooide class p.p. (Gallego and Distel 2004; Gallego *et al.* 2004); Festucoid p.p (Marx *et al.* 2004; Lu *et al.* 2006); Pooideae p.p. (Piperno 2006).

ID	Acronym	Brief description
91	Rn01	Short cylindrical element or narrowly conical and transverse section slightly elliptical.
92	Rn02	Short cylindrical element or narrowly conical and circular transverse section.

Triangular (= Triangular) Tr

Equivalent names

Triangle (Mulholland and Rapp 1992); Triangular (Zucol and Bonomo 2008; Bonomo *et al.* 2009); Triangulita (Parra and Flórez 2001).

ID	Acronym	Brief description
93	Tr01	Element with triangular contour, generally small, less 10–15 μm length of each side.

Structured lobular (= Jigsaw ?) sLb

ID	Acronym	Brief description
94	sLb02	Element with lobular rounded projection, arranged in different angles in reference to their axes.

Short tracheid (= Irregular body with striate ornamentation) Ta

Equivalent names

Tracheid p.p. (Kondo *et al.* 1994); Irregular body with striate ornamentation (Wallis 2003); MT13 Cuerpos irregulares con ornamentaciones estriadas (Brea *et al.* 2008).

ID	Acronym	Brief description
95	Ta01	Element with irregular body and striate surface ornamentation.

Polygonal plate (= Polygonal plate) Pg

Equivalent name

Pentagon (Mulholland and Rapp 1992).

ID	Acronym	Brief description
96	Pg01	Flat element with straight sides and pentagonal or hexagonal ontline, smooth homogeneous body or slightly slimmed centrally.

Calyptras shaped (= Root cap shaped) Ca

ID	Acronym	Brief description
97	Ca01	Element that looks like a root cap; or calyptra-shaped.

Plate papillose (= Laminar papillae plate) Lp

ID	Acronym	Brief description
98	Lp01	Laminar element with cylindrical protuberance (papillose type) and thin body.

Elongate papillose (= Prismatic papillae) Pp

Equivalent names

VI.I. Epidermal with a faintly stippled surface and small protuberances, Echinate platelets (Bozarth 1992): MT12 Cuerpos prismáticos laminares de superficie columelada o equinada (Brea *et al.* 2008).

ID	Acronym	Brief description
99	Pp01	Elongate element with cylindrical protuberance (pit refilled aspect). Difficult to define their prismatic body.

REFERENCES

Alexandre, A., J. D. Meunier, A. M. Lezine, A. Vicens, and D. Schwartz 1997. Phytoliths indicators of grasslands dynamics during the late Holocene in intertropical Africa. *Palaeogeography, Palaeoclimatology, Palaeoecology,* **136**, 213–229.

Andreis, R. R. 1972. Paleosuelos de la Formación Musters (Eoceno Medio), Laguna del Mate, Prov. de Chubut. Rep. Argentina. *Revista Asociación Argentina de Mineralogía, Petrología y Sedimentología,* **3**, 91–97.

Archangelsky, S. 1973. Palinología del Paleoceno de Chubut. I. Descripciones sistemáticas. *Ameghiniana,* **10**, 339–399.

Baker, G. 1960. Hook-shaped opal phytoliths in the epidermal cells of oats. *Australian Journal of Botany*, **8**, 69–74.

Barboni, D., R. Bonnfille, A. Alexandre, and J. D. Meunier 1999. Phytoliths as paleoenvironmental indicators, West Side Middle Awash Valley, Ethiopia. *Palaeogeography, Palaeoclimatology, Palaeoecology*, **152**, 87–100.

Barboni, D., L. Bremond, and R. Bonnefille 2007. Comparative study of modern phytoliths assemblages from inter-tropical Africa. *Palaeogeography, Palaeoclimatology, Palaeoecology*, **246**, 454–470.

Barreda, V. and L. Palazzesi 2007. Patagonian vegetation turnovers during the Paleogene–early Neogene: origin of arid-adapted floras. *Botanical Review*, **73**, 31–50.

Bertoldi de Pomar, H. 1971. Ensayo de clasificación morfológica de los silicofitolitos. *Ameghiniana*, **8**, 317–328.

Bonomo, M., A. F. Zucol, B. Gutiérrez Téllez, A. Coradeghini, and M. S. Vigna 2009. Late Holocene palaeoenvironments of the Nutria Mansa 1 archaeological site, Argentina. *Journal of Paleolimnology*, **41**, 273–296.

Bozarth, S. R. 1987. Diagnostic opal phytolith from rinds of selected *Cucurbita* species. *American Antiquity*, **52**, 607–615.

Bozarth, S. R. 1992. Classification of opal phytoliths formed in selected dicotyledons native to the Great Plains. In Rapp, G. and Mulholland, S. (eds.), *Phytolith Systematics: Emerging Issues*. New York: Plenum Press, pp. 193–214.

Brea, M., A. F. Zucol, M. S. Raigemborn, and S. Matheos 2008. Reconstrucción de paleocomunidades arbóreas mediante análisis fitolíticos en sedimentos del Paleoceno superior-Eoceno? (Formación Las Flores), Chubut, Argentina. In Korstanje, A. and Babot, P. (eds.), *Interdisciplinary Nuances in Phytolith and Other Microfossil Studies*. 91–108. Oxford, UK: British Archaeological Reports.

Bremond, L., A. Alexandre, E. Véla, and J. Guiot 2004. Advantages and disadvantages of phytolith analysis for the reconstruction of Mediterranean vegetation: an assessment based on modern phytolith, pollen and botanical data (Luberon, France). *Review of Paleobotany and Palynology*, **129**, 213–228.

Bremond, L., A. Alexandre, C. Hély, and J. Guiot 2005. A phytoliths index as a proxy of tree cover density in tropical areas: calibration with leaf area index along a forest–savanna transect in southeastern Cameroon. *Global and Planetary Changes*, **45**, 277–293.

Brown, D. A. 1984. Prospects and limits of a phytolith key for grasses in the Central United States. *Journal of Archaeological Science*, **11**, 345–368.

Brown, D. A. 1986. Taxonomy of Midcontinent grasslands phytolith key. In Rovner, I. (ed.), *Plant Opal Phytolith Analysis in Archaeology and Paleoecology*, Occasional Paper no.1. Raleigh, NC: North Carolina State University Press, pp. 67–85.

Carter, J. A. 2000. Phytoliths from loess in Southland, New Zealand. *New Zealand Journal of Botany*, **38**, 325–332.

Christin, P.-A., G. Besnard, E. Samaritani, M. R. Duvall, T. R. Hodkinson, V. Savolainen, and N. Salamin 2008. Oligocene CO_2 decline promoted C_4 photosynthesis in grasses. *Current Biology*, **18**, 37–43.

Cifelli, R. L. 1985. South American ungulate evolution and extinction. In Stehli, F. G. and Webb, S. D. (eds.), *The Great American Biotic Interchange*. New York: Plenum Press, pp. 249–266.

Clayton, W. D. 1981. Evolution and distribution of grasses. *Annals of the Missouri Botanical Garden*, **68**, 5–14.

Coradeghini, A. and M. S. Vigna 2001. Flora de quistes crisofíceos fósiles en sedimentos recientes de Mallín Book, Río Negro (Argentina). *Revista Española de Micropaleontología*, **33**, 163–181.

Crepet, W. L. and G. D. Feldman 1991. The earliest remains of grasses in the fossil record. *American Journal of Botany*, **78**, 1010–1014.

Deflandre, G. 1963. Les phytolithaires (Ehrenberg): nature et signification micropaléontologique, pédologique et géologique. *Protoplasma*, **57**, 234–259.

Dumitrica, P. 1973. Phytolitharia. *Initial Reports of the Deep Sea Drilling Project*, **13**, 940–943.

Ehrenberg, C. G. 1841. Über verbreitung und einfluss des mikroskopischen lebens in Süd und Nordamerika. *Monatsbericht der Koiglich Preussischen Akademie der Wissenschaften*, 139–144.

Ehrenberg, C. G. 1854. *Mikrogeologie*, vol. I, text. vol. II, atlas. Leipzig: Leopold Voss.

Elias, M. K. 1942. *Tertiary Prairie Grasses and Other Herbs from the High Plains*, Special Paper no. 41. Boulder, CO: Geological Society of America.

Fredlund, G. G. and L. T. Tieszen 1994. Modern phytoliths assemblages from North American Great Plains. *Journal of Biogeography*, **21**, 321–335.

Fujiwara, H. 1993. Research into the history of rice cultivation using plant opal analysis. In Pearsall, D. M. and Piperno, D. R. (eds.), *Current Research in Phytolith Analysis: Applications in Archaeology and Paleoecology*, Research Paper no. 10. Philadelphia, PA: Museum Applied Science Center for Archaeology, pp. 147–158.

Gallego, L. and R. A. Distel 2004. Phytolith assemblages in grasses native to central Argentina. *Annals of Botany*, **94**, 1–10.

Gallego, L., R. A. Distel, R. Camina, and R. M. Rodríguez Iglesias 2004. Soil phytoliths as evidence for species replacement in grazed rangelands of central Argentina. *Ecography*, **27**, 1–8.

Gordon-Gray, K. D., L. van Laren, and V. Bandu 1978. Silica deposits in *Rhynchospora* species (Cyperaceae). *Proceedings of the Electron Microscopy Society of Southern Africa*, **8**, 83–84.

Grass Phylogeny Working Group 2001. Phylogeny and subfamilial classification of the grasses (Poaceae). *Annals of the Missouri Botanical Garden*, **88**, 373–457.

Hammer, Ø., D. A. T. Harper, and P. D. Ryan 2007. PAST - PAlaeontological STatistics, version 1.75. Available at http://palaeo-electronica.org/2001_1/past/

Hart, D. M., C. Lentfer, L. A. Wallis, and D. Bowdery 2000. A universal phytolith key: point class. *Abstracts III International Meeting on Phytolith Research*, 13–14.

ICPNWG. 2005. International Code for Phytolith Nomenclature 1.0. *Annals of Botany*, **96**, 253–260. doi:10.1093/aob/mci172.

Jacobs, B. F., J. D. Kingston, and L. L. Jacobs 1999. The origin of grass-dominated ecosystems. *Annals of the Missouri Botanical Garden*, **86**, 590–643.

Kay, R. F., R. H. Madden, M. G. Vucetich, A. A. Carlini, M. M. Mazzoni, G. H. Ré, M. Heizler, and H. Sandeman 1999. Revised geochronology of the Casamayoran South American Land Mammals Age: climatic and biotic implications. *Proceeding of the National Academy of Sciences USA*, **96**, 13 235–13 240.

Kealhofer, L. and D. R. Piperno 1998. *Opal Phytoliths in South-East Asian Flora*, Smithsonian Contributions to Botany no. 88. Washington, DC: Smithsonian Institution Press.

Kondo, R. 1977. Opal phytoliths: inorganic, biogenic particles in plants and soils. *Japan Agricultural Research Quarterly*, **11**, 198–203.

Kondo, R. and T. Pearson 1981. Opal phytoliths in tree leaves. II. Opal phytolith in dicotyledon angiosperm tree leaves. *Research Bulletin of Obihiro University Series*, 1, **12**, 217–230.

Kondo, R. and T. Sase 1986. Opal phytoliths, their nature and application. *Daiyonki Kenkyu*, **25**, 31–63.

Kondo, R., T. Sase, and Y. Kato 1987. Opal phytolith analysis of andisols with regard to interpretation of paleovegetation. *Proceedings IX International Soil Classification Workshop*, 520–534.

Kondo, R., C. Childs, and I. Atkinson 1994. *Opal Phytoliths of New Zealand*. Lincoln, NZ: Manaaki Whenua Press.

Linder, H. P. 1986. The evolutionary history of the Poales/Restionales: a hypothesis. *Kew Bulletin*, **42**, 297–318.

Locker, S. and E. Martini 1986. Phytoliths from the Southwest Pacific, Site 591. *Initial Reports of the Deep Sea Drilling Project*, **90**, 1079–1084.

Locker, S. and E. Martini 1989. Phytoliths at DSDP Site 591 in the southwest Pacific and the aridification of Australia. *Geologische Rundschau*, **78**, 1165–1172.

Lu, H. Y., N. Q. Wu, X. D. Yang, H. Jiang, K. Liu, and T. S. Liu 2006. Phytoliths as quantitative indicators for the reconstruction of past environmental conditions in China. I. Phytolith-based transfer functions. *Quaternary Science Reviews*, **25**, 945–959.

Marx, R., E. L. Daphne, K. M. Lloyd, and W. G. Lee 2004. Phytolith morphology and biogenic silica concentrations and abundance in leaves of *Chionochloa* (Danthonieae) and *Festuca* (Poeae) in New Zealand. *New Zealand Journal of Botany*, **442**, 677–691.

Mazzoni, M. M. 1979. Contribución al conocimiento petrográfico de la Formación Sarmiento, barranca S del lago Colhué Huapí, provincia del Chubut. *Revista Asociación Argentina de Mineralogía, Petrología y Sedimentología*, **10**(3–4), 33–53.

Mehra, P. N. and O. P. Sharma 1965. Epidermal silica cells in the Cyperaceae. *Botanical Gazette*, **126**, 53–58.

Metcalfe, C. R. 1971. *Anatomy of Monocotyledons* vol. V, *Cyperaceae*. Oxford, UK: Clarendon Press.

Moreno, C. E. 2001. *Métodos para medir la biodiversidad*, Manuales y Tesis no. 1. Zaragoza, Spain: Sociedad Entomológica Aragonesa.

Mulholland, S. M. and G. Rapp 1989. Characterization of grass phytoliths for archaeological analysis. *Materials Research Society Bulletin*, **14**(3), 36–39.

Mulholland, S. M. and G. Rapp 1992. A morphological classification of grass silica-bodies. In Rapp, G. and Mulholland, S. (eds.), *Phytolith Systematics: Emerging Issues*. New York: Plenum Press, pp. 65–89.

Muller, J. 1981. Fossil pollen record extant angiosperms. *Botanical Review*, **47**, 1–142.

Muller, J. 1984. Significance of fossil pollen for angiosperm history. *Annals of the Missouri Botanical Garden*, **71**, 419–443.

Nalepka, D. and A. Walanus 2003. Data processing in pollen analysis. *Acta Palaeobotanica*, **43**, 125–134.

Ollendorf, A. L. 1992. Toward a classification scheme of sedge (Cyperaceae) phytoliths. In Rapp, G. and Mulholland, S. (eds.), *Phytolith Systematics: Emerging Issues*. New York: Plenum Press, pp. 91–111.

Pagani, M., J. C. Zachos, K. H. Freeman, B. Tipple, and S. Bohaty 2005. Marked decline in atmospheric carbon dioxide concentrations during the Paleocene. *Science*, **309**, 600–603.

Palazzesi, L. and V. Barreda 2007. Major vegetation trends in the Tertiary of Patagonia (Argentina): a qualitative paleoclimatic approach based on palynological evidence. *Flora*, **202**, 328–337.

Parra, L. N. S. and M. T. M. Flórez 2001. Propuesta de clasificación morfológica para los Fitolitos altoandinos colombianos. *Crónica forestal y del medio ambiente*, **16**, 35–66.

Parry, D. W. and F. Smithson 1958. Silicification of bulliform cells in grasses. *Nature*, **181**, 1549–1550.

Parry, D. W. and F. Smithson 1964. Types of opaline silica depositions in the leaves of British grasses. *Annals of Botany*, **28**, 169–185.

Parry, D. W. and F. Smithson 1966. Opaline silica in the inflorescences of some British grasses and cereals. *Annals of Botany*, **30**, 524–538.

Pearsall, D. M. 1993. Contributions of phytolith analysis for reconstructing subsistence: examples from research in Ecuador. In Pearsall, D. M. and Piperno, D. R. (eds.), *Current Research in Phytolith Analysis: Applications in Archaeology and Paleoecology*, Research Paper no. 10. Philadelphia, PA: Museum Applied Science Center for Archaeology, pp. 109–122.

Pearsall, D. M. 2000. *Paleoethnobotany: A Handbook of Procedures*, 2nd edn. San Diego, CA: Academic Press.

Pearsall, D. M. 2006. *Phytoliths in the Flora of Ecuador: The University of Missouri Online Phytolith Database*. Available at www.missouri.edu/~phyto/. [With contributions by A. Biddle, Dr. K. Chandler-Ezell, S. Collins, N. Duncan, S. Stewart, C. Vientimilla, Zhijun Zhao, and B. Grimm, page designer and editor.]

Pearsall, D. M. and E. H. Dinan 1992. Developing a phytolith classification system. In Rapp, G. and Mulholland, S. (eds.), *Phytolith Systematics: Emerging Issues*. New York: Plenum Press, pp. 37–64.

Piperno, D. R. 1985. Phytolith analysis and tropical paleo-ecology: production and taxonomic significance of siliceous form in

New World plants domesticated and wild species. *Review of Palaeobotany and Palynology*, **45**, 185–228.

Piperno, D. R. 1988. *Phytolith Analysis: An Archeological and Geological Perspective*. San Diego, CA: Academic Press.

Piperno, D. R. 1989. The occurrence of phytoliths in the reproductive structures of selected topical angiosperms and their significance in tropical paleoecology, paleoethnobotany and systematics. *Review of Palaeobotany and Palynology*, **61**, 147–173.

Piperno, D. R. 2006. *Phytoliths: A Comprehensive Guide for Archaeologists and Paleoecologists*. Lanham, MD: AltaMira Press.

Piperno, D. R. and P. Becker 1996. Vegetational history of a site in the central Amazon basin derived from phytolith and charcoal records from natural soils. *Quaternary Research*, **45**, 202–209.

Piperno, D. R. and D. M. Pearsall 1998. *The Silica Bodies of Tropical American Grasses: Morphology, Taxonomy and Implications for Grass Systematics and Fossil Phytolith Identification*, Smithsonian Contributions to Botany no. 85. Washington, DC: Smithsonian Institution Press.

Powers-Jones, A. and J. Padmore 1993. The use of quantitative methods and statistical analyses in the study of opal phytoliths. In Pearsall, D. M. and Piperno, D. R. (eds.), *Current Research in Phytolith Analysis: Applications in Archaeology and Paleoecology*, Research Paper no.10. Philadelphia, PA: Museum Applied Science Center for Archaeology, pp. 47–56.

Prasad, V., C. A. E. Strömberg, H. Alimohammadian, and A. Sahni 2005. Dinosaur coprolites and the early evolution of grasses and grazers. *Science*, **310**, 1177–1180.

Prat, H. 1936. La systématique des Graminées. *Annales des Sciences Naturelles (Botanique), Séries 10*, **18**, 165–258.

Prat, H. 1948. General features of the epidermis in *Zea mays*. *Annals of the Missouri Botanical Garden*, **35**, 341–351.

Raigemborn, M., M. Brea, A. Zucol, and S. Matheos 2006. Fossil wood and phytolith assemblages from the upper Paleocene–Eocene? of Central Patagonia, Argentina. *Abstracts Conference on Climate and Biota of the Early Paleogene*, 108.

Raigemborn, M., M. Brea, A. Zucol, and S. Matheos 2009. Early Paleogene climatic conditions at mid latitude Southern Hemisphere: mineralogical and paleobotanical proxies from continental sequences in Golfo San Jorge basin (Chubut, Patagonia, Argentina). *Geologica Acta*, **7**, 125–145.

Rapp, G. and Mulholland, S. (eds.) 1992. *Phytolith Systematics: Emerging Issues*. New York: Plenum Press.

Retallack, G. J. 2001. Cenozoic expansion of grasslands and climatic cooling. *Journal of Geology*, **109**, 407–426.

Retallack, G. J. 2004. Late Oligocene bunch grassland and early Miocene sod grassland paleosols from central Oregon, USA. *Palaeogeography, Palaeoclimatology, Palaeoecology*, **207**, 203– 237.

Rull, V. and T. Vegas-Vilarrúbia 2000. Chrysophycean stomacysts in a Caribbean mangrove. *Hydrobiologia*, **428**, 145–150.

Runge, F. 1999. The opal phytolith inventory of soils in central Africa: quantities, shapes, classification, and spectra. *Review of Palaeobotany and Palynology*, **107**, 23–53.

Shulmeister, J., J. M. Soons, G. W. Berger, M. Harper, S. Holt, N. Moar, and J. A. Carter 1999. Environmental and sea-level changes on Banks Peninsula (Canterbury, New Zealand) through three glaciation–interglaciation cycles. *Palaeogeography, Palaeoclimatology, Palaeoecology*, **152**, 101–127.

Spalletti, L. A. and M. M. Mazzoni 1979. Estratigrafía de la Formación Sarmiento en la barranca sur del Lago Colhué Huapí, provincia del Chubut. *Revista de la Asociación Geológica Argentina*, **34**, 271–281.

Stevens, P. F. 2001 onwards. *Angiosperm Phylogeny Website*. Version 9, June 2008 [and more or less continuously updated since].

Strömberg, C. A. E. 2004. Using phytolith assemblages to reconstruct the origin and spread of grass-dominated habitats in the Great Plains of North America during the Late Eocene to Early Miocene. *Palaeogeography, Palaeoclimatology, Palaeoecology*, **207**, 239–275.

Strömberg, C. A. E. 2005. Decoupled taxonomic radiation and ecological expansion of open habitat grasses in the Cenozoic of North America. *Proceedings of the National Academy of Sciences USA*, **102**, 11 980–11 984.

Strömberg, C. A. E., E. M. Friis, M. M. Liang, L. Werdelin, and Y. L. Zhang 2007a. Paleaeoecology of an Early–Middle Miocene lake in China: preliminary interpretation based on phytoliths from the Shanwang Basin. *Vertebrata Palasiatica*, **45**, 145–160.

Strömberg, C. A. E., L. Werdelin, E. M. Friis, and G. Saraç 2007b. The spread of grass-dominated habitats in Turkey and surrounding areas during the Cenozoic: Phytolith evidence. *Palaeogeography, Palaeoclimatology, Palaeoecology*, **250**, 18–49.

Taugourdeau-Lantz, J., J. Laroche, G. Lachkar, and D. Pons 1976. La silice chez les vegetaux: problème des phytolithaires. I. *Travaux du Laboratoire de Micropalontologie, Université Pierre-et-Marie Curie, Paris*, **5**, 255–303.

Thomasson J. R. 1978. Observations on the characteristics of the lemma and palea of the late Cenozoic grass *Panicum elegans*. *American Journal of Botany*, **65**, 34–39.

Tomlinson, P. B. 1961. *Anatomy of the Monocotyledons, vol. II, Palmae*. Oxford, UK: Clarendon Press.

Twiss, P. C. 1992. Predicted world distribution of C_3 and C_4 grass phytoliths. In Rapp, G. and Mulholland, S. (eds.), *Phytolith Systematics: Emerging Issues*. New York: Plenum Press, pp. 113–128.

Twiss, P. C., E. Suess, and R. M. Smith 1969. Morphological classification of grass phytoliths. *Soil Science of America Proceeding*, **33**, 109–115.

Utescher, T. and V. Mosbrugger 2007. Eocene vegetation patterns reconstructed from plant diversity: a global perspective. *Palaeogeography, Palaeoclimatology, Palaeoecology*, **247**, 243–271.

Vicentini, A., J. C. Barber, S. S. Aliscioni, L. M. Giussani, and E. A. Kellogg 2008. The age of the grasses and clusters of origins of C_4 photosynthesis. *Global Change Biology*, **14**, 2963–2977.

Walanus, A. and D. Nalepka 1999a. POLPAL: program for counting pollen grains, diagrams plotting and numerical analysis. *Acta Palaeobotanica* (Suppl.), **2**, 659–661.

Walanus, A. and D. Nalepka 1999b. *POLPAL: Numerical Analysis*. Warsaw: W. Szafer Institute of Botany, Polish Academy of Sciences.

Walanus, A. and D. Nalepka 2002. POLPAL: *System Manual*. Warsaw: W. Szafer Institute of Botany, Polish Academy of Sciences.

Wallis, L. 2003. An overview of leaf phytolith production patterns in selected northwest Australian flora. *Review of Palaeobotany and Palynology*, **125**, 201–248.

Zucol, A. F. 1996. Estudios morfológicos comparativos de especies de los géneros *Stipa, Panicum y Paspalum* (Poaceae), de la Provincia de Entre Ríos. Ph.D, thesis, Universidad Nacional de La Plata.

Zucol, A. F. 1999. Fitolitos: hacia un sistema clasificatorio. *Actas I Encuentro de Investigaciones Fitolíticas*, 9–10.

Zucol, A. F. 2005. Estudios paleoagrostológicos en el Cenozoico de Patagonia, su importancia en el conocimiento de la historia evolutiva de los paleopastizales. *Resúmenes XVI Congreso Geológico Argentino*, 123–129.

Zucol, A. F. and M. Bonomo 2008. Estudios arqueobotánicos del sitio Nutria Mansa 1 (partido de General Alvarado, provincia de Buenos Aires). II. Análisis fitolíticos comparativos de artefactos de molienda. In Korstanje, A. and Babot, P. (eds.), *Interdisciplinary Nuances in Phytolith and Other Microfossil Studies*. 173–185. Oxford, UK: British Archaeological Reports.

Zucol, A. F. and M. Brea 2005. Sistemática de fitolitos, pautas para un sistema clasificatorio: un caso en estudio en la Formación Alvear (Pleistoceno inferior). *Ameghiniana*, **42**, 685–704.

Zucol, A. F., M. M. Mazzoni, and R. H. Madden 1999. Análisis fitolíticos en la secuencia sedimentaria de Gran Barranca, Chubut. *Ameghiniana*, **36**, (supl. 4), 43R.

Zucol, A. F., M. Brea, E. Bellosi, A. A. Carlini, and G. Vucetich 2007. Preliminary phytolith analysis of Sarmiento Formation in the Gran Barranca (Central Patagonia, Argentina). In Madella, M. and Zurro, D. (eds.), *Plants, People and Places. Recent Studies in Phytolith Analysis*. Oxford, UK: Oxbow Books, pp. 189–195.

23 Stable isotopes of fossil teeth and bones at Gran Barranca as monitors of climate change and tectonics

Matthew J. Kohn, Alessandro Zanazzi, and Jennifer A. Josef

Abstract

Fossiliferous sediments at Gran Barranca span a remarkable range of ages from the middle Eocene through the early Miocene, including the paleoceanographically significant Eocene–Oligocene transition (EOT). These strata provide an unparalleled opportunity to examine terrestrial isotope systematics attending global climate change in the southern hemisphere, as well as possible effects of Andean orogenesis on regional climate. Fossil tooth enamel shows minimal changes in oxygen isotope composition throughout the sequence, including a close bracket to the EOT. Variances in $\delta^{18}O$ and $\delta^{13}C$ from tooth enamel in all time slices near the EOT are statistically indistinguishable, further suggesting minimal changes to climate seasonality, e.g. mean annual range of temperature. The presence of a transcontinental waterway west of Gran Barranca apparently buffered temperatures and isotope compositions through at least the mid-Oligocene, making these fossils sensitive to changes in global climate, but not to regional factors such as Andean uplift. The absence of an isotopic shift across the EOT is consistent with a small decrease in temperature (2–3 °C), implying small changes to sea-surface temperatures for the Southern Ocean at this latitude.

At 28–20 Ma, enamel $\delta^{18}O$ values drop compared to predicted compositions, whereas preliminary $\delta^{18}O$ values of bone at 20 Ma rise, suggesting lower temperatures. Possibly cooling was caused by changes to ocean circulation, such as strengthening of the Antarctic Circumpolar Current (ACC). If so, the most likely timing of cooling (between 30 and 28 Ma) may correspond with isotope events Oi2, Oi2a, and/or Oi2b. Alternatively the largest decrease to $\delta^{18}O$ at ~29 Ma also correlates with expected regression of a seaway intervening between the Andes and Gran Barranca, perhaps permitting an Andean isotopic rainshadow to manifest itself eastward. If so, then the Andes likely reached quasi-modern elevations by then and maintained those elevations to the present. Consistently low $\delta^{18}O$ at ~12 Ma at Cerro Guenguel (~100 km from Gran Barranca) and for modern sheep teeth near Gran Barranca are attributable to a combination of low temperatures and presence of an Andean rainshadow.

Resumen

Los sedimentos fosilíferos de Gran Barranca abarcan un importante rango temporal, desde el Eoceno medio hasta el Mioceno temprano, incluyendo la transición Eoceno–Oligoceno (TEO), tan significativa desde un punto de vista paleo-oceanográfico. Estos niveles proveen una oportunidad única para estudiar los isótopos continentales, para comprender el cambio climático global en el Hemisferio Sur así como posibles efectos de la orogenia andina en el clima de la región. El esmalte de los dientes fósiles muestra mínimos cambios en la composición de isótopos de oxigeno a lo largo de la secuencia, aun en las cercanías del TEO. Los cambios de $\delta^{18}O$ y $\delta^{13}C$ del esmalte dentario en todos los cortes temporales cercanos al TEO son estadísticamente indistinguibles, lo que sugiere cambios mínimos en estacionalidad e.g. MART. La presencia de una corredor marino transcontinental al Oeste de Gran Barranca aparentemente amortiguó las temperaturas y la composición de isótopos al menos en el Oligoceno medio, haciendo a estas faunas sensibles a cambios climáticos globales, pero no a factores regionales como el levantamiento de los Andes. La ausencia de un cambio isotópico a través del TEO es consistente con una pequeña disminución de temperatura (2–3 °C), que implica pequeños cambios en las temperaturas superficiales marinas para los océanos australes en esta latitud.

Entre 28 y 20 Ma, los valores de $\delta^{18}O$ del esmalte cayeron en comparación a las composiciones previstas, mientras los valores preliminares de $\delta^{18}O$ de huesos se elevaron a los 20 Ma, sugiriendo temperaturas menores. Posiblemente el enfriamiento fue causado por cambios de circulación oceánica, como el fortalecimiento de la Corriente Circumpolar Antártica (CCA). En este caso, la mejor correlación para el enfriamiento (entre 3028 Ma) podría corresponder a los eventos isotópicos Oi2, Oi2a, y/o Oi2b. Alternativamente, el mayor descenso de $\delta^{18}O$ hacia los ~29 Ma también podría correlacionarse con la supuesta regresión del corredor marino entre los Andes y la Gran Barranca, permitiendo que la sombra de lluvias causada por los Andes se manifestara hacia el Este. Si esto

The Paleontology of Gran Barranca: Evolution and Environmental Change through the Middle Cenozoic of Patagonia, eds. R. H. Madden, A. A. Carlini, M. G. Vucetich, and R. F. Kay. Published by Cambridge University Press. © Cambridge University Press 2010.

fue así, los Andes ya habrían adquirido una elevación cercana a la actual en esos tiempos. Valores de $\delta^{18}O$ consistentemente bajos a ~12 Ma en Cerro Guenguel (~100 km de Gran Barranca) y en dientes de ovejas actuales cerca de Gran Barranca pueden atribuirse a una combinación de bajas temperaturas y la presencia de una sombra de lluvias originada por los Andes.

Introduction

Most paleoclimatic studies focus on the marine record because isotopic compositions of benthic and planktonic foraminifera provide a continuous record of ocean temperatures and compositions over many tens of millions of years (e.g. Miller *et al.* 1987; Zachos *et al.* 2001). In contrast, continental climate records are inferred with more difficulty, both because a direct and continuous paleoclimate proxy is generally lacking, and because most marine records that could be used to infer a continental history are from sites distant from the continents. In some areas, soil carbonates and paleoflora can be used for continental climate studies, but in southern South America preservation is temporally and geographically sporadic (Romero 1986). Moreover, older paleofloral studies have sometimes yielded coexisting flora from mutually exclusive environments, in comparison to modern ecosystems (Berry 1938; Romero 1986). Fortunately, fossil vertebrates can be common in some continental settings, and their stable isotope compositions encode continental paleoclimate (e.g. see summaries of Koch 1998; Kohn and Cerling 2002). Thus, fossiliferous sediments at Gran Barranca provide an important record of paleoclimate at intermediate southern latitudes (~45° S). Our $\delta^{18}O$ and $\delta^{13}C$ data from fossil teeth not only inform geologic processes specific to southern South America, but also provide a geographically important reference for global paleoclimate.

Gran Barranca occupies a geologically complex area in South America (Fig. 23.1). To the west the high Andean mountains rise, to the east the South Atlantic and Southern Ocean currents converge, and to the south inland seaways have advanced and retreated. These three physical conditions potentially permit investigation of the impact on terrestrial climate of global cooling and ice-cap development, ocean circulation, and regional orogenesis. Fossil teeth at Gran Barranca span the Eocene–Oligocene transition (EOT) and continue into the early Miocene (Kay *et al.* 1999). Thus fossils from Gran Barranca should illuminate climate change in a continental setting with both mountain uplift and ice cap formation in close proximity.

Our main goals in this study were (1) to develop a baseline record of $\delta^{18}O$ in tooth enamel for the southern continental hemisphere from ~40Ma to the present; (2) to compare this continental record with the foraminiferal $\delta^{18}O$ record; and (3) to examine climate and ecological changes attending

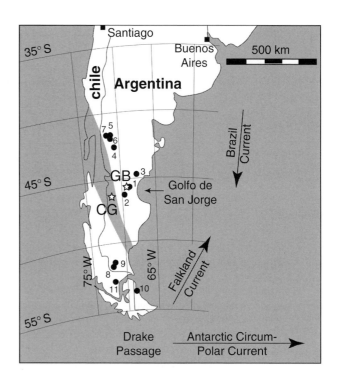

Fig. 23.1. Map of southern South America, showing sample sites (stars: GB, Gran Barranca; CG, Cerro Guenguel). Gray region within continent shows a transcontinental inland seaway proximal to the study site, and present throughout the Eocene and until at least the early Oligocene (~30 Ma: Smith *et al.* 1994), and possibly the mid-Miocene (Romero 1986). Main ocean currents in the vicinity are the Antarctic Circumpolar Current (ACC), the Falkland (or Malvinas) Current, and the Brazil Current. Dots are paleoflora locations – see Table 23.1 for ages and references.

the switch from "greenhouse" to "icehouse" conditions across the EOT. We additionally collected preliminary stable isotope data from Eocene and Oligocene fossil bone to help constrain paleoecological and temperature changes.

Background

Study area

The majority of fossils for this study were collected at Gran Barranca, an escarpment of volcanogenic sediments located south of Lake Colhue-Huapi (S45° 42' 49", W68° 44' 16") in Chubut Province, Argentina (Fig 23.1). To investigate mid-Miocene climate, we also analyzed one 11.8 Ma tooth from the Cerro Guenguel North Amphitheatre locality in the Chubut Province of Argentina (S46° 04' 58", W71° 40' 01", or about 100 km to the SW; Fig. 23.1). The modern sheep teeth are from skeletons at El Criado, Argentina, near Gran Barranca. The chronologic, depositional, and stratigraphic framework of the region is outlined in other chapters (e.g. see Ré *et al.* Chapter 4 and Bellosi Chapter 19, this book).

Isotope compositions of biogenic phosphates

Oxygen isotope compositions of modern biogenic phosphates record the isotopic composition of body water, which correlates strongly and positively with meteoric water (Longinelli 1984; Luz *et al.* 1984), and weakly and negatively with relative humidity (Ayliffe and Chivas 1990; Luz *et al.* 1990), i.e.:

$$\delta^{18}O(PO_4) = m * \delta^{18}O(\text{precipitation}) + r * h(\text{relative humidity}) + c \qquad (23.1)$$

where *m*, *r*, and *c* are taxon-specific constants. $\delta^{18}O$ is defined by:

$$\delta^{18}O = 1000 * [(^{18}O/^{16}O)_{Smp}/(^{18}O/^{16}O)_{Std} - 1] \qquad (23.2)$$

where $(^{18}O/^{16}O)_{Smp}$ is the isotope ratio in the sample and $(^{18}O/^{16}O)_{Std}$ is the isotope ratio in a standard. Just as for-aminifera $\delta^{18}O$ values record a combination of changing temperature and ocean water composition (Miller *et al.* 1987), so too do modern mammalian phosphates chronicle combined changes in meteoric water compositions and humidity. Temperature and precipitation amount, as well as the source and movement of moisture, can modify meteoric water isotopic compositions (Dansgaard 1964; Rozanski *et al.* 1993) and relative humidity. These trends in turn reflect rainout (distillation) processes, which lower moisture content and $\delta^{18}O$ values. Distillation of moisture occurs either during transport from low latitudes to high latitudes, or from low elevations to high elevations.

Carbon isotope compositions of modern biogenic phosphates record the diet of the animal during the time the tooth formed. Carbon isotope compositions of plants (diet) reflect photosynthetic mechanism – in modern settings primarily distinguishing C_3 plants (trees, shrubs, herbs, and cool-climate grasses) from C_4 plants (warm-climate grasses plus a few herbs and desert-adapted shrubs) – as well as light level, leaf litter recycling, and aridity (Koch 1998; Cerling and Harris 1999; Kohn and Cerling 2002). There is no evidence for C_4 plants in southern Argentina for the time periods considered, so changes to $\delta^{13}C$ at Gran Barranca must reflect changes to plant cover and aridity, combined with any secular changes to atmospheric CO_2 compositions (note that $\delta^{13}C$ is defined analogously to $\delta^{18}O$ in Eq. 23.2).

The two main sources of isotopic data in fossil vertebrates are bone and enamel phosphate. The mineral in phosphatic tissues is similar to geological apatite but contains significant CO_3. Enamel has the general formula $\sim Ca_{4.3}(Na,Mg)_{0.2}[(PO_4)_{2.5}(HPO_4)_{0.2}(CO_3)_{0.3}](OH)_{0.5}$ (e.g. Driessens and Verbeeck 1990, pp. 107 and 128) where CO_3 and HPO_4 substitution for PO_4 is charge-balanced by vacancies in the Ca and OH sites. Bone mineral compositions are less well known, but include much more CO_3 (Driessens and Verbeeck 1990) and much less OH, possibly

none (Pasteris *et al.* 2004). For all biogenic phosphates, oxygen isotope compositions are measured for either the PO_4 or CO_3 components, whereas carbon isotope compositions are measured for the CO_3 component.

Enamel is commonly preferred to bone for paleoclimate studies because its lower organic content, higher density, and larger apatite crystals make it less susceptible to diagenetic alteration (Ayliffe *et al.* 1994; see review of Kohn and Cerling 2002). Thus, enamel records the original isotope composition of the animal in which it was precipitated, in turn dependent on water compositions and humidity (for oxygen isotopes), and original diet (for carbon isotopes). For these reasons, we focused principally on enamel in this study.

Tooth enamel grows progressively from the occlusal surface towards the root, which causes complications in interpreting isotope compositions. Initial mineralization produces a thin shell of enamel at a vertical rate of ~ 1 mm per 5–15 days (Fricke and O'Neil 1996; Fricke *et al.* 1998; Kohn *et al.* 1998; Kohn 2004). The thickness of enamel takes longer to develop, at a rate of ~ 1 mm/month (Passey and Cerling 2002). Large herbivore teeth require a few months to more than a year to form depending generally upon the length and thickness of enamel. Because humidity and meteoric water $\delta^{18}O$ vary during the year, an animal's isotope composition also varies, and is recorded as isotope zoning along the length of the tooth. This isotope zoning is ordinarily expressed as high $\delta^{18}O$ reflecting a summer/dry season and low $\delta^{18}O$ reflecting a winter/wet season. Zoning in carbon isotopes can reflect seasonal changes in diet and environment.

Investigating isotope zoning benefits terrestrial climate research in two ways. Firstly, data from the entire length of the tooth or from multiple teeth can be used to estimate average yearly composition. A single analysis from a tooth reveals little in terms of average yearly climate because it may be uncertain what part of the year it represents. Only by examining the full range of values from one tooth or a group of teeth can any confidence be placed upon yearly estimates of $\delta^{18}O$ of tooth enamel, meteoric water composition, temperature, or humidity. Secondly, zoning permits identification of changes to seasonal highs and lows through time. With small teeth, multiple teeth at the same locality are needed because not all teeth form at the same time of the year, and therefore different teeth record varying portions of the yearly climate (Kohn *et al.* 1998). With some longer teeth, multiple year records of climatic changes can be investigated.

Oxygen isotope variations in enamel can be used in principle to infer paleoseasonality, but with some limitations. Several factors dampen the isotope zoning preserved in enamel compared to seasonal variations in meteoric water compositions, including (1) animals derive a significant fraction of their oxygen from air O_2, whose composition is seasonally invariant (Kohn 1996), (2) the residence time of water in most large herbivores is several weeks to ~1 month (Kohn *et al.* 2002; Kohn 2004), and (3) enamel at any point on a tooth precipitates over a significant period of time, usually between a week and several months (Balasse 2002; Passey and Cerling 2002; Kohn 2004). Models of oxygen turnover can be used to estimate damping factors for air O_2 and residence times. However, enamel precipitation times cannot be directly estimated from fossils, other than qualitatively from enamel thickness and body mass.

It is tempting to use the range of isotope compositions, plus temperature coefficients of seasonal $\delta^{18}O$ to infer changes to mean annual range of temperature (MART). Such estimates would be smaller than true MART because of the uncorrectable damping from enamel precipitation. However, these estimates could also be overestimated because animals record specific seasonal compositions, i.e. we measure a total range of compositions over many years, not the mean of the ranges for those years. Modern monthly temperature records allow correction for this effect, and we have found that taking 95% limits to an array of data closely approximates mean annual ranges (Zanazzi *et al.* 2007).

Unlike enamel, bone undergoes extensive alteration after burial, and appears to adopt a wholly pedogenic composition (Kohn and Law 2006). This alteration means the isotope compositions do not reflect original biogenic compositions (dependent on body water composition and diet), but rather soil water and soil CO_2 compositions plus temperature. Like pedogenic carbonate, these compositions can be used effectively as a climate proxy (Kohn and Law 2006; Zanazzi *et al.* 2007). In general soil water $\delta^{18}O$ reflects local water compositions, whereas soil CO_2 $\delta^{13}C$ reflects overall plant compositions. Both $\delta^{18}O$ and $\delta^{13}C$ values can be affected by evaporative enrichment in semi-arid to arid settings, and in favorable cases the $\delta^{13}C$ value can be used to infer the unevaporated $\delta^{18}O$ value (Kohn and Law 2006). For paleoclimate studies, we generally prefer numerous (hundreds) of stratigraphically distributed bone fragments. Only 25 bone samples were made available for this study, which nonetheless did provide some information regarding climate change across the EOT and into the earliest Miocene.

Marine climate overview

Ultimately, we wish to link terrestrial and marine climate, so we briefly review the marine record here, with special emphasis on southern South America and neighboring Antarctica. The following discussion is based primarily on Kennett (1977), Miller *et al.* (1987, 1991), Frakes *et al.* (1992), and Zachos *et al.* (2001).

The past ~50 m.y. have been dominated by global cooling, with periodic cooling and warming steps. Peak Cenozoic warmth at ~53 Ma was succeeded by synchronous cooling of surface and deep ocean waters. A marked increase in $\delta^{18}O$ values of benthic foraminifera at the EOT (33.7–33.4 Ma) was accompanied by major ice-sheet growth on Antarctica. Ocean currents underwent reorganization as Antarctic bottomwater began to form and circulate. The causes of this change are debated, but generally viewed as resulting from one of two mechanisms: (a) formation of the Antarctic Circumpolar Current (ACC) (Kennett 1977) as the Tasmanian gateway opened between Antarctica and Tasmania, and possibly the Drake Passage widened beyond a critical threshold between Antarctica and South America, or (b) a drop in atmospheric CO_2 levels (DeConto and Pollard 2003). Although the timing of opening of the Tasmanian gateway (35.5–30.2 Ma: Exon *et al.* 2001; Stickley *et al.* 2004) includes the EOT, different studies conflict on the timing of the opening of the Drake Passage and the development of the ACC (Fig. 23.2), and they may be temporally unrelated to the EOT. For example, some studies suggest that the opening of the Drake Passage corresponds to an intense but brief cooling event in the earliest Miocene (Mi1: Barker and Burrell 1977, 1982; Pfuhl and McCave 2005). Opening of the Drake Passage is predicted to have caused cooling of southern South America by at least 3–4° C (Nong *et al.* 2000; Toggweiler and Bjornsson 2000).

Sea-surface temperatures (SSTs) are more relevant to questions of continental climate, but again different studies of the EOT conflict. At lower latitudes (<60°), SSTs appear to have changed little (<2 °C: Zachos *et al.* 1994; Ivany *et al.* 2000; Lear *et al.* 2000, 2004), and Lear *et al.* (2000) argued that a ~1‰ increase in benthic foraminifera $\delta^{18}O$ could have resulted from polar ice buildup without high-latitude cooling. However, a recent high-resolution record from the tropical Pacific (Coxall *et al.* 2005) shows an even larger increase in $\delta^{18}O$ across the EOT, 1.5 ± 0.1‰, which is not reasonably supportable by ice volume alone. Likely, a combination of ice volume (0.5–1.0‰ effect) plus high-latitude cooling (0.5–1.0‰ effect or ~2–4 °C: Zachos *et al.* 1994; Bohaty and Zachos 2003) was responsible. Faunal studies qualitatively suggest some high-latitude cooling across the EOT (e.g. see summary of Schumacher and Lazarus 2004), consistent with Coxall *et al.*'s interpretation.

After the EOT, the early–mid-Oligocene was fairly stable climatically with little fluctuation in the $\delta^{18}O$ of foraminifera, and a presumed cooler climate compared to the Eocene. Short-term cooling events termed Oi2, Oi2a, and Oi2b (Miller *et al.* 1991; Pekar and Miller 1996; Pekar *et al.* 2002) occurred between ~30 and ~27 Ma; their causes are unknown. Fossil Oligocene plants indicate that Antarctica was not always completely covered by ice (Romero 1986). Global warming in the late Oligocene returned ocean $\delta^{18}O$ to values similar to the late Eocene, albeit punctuated in the earliest Miocene by the Mi1 brief cooling event (Miller *et al.* 1987). Early Miocene warming reduced or

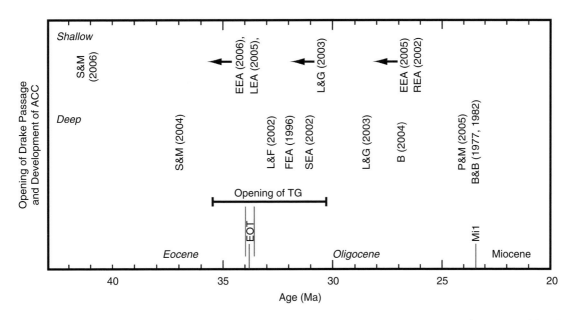

Fig. 23.2. Summary of age estimates for opening of Drake Passage, either as a shallow or deep passageway. References are: S&M, Scher and Martin (2004, 2006); EEA, Eagles *et al.* (2005, 2006); LEA, Livermore *et al.* (2005); L&F, Latimer and Filippelli (2002); FEA, Fulthorpe *et al.* (1996); SEA, Schut *et al.* (2002); L&G, Lawver and Gahagan (2003); B, Barker (2004); REA, Robert *et al.* (2002); P&M, Pfuhl and McCave (2005); B&B, Barker and Burrell (1977, 1982). EOT is the Eocene–Oligocene transition; Mi1 is an early Miocene glaciation event. Opening of the Tasmanian Gateway (TG) from Stickley *et al.* (2004).

eliminated Oligocene ice volume, culminating in the mid-Miocene climatic optimum (MMCO) at 14–16 Ma, as indicated by low $\delta^{18}O$ values of foraminifera, warmer ocean temperatures, and the melting of the Antarctic ice caps. Subsequent late Miocene through Pleistocene global cooling and growth of the major East Antarctic ice-sheet was generally accompanied by increased aridity (e.g. Rea *et al.* 1994) and cyclic glaciations.

Southern South American paleoflora

Southern South America has several Cenozoic paleoflora localities that are key to estimating past ecological environments (Table 23.1). During the Paleocene, the Golfo de San Jorge region, east of Gran Barranca, had a paleofloral assembly that was similar to that of present-day southern Brazil (∼26° S), including mangrove swamp, tropical rainforest, mountain rainforest, highland and sclerophyllous forest, and savannas, i.e. a subtropically humid climate (Wolfe 1971; Romero 1986). Early Eocene paleofloras are interpreted as neotropical (Romero 1986). The Eocene sequences in northwest Patagonia indicate tropical and subtropical, temperate, semi-arid, and rainforest environments (Romero 1986). It is unclear why flora indicate such disparate conditions, but considering Gran Barranca's low elevation, we assume the environment was warm and likely humid. Pollen and leaf species for the mid-Eocene indicate a subtropical rainforest (Romero 1986). During the late Eocene northwest Patagonia

changed from paratropical rainforest to mixed mesophytic forest (Wolfe 1971; Romero 1986). Phytoliths at Gran Barranca from before and after the EOT are dominated by palms, indicative of warm and moderately humid conditions (see discussion of Kohn *et al.* 2004). Oligocene paleofloras have been found only at the southern tip of South America, not from central Patagonia, but these indicate conditions that are comparable to localities now 1000 km further north (Romero 1986), which are 5–10 °C warmer. Thus, the Oligocene appears to have been warmer than at present. Miocene paleofloras of Patagonia are unfortunately absent, despite occurrences elsewhere in South America. The lack of preserved leaves and pollen has been viewed as the result of drying due to development of the Andean rainshadow (Romero 1986). Paleosol evidence from the Argentine Andes margin (∼26° S) indicates a hot climate with seasonal shifts between humid and dry conditions by 10.7–12.0 Ma (Kleinert and Strecker 2001). The modern environment is a semi-arid scrubland.

Paleocoastlines

The size and existence of bodies of water near Gran Barranca likely played a major role in mediating regional climate by altering ocean circulation or regional weather patterns. The following description is based principally on Smith *et al.* (1994), who used marine vs. non-marine facies interpretations to infer the locations of paleocoastlines.

Table 23.1. *Paleoflora estimates of climate in Argentina*

Time	Locality[a]	Conditions	Estimated relative humidity	Reference[b]
	Golfo de San Jorge			
Paleocene	Funes (1), Sur del Río Deseado (2)	Subtropical humid	75%	1,2
Early Eocene	Cañadon Hondo (3)	Neotropical	75%	1,2
Modern	Las Heras (12)	Semi-arid	55%	5
	Northwest Patagonia			
Paleocene	Laguna del Hunco (4)	Paratropical	75%	1,2
Early Eocene	Río Pichileufu (5), Río Chenqueniyeu (6)	Humid and warm	75%	1,2
Late Eocene	Ñirihuau (7)	Mixed mesophytic	70%	1,2,3
	Southernmost SA			
Mid Eocene	Río Turbio (8)	Subtropical humid	75%	1,2,3
Oligocene	Río Guillermo (9), Río Leona (10), Loreto Chile (11)	Mixed mesophytic	70%	1,2

Notes: [a]Numbers in parentheses for locality refer to locations in Fig. 23.1.
[b]References are: 1, Romero (1986); 2, Wolfe (1971); 3, Wolfe and Poore (1982); 4, Kleinert and Strecker (2001); 5, Pearce and Smith (1984).

In the mid-Eocene (~45 Ma) Golfo de San Jorge was slightly larger than the modern gulf, and a narrow transcontinental seaway extended from northwest of Gran Barranca on the Pacific coast to the southeast corner of South America (Fig. 23.1). This basic configuration persisted through the late Eocene and early Oligocene. A marine regression in the late Oligocene may have eliminated the inland sea. However, based on paleofloral evidence, Romero (1986) inferred a large Oligocene basin covering the southeastern section of southern South America. Thus it is possible that the inland seaway remained open near Gran Barranca throughout the Oligocene. The sea-level curve of Haq *et al.* (1987) suggests that any Oligocene disappearance of the inland sea would have occurred during the late Oligocene, i.e. post- ~30 Ma. The inland sea was again present in the early Miocene (~20 Ma), and Golfo de San Jorge may have connected the inland sea to the Atlantic Ocean, resulting in two mouths on the eastern side of South America. Sometime between 20 and 12 Ma, Golfo de San Jorge disconnected from the transcontinental seaway, and the seaway likely closed.

Tectonic setting

Because northwestward movement of South America over the past 40 m.y. has been so minimal (a few degrees), the only tectonic factors that could have influenced precipitation

$\delta^{18}O$ at Gran Barranca are Andean uplift and opening of the Drake Passage. An increase in Andean mean elevation could have caused regression of the marine seaway and would have disrupted atmospheric circulation, which would have affected the moisture/precipitation source for Gran Barranca. Today at 46 ° S winds and moisture principally traverse South America from the west to east, across the Andes, where a majority of the moisture rains out on western slopes. This rainout should yield lower $\delta^{18}O$ precipitation and tooth enamel to the east. However, rainout also decreases humidity in the eastern rainshadow, which should yield an increase in $\delta^{18}O$ of surface water and tooth enamel. An absence of independent estimators of humidity (e.g. from floristics) could compromise deconvolution of these two effects.

Andean volcanic activity began as early as the Late Jurassic (Coira *et al.* 1982; Ramos *et al.* 1982), and has continued in various forms to the present day. At 45 ° S to 52 ° S latitude, major pulses of magmatism include the mid-Cretaceous Patagonian Batholith (98 ± 4 Ma), as well as several pulses of basaltic activity – during the Paleocene to late Eocene (57–36 Ma), just prior to the EOT (35–34 Ma), during the mid- to late-Oligocene (29–25 Ma), and post-early Miocene (16–3 Ma) (Ramos *et al.* 1982). Volcanic ash is a major constituent of sediments at Gran Barranca, but the correlation between volcanic history and Andean paleoelevation at this latitude is unknown. Stable isotope compositional trends of sedimentary carbonates have been interpreted to reflect abrupt surface uplift of the Andes at 16.5 Ma at approximately this latitude (47.5 ° S) (Blisniuk *et al.* 2005). However the isotopic impact of the intervening seaway was not considered, particularly its expansion and elimination in the Miocene, potentially compromising conclusions regarding Andean surface elevations. Rather than reflecting range heights, the isotope trends might instead reflect closure of the seaway at 16.5 Ma, albeit perhaps related to Andean orogenesis.

Samples

Samples were provided via the Universidad Nacional de La Plata, La Plata, Argentina, from well-located strata. Ages are based principally on magnetostratigraphy, as anchored with ^{40}Ar/^{39}Ar ages from intercalated tuffs (Kay *et al.* 1999; Ré *et al.* Chapter 4, this book). All samples, whether individual teeth or tooth fragments, are from large (>100 kg) notoungulates, from the families Astrapotheriidae, Isotemnidae, and Leontiniidae. Large notoungulates were targeted both because some studies suggest that large mammals should have higher water turnover relative to energy expenditure (Bryant and Froelich 1995), and because recent morphological observations by Madden (unpubl. data) suggest that notoungulates employed hind-gut fermentation, which requires higher water turnover (McNab 2002; see also discussion in Kohn and Fremd 2007). Higher water turnover confers a greater sensitivity of tooth δ^{18}O to water composition rather than other climatic or physiological factors, such as relative humidity or heat loss mechanisms (Kohn 1996).

Some materials were previously analyzed for the δ^{18}O of the PO$_4$ component (Kohn *et al.* 2004). In this study we expand the earlier dataset by including three new data sources: (1) the PO$_4$ component of late Oligocene to Recent samples, which interrogates possible Andean uplift to the west; (2) the CO$_3$ component of samples spanning the EOT, which allows comparison of the δ^{18}O of coexisting components (PO$_4$ vs. CO$_3$), as well as an evaluation of any changes to δ^{13}C; and (3) the CO$_3$ component of fossil bone from select horizons. Kohn *et al.* (2004) had to assume changes to relative humidity across the EOT based on floristics, whereas δ^{13}C values can identify major changes in aridity and ecosystem structure, in addition to any secular changes to atmospheric CO$_2$ compositions, as determined via comparison to the marine foraminiferal record.

To measure tooth compositions, a 2–5-mm-wide strip of enamel was first cut lengthwise from each tooth and sectioned every 1–2 mm along the length. Sampling the entire thickness of enamel in large herbivores likely integrates one to several months' time required for the outward growth and maturation of enamel (Kohn and Cerling 2002; Passey and Cerling 2002; Kohn 2004), although enamel mineralization rates are as yet uncharacterized for the notoungulates analyzed here. Isotope variations along each tooth were not sufficiently diagnostic to refine mineralization rate estimates. Adhering dentine was removed from subsamples by using a dental drill. To measure bone compositions, 2–3 mg of bone was cut from each bone fragment.

For analysis of the PO$_4$ component, ~10 mg of cleaned enamel was ground with a mortar and pestle, then processed according to standard protocols (Dettman *et al.* 2001; Kohn *et al.* 2002). Briefly, this included dissolution in HF to produce CaF$_2$, which was discarded, neutralization in NH$_4$OH, and addition of excess AgNO$_3$ to precipitate Ag$_3$PO$_4$, which was analyzed using the O'Neil *et al.* (1994) method. This approach yields an isotopic scale compression (Vennemann *et al.* 2002), which we corrected by assuming a value for NBS-120c and SP3–3 of 22.5‰ (Vennemann *et al.* 2002) and 7.3‰ (Farquhar *et al.* 1993) respectively. Reproducibility was ~±0.3‰ (±1 σ).

For analysis of the CO$_3$ component, cleaned enamel or bone was split to 2–3 mg, ground with a mortar and pestle, then chemically processed using H$_2$O$_2$ and Ca-acetate buffer pretreatments, and analyzed for the δ^{13}C and δ^{18}O of the CO$_3$ component in an automatic carbonate device at 90 °C (Koch *et al.* 1997). The δ^{18}O (Vienna standard mean ocean water [V-SMOW]) and δ^{13}C (V-PDB) compositions of phosphate standard NBS-120c were 29.0±0.1‰ and −6.3±0.1‰ (±1 σ), respectively. All compositions are reported with respect to V-SMOW for δ^{18}O, and V-PDB for δ^{13}C.

Statistical analysis (Tables 23.3 and 23.4 below) of mean δ^{18}O and δ^{13}C values was based on 2-sided *t*-tests and Mann–Whitney tests; differences in variances were assessed by using *F*-tests.

Isotope results

Most teeth show moderate variability in δ^{13}C values, between ~−13 and ~−9‰ (Figs. 23.3, 23.4; Table 23.2), indicating both dietary and ecosystem variation. *F*-tests show few significant differences in variability for different time slices at 95% confidence (Tables 23.3 and 23.4). Compositions average ~−11.5‰, consistent with C$_3$ ecosystems. Values for δ^{13}C at 33.5 Ma are ~1.5‰ higher than times immediately before and after. About 0.5‰ is attributable to secular changes in atmospheric CO$_2$ compositions, as inferred from the benthic marine record, leaving a ~1‰ discrepancy to be explained. Even correcting for the 0.5‰ secular shift, 2-sided *t*-tests and Mann–Whitney tests indicate this difference is significant at >95% confidence compared to times immediately before and after. However, the paucity of data at 33.5 Ma and overlap with the Eocene data argue for extreme caution in interpreting this difference.

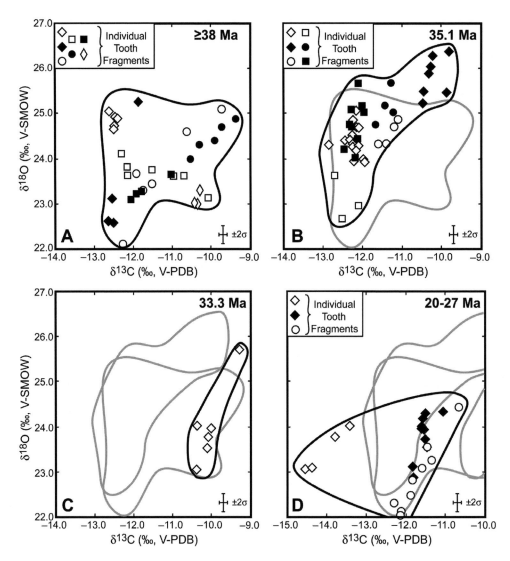

Fig. 23.3. Plots of $\delta^{18}O$ vs. $\delta^{13}C$ for four time slices – two before the EOT, and two after. By 38 Ma, a broad positive correlation between $\delta^{18}O$ and $\delta^{13}C$ for each tooth may reflect consistent changes in diet attending summer/dry vs. winter/wet seasonality. Note change of scale in D.

Bone $\delta^{13}C$ values (Fig. 23.5, Table 23.5) average ~9.5‰, or ~2‰ higher than in teeth. Higher values are expected for bone CO_3 for two reasons. Firstly, it equilibrates at lower temperature (i.e. it is pedogenic rather than biogenic), and $\delta^{13}C$ values of carbonates increase with decreasing temperature relative to the fluid from which they precipitate (Romanek *et al.* 1992). Secondly, soil CO_2 should be slightly enriched in ^{13}C relative to plant biomass (Cerling and Quade 1993). Bone $\delta^{13}C$ trends contrast with teeth in showing an apparently large negative excursion to −11‰ by 30 Ma. Because there are so few bone data at 30 Ma, and data scatter for bone is typically large (Kohn and Law 2006; Zanazzi *et al.* 2007), we cannot yet evaluate whether this represents a single anomaly at 30 Ma, or a systematic trend from 38 to 30 Ma.

Variations in tooth enamel $\delta^{18}O$ are moderate, between ~22 and ~26.5‰ for CO_3 analyses, and ~13 to ~18‰ for PO_4 analyses (Figs. 23.3, 23.4, 23.6; Table 23.2), which we interpret as climate seasonality. Consistent positive correlations between $\delta^{18}O$ vs. $\delta^{13}C$ (Fig. 23.3) by 38 Ma suggest seasonal co-variation between meteoric water composition and diet. The true seasonal variation in meteoric water compositions may be somewhat different because enamel precipitation and water turnover in animals dampens primary ecological signals (Passey and Cerling 2002; Kohn 2004), and because our samples presumably span multiple years. However, if mineralization and water turnover rates in different notoungulates were not drastically different, relative changes to paleoseasonality may be inferred. *F*-tests indicate no

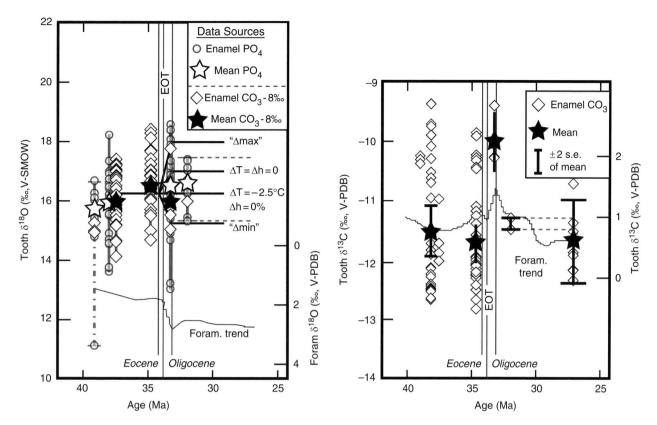

Fig. 23.4. Plots of $\delta^{18}O$ and $\delta^{13}C$ vs. time for tooth enamel data spanning the EOT, including oxygen isotope data from tooth enamel PO_4 (Kohn *et al.* 2004). Small symbols are individual measured compositions. Vertical bars span composition ranges; dot–dash bar at 40.3 Ma is extension to single anomalous datum. Horizontal bars are age constraints for specific levels; for some times, the age uncertainty is smaller than the symbol. Oxygen isotope data at 37.8 Ma are shifted slightly for the CO_3 component to facilitate comparison with PO_4 data. Similarly, data for ≤ 37.8 Ma are offset from 37.8 Ma data to facilitate comparison. For oxygen, compositions are most consistent with relatively small changes to climate across the EOT; models show range of predicted isotope trends for mean compositions relative to the late Eocene. See Table 23.6 for parameters in models. The change to $\delta^{13}C$ is consistent with secular changes to atmospheric CO_2 compositions (as indicated by the marine foraminiferal curve from Zachos *et al.* 2001), except sparse data at 33.5 Ma.

significant differences in compositional variability for teeth from any of the fossil time slices, indicating no resolvable change to seasonality from $\sim$40 to $\sim$12 Ma (Table 23.4).

For comparing average climate at different times, we use mean values (Figs. 23.4–23.6), although use of medians does not yield substantially different results. Mean compositional differences between PO_4 and CO_3 components of teeth from the same time periods are $\sim$8‰, which is consistent with preservation of original biogenic fractionations (Bryant *et al.* 1996; Iacumin *et al.* 1996). Mean compositions for the PO_4 component in enamel are surprisingly uniform throughout the study period, and are generally within 1‰ of $\sim$15‰ from the late Eocene to the present. The biggest compositional shifts are within the late Eocene, and at 27.8 and 11.8 Ma. Interestingly data for 37.8 vs. 33.5 Ma (i.e. across the EOT) exhibit no statistically significant differences (Table 23.4). Fossil bone compositions show similarly small changes, with nearly

constant mean compositions from 40 to 30 Ma, including the EOT, and a possible 0.5–1‰ increase at $\sim$20 Ma. We interpret changes to mean $\delta^{18}O$ values in tooth enamel as reflecting changes in the yearly average climate, either mean annual humidity (MAH), mean annual temperature (MAT), or both. We interpret bone changes to $\delta^{18}O$ values as reflecting changes to soil water composition and temperature. Interpretation requires consideration of several influential geologic and climatic factors, and is best facilitated through compositional modeling.

Models of oxygen isotopes

Modeling isotope compositions of tooth enamel requires constraining terms in Eq. (23.1): precipitation $\delta^{18}O$, relative humidity, and taxon-specific constants. Typical values for large herbivores ($m = 0.85$, $r = 0.15$), and for climatic

Table 23.2. *Tooth $\delta^{18}O$ and $\delta^{13}C$ compositions of fossil notoungulate teeth from Gran Barranca and Cerro Guenguel, Argentina*

Age: 40.3 ± 1.3 Ma

Sample	$\delta^{13}C$	$\delta^{18}O$	Sample	$\delta^{13}C$	$\delta^{18}O$
Ar99-269 C	−10.27	23.30	Ar99-269I	−10.32	23.00
Ar99-269L	−10.39	23.02			

Age: 37.8 ± 0.8 Ma

Sample	$\delta^{13}C$	$\delta^{18}O$	Sample	$\delta^{13}C$	$\delta^{18}O$
EN-1a A	−10.09	23.14	EN-1a D	−10.71	23.62
EN-1a G	−10.97	23.61	EN-1a J	−11.51	23.76
EN-1a M	−12.17	23.83	EN-1a P	−12.15	23.83
EN-1a S	−12.30	24.12			
EN-1b B	−12.50	22.59	EN-1b H	−12.67	22.59
EN-1b L	−12.55	23.12	EN-1b O	−11.86	25.28
EN-2b A	−12.47	24.92	EN-2b D	−12.63	25.05
EN-2b G	−12.50	24.69	EN-2b J	−12.50	24.74
EN-2b M	−12.44	24.91			
EN-2a A	−10.54	23.99	EN-2a J	−9.91	24.40
EN-2a M	−10.31	24.31	EN-2a P	−9.74	24.31
EN-2a T	−9.38	24.88			
Ar03-55-1 D	−9.75	25.10	Ar03-55-1 G	−10.63	24.62
Ar03-55-2 B	−11.74	23.31	Ar03-55-2 E	−11.51	23.45
Ar03-55-2 H	−11.92	23.67	Ar03-55-2 K	−12.27	22.11
ERGB-D	−11.90	23.32	ERGB-F	−11.74	23.26
ERGB-H	−12.04	23.10	ERGB-J	−10.99	23.64
			Average	**−11.39**	**23.93**

Age: $\leq 37.8 \pm 0.8$ Ma

Sample	$\delta^{13}C$	$\delta^{18}O$	Sample	$\delta^{13}C$	$\delta^{18}O$
ENN-1 A	−11.61	24.33	ENN-1 D	−11.41	24.34
ENN-1 G	−11.22	24.71	ENN-1 M	−11.12	24.71
ENN-2a G	−12.70	23.64	ENN-2a J	−12.10	22.96
ENN-2a M	−12.52	22.68			
ENN-2b A	−10.50	25.24	ENN-2b D	−10.49	25.51
ENN-2b G	−10.35	25.90	ENN-2b I	−10.29	25.90
ENN-2b L	−10.25	26.30	ENN-2b P	−9.84	26.39
ENN-2b S	−9.91	25.48			
ELNNR-A	−12.85	23.95	ELNNR-B	−12.25	24.71
ELNNR-C	−12.16	24.01	ELNNR-D	−12.24	24.43
ELNNR-E	−11.94	25.09	ELNNR-F	−12.09	24.71
ELNNR-G	−11.99	24.20	ELNNR-H	−12.31	24.30
ELNNR-I	−12.17	24.55	ELNNR-J	−12.15	23.97
ELNNR-K	−12.17	24.43	ELNNR-L	−12.11	24.29
ELNNR-M	−12.26	24.16	ELNNR-N	−12.23	24.36
ELNNR-O	−12.45	23.88	ELNNR-P	−12.25	23.55
PROF M-C	−11.20	25.03	PROF M-F	−11.26	25.69
PROF M-I	−11.41	25.18	PROF M-L	−11.65	24.76
			Average	**−11.63**	**24.62**

Age: 33.5 ± 0.2 Ma

Sample	$\delta^{13}C$	$\delta^{18}O$	Sample	$\delta^{13}C$	$\delta^{18}O$
Ar00-115-B	−10.05	23.98	Ar00-115-C	−10.12	23.79
Ar00-115-F	−10.15	23.54	Ar00-115-H	−10.41	24.03
Ar00-115-I	−10.41	23.07	Ar00-115-L	−9.33	25.72
			Average	**−10.08**	**24.65**

Table 23.2. (*cont.*)

Age: 30.3 ± 0.8 Ma

Sample	$\delta^{13}C$	$\delta^{18}O$	Sample	$\delta^{13}C$	$\delta^{18}O$
Ar03-144-2 B	−11.51	24.68	Ar03-144-2 E	−11.30	24.02
Ar99-107a		17.36	Ar99-107d		16.22
Ar99-107g		17.08	Ar99-107j		15.42
Ar99-107m		16.29			
Ar99-106h		16.98			
			Average	**−11.40**	**16.51**

Age: 27.8 ± 1.4 Ma

Sample	$\delta^{13}C$	$\delta^{18}O$	Sample	$\delta^{13}C$	$\delta^{18}O$
Ar00-03#1a		13.94	Ar00-03#1j		13.70
Ar00-03#1p		12.64	Ar00-03#2a		14.02
Ar93-27a		14.37	Ar93-27d		14.39
Ar93-27j		14.14	Ar93-27m		15.16
Ar93-27o		14.88			
Ar93-53d		14.18	Ar93-53g		14.03
Ar93-53p		14.72	Ar93-53s		14.02
Ar93-53v		13.87			
Ar93-60c		15.08	Ar93-60c	−12.31	22.26
Ar93-60f		14.05	Ar93-60i	−11.61	23.06
Ar93-60i		12.10	Ar93-60l	−10.67	24.41
Ar93-60o	−11.36	23.23	Ar93-60r	−11.84	22.77
Ar93-60u	−11.88	22.44	Ar93-60v		13.56
Ar93-60x	−12.14	22.08	Ar93-60aa	−12.15	21.98
Ar93-60ad	−11.47	23.48			
			Average	**−11.71**	**14.32**

Age: 20.1 ± 0.1 Ma

Sample	$\delta^{13}C$	$\delta^{18}O$	Sample	$\delta^{13}C$	$\delta^{18}O$
Ar 99-126a		17.93	Ar 99-126d		15.16
Ar 99-126g		15.06	Ar 99-126j		16.74
Ar 99-126m		17.51	Ar 99-126p		16.93
Ar 99-126s		17.71	Ar 99-126v		15.81
Ar 99-126y		15.64			
Ar 01-545#1a		15.90	Ar 01-545#1d		15.49
Ar 01-545#1m		14.88	Ar 01-545#1p		15.00
Ar 01-545#1s		16.07	Ar 01-545#1v		15.46
Ar 01-545#2a		14.70	Ar 01-545#2d		14.55
Ar 01-545#2g		13.93	Ar 01-545#2m		15.06
Ar 01-545#2p		15.05	Ar 01-545#2s		14.39
Ar 01-545#2v		14.29			
Ar 99-256a		17.29			
Ar00-210#2d		16.72	Ar00-210#2e		15.33
Ar00-213b	−14.41	23.07	Ar00-213d	−14.57	23.05
Ar00-213f	−13.81	23.75	Ar00-213h	−13.43	23.99
Ar93-34c	−11.48	23.53	Ar93-34f	−11.07	24.32
Ar93-34i	−11.56	23.92	Ar93-34l	−11.62	23.94
Ar93-34o	−11.59	24.16	Ar93-34r	−11.52	24.29
Ar93-34u	−11.59	24.01	Ar93-34x	−11.53	23.71
Ar93-34aa	−11.82	22.85	Ar93-34ad	−11.84	23.10
			Average	**−12.27**	**15.70**

Table 23.2. (*cont.*)

Age: 11.8 ± 0.2 Ma (Cerro Guenguel)

Sample	$\delta^{18}O$	Sample	$\delta^{18}O$	Sample	$\delta^{18}O$
Ar 91-487b	15.18	Ar 91-487c	14.81	Ar 91-487d	14.40
Ar 91-487g	14.94	Ar 91-487h	15.36	Ar 91-487i	15.33
Ar 91-487j	14.60	Ar 91-487k	15.08	Ar 91-487l	13.76
Ar 91-487m	12.76	Ar 91-487n	15.55	Ar 91-487o	14.64
Ar 91-487p	14.70	Ar 91-487q	13.44	Ar 91-487s	15.71
Ar 91-487t	16.64	Ar 91-487v	12.77	Ar 91-487w	15.94
Ar 91-487x	15.44	Ar 91-487y	14.17	Ar 91-487z	15.16
Ar 91-487aa	15.63	Ar 91-487bb	14.64	Ar 91-487cc	13.52
Ar 91-487dd	15.15	Ar 91-487ee	14.44	Ar 91-487ff	15.64
Ar 91-487hh	15.02	Ar 91-487ii	14.04	Ar 91-487jj	14.01
Ar 91-487kk	15.00	Ar 91-487ll	13.78	Ar 91-487mm	14.84
Ar 91-487nn	14.12			**Average**	**14.71**

Age: modern sheep (near Gran Barranca)

Sample	$\delta^{18}O$	Distance	Sample	$\delta^{18}O$	Distance
EC#2M1a	11.42	16	EC#2M3h	19.59	26
EC#2M1d	15.36	10	EC#2M3k	13.28	20
EC#2M1g	18.76	4	EC#2M3n	14.11	14.5
EC#2M2b	17.16	21.5	EC#2M3q	15.34	8.5
EC#2M2e	16.35	16	EC#2P3a	14.49	1.5
EC#2M2h	13.03	10	EC#2P4a	14.60	1
EC#2M2k	13.32	4.5	EC#2P4e	15.00	9
EC#2M2m	14.50	1	EC#2P4g	17.45	13
EC#2M3b	13.55	38	**Average**	**15.80**	

Notes: Data are reported to the nearest 0.01 (the internal precision of the mass spectrometer), but external reproducibility is considerably poorer. See text for discussion. Except for modern sheep, letter "A" or "a" corresponds to occlusal surface, and each letter designation is ~1.5 mm from previous analysis (i.e. a is the first 1.5 mm including occlusal surface, b spans 1.5 to 3 mm from occlusal surface, etc.). Double letters (e.g., aa, ab, etc.) designate next samples below letter z. Values for $\delta^{18}O$ and $\delta^{13}C$ are in ‰, relative to V-SMOW and V-PDB respectively. Samples with a 1a vs. 1b, 2a vs. 2b, or #1 vs. #2 designation are from composites – these different teeth may have come from one individual, or from different individuals. Averages for combined CO_3 and PO_4 data subtract 8‰ from CO_3 analyses to correct for *in vivo* fractionation relative to PO_4 (Bryant *et al.* 1996; Iacumin *et al.*, 1996) and include data from Kohn *et al.* (2004) for samples older than 30 Ma. Distances for modern sheep are in mm below the occlusal surface. Samples with age "≤37.8 ± 0.8" were collected above level of samples with age 37.8 ± 0.8.

controls on meteoric water compositions ($\Delta\delta^{18}O = 0.35‰/°C * \Delta T$) yield the following equation (Kohn *et al.* 2004):

$$\delta^{18}O(PO_4) = \delta^{18}O(SW) - 0.15‰/‰\,r.h. + 0.3‰/°C * T + c \tag{23.3}$$

The taxon-specific constant, c, has the largest uncertainty, so we instead recast Eq. (23.3) in differential form:

$$\Delta^{18}O(PO_4) = \Delta^{18}O(SW) - 0.15‰/‰\Delta r.h. + 0.3‰/°C\,\Delta T \tag{23.4}$$

Equation (23.4) predicts compositional changes relative to a reference climate state. For example, an increase of 1.2‰ in the $\delta^{18}O$ of tooth enamel would reflect a ~4 °C increase in temperature (1.4‰ increase in $\delta^{18}O$ of local water), or a ~8% decrease in humidity, or a combined shift in both temperature and humidity. In considering fossil species, we assume that notoungulates in the lineages investigated had sufficiently similar behavior and physiology that their values of m, r, and c are comparable, permitting comparison of their isotope compositions via Eq. (23.4). For temperate

Table 23.3. *Statistical tests of oxygen isotope compositions of fossil teeth from southern Argentina*

				t-tests					
Age (Ma)	40	38	≤38	33.5	30	28	20	12 Ma	0.0 Ma
40		0.2421	0.0010	0.0017	0.0060	0.0002	0.6761	0.0061	0.7218
38	0.1866		0.0008	0.0007	0.0941	0.0000	0.2478	0.0000	0.7250
≤38	0.2614	0.7522		0.9131	0.7260	0.0000	0.0001	0.0000	0.0499
33.5	0.1587	0.8026	0.6104		0.6951	0.0000	0.0001	0.0000	0.0447
30	0.8961	0.3022	0.3883	0.2626		0.0000	0.0261	0.0000	0.3490
28	0.3147	0.6363	0.8695	0.5194	0.4466		0.0000	0.0812	0.0015
20	0.1718	0.8847	0.6724	0.9190	0.2807	0.5716		0.0000	0.7919
12	0.2673	0.7257	0.9771	0.5873	0.3955	0.8900	0.6481		0.0091
0	0.0011	0.0000	0.0001	0.0003	0.0050	0.0001	0.0001	0.0001	
				F-tests					
				Mann–Whitney tests					
Age (Ma)	40	38	≤38	33.5	30	28	20	12 Ma	0.0 Ma
40		0.1829	0.0009	0.0015	0.0048	0.0003	0.7601	0.0061	0.8630
38			0.0022	0.0034	0.0832	0.0000	0.1538	0.0000	0.4500
≤38				0.9514	0.7592	0.0000	0.0000	0.0000	0.0893
33.5					0.9634	0.0000	0.0000	0.0000	0.0694
30						0.0000	0.0122	0.0000	0.0182
28							0.0000	0.0544	0.0182
20								0.0000	0.6718
12									0.1560

Note: comparisons to 40 Ma omit one anomalously low $\delta^{18}O$ value for that time.

Table 23.4. *Statistical tests of carbon isotope compositions of fossil teeth from southern Argentina*

			t-tests			
Age (Ma)	38–40	≤38	33.5	30	28	20
38–40		0.2990	0.0645	0.9866	0.3713	0.0135
≤38	0.2653		0.0050	0.7109	0.7815	0.0404
33.5	0.0423	0.0980		0.0327	0.0005	0.0038
30	0.2264	0.2750	0.5488		0.4330	0.3404
28	0.0424	0.1340	0.6079	0.4417		0.2039
20	0.4334	0.0944	0.0226	0.1901	0.0193	
			F-tests			
			Mann–Whitney tests			
Age (Ma)	38–40	≤38	33.5	30	28	20
38–40		0.5038	0.1001	0.8095	0.7997	0.1921
≤38			0.0090	0.5369	0.7805	0.5521
33.5				0.0000	0.0016	0.0000
30					0.3273	0.1000
28						0.7813

Note: *t*-tests for 33.5 Ma account for a 0.5‰ increase to $\delta^{13}C$ at that time, as indicated by compositions of benthic foraminifera.

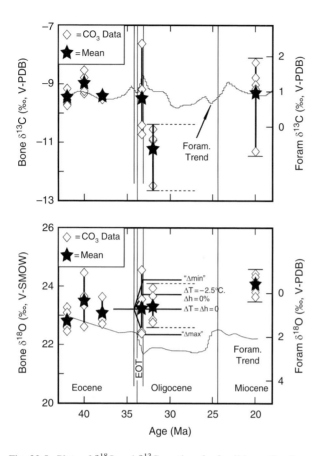

Fig. 23.5. Plots of $\delta^{18}O$ and $\delta^{13}C$ vs. time for fossil bone. Small symbols are individual measured compositions. Vertical bars span composition ranges. Horizontal black bars are age constraints for specific levels; for some times, the age uncertainty is smaller than the symbol. Data $\geq$40.3 Ma are plotted slightly offset from 40.3 Ma data for clarity. The change to $\delta^{13}C$ can be reconciled with secular changes to atmospheric CO_2 compositions (as indicated by the marine foraminiferal curve from Zachos *et al.* 2001), except data at 30 Ma. Possibly, $\delta^{13}C$ values show a systematic decrease from late Eocene through early Oligocene, with rebound in early Miocene. For oxygen, models show range of predicted isotope trends for mean compositions relative to the late Eocene. See Table 23.6 for parameters in models.

conditions, disparate modern herbivores have similar compositions (Kohn and Cerling 2002), so this assumption is likely problematic only for low relative humidities, or for comparing to the modern sheep data. In that context, we chose not to use modern conditions and sheep compositions as our baseline, but instead used the late Eocene because it is well characterized florally, and corresponds to a humid period when isotope compositions of disparate taxa should be most similar (Kohn and Cerling 2002). An alternative baseline would simply shift predicted compositions uniformly, but not their predicted relative differences.

We modeled bone $\delta^{18}O$ changes (EOT models only) based on possible changes to precipitation $\delta^{18}O$, and temperature. In integrated form, this expression is:

Table 23.5. *Bone $\delta^{18}O$ and $\delta^{13}C$ compositions from Gran Barranca*

Age: $\geq$40.3 $\pm$ 1.3 Ma

Sample	$\delta^{13}C$	$\delta^{18}O$	Sample	$\delta^{13}C$	$\delta^{18}O$
Ar99-308	−9.17	22.46	Ar99-309	−9.37	22.60
Ar99-311	−9.54	23.09	Ar99-317	−9.70	23.27
Ar99-448	−9.73	22.63	**Average**	**−9.50**	**22.81**

Age: 40.3 $\pm$ 1.3 Ma

Sample	$\delta^{13}C$	$\delta^{18}O$	Sample	$\delta^{13}C$	$\delta^{18}O$
Ar00-346	−9.33	22.61	Ar00-42	−8.78	23.65
Ar00-45	−9.24	24.46	Ar99-162	−8.53	23.34
			Average	**−8.97**	**23.52**

Age: 37.8 $\pm$ 0.8 Ma

Sample	$\delta^{13}C$	$\delta^{18}O$	Sample	$\delta^{13}C$	$\delta^{18}O$
Ar00-40	−9.40	23.63	Ar99-06	−9.34	22.86
Ar99-119	−9.52	22.71	Ar99-329	−7.61	22.38
			Average	**−8.97**	**22.89**

Age: 33.5 $\pm$ 0.2 Ma

Sample	$\delta^{13}C$	$\delta^{18}O$	Sample	$\delta^{13}C$	$\delta^{18}O$
Ar00-123	−9.19	22.44	Ar99-32	−10.72	23.55
Ar99-33	−10.41	24.56	**Average**	**−10.11**	**23.52**

Age: 30.3 $\pm$ 0.8 Ma

Sample	$\delta^{13}C$	$\delta^{18}O$	Sample	$\delta^{13}C$	$\delta^{18}O$
Ar00-294	−10.55	23.66	Ar99-107	−10.93	22.75
Ar99-347	−10.90	23.93	Ar99-392	−12.49	22.85
			Average	**−11.22**	**23.30**

Age: 20.1$\pm$0.1 Ma

Sample	$\delta^{13}C$	$\delta^{18}O$	Sample	$\delta^{13}C$	$\delta^{18}O$
Ar00-313	−8.26	23.80	Ar00-356	−8.78	24.14
Ar00-359	−11.31	24.43	Ar99-431	−9.14	24.31
Ar99-97	−9.02	23.64	**Average**	**−9.30**	**24.06**

Notes: Data are reported to the nearest 0.01 (the internal precision of the mass spectrometer), but external reproducibility is considerably poorer. See text for discussion. Values for $\delta^{18}O$ and $\delta^{13}C$ are in ‰, relative to V-SMOW and V-PDB respectively. Samples with ages "$\geq$40.3 $\pm$ 1.3 Ma" were collected below level of samples with age 40.3 $\pm$ 1.3 Ma.

$$\delta^{18}O(\text{bone } CO_3) = \delta^{18}O(\text{SW}) + (0.35 - 0.21)‰/°C * T + c \tag{23.5}$$

Equation (23.5) includes a term for seawater composition, the temperature dependence of isotopes in precipitation (0.35‰/°C), the temperature dependence of the fractionation between bone carbonate and soil water between 5 and 25 °C (−0.21‰/°C: Kim and O'Neil 1997), and a constant, c, that accounts for systematic fractionations between precipitation and soil water, and between soil

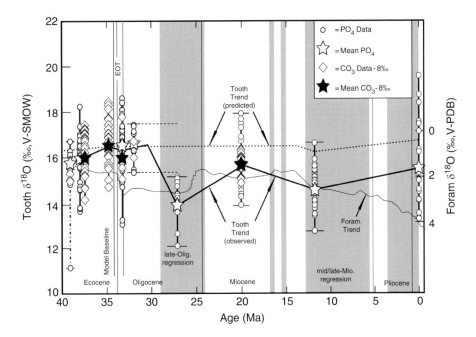

Fig. 23.6. Summary of isotope ranges and values through time for all data. Small symbols are individual measured compositions. Vertical gray lines separate geologic epochs. Vertical gray bands demarcate major marine regressions (Haq *et al.* 1987). Vertical black bars span composition ranges; dot–dash bar at 40.3 Ma is extension to single anomalous datum. Dashed line is predicted $\delta^{18}O$ values of tooth enamel (relative to average late Eocene compositions), and shows good correspondence through the early Oligocene, but values that are too high from the late Oligocene to the present. Predicted compositions are based on regional floral and global marine records; see Table 23.7 for model parameters.

water and bone CO_3. Again, recasting in differential form yields the simple expression:

$$\Delta^{18}O(\text{bone } CO_3) = \Delta^{18}O(SW) + 0.14‰/°C \Delta T \quad (23.6)$$

For modeling enamel compositional changes across the EOT (Table 23.6), we directly input changes to temperature, relative humidity, and ice volume effects (i.e. Eqs. 23.4 and 23.6). For modeling the entire enamel record, we estimated changes to temperature and humidity from paleofloral and global climate records (Table 23.7). In making these predictions, we assumed:

(1) Relative humidities were at least 70% from the Eocene through the Oligocene, as indicated by paleoflora. Because there are so few Miocene and younger paleoflora, patterns of post-Oligocene changes in humidity are less clear, although modern conditions are obviously much drier (r.h. = 55%), so a net post-Oligocene decrease in relative humidity must have occurred. We assume this resulted from a combination of effects – post-middle-Miocene cooling, closure of the interior seaway west of the study area, and development of the Andean rainshadow.

(2) Temperatures relative to the late Eocene can be estimated from paleoflora, the global marine isotope record, and general circulation models (GCMs). There are very few

Table 23.6. *Models of isotope changes across the Eocene–Oligocene transition*

Model	ΔT (°C)	Δh	ΔOcean	$\Delta\delta^{18}O$ (enamel, ‰)	$\Delta\delta^{18}O$ (bone, ‰)
Preferred	−2.5	0	0.75	0	+0.4
Zero	0	0	0.75	+0.75	+0.75
ΔMax	0	−5	1.0	+1.75	+1.0
ΔMin	−5	0	0.50	−1.0	−0.2

Notes: Predicted composition changes are differences relative to samples ≤38 Ma and are rounded to nearest 0.2–0.25‰. ΔOcean is the change to seawater $\delta^{18}O$ resulting from ice volume. Bone compositions are based on observed, constant enamel composition across the EOT, plus assumed changes to relative humidity (which affects estimated water composition: Eq. 23.5) and temperature. The temperature-dependence of bone CO_3 compositions mitigates isotopic shifts in the "ΔMin" model.

direct estimates of paleotemperature in southern South America, particularly near our study area. Essentially, we anchored estimates of temperature in the Eocene and Oligocene from paleoflora, and scaled expected

Table 23.7. *Assumed changes in temperature and humidity, and predicted* $\Delta\delta^{18}O$

Time (Ma)	Δh	ΔT (°C)	ΔOcean	$\Delta\delta^{18}O$ (enamel, ‰)	Comments
38	5 ($\sim$75%)	+2 ($\sim$22)	0	−0.15	Mid-late Eocene drying
35	0 ($\sim$70%)	0 ($\sim$20)	0	0	Late Eocene baseline
34–33	0 ($\sim$70%)	−2.5 ($\sim$17.5)	+0.75	0	Eocene–Oligocene transition
33–26	0 ($\sim$70%)	−2.5 ($\sim$17.5)	+0.75	0	Oligocene cool period
26–25	0 ($\sim$70%)	0 ($\sim$20)	0	0	Late Oligocene warming
25–14	0? ($\sim$70%)	0 ($\sim$20)	0	0	Miocene warm period
14–12	−5? ($\sim$65%)	−5 ($\sim$15)	+0.25	−0.5	Mid-Miocene cooling
12–0	−15 (55%)	−10 (10)	+1.0	0.25 ‰	Post-mid-Miocene cooling

Notes: Predicted composition changes are differences relative to 35.1 Ma. ΔOcean is the change to seawater $\delta^{18}O$ resulting from ice volume.

temperature changes for the Miocene according to the global marine oxygen isotope curve (Zachos *et al.* 2001). Several possible temperature shifts are considered for the EOT because climate change then is unclear and controversial. General circulation models for the opening of the Drake Passage (Nong *et al.* 2000; Toggweiler and Bjornsson 2000), suggest that temperature should have decreased across the EOT by several degrees at 45 ° S latitude. We consider a maximum temperature decrease up to 5 °C. In comparison, Drake Passage GCMs suggest that temperatures in continental interiors should have increased by several degrees at 45 ° N latitude (Toggweiler and Bjornsson 2000), whereas it dropped by 5–10 °C (Zanazzi *et al.* 2007), implicating a decrease in atmospheric CO_2 concentrations rather than opening of the Drake Passage alone. For the Miocene, the marine record suggests conditions should have been similar to the Eocene (*c.* 20 ° C).

(3) Changes in ice volume changed the $\delta^{18}O$ of seawater, which in turn changed the $\delta^{18}O$ of meteoric waters and O_2 (because of atmosphere–ocean coupling: Bender *et al.* 1985). Based on the benthic foraminiferal record (Zachos *et al.* 2001; Coxall *et al.* 2005), our preferred model assumes that (a) Antarctic glaciation during the EOT caused an increase in world ocean $\delta^{18}O$ of 0.75‰ (we consider a range of 0.5 to 1.0‰), (b) no isotopic shift occurred during the Oligocene cool period, (c) most glacial ice melted during the late Oligocene, causing a 0.75‰ decrease to world ocean $\delta^{18}O$, and (d) cooling since the MMCO has caused an increase in world ocean $\delta^{18}O$ of 1.0‰.

This approach (Tables 23.6, 23.7) yields predictions that correspond to global climate effects, as modified by local humidity, and permits comparison of our regional isotope patterns with global climate patterns. Similarities are explainable in terms of global processes, whereas

disparities presumably result from regional factors, which could affect both temperature and humidity.

Interpretations

Carbon isotope compositions that bracket the EOT are consistent with C_3 ecosystems, and a complete absence of C_4 grasses. Phytoliths indicate the presence of some grasses through this time, so these grasses were likely C_3. Values of $\delta^{13}C$ are not particularly high, even at 33.5 Ma; for example, an "average" C_3 diet at that time would confer $\delta^{13}C$ values of $\sim$−11‰ to tooth enamel (Passey *et al.* 2002), and a nominal ±2‰ variation about the mean for modern C_3 plants (e.g. Cerling *et al.* 1997) encompasses all observations. Thus, we do not infer drought stress, which would require values above $\sim$−9‰. In considering $\delta^{13}C$ trends, we note the general paucity of data for teeth between 28 and 38 Ma, and for bone in general. Thus, although at $\sim$33.5 Ma tooth enamel $\delta^{13}C$ appears to increase by $\sim$1‰ relative to marine isotope trend, we believe we have not yet collected sufficient data to make robust specific conclusions. This caution is underscored by the irreconcilable differences in enamel vs. bone $\delta^{13}C$ trends.

For interpreting oxygen isotopes, our preferred input parameters indicate a surprising insensitivity of enamel compositions to general climate trends (Figs. 23.4–23.6). This insensitivity occurs because decreased temperature (which causes a decrease in modeled $\delta^{18}O$) generally corresponds with increased ice volume and decreased relative humidity (which cause an increase in modeled $\delta^{18}O$). The predicted vs. observed $\delta^{18}O$ curves yield some obvious similarities and discrepancies, both for enamel and bone (Figs. 23.4–23.6). To facilitate discussion, we focus on three major time periods: (1) The EOT, when global climate rapidly and dramatically cooled, (2) the Oligocene through early Miocene, when global climate significantly warmed, and (3) the MMCO and subsequent global cooling and drying.

Eocene–Oligocene transition

Prior to the EOT, the small increase in $\delta^{18}O$ in the late Eocene probably reflects either a small increase in temperature ($\leq 2\,°C$) or decrease in relative humidity ($\sim 3\%$), or both. Paleoflora suggest temperature shifts were minimal, and perhaps decreased slightly ($-2\,°C$), so we instead ascribe the composition shift to decreased humidity (Kohn *et al.* 2004). For the EOT, enamel data are best matched by a small decrease to temperature ($2–3\,°C$), assuming no change to relative humidity, and a preferred 0.75‰ increase in global $\delta^{18}O$ due to ice volume. The persistence of palms across the EOT was interpreted to reflect unchanged humidities of $\sim 70\%$ (Kohn *et al.* 2004). The preferred, small temperature decrease contrasts slightly with Kohn *et al.*'s interpretations because we now recognize a small increase in $\delta^{18}O$ *prior* to the EOT. *F*-tests suggest no significant change to isotope variability, suggesting that MART did not change much across the EOT.

To maintain constant enamel compositions, a maximal decrease in relative humidity of 5% across the EOT would necessitate a temperature drop of $5\,°C$. Such a model would still be consistent with bone compositions. A ± 0.25‰ uncertainty in seawater composition (ice volume) contributes an additional uncertainty of about $\pm 1\,°C$. Larger temperature drops are inconsistent with both the enamel and bone compositions. Temperature shifts $<5\,°C$ at Gran Barranca appear smaller than evident from enamel and bone compositions in central North America ($-8\,°C$: Zanazzi *et al.* 2007). Whereas continental interiors may have been more sensitive to climate change, Gran Barranca's position within a narrow peninsula must have buffered temperatures to local sea-surface temperatures. If sea-surface temperatures did not change much across the EOT at mid and low latitudes (Zachos *et al.* 1994; Lear *et al.* 2000, 2004; Billups and Schrag 2003), then no major temperature and $\delta^{18}O$ shifts could have occurred at Gran Barranca. Instead, the clearest decrease in $\delta^{18}O$ occurred later, by ~ 28 Ma.

Oligocene to early Miocene

The biggest isotopic shift in our enamel record (>2‰ from mean or modeled values) occurs between 30 and 28 Ma, when the marine record suggests global climate was generally stable. The 28 Ma data are unquestionably different from 30 Ma (a 2-sided *t*-test yields $p < 1 \times 10^{-6}$) and from 33.5 Ma (a 2-sided *t*-test yields $p < 1 \times 10^{-12}$). One possibility is that temperature dropped dramatically ($\geq 10\,°C$), but due to regional rather than global reasons. At 28 Ma, an absence of bone data precludes direct temperature estimates, but at 20 Ma, the difference between enamel PO_4 and bone CO_3 compositions increases to ~ 8.5‰ compared

to ~ 7‰ for late Eocene and early Oligocene differences. This increase in fractionation translates into a $\sim 7\,°C$ temperature decrease, assuming no change to relative humidity. A decrease in relative humidity would increase the estimated temperature drop. Considering how low enamel compositions are at 28 Ma compared to 20 Ma, an even larger temperature drop may be evident, although the lack of bone data precludes definitive interpretation. Because we have few bone isotope data, the timing of such a temperature decrease is poorly known, bracketed only between 30 and 20 Ma, but possibly between 30 and 28 Ma. This time corresponds with marine isotope anomalies Oi2, Oi2a, and Oi2b (30.5 to 27 Ma: Miller *et al.* 1991; Pekar and Miller 1996; Pekar *et al.* 2002). Cooling at Gran Barranca, if it occurred during this time, may reflect one (or more) of these events.

An alternative possibility notes the correspondence in the early late-Oligocene between decreased $\delta^{18}O$ and decreased sea level (Haq *et al.* 1987). Smith *et al.* (1994) argued for regression of the marine seaway west of Gran Barranca, essentially coincident with this event. If the Patagonian Andes were already at significant elevation, then the appearance of an Andean rainshadow would yield a corresponding reduction in $\delta^{18}O$ of local surface waters and tooth enamel. The subsequent small increase in $\delta^{18}O$ in the early Miocene then reflects reappearance of the seaway, and reduction of a rainshadow effect. Although we acknowledge some controversy regarding the extent of land and sea areas in the Oligocene (Romero 1986), changes in global sea level in this period (Haq *et al.* 1987) must have affected the geometry of the seaway, and hence the influence of an Andean rainshadow if one was present.

Mid-Miocene to Recent

In this interval, $\delta^{18}O$ values remain below predictions, despite large decreases in relative humidity. Because the marine seaway had certainly closed by the middle Miocene (Smith *et al.* 1994), these relatively low $\delta^{18}O$ values likely reflect lower temperatures and possibly an Andean influence. Blisniuk *et al.* (2004) argue for development of a rainshadow at ~ 16.5 Ma. In comparison to the late Eocene and early Oligocene, when moisture sources were directly adjacent to Gran Barranca, orographic depletion of ^{18}O in precipitation by the mid-Miocene would unquestionably yield compositions below predictions.

Development of the Antarctic Circumpolar Current

We found no obvious isotope anomalies either in the late Eocene or across the EOT that would implicate formation of the ACC. However, the drop in enamel $\delta^{18}O$ at ~ 28 Ma could have resulted from decreased temperature associated with widening of the Drake Passage and further strengthening of the ACC. Few marine data directly support this hypothesis, but we do note the occurrence of global marine isotope anomalies Oi2, Oi2a, and Oi2b (Miller *et al.* 1991; Pekar and Miller 1996; Pekar *et al.* 2002) between 30.5 and 27 Ma. Possibly one or more of these anomalies resulted from changes to the character of the ACC (depth of circulation, strength, etc.) and south Atlantic currents, and resulted in cooling of southern Patagonia.

Andean elevations

If instead the drop in enamel δ^{18}O at ~28 Ma was produced orographically, then the Andes must have achieved significant elevations by 28 Ma, and maintained them until the present. The paleoelevation of the central Andes has been closely scrutinized (e.g. Garzione *et al.* 2006; Ghosh *et al.* 2006), and major increases in elevation there did not occur until the late Miocene. However, we know of only one study of southern Andes paleoelevations (Blisniuk *et al.* 2005; see also Kleinert and Strecker 1997), and it does not address this time interval. We see little evidence for increased elevations at 12 or 0 Ma relative to 28 Ma, as determined by the deviation between mean and modeled compositions. One scenario is that modern Andean elevations were established by the late Oligocene and remained essentially static since. For example, tectonic changes at ~28 Ma occurred both in volcanic activity (Ramos *et al.* 1982) and in plate convergence direction (Somoza and Ghidella 2005), possibly leading to uplift. The isotopic data collected by Blisniuk *et al.* (2005) were not evaluated within the context of an intervening seaway and local water sources. Scenarios involving changes to sea level could alternatively explain their data. Data from Kleinert and Strecker (1997) suggest more strongly that elevations increased significantly after ~12 Ma, but their study site is considerably farther north where tectonics may be distinct from the southern Andes. Because of the intervening marine seaway, our data cannot address whether modern elevations could have been attained earlier, for example as early as the formation of the Patagonian batholith in the late Cretaceous.

Conclusions

The global switch from "greenhouse" to "icehouse" conditions at the EOT had a small, perhaps even negligible, impact at Gran Barranca, despite its proximity to Antarctica and the developing ice cap. We estimate a temperature decrease of 2–3 °C, and no more than ~5 °C, consistent with data indicating that sea-surface temperatures at latitudes <60 ° changed very little (1–2 °C) across the EOT (Zachos *et al.* 1994; Lear *et al.* 2000; 2004; Billups and Schrag 2003). One of two mechanisms likely caused the discordant drop in enamel δ^{18}O between 30 and 28 Ma. If the Andes were significantly elevated by 28 Ma, then

regression of a seaway that intervened between Gran Barranca and the Andes would have allowed a rainshadow to be manifest, lowering δ^{18}O values. This model implies that the Andes had already achieved quasi-modern elevations by the late Oligocene, and their isotopic influence resulted from changes to seaway presence and/or geometry. Alternatively, changes to the ACC could have caused large permanent temperature decreases by 28 Ma, as implied by lower apparent temperatures at 20 Ma compared to ≥30 Ma samples. In this model, marine isotope anomalies Oi2, Oi2a, and/or Oi2b could reflect changes to the ACC and marine circulation in general.

ACKNOWLEDGEMENTS

This chapter constitutes a portion of the MS work of J. Josef and PhD work of A. Zanazzi. This material is based upon work supported by the National Science Foundation under Grant Nos. EAR-9909568, EAR-0304181, and ATM-0400532 (to MJK). Field research was supported by US National Science Foundation grants EAR-0087636, BCS-0090255, and DEB-9907985 to Richard F. Kay and Richard H. Madden. We thank R. Madden, G. Vucetich, R. Kay, and A. Carlini for providing samples, Bruce MacFadden and an anonymous reviewer for their perceptive comments that improved the script, and the Government of the Province of Chubut for authorization to make the collections that yielded the fossils analyzed in this study.

REFERENCES

Ayliffe, L. K. and A. R. Chivas 1990. Oxygen isotope composition of the bone phosphate of Australian kangaroos: potential as a palaeoenvironmental recorder. *Geochimica Cosmochimica Acta*, **54**, 2603–2609.

Ayliffe, L. K., A. R. Chivas, and M. G. Leakey 1994. The retention of primary oxygen isotope compositions of fossil elephant skeletal phosphate. *Geochimica Cosmochimica Acta*, **58**, 5291–5298.

Balasse, M. 2002. Reconstructing dietary and environmental history from enamel isotopic analysis: time resolution of intra-tooth sequential sampling. *International Journal of Osteoarchaeology*, **12**, 155–165.

Barker, P. F. 2004. Origin signature and palaeoclimatic influence of the Antarctic Circumpolar Current. *Earth-Science Reviews*, **66**, 143–162.

Barker, P. F. and J. Burrell 1977. The opening of the Drake Passage. *Marine Geology*, **25**, 15–34.

Barker, P. F. and J. Burrell 1982. The influence on Southern Ocean circulation, sedimentation and climate of the opening of Drake Passage. In Craddock, C. (ed.), *Antarctic Geoscience*. Madison, WI: University of Wisconsin Press, pp. 377–385.

Bender, M., L. D. Labeyrie, D. Raynaud, and C. Lorius 1985. Isotopic composition of atmospheric O_2 in ice linked with deglaciation and global primary productivity. *Nature,* **318,** 349–352.

Berry, E. W. 1938. *Tertiary Flora from the Rio Pichileufu, Argentina,* Special Paper no. 12. Boulder, CO: Geological Society of America.

Billups, K. and D. P. Schrag 2003. Application of benthic foraminiferal Mg/Ca ratios to questions of Cenozoic climate change. *Earth and Planetary Science Letters,* **209,** 181–195.

Blisniuk, P. J., L. A. Stern, C. P. Chamberlain, B. Idleman, and P. K. Zeitler 2005. Climatic and ecologic changes during Miocene surface uplift in the Southern Patagonian Andes. *Earth and Planetary Science Letters,* **230,** 125–142.

Bohaty, S. M. and J. C. Zachos 2003. Significant Southern Ocean warming event in the late middle Eocene. *Geology,* **31,** 1017–1020.

Bryant, J. D. and P. N. Froelich 1995. A model of oxygen isotope fractionation in body water of large mammals. *Geochimica Cosmochimica Acta,* **59,** 4523–4537.

Bryant, J. D., P. L. Koch, P. N. Froelich, W. J. Showers, and B. J. Genna 1996. Oxygen isotope partitioning between phosphate and carbonate in mammalian apatite. *Geochimica Cosmochimica Acta,* **60,** 5145–5148.

Cerling, T. E. and J. M. Harris 1999. Carbon isotope fractionation between diet and bioapatite in ungulate mammals and implications for ecological and paleoecological studies. *Oecologia,* **120,** 347–363.

Cerling, T. E. and J. Quade 1993. Stable carbon and oxygen isotopes in soil carbonates. In Swart, P. K., Lohmann, K. C., McKenzie, J. A., and Savin, S. (eds.), *Climate Change in Continental Isotopic Records.* Washington, DC: American Geophysical Union, pp. 217–231.

Cerling, T. E., J. M. Harris, B. J. MacFadden, M. G. Leakey, J. Quade, V. Eisenmann, and J. R. Ehleringer 1997. Global vegetation change through the Miocene/Pliocene boundary. *Nature,* **389,** 153–158.

Coira, B., J. Davidson, C. Mpodozis, and V. Ramos 1982. Tectonic and magmatic evolution of the Andes of northern Argentina and Chile. *Earth-Science Reviews,* **18,** 303–332.

Coxall, H. K., P. A. Wilson, H. Palike, C. H. Lear, and J. Backman 2005. Rapid stepwise onset of Antarctic glaciation and deeper calcite compensation in the Pacific Ocean. *Nature,* **433,** 53–57.

Dansgaard, W. 1964. Stable isotopes in precipitation. *Tellus,* **16,** 436–468.

DeConto, R. J. and D. Pollard 2003. Rapid Cenozoic glaciation of Antarctica triggered by declining atmospheric CO_2. *Nature,* **421,** 245–249.

Dettman, D. L., M. J. Kohn, J. Quade, F. J. Ryerson, T. P. Ojha, and S. Hamidullah 2001. Seasonal stable isotope evidence for a strong Asian monsoon throughout the past 10.7 m.y. *Geology,* **29,** 31–34.

Driessens, F. C. M. and R. M. H. Verbeeck 1990. *Biominerals.* Boca Raton, FL: CRC Press.

Eagles, G., R. A. Livermore, J. D. Fairhead, and P. Morris 2005. Tectonic evolution of the west Scotia Sea. *Journal of Geophys Research,* **110,** B02401. doi:10.1029/2004JB003154.

Eagles, G., R. A. Livermore, and P. Morris 2006. Small basins in the Scotia Sea: the Eocene Drake Passage gateway. *Earth and Planetary Science Letters,* **242,** 343–353.

Exon, N., J. Kennett, M. Malone, H. Brinkhuis, G. Chaproniere, A. Ennyu, P. Fothergill, M. Fuller, M. Grauert, P. Hill, T. Janecek, C. Kelly, J. Latimer, K. McGonigal, S. Nees, U. Ninnemann, D. Nuernberg, S. Pekar, C. Pellaton, H. Pfuhl, C. Robert, U. Röhl, S. Schellenberg, A. Shevenell, C. Stickley, N. Suzuki, T. Yannick, W. Wei, and T. White 2002. Drilling reveals climatic consequences of Tasmanian gateway opening. *EOS, Transactions of the American Geophysical Union,* **83,** 253–259.

Farquhar, J., T. Chacko, and B. R. Frost 1993. Strategies for high-temperature oxygen isotope thermometry: a worked example from the Laramie Anorthosite Complex, Wyoming, USA. *Earth and Planetary Science Letters,* **117,** 407–422.

Frakes, L. A., J. E. Francis, and J. I. Syktus 1992. *Climate Modes of the Phanerozoic: The History of Earth's Climate over the past 600 Million Years.* Cambridge, UK: Cambridge University Press.

Fricke, H. C. and J. R. O'Neil 1996. Inter-and intra-tooth variation in the oxygen isotope composition of mammalian tooth enamel phosphate: implications for palaeoclimatological and palaeobiological research. *Palaeogeography, Palaeoclimatology, Palaeoecology,* **126,** 91–99.

Fricke, H. C., W. C. Clyde, and J. R. O'Neil 1998. Intra-tooth variations in $\delta^{18}O(PO_4)$ of mammalian tooth enamel as a record of seasonal variations in continental climate variables. *Geochimica Cosmochimica Acta,* **62,** 1839–1850.

Fulthorpe, C. S., R. M. Carter, K. G. Miller, and J. Wilson 1996. Marshall Paraconformity: a mid-Oligocene record of inception of the Antarctic Circumpolar Current and coeval glacioeustatic lowstand? *Marine and Petroleum Geology,* **13,** 61–77.

Garzione, C. N., P. Molnar, J. C. Libarkin, and B. J. MacFadden 2006. Rapid late Miocene rise of the Bolivian Altiplano: evidence for removal of mantle lithosphere. *Earth and Planetary Science Letters,* **241,** 543–556.

Ghosh, P., C. N. Garzione, and J. M. Eiler 2006. Rapid uplift of the Altiplano revealed through $^{13}C-^{18}O$ bonds in paleosol carbonates. *Science,* **311,** 511–515.

Haq, B. J., J. Hardenbol, and P. R. Vail 1987. Chronology of fluctuating sea levels since the Triassic. *Science,* **235,** 1156–1167.

Iacumin, P., H. Bocherens, A. Mariotti, and A. Longinelli 1996. Oxygen isotope analyses of co-existing carbonate and phosphate in biogenic apatite: a way to monitor diagenetic alteration of bone phosphate? *Earth and Planetary Science Letters,* **142,** 1–6.

Ivany, L. C., W. P. Patterson, and K. C. Lohmann 2000. Cooler winters as a possible cause of mass extinctions at the Eocene/Oligocene boundary. *Nature,* **407,** 887–890.

Kay, R. F., R. H. Madden, G. M. Vucetich, A. A. Carlini, M. M. Mazzoni, G. H. Ré, M. Heizler, and H. Sandeman 1999. Revised geochronology of the Casamayoran South American Land Mammal Age: Climatic and biotic implications. *Proceedings of the National Academy of Sciences USA*, **96**, 13 235–13 240.

Kennett, J. P. 1977. Cenozoic evolution of Antarctic glaciation, the circum-Antarctic Ocean, and their impact on global paleoceanography. *Journal of Geophysical Research*, **82**, 3843–3860.

Kim, S. -T. and J. R. O'Neil 1997. Equilibrium and nonequilibrium oxygen isotope effects in synthetic carbonates. *Geochimica Cosmochimica Acta*, **61**, 3461–3475.

Kleinert, K. and M. R. Strecker 2001. Climate change in response to orographic barrier uplift: paleosol and stable isotope evidence from the late Neogene Santa Maria basin, northwestern Argentina. *Geological Society of America Bulletin*, **113**, 728–742.

Koch, P. L. 1998. Isotopic reconstruction of past continental environments. *Annual Review of Earth and Planetary Sciences*, **26**, 573–613.

Koch, P. L., N. Tuross, and M. L. Fogel 1997. The effects of sample treatment and diagenesis on the isotopic integrity of carbonate in biogenic hydroxylapatite. *Journal of Archaeological Science*, **24**, 417–429.

Kohn, M. J. 1996. Predicting animal $\delta^{18}O$: accounting for diet and physiological adaptation. *Geochimica Cosmochimica Acta*, **60**, 4811–4829.

Kohn, M. 2004. Comment: Tooth enamel mineralization in ungulates: implications for recovering a primary isotopic time-series, by B. H. Passey and T. E. Cerling (2002). *Geochimica Cosmochimica Acta*, **68**, 403–405.

Kohn, M. J. and T. E. Cerling 2002. Stable isotope compositions of biological apatite. *Reviews in Mineralogy and Geochemistry*, **48**, 455–488.

Kohn, M. and T. Fremd 2007. Tectonic controls on isotope compositions and species diversification, John Day Basin, central Oregon. *PaleoBios*, **27**, 48–61.

Kohn, M. and J. Law 2006. Stable isotope chemistry of fossil bone as a new paleoclimate indicator. *Geochimica Cosmochimica Acta*, **70**, 931–946.

Kohn, M. J., M. J. Schoeninger, and J. W. Valley 1998. Variability in herbivore tooth oxygen isotope compositions: reflections of seasonality or developmental physiology? *Chemical Geology*, **152**, 97–112.

Kohn, M. J., J. L. Miselis, and T. J. Fremd 2002. Oxygen isotope evidence for progressive uplift of the Cascade Range, Oregon. *Earth and Planetary Science Letters*, **204**, 151–165.

Kohn, M. J., J. A. Josef, R. Madden, R. Kay, G. Vucetich, and A. A. Carlini 2004. Climate stability across the Eocene–Oligocene transition, southern Argentina. *Geology*, **32**, 621–624.

Latimer, J. C. and G. M. Filippelli 2002. Eocene to Miocene terrigenous inputs and export production: geochemical evidence from ODP Leg 177, Site 1090. *Palaeogeography, Palaeoclimatology, Palaeoecology*, **182**, 151–164.

Lawver, L. A. and L. M. Gahagan 2003. Evolution of Cenozoic seaways in the circum-Antarctic region. *Palaeogeography, Palaeoclimatology, Palaeoecology*, **198**, 11–37.

Lear, C. H., H. Elderfield, and P. A. Wilson 2000. Cenozoic deep-sea temperatures and global ice volumes from Mg/Ca in benthic foraminiferal calcite. *Science*, **287**, 269–272.

Lear, C. H., Y. Rosenthal, H. K. Coxall, and P. A. Wilson 2004. A Late Eocene to early Miocene ice-sheet dynamics and the global carbon cycle. *Paleoceanography*, **19**, PA4015. doi:10.1029/2004PA001039.

Livermore, R., A. Nankivell, G. Eagles, and P. Morris 2005. Paleogene opening of Drake Passage. *Earth and Planetary Science Letters*, **236**, 459–470.

Longinelli, A. 1984. Oxygen isotopes in mammal bone phosphate: a new tool for paleohydrological and paleoclimatological research? *Geochimica Cosmochimica Acta*, **48**, 385–390.

Luz, B., Y. Kolodny, and M. Horowitz 1984. Fractionation of oxygen isotopes between mammalian bone-phosphate and environmental drinking water. *Geochimica Cosmochimica Acta*, **48**, 1689–1693.

Luz, B., A. B. Cormie, and H. P. Schwarcz 1990. Oxygen isotope variations in phosphate of deer bones. *Geochimica Cosmochimica Acta*, **54**, 1723–1728.

McNab, B. K. 2002. *The Physiological Ecology of Vertebrates: A View from Energetics*. Ithaca, NY: Cornell University Press.

Miller, K. G., R. G. Fairbanks, and G. S. Mountain 1987. Tertiary oxygen isotope synthesis, sea level history, and continental margin erosion. *Paleoceanography*, **2**, 1–19.

Miller, K. G., J. D. Wright, and R. G. Fairbanks 1991. Unlocking the ice house: Oligocene–Miocene oxygen isotopes, eustasy, and margin erosion. *Journal of Geophysical Research*, **96**, 6829–6848.

Nong, G. T., R. G. Najjar, D. Seidov, and W. H. Peterson 2000. Simulation of ocean temperature change due to the opening of Drake Passage. *Geophysical Research Letters*, **27**, 2689–2692.

O'Neil, J. R., L. J. Roe, E. Reinhard, and R. E. Blake 1994. A rapid and precise method of oxygen isotope analysis of biogenic phosphate. *Israel Journal of Earth Sciences*, **43**, 203–212.

Passey, B. H. and T. E. Cerling 2002. Tooth enamel mineralization in ungulates: implications for recovering a primary isotopic time-series. *Geochimica Cosmochimica Acta*, **66**, 3225–3234.

Passey, B. H., T. E. Cerling, M. E. Perkins, M. R. Voorhies, J. M. Harris, and S. T. Tucker 2002. Environmental change in the Great Plains: an isotopic record from fossil horses. *Journal of Geology*, **110**, 123–140.

Pasteris, J. D., B. Wopenka, J. J. Freeman, K. Rogers, E. Valsami-Jones, J. A. M. van der Houwen, and M. J. Silva 2004. Lack of OH in nanocrystalline apatite as a function of degree of atomic order: implications for bone and biomaterials. *Biomaterials*, **25**, 229–238.

Pearce, E. A. and C. G. Smith 1984. *The Times Books World Weather Guide*. London: Times Books.

Pekar, S. F. and K. G. Miller 1996. New Jersey Oligocene "Icehouse" sequences (ODP Leg 150X) correlated with global $\delta^{18}O$ and Exxon eustatic records. *Geology*, **24**, 567–570.

Pekar, S. F., N. Christie-Blick, M. A. Kominz, and K. G. Miller 2002. Calibration between eustatic estimates from backstripping and oxygen isotopic records for the Oligocene. *Geology*, **30**, 903–906.

Pfuhl, H. A. and I. N. McCave 2005. Evidence for late Oligocene stablishment of the Antarctic Circumpolar Current. *Earth and Planetary Science Letters*, **235**, 715–728.

Ramos, V. A., H. Niemeyer, J. Skarmeta, and J. Munoz 1982. Magmatic evolution of the Austral Patagonian Andes. *Earth-Science Reviews*, **18**, 411–443.

Rea, D. K. 1994. The paleoclimatic record provided by eoloian deposition in the deep sea: the geologic history of wind. *Reviews of Geophysics*, **32**, 159–195.

Robert, C. M., L. Dister-Haass, and H. Chamley 2002. Late Eocene–Oligocene oceanographic development at southern high latitudes, from terrigenous and biogenic particles: a comparison of Kerguelen Plateau and Maud Rise, ODP Sites 744 and 689. *Marine Geology*, **191**, 37–54.

Romanek, C. S., E. L. Grossman, and J. W. Morse 1992. Carbon isotopic fractionation in synthetic aragonite and calcite: effects of temperature and precipitation rate. *Geochimica Cosmochimica Acta*, **56**, 419–430.

Romero, E. J. 1986. Paleogene phytogeography and climatology of South America. *Annals of the Missouri Botanical Garden*, **73**, 449–461.

Rozanski, K., L. Araguas-Araguas, and R. Gonfiantini 1993. Isotopic patterns in modern global precipitation. In Swart, P. K., Lohmann, K. C., McKenzie, J. A., and Savin, S. (eds.), *Climate Change in Continental Isotopic Records*. Washington, DC: American Geophysical Union, pp. 1–36.

Scher, H. D. and E. E. Martin 2004. Circulation in the Southern Ocean during the Paleogene inferred from neodymium isotopes. *Earth and Planetary Science Letters*, **228**, 391–405.

Scher, H. D. and E. E. Martin 2006. Timing and climatic consequences of the opening of Drake Passage. *Science*, **312**, 428–430.

Schumacher, S. and D. Lazarus 2004. Regional differences in pelagic productivity in the late Eocene to early Oligocene: a comparison of southern high latitudes and lower latitudes. *Palaeogeography, Palaeoclimatology, Palaeoecology*, **214**, 243–263.

Schut, E. W., G. Uenzelmann-Neben, and R. Gersonde 2002. Seismic evidence for bottom current activity at the Agulhas Ridge. *Global Planetary Change*, **34**, 185–198.

Smith, A. G., D. G. Smith, and B. M. Funnell 1994. *Atlas of Mesozoic and Cenozoic Coastlines*. Cambridge, UK: Cambridge University Press.

Somoza, R. and M. E. Ghidella 2005. Convergencia en el margen occidental de América del Sur durante el Cenozoico: subducción de las placas de Nazca, Farallón y Aluk. *Revista de Asociación Geológica Argentina*, **60**, 797–809.

Stickley, C. E., H. Brinkhuis, S. A. Schellenberg, A. Sluijs, U. Rohl, M. D. Fuller, M. Grauert, M. Huber, J. Warnaar, and G. L. Williams 2004. Timing and nature of the deepening of the Tasmanian Gateway. *Paleoceanography*, **19**, 1–18.

Toggweiler, J. R. and H. Bjornsson 2000. Drake Passage and paleoclimate. *Journal of Quaternary Science*, **15**, 319–328.

Vennemann, T. W., H. C. Fricke, R. E. Blake, J. R. O'Neil, and A. Colman 2002. Oxygen isotope analysis of phosphates: a comparison of techniques for analysis of Ag_3PO_4. *Chemical Geology*, **185**, 321–336.

Wolfe, J. A. 1971. Tertiary climatic fluctuations and methods of analysis of Tertiary floras. *Palaeogeography, Palaeoclimatology, Palaeoecology*, **9**, 27–57.

Wolfe, J. A. 1978. A paleobotanical interpretation of Tertiary climates in the Northern Hemisphere. *American Scientist*, **66**, 694–703.

Wolfe, J. A. 1992. Climatic, floristic, and vegetational changes near the Eocene/Oligocene boundary in North America. In Prothero, D. R. and Berggren, W. A. (eds.), *Eocene–Oligocene Climatic and Biotic Evolution*. Princeton, NJ: Princeton University Press, pp. 421–436.

Wolfe, J. A. and R. Z. Poore 1982. Tertiary marine and nonmarine climatic trends. In Berggren, W. A. and Crowell, J. A. (eds.), *Climate in Earth History*. Washington, DC: National Academy Press, pp. 154–158.

Zachos, J. C., L. D. Stott, and K. C. Lohmann 1994. Evolution of early Cenozoic marine temperatures. *Paleoceanography*, **9**, 353–387.

Zachos, J., M. Pagani, L. Sloan, E. Thomas, and K. Billups 2001. Trends, rhythms, and aberrations in global climate 65 Ma to present. *Science*, **292**, 686–693.

Zanazzi, A., M. J. Kohn, B. J. MacFadden, and D. O. Terry Jr. 2007. Large temperature drop across the Eocene–Oligocene transition in central North America. *Nature*, **445**, 639–642.

24 Hypsodonty and body size in rodent-like notoungulates

Marcelo A. Reguero, Adriana M. Candela, and Guillermo H. Cassini

Abstract

Four families of endemic rodent-like notoungulates of the Suborder Typotheria ("Archaeohyracidae," Interatheriidae, Hegetotheriidae, and Mesotheriidae) show a high degree of hypsodonty. We evaluate the evolutionary pattern expressed by the variability of hypsodonty and body mass in this group, and consider whether the traits are correlated. We used hypsodonty and body mass data for 36 species of Typotheria to test for an association between these features. Major evolutionary changes in hypsodonty and body size in Typotheria occurred during the Eocene–Oligocene transition (EOT) and it is well documented in the Patagonian fossil record at Gran Barranca. At least two evolutionary trends are recognized in two clades of Typotheria: (1) in Mesotheriidae an increasing of the hypsodonty, associated with increasing of the body mass in species with complex occlusal designs, and (2) the Hegetotheriidae are characterized by possessing high crown teeth with simplified occlusal designs and small body sizes (Hegetotheriidae Pachyrukhinae). The Hegetotheriidae Pachyrukhinae show the highest indices of hypsodonty and were the smallest typotheres.

Resumen

Cuatro familias de notoungulados rodentiformes del Suborden Typotheria ("Archaeohyracidae," Interatheriidae, Hegetotheriidae y Mesotheriidae) muestran un alto grado de hipsodoncia. Nosotros evaluamos el patrón evolutivo expresado por la variabilidad de la hipsodoncia y la masa corporal en ese grupo. Datos de hipsodoncia y masa corporal de 36 especies de Typotheria fueron utilizados para explorar la asociación entre estos caracteres. Los mayores cambios evolutivos en hipsodoncia y tamaño corporal ocurren durante la transición Eoceno–Oligoceno (EOT) y están bien documentados en el registro patagónico de Gran Barranca. Al menos dos tendencias evolutivas fueron reconocidas en dos clados de Typotheria: 1) en los Mesotheriidae hay un incremento de la hipsodoncia, asociado a un incremento de la masa corporal en especies con diseños oclusales complejos, y 2) los Hegetotheriidae se caracterizan por poseer molariformes de coronas muy altas con diseños oclusales muy simplificados y tamaño corporal pequeño (Hegetotheriidae Pachyrukhinae).

The Paleontology of Gran Barranca: Evolution and Environmental Change through the Middle Cenozoic of Patagonia, eds. R. H. Madden, A. A. Carlini, M. G. Vucetich, and R. F. Kay. Published by Cambridge University Press. © Cambridge University Press 2010.

Los Hegetotheriidae Pachyrukhinae muestran los índices de hipsodoncia más altos y fueron los tipoterios más pequeños.

Introduction

Notoungulates are unique in having attained impressive morphologic and taxonomic diversity (13 or so families, at least 140 genera: Croft 1999) without leaving a single living representative. Among them, the Typotheria, so-called "rodent-like" notoungulates, experienced a broad adaptive radiation including small rodent-like (Interatheriidae), rabbit-like (Hegetotheriidae), and sheep-like (Mesotheriidae) forms.

The most noteworthy period of concentrated evolutionary change in morphological enhancements to tooth functional longevity (hypsodonty) and body size in Typotheria occurred during the Eocene–Oligocene transition (EOT) and it is well documented in the Sarmiento Formation at Gran Barranca (Chubut). Gran Barranca contains the most complete sequences of Paleogene continental sediments known in South America. A refined local biostratigraphy has allowed assignment of most mammalian-bearing horizons to specific biostratigraphic zones. Large samples of specimens of Typotheria are available from stratigraphically controlled localities throughout this sequence.

The rodent-like notoungulates were grouped by Simpson (1967) in the suborder Typotheria, and defined by him as notoungulates with "cheek teeth brachydont in earliest and most primitive forms, but with accelerated tendency toward hypsodonty even in the early faunas and by Deseadan time all hypsodont and most rootless" (Simpson 1967, p. 15). His concept of the Typotheria included such primitive and brachyodont families as Oldfieldthomasiidae and Archaeopithecidae together with Interatheriidae. The latest systematic and phylogenetic revision of Typotheria (Reguero and Prevosti this book) groups in this suborder only the hypsodont, small to medium-size notoungulates: Campanorcidae, "Archaeohyracidae," Interatheriidae, Hegetotheriidae, and Mesotheriidae.

During the EOT vertebrate faunas from Gran Barranca show a great radiation of two families of Typotheria, "Archaeohyracidae" and Interatheriidae, possessing high-crowned incisors and cheek teeth that we today associate with life in

362

"open" habitats (Simpson 1951; Webb 1983; MacFadden 2000a, 2000b; Janis *et al.* 2000). The oldest example of a hypsodont taxon (*Acropithecus* in the Vacan, early Eocene) of Patagonia antedates the first appearance of members of hypsodont clades in the North American fossil record (Oreodontidae, Leporidae, Castoridae, Geomyidae, Heteromyidae).

Patterson and Pascual (1968) highlighted "the precocity shown by certain ungulates in the acquisition of high-crowned, or hypsodont, and rootless, or hypselodont, teeth. Notohippids by mid-Eocene time were considered as advanced in this respect as equines were in the early Pliocene, about 30 million years later."

The South American continental biotic record was temporally calibrated against the record of climatic change by Kay *et al.* (1999) in Gran Barranca (for a revision, see Ré *et al.* Chapter 4, this book).

Although the dentitions of these notoungulates are well represented in the fossil record, and many other aspects of the typothere evolutionary history are well studied, their dietary habitats have been described only in the broadest terms.

In ungulates hypsodonty is widely considered an adaptation for coping with the high rate of dental wear caused by abrasive diets, such as grazing, due to high amounts of fiber, silica, or exogenous grit (dust, sand, ash) on the surface of ingested food (Simpson 1944; Fortelius 1985; Janis 1988; Pérez Barbería and Gordon 2001; Williams and Kay 2001). This evolutionary hypothesis is supported by an apparent correlation of crown height with ecological factors (e.g. diets, habitats) among modern ungulates. Janis (1988) found that crown height was associated with open habitats, but not directly related to the plants ingested. Recently, Williams and Kay (2001) examined several factors, i.e. coevolution of climate, diet, and feeding behavior, with crown height after removing phylogenetic effects; they did not find a relationship between crown height and climate. Williams and Kay (2001) found that tooth crown height was positively correlated with the proportion of monocots (primarily grass) in the diet and was negatively correlated with foraging height preferences.

The evolutionary trend of hypsodonty seen in different early and middle Miocene North American ungulate lineages is often considered as an indication of coevolution with a savanna grassland vegetation (Simpson 1944; Webb 1977; Fortelius 1985; Janis 1988, 1993; Pérez-Barbería and Gordon 1998a, 1998b). However, Mihlbachler and Solounias (2006) in a mesowear study of the North American Artiodactyla Merycoidodontidae found no clear indication that the overall trend of increasing dietary abrasion imposed sufficient selection to drive crown height evolution. In another recent report (Strömberg 2005) demonstrates that habitats dominated by C_3 grasses in the Central Great Plains were established at least 4 m.y. prior to the emergence of hypsodonty among Holarctic equids (Strömberg 2005). This retardation in the evolution of full hypsodonty was suggested as a result of a weak or fluctuating selection pressures and/or of the phylogenetic inertia (Strömberg 2005).

The hypsodonty exhibited in several lineages of South American ungulates, especially during the EOT, has been considered as associated to environmental changes (Pascual *et al.* 1996), but there is yet no precise analysis of the time of the first appearance and history of change of selective forces that might have acted on hypsodonty. In this chapter we examine the timing of the appearance of hypsodonty. Additionally, we explore the potential relationships between the evolutionary acquisition of hypsodonty and other morphological features, such as body size and occlusal designs, currently considered as very important to understand the biology of organisms.

Materials and methods

Taxa

"Archaeohyracids" are small, somewhat gliriform typotheres, defined by Simpson (1967, p. 104) as "Early notoungulates with accelerated hypsodonty (not reaching continuous growth)" (Fig. 24.1A).

Interatheriids are small, somewhat gliriform typotheres, most of which approximated the size of a house cat (*Felis sylvestris*). They are often very common in the faunas in which they occur but are rare or absent in some mid-latitude faunas (i.e. those between 10° and 23° S latitude) of Chile and Bolivia (Croft *et al.* 2004) (Fig. 24.1B).

Mesotheriids are small to medium typotheres (mostly between 5 kg and 50 kg: Croft *et al.* 2004) that were likely adapted for a fossorial lifestyle (*sensu* Hildebrand 1985; Shockey *et al.* 2004). They are very common in some Oligocene and early Miocene faunas and are moderately abundant in most middle Miocene to Pleistocene faunas, but have not been recorded in any low-latitude faunas (i.e. those north of 10° S: Croft *et al.* 2003, 2004; Reguero and Castro 2004) (Fig. 24.1C).

Hegetotheriids are similar in size to interatheriids, and some of the later representatives were very similar in overall morphology to modern rabbits (leporids) or certain caviomorph rodents (e.g. *Dolichotis*) (Elissamburu 2004; Reguero *et al.* 2007). Two hegetotheriid subfamilies are generally recognized: Hegetotheriinae and Pachyrukhinae. Hegetotheriinae is likely paraphyletic, though it may include a monophyletic subset of Miocene taxa (Cifelli 1993; Croft 2000; Croft and Anaya 2004; Croft *et al.* 2004; Reguero and Prevosti this book). Pachyrukhinae is universally considered monophyletic (Cerdeño and Bond 1998; Reguero and Prevosti this book) and the clade is certainly recognizable as early as the Deseadan (Dozo *et al.* 2000) and potentially as early as the Tinguirirican (Reguero 1993) (Fig. 24.1D).

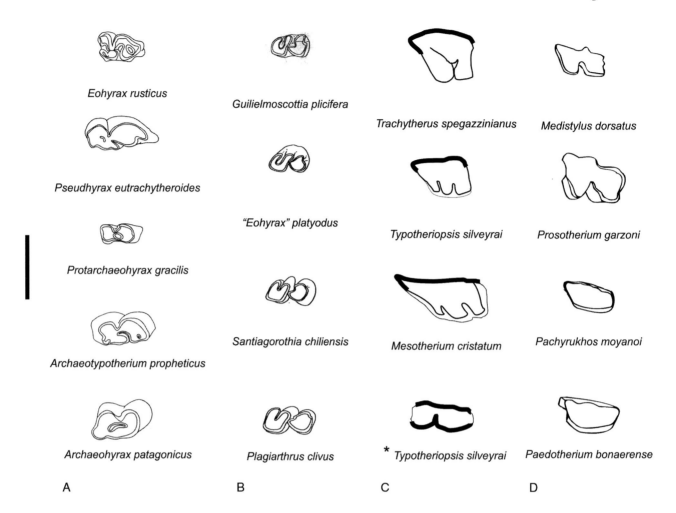

Fig. 24.1. Occlusal schematic views of left m1 (A–B) and left M1 (C–D, except tagged * left m1). (A) "Archaeohyracidae";
(B) Interatheriidae; (C) Mesotheriidae; (D) Pachyrukhinae. Scale bar 1 cm.

Data

Specimens analyzed are housed at Museo Provincial "Egidio Feruglio," Trelew, Argentina, Museo de La Plata, La Plata, Argentina, and Museo de Ciencias Naturales "Bernardino Rivadavia," Buenos Aires, Argentina.

Estimation of body size is a controversial issue (Fortelius 1990; Smith 2002). We use specific body masses (g) for a total of 36 species of Typotheria. Although estimating body mass using only dental remains is more error-prone than if the estimates were based on skeletons, most of the species in Typotheria are known by dental remains, so we used dental measurements for estimating body mass.

The masses were predicted using regression equation of Damuth (1990) of body mass, based on first molar length calculated from recent non-selenodont ungulates:

$$\log_{10} \text{ body mass} = 3.17 * \log 10 \text{ (m1 length)} + 1.04.$$

Molar length was used instead of area because length values tend to be more highly correlated with body mass in living ungulates (Janis 1990). Accurate molar crown height data are very difficult to obtain, particularly among hypsodont ungulates, because it is difficult to find specimens at the precise age where a tooth crown is fully formed yet unworn. Among the material examined, m1 was generally unmeasurable, so we used m2, the only tooth for which we were able consistently to find examples with a measurable unworn tooth for each species. Samples used for the study are listed in Appendix 24.1. We followed Janis (1988) by calculating a crown height ratio (m2 ratio) as the crown height divided by the labiolingual width of the tooth. Crown height was measured on the labial side of the tooth at the hypoflexid.

Statistical analyses

The comparison of body size with hypsodonty using a reduced major axis regression analysis was performed using PAST software version 1.86d Hammer *et al.* (2001). Because the body mass variable departures from normality,

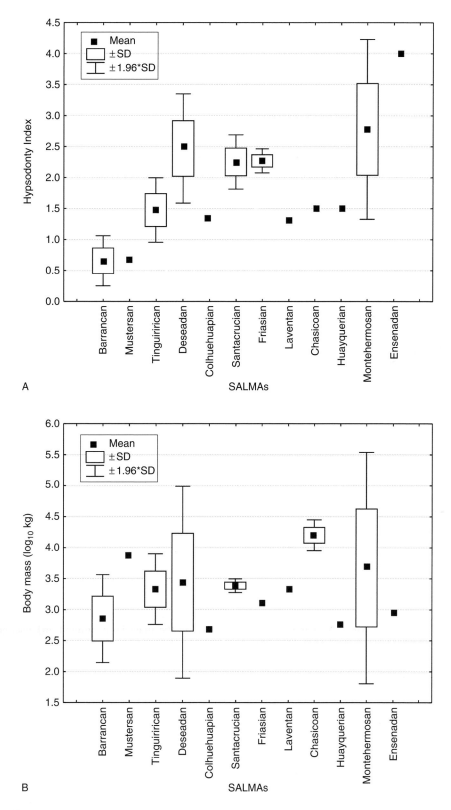

Fig. 24.2. Box plot of hypsodonty index (A) and body mass, $\log_{10}$ transformed (B) in Typotheria through time (represented by SALMAS).

Table 24.1. *Results of simple linear regression for each group*

Taxon	Range of W (g)	n	R^2	Intercept at $W = 1$ g	Standard error	Slope	Standard error	Significance p value (*d.f.*)
Typotheria	224–73435	36	$1.6E^{-04}$	−1.501	0.6261	1.045	0.1792	0.941 (*1, 34*)
Mesotheriidae	2358–73435	11	0.337	−0.0919	0.5822	0.5366	0.1456	0.0611 (*1, 9*)
Hegetotheriidae	224–13681	11	0.543	6.748	0.8349	−1.289	0.2904	<0.01* (*1, 9*)
Interatheriidae	283–2817	7	0.182	−3.426	1.8726	1.797	0.7268	0.340 (*1, 5*)
Typotheria (-H)	224–73435	25	0.255	−1.416	0.5913	0.911	0.1639	<0.01* (*1, 23*)

Note: The regressions were not significant in all groups at p level of 0.05 except for Hegetotheriidae and Typotheria (without Hegetotheriidae) tagged by an asterisk ($p = 0.01$).

this variable was $\log_{10}$ transformed. The hypsonty index (HI) and $\log_{10}$–transformed body mass are normally distributed (Shapiro Wilk: *W*: 0.968, *p* 0.375; and Shapiro–Wilk: *W*: 0.944, *p* 0.066, respectively). When the regression was significant, outliers and residual distribution were examined.

Results

Species of Typotheria increase in hypsodonty through time (Fig. 24.2A), whereas the body mass shows no identifiable pattern (Fig. 24.2B). This may be the result of processes of change of body mass and hypsodonty operating differently among the diverse lineages of Typotheria.

The initiation and increase of hypsodonty among typotheres is detected in Gran Barranca in the Tinguirirican and Deseadan SALMAs (Fig. 24.2A). However, Kohn *et al.* (2004) considered that this phenomenon had begun previously, between 39.3 and 38.0 Ma, ~5 m.y. prior to the EOT. The increase in grass abundance (Mazzoni 1979) and possible small increase in $\delta^{18}O$ values in teeth at that time suggest that a small decrease in relative humidity (~5%) could have driven hypsodonty (Kohn *et al.* 2004). Thus, the rise of hypsodonty in South America could reflect climate change and its impact on floras, but probably started prior to the EOT, which seems insignificant climatically. As is depicted in Fig. 24.2A the increase of the hypsodonty is continuous through the Oligocene in Patagonia when the highest diversity at the family level was recorded. Outside Patagonia, high values of hypsodonty are recorded in the Montehermosan when Typotheria show a decline in its diversity, and one of the clades, Hegetotheriidae, experienced a radiation of the smallest representantives (Pachyrukhinae).

Figure 24.2B shows no clear tendency of change in body mass from the late Eocene to middle Miocene in Patagonia, a period when the hypsodonty increases and the highest

diversity of Typotheria is recorded. A possible increase of body mass seems to appear from the Colhuehuapian to the Chasicoan, but the few available individuals make this possible tendency uncertain.

Regression analyses for Typotheria indicate that there was not a significant relation between hypsodonty and body mass (see Table 24.1, Fig. 24.3A), so the body mass would have no influence on variation of hypsodonty. However, it is possible that this result, obtained through all the species of Typotheria (which pertain in turn to more than one clade; see Reguero and Prevosti this book), express variation due to both within and among taxa (Harvey and Page 1991). Therefore, the relationships between both characters were also analyzed separately at different taxonomic groups. The regression of hypsodonty on $\log_{10}$ body mass considering only Hegetotheriidae is significant, showing a negative slope (Table 24.1, Fig. 24.3C), whereas for Mesotheriidae and Interatheriidae the regressions were not significant (Table 24.1, Fig. 24.3B). However, when the analysis was performed without Hegetotheriidae the regression resulted as significant with a positive slope (Table 24.1, Fig. 24.3D). This suggests the existence of different relationships between these characters within different clades.

Discussion

As interpreted for all ungulates, hypsodonty is thought to arise as a consequence of various dietary and environmental factors: (1) high silica phytolith abundance in especially coarse grasses, (2) prevalence of grass life forms with areas of exposed soil around them, (3) high levels of soil disturbance or soil mineral mobility, (4) large areas of continuously available accumulations of volcanic ash (and other potential sources of mineral dust) subject to erosion–entrainment–transport–deposition cycles extending over evolutionary timescales. The advantage of hypsodonty is to increase the useful life of the dental battery and allow

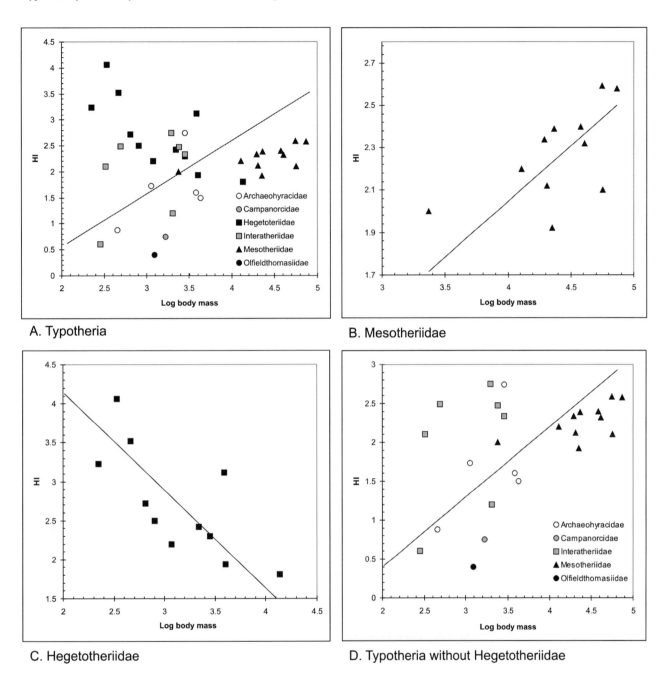

A. Typotheria

B. Mesotheriidae

C. Hegetotheriidae

D. Typotheria without Hegetotheriidae

Fig. 24.3. Relationship between hypsodonty index (HI) and body mass ($\log_{10}$ transformed).

more abrasive material to be processed without shortening the functional life of the teeth (Radinsky 1984). McNaughton *et al.* (1985) observed that herbivores that feed close to the ground ingest significant quantities of gritty dust along with their plant food. In extant ungulates there is correlation between hypsodonty and exogenous grit incorporation, when feeding close to the ground in open habitats (Williams and Kay 2001; Mendoza and Palmqvist 2008). Abrasive materials

(whether from phytoliths in grass or from exogenous grit) associated with open areas may have been important selective factors acting on hypsodonty in typotherian notoungulates. At Gran Barranca, the specimens of Typotheria recorded before and near the time of the EOT document increasing hypsodonty (Fig. 24.2A). The initiation of an evolutionary response to increased tooth abrasion seems to be related to local change in vegetation and eolian deposition.

368

M. A. Reguero et al.

For some hypsodont mammals, body size has been proposed to be correlated with increased hypsodonty, a result that would be manifested by a positive allometric scaling to accommodate the metabolic requirements of mammals with larger body size or to permit extend tooth use in mammals with longer lifespans (Simpson 1944; Radinsky 1984). But studies of extant and fossil Holarctic ungulates have rejected these hypotheses (Fortelius 1985; Janis 1988; Solounias *et al.* 1994).

In at least some clades of Typotheria the variation of hypsodonty may be reflecting correlated changes due to changes in body size plus a variation due to directional selection, independent of body size. Our observations indicate that at least two evolutionary trends can be identified in Typotheria: (1) one characterized by increasing hypsodonty, associated with increasing body size in species with complex occlusal designs (*Mesotherium cristatum*: Figs. 24.1C, 24.3B), and (2) another characterized by increasing hypsodonty associated with decreasing body mass in species with simplified occlusal designs (*Paedotherium bonaerense*: Figs. 24.1D, 24.3C). The highest indexes of hypsodonty are seen in species with the most simple designs (Hegetotheriidae).

Occlusal designs seem to play an important role in the relationship between hypsodonty and body size, and these three factors would form a complex of characters, all being targets selection. In the case of larger species, complex occlusal designs would increase the effectiveness of chewing (Strömberg 2005).

This understanding of selective factors and adaptive response of hypsodonty in association with other morphological features, such as body mass and occlusal designs, in Typotheria, can only be tested with a higher-resolution phytolith record at Gran Barranca, or by other plant fossil material allowing a detailed reconstruction of vegetation changes during Eocene to Miocene in southern South America. Whether hypsodonty arose coincidently with or just after the spread of open grass-dominated habitats in South America remains to be tested.

We hope that the understanding of the variation of these features may be useful to test hypotheses regarding the adaptive nature of hypsodonty in South American ungulates, and its value as an indicative feature of environmental changes during South American Paleogene.

Appendix 24.1 Chronological distribution, number of specimens (n), body mass (in g and $\log_{10}$ transformed), and hypsodonty index (HI) for the species of Typotheria studied

Taxa	Family	SALMA[a]	n	Body mass (g) Mean	Body mass (g) $\log_{10}$	HI Mean
Oldfieldthomasia debilitata	Oldfieldthomasiidae	BAR	3	1 238	3.093	0.40
Punahyrax bondesioi	"Archaeohyracidae"	BAR	2	453	2.656	0.88
Notopithecus adapinus	Interatheriidae	BAR	3	283	2.452	0.60
Typotheriopsis silveyrai	Mesotheriidae	CHA	2	23 244	4.366	1.93
Cochilius volvens	Interatheriidae	COL	2	490	2.691	2.49
Archaeohyrax patagonicus	"Archaeohyracidae"	DES	3	2 817	3.450	2.74
Sallatherium altiplanense	Hegetotheriidae	DES	2	2 817	3.450	2.30
Propachyrucos smithwoodwardii	Hegetotheriidae	DES	1	798	2.902	2.50
Prosotherium garzoni	Hegetotheriidae	DES	2	644	2.809	2.72
Pachyrukhos moyanoi	Hegetotheriidae	DES	3	461	2.664	3.52
Prohegetotherium schiaffinoi	Hegetotheriidae	DES	3	1 178	3.071	2.20
Anatrachytherus soriai	Mesotheriidae	DES	1	25 025	4.398	1.92
Trachytherus spegazzinianus	Mesotheriidae	DES	3	40 752	4.610	2.32
Trachytherus curuzucuatiense	Mesotheriidae	DES	1	56 694	4.754	1.85
Plagiarthrus clivus	Interatheriidae	DES	3	2 063	3.314	2.75
Archaeophylus patrius	Interatheriidae	DES	2	333	2.522	2.10
Mesotherium cristatum	Mesotheriidae	ENS	3	73 436	4.866	2.02
Microtypotherium choquecotense	Mesotheriidae	FRI	1	12 927	4.112	2.20
Eutypotherium lehmannitschei	Mesotheriidae	FRI	2	19 551	4.291	2.03
Hemihegetotherium achataleptum	Hegetotheriidae	HUA	2	13 681	4.136	1.81
Miocochilius anomopodus	Interatheriidae	LAV	2	2 393	3.379	2.47
Tremacyllus impressus	Hegetotheriidae	MAR	3	224	2.351	3.23
Paedotherium bonaerense	Hegetotheriidae	MAR	3	340	2.531	4.06
Hemihegetotherium sp. nov.	Hegetotheriidae	MON	1	3 909	3.592	3.12
Hemihegetotherium torresi	Hegetotheriidae	MON	2	4 032	3.606	3.61
Plesiotypotherium achirense	Mesotheriidae	MON	1	20 359	4.309	2.12
Pseudotypotherium hystatum	Mesotheriidae	MON	2	38 014	4.580	2.10
Mesotherium maendrum	Mesotheriidae	MON	2	55 656	4.746	2.00
Pseudhyrax eutrachytheroides	"Archaeohyracidae"	MUS	2	3 825	3.583	1.60
Hegetotherium mirabile	Hegetotheriidae	SAN	2	2 190	3.340	2.42
Eotypotherium chico	Mesotheriidae	SAN	1	2 358	3.373	2.00
Protypotherium australe	Interatheriidae	SAN	4	2 817	3.450	2.34
Archaeotypotherium propheticus	"Archaeohyracidae"	TIN	3	4 285	3.632	1.50
Protarchaeohyra × gracilis	"Archaeohyracidae"	TIN	3	1 125	3.051	1.73
Santiagorothia chiliensis	Interatheriidae	TIN	2	2 030	3.307	1.20
Campanorco inauguralis	Campanorcidae	VAC	1	1 665	3.221	0.75

Note: [a]*SALMA abbreviations (ordered temporally): VAC, Vacan; BAR, Barrancan; MUS, Mustersan; TIN, Tinguirirican; DES, Deseadan; COL, Colhuehuapian; SAN, Santacrucian; FRI, Friasian; LAV, Laventan; CHAS, Chasicoan; HUA, Huayquerian; MON, Montehermosan; MAR, Marplatan; ENS, Ensenadan.*

REFERENCES

Cerdeño, E. and M. Bond 1998. Taxonomic revision and phylogeny of *Paedotherium* and *Tremacyllus* (Pachyrukhinae, Hegetotheriidae, Notoungulata) from the late Miocene to Pleistocene of Argentina. *Journal of Vertebrate Paleontology*, **18**, 799–811.

Cifelli, R. L. 1993. The phylogeny of the native South American ungulates. In Szalay, F. S., Novacek, M. J., and McKenna, M. C. (eds.), *Mammal Phylogeny: Placentals*. New York: Springer-Verlag, pp. 195–216.

Croft, D. A. 1999. Placentals: South American ungulates. In Singer, R. (ed.), *The Encyclopedia of Paleontology*. Chicago, II: Fitzroy-Dearborn, pp. 890–906.

Croft, D. A. 2000. Archaeohyracidae (Mammalia, Notoungulata) from the Tinguiririca Fauna, central Chile, and the evolution and paleoecology of South American mammalian herbivores. Ph.D. thesis, University of Chicago.

Croft, D. A. and F. Anaya 2004. A new hegetotheriid from the middle Miocene of Quebrada Honda, Bolivia, and a phylogeny of the Hegetotheriidae. *Journal of Vertebrate Paleontology*, **24**(3 suppl.), 48–49A.

Croft, D. A., J. J. Flynn, and A. R. Wyss 2003. Diversification of mesotheriids (Mammalia: Notoungulata: Typotheria) in the middle latitudes of South America. *Journal of Vertebrate Paleontology*, **23**(3 suppl.), 43A.

Croft, D. A., J. J. Flynn, and A. Wyss 2004. Notoungulata and Litopterna of the early Miocene Chucal Fauna, northern Chile. *Fieldiana, Geology*, n.s., **50**, 1–52.

Damuth, J. 1990. Problems in estimating body masses of archaic ungulates using dental measurements. In Damuth, J. and MacFadden, B. J. (eds.), *Body Size in Mammalian Paleobiology: Estimation and Biological Implications*. Cambridge, UK: Cambridge University Press, pp. 229–253.

Dozo, M. T., M. A. Reguero, and E. Cerdeño 2000. *Medistylus dorsatus* (Ameghino, 1903), un Hegetotheriidae Pachyrukhinae (Mammalia, Notoungulata) del Deseadense de la provincia de Chubut, Argentina. *Ameghiniana*, **37**, 24R.

Elissamburu, A. 2004. Morphometric and morphofunctional analysis of the appendicular skeleton of *Paedotherium* (Mammalia, Notoungulata). *Ameghiniana*, **41**, 363–380.

Fortelius, M. 1985. Ungulate cheek teeth: developmental, functional, and evolutionary interrelations. *Acta Zoologica Fennica*, **180**, 1–76.

Fortelius, M. 1990. The mammalian dentition: a "tangle" view. *Netherland Journal of Zoology*, **40**, 312–328.

Hammer, O., D. A. T. Harper, and P. D. Ryan 2001. PAST: PAleontological STatistics software package for education and data analysis. *Palaeontologia Electronica*, **4**(1), 1–9.

Harvey, P. H. and M. D. Pagel 1991. *The Comparative Method in Evolutionary Biology*. Oxford, UK: Oxford University Press.

Hildebrand, M. 1985. Digging of quadrupeds. In Hildebrand, M. Bramble, D. M., Liem, K. F., and Wake, D. B. (eds.), *Functional Vertebrate Morphology*. Cambridge, MA: Belknap Press of Harvard University Press, pp. 90–108.

Janis, C. M. 1988. Estimation of tooth volume and hypsodonty indices in ungulate mammals, and the correlation of these factors with dietary preference. In Russell, E., Santoro, J. P., and Signogneau-Russell, D. (eds.), *Teeth Revisited: Proceedings VII International Symposium on Dental Morphology, Mémoires du Muséum National d'Histoire Naturelle*, Serie C, **53**, 367–387.

Janis, C. M. 1990. The correlation between diet and dental wear in herbivorous mammals, and its dental relationship to the determination of diets in extinct species. In Boucot, J. (ed.), *Evolutionary Paleobiology of Behavior and Coevolution*. Amsterdam: Elesvier, pp. 241–260.

Janis, C. M. 1993. Tertiary mammal evolution in the context of changing climates, vegetation, and tectonic events. *Annual Reviews of Ecology and Systematics*, **24**, 467–500.

Janis, C. M., J. Damuth, and J. M. Theodor 2000. Miocene ungulates and terrestrial primary productivity: where have all the browsers gone? *Proceedings of the National Academy of Sciences USA*, **97**, 7899–7904.

Kay, R., R. H. Madden, M. G. Vucetich, A. A. Carlini, M. M. Mazzoni, G. H. Ré, M. Heizler, and H. Sandeman 1999. Revised geochronology of the Casamayoran South American Land Mammal Age: climatic and biotic implications. *Proceedings of the National Academy of Sciences USA* **96**, 13 235–13 240.

Kohn, M. J., J. A. Josef, R. Madden, R. Kay, M. G. Vucetich, and A. A. Carlini 2004. Climate stability across the Eocene–Oligocene transition, southern Argentina. *Geology*, **32**, 621–624.

MacFadden, B. J. 2000a. Origin and evolution of the grazing guild in Cenozoic New World terrestrial mammals. In Sues, H. D. (ed.), *Evolution of Herbivory in Terrestrial Vertebrates*. Cambridge, UK: Cambridge University Press, pp. 223–244.

MacFadden, B. J. 2000b. Cenozoic mammalian herbivores from the Americas: reconstructing ancient diets and terrestrial communities. *Annual Reviews of Ecology and Systematics*, **31**, 33–59.

Mazzoni, M. M. 1979. Contribución al conocimiento petrográfico de la Formación Sarmiento, barranca sur del lago Colhue-Huapi, provincia de Chubut. *Revista de la Asociación Argentina de Mineralogía Petrología y Sedimentología*, **10**, 33–54.

McNaughton, S. J., J. L. Tarrants, M. M. McNaughton, and R. H. Davis 1985. Silica as the defense against herbivory and a growth promoter in African grasses. *Ecology*, **66**, 528–535.

Mendoza, M. and P. Palmqvist 2008. Hypsodonty in ungulates: an adaptation for grass consumption or for foraging in open habitat? *Journal of Zoology*, **274**, 134–142.

Mihlbachler, M. C. and N. Solounias 2006. Coevolution of tooth crown height and diet in oreodonts (Merycoidodontidae, Artiodactyla) examined with phylogenetically independent contrasts. *Journal of Mammalian Evolution*, **13**, 11–36.

Pascual, R., E. Ortíz-Jaureguízar, and J. L. Prado 1996. Land-mammals: paradigm for Cenozoic South American geobiotic evolution. *Münchner Geowissenschaftliche Abhandlungen*, **30**(A), 265–319.

Patterson, B. and R. Pascual 1968. The fossil mammal fauna of South America. *Quarterly Review of Biology*, **43**, 409–451.

Pérez-Barbería, F. J. and I. J. Gordon 1998a. Factors affecting food comminution during mastication in herbivorous mammals: a review. *Biological Journal of the Linnean Society*, **63**, 233–256.

Pérez-Barbería, F. J. and I. J. Gordon 1998b. The influence of sexual dimorphism in body size and mouth morphology on diet selection and sexual segregation in cervids. *Acta Veterinaria Hungarica*, **46**, 357–367.

Pérez-Barbería, F. J., I. J. Gordon, and C. Nores 2001. Evolutionary transitions among feeding styles and habitats in ungulates. *Evolutionary Ecology Research*, **3**, 221–230.

Radinsky, L. B. 1984. Ontogeny and phylogeny in horse skull evolution. *Evolution*, **38**, 1–15.

Reguero, M. A. 1993. Los Typotheria y Hegetotheria (Mammalia: Notoungulata) eocenos de la localidad Cañadón Blanco, Chubut. *Ameghiniana*, **30**, 336.

Reguero, M. A. and P. V. Castro 2004. Un nuevo Trachytheriinae (Mammalia, †Notoungulata) del Deseadense (Oligoceno tardío) de Patagonia, Argentina: implicancias en la filogenia, biogeografía y bioestratigrafía de los Mesotheriidae. *Revista Geológica de Chile*, **31**, 45–64.

Reguero, M. A., M. T. Dozo, and E. Cerdeño 2007. *Medistylus dorsatus* (Ameghino, 1903), an enigmatic Pachyrukhinae (Hegetotheriidae, Notoungulata) from the Deseadan of the Chubut province, Argentina: systematics and paleoecology. *Journal of Paleontology*, **81**, 1301–1307.

Shockey, B. J., D. A. Croft, and F. Anaya 2004. Distinctive fossorial adaptations in mesotheriids (Mammalia: Notoungulata). *Journal of Vertebrate Paleontology*, **24**(3 suppl.), 112A.

Simpson, G. G. 1944. *Tempo and Mode in Evolution*. New York: Columbia University Press.

Simpson, G. G. 1951. *Horses: The story of the Horse Family in the Modern World and through Sixty Million Years of History*. Oxford, UK: Oxford University Press.

Simpson, G. G. 1967. The beginning of the Age of Mammals in South America. II. *Bulletin of the American Museum of Natural History*, **137**, 1–260.

Smith, R. J. 2002. Estimation of body mass in paleontology. *Journal of Human Evolution*, **43**, 271–287.

Solounias, N., M. Fortelius, and P. Freeman 1994. Molar wear rates in ruminants: a new approach. *Annales Fennici Zoologici*, **31**, 219–227.

Strömberg, C. A. E. 2005. Decoupled taxonomic radiation and ecological expansion of open-habitat grasses in the Cenozoic of North America. *Proceedings of the National Academy of Sciences USA*, **102**, 11 980–11 984.

Webb, S. D. 1977. A history of savanna vertebrates in the New World. I. North America. *Annual Reviews of Ecology and Systematics*, **8**, 355–380.

Webb, S. D. 1983. The rise and fall of the late Miocene ungulate fauna in North America. In Nitecki, M. H. (ed.), *Coevolution*. Chicago, IL: University of Chicago Press, pp. 267–306.

Williams, H. S. H. and R. Kay 2001. A comparative test of adaptive explanations for hypsodonty in ungulates and rodents. *Journal of Mammalian Evolution*, **8**, 207–229.

PART IV REGIONAL APPLICATIONS

25 Vegetation during the Eocene–Miocene interval in central Patagonia: a context of mammal evolution

Viviana Barreda and Luis Palazzesi

Abstract

The major vegetation trends in the context of the mammal evolution, during the Eocene–Miocene in Patagonia, are analyzed on the basis of the paleobotanical and palynological records. The long-term dynamics of the vegetation were most likely controlled by changing climatic conditions, which was in turn linked to important paleogeographic and tectonic events. The major patterns of change in the vegetation were related to: (1) the shift from megatherm (palms, tropical vines, tree ferns) to micromesotherm (southern beech, gymnosperm) communities, from the Barrancan towards the Tinguiririran, and (2) the rise to importance of shrubby and herbaceous communities (grasses, sedges, and bushes) in extra-Andean Patagonia, from the Deseadan towards the Colhuehuapian. Even when the major trends in the floras were linked to the establishment of communities progressively adapted to cooler and drier conditions, some megatherm and hydrophilic plants persisted, mainly during the Deseadan and Pinturan time intervals.

Resumen

Las principales tendencias de la vegetación como marco de la evolución de los mamíferos, durante el Eoceno-Mioceno de Patagonia, son analizadas sobre la base de los registros paleobotánicos y palinológicos. Los grandes cambios en la flora fueron controlados por las variables condiciones paleoclimáticas imperantes, producto de los eventos paleogeográficos y tectónicos del momento. Los principales patrones de cambio observados en la vegetación se vinculan con: (1) el progresivo reemplazo de comunidades megatérmicas (con palmeras, lianas, helechos arborescentes) por otras micro-mesotérmicas (con *Nothofagus* y gimnospermas), particularmente evidente desde el SALMA Barranquense a la Tinguiririquense, y (2) el surgimiento y la expansión de comunidades herbáceo arbustivas (con pastos, juncos y arbustos) en la Patagonia extraandina, documentado en el intervalo Deseadense-Colhuehuapense. Si bien, las tendencias más importantes estuvieron vinculadas al establecimiento de comunidades progresivamente más adaptadas al frío y la aridez, en algunos momentos, principalmente en el Deseadense y en el Pinturense, todavía pudieron reconocerse elementos megatérmicos de preferencias húmedas.

Introduction

The extant Patagonian flora is a result of several past events. Prominent external forcing factors were responsible of the long-term paleoclimatic trends that led climate to shift from warm and humid in the early Paleogene to semi-arid and cold–temperate by the late Neogene. Such major patterns of change were linked, in part, to the fragmentation of Gondwana (split of Antarctica from Australia and South America), the thermal isolation of Antarctica, mountain-building processes (Zachos *et al.* 2001; Blisniuk *et al.* 2005), and changes in atmospheric CO_2 concentration. Vegetation communities experienced marked variations in composition, structure, distribution, and abundance throughout the Cenozoic, controlled by these paleoclimatic, paleogeographic, and paleoaltitudinal events. Fossil pollen and leaf remains, preserved at several sites in Patagonia, are useful tools to interpret these modifications.

The first attempts to analyze the evolution of the Cenozoic vegetation in southern South America were carried out by Menéndez (1971), Volkheimer (1971), and Romero (1978, 1986, 1993), but relevant data were relatively scant and the ages of the fossil-bearing deposits uncertain or not well established. During last few decades a progressive increase in paleobotanical (e.g. Romero and Zamaloa 1985; Barreda 1996; Barreda *et al.* 2003; Melendi *et al.* 2003; Wilf *et al.* 2003, 2005; Guerstein *et al.* 2004; Zamaloa and Romero 2005), stratigraphical, and chronological studies (Fleagle *et al.* 1995; Palamarczuk and Barreda 1998; Casadío *et al.* 2001; Wilf *et al.* 2005; Parras and Casadío 2006) has allowed tracing the floristic evolution of the Patagonian vegetation with more confidence.

Plants are important in providing habitat, refugia, and food for animal populations. However, a comprehensive comparison between the major stages in mammal and

The Paleontology of Gran Barranca: Evolution and Environmental Change through the Middle Cenozoic of Patagonia, eds. R. H. Madden, A. A. Carlini, M. G. Vucetich, and R. F. Kay. Published by Cambridge University Press. © Cambridge University Press 2010.

vegetation evolution from southern South America has never been published to date. In this chapter we attempt to summarize Patagonian vegetation trends during the Eocene–Miocene interval as a framework for mammal evolution.

Materials and methods

Plant fossil records are biased by the paleoenvironmental conditions in which the organisms grew and were deposited. In Patagonia, most paleobotanical and palynological assemblages are recovered from fine-grained layers deposited in near-shore marine environments. Gran Barranca, the major paleontological site that this book is about, is the type area of the Sarmiento Formation. This unit is dominated by massive fine tuffs formed from intense volcanism. Such environments were not suitable for plant preservation; hence floristic inferences are mostly based on plant fossil records coming from regional Patagonian outcrops (Fig. 25.1).

The Gran Barranca is the reference section for the continental middle Cenozoic sequences of South American Land Mammal Ages (SALMA): Barrancan, Mustersan, Tinguiririran, Deseadan, Colhuehuapian, and Pinturan. New Ar/Ar determinations, magnetic polarity studies, increased fossil mammal collections, and detailed stratigraphic analyses have been crucial in providing a more accurate age control for the SALMAs (Ré *et al.* Chapter 4, this book). At Gran Barranca, the Barrancan appears to represent an interval between 41.5 and 38.4 Ma (middle Eocene); the Mustersan between 38.4 and 36.6 Ma (middle–late Eocene); the Tinguiririran between 33.7 and 33.2 Ma; a pre-Deseadan fauna called "La Cantera" is between 31.1 and 29.5 Ma (late Eocene – early Oligocene); the Deseadan between 29.2 and 26.3 Ma (Oligocene); the Colhuehuapian between 20.2 and 20.0 Ma (early Miocene); and the Pinturan between 19.7 and 18.7 Ma (early Miocene).

Detailed paleoclimatic determinations of mean annual temperature (MAT), and mean annual precipitation (MAP) are available for only a few localities in Patagonia (Laguna del Hunco, Río Pichileufú). Most investigations do not include this specific analysis, so in this study only broad paleoclimatic inferences are presented. The terms megatherm (>24 °C), mesotherm (>14 °C, <20 °C), and microtherm (<12 °C), with two inter-zones, were used to express the plant response to major environmental variables such as light, temperature, and precipitation, following the classification of Nix (1982). Although these terms were based on the Australian biota, Paleogene and early Neogene communities from Patagonia shared several taxa with those of Australia, so they should be applicable.

Reconstructions were based on analogies with modern taxa. That means it is assumed that fossil taxa have ecological requirements similar to those of their closest living relatives.

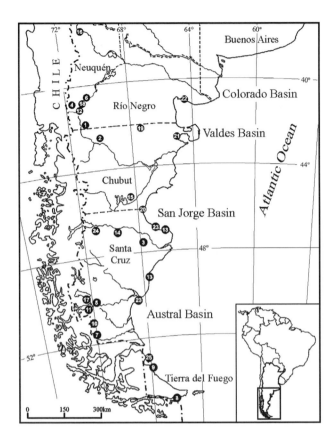

Fig. 25.1. Location map showing paleobotanical and palynological sites. 1, Laguna del Hunco Flora (Wilf *et al.* 2003, 2005, and references therein); 2, Río Pichileufú Flora (Wilf *et al.* 2005, and references therein); 3, Estancia Laguna Los Manantiales (Zamaloa and Andreis 1995); 4, La Huitrera Formation (lower section) (Melendi *et al.* 2003); 5, Calafate Formation (Sepúlveda and Norris 1982); 6, La Huitrera Formation (upper section) (Báez *et al.* 1990; Melendi *et al.* 2003); 7, Río Turbio Formation (Berry 1937; Hünicken 1955; Archangelsky 1972; Romero 1977; Romero and Zamaloa 1985; Romero and Castro 1986; Ancibor 1988, 1990; Brea 1993); 8, Sloggett Formation (Olivero *et al.* 1998); 9, Estancia La Sara (Menéndez and Caccavari 1975); 10, Río Guillermo Formation (Dusén 1907; Frenguelli 1941; Brandmayr 1945; Hünicken 1955); 11, Río Leona Formation (Dusén 1907; Frenguelli 1941; Brandmayr 1945; Hünicken 1955; Barreda *et al.* 2004; 2009; Pujana 2007); 12, Río Foyel Formation (Pöthe de Baldis 1984; Barreda *et al.* 2003); 13 (two sites), San Julián Formation (Pöthe de Baldis 1974; Barreda 1997; Barreda and Palamarczuk 2000a); 14, Deseado River, Sarmiento Formation (Spegazzini 1924); 15, Cañadón Hondo, Sarmiento Formation (Berry 1932); 16, Lileo Formation (Leanza *et al.* 2002); 17, Centinela Formation (Guerstein *et al.* 2004); 18, Ñirihuau Formation (Fiori 1939); 19, Sierra La Colonia locality (Archangelsky and Zamaloa 2003); 20, Chenque Formation (Romero 1970; Barreda 1993, 1996; Palamarczuk and Barreda 1998); 21, Gaiman Formation (Palazzesi and Barreda 2005); 22, Barranca Final Formation (Gamerro and Archangelsky 1981; Guerstein 1990a, 1990b; Guerstein and Quattrocchio 1988); 23, Monte León Formation (Barreda and Palamarczuk 2000b, 2000c); 24, Pinturas Formation (Zamaloa 1993); 25, Cullen Formation (Vergel and Durango de Cabrera 1988; Zamaloa and Romero 1990; Zetter *et al.* 1999; Zamaloa 2000, 2004; Zamaloa and Romero 2005).

		Coeval Mammal Faunas (SALMAs)	Botanical information	Inferred vegetation	% entire-margin leaves	Inferred paleoclimate
Miocene	M	"Pinturan." Colhuehuapian	20, 24, 25	New increase of megatherm elements – swamp communities and humid forests	No record	Warm–temperate Humid to subhumid
	E / Bu.		20, 21, 22, 23	Arid-adapted vegetation expanded	No record	Temperate – Subhumid to semi-arid
	Aq.		16, 17, 18, 19, 20	Patches of seasonally dry forests	No record	Temperate to warm–temperate–subhumid
Oligocene	L	Deseadan	11, 12, 13, 14, 15	Meso-megatherm forests –first occurrences of modern shrubby and herbaceous lineages – first Asteracean record (sunflower family)	No record	Warm–temperate Humid
	E	"La Cantera" Tinguiririan	8, 9, 10	Spreading of microtherm *Nothofagus* gymnosperm dominated forests	~27%	Temperate to cold–temperate Humid
Eocene	L	Mustersan Barrancan	5, 6, 7	First irruption of micro-mesotherm *Nothofagus* forests – some megatherm lineages still prevail – first poaceae records (grass family)	~40%	Temperate to Warm–temperate Humid
	M		1, 2, 3, 4	Megatherm communities	~69%	Warm–temperate Humid MAT 14-18°C MAP 1050-1250 mm
	E					

Fig. 25.2. Inferred plant communities and paleoclimatic trends from the Eocene–Miocene in Patagonia contrasted with the SALMAs. References for paleobotanical and palynological sites 1–25 are given in Fig. 25.1 caption. Estimated ages of floras 1–4 are based on: Melendi *et al.* 2003, Wilf *et al.* 2005; floras 5–7 on: Archangelsky 1968, Malumián 1999, Melendi *et al.* 2003; floras 8–10 on: Menéndez and Caccavari 1975, Olivero *et al.* 1998, Malumián 1999, Marenssi *et al.* 2005; floras 11–15 on: Casadío *et al.* 2001, Parras *et al.* 2008, Barreda *et al.* 2009; floras 16–20 on: Palamarczuk and Barreda 1998, Barreda and Palamarczuk 2000c, Guerstein *et al.* 2004, Parras and Casadío 2006; floras 21–23 on: Fleagle *et al.* 1995, Barreda and Palamarczuk 2000c; floras 24–25 on: Fleagle *et al.* 1995, Zamaloa 2000. Estimated ages of SALMAs are based on: Ré *et al.* Chapter 4 this book. Data on % of entire-margin leaves, are after Romero (1986).

Results

The plant communities inferred are shown in Fig. 25.2. These are presented against previous published paleoclimatic data, and the SALMAs from the Gran Barranca area. A brief analysis of the paleobotanical and palynological data is provided in the following paragraphs.

Paleobotanical records

The most abundant and diverse floras from the Cenozoic of southern South America are those from the early–middle Eocene Laguna del Hunco and Río Pichileufú floras (northwestern Patagonia). They have been investigated since the beginning of the last Century, and most recently summarized by Wilf *et al.* (2003, 2005). Isotopic data from $^{40}Ar/^{39}Ar$ indicates an age of 51.91 ± 0.22 Ma (early Eocene) for the Laguna del Hunco and 47.46 ± 0.05 Ma (middle Eocene) for the Río Pichileufú floras (Wilf *et al.* 2005). They present a high plant diversity (Wilf *et al.* 2003, 2005), with the presence of abundant megatherm families (palms, Myrtaceae *Myrcia*, Sapindaceae *Schmidelia* and

Cupania, Lauraceae, Rubiaceae *Coprosma*, Casuarinaceae *Gymnostoma*). Other taxa with broader climatic requirements include conifers, cycads, and ginkgoales (among gymnosperms), along with Proteaceae *Lomatia* and Malvales (among angiosperms). Arid-adapted elements represented by shrubs or small trees of Anacardiaceae (*Schinopsis*), Celtidaceae (*Celtis*), and Fabaceae (Caesalpinoideae, *Cassia*) are reported in some geographic areas (Hünicken 1955; Wilf *et al.* 2003, 2005). The first Poaceae remains, similar to modern *Chusquea* (tall Bambusoideae grasses) are also recorded (Frenguelli and Parodi 1941).

Middle to late Eocene data come from the Río Turbio Formation, southwestern Patagonia (Berry 1937; Hünicken 1955; Ancibor 1988, 1990; Brea 1993). An important change in vegetation is suggested by the invasion of *Nothofagus* forests, coinciding with the occurrence of other microtherm to mesotherm families such as Podocarpaceae, Araucariaceae, Myrtaceae, Cunoniaceae, Gunneraceae, Caryophyllaceae, and Proteaceae. Megatherm Lauraceae, Tiliaceae–Bombacaceae, Malpighiaceae, Sapindaceae, and Rubiaceae are documented as well (Hünicken 1955).

Poaceae is also reported (Berry 1937). Although this assignment needs further investigation, the coeval presence of phytolith remains related to grasses (Andreis 1972; Zucol and Brea 2005) would support Berry's determination.

Data on the early Oligocene come from the Río Guillermo Formation of southwestern Patagonia (Dusén 1907; Frenguelli 1941; Brandmayr 1945; Hünicken 1955). Assemblages are dominated by austral families such as Araucariaceae, Nothofagaceae, and Podocarpaceae. Anacardiaceae and Myrtaceae are also present. Leaves are smaller than those reported from the early Eocene suggesting cooler conditions (Romero 1986). There are no recorded megatherm elements.

Late Oligocene to early Miocene paleobotanical data are scanty, restricted to leaf imprints and woody remains from Cañadón Hondo (Berry 1932) and Deseado River (Spegazzini 1924) localities (assigned to the Sarmiento Formation, *Pyrotherium* Zone of Deseadan SALMA); from Río Leona Formation at southwestern Patagonia (Barreda *et al.* 2004; Pujana 2007); and from Ñirihuau and Chenque formations (Fiori 1939; Romero 1970). Forests were dominated by southern elements of the Podocarpaceae, Nothofagaceae, Araucariaceae, and Proteaceae, along with fern communities. Megafossils of Nothofagaceae are also reported from the Sarmiento Formation at Cañadón Hondo and Deseado river areas (Spegazzini 1924; Berry 1932). Megatherm Lauraceae (*Ulminium atlanticum*) were still present in central Patagonia (Romero 1970). Fossil Poaceae leaves are recorded from the Ñirihuau Formation (Fiori 1939).

Palynological records

Early to middle Eocene palynological records come from two localities. One is in northwestern Patagonia (La Huitrera Formation, lower section, Nahuel Huapi Este locality), and the other is in southeastern Patagonia (Estancia Laguna Manantiales area) (Zamaloa and Andreis 1995; Melendi *et al.* 2003). Assemblages are dominated by gymnosperms (Podocarpaceae, Araucariaceae). Angiosperms with megatherm characters are abundant, particularly palms and Tiliaceae–Bombacaceae, while *Nothofagus* pollen grains are absent or very scarce.

Middle to late Eocene data come from outcrops of the La Huitrera (northwest Patagonia, near Confluencia), Río Turbio and Calafate (southwest Patagonia) Formations, and core samples from the Estancia La Sara as well (southernmost Patagonia) (Archangelsky 1972; Menéndez and Caccavari 1975; Romero 1977; Romero and Zamaloa 1985; Romero and Castro 1986; Sepúlveda and Norris 1982; Báez *et al.* 1990; Melendi *et al.* 2003). Abundant *Nothofagidites* pollen grains were recorded in these assemblages associated with other taxa related to Gondwanan lineages (Podocarpaceae, Araucariaceae, Myrtaceae, Cunoniaceae, Gunneraceae, Caryophyllaceae, and Proteaceae). Megatherm elements of

the Tiliaceae–Bombacaceae and Aquifoliaceae (*Ilex*) are reported (Romero and Zamaloa 1985; Romero and Castro 1986; Melendi *et al.* 2003).

The early Oligocene is represented by the Sloggett Formation, with outcrops in southern Patagonia (Olivero *et al.* 1998). Assemblages are dominated by temperate forest elements of the Nothofagaceae, Podocarpaceae, Araucariaceae, and Proteaceae (Olivero *et al.* 1998). No megatherm components have been recorded so far.

Late Oligocene palynological data come from the San Julián (southeastern Patagonia), Río Leona (southwestern Patagonia), and Río Foyel (northwestern Patagonia) Formations (Pöthe de Baldis 1974, 1984; Barreda 1997; Barreda and Palamarczuk 2000a; Barreda *et al.* 2003, 2004, 2009).

Late Oligocene assemblages are highly diverse. Podocarpaceae, Nothofagaceae, and Araucariaceae are abundant, but there are large numbers of megatherm trees, shrubs, vines, and herbs such as Arecaceae, Malpighiaceae, Rubiaceae (*Gardenia*), Combretaceae (*Combretum/Terminalia*), Sapindaceae (*Cupania*), and Chloranthaceae, among others. Shoreline communities include herbs and low shrubs with the first records of the sunflower family (Asteraceae, Mutisiae) and the Rosaceae.

Early to middle Miocene palynological data are abundant, with records from the Lileo, Barranca Final, Gaiman, Chenque, Monte León, Centinela, Pinturas, and Cullén Formations (Fig. 25.1), and the unnamed deposits of Sierra La Colonia locality (Guerstein 1990a, 1990b; Zamaloa 1993, 2004; Barreda 1996; Barreda and Palamarczuk 2000a, 2000b, 2000c; Leanza *et al.* 2002; Archangelsky and Zamaloa 2003; Guerstein *et al.* 2004; Palazzesi and Barreda 2005).

Early Miocene (Aquitanian) assemblages were characterized by the development of xerophytic and halophytic herbs and shrubs (Chenopodiaceae, Ephedraceae, and Convolvulaceae *Cressa/Wilsonia*) which could have been spread in some areas of central Patagonia (e.g. Sierra La Colonia locality). Grasses were present, but in low amounts. Forests were still widespread dominated by southern lineages (Podocarpaceae, Nothofagaceae, Araucariaceae), with some megatherm elements like Arecaceae and Symplocaceae (*Symplocos*) (Barreda 1993, 1996). During the latest Aquitanian there was an increase of arid-loving taxa such as Ephedraceae, Chenopodiaceae, Convolvulaceae (*Cressa/Wilsonia*), Calyceraceae, and Asteraceae (Barnadesioideae *Chuquiraga*, Nassauvieae), and some sclerophyll trees were present such as Casuarinaceae (*Casuarina*), Fabaceae (Mimosoideae, *Acacia = Acaciapollenites myriosporites*, *Anadenanthera = Polyadopollenites* sp.), (Caesalpinoidea, *Caesalpinia = Margocolporites vanwijhei*), and Proteaceae. Rainforest trees contributed to the pollen assemblages and may have been from riparian or gallery forests.

Latest early Miocene (Burdigalian) assemblages show a new increase in megatherm angiosperms of the Sapindaceae

(*Cupania*) and Euphorbiaceae (*Alchornea*). Tree ferns and aquatic herbs and hydrophytes (Cyperaceae, Sparganiaceae, Restionaceae) were widespread in central Patagonia. Forests were still widespread, but dry pockets of Chenopodiaceae, Convolvulaceae (*Cressa/Wilsonia*), and Asteraceae (Barnadesioideae *Chuquiraga*, Nassauviieae) developed in coastal salt marshes.

Discussion and conclusions

Patagonia occupies a region important for understanding southern South American extant vegetation and the events that influenced its history. The most important trends in the evolution of the Patagonian vegetation during the Eocene–Miocene interval were related to the shift from megatherm to microtherm–mesotherm communities, and the first radiation of arid-loving taxa in eastern areas.

The presumptive coeval floras with the SALMAs recognized in the Gran Barranca area are presented in this overview in order to show the broad floristic context within which the mammals spread and evolved.

Before spread of the Barrancan fauna during the early–middle Eocene (~52–47 Ma), the maximum southward dispersal of megatherm elements had occurred. The highly diverse floras recovered in central Patagonia are representative of this migration. These Eocene records contain a broadleaved vegetation (with *Nothofagus* absent or very scarce) growing under a MAT of about 14–18 °C and abundant rainfall (MAP 1050–1250 mm: Wilf *et al.* 2005). Aridity in northern Patagonia would have developed, though, supported by the presence of evaporites and other sedimentological evidences (Melchor *et al.* 2002; Ziegler *et al.* 2003).

Paleofloras coeval with the Barrancan and Mustersan Ages (middle to late Eocene) are marked by the spread of Nothofagaceae forests, supporting the gradual cooling conditions suggested by isotopic data (Zachos *et al.* 2001). Some megatherm lineages still persisted even in southern Patagonia, and coastal regions might have acted as refugia. A seasonal climate was suggested by the occurrence of distinct growth rings in fossil Nothofagaceae wood (Brea 1993). The first evidence for Poaceae, from both megafossils and phytolitic remains, is recorded during this time interval. Grass pollen grains have not been reported from the Eocene so far in Patagonia.

Paleofloras equivalent to the Tinguirirican SALMA, and the pre-Deseadan "La Cantera" fauna (latest Eocene to early Oligocene) are characterized by spread of microtherm *Nothofagus*–gymnosperm dominated forests. The understory would have been composed of ferns (Lophosoriaceae, Pteridaceae). These assemblages suggest a temperate to cold–temperate climate with abundant rainfall. The extinction of Juglandaceae and Aquifoliaceae and the northward migration of megathermal elements such as *Cupania* and

Bombacaceae are in agreement with the cooling trend observed by this time (Zachos *et al.* 2001).

The floristic composition of the Deseadan SALMA (Oligocene) is vital to understanding the one present today. Its early stage (early Oligocene) is similar to that developed in the Tinguirirican and "La Cantera" SALMAs; however, some openings in the forest canopy, probably related to volcanic activity, might have developed. The late Deseadan stage (late Oligocene) was characterized by a new southward migration of megatherm elements. Closed forests would have spread in eastern Patagonia, although the first occurrences of modern shrubby and herbaceous lineages (Asteraceae, Goodeniaceae) also took place.

The earliest Miocene floras lack strictly coeval mammal faunas. Botanical information indicates that during this time many modern taxa appeared (Barnadesioideae *Chuquiraga*, and Nassauviieae of the Asteraceae) and diversified (Leguminosae; *Prosopis* type), contributing to a more complex vegetation structure with patches of open and dry-tolerant taxa.

Colhuehuapian coeval floras (early Miocene, earliest Burdigalian) are characterized by the maximum expansion of arid vegetation. Important xerophitic elements first appear at this time in Patagonia (Leguminosae as *Acacia*, and *Anadenanthera* types), some of them endemic to South America (Calyceraceae). Other previously reported sclerophyllous elements (Anacardiacea, Gramineae, Ephedraceae, Leguminosae [Caesalpinoidea and *Prosopis* types] and Chenopodiaceae) increase in abundance and diversity. Forests were reduced in extent and would have been restricted to gallery forests habitats beside rivers and streams.

By the Pinturan SALMA (early Miocene, late Burdigalian) a decline of arid-adapted taxa, particularly Ephedraceae, was matched by an expansion of hydrophytic herbs (Restionaceae, Sparganiaceae/Typhaceae, Cyperaceae) and aquatic ferns (*Azolla*). Moreover, a new increase in megatherm elements took place and mixed Neotropical and Antarctic lineages coexisted, with an important understory of vines, ferns, and some herbaceous angiosperms. Floras equivalent to the Colhuehuapian and Pinturan SALMAs developed adaptive traits that allowed the mammals to spread under a changing scenario, with the radiation of modern taxa.

The long-term effect of mammalian herbivores on vegetation has been noted by several authors (e.g. Stebbins 1981; Owen-Smith 1987). Unfortunately, interdisciplinary studies in Patagonia are scant and further work on this subject is needed.

ACKNOWLEDGEMENTS
Eduardo Bellosi is acknowledged for useful discussions. Special thanks are extended to Alan Graham for his critical reading of the manuscript as reviewer. Support was provided by Consejo Nacional de Investigaciones Científicas y

Técnicas, Argentina PIP-CONICET 5001, and Agencia Nacional de Promoción Científica y Tecnológica, Argentina, PICT 32344.

REFERENCES

Ancibor, E. 1988. Determinación xilológica de una raíz petrificada de Proteaceae de la Formación Río Turbio (Eoceno), Santa Cruz, Argentina. *Ameghiniana*, **25**, 289–295.

Ancibor, E. 1990. Determinación xilológica de la madera fósil de una fagácea de la Formación Río Turbio (Eoceno), Santa Cruz. Argentina. *Ameghiniana*, **27**, 179–184.

Andreis, R. R. 1972. Paleosuelos de la Formación Musters (Eoceno medio), Laguna del Mate, provincia de Chubut, República Argentina. *Revista de la Asociación Argentina de Mineralogía, Petrología y Sedimentología*, **3**, 91–97.

Archangelsky, A., and M. C. Zamaloa 2003. Primeros resultados palinológicos del Paleógeno del sector oriental de la Sierra La Colonia, provincia del Chubut, Argentina. *Revista del Museo Argentino de Ciencias Naturales*, n.s., **5**, 119–124.

Archangelsky, S. 1968. Sobre el paleomicroplancton del Terciario Inferior de Río Turbio, provincia de Santa Cruz. *Ameghiniana*, **5**, 406–416.

Archangelsky, S. 1972. Esporas de la Formación Río Turbio (Eoceno), Provincia de Santa Cruz. *Revista del Museo de La Plata*, n.s., *Sección Paleontología*, **6**, 65–100.

Báez, A. M., M. C. Zamaloa, and E. J. Romero 1990. Nuevos hallazgos de microfloras y anuros paleógenos en el noroeste de Patagonia: implicancias paleoambientales y paleobiogeográficas. *Ameghiniana*, **27**, 83–94.

Barreda, V. D. 1993. Late Oligocene?–Miocene pollen of the families Compositae, Malvaceae and Polygonaceae from the Chenque Formation, Golfo San Jorge basin, southeastern Argentina. *Palynology*, **17**, 169–186.

Barreda, V. D. 1996. Bioestratigrafía de polen y esporas de la Formación Chenque, Oligoceno tardío?–Mioceno de las provincias de Chubut y Santa Cruz, Patagonia, Argentina. *Ameghiniana*, **33**, 35–56.

Barreda, V. D. 1997. Palinoestratigrafía de la Formación San Julián en el área de Playa La Mina (Provincia de Santa Cruz), Oligoceno de la Cuenca Austral. *Ameghiniana*, **34**, 283–294.

Barreda, V. D. and S. Palamarczuk 2000a. Palinoestratigrafía de depósitos del Oligoceno tardío–Mioceno, en el área sur del Golfo San Jorge, provincia de Santa Cruz, Argentina. *Ameghiniana*, **37**, 103–117.

Barreda, V. D. and S. Palamarczuk 2000b. Palinomorfos continentales y marinos de la Formación Monte León en su área tipo, provincia de Santa Cruz, Argentina. *Ameghiniana*, **37**, 3–12.

Barreda, V. D. and S. Palamarczuk 2000c. Estudio palinoestratigráfico integrado del entorno Oligoceno tardío–Mioceno en secciones de la costa patagónica y plataforma continental argentina. In Aceñolaza, F. G. and Herbst, R.(eds.), *El Neogeno de Argentina*. Tucumán: Instituto Superior de Correlación Geológica, pp. 103–138.

Barreda, V. D., V. García, M. E. Quattrocchio, and W. Volkheimer 2003. Palynostratigraphic analysis of the Río Foyel

Formation (Latest Oligocene–Early Miocene), northwestern Patagonia, Argentina. *Revista Española de Micropaleontología*, **35**, 229–240.

Barreda, V. D., S. Césari, S. Marensi, and L. Palazzesi 2004. The Río Leona Formation: a key record of the Oligocene flora in Patagonia. *Abstracts VII International Organization of Paleobotany Conference*, 9–10.

Barreda, V. D., L. Palazzesi, and S. Marenssi 2009. Palynological record of the Paleogene Río Leona Formation (southernmost South America): stratigraphical and paleoenvironmental implications. *Review of Palaeobotany and Palynology*, **154**, 22–33.

Berry, E. W. 1932. Fossil plants from Chubut Territory collected by the Scarritt Patagonian Expedition. *American Museum Novitates*, **536**, 1–10.

Berry, E. W. 1937. Eogene plants from Río Turbio, in the Territory of Santa Cruz, Patagonia. *Johns Hopkins University Studies in Geology*, **12**, 91–98.

Blisniuk, P. M., L. A. Stern, C. P. Chamberlain, B. Idelman, and K. Zeitler 2005. Climatic and ecologic changes during Miocene surface uplift in the Southern Patagonian Andes. *Earth and Planetary Science Letters*, **230**, 125–142.

Brandmayr, J. 1945. Contribución al conocimiento geológico del extremo Sud Sud-Oeste del Territorio de Santa Cruz (Región Cerro Cazador-Alto Río Turbio). *Boletín de Informaciones Petroleras*, **22**, 415–442.

Brea, M. 1993. Inferencias paleoclimáticas a partir del estudio de los anillos de crecimiento de leños fósiles de la Formación Río Turbio, Santa Cruz, Argentina. I. *Nothofagoxylon paraprocera* Ancibor, 1990. *Ameghiniana*, **30**, 135–141.

Casadío, S., A. Parras, S. Marensi, and M. Griffin 2001. Edades $^{87}Sr/^{86}Sr$ de *Crassostrea? hatcheri* (Ortmann) – Bivalvia, Ostreoidea – en el 'Patagoniano' de Santa Cruz, Argentina. *Ameghiniana*, **38** (4) Supl. 30R.

Dusén, P. 1907. Über die tertiäre Flora der Magellansländer. In *Wissenschaften Ergebnisse Schwedische Expedizion nach den Magellansländer, 1895–1897*, vol. **I**, *Geologische und Paläontologische*. Uppsala: Archiv für Botanik, pp. 1–27.

Fiori, A. 1939. Filliti Terziarie della Patagonia. II. Fillite del Río Ñirihuau: giornale di geologia. *Annali del Reale Museo Geologico di Bologna,* Serie 2, **13**, 41–67.

Fleagle, J. G., T. M. Bown, C. C. Swisher, and G. Buckley 1995. Age of the Pinturas and Santa Cruz formations. *Actas IV Congreso Argentino de Paleontología y Bioestratigrafía*, 129–135.

Frengüelli, J. 1941. Nuevos elementos florísticos del Magellaniano de Patagonia austral. *Notas del Museo de La Plata*, **6**, 173–202.

Frengüelli, J. and L. R. Parodi 1941. Una *Chusquea* fósil de El Mirador (Chubut). *Notas del Museo de La Plata*, **6**, 235–238.

Gamerro, J. C. and S. Archangelsky 1981. Palinozonas neocretácicas y terciarias de la Plataforma Continental Argentina de la Cuenca del Colorado. *Revista Española de Micropaleontología*, **13**, 110–140.

Guerstein, G. R. 1990a Palinología estratigráfica del Terciario de la cuenca del Colorado. República Argentina. I. Especies terrestres de la perforación Nadir N° 1. *Revista Española de Micropaleontología*, **22**, 33–61.

Guerstein, G. R. 1990b. Palinología estratigráfica del Terciario de la cuenca del Colorado. República Argentina. III. Estudio sistemático y estadístico de la perforación Puerto Belgrano N° 20. *Revista Española de Micropaleontología*, **22**, 459–480.

Guerstein, G. R., M. V. Guler, and S. Casadío 2004. Palynostratigraphy and palaeoenvironments across the Oligocene–Miocene boundary within the Centinela Formation, southwestern Argentina. In Beaudoin A. B., and Head, M. J. (eds.), *The Palynology and Micropalaeontology of Boundaries*, Special Publication no. 230. London: Geological Society of London, pp. 325–343.

Guerstein, G. R. and M. Quattrocchio 1988. Palinozonas e interpretación estratigráfica mediante análisis de agrupamiento del Terciario de la Cuenca del Colorado, República Argentina. *Actas II Jornadas Geológicas Bonaerenses*, **1**, 27–35.

Hünicken, M. 1955. Depositos Neocretácicos y Terciarios del extremo SSW de Santa Cruz. *Revista del Instituto Nacional de Investigación de las Ciencias Naturales y Museo Argentino de Ciencias Naturales "Bernardino Rivadavia,"* **4**, 1–161.

Leanza, H. A., W. Volkheimer, C. A. Hugo, D. L. Melendi, and E. I. Rovere 2002. Lutitas negras lacustres cercanas al límite Paleógeno–Neógeno en la región nor-occidental de la provincia del Neuquén: evidencias palinológicas. *Revista de la Asociación Geológica Argentina*, **57**, 280–288.

Malumián, N. 1999. La sedimentación y el volcanismo terciarios en la Patagonia Extraandina. *Anales del Instituto de Geologia y Recursos Minerales*, **29**, 557–612.

Marenssi, S. A, C. O. Limarino, A. Tripaldi, and L. I. Net 2005. Fluvial systems variations in the Río Leona Formation: tectonic and eustatic controls on the Oligocene evolution of the Austral (Magallanes) Basin, southernmost Argentina. *Journal of South American Earth Sciences*, **19**, 359–372.

Melchor, R., J. Genise, and S. E. Miquel 2002. Ichnology, sedimentology and paleontology of Eocene calcareous paleosols from a palustrine sequence, Argentina. *Palaios*, **17**, 16–35.

Melendi, D. L., L. H. Scafati, and W. Volkheimer 2003. Palynostratigraphy of the Paleogene Huitrera Formation in N-W Patagonia, Argentina. *Neues Jahrbuch für Geologie und Paläontologie, Abhandlungen*, **228**, 205–273.

Menéndez, C. 1971. Floras Terciarias de la Argentina. *Ameghiniana*, **8**, 357–371.

Menéndez, C. and M. A. Caccavari de Filice 1975. Las especies de *Nothofagidites* (polen fósil de *Nothofagus*) de sedimentos Terciarios y Cretácicos de Estancia La Sara, Norte de Tierra del Fuego, Argentina. *Ameghiniana*, **12**, 165–183.

Nix, H. 1982. Environmental determinants of biogeography and evolution in Terra Australis. In Barker, W. R. and Greenslade, P. J. M. (eds.), *Evolution of the Flora and Fauna of Arid Australia*. Adelaide, SA: Peacock Publications, pp. 47–66.

Olivero, E. B., V. Barreda, S. Marenssi, S. Santillana, and D. Martinioni 1998. Estratigrafía, sedimentología y palinología de la Formación Sloggett (Paleogeno continental), Tierra del Fuego. *Revista de la Asociación Geológica Argentina*, **53**, 504–516.

Owen-Smith, N. 1987. Pleistocene extinctions: the pivotal role of mega-herbivores. *Paleobiology*, **13**, 351–362.

Palamarczuk, S. and V. Barreda 1998. Bioestratigrafía de dinoflagelados de la Formación Chenque (Mioceno), provincia del Chubut. *Ameghiniana*, **35**, 415–426.

Palazzesi, L. and V. Barreda 2005. Comunidades florísticas miocenas de Península Valdés: evidencias palinológicas. *Ameghiniana*, **42** supl., 9R.

Parras, A. and S. Casadío 2006. The oyster *Crassotrea? hatcheri* (Ortmann, 1987), a physical ecosystem engineer from the upper Oligocene – lower Miocene of Patagonia, Southern Argentina. *Palaios*, **21**, 168–186

Parras, A., M. Griffin, R. Feldmann, S. Casadío, C. Schweitzer, and S. Marenssi 2008. Isotopic ratios and faunal affinities in the correlation of marine beds across the Paleogene/ Neogene boundary in southern Patagonia, Argentina. *Journal of South American Earth Science*, **26**, 204–216.

Pöthe de Baldis, E. D. 1974. La microflora del carbón de Cabo Curioso (Eoceno Superior-Oligoceno Inferior), provincia de Santa Cruz. *Resumenes I Congreso Argentino de Paleontología y Bioestratigrafía*, 30.

Pöthe de Baldis, E. D. 1984. Microfloras Cenozoicas. *Relatorio IX Congreso Geológico Argentino*, **2**, 393–411.

Pujana, R. R. 2007. New fossil woods of Proteaceae from the Oligocene of southern Patagonia. *Australian Systematic Botany*, **20**, 119–125.

Romero, E. J. 1970. *Ulminium atlanticum* n. sp. tronco petrificado de Lauraceae del Eoceno de Bahía Solano, Chubut, Argentina. *Ameghiniana*, **7**, 205–224.

Romero, E. J. 1977. *Polen de gimnospermas y fagáceas de la Formación Río Turbio (Eoceno), Santa Cruz, Argentina*. Buenos Aires: Fundación para la Educación, la Ciencia y la Cultura.

Romero, E. J. 1978. Paleoecología y paleofitogeografía de las Tafofloras del Cenofítico de Argentina y áreas vecinas. *Ameghiniana*, **15**, 209–227.

Romero, E. J. 1986. Paleogene phytogeography and climatology of South America. *Annals of the Missouri Botanical Garden*, **73**, 449–461.

Romero, E. J. 1993. South American paleofloras. In Goldblatt, P. (ed.), *Biological Relationships between Africa and South America*. New Haven, CT: Yale University Press, pp. 62–85.

Romero, E. J. and M. T. Castro 1986. Material fúngico y granos de polen de angiospermas de la Formación Río Turbio (Eoceno), provincia de Santa Cruz, República Argentina. *Ameghiniana*, **23**, 101–118.

Romero, E. J. and M. C. Zamaloa 1985. Polen de angiospermas de la Formación Río Turbio (Eoceno), provincia de Santa Cruz, Argentina. *Ameghiniana*, **22**, 101–118.

Sepúlveda, E. G. and G. Norris 1982. A comparison of some Paleogene fungal palynomorphs from Arctic Canada and from Patagonia, Southern Argentina, *Ameghiniana*, **19**, 319–334.

Speggazzini, C. 1924. Sobre algunas impresiones vegetales eocénicas de Patagonia. *Comunicación del Museo Nacional de Historia Natural de Buenos Aires*, **2**(10), 95–107.

Stebbins, G. L. 1981. Coevolution of grasses and herbivores. *Annals of the Missouri Botanical Garden*, **68**, 5–86.

Vergel, M. M., and J. Durango de Cabrera 1988. Palinología de la Formación Cullén (Terciario) de las inmediaciones de Cañadón Beta, Tierra del Fuego, República Argentina. *Actas V Congreso Geológico Chileno*, **2**, 227–245.

Volkheimer, W. 1971. Aspectos paleoclimatológicos del Terciario Argentino. *Revista del Museo Argentino de Ciencias Naturales* "Bernardino Rivadavia," **1**, 243–262.

Wilf, P., N. R. Cúneo, K. R. Johnson, J. F. Hicks, S. L. Wing, and J. D. Obradovich 2003. High plant diversity in Eocene South America: evidence from Patagonia. *Science*, **300**, 122–125.

Wilf, P., K. R. Johnson, N. R. Cúneo, M. E. Smith, B. S. Singer, and M. A. Gandolfo 2005. Eocene plant diversity at Laguna del Hunco and Río Pichileufú, Patagonia, Argentina. *American Naturalist*, **165**, 634–650.

Zachos, J., M. Pagani, L. Sloan, E. Thomas, and K. Billups 2001. Trends, rhythms and aberrations in global climate 65 Ma to Present. *Science*, **292**, 686–693.

Zamaloa, M. C. 1993. Hallazgos palinológicos en la Formación Pinturas, sección cerro Los Monos (Mioceno Inferior),

provincia de Santa Cruz, Argentina. *Ameghiniana*, **30**, 353.

Zamaloa, M. C. 2000. Palinoflora y ambiente en el Terciario del nordeste de Tierra del Fuego, Argentina. *Revista del Museo Argentino de Ciencias Naturales*, n.s., **2**, 43–51.

Zamaloa, M. C. 2004. Miocene algae and spores from Tierra del Fuego, Argentina. *Alcheringa*, **28**, 205–227.

Zamaloa, M. C. and R. R. Andreis 1995. Asociación palinológica del Paleoceno temprano (Formación Salamanca) en Ea. Laguna Manantiales, Santa Cruz, Argentina. *Actas VI Congreso Argentino de Paleontología y Bioestratigrafía*, **1**, 301–305.

Zamaloa, M. C. and E. J. Romero 2005. Neogene palynology of Tierra del Fuego, Argentina: conifers. *Alcheringa*, **29**, 113–121.

Zetter, R., C. C. Hofmann, I. Draxler, J. Durango de Cabrera, M. M. Vergel, and F. Vervoorst 1999. A rich Middle Eocene microflora at Arroyo de los Mineros, near Cañadón Beta, NE Tierra del Fuego Province, Argentina. *Abhandlungen der Geologischen Bundesanstalt*, **56**, 436–460.

Ziegler, A. M., G. Eshel, P. M. Rees, T. A. Rothfus, D. B. Rowley, and D. Sunderlin 2003. Tracing the tropics across land and sea: Permian to present. *Lethaia*, **36**, 227–254.

Zucol, A. F. and M. Brea 2005. El registro fitolítico de los sedimentos cenozoicos de la localidad de Gran Barranca: su aporte a la reconstrucción paleoecológica. *Actas XVI Congreso Geológico Argentino*, **4**, 395–402.

26 Paleogene climatic and biotic events in the terrestrial record of the Antarctic Peninsula: an overview

Marcelo A. Reguero and Sergio A. Marenssi

Abstract

The James Ross Basin in the Antarctic Peninsula contains a terrestrial Eocene fauna that predates the establishment of the permanent ice-sheets at the Eocene–Oligocene boundary in Antarctica. Mammalian fossil specimens are available from three stratigraphically controlled localities throughout this sequence.

The Seymour Island La Meseta Fauna (La Meseta Alloformation, *Cucullaea* I Allomember, middle Eocene) contains at least 12 mammal taxa, predominantly tiny marsupials (mostly endemic and new taxa). The endemism of the marsupials suggests the existence of some form of isolating barrier (climatic and/or geographic) since the early Eocene. Faunal similarity between the La Meseta Fauna and the Paso del Sapo fauna assigned to the Vacan (?late Paleocene – ?early Eocene) of Patagonia strongly suggests that the former is more derived from the latter. The occurrence on Seymour Island of the sudamericid *Sudamerica ameghinoi*, which had become extinct in South America in the Paleocene, also indicates that isolation may have allowed extended survival of this Gondwanan group in the Eocene of Antarctica and that the factors that caused their extirpation in South America did not affect Antarctica. The faunal evidence indicates that the La Meseta mammalian fauna was derived from late Paleocene – early Eocene Riochican/Vacan faunas. The dispersal and vicariance events may have occurred during the onset of the late Paleocene – early Eocene climatic optimum between 58.5 and 56.5 Ma, when major regressive events are recorded either in the northern Antarctic Peninsula and southernmost Patagonia. The absence of notoungulates in the La Meseta fauna is noteworthy. We speculate that the notoungulates could have passed into Antarctica during the latest part of the Paleocene or early Eocene when the environmental conditions were warmer, and then became extinct at the onset of the climatic deterioration during the early Eocene.

Resumen

La Cuenca de James Ross en la Península Antártica contiene una fauna terrestre del Eoceno que predata el

establecimiento de la cubierta de hielo permanente en el límite Eoceno/Oligoceno. Los mamíferos terrestres provienen de tres niveles fosilíferos de la Aloformación La Meseta en la Isla Seymour (=Marambio).

La fauna de la Isla Seymour (Aloformación La Meseta, Alomiembro *Cucullaea* I, Eoceno medio) contiene al menos doce taxones de mamíferos, predominantemente pequeños marsupiales (mayormente endémicos y nuevos taxones). El endemismo de estos marsupiales sugiere la existencia de alguna forma de barrera de aislamiento (climática y/o geográfica) desde el Eoceno temprano. Las similitudes faunísticas entre la fauna de La Meseta y la de Paso del Sapo, asignada al Vaquense (? Paleoceno tardío - ?Eoceno temprano), de Patagonia sugieren fuertemente que la primera derivó de la última. La presencia en la Isla Seymour del sudamericido *Sudamerica ameghinoi*, extinguido en el Paleoceno de América del Sur, también indica que el aislamiento pudo haber permitido la supervivencia de este grupo gondwánico en el Eoceno de Antártida y que los factores que causaron su seudoextinción en América del Sur no afectaron a este continente. Las evidencias faunísticas indican que la fauna de La Meseta derivó de la fauna Vaquense de probable edad Paleoceno tardío/Eoceno temprano. Los eventos de dispersión y vicarianza pudieron haber ocurrido durante el establecimiento del optimun climático del Cenozoico en el límite Paleoceno/Eoceno, cuando la mayor fase regresiva se registra tanto en el norte de la Península Antártica el la parte sur de Patagonia (entre los 58.5 y 56.5 Ma). Inferimos que los notoungulados podrían haber pasado a Antártida en la última parte del Paleoceno o el Eoceno temprano, cuando las condiciones ambientales fueron más cálidas, y luego se extinguieron como consecuencia del deterioro climático registrado en el Eoceno de la Península Antártica.

Introduction

Since the early Cretaceous the Antarctic Peninsula has been located at almost the same paleolatitude (60–65° S) (Lawver *et al.* 1992), but became glaciated only more recently: a cool but not glacial early Cenozoic Antarctic climate is well known (see, for example, Dingle *et al.* 1998; Stilwell and Feldman 2000; Dutton *et al.* 2002). Evidence from an ever-increasing range of sources (ice-rafted debris,

The Paleontology of Gran Barranca: Evolution and Environmental Change through the Middle Cenozoic of Patagonia, eds. R. H. Madden, A. A. Carlini, M. G. Vucetich, and R. F. Kay. Published by Cambridge University Press. © Cambridge University Press 2010.

Antarctic continental shelf drilling, marine benthic oxygen isotopes, clay mineralogy, deep-sea biotic changes, and hiatuses indicating Southern-Origin Bottom Water [SOBW] onset) points to the initiation of Antarctic glaciation at sea level close to the Eocene–Oligocene boundary (∼34 Ma). Before that, the northern tip of the Antarctic Peninsula and southernmost South America were physically connected allowing the dispersal of plants and animals between the two (Olivero *et al.* 1990; Marenssi *et al.* 1994; Shen 1995; Reguero *et al.* 1998, 2002).

Evidence of Cretaceous and Paleogene Antarctic terrestrial faunas comes almost exclusively from the James Ross Basin, Antarctic Peninsula (Reguero and Gasparini 2006), and the knowledge of the Paleogene terrestrial vertebrates in Antarctica has been based almost exclusively on fossils from several horizons of the Eocene La Meseta Formation in Seymour (= Marambio) Island, east of the Antarctic Peninsula (Fig. 26.1), and secondly from the Eocene of the Fildes Peninsula, King George (25 de Mayo) Island (Covacevich and Rich 1982; Jianjun and Shuonan 1994).

In the last two decades, the James Ross Basin has been the scene of concentrated paleontological and geological fieldwork. Our knowledge of Paleogene mammalian fauna has increased, but data are still sparse for the late Cretaceous and Paleocene history of mammals in Antarctica. In contrast, major progress and results have been obtained for the Eocene. Thus, the overview proposed here only concerns the Eocene, because it is that interval that is most relevant for the Gran Barranca.

The discovery of terrestrial fossil mammals from Antarctica has been recorded from the northwest portion of Seymour (= Marambio) Island, Antarctic Peninsula (Fig. 26.1). The terrestrial mammal-bearing unit, La Meseta Formation, represents sedimentation in coastal and shallow-marine environments. Fossiliferous rocks there have generally been viewed as spanning a time period from early Eocene to early Oligocene (Ivany *et al.* 2006).

These terrestrial mammals suggest close biogeographic links with Paleogene faunas of Patagonia (Bond *et al.* 1993; Marenssi *et al.* 1994; Reguero *et al.* 1998, 2002). Even as knowledge of the Seymour Island fauna and geology has grown, it has proven difficult to establish precise biostratigraphic correlations between the terrestrial Seymour Island fauna and those elsewhere from South America, e.g. Patagonia. Correlation of Eocene Antarctic and Patagonian fossil assemblages has been a long-standing problem for several reasons: (1) the high proportion of mammals and birds (penguins) that were endemic to Antarctica or had distant relatives elsewhere; (2) floras and marine vertebrate faunas characterized by long-ranging taxa and taxa of unknown range; (3) lack of radiometrically datable rocks at appropriate stratigraphic position, and (4) the inability to

correlate directly Antarctic Eocene rocks with sections in Patagonia that have reliable biostratigraphic or radiometric dates. Despite the increasing number of discoveries, the Eocene record of Antarctic terrestrial mammals is regrettably sparse.

This chapter summarizes information documenting major changes in the terrestrial biota in the Eocene of the James Ross Basin before significant permanent ice-sheets first appeared at the Eocene–Oligocene boundary in Antarctica. We present faunal data supporting the existence of a high-latitude and high-altitude land biota with differences from the contemporaneous or nearly so faunas of Patagonia. Biostratigraphic range data and taxonomic composition of terrestrial mammalian faunas from the middle through late Eocene sequence at Seymour Island are examined.

Geological and tectonic setting of the Paleogene in the Antarctic Peninsula

Antarctic Paleogene sedimentary rocks are only exposed around the northern part of the Antarctic Peninsula, on the South Shetland Islands, and on the James Ross Island Group (Fig. 26.1). They were deposited in very different tectonic settings and environments. The South Shetland sequence represents an outer-arc (Birkenmajer 1995) or fore-arc (Elliot 1988) succession composed mainly of terrestrial volcanic and sedimentary deposits. The James Ross Island sequence is made up of marine clastic rocks deposited within a back-arc basin (Elliot 1988; Hathway 2000) and representing the topmost beds of a regressive megasequence (Pirrie *et al.* 1991).

Other Paleogene successions are only documented in drill sites. They are all around Antarctica, in the Ross Sea basins, Weddell Sea, Kerguelen Plateau, and off Prydz Bay (Webb 1991). Evidence from the East Antarctica sites have been interpreted as indicating the existence of relatively warm to cold–temperate marine waters through the Paleocene and Eocene with a strong cooling near the Eocene–Oligocene boundary (Robert *et al.* 2002). Ice-rafted detritus off Dronning Maud Land (Site 693) and in the southern Kerguelen Plateau (Site 747) indicate ice accumulations on East Antarctica in the early Oligocene. Glacial diamictites recorded off Prydz Bay have been assigned a maximum age of late middle Eocene (Webb 1991). The closest drill-hole record to the Antarctic Peninsula outcrops come from the ODP Leg 113 (Site 696) on the South Orkney microcontinent. Hole 696B drilled into glauconitic sandy mudstones yielding dinoflagellate cysts, sporomorphs, bentic foraminifers, and calcareous nannofossils indicating middle Eocene to earliest Oligocene ages.

King George (25 de Mayo) Island is located at the southern end of the South Scotia Ridge, within the South Shetland

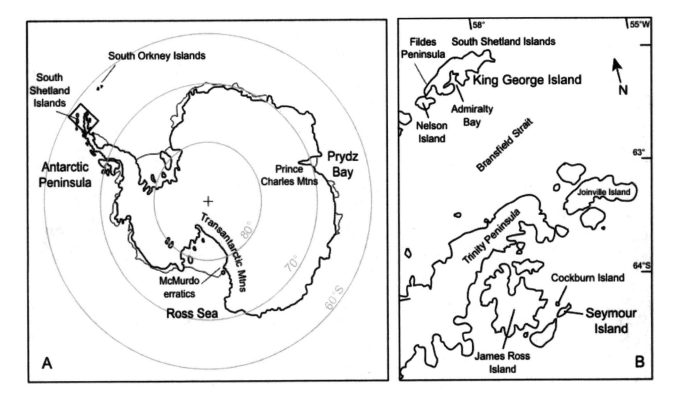

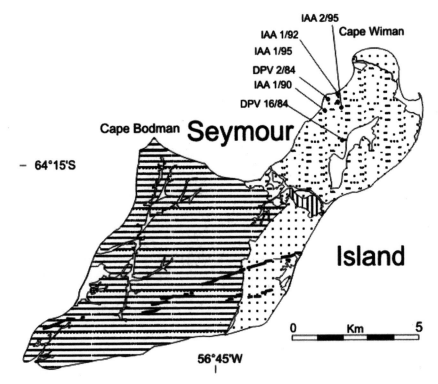

Fig. 26.1. Location map showing the location of the Seymour (Marambio) Island to the east of the Antarctic Peninsula and location of the main geological groups from which the fossil vertebrates were recovered. The mammals come from various localities in the northern part of Seymour Island (IAA and DPV localities mentioned in the text).

Island Arc (Fig. 26.1). It is bounded to the north by the South Shetland subduction trench and to the south is separated from the Antarctic Peninsula crustal block by the Bransfield Rift, a late Cenozoic structure situated off the axis within the marine back-arc Bransfield Basin (Birkenmajer 1995). This basin separates the South Shetland Islands outer arc from the Antarctic Peninsula inner arc and it is the site of Pleistocene to Recent volcanic activity.

King George Island and the neighboring Nelson Island consist of several tectonic blocks bounded by two systems of strike–slip faults of Tertiary (54–21 Ma) age (Birkenmajer 1989). Thus, considerable differences in stratigraphic succession, age, and character of the rocks occur between particular blocks.

The stratigraphic sequence includes mainly Upper Cretaceous to Lower Miocene island-arc extrusive and intrusive rocks comprising mainly terrestrial lavas, pyroclastic, and volcaniclastic sediments often with terrestrial plant fossils. Hypabyssal dykes and plutons intrude the latter. Fossiliferous marine and glaciomarine sediments are also represented.

Paleogene rocks of the Fildes Peninsula area were grouped into the Fildes Peninsula Group (Hawkes 1961; Barton 1965) composed of Campanian to uppermost Oligocene or earliest Miocene subalkaline volcanic rocks and plant-bearing volcaniclastics. Based on lithologic, biotic, and volcanic characteristics Shen (1995) subdivided this Group into five formational units named Half Three Point (late Cretaceous), Jasper (Paleocene), Agate Beach (Paleocene), Fossil Hill (Eocene), and Block Hill (Eocene) Formations and Suffield Point Volcanics (Miocene).

Birkenmajer (1988), on the basis of radiometric, sedimentological and paleontological data, proposed for the Cenozoic of the Antarctic Peninsula four glaciations separated by three interglacials. The first undisputed, extensive glaciation in West Antarctica, called Polonez, took place at the early–late Oligocene boundary (29.3 Ma: Dingle et al. 1997).

The Paleogene back-arc deposits comprise more than 1000 meters of shallow marine to coastal fossiliferous clastic sedimentary rocks mainly of Paleocene and Eocene ages (Elliot 1988; Sadler 1988; Marenssi et al. 1998a). They are exposed on Seymour and Cockburn Islands approximately 100 km southeast of the northern tip of the Antarctic Peninsula representing the uppermost part of the James Ross Basin succession (del Valle et al. 1992) (Fig. 26.1).

The early Eocene to latest late Eocene La Meseta Formation (Elliot and Trautman 1982) is an unconformity-bounded unit (La Meseta Alloformation of Marenssi et al. 1998a). This unit has a maximum composite thickness of 720 meters filling up a 7-km-wide valley cut down into the older sedimentary rocks of the island after the regional uplift and tilting of the Paleocene and Marambio Group

beds. The La Meseta Formation comprises mostly poorly consolidated siliciclastic fine-grained sediments deposited in deltaic, estuarine, and shallow marine environments as a part of a tectonically controlled incised valley system (Marenssi 1995; Marenssi et al. 1998b), which spans nearly all of the Eocene, and for some authors includes the Eocene–Oligocene boundary (Ivany et al. 2006). This unit is made up of several lens-shaped unconformity-bounded members representing different sedimentation stages (Marenssi et al. 2002) (Fig. 26.2). The richly fossiliferous Eocene sediments have yielded the first land mammal (Woodburne and Zinsmeister 1984), the first placental land mammal (Carlini et al. 1990), and the first record of "extinct South American ungulates" (Bond et al. 1990) from Antarctica among other terrestrial vertebrates (Reguero et al. 2002). It has also yielded dung beetle brood balls (Laza and Reguero 1993), leaves, tree trunks, a flower, and diverse marine vertebrate and invertebrate faunas (Stilwell and Zinsmeister 1992; Torres et al. 1994; Gandolfo et al. 1998a, 1998b). The oldest dates in the formation, 52–54 Ma (based on $^{87}Sr/^{86}Sr$ isotopic ratios after Reguero et al. 2002), come from the 150-m level in the boundary of the Valle de las Focas and Acantilados allomembers of Marenssi and Santillana (1994). Consequently, the base of the La Meseta should be close to the beginning of the Eocene at 54.8 Ma (Gradstein et al. 2004).

Eocene/Oligocene Antarctic glaciation

The glaciation of East Antarctica is often attributed to the tectonic opening of the ocean gateways between Antarctica and Australia (Tasmanian Gateway), and Antarctica and South America (Drake Passage), leading to the organization of the Antarctic Circumpolar Current (ACC) and the "thermal isolation" of Antarctica (Kennett 1977; Exon et al. 2001). While most tectonic reconstructions place the opening of the Tasmanian Gateway close to the Eocene–Oligocene boundary, Drake Passage did not open until several million years later (Lawver and Gahagan 1998), and may not have provided a significant deep-water passage until the Miocene (Barker and Burrell 1977). Most recently, once more using mainly the major plate separations, Lawver and Gahagan (2003) argued that Drake Passage opened for deep-water circulation via the Powell Basin before 28 Ma and proposed a 31-Ma onset for the ACC. Evidence from the La Meseta Formation, Seymour Island shows no indication of any glacial deposition during the early Paleocene or latest Eocene cold episodes and the region remained well-vegetated and inhabited by terrestrial faunas for the entire Eocene (Marenssi et al. 1994; Dingle et al. 1998; Reguero et al. 1998, 2002).

Recently Ivany et al. (2006) described rock units from the topmost levels of the La Meseta Formation and overlying horizons, which indicated ice-sheet formation in this

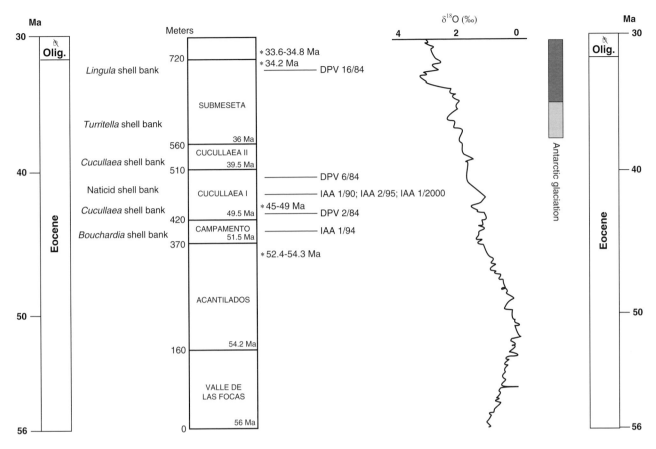

Fig. 26.2. Stratigraphy for the Eocene La Meseta Formation, Seymour Island after Marenssi and Santillana (1994). Radiometric dates are based on $^{87}Sr/^{86}Sr$ isotopic ratios after Reguero *et al.* (2002), Dutton *et al.* (2002), and Dingle and Lavelle (1998b). Comparison between global climate change curve over the past 60 million years (according to Zachos *et al.* 2001), showing a mainly cooling trend, is inferred from foraminiferal oxygen isotope records from all major ocean basins with $^{18}O/^{16}O$ ratios plotted as per mil deviation from a standard ($\delta^{18}O$). The vertical gray bars indicate the relative extent of polar ice sheets: light gray, ice volumes less than half of the maximum extent; dark gray, ice volumes close to the maximum extent (after Zachos *et al.* 2001), with La Meseta Formation unconformity-bounded internal units and ages (after Marenssi 2006).

area of the Antarctic Peninsula at the Eocene–Oligocene boundary. Dingle and Lavelle (1998a) found the earliest observed glacial event on the Antarctic Peninsula to be 29.8 ± 0.6 Ma or at least 4 m.y. later than the first Cenozoic glacial event observed in East Antarctica.

Eocene Antarctic paleoclimate and paleoenvironment

The climatic evolution of the Southern Ocean was in part brought about by rearrangement of the southern hemisphere land masses, and is related to circulation changes affecting global heat transport (Stott *et al.* 1990). The separation of South America from Antarctica cleared the way for a free ACC which produced the thermal and physical isolation of Antarctica. It developed at some time during the Cenozoic, as a series of deep-water gaps opened around Antarctica, and has been widely viewed as having reduced meridional

heat transport, isolating the continent within an annulus of cold water and thus being at least partly responsible for Antarctic glaciation (e.g. Kennett 1977). Recent articles provide data which indicate a time frame for opening of the Drake Passage resulting in the development of ACC and subsequent onset of Antarctic climatic cooling, leading to the development of an earliest Oligocene ice-sheet on the Antarctic Peninsula. Scher and Martin (2006) utilized a rare earth element neodymium (Nd) contained in fish teeth extracted from deep-sea cores taken in the south Atlantic Ocean between southern Africa and Queen Maud Land on the Antarctic craton to provide data on a deep-water opening of the Drake Passage. They used ratios of $^{143}Nd/^{144}Nd$ to determine the transition from non-radiogenic to radiogenic Nd values which would mark the flow of the more radiogenic Pacific waters into the Atlantic Ocean as the signal for the deep-water opening of the Drake Passage in the late Eocene (*c.* 41 Ma).

Livermore *et al.* (2005) propose a shallow-water opening (<1000 m deep) as early as 50 Ma based on tectonic evidence in the Weddell Sea. They correlated the formation of new tectonic basins in the area that would later become the Scotia Arc and the northern Antarctic Peninsula with a drop in Southern Ocean temperatures based on oxygen isotope data from benthic foraminifera to represent the initial shallow-water opening between South America and Antarctica.

Although from indirect sources, terrestrial environments and biota have also been interpreted. Reguero *et al.* (1998, 2002) proposed for the Eocene a landscape with a mountainous or highland area located to the west, possibly with sporadically active volcanoes, and a lowland irregular coastal area bounded by a cool–temperate sea to the east. The presence of *Nothofagus*-dominated forest bounded by grassland areas has also been proposed using the paleoflora found in the La Meseta Formation.

Paleoclimatic interpretations of the Paleogene of the James Ross Basin were proposed using different lines of evidence (sediments, fossils, and stable isotope analysis). The idea of a cool–temperate marine environment from the late Cretaceous through the late Eocene has been proposed some time ago based on the presence of particular taxa of marine mollusks (Weddellian Province of Zinsmeister 1982). It has been confirmed by stable isotope analysis (Gazdzicki *et al.* 1992; Dichfield *et al.* 1994; Pirrie *et al.* 1998; Dutton *et al.* 2002). From the terrestrial environments, Francis (1991) interpreted a marked cooling of the climate during the Cretaceous–Paleocene transition on Seymour Island based on the study of growth ring patterns of tree trunks collected from the López de Bertodano and Sobral Formations. By the time of deposition of the Cross Valley Formation (Upper Paleocene) the climate had returned to a warmer level (Francis 1991). The late Paleocene paleoflora from Seymour Island is dominated by angiosperm taxa with some 36 species present. CLAMP analysis of the angiosperm leaves suggest that the mean annual temperature (MAT) was 13.5 °C for this time period in the Antarctic Peninsula (Francis *et al.* 2003).

Gandolfo *et al.* (1996, 1998a) indicate the presence of subtropical forests based on the percentage of entire-margin leaves (43%) from the middle Eocene paleofloras of the La Meseta Formation (where the La Meseta fauna has been recovered). Likewise, several lines of evidence point to indicate a seasonal cool–temperate rainy climate for the middle Eocene of the Seymour Island with a cooling trend through the Eocene until the earliest Oligocene (Case 1988; Francis 1991; Torres *et al.* 1994; Gandolfo *et al.* 1996, 1998a). Case (1988) suggested that the nothophyllous leaves of the lower part of the La Meseta Formation indicate a period of climatic amelioration with respect to the middle and upper part of the unit where microphyllous leaves were recovered. This Eocene flora is dominated by *Nothofagus*

species (Case 1988; Francis *et al.* 2003) as a response to the cooler temperatures (MAT 10.8 °C) and lower mean annual precipitation (MAP) (1534 mm) in the middle Eocene (Francis *et al.* 2003). The palynoflora presents austral characteristics (Zamaloa *et al.* 1987) while taxonomic analysis indicates a mixture of Antarctic (dominant) and Neotropical megafloral elements (Gandolfo *et al.* 1996, 1998a). Morphological analysis of the leaves led Gandolfo *et al.* (1996, 1998a) to interpret the presence of a mixed mesophitic forest (32% of entire-margin leaves) developed under a seasonal cool–temperate climate with dry and rainy seasons. Mean annual paleotemperatures were calculated to be between 11 °C and 13 °C and the mean of the coldest months between −3 °C and 2 °C.

That the pre-glacial climatic phase (Dingle and Lavelle 1998a) lasted until about 30 Ma is indicated by two floras. The older one, Petrified Forest Creek flora (Fig. 26.1) from the Arctowski Cove Formation, Ezcurra Inlet Group, is a *Nothofagus*–pteridophyte pollen-spore assemblage not older than late Eocene – early Oligocene. It gives evidence of a *Nothofagus* forest with well-developed undergrowth developed in moist climatic conditions comparable with the present-day Auckland Province, New Zealand (Birkenmajer and Zastawniak 1989). Dingle and Lavelle (1998b) interpret a cool–cold climate with high rainfall for this interval. The early Oligocene *Cytadella* flora from the Point Thomas Formation, Ezcurra Inlet Group, suggests the presence of broadleaved *Nothofagus* forest with ferns in the undergrowth similar to the Petrified Forest Creek Flora. The *Cytadella* flora resembles communities living with MAT between 11 °C and 15 °C, and MAP between 1200 and 3200 mm.

Additionally, Dingle *et al.* (1998), on the basis of chemical analysis and clay mineralogy, have recorded a climatic deterioration from very warm, non-seasonally wet conditions at the end of the Paleogene global optimum (early middle Eocene) to a latest Eocene cold, frost-prone, and relatively dry regime.

During Paleogene times a dramatic climatic change is recorded at southern high latitudes either in the marine and the terrestrial realms. Maximum marine paleotemperatures are recorded in ODP sediments of late Paleocene to early Eocene age while at the same time paleofloral evidence from Seymour Island indicates the existence of subtropical forests at 62° S. The rest of the Eocene and early Oligocene was a period of stepwise climatic deterioration culminating with the first extensive Antarctic glaciation at about 29 Ma. Warm to cool temperate climate with high rainfall prevailed. A mixed (austral and tropical) paleoflora developed during that time. These southern high-latitude forests experienced a growth regime with long darkness winters and near-continual low-angle incident sunlight summers showing both latitudinal and altitudinal gradients.

Table 26.1. *Taxonomic list, stratigraphy, and references for the terrestrial mammals from the Eocene (La Meseta Formation) of Seymour Island, Antarctic Peninsula*

Taxon	Stratigraphy (Allomember)	References
MAMMALS		
MARSUPIALIA		
POLYDOLOPIDAE		
Polydolops dailyi	Cucullaea I	Woodburne and Zinsmeister 1982; Candela and Goin 1995
Polydolops seymouriensis	Cucullaea I	Case *et al.* 1988; Candela and Goin 1995
MICROBIOTHERIIDAE		
Gen. et sp. indet.	Cucullaea I	Goin and Carlini 1995
Marambiotherium glacialis	Cucullaea I	Goin *et al.* 1999
Woodburnodon casei	Cucullaea I	Chornogubsky *et al.* 2009
DERORHYNCHIDAE		
Derorhynchus minutus	Cucullaea I	Goin *et al.* 1999
Pauladelphys juanjoi	Cucullaea I	Goin *et al.* 1999
Xenostylus peninsularis	Cucullaea I	Goin *et al.* 1999
Gen. et sp. indet. 1	Cucullaea I	Goin *et al.* 1999
PREPIDOLOPIDAE		
Perrodelphys coquinense	Cucullaea I	Goin *et al.* 1999
GONDWANATHERIA		
SUDAMERICIDAE		
cf. *Sudamerica ameghinoi*	Cucullaea I	Goin *et al.* 2006
XENARTHRA		
Tardigrada indet.	Cucullaea I	Carlini *et al.* 1990; Vizcaíno and Scillato Yané 1995
LITOPTERNA		
SPARNOTHERIODONTIDAE		
Notolophus arquinotiensis	Cucullaea I, Submeseta	Bond *et al.* 1990, 2006
Sparnotheriodontidae indet.	Acantilados	Reguero *et al.* 2002
ASTRAPOTHERIA		
TRIGONOSTYLOPIDAE		
Trigonostylops sp.	Cucullaea I	Hooker 1992; Marenssi *et al.* 1994
MAMMALIA *INCERTAE SEDIS*		
Zalambdodont mammal	Cucullaea I	Goin and Reguero 1993; MacPhee *et al.* 2008
Gen. et sp. indet. 2	Acantilados	Vizcaíno *et al.* 1997

Seymour Island terrestrial vertebrates

Seymour Island (= Marambio) contains the only Eocene land vertebrate fauna known in Antarctica (Table 26.1), and represents the southernmost part of the distribution of some Paleogene South American (Patagonian) land mammal lineages. This was especially so because paleogeographic reconstructions of the Antarctic Peninsula during the Eocene, utilizing paleomagnetic data collected on the continent itself, indicate a paleolatitude perhaps as far south as 63° (Lawver *et al.* 1992).

Among the Eocene terrestrial mammals from the La Meseta Formation, Seymour Island, the Sparnotheriodontidae, a family of the order Litopterna, an extinct South American ungulate group, and the marsupial family Polydolopidae were the dominant taxa. They were not usually dominant elements in much the larger Paleogene associations elsewhere in South America (Reguero *et al.* 1998, 2002). The most abundant group in the fauna is a suite of "opposum-like" marsupials (Goin *et al.* 1999).

The Antarctic endemic polydolopine *Polydolops dailyi*, a species closely related to the species *Polydolops thomasi* (Candela and Goin 1995) present in the Vacan (?late Paleocene – ?early Eocene) and Barrancan (late Eocene) "subages" (Fig. 26.3) of Patagonia. Their close relationships suggest a short interval of differentiation between both species. Recently Chornogubsky *et al.* (2009) revalidated the genus *Antarctodolops* and recognized a new species of this same genus.

The Antarctic derorhynchid marsupials *Derorhynchus minutus* and *Pauladelphys juanjoi* show close similarities

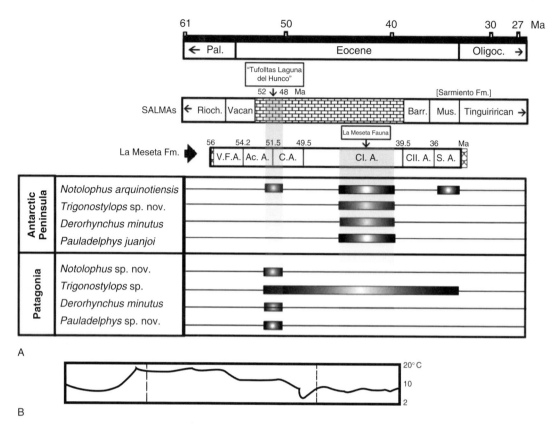

Fig. 26.3. Relationships between the evolutionary history of Antarctic and Patagonian fossil mammals, stratigraphy, temperature (A), and sea level at high latitudes during the Cenozoic (B). The temperature curve is for surface waters. Timescale is according to Berggren *et al.* (1995). Abbreviations: Ac. A., Acantilados Allomember; C. A., Campamento Allomember; CI. A., *Cucullaea* I Allomember; CII. A., *Cucullaea* II Allomember; S. A., Submeseta Allomember; V. F. A., Valle de las Focas Allomember. Heavy lines to the right of mammal names indicate known stratigraphic range. Data from eustatic sea levels after Haq *et al.* (1987).

with two new, unpublished, species of *Derorhynchus* from Las Flores, Chubut Province, Patagonia, of Itaboraian age (Goin *et al.* 1999).

The Antarctic microbiotheriid *Marambiotherium glacialis* closely resembles the Itaboraian species *Mirandatherium alipioi* of Patagonia (Goin *et al.* 1999); the Antarctic microbiotheriid *Woodburnodon casei* seems to be plesiomorphic to all known microbiotherians and is the largest known microbiotheriid marsupial (Goin *et al.* 2007).

The La Meseta land vertebrate fauna (LMF) is unusual in being dominated by the large-sized and bizarre litoptern ungulate *Notiolofos arquinotiensis* (Bond *et al.* 2006, 2009). This is not the case in the Eocene Patagonian fossil record where the Eocene faunas known from Vacan (?late Paleocene – ?early Eocene), Barrancan (late Eocene), and Mustersan (late Eocene) are dominated by notoungulates. The most unexpected circumstance in the LMF is the apparent lack of notoungulates and other ungulate groups such as primitive Condylarthra and non-sparnotheriodontid Litopterna. Notoungulata were the most diverse (morphologically as well as taxonomically) and successful of the

South American ungulate groups. One of the most important radiations of notoungulates in Patagonia occurred during the late Paleocene – early Eocene. As we suppose that no barrier to dispersal existed between Patagonia and Antarctic Peninsula during the Paleocene, the absence of this group in Antarctica could be explained by suggesting that the LMF is composed only of those taxa that were able to adapt to cooler conditions. If the notoungulates migrated southward into the Antarctic Peninsula during the late Paleocene, they presumably became extinct prior the deposition of the La Meseta Formation (late early Eocene – late Eocene). Although it is certainly possible that notoungulates were present at LMF but remain unsampled, the lack of specimens referable to this clade from among more than 80 identified specimens at least speaks to the rarity of this group, if they existed in this region at that time; it is difficult to envision a taphonomic bias that would preserve other closely related and similarly sized ungulates (i.e. sparnotheriodontids and trigonostylopids) to the exclusion of notoungulates.

Bond (1999) remarks the noteworthy difference in the geographic distribution between Notoungulata and

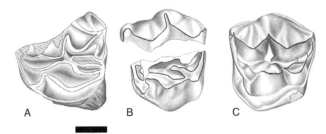

Fig. 26.4. Occlusal views of upper molars of *Notiolofos* spp. from Antarctica and Patagonia. (A) *Notiolofos arquinotiensis*, MLP 95-I-10–6, left M3, holotype from Seymour Island, Antarctica, La Meseta Formation, Submeseta Allomember; (B) *Notiolofos arquinotiensis*, MLP 90-I-20–1, left M1 or M2 from Seymour Island, Antarctica, La Meseta Formation, *Cucullaea* I Allomember; (C) *Notiolofos* sp. nov., MLP 66-V-12–2, right M1 (reversed) from Laguna Fría, Chubut, Tufolitas Laguna del Hunco Formation. Scale bar 5 mm (drawing by A. Viñas).

Litopterna in the late Pleistocene (Lujanian) in Patagonia; whereas the litoptern *Macrauchenia patachonica* has a wide range of distribution southward (Santa Cruz Province) in this age, the notoungulate *Toxodon* reached only Bahia Blanca (38° 45′ S) in the Buenos Aires Province. This fact suggests that some factor (geographic or environmental) affected the dispersal of notoungulates to southern latitudes.

The Antarctic sparnotheriodontid litoptern *Notiolofos arquinotiensis* (Bond *et al.* 2006), previously referred to *Victorlemoinea* (= *Sparnotheriodon*) by Bond *et al.* 1990, has close affinity with an undescribed species from Patagonia (Goin *et al.* 2000). The new taxon from Laguna Fría, Paso del Sapo, Chubut Province, Patagonia, is more advanced than the primitive *Victorlemoinea prototypica* from the Itaboraian of Brazil, and more similar in size to *V. labyrinthica* from Cañadón Vaca (Vacan Age). However, the Laguna Fría taxon exhibits intermediate character states, i.e. more reduced hypocone in the first molars and shorter lingual crest of the metaconid (Fig. 26.4), that show a close affinity with *Notiolofos arquinotiensis* (Bond *et. al.* 2006). Based on these differences we tentatively refer it to a new species of this genus. The record of *Notiolofos arquinotiensis* in the La Meseta Formation can be traced from the Acantilados allomember (early middle Eocene) to the Submeseta Allomember (latest Eocene) (Fig. 26.4). In the *Cucullaea* Allomember (middle Eocene), when the climatic conditions were cool–temperate, this species was common. Its last record occurs in the highest horizon of the Submeseta Allomember, dated by Dingle and Lavelle (1998a) at ~34.2 Ma (Fig. 26.2).

The Antarctic astrapothere is reminiscent of the trigonostylopid *Trigonostylops* (Bond *et al.* 1990) but represents a different new taxon (Bond *et al.* 2008). The species of *Trigonostylops* are known from the Riochican through the Mustersan of Patagonia.

The high proportion of endemic taxa, mainly tiny marsupials in LMF fauna, together with relicts such as protodidelphid and derorhynchine marsupials, and gondwanatherians, give it a very distinctive Paleocene Patagonian appearance, indicating that some form of isolating barrier – climatic, geographic, or topographic – was in effect. Several types of environmental factors could result from the high latitude, of which temperature and geographic may be the most important. The relatively low temperatures of the Antarctic regions during the middle–late Eocene appear to have been matched by the development of a characteristic Antarctic biota (Marenssi *et al.* 1994).

The Antarctic sudamericid represents the youngest record of the group and is very closely related to *Sudamerica ameghinoi* from the Paleocene of Punta Peligro in Patagonia, but is more derived than the latter in the microstructure of the enamel (Goin *et al.* 2006). The Gondwanatheria is a peculiar mammalian order with a widespread Gondwanan distribution in the late Cretaceous of Patagonia (Bonaparte 1986), Madagascar, India (Krause *et al.* 1997), probably Africa (Krause *et al.* 2003), and in the early Paleocene of Patagonia (Scillato Yané and Pascual 1984). These mammals have gliriform incisors, and were the earliest South American mammals to develop hypsodont cheek teeth with thick cementum. Krause *et al.* (1997) suggested that Antarctica might have served as an important Cretaceous biogeographical link between South America and Indo-Madagascar. In recognition of the derived enamel structure of the Antarctic sudamericid (Goin *et al.* 2006) and the endemism of the contemporaneous ungulate mammals with respect to South American relatives (Bond *et al.* 1995), the most parsimonious hypothesis is that the La Meseta Formation mammal fauna was relictual in the Antarctic Peninsula relative to a biota shared between the Antarctic Peninsula and South America in the early Paleocene. This vicariant hypothesis appears to reflect the subsequent separation of the Antarctic and South American continents. The remarkably good fossil record of late Paleocene (Itaboraian SALMA) South American mammals, both in Patagonia and Brazil (see, for example, Pascual *et al.* 1996), adds support to this hypothesis: no gondwanatheres have been recorded in Itaboraian beds, or in the subsequent Tertiary land mammal-bearing beds.

The Antarctic zalambdodont mammal (MacPhee *et al.* 2008) might shed some additional biogeographical light on the origins of the LMF.

The Antarctic Pilosa shares plesiomorphic features with primitive Tardigrada and Vermilingua. Despite some dubious Eocene records (Simpson 1948) the Antarctic form represents the earliest unquestionable record of Pilosa (Vizcaíno and Scillato Yané 1995).

Age of the vertebrate-bearing horizons of the La Meseta Formation

The La Meseta Formation should contain a near complete record of the transition from the warm early Eocene conditions, through the transitional cooling middle Eocene to the greater cooling in the late Eocene with the later opening of the Drake Passage and the onset of early Oligocene ice-sheet development. Although the overall age of the La Meseta Formation may span much of the Eocene, the age of the three principal terrestrial vertebrate-bearing horizons (Acantilados, *Cucullaea* I, and Submeseta allomembers) can be more tightly constrained (Fig. 26.4).

Cocozza and Clarke (1992) and later Askin (1997) indicated that the lower third of this unit (Valle de las Focas, Acantilados, and the lower part of Campamento allomembers) is late early Eocene in age. One biogenic carbonate sample from shells of *Ostrea antarctica* from the lower part of the Acantilados Allomember yield an $^{87}Sr/^{86}Sr$ age between 52.4 and 54.3 Ma (Reguero *et al.* 2002). Therefore, the erosional event at the base of La Meseta Formation has to be older than the peak of the late Ypresian lowstand.

The terrestrial vertebrates from *Cucullaea* I Allomember, though numerically small, suggest a late middle Eocene age (Bartonian, ~37 to ~41 Ma: Goin *et al.* 1999). Based on the study of palynofloras, Askin (1997) considered the middle part of the La Meseta Formation (*Cucullaea* I and *Cucullaea* II Allomembers) to be middle Eocene. This age corresponds well with the $^{87}Sr/^{86}Sr$-derived ages of 44.54 or 47.35 Ma reported by Dutton *et al.* (2002) for Telm 5 (*Cucullaea* I or *Cucullaea* II Allomembers). This temporal assignment is also consistent with the middle Eocene age assigned to the ichthyofauna found in the same depositional horizon (Cione and Reguero 1994, 1998). Therefore, the LMF would partly fill this considerable temporal gap in the Eocene record of the mammalian evolution in South America (see below).

The upper third of the unit (Submeseta Allomember) is regarded as late Eocene – earliest Oligocene by Askin (1997). However, Dingle and Lavelle (1998a) reported a $^{87}Sr/^{86}Sr$-derived age of 34.2 Ma (latest Eocene) for the topmost few meters of the unit, bracketing it into the Eocene. In addition, Dutton *et al.* (2002) presented $^{87}Sr/^{86}Sr$-derived ages of 36.13, 34.96, and 34.69 Ma for Telm 7 (Submeseta Allomember).

Correlation between Antarctic Peninsula and Patagonia

Very few genera and only one species are in common between the Eocene of Antarctica and any fauna in the Eocene sequence in Patagonia. Thus, this justifies only a preliminary mammal-based faunal correlation between these two continents (Antarctica and South America).

After the discovery of two species of polydolopid marsupials and the two ungulates, the LMF was initially considered to be a late Eocene representative of a Casamayoran mammalian fauna of Patagonia (Marenssi *et al.* 1994). More recent mammalian discoveries provide a different interpretation as to the origins of this now middle Eocene Antarctic paleofauna or even that the LMF had its origin in times that dates from the early Eocene. The Vacan fauna (early Casamayoran) at Cañadón Vaca, Chubut, contains archaic notoungulate families (e.g. Henricosborniidae, Isotemnidae), and the relative primitiveness of this fauna documents a great faunal difference from the subsequent Barrancan "subage" (Cifelli 1985). The middle Eocene for the Barrancan (late Casamayoran) dated in Gran Barranca, Patagonia and the refined ages for the Paleocene and early Eocene faunas in the San Jorge Basin (Bond *et al.* 1995; Marshall *et al.* 1997) seem to be more consistent with the observed taxonomic similarities and differences between the land mammal fauna from Seymour Island and those from the Paleocene and Eocene of Patagonia, e.g. Gran Barranca. Reassessment of the age of the taxa from the middle levels of the La Meseta Formation at Seymour Island shows that they are middle Eocene and not late Eocene as believed earlier by several authors (Zinsmeister 1978; Woodburne and Zinsmeister 1982; Wrenn and Hart 1988). Consequently the LMF from *Cucullaea* I Allomember is a unique fauna that may document the evolution of mammals in the southern area of South America in the middle Eocene (~49 and ~37 Ma).

Fossil mammals originally referred to the Vacan "Subage" of the Casamayoran SALMA have been recovered from near Laguna Fría, Paso del Sapo, Chubut from the same stratigraphic unit, Tufolitas Laguna del Hunco Formation, as rich fossil leaf floras at nearby Laguna del Hunco, Chubut. The Tufolitas Laguna del Hunco Formation is the same unit that yields leaves, so the age of the leaves is probably similar to the age of these mammals. Based on new Ar/Ar dates from the fossiliferous exposures themselves, deposition of the fossiliferous beds yielding the macroflora at Laguna del Hunco occurred from 52.4–51.6 Ma, near the base of Chron 23 (Wilf *et al.* 2003). The species-level morphotypes and the floral composition suggest a rich subtropical forest.

The mammal fauna is in the Laguna Fría tuffs overlies unconformably the Barda Colorada ignimbrites dated at 49.51 ± 0.32 and 52.05 ± 0.23 Ma. The fauna is covered unconformably by the Huancache andesitic basalts dated at 47.89 ± 1.21 Ma. These dates establish an early Eocene age for the Vacan SALMA at this locality (Tejedor *et al.* 2009). As we mentioned above, the LMF has similarities with that of Laguna Fría such as the derorhynchid marsupials and sparnotheriodontid litopterns (Fig. 26.3). The taxonomic and potentially biogeographic relations between Laguna Fría and the Eocene fauna from Antarctica led Goin *et al.* (2000) to support affinities also as a biochronologic unit.

Discussion

The late Paleocene – early Eocene was the apogee of Cenozoic warmth. During this interval, the tropics extended between 10–15° poleward, and both polar regions were populated with temperate forests (Frakes *et al.* 1992). The late Paleocene subtropical Cross Valley Flora on Seymour Island, which documents the warmest climatic conditions in the Paleogene of this continent, presents strong evidence for this warming period in Antarctica. Based on the record of South American ungulates, one of the most probable dispersal events between Patagonia and the Antarctic Peninsula could have occurred during the late Paleocene and it was probably enhanced by the beginning of the "climatic optimum" period (late Paleocene – early Eocene) and with the sea-level lowstand identified between 58.5 and 56.5 Ma (Haq *et al.* 1987). Lowering of the sea level might have increased the extension of low-lying coastal areas, providing an easier route than crossing rough mountainous terrain (Antarctic Peninsula) by leaving a long, continuous coastal region bordered by shallow seas and high mountains. The actual placement of the coastline during the latest Paleocene remains speculative, but the overall physical consequences are not. The geological record shows a decrease in marine rocks (Cross Valley Formation) before the Paleocene–Eocene boundary.

Perhaps the most striking aspect of the LMF, especially given its age, is the abundance of a primitive litoptern and the absence of notoungulates. In terms of abundance, LMF is dissimilar to other early to middle Eocene faunas of Patagonia, where notoungulates constitute a significant proportion of identified specimens.

Despite several shared taxa, there are profound differences between the ungulate fauna of Seymour Island and those from Vacan localities in Argentina.

The presence of the earliest members of the marsupials and edentates in Antarctic Peninsula, including early diverging species, suggests that the southern latitudes of South America served as a center of diversification for these clades.

ACKNOWLEDGEMENTS
We especially acknowledge the Instituto Antártico Argentino and Fuerza Aérea Argentina, which provided logistical support for our participation in the Antarctic field work. We also have benefited from collaborative effort in the field (prospecting and picking) of Juan José Moly, Sergio Santillana, Sergio F. Vizcaíno, Cecilia Besendjak, Laura Net, Hugo de Vido, Andrea Concheyro, and Rolando Maidana. Part of this study was funded by the National Geographic Society (Grants 6615–99; 7125–01 and 7599–04), Instituto Antártico Argentino (PICTA 23–2004 and 1–2008), and Consejo de Investigaciones Científicas y Técnicas (CONICET).

REFERENCES

Askin, R. A. 1997. Eocene–?earliest Oligocene terrestrial palynology of Seymour Island, Antarctica. In Ricci, C. A. (ed.), *The Antarctic Region: Geological Evolution and Processes*. Siena: Museo Nazionale Antartide, pp. 993–996.

Barker, P. F. and J. Burrell 1977. The opening of Drake Passage. *Marine Geology*, **25**, 15–34.

Berggren, W. A., D. V. Kent, C. C. Swisher III, and M. - P. Aubry 1995. A revised Cenozoic geochronology and chronostratigraphy. In Berggren, W. A., Kent, D. V., Aubry, M. -P., and Hardenbol, J. (eds.), *Geochronology, Time Scales and Global Stratigraphic Correlation,* Special Publication no. 54. Tulsa, OK: Society for Sedimentary Geology, pp. 129–212.

Barton, C. M. 1965. The Geology of South Shetland Islands. III. The stratigraphy of King George Island. *Scientific Reports British Antarctic Survey*, **44**, 1–33.

Birkenmajer, K. 1988. Tertiary glacial and interglacial deposits, South Shetland Islands, Antarctica: geochronology vs. biostratigraphy (a progress report). *Bulletin of the Polish Academy of Sciences, Earth Sciences*, **36**, 133–145.

Birkenmajer, K. 1989. A guide to Tertiary geochronology of King George Island, West Antarctica. *Polish Polar Research*, **10**, 555–579.

Birkenmajer, K. 1995. Mesozoic–Cenozoic magmatic arcs of Northern Antarctic Peninsula: subduction, rifting and structural Evolution. In Srivastava, R. K. and Chandra, R. (eds.), *Magmatism in Relation to Diverse Tectonic Settings.* New Delhi: IBH Publishing Co., pp. 329–344.

Birkenmajer, K. and E. Zastawniak 1989. Late Cretaceous–Early Neogene vegetation history of the Antarctic Peninsula sector, Gondwana breakup and Tertiary glaciations. *Bulletin of the Polish Academy of Sciences, Earth Sciences,* **37**, 63–88.

Bonaparte, J. F. 1986. History of the terrestrial Cretaceous vertebrates of Gondwana. *Actas IV Congreso Argentino de Paleontología y Bioestratigrafía*, **2**, 63–95.

Bond, M. 1999. Quaternary native ungulates of Southern South America. A synthesis. In Rabassa, J. and Salemme, M. (eds.), *Quaternary of South America and Antarctic Peninsula*. Rotterdam: A. A. Balkema, pp. 177–205.

Bond, M., R. Pascual, M. A. Reguero, S. N. Santillana, and S. A. Marenssi 1990. Los primeros ungulados extinguidos sudamericanos de la Antártida. *Ameghiniana*, **16**, 240.

Bond, M., M. A. Reguero, and S. F. Vizcaíno 1993. Mamíferos continentales de la Formación La Meseta (Terciario, Antártida): biocronología. *Actas XIII Congreso Brasileiro de Paleontologia e I Simpósio Paleontológico do Cone Sul*, 93.

Bond, M., A. A. Carlini, F. J. Goin, L. Legarreta,
E. Ortíz-Jaureguízar, R. Pascual, and M. A. Uliana 1995.
Episodes in South American land mammal evolution and
sedimentation: testing their apparent concurrence in a
Paleocene succession from central Patagonia. *Actas VI
Congreso de Paleontología y Bioestratigrafía*, **2**, 47–58.

Bond, M., M. A. Reguero, S. F. Vizcaíno, and S. A. Marenssi 2006.
A new "South American ungulate" (Mammalia: Litopterna)
from the Eocene of the Antarctic Peninsula. In Francis,
J. E., Pirrie, D., and Crame, J. A. (eds.), *Cretaceous–
Tertiary High Latitude Palaeonvironments, James Ross
Basin, Antarctica,* Special Publication no. 258. London:
Geological Society of London, pp. 163–176.

Bond, M., M. A. Reguero, A. Kramarz, J. J. Moly, S. N. Santillana,
and S. A. Marenssi 2008. Un Astrapotheria (Mammalia) del
Eoceno de la Formación La Meseta, Isla Marambio
(Seymour), Península Antártica. *Resúmenes IV Simposio
Latinoamericano sobre Investigaciones Antárticas y VII
Reunión Chilena de Investigación Antártica*, 409.

Bond, M., M. A. Reguero, S. F. Vizcaíno, S. A. Marenssi, and
E. Ortíz-Jaureguízar 2009. *Notiolofos,* a replacement name
for *Notolophus* Bond, Reguero, Vizcaíno and Marenssi,
2006, a preoccupied name. *Journal of Vertebrate
Paleontology*, **29**, 979.

Candela, A. and F. J. Goin 1995. Revisión de las especies
antárticas de marsupiales polidolopinos
(Polydolopimorphia, Polydolopidae). *Resúmenes III
Jornadas de Comunicaciones sobre Investigaciones
Antárticas*, 55–58.

Carlini, A. A., R. Pascual, M. A. Reguero, G. J. Scillato-Yané,
E. P. Tonni, and S. F. Vizcaíno 1990. The first Paleogene
land placental mammal from Antarctica: its paleoclimatic
and paleobiogeographical bearings. *Abstracts IV
International Congress of Systematic and Evolutionary
Biology*, 325.

Case, J. 1988. Paleogene floras from Seymour Island, Antarctic
Peninsula. In Feldmann, R. M. and Woodburne, M. O.
(eds.), *Geology and Paleontology of Seymour Island,
Antarctic Peninsula,* Memoir no. 169. Boulder, CO:
Geological Society of America, pp. 523–530.

Case, J. A., M. O. Woodburne, and D. S. Chaney 1988. A new
genus and species of polydolopid marsupial from the La
Meseta Formation, late Eocene, Seymour Island, Antarctic
Peninsula. In Feldmann, R. M. and Woodburne, M. O.
(eds.), *Geology and Paleontology of Seymour Island,
Antarctic Peninsula,* Memoir no. 169. Boulder, CO:
Geological Society of America, pp. 505–521.

Chornogubsky, L., F. J. Goin, and M. A. Reguero 2009.
A reassessment of Antarctic polydolopid marsupials
(Middle Eocene, La Meseta Formation). *Antarctic Science*,
21, 285–297.

Cifelli, R. L. 1985. Biostratigraphy of the Casamayoran, early
Eocene, of Patagonia. *American Museum Novitates*,
2820, 1–26.

Cione, A. L. and M. A. Reguero 1994. New records of the sharks *Isurus*
and *Hexanchus* from the Eocene of Seymour Island, Antarctica.
Proceedings of the Geologists Association, **105**, 1–14.

Cione, A. L., M. A. Reguero, and C. Acosta Hospitaleche 2007.
Did the continent and sea have different temperatures
in the northern Antarctic Peninsula during the Middle
Eocene? *Revista de la Asociación Geológica Argentina*,
62, 586–596.

Coccozza, C. and C. Clarke 1992. Eocene microplankton from
La Meseta Formation. *Antarctic Science*, **4**, 355–362.

Covacevich, V. and P. V. Rich 1982. New birds ichnites from
Fildes Peninsula, King George Island, West Antarctica.
In Craddock, C. (ed.), *Antarctic Geoscience*. Madison, WI:
University of Wisconsin Press, pp. 245–254.

del Valle, R. A., D. H. Elliot, and D. I. M. Macdonald 1992.
Sedimentary basins on the east flank of the Antarctic
Peninsula: proposed nomenclature. *Antarctic Science*,
4, 477–478.

Dingle, R. and M. Lavelle 1998a. Antarctic Peninsula cryosphere:
early Oligocene (c. 30 Ma) initiation and a revised glacial
chronology. *Journal of the Geological Society of London*,
155, 433–437.

Dingle, R. and M. Lavelle 1998b. Late Cretaceous–Cenozoic
climatic variations of the northern Antarctic Peninsula:
new geochemical evidence and review. *Palaeogeography,
Palaeoclimatology, Palaeoecology*, **107**, 79–101.

Dingle, R., J. McArthur, and P. Vroon 1997. Oligocene and
Pliocene interglacial events in the Antarctic Peninsula dated
using strontium isotope stratigraphy. *Journal of the
Geological Society of London*, **154**, 257–264.

Dingle, R., S. Marenssi, and M. Lavelle 1998. High latitude
Eocene climate deterioration: evidence from the northern
Antarctic Peninsula. *Journal of South American Earth
Sciences*, **11**, 571–579.

Ditchfield, P. W., J. D. Marshall, and D. Pirrie 1994. High
latitude palaeotemperature variation: new data from the
Tithonian to Eocene of James Ross Island, Antarctica.
Palaeogeography, Palaeoclimatology, Palaeoecology,
107, 79–101.

Dutton, A. L., K. Lohmann, and W. J. Zinsmeister 2002. Stable
isotope and minor element proxies for Eocene climate
of Seymour Island, Antarctica. *Paleoceanography*,
17(2), 1–13.

Elliot, D. H. 1988. Tectonic setting and evolution of the
James Ross Basin, northern Antarctic Peninsula. In
Feldmann, R. M. and Woodburne, M. O. (eds.), *Geology
and Paleontology of Seymour Island, Antarctic Peninsula,*
Memoir no. 169. Boulder, CO: Geological Society of
America, pp. 541–555.

Elliot, D. H. and T. A. Trautman 1982. Lower Tertiary strata
on Seymour Island, Antarctic Peninsula. In Craddock, C.
(ed.), *Antarctic Geoscience*. Madison, WI: University of
Wisconsin Press, pp. 287–297.

Exon, N. F., J. P. Kennett, and M. J. Malone (eds.) 2001.
*Proceedings of the Ocean Drilling Project, Initial Reports
189.* Available at www.odp.tamu.edu/publications/189_IR/
189ir.htm.

Frakes, L. A., J. E. Francis, and J. I. Syktus 1992. *Climate Modes
of the Phanerozoic*. Cambridge, UK: Cambridge
University Press.

Francis, J. E, 1991. Palaeoclimatic significance of Cretaceous/ early Tertiary fossil forests of the Antarctic Peninsula. In Thomson, M. R. A., Crame, J. A., and Thomson, J. W. (eds.), *Geological Evolution of Antarctica*. Cambridge, UK: Cambridge University Press, pp. 623–627.

Francis, J. E., A. M. Tosolini, and D. J. Cantrill 2003. Biodiversity and climatic change in Antarctic Paleogene floras. *Antarctic Contributions to Global Earth Sciences, IX International Symposium on Antarctic Earth Sciences*, 107.

Gandolfo, M. A., S. A. Marenssi, and S. N. Santillana 1996. Flora y Paleoclima de la Formación La Meseta (Eoceno–Oligoceno inferior?), isla Marambio (Seymour), Antártida. *Actas I Congreso del Paleógeno de América del Sur*, 31–32.

Gandolfo, M. A., S. A. Marenssi, and S. N. Santillana 1998a. Flora y paleoclima de la Formación La Meseta (Eoceno medio), isla Marambio (Seymour), Antártida. In Casadio, S. (ed.), *Paleógeno de América del Sur y de la Península Antártica*. Publicatíon Especial de la Asociación Paleontológica Argentina, 5, 155–162.

Gandolfo, M. A., P. Hoc, S. Santillana, and S. Marenssi 1998b. Una flor fósil morfológicamente afín a las Grossulariaceae (Orden Rosales) de la Formación La Meseta (Eoceno medio), Isla Marambio, Antártida. In Casadio, S. (ed.), *Paleógeno de América del Sur y de la Península Antártica*. Publicatíon Especial de la, Asociación Paleontológica Argentina, 5, 147–153.

Gazdzicki, A. J., M. Gruszczynski, A. Hoffman, K. Malkowski, S. A. Marenssi, S. Halas, and A. Tatur 1992. Stable carbon and oxygen isotope record in the Paleogene La Meseta Formation, Seymour Island, Antarctica. *Antarctic Science*, 4, 461–468.

Goin, F. and A. Carlini 1995. An early Tertiary microbiotheriid marsupial from Antarctica. *Journal of Vertebrate Paleontology*, 15, 205–207.

Goin, F. and M. Reguero 1993. Un "enigmático insectívoro" del Eoceno de Antártida. *Ameghiniana*, 30, 108.

Goin, F. J., M. A. Reguero, and S. F. Vizcaíno 1995. Novedosos hallazgos de "comadrejas" (Marsupialia) del Eoceno medio de Antártida. *Resúmenes III Jornadas de Comunicaciones sobre Investigaciones Antárticas*, 59–62.

Goin, F. J., J. A. Case, M. O. Woodburne, S. F. Vizcaíno, and M. A. Reguero 1999. New discoveries of "opossum-like" marsupials from Antarctica (Seymour Island, Medial Eocene). *Journal of Mammalian Evolution*, 6, 335–365.

Goin, F., M. Tejedor, M. Bond, G. López, and M. Reguero 2000. Mamíferos Eocenos de Paso del Sapo, Chubut. *Ameghiniana*, 37, 25R.

Goin, F. J., M. A. Reguero, R. Pascual, W. von Koenigswald, M. O. Woodburne, J. A. Case, C. Vieytes, S. A. Marenssi, and S. F. Vizcaíno 2006. First gondwanatherian mammal from Antarctica. In Francis, J. E., Pirrie, D., and Crame, J. A. (eds.), *Cretaceous–Tertiary High-Latitude Palaeoenvironments, James Ross Basin, Antarctica,* Special Publication no. 258. London: Geological Society of London, pp. 135–144.

Goin, F. J., N. Zimicz, M. A. Reguero, S. N. Santillana, S. A. Marenssi, and J. J. Moly 2007. New mammal from the Eocene of Antarctica, and the origins of the Microbiotheria. *Revista de la Asociación Geológica Argentina*, 62, 597–603.

Gradstein, F., J. Ogg, and A. Smith (eds.) 2004. *A Geologic Time Scale 2004*. Cambridge, UK: Cambridge University Press.

Haq, B. U., J. Hardenbol, and P. R. Vail 1987. Chronology of fluctuating sea levels since the Triassic. *Science*, 235, 1156–1167.

Hathway, B. 2000. Continental rift to back-arc basin: Jurassic–Cretaceous stratigraphical and structural evolution of the Larsen Basin, Antarctic Peninsula. *Journal of the Geological Society of London*, 157, 417–432.

Hawkes, D. D. 1961. The Geology of South Shetland Islands. I. *The Petrography of King George Island. Scientific Reports of the Falkland Islands Dependencies Survey*, 26, 1–28.

Hooker, J. J. 1992. An additional record of a placental mammal (Order Astrapotheria) from the Eocene of Western Antarctica. *Antarctic Science*, 4, 107–108.

Ivany, L. C., S. Van Simaeys, E. W. Domack, and S. D. Samson 2006. Evidence for an earliest Oligocene ice sheet on the Antarctic Peninsula. *Geology*, 34, 377–380.

Jianjun, L. and Z. Shuonan 1994. New Materials of bird ichnites from Fildes Peninsula, King George Island of Antarctica and their biogeographic significance. In Shen, Y. (ed.), *Stratigraphy and Palaeontology of Fildes Peninsula, King George Island, Antarctica,* State Antarctic Commitee, Monograph no. 3. Beijing: Science Press, pp. 239–249.

Kennett, J. P. 1977. Cenozoic evolution of Antarctic glaciation, the circum-Antarctic oceans and their impact on global paleoceanography. *Journal of Geophysical Research*, 82, 3843–3859.

Krause, D. W., G. V. R. Prasad, W. von Koenigswald, A. Sahni, and F. E. Grine 1997. Cosmopolitanism among Gondwana Late Cretaceous mammals. *Nature*, 390, 504–507.

Krause, D. W., M. D. Gottfried, P. M. O'Connor, and E. M. Roberts 2003. A Cretaceous mammal from Tanzania. *Acta Palaeontologica Polonica*, 48, 321–330.

Lawver, L. A. and L. M. Gahagan 1998. Opening of Drake Passage and its impact on Cenozoic ocean circulation. In Crowley, T. J. and Burke, K. C. (eds.), *Tectonic Boundary Conditions for Climate Reconstructions*. Oxford, UK: Oxford University Press, pp. 212–223.

Lawver, L. A. and L. Gahagan 2003. Evolution of Cenozoic seaways in the circum-Antarctic region. *Palaeogeography, Palaeoclimatology, Palaeoecology*, 198, 11–37.

Lawver, L. A., L. M. Gahagan, and F. M. Coffin 1992. The development of paleoseaway around Antarctica. In Kennett, J. P. and Warnke, D. A. (eds.), *The Antarctic Paleoenvironment: A Perspective on Global Change,* Antarctic Research Series no. 65. Washington, DC: American Geophysical Union, pp. 7–30.

Laza, J. H. and M. A. Reguero 1990. Extensión faunística de la antigua región neotropical en la Península Antártica durante el Eoceno. *Ameghiniana*, 26, 245.

Livermore, R., A. Nankivel, G. Eagles, and P. Morris 2005. Paleogene opening of Drake Passage. *Earth and Planetary Science Letters*, 236, 459–470.

MacPhee, R., M. A. Reguero, P. Strganac, M. Nishida, and P. Jacobs 2008. Out of Antarctica: Paleontological reconnaissance of Livingston Island (South Shetlands) and Seymour Island (James Ross Group). *Journal of Vertebrate Paleontology*, **28** (3 Suppl.), 64A.

Marenssi, S. A. 1995. Sedimentología y paleoambientes de sedimentación de la Formación La Meseta, isla Marambio, Antártida. Ph.D. thesis, Universidad de Buenos Aires.

Marenssi, S. A. 2006. Eustatically controlled sedimentation recorded by Eocene strata of the James Ross Basin. In Francis, J. E., Pirrie, D., and Crame, J. A. (eds.), *Cretaceous–Tertiary High-Latitude Environments, James Ross Basin, Autarctia*, Special Publication no. 258. London: Geological Society of London, pp. 125–133.

Marenssi, S. A. and S. N. Santillana 1994. Unconformity-bounded units within the La Meseta Formation, Seymour Island, Antarctica: a preliminary approach. *Abstracts XXI Polar Symposium, Warsaw, 33–37.*

Marenssi, S. A., M. A. Reguero, S. N. Santillana, and S. F. Vizcaíno 1994. Eocene land mammals from Seymour Island, Antarctica: palaeobiogeographical implications. *Antarctic Science*, **6**, 3–15.

Marenssi, S. A., S. N. Santillana, and C. A. Rinaldi 1998a. *Paleoambientes sedimentarios de la Aloformación La Meseta (Eoceno), Isla Marambio (Seymour), Antártida.* Buenos Aires: Instituto Antártico Argentino.

Marenssi, S. A., S. N. Santillana, and C. A. Rinaldi 1998b. Stratigraphy of the La Meseta Formation (Eocene), Marambio (Seymour) Island, Antarctica. In Casadio, S. (ed.), *Paleógeno de América del Sur y de la Península Antártica.* Publicacíon Especial de la Asociación Paleontológica Argentina, **5**, 137–146.

Marenssi, S. A, L. I. Net, and S. N. Santillana 2002. Provenance, depositional and paleogeographic controls on sandstone composition in an incised valley system: the Eocene La Meseta Formation, Seymour Island, Antarctica. *Sedimentary Geology*, **150**, 301–321.

Marshall, L. G., T. Sempere, and R. F. Butler 1997. Chronostratigraphy of the mammal-bearing Paleocene of South America. *Journal of South American Earth Sciences*, **10**, 49–70.

Mazzoni, M. M., K. Kawashita, S. Harrison, and E. Aragón 1991. Edades radiométricas eocenas en el borde occidental del Macizo Norpatagónico. *Revista de la Asociación Geológica Argentina*, **46**, 150–158.

Olivero, E., Z. Gasparini, C. Rinaldi, and R. Scasso 1990. First record of dinosaurs in Antarctica (Upper Cretaceous, James Ross Island): palaeogeographical implications. In Thomson, M. R. A., Crame, J. A., and Thomson, J. W. (eds.), *Geological Evolution of Antarctica*. Cambridge, UK: Cambridge University Press, pp. 617–622.

Pascual, R., E. Ortíz-Jaureguízar, and J. L. Prado 1996. Land mammals: paradigm for Cenozoic South American geobiotic evolution. *Münchner Geowissenschaftliche Abhandlungen A*, **30**, 265–319.

Pirrie, D., J. D. Marshall, and J. A. Crame 1998. Marine high Mg calcite cements in Teredolites bored fossil wood: evidence for cool palaeoclimates in the Eocene La Meseta Formation, Seymour Island, Antarctica. *Palaios*, **13**, 276–286.

Reguero, M. A. and Z. Gasparini 2006. Late Cretaceous–Early Tertiary marine and terrestrial vertebrates from James Ross Basin, Antarctic Peninsula: a review. In Rabassa, J. and Borla, M. L. (eds.), *Antarctic Peninsula and Tierra del Fuego: 100 years of Swedish–Argentine Scientific Cooperation at the End of the World*. London: Taylor & Francis, pp. 55–76.

Reguero, M. A., S. F. Vizcaíno, F. J. Goin, S. A. Marenssi, and S. N. Santillana 1998. Eocene high-latitude terrestrial vertebrates from Antarctica as biogeographic evidence. In Casadio, S. (ed.), *Paleógeno de América del Sur y de la Península Antártica*. Publicacíon Especial de la Asociación Paleontológica Argentina, **5**, 185–198.

Reguero, M. A., S. A. Marenssi, and S. N. Santillana 2002. Antarctic Peninsula and Patagonia Paleogene terrestrial environments: biotic and biogeographic relationships. *Palaeogeography, Palaeoclimatology, Palaeoecology*, **277**, 1–22.

Robert, C., L. Diester-Haass, and H. Chamley 2002. Late Eocene–Oligocene oceanographic development at southern high latitudes, from terrigenous and biogenic particles: a comparison of Kerguelen Plateau and Maud Rise, ODP Sites 744 and 689. *Marine Geology*, **191**, 37–54.

Sadler, P. 1988. Geometry and stratification of uppermost Cretaceous and Paleogene units on Seymour Island, northern Antarctic Peninsula. In Feldmann, R. M. and Woodburne, M. O. (eds.), *Geology and Paleontology of Seymour Island, Antarctic Peninsula*, Memoir no. 169. Boulder, CO: Geological Society of America, pp. 303–320.

Scher, H. D. and E. E. Martin 2006. The timing and climatic influence of the opening of Drake Passage. *Science*, **312**, 428–430.

Scillato Yané, G. J. and R. Pascual 1984. Un peculiar Paratheria, Edentata (Mammalia) del Paleoceno medio de Patagonia. *Resúmenes I Jornadas Argentinas de Paleontología de Vertebrados*, 15.

Shen, Y. 1995. Subdivision and correlation of Cretaceous to Paleogene volcano-sedimentary sequence from Fildes Peninsula, King George Island, Antarctica. In Shen, Y. (ed.), *Stratigraphy and Palaeontology of Fildes Peninsula, King George Island, Antarctica*. State Antarctic Committee Monograph no. 3. Beijing: Science Press, pp. 1–36.

Stilwell, J. D. and W. J. Zinsmeister 1992. *Molluscan Systematics and Biostratigraphy: Lower Tertiary La Meseta Formation, Seymour Island, Antarctic Peninsula*, Antarctic Research Series no. 55. Washington, DC: American Geophysical Union.

Stilwell, J. D. and R. M. Feldmann (eds.) 2000. *Paleobiology and Paleoenvironments of Eocene Rocks, McMurdo Sound, East Antarctica*, Antarctic Research Series no. 76. Washington, DC: American Geophysical Union.

Tejedor, M. F., F. J. Goin, J. N. Gelfo, G. M. López, M. Bond, A. A. Carlini, G. J. Scillato-Yané, M. O. Woodburne, L. Chornogubsky, E. Aragón, M. A. Reguero, N. Czaplewski, S. Vincon, G. Martin, and M. Ciancio 2009. New Early Eocene mammalian fauna from Western

Patagonia, Argentina. *American Museum Novitates*, **3638**, 1–43.

Torres, T., S. A. Marenssi, and S. N. Santillana 1994. Maderas fósiles de la isla Seymour, Formación La Meseta, Antártica. *Serie Científica del Instituto Antartico Chileno*, **44**, 17–38.

Vizcaíno, S. F. and G. J. Scillato-Yané 1995. An Eocene tardigrade (Mammalia, Xenarthra) from Seymour Island, West Antarctica. *Antarctic Science*, **7**, 407–408.

Vizcaíno, S. F., M. Bond, M. A. Reguero, and R. Pascual 1997. The youngest record of fossil land mammals from Antarctica, its significance on the evolution of the terrestrial environment of the Antarctic Peninsula during the late Eocene. *Journal of Paleontology*, **71**, 348–350.

Vizcaíno, S. F., M. A. Reguero, F. J. Goin, C. P. Tambussi and J. I. Noriega 1998. Community structure of Eocene terrestrial vertebrates from Antarctic Peninsula. In Casadio, S. (ed.), *Paleógeno de América del Sur y de la Peninsula Antártica*. Publicacíon Especial de la Asociación Paleontológica Argentina, **5**, 177–183.

Webb, P. N. 1991. Evolution of Cenozoic paleoenvironments. In Thomson, M. R. A., Crame, J. A., and Thomson, J. W. (eds.), *Geological Evolution of Antarctica*. Cambridge, UK: Cambridge University Press, pp. 599–607.

Wilf, P., N. R. Cúneo, K. R. Johnson, J. F. Hicks, S. L. Wing, and J. D. Obradovich 2003. High plant diversity in Eocene South America: evidence from Patagonia. *Science*, **300**, 122–125.

Woodburne, M. O. and W. J. Zinsmeister 1982. Fossil land mammals from Antarctica. *Science*, **218**, 284–286.

Woodburne, M. O. and W. J. Zinsmeister 1984. The first land mammal from Antarctica and its biogeographic implications. *Journal of Paleontology*, **58**, 913–948.

Wrenn, J. H. and G. F. Hart 1988. Paleogene dinoflagellates cyst biostratigraphy of Seymour Island, Antarctica. In Feldmann, R. M. and Woodburne, M. O. (eds.), *Geology and Paleontology of Seymour Island, Antarctic Peninsula*, Memoir no. 169. Boulder, CO: Geological Society of America, pp. 321–447.

Zachos, J., M. Pagani, L. Sloan, E. Thomas, and K. Billups 2001. Trends, rhythms, and aberrations in global climate 65 Ma to present. *Science*, **292**, 686–693.

Zamaloa, M. C., E. J. Romero, and L. Stinco 1987. Polen y esporas de la Formación La Meseta (Eoceno superior–Oligoceno) de la isla Marambio (Seymour), Antártida. *Actas VII Congreso Argentino de Paleobotánica y Palinología*, 199–203.

Zinsmeister, W. J. 1978. Eocene nautiloid fauna from the La Meseta Formation of Seymour Island, Antarctic Peninsula. *Antarctic Journal of United States*, **13**, 24–25.

Zinsmeister, W. J. 1982. Late Cretaceous–early Tertiary molluscan biogeography of southern Circum-Pacific. *Journal of Paleontology*, **56**, 84–102.

27 Mid-Cenozoic paleoclimatic and paleoceanographic trends in the southwestern Atlantic Basins: a dinoflagellate view

G. Raquel Guerstein, M. Verónica Guler, Henk Brinkhuis, and Jeroen Warnaar

Abstract

Middle Eocene to Miocene organic-walled dinoflagellate cysts (dinocysts) from marine deposits in southwest Atlantic sedimentary basins in Patagonia (Austral, Golfo San Jorge, Colorado, and Punta del Este, from south to north) reveal details of the paleoclimatic and paleoceanographic evolution of the region. There is evidence for four transgressive events during this interval, viz. in the middle Middle Eocene, Late Eocene, Late Oligocene – earliest Miocene, and Middle to Late Miocene. The Middle Eocene dinocyst assemblages are dominated by typical high-latitude, sub-Antarctic endemic floras recorded from many different sites from around the Southern Ocean. During the Late Eocene, these endemic species became less abundant and were partially replaced by more diverse assemblages with markers of cooler, more offshore conditions and an increased number of heterotrophic Protoperidiniaceae. In contrast, the Late Oligocene and Middle Miocene assemblages are often dominated by more cosmopolitan species. Hence, material from the last two transgressive events suggest more temperate to warm surface water conditions. The trends recorded from the Southwest Atlantic basins coincide with the large-scale paleoclimatic and paleoceanographic trends previously proposed by other authors on the basis of analysis of sites in the Southern Ocean.

Resumen

Los quistes de dinoflagelados de pared orgánica de depósitos marinos del Eoceno Medio al Mioceno de diferentes cuencas sedimentarias del Atlántico Sudoccidental (Austral, San Jorge, Colorado y Punta del Este), revelan detalles sobre la la evolución paleoclimática y paleoceanográfica de la región. Existen evidencias de cuatro eventos transgresivos, en el Eoceno Medio, Eoceno Tardío, Oligoceno Tardío – inicios del Mioceno y Mioceno Medio a Tardío. Las asociaciones del Eoceno Medio están dominadas por floras subantárticas endémicas, típicas de altas latitudes, registradas en diferentes sitios del Océano

Austral. Durante el Eoceno Tardío estas especies endémicas se hicieron menos abundantes y fueron parcialmente reemplazadas por asociaciones más diversas con indicadores de aguas frías, especies oceánicas y un mayor número de quistes de dinoflagelados heterotróficos protoperidiniaceos. En contraste, las asociaciones del Oligoceno Tardío y Mioceno Medio están generalmente dominadas por especies cosmopolitas. Estos dos últimos eventos transgresivos sugieren condiciones de aguas superficiales templadas a cálidas. Las tendencias registradas en las cuencas del Atlántico Sudoccidental coinciden con las tendencias a gran escala propuestas previamente por otros autores sobre la base del análisis de sitios en el Océano Austral.

Introduction

Dinoflagellates constitute an important component of the eukaryotic marine plankton. Many produce resistant organic-walled cysts (dinocysts) typically encountered in the fossil record. Their analysis contributes to the assessment of physical and chemical conditions of ancient surface waters. From a micropaleontological perspective these microfossils are very important in the studies of the Paleogene and early Neogene, as they include biostratigraphic and paleoenvironmental index species. The spatial qualitative and quantitative distribution of dinocysts thus helps to portray the paleoclimatic and paleoceanographic evolution of, for example, the southwest Atlantic basins.

During the Cenozoic the extra-Andean Patagonian regions were affected by several trangressions that flooded wide areas. Malumián (1999) described five Cenozoic sedimentary cycles that were related mainly to the subsidence of the Atlantic margin of Patagonia, suggesting that the cycles were controlled by tectonic events and climatic changes. The first cycle began during the Maastrichtian with a transgressive event represented in all the southwest Atlantic basins. The second event was during the Middle Eocene and was best documented south of 40° S, though it was also recorded from deep wells in the Punta del Este and Colorado Basins. The third cycle started with widespread Late Eocene marine units, and extended during the Early

The Paleontology of Gran Barranca: Evolution and Environmental Change through the Middle Cenozoic of Patagonia, eds. R. H. Madden, A. A. Carlini, M. G. Vucetich, and R. F. Kay. Published by Cambridge University Press. © Cambridge University Press 2010.

Oligocene with continental deposits. The fourth cycle is characterized by two important transgressive events in terms of the covered area. The first event, informally named "Patagoniano," occurred during the Late Oligocene – Early Miocene and the second, originally referenced as "Entrerriense," has been dated as Middle to early Late Miocene. The fifth and last sedimentary cycle defined by Malumián (1999) is related to an orogenic phase and uplifting of the Patagonian Andes. The most important sedimentary unit is considered the regressive phase of the "Entrerriense" transgression.

During the depositional time of the Sarmiento Formation at Gran Barranca (Middle Eocene to Middle Miocene) there is evidence for four main marine transgressions at the southwest Atlantic basins. The transgressive events were dated mainly by calcareous microfossils and palynomorphs as Middle Eocene, Late Eocene, Late Oligocene – earliest Miocene, and Middle to Late Miocene respectively (Malumián 1999 and references therein; Guler *et al.* 2002; Guler and Guerstein 2002; Guerstein *et al.* 2004, 2008).

These sea-level changes that occurred from the Middle Eocene to the Middle Miocene should partially reflect the global transition from Early Eocene "greenhouse" conditions to the "icehouse" conditions typical for the latter part of the Paleogene through the Neogene (Zachos *et al.* 2001, 2008). The aim of this contribution is to analyze the broad environmental changes as reflected by dinocyst assemblages from the Austral, San Jorge, Colorado, and Punta del Este Basins (Fig. 27.1). These will be compared to, and integrated with, existing records from: (1) Patagonia, (2) other South Atlantic basins, and (3) Southern Ocean sites. Subsequently, possible relationships with large-scale tectonic processes (e.g. the opening and deepening of the Drake Passage and the Tasman Gateway) that occurred during this time interval are investigated.

Our results, based on compiled dinoflagellate data from southwestern Atlantic basins, attempt to reconstruct the paleoclimatic and paleoceanographic conditions in Patagonia from the Middle Eocene to the Middle Miocene. Therefore, we believe that this chapter will contribute to a better understanding of the paleontology of Gran Barranca in a regional scenario.

Materials and methods

The data-set used in this study was compiled from published information on Middle Eocene to Late Miocene dinocyst assemblages from five Southwest Atlantic basins: Austral, San Jorge, Valdés, Colorado, and Punta del Este (Fig. 27.1). Table 27.1 shows the sources of information, including location, type and number of samples, repository, and references.

Fig. 27.1. Distribution map of compiled sites and localities.

Age calibration of these records is mostly based on calcareous microfossil and/or palynomorph stratigraphies and to a lesser extends on radiometric dating. Figure 27.2 shows the stratigraphic position of the units discussed in this study.

The taxonomy and nomenclature of the dinocysts, unless indicated otherwise, are as cited in Fensome and Williams (2004) and Fensome *et al.* (2007). The timescale used is from Gradstein *et al.* (2004). The tectonic reconstructions derived from Plate Tectonic Reconstructions Online Paleogeographic Mapper (http://www.serg.unicam.it/Reconstructions.htm), after Schettino and Scotese (2005).

Dinocyst assemblages from southwest Atlantic basins

Middle Eocene

The marine Middle Eocene deposits extend into the Austral Basin, cropping out both in Tierra del Fuego and Santa Cruz Provinces they also crop out in the Valdés Basin. Towards the north, they were only recorded in the offshore Colorado and Punta del Este Basins (Fig. 27.3A). At the four basins this marine episode is very well documented through palynological data, with rich and diverse dinocyst assemblages (Archangelsky 1969; Guerstein *et al.* 2003, 2008).

Dinocyst assemblages from northeastern Tierra del Fuego, at the Río de la Turba section (Leticia Formation) are dominated by the so-called circum-Antarctic, endemic

Table 27.1. *Site locations, type of samples, number of samples palynologically productive*

Basin	Site	Location	Type/number of samples	Repository[a]	References
Austral	Río de la Turba	Northeast Tierra del Fuego Province	outcr: 10	UNSLP	Guerstein *et al.* 2008
	Cabo Peñas – Cerro Águila – Río Candelaria	Northeast Tierra del Fuego Province	outcr: 20	UNSLP	Guerstein *et al.* 2002, 2008
	Cabo Campo del Medio	East Tierra del Fuego Province	outcr: 9	UNSLP	Guerstein *et al.* 2003, R. G. (pers. obs.)
	Río Turbio: D-15, D-16, T-134	Southwest Santa Cruz Province	core: 22	UNLP	Archangelsky 1969
	Ea. 25 de mayo	Southwest Santa Cruz Province	outcr: 23	UNSLP	Guerstein *et al.* 2004
	Monte Entrada – Monte Observación	East Santa Cruz Province	outcr: 20	MACN (CIRGEO)	Barreda and Palamarczuk 2000b
San Jorge	Mazarredo – Punta Nava – El Faro	Northeast Santa Cruz Province	outcr: 12	MACN (CIRGEO)	Barreda and Palamarczuk 2000a
	Cerro Chenque	Southeast Chubut Province	outcr: 13	MACN (CIRGEO)	Palamarczuk and Barreda 1998
Valdés	Puerto Pirámide	Northeast Chubut Prov.	outcr: 1	MACN - UNSLP	Palazzesi and Barreda 2003
Colorado	Ombucta x-1	Southwest Buenos Aires Province	drill: 31	UNSLP	Guerstein and Guler 2000
	Pejerrey x-1	Offshore Buenos Aires Province	drill: 26	UNSLP	Guerstein *et al.* 2003
	Cx-1	Offshore Buenos Aires Province	drill: 40	UNSLP	Guerstein and Junciel 2001; Guler *et al.* 2001
	Barranca Final	East Río Negro Province	outcr: 10	UNSLP	Guler *et al.* 2002
	Fx-1, Dx-1	Offshore Buenos Aires Province	drill: 33	UNSLP	Guler and Guerstein 2002
Punta del Este	Gaviotín x-1	Southern Uruguay	drill: 5	Fac. Ciencias, Univ. de la República – Uruguay	Daners *et al.* 2004; Guerstein *et al.* 2003

Notes: [a]UNSLP, Laboratorio de Palinología de la Universidad Nacional del Sur; UNLP, Universidad Nacional de La Plata; MACN, Museo Argentino de Ciencias Naturales; (CIRGEO): originally stored at CIRGEO (Centro de Investigaciones en Recursos Geológicos).

Transantarctic Flora of Wrenn and Beckman (1982). The Transantarctic Flora is also recorded from the Leticia Formation in the Cabo Campo del Medio section, southeastern Tierra del Fuego (R. G. pers. obs.), and from southwestern Santa Cruz within the Río Turbio Formation (Archangelsky 1969).

Wrenn and Beckmann (1982) considered that the Transantarctic Flora was primarily represented by *Alterbidinium*? *distinctum*, *Arachnodinium antarcticum*, *Deflandrea antarctica*, *D. cygniformis*, *D. granulata*, *D. oebisfeldensis sensu* Cookson and Cranwell, *Spinidinium macmurdoense*,

Vozzhenikovia apertura, and *Wilsonidinium echinosuturatum*. However, in the Argentinian sections, these elements are commonly joined by *Thalassiphora pelagica* and *Enneadocysta dictyostila* (often as *Enneadocysta partridgei*: see Fensome *et al.* 2007). Similar situations have been reported by, for example, Truswell (1997) from the west Tasman continental margin (see also Brinkhuis *et al.* 2003). However, Brinkhuis *et al.* remarked that both species were also recorded at somewhat lower latitudes. Many experts, including Pross (2001) and Pross and Brinkhuis (2005), have considered *Thalassiphora pelagica* to be a typical cosmopolitan taxon

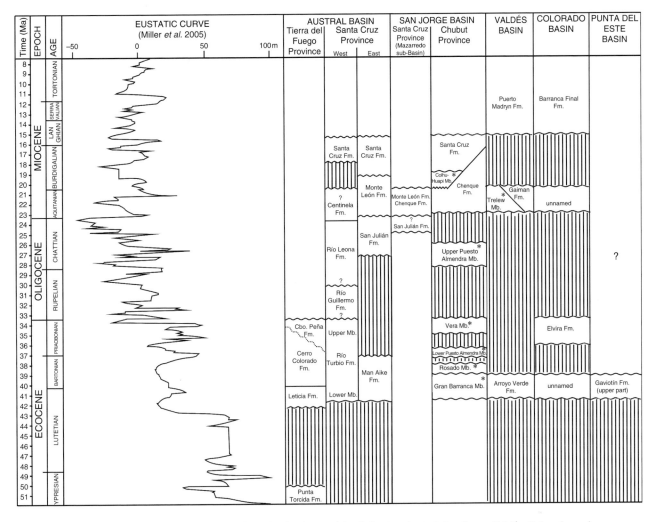

Fig. 27.2. Correlation of lithostratigraphical units from the Southwest Atlantic basins: Austral, San Jorge, Valdés, Colorado, and Punta del Este. Compiled and modified from Malumián (1999), Guler *et al.* (2002), Guerstein *et al.* (2004), Daners *et al.* (2004), Kay *et al.* (1996 a, 1996b), Parras *et al.* (2008), Ré *et al.* (Chapter 4, this book). Units of the Sarmiento Formation with asterisk.

in Paleogene and early Neogene dinocyst assemblages, in the southwest Atlantic basins at least in low abundances at all localities studied. Abundant *Thalassiphora pelagica* have been recorded from the Middle Eocene of the Río Turbio Formation (Archangelsky 1969) and from the Pejerrey x-1 and Gaviotín wells (Colorado and Punta del Este Basins) (Guerstein *et al.* 2003).

Enneadocysta dictyostila (as *Enneadocysta partridgei*) was also considered a member of the Transantarctic Flora by Brinkhuis *et al.* (2003), Sluijs *et al.* (2003), and Warnaar (2006). Indeed, the abundant occurrence of *Enneadocysta dictyostila* constitutes an important biostratigraphic event within the middle Eocene all over the Southern Ocean (see also Raine *et al.* 1997; Röhl *et al.* 2004; Warnaar 2006). Moreover, based on personal observations (R. G.), a bloom of this species in the middle Member of the Leticia Formation (southeastern Tierra del Fuego) coincides with the unique

interval bearing nannofossils throughout the section (Malumián *et al.* 1994; Olivero pers. comm. to R. G. 2004).

Late Eocene

Rocks of the Late Eocene transgression are recorded in northern Tierra del Fuego, eastern Santa Cruz, and offshore in the Colorado Basin (Fig. 27.3B). In the three sections of the type area of the Cabo Peña Formation in northeast Tierra del Fuego, the Transantarctic Flora is still represented, but differs from Middle Eocene assemblages by containing more typical offshore species along with the cool-water marker *Gelatia inflata*, as well as common heterotrophic Proto-peridiniaceans (Guerstein *et al.* 2008). In the offshore Colorado Basin, this marine episode is represented by the Elvira Formation and was recorded in the Pejerrey x-1 and Cx-1 wells (Guerstein and Junciel 2001; Guerstein *et al.* 2003). An

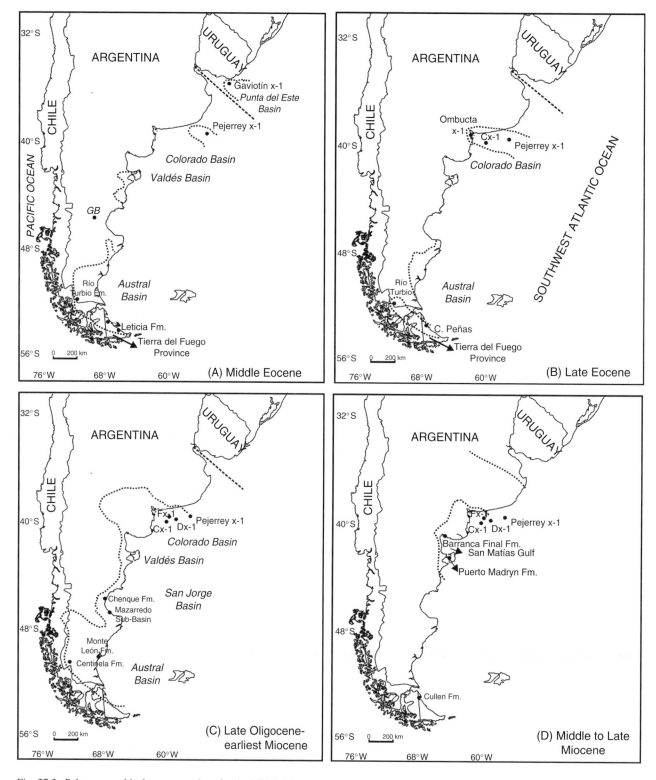

Fig. 27.3. Paleogeographical reconstructions for the Middle Eocene to the Late Miocene in the southwest Atlantic basins. Dashed lines indicate paleo-shorelines. GB, Gran Barranca. (Modified from Malumián 1999.)

increase in protoperidiniaceans (*Lejeunecysta* spp.) characterizes the top of this interval in the Cx-1 well.

Most extant protoperidiniaceans are heterotrophic, and commonly inhabit cool (upwelled) surface waters rich in dissolved nutrients and characterized by high primary productivity (e.g. Reichart and Brinkhuis 2003; Sluijs *et al.* 2005). The appearance of these components in the Late Eocene deposits may be due to the influence of increased eukaryotic productivity and the intensification of ocean circulation and upwelling, possibly related to the shallow opening of the Drake Passage during the late Middle Eocene (Scher and Martin 2006).

Late Oligocene – earliest Miocene
The deposits from the Late Oligocene – earliest Miocene transgression have widespread records throughout Patagonia (Fig. 27.3C). It was one of the most important marine transgressions that occurred during the Cenozoic in the southernmost part of South America. The sediments deposited during this event, informally known as "Patagoniano," are an important component of the sedimentary succession of the Colorado, San Jorge, and Austral Basins.

The Centinela Formation, cropping out in southwestern Santa Cruz Province, represents the maximum inland shoreline extension to the west near the Oligocene–Miocene boundary. Guerstein *et al.* (2004) assigned a Late Oligocene – earliest Miocene age to this formation, which agrees with the ^{87}Sr/^{86}Sr ages of about 23 Ma indicated for the lower part of the section by Casadío *et al.* (2000).

To the east, this marine episode is recorded at several sections exposed along the Atlantic coast, which yield rich and diverse dinocysts assemblages: Chenque Formation (Palamarczuk and Barreda 1998), Monte León Formation (Barreda and Palamarczuk 2000a), and coeval deposits from the Mazarredo sub-Basin in northern Santa Cruz Province (Barreda and Palamarczuk 2000b).

All sections from southern Patagonia show an increasing marine influence at the base of the Miocene, with oceanic and outer neritic species, and a subsequent regressive trend towards the top. Unlike the underlying units, the endemic Transantarctic Flora species are almost absent and cosmopolitan taxa are dominant (*Operculodinium israelianum-centrocarpum, Nematosphaeropsis rigida, Lingulodinium machaerophorum, L. hemicystum, Spiniferites* spp.). The presence of *Tuberculodinium vancampoae* in assemblages from the San Jorge Basin indicates temperate to warm surface waters (Wall *et al.* 1977; Edwards and Andrle 1992). These data are in accordance with results from recent general circulation model experiments that indicate a flow of warm waters from the Pacific Ocean through the Drake Passage into the southwest Atlantic Ocean (Sijp and England 2004; Warnaar 2006).

Middle to early Late Miocene
According to Malumián (1999), the Middle Miocene transgression is the most widespread Cenozoic transgression, covering the central part of Argentina. In Patagonia it is restricted to the present coastal areas in Valdés Basin and San Matías Gulf, Colorado Basin (Fig. 27.3D).

In the Colorado Basin, this marine episode is represented by the upper part of Barranca Final Formation, the type section of which crops out on the San Matías Gulf coast. The dinocyst assemblages are characterized by a scarcity of oceanic markers and high abundances of protoperidiniaceans, suggesting nutrient-rich shallow waters. They reflect warm–temperate to warm surface waters, mainly based on the presence of *Tuberculodinium vancampoae* and *Lingulodinium machaerophorum* (Guler *et al.* 2002; Guler 2003). Preliminary records from Puerto Pirámide section (Valdés Basin) show similar assemblages bearing *Tuberculodinium vancampoae* and *Lingulodinium hemicystum* (Palazzesi and Barreda 2004). These climatic conditions coincide with those established during the Neogene Climatic Optimum, at about 16 Ma (Flower and Kennett 1994).

The Middle to Late Miocene interval has been recorded in different wells drilled in the offshore Colorado Basin (Gamerro and Archangelsky 1981; Guerstein and Junciel 2001; Guler and Guerstein 2002). The transgressive interval correlates with the palynozone B of Gamerro and Archangelsky (1981). Upwards, palynological assemblages are dominated by sporomorphs reflecting a regressive trend, possibly related to a cooling climatic event (Guler *et al.* 2001; Guler and Guerstein 2002).

Comparison with other sites from the Southern Ocean

The Middle Eocene dinocyst assemblages from the onshore Tierra del Fuego and Santa Cruz Basins and the offshore Punta del Este and Colorado Basins closely resemble those from other sites in the Southern Ocean. They are all dominated by the Transantarctic Flora and in most of the records there was at least one peak of *Enneadocysta dictyostila*, for example: off southeastern Australia and western New Zealand (Haskell and Wilson 1975, sites 280–283); from the upper part of La Meseta Formation in northern Seymour Island (Cocozza and Clarke 1992, site DJ277) and from Brunce Bank in the Scotia Sea, off Antarctica (Mao Shaozhi and Mohr 1995, site 696) (Fig. 27.4A).

In a study of ODP Site 1172 (1168–1172) on the East Tasman Plateau (Fig 27.4A), Röhl *et al.* (2004) recorded alternating dominances of *Deflandrea* spp. and *Enneadocysta* spp. during part of the Middle Eocene. These authors correlated peaks of *Enneadocysta* spp. with maxima in nannoplankton occurrences, whereas maxima of *Deflandrea* spp.

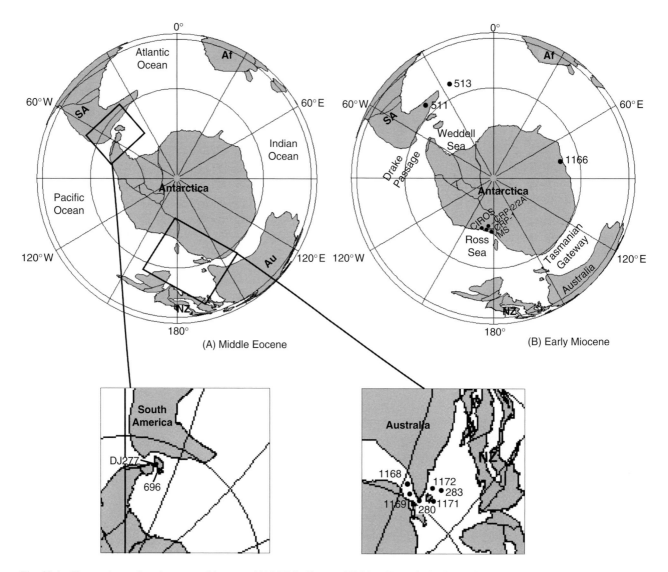

Fig. 27.4. Circum-Antarctic paleogeographic maps. (A) Middle Eocene (40 Ma – Bartonian); (B) Early Miocene (20 Ma – Burdigalian). Gray areas indicate continental crust. For locality codes see text. Maps derived from Plate Tectonic Reconstructions On-line Paleogeographic Mapper.

were related to diatom blooms and nutrient enrichment intervals. Based on this pattern, they proposed that abundances of *Enneadocysta* spp. might reflect more offshore, oligotrophic conditions, whereas *Deflandrea* peaks could indicate more inshore, eutrophic conditions.

Fensome *et al.* (2007) considered the genus *Enneadocysta* to be a member of the gonyaulacacean family Areoligeraceae. Based on this interpretation, Guerstein *et al.* (2008) suggested that *Enneadocysta dictyostila* was possibly an autotrophic species that preferred well lit, surface waters. Conspicuously, this member of the Transantarctic Flora is only consistently recorded eastern Tasman region, precisely where the general circulation model experiments by Huber *et al.* (2004) showed lower temperatures compared with other circum-Antarctic

sites. Thus, high abundances of *Enneadocysta dictyostila* may reflect relative cooling and deepening episodes related to the setting of the "refrigerator trap" system in the Southern Ocean that governed gyres in the South Pacific and South Indian–South Atlantic oceans (Warnaar 2006).

At ODP Site 511, eastern Malvinas Plateau (Fig. 27.4B), the Eocene–Oligocene boundary is well determined by independent dating, mainly based on planktonic foraminifers, calcareous nannoplankton, and radiolarians (Basov *et al.* 1983). Below the Eocene–Oligocene boundary, dinocyst assemblages are dominated by a few species of the Transantarctic Flora, such as *Alterbidinium*? *distinctum*, *Deflandrea antarctica*, and *Vozzhennikovia* sp., the latter being dominant close to the boundary (Goodman and Ford 1983; Guerstein

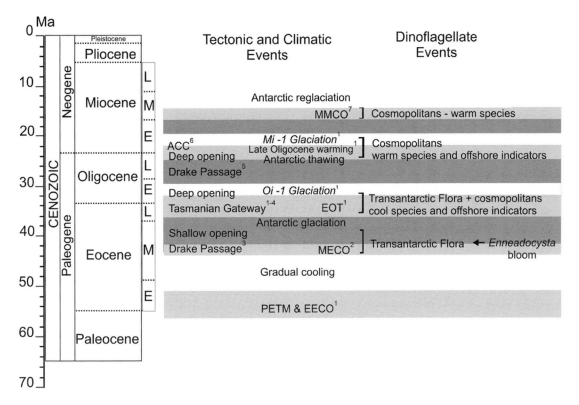

Fig. 27.5. Tectonic, climatic, and dinoflagellate events during the Mid Cenozoic in Patagonia. References: 1, Zachos *et al.* (2001); 2, Bohaty and Zachos (2003); 3, Scher and Martin (2006); 4, Stickley *et al.* (2006); 5, Pekar *et al.* (2006); 6, Warnaar (2006); 7, Flower and Kennett (1994). ACC, Antarctic Cirumpolar Current; EECO, Early Eocene Climatic Optimum; EOT, Eocene–Oligocene transition; PETM, Late Paleocene Thermal Maximum; MECO, Middle Eocene Climatic Optimum; MMCO, Middle Miocene Climatic Optimum.

et al. 2002). The oxygen isotopic data associated with the presence of *Gelatia inflata* may reflect cool surface-water conditions (Muza *et al.* 1983; Guerstein *et al.* 2002).

In the Late Eocene – Early Oligocene from CIROS-1 Drillhole, McMurdo Sound, East Antarctica (Fig. 27.4B), Hannah (1997) observed a decrease in the number of dinocysts across the Eocene–Oligocene boundary; the author related this change to cold conditions produced by ice-sheet development near the depositional area. The Transantarctic Flora was also recorded by Levy and Hardwood (2000) in the erratics from McMurdo Sound (Fig. 27.4B, MS). They pointed out the decreasing species richness towards the Eocene–Oligocene boundary, the earliest Oligocene assemblages being very poor in dinocysts. These trends are compatible with those indicated by Macphail and Truswell (2004) at ODP Site 1166, eastern Antarctica and Goodman and Ford 1983 at ODP Site 513, Malvinas Plateau (Fig. 27.4B). We consider that decreasing abundance and species richness of dinocysts may result from poorer preservation from most circum-Antarctic sites, a consequence of increasing circulation, oxidation, and winnowing. Alternatively, and in agreement with Hannah (1997), the decreasing diversity may also be the result of a higher sediment supply through land-ice erosion.

Oligocene and Miocene marine assemblages from CRP-1 and CRP-2 (Fig. 27.4B), two sites from McMurdo Sound and Victoria Land Basin respectively, are mainly composed of previously undescribed taxa (Hannah *et al.* 1998, 2000). At both sites it is remarkable that many dinocysts are reworked species of the Paleogene Transantarctic Flora. This aspect throughout the Oligocene–Miocene deposits reflects stronger ice-rafting and erosion around the Antarctic continent that was much stronger than in Patagonia.

Sluijs *et al.* (2003) studied the Eocene–Oligocene transition at four sites from ODP Leg 189 (1168–1172) around the Tasman Rise (Fig. 27.4A) These authors pointed out a drastic replacement of the Middle Eocene Transantarctic Flora components by more cosmopolitan and offshore species at about 35.5 Ma, relating oceanographic and environmental changes to the deepening of the Tasmanian Gateway. This replacement is almost synchronous throughout the Tasmanian region and coincides with the change observed in Tierra del Fuego, though in our sections the dinocyst assemblages indicate a stratigraphic gap between the late Middle Eocene and the latest Eocene. This hiatus may indicate the effect of current activities not evidenced in the Tasman Sea, possibly because the latter sites may represent sheltered areas (see also Warnaar 2006).

Huber *et al.* (2004) reconstructed the paleoceanographic circulation pattern from the Tasmanian region during the Eocene based on phytoplankton data from ODP Leg 189 and fully coupled climate model simulations. Their results indicated clockwise subpolar gyres dominating the Southern Ocean circulation. These gyres would have produced a relatively cool, northward-flowing current along the eastern margin of Australia (Tasman Current) and the eastern margin of South America. Stickley *et al.* (2004) suggested that the deepening of the Tasmanian Gateway led to a change in the circulation pattern around Antarctica that caused warming of surface waters off southern and eastern Tasmania and the disappearance of the endemic Transantarctic Flora during the Early Oligocene.

Late Oligocene to Early Miocene sections from southern Patagonia show an increase in cosmopolitan taxa, with endemic species almost absent. These elements, along with the consistent presence of *Tuberculodinium vancampoae*, may reflect warm surface waters from the Pacific due to an increased flow through the Drake Passage before the installment of the Antarctic Circumpolar Current (Sijp and England 2006; Pekar *et al.* 2006; Warnaar 2006).

Conclusions

In this chapter, we have summarized the dinoflagellate cyst events related to the large-scale tectonic changes and global climatic trends during the Early Cenozoic (Fig. 27.5). Assemblages from the Middle Eocene to the earliest Oligocene indicate progressive cooling and enhanced productivity, with the best representation of the Transantarctic Flora in the Middle Eocene. During the Late Eocene, endemic components were still present, but now co-occur with bipolar cool markers such as *Gelatia inflata*, typical oceanic species, and increasingly abundant heterotrophic protoperidiniaceans. Thus, cool–temperate conditions may have prevailed until at least the earliest Oligocene.

Late Oligocene to Early Miocene assemblages from southern Patagonia show that cosmopolitan taxa increasingly replace the endemic Eocene Transantarctic Flora. Species typical for temperate to warm surface waters are common and may indicate the throughflow of warm waters from the Pacific via the Drake Passage before the development of the Antarctic Circumpolar Current. Dinocyst assemblages from the Middle Miocene suggest temperate warm to warm surface waters. These data agree with the climatic conditions established since the Neogene Climatic Optimum occurred close to the Early–Middle Miocene boundary.

ACKNOWLEDGEMENTS
The authors thank the reviewers J. Pross and R. Fensome whose constructive comments considerably improved the manuscript. This study was partially supported by grants from the Universidad Nacional Sur, Consejo Nacional de Investigaciones Científicas y Técnicas, Argentina (PIP 6416) and Fondo Nacional para la Investigación Científica y Tecnológica (PICT 07–26057).

Appendix 27.1   List of species cited in the text

Alterbidinium? distinctum (Wilson 1967a) Lentin and Williams 1985

Arachnodinium antarcticum Wilson and Clowes 1982

Deflandrea antarctica Wilson 1967

Deflandrea cygniformis Pöthe de Baldis 1966

Deflandrea granulata Menéndez 1965

Deflandrea oebisfeldensis sensu Cookson and Cranwell 1967

Enneadocysta dictyostila (Menéndez 1965) Stover and Williams 1995 emend. Fensome *et al.* 2007

Enneadocysta partridgei Stover and Williams 1995

Gelatia inflata Bujak 1984

Lingulodinium hemicystum McMinn 1991

Lingulodinium machaerophorum (Deflandre and Cookson 1955) Wall 1967

Nematosphaeropsis rigida Wrenn 1988

Operculodinium centrocarpum (Deflandre and Cookson 1955) Wall 1967

Operculodinium israelianum (Rossignol 1962) Wall 1967

Spinidinium macmurdoense (Wilson 1967a) Lentin and Williams 1976

Thalassiphora pelagica (Eisenack 1954) Eisenack and Gocht 1960 emend. Benedek and Gocht 1981

Tuberculodinium vancampoae Rossignol 1962

Vozzhenikovia apertura (Wilson 1967a) Lentin and Williams 1976

Wilsonidinium echinosuturatum (Wilson 1967a) Lentin and Williams 1976

REFERENCES

Archangelsky, S. 1969. Estudio del paleomicroplancton de la Formación Río Turbio (Eoceno), Provincia de Santa Cruz. *Ameghiniana*, **3**, 181–218.

Barreda, V. D. and S. Palamarczuk 2000a. Palinoestratigrafía de depósitos del Oligoceno tardío-Mioceno en el área sur del Golfo San Jorge, provincia de Santa Cruz, Argentina. *Ameghiniana*, **37**, 103–117.

Barreda, V. D. and S. Palamarczuk 2000b. Palinomorfos continentales y marinos de la Formación Monte León en su área tipo, provincia de Santa Cruz, Argentina. *Ameghiniana*, **37**, 3–12.

Basov, I. A., P. F. Ciesielski, V. A. Krasheninnikov, F. M. Weaver, and S. W. Wise Jr. 1983. Biostratigraphic and paleontologic synthesis: Deep Sea Drilling Project Leg 71, Falkland Plateau and Argentine Basin. *Initial Reports of the Deep Sea Drilling Project*, **71**, 445–460.

Bohaty, S. M. and J. C. Zachos 2003. Significant Southern Ocean warming event in the late middle Eocene. *Geology*, **31**, 1017–1020.

Brinkhuis, H., S. Sengers, A. Sluijs, J. Warnaar, and G. L. Williams 2003. Latest Cretaceous to earliest Oligocene, and Quaternary dinoflagellate cysts from ODP Site 1172, East Tasman Plateau. In Exon, N. F., Kennett, J. P., and Malone, M. (eds.), *Proceedings of the Ocean Drilling Program, Scientific Results* 189, pp. 1–48. Available at www-odp.tamu.edu/publications/189_SR/1

Casadío, S., G. R. Guerstein, S. Marenssi, S. Santillana, R. Feldmann, A. Parras, and C. Montalvo 2000. Evidencias para una edad oligocena de la Formación Centinela, suroeste de Santa Cruz, Argentina. *Ameghiniana*, 37, **71R**.

Cocozza, C. D. and C. M. Clarke 1992. Eocene microplankton from La Meseta Formation, northern Seymour Island. *Antarctic Science*, **4**, 355–362.

Daners, G., G. R. Guerstein, M. V. Guler, and G. Veroslavsky 2004. Las transgresiones del Maastrichtiense – Daniense y Eoceno medio en la cuenca Punta del Este y su correlación regional basada en dinoflagelados. *Actas IV Congreso Uruguayo de Geología y II Reunión de Geología Ambiental y Ordenamiento Territorial del Mercosur*, 9.

Edwards, L. E. and V. A. S. Andrle 1992. Distribution of selected dinoflagellate cysts in modern marine sediments. In Head, M. J. and Wrenn, J. H. (eds.), *Neogene and Quaternary Dinoflagellate Cysts and Acritarchs*. Dallas, TX: American Association of Stratigraphic Palynologists, pp. 259–288.

Fensome, R. A. and G. L. Willliams 2004. *The Lentin and Williams Index of Fossil Dinoflagellates*, 2004 edn. Dallas, TX: American Association of Stratigraphic Palynologists.

Fensome, R. A., G. R. Guerstein, and G. L. Williams 2007. The Paleogene dinoflagellate cyst genera *Enneadocysta* and *Licracysta* gen. nov.: new insights based on material from offshore eastern Canada and southern Argentina. *Micropaleontology*, **52**, 385–410.

Flower, B. P. and J. P. Kennett 1994. The Middle Miocene climatic transition: East Antarctic ice sheet development, deep ocean circulation and global carbon cycling. *Palaeogeography, Palaeoclimatology, Palaeoecology*, **108**, 537–555.

Gamerro, J. C. and S. Archangelsky 1981. Palinozonas Neocretácicas y Terciarias de la plataforma continental Argentina en la Cuenca del Colorado. *Revista Española de Micropaleontología*, **13**, 119–140.

Goodman, D. K. and L. N. Ford Jr. 1983. Preliminary dinoflagellate biostratigraphy for the middle Eocene to lower Oligocene from the southwest Atlantic Ocean. *Initial Reports of the Deep Sea Drilling Project*, **71**, 859–877.

Gradstein, F., J. Ogg, and A. Smith, (eds.) 2004. *A Geologic Time Scale 2004*. Cambridge, UK: Cambridge University Press.

Guerstein, G. R. and M. V. Guler 2000. Bioestratigrafía basada en quistes de dinoflagelados del Eoceno tardío – Mioceno del pozo Ombucta x-1, Cuenca del Colorado, Argentina. *Ameghiniana*, **37**, 81–90.

Guerstein, G. R and G. L. Junciel 2001. Quistes de dinoflagelados del Cenozoico de la Cuenca del Colorado, Argentina. *Ameghiniana*, **38**, 299–316.

Guerstein, G. R., C. Cayulef, and M. V. Guler 2002. Dinoflagellate cysts from Eocene/Oligocene boundary beds in the western South Atlantic. *Geological Association of Canada 2000, Session SS22 – The Palynology and Micropalaeontology of Boundaries*, Abstract 54.

Guerstein, G. R., M. V. Guler, G. Daners, and S. Archangelsky 2003. Quistes de dinoflagelados del Eoceno Medio del Atlántico Sudoccidental y su correlación con otros sitios del Hemisferio Sur. *Ameghiniana*, **40**, 87R.

Guerstein, G. R., M. V. Guler, and S. Casadío 2004. Palynostratigraphy and Palaeoenvironments across the Oligocene–Miocene boundary within the Centinela Formation, Southwestern Argentina. In Head, M. and Beaudoin, A. (eds.), *The Palynology and Micropalaeontology of Boundaries*, Special Publication no. 230. London: Geological Society of London, pp. 325–343.

Guerstein, G. R., M. V. Guler, G. L. Williams, R. A. Fensome, and J. O. Chiesa 2008. Mid Palaeogene dinoflagellate cysts from Tierra del Fuego, Argentina: biostratigraphy and palaeoenvironments. *Journal of Micropalaeontology*, **27**, 75–94.

Guler, M. V. 2003. Quistes de dinoflagelados de la familia Protoperidiniaceae del Neógeno de la cuenca del Colorado, Argentina. *Ameghiniana*, **40**, 457–467.

Guler, M. V. and G. R. Guerstein 2002. Bioestratigrafía del Oligoceno–Plioceno Temprano de la Cuenca del Colorado (Argentina), basada en quistes de dinoflagelados. *Revista Española de Micropaleontología*, **34**, 359–371.

Guler, M. V., G. R. Guerstein, and N. Malumián 2002. Bioestratigrafía de la Formación Barranca Final, Neógeno de la Cuenca del Colorado, Argentina. *Ameghiniana*, **39**, 103–110.

Guler, M. V., G. R. Guerstein, and M. E. Quattrocchio 2001. Palinología del Neógeno de la perforación Cx-1, cuenca del Colorado, Argentina. *Revista Española de Micropaleontología*, **33**, 183–204.

Hannah, M. J. 1997. Climate controlled dinoflagellate distribution in late Eocene–earliest Oligocene strata from CIROS-1 drillhole, McMurdo Sound, Antarctica. *Terra Antartica*, **4**, 73–78.

Hannah, M. J., J. H. Wrenn, and G. J. Wilson 1998. Early Miocene and Quaternary marine palynomorphs from CRP-1, McMurdo Sound. *Terra Antartica*, **5**, 527–538.

Hannah, M. J., G. J. Wilson, and J. H. Wrenn 2000. Oligocene and Miocene Marine Palynomorphs from CRP-2/2A, Victoria Land Basin, Antarctica. *Terra Antartica*, **7**, 503–511.

Haq, B. U., J. Hardenbol, and P. R. Vail 1987. Chronology of fluctuating sea levels since the Triassic. *Science*, **235**, 1156–1167.

Haskell, T. R. and G. J. Wilson 1975. Palynology of sites 280–284, Deep Sea Drilling Project Leg 29, off southeastern Australia and western New Zealand. *Initial Reports of the Deep Sea Drilling Project*, **29**, 723–741.

Huber, M., H. Brinkhuis, C. E. Stickley, K. Döös, A. Sluijs, J. Warnaar, S. A. Schellenberg, and G. L. Williams 2004. Eocene circulation of the Southern Ocean: was Antarctica kept warm by subtropical waters? *Paleoceanography*, **19**, PA4026, doi: 10.1029/2004PA001014.

Kay, R. F., R. M. Madden, M. G. Vucetich, A. A. Carlini, M. M. Mazzoni, G. H. Ré, M. Heizler, and H. Sandeman 1999a. Revised age of the Casamayoran South American Land Mammal "Age": climatic and biotic implications. *Proceedings of the National Academy of Sciences USA*, **96**, 13 235–13 240.

Kay, R. F., R. H. Madden, M. Mazzoni, M. G. Vucetich, G. Ré, M. Heizler, and H. Sandeman 1999b. The oldest Argentine primates: first age determinations for the Colhuehuapian South American Land Mammal 'Age.' *American Journal of Physical Anthropology*, **28**(Suppl.), 166.

Levy, R. H. and D. M. Hardwood 2000. Tertiary marine palynomorphs from the McMurdo Sound erratics, Antarctica. In Stilwell, J. D. and Feldmann, R. M. (eds.), *Paleobiology and Paleoenvironments of Eocene Rocks, McMurdo Sound, East Antarctica*, Antarctic Research Series no. 76. Washington, DC: American Geophysical Union, pp. 183–242.

Macphail, M. K. and E. M. Truswell 2004. Palynology of Site 1166, Prydz Bay, East Antarctica. *Proceeding of the Ocean Drilling Program, Scientific Results*, **188**, 1–43.

Malumián, N. 1999. La sedimentación y el volcanismo terciarios en la Patagonia Extraandina. *Anales del Instituto de Geología y Recursos Minerales*, **29**, 557–612.

Malumián, N., E. B. Olivero, and A. Concheyro 1994. Eocene microfossils from the Leticia Formation, Tierra del Fuego Island, Argentina. *Ameghiniana*, **31**, 398.

Mao, S. and B. A. R. Mohr 1995. Middle Eocene dinocysts from Bruce Bank (Scotia Sea, Antarctica) and their paleo environmental and paleogeographic implications. *Review of Palaeobotany and Palynology*, **86**, 235–263.

Miller, K. G., M. A. Kominz, J. V. Browning, J. D. Wright, G. S. Mountain, M. E. Katz, P. J. Sugarman, B. S. Cramer, N. Christie-Blick, and S. F. Pekar 2005. The Phanerozoic record of global sea-level change. *Science*, **310**, 1293–1298.

Muza, J. P., D. F. Williams, and S. W. Wise 2004. Paleogene oxygen isotope record for deep sea drilling sites 511 and 512, subantarctic South Atlantic Ocean: paleotemperatures, paleoceanographic changes, and the Eocene/Oligocene boundary event. *Initial Reports of the Deep Sea Drilling Project*, **71**, 409–422.

Palamarczuk, S. and V. D. Barreda 1998. Bioestratigrafía en base a quistes de dinoflagelados de la Formación Chenque (Mioceno), provincia del Chubut, Argentina. *Ameghiniana*, **35**, 415–426.

Palazzesi, L. and V. D. Barreda 2003. Primer registro palinológico de la Formación Puerto Madryn, Mioceno de la provincia de Chubut, Argentina. *Ameghiniana*, **41**, 355–362.

Parras, A., M. Griffin, R. Feldmann, S. Casadío, C. Schweitzer, and S. Marenssi 2008. Isotopic ratios and faunal affinities in the correlation of marine beds across the Paleogene/Neogene boundary in southern Patagonia, Argentina. *Journal of South American Earth Sciences*, **26**, 204–216.

Pekar, S. F., R. M. DeConto, and D. M. Harwood 2006. Resolving a late Oligocene conundrum: deep-sea warming and Antarctic glaciation. *Palaeogeography, Palaeoclimatology, Palaeoecology*, **231**, 29–40.

Pross, J. 2001. Paleo-oxygenation in Tertiary epeiric seas: evidence from dinoflagellate cysts. *Palaeogeography, Palaeoclimatology, Palaeoecology*, **166**, 369–381.

Pross, J. and H. Brinkhuis 2005. Organic-walled dinoflagellate cysts as paleoenvironmental indicators in the Paleogene: a synopsis of concepts. *Paläontologische Zeitschrift*, **79**, 53–59.

Raine, J. I., R. A. Askin, E. M. Crouch, M. J. Hannah, R. H. Levy, and J. H. Wrenn 1997. Palynomorphs. *Institute of Geological and Nuclear Sciences, Science Reports*, **97**, 25–33.

Reichart, G.-J. and H. Brinkhuis 2003. Late Quaternary *Protoperidinium* cysts as indicators of paleoproductivity in the northern Arabian Sea. *Marine Micropaleontology*, **49**, 303–370.

Röhl, U., H. Brinkhuis, C. E. Stickley, M. Fuller, S. A. Schellenberg, G. Wefer, and G. L. Williams 2004. Sea level and astronomically induced environmental changes in middle and late Eocene sediments from the East Tasman Plateau. In Exon, N. F., Malone, M., and Kennett, J. P. (eds.), *Climate Evolution of the Southern Ocean and Australia's Northward Flight from Antarctica*. Washington, DC: American Geophysical Union, pp. 127–151.

Scher, H. D. and E. E. Martin 2006. Timing and climatic consequences of the opening of Drake Passage. *Science*, **312**, 428–430.

Schettino, A. and C. R. Scotese 2005. Apparent polar wander paths for the major continents (200 Ma – Present Day): a paleomagnetic reference frame for global plate tectonic reconstructions. *Geophysical Journal International*, **163**, 727–759.

Sijp, W. P. and M. H. England 2004. Effect of the Drake Passage through flow on Global Climate. *Journal of Physical Oceanography*, **34**, 1254–1266.

Sluijs, A., H. Brinkhuis, C. E. Stickley, J. Warnaar, G. L. Williams and M. Fuller 2003. Dinoflagellate cysts from the Eocene/Oligocene transition in the Southern Ocean; results from ODP Leg 189. In: Exon, N. F., Kennett, J. P. and Malone, M. J., (eds.), *Proceedings of the Ocean Drilling Program, Scientific Results*, **189**. Available at www-odp.tamu.edu/ publications/189_SR/104/104.htm.

Sluijs, A., J. Pross, and H. Brinkhuis 2005. From greenhouse to icehouse: organic-walled dinoflagellate cysts as paleoenvironmental indicators in the Paleogene. *Earth-Science Reviews*, **68**, 281–315.

Stickley, C. E., H. Brinkhuis, S. A. Schellenberg, A. Sluijs, U. Röhl, M. Fuller, M. Grauert, M. Huber, J. Warnaar, and G. L. Williams 2004. Timing and nature of the deepening of the Tasmanian Gateway. *Paleoceanography*, **19**, 1–18.

Truswell, E. M. 1997. Palynomorph assemblages from marine Eocene sediments on the West Tasmanian continental margin and the South Tasman Rise. *Australian Journal of Earth Sciences*, **4**, 633–654.

Wall, D., B. Dale, G. Lohmann, and W. Smith 1977. The environmental and climatic distribution of dinoflagellate cysts in modern marine sediments from regions in the North and South Atlantic Oceans and adjacent seas. *Marine Micropaleontology*, **2**, 121–200.

Warnaar, J. 2006. Climatological implications of Australian–Antarctic separation. Ph.D. thesis, University of Utrecht.

Wrenn, J. H. and S. W. Beckman 1982. Maceral, total organic carbon, and palynological analyses of Ross Ice Shelf Project site J9 cores. *Science*, **216**, 187–189.

Zachos, J. C., M. Pagani, L. C. Sloan, E. Thomas, and K. Billups 2001. Trends, rhythms, and aberrations in global climate 65 Ma to present. *Science*, **292**, 686–693.

Zachos, J. C., G. R. Dickens, and R. E. Zeebe 2008. An early Cenozoic perspective on greenhouse warming and carbon-cycle dynamics. *Nature*, **451**, 279–283.

28 Divisaderan: Land Mammal Age or local fauna?

Guillermo Marcos López

Abstract

The mammalian fauna from the Divisadero Largo Formation type locality (Mendoza, Argentina) was originally characterized by the coexistence of taxa with generalized morphology along with some of much more modern aspect. This unique composition led to the proposition that the assemblage represented a Divisaderan South American Land Mammal Age (SALMA). The notoungulates *Ethegotherium carettei* (Hegetotheriidae) and "*Trachytherus? mendocensis*" (Mesotheriidae) are the two species with most derived characteristics, comparable to closely related taxa in Deseadan and even post-Deseadan faunas. Based on arguments presented here, these two species probably do not come from the Divisadero Largo Formation but rather from the overlying Mariño Formation, and the systematic position of many other elements of this fauna are reinterpreted. This revision leads to a new proposition that the Divisaderan should not be considered a SALMA, but a local fauna with taxa that are much older than previously considered. The antiquity of the Divisadero Largo fauna is best established with reference to the sequence at Gran Barranca, and provides an example of the significance of this sequence for correlation.

Resumen

La fauna exhumada en la localidad tipo de la Formación Divisadero Largo (Mendoza, Argentina) se caracterizó por la coexistencia de taxa de características generalizadas junto a otros de aspecto mucho más moderno. Esto permitió sustentar la "Edad Mamífero" Divisaderense, la cual está representada exclusivamente por esta fauna. Los notoungulados *Ethegotherium carettei* (Hegetotheriidae) y "*Trachytherus? mendocensis*" (Mesotheriidae) son las dos especies de características más derivadas, comparables a las presentes en especies deseadenses y aún post-deseadenses. En este trabajo se certifica que estas dos especies no provendrían de niveles de la Formación Divisadero Largo sino de la suprayacente Formación Mariño y se reinterpreta la posición sistemática de muchos otros integrantes de esta fauna. Esto permite establecer que la "Edad Mamífero" Divisaderense carece de identidad y que no debería ser considerada como tal, sino como una Fauna Local. Por otro lado, el análisis de los taxa que conforman la fauna Divisaderense sugiere considerarla como mucho más antigua de lo que, hasta el momento, se consideraba. La antiguedad de la fauna de Divisadero Largo se establece con referencia a la secuencia de Gran Barranca, un ejemplo de la utilidad de la secuencia para fines de correlación.

The Paleontology of Gran Barranca: Evolution and Environmental Change through the Middle Cenozoic of Patagonia, eds. R. H. Madden, A. A. Carlini, M. G. Vucetich, and R. F. Kay. Published by Cambridge University Press. © Cambridge University Press 2010.

Introduction

The "**magic slate**" was a very popular toy in the seventies; it consisted of a small board with a dark waxed cardboard background overlaid by a clear celluloid sheet. A pointed stylus was used to draw or write, and the strokes could be immediately erased by lifting the celluloid sheet. As early as 1924 Sigmund Freud expressed his interest in this toy which provided him a metaphor for the way memory works. Freud observed that while lines were immediately erased, more intense strokes remained impressed in the dark background for ever.

Since its discovery, the fauna from the Divisadero Largo Formation in Mendoza Province has been controversial, not only with respect to the taxonomic identity and mixed evolutionary stage of its components, but also its age, correlation, and validity as a land mammal age. However, like a "magic slate," careful scrutiny reveals a few indelible strokes that further clarify its relationships.

When referring to the fossil mammals in this fauna, Simpson *et al.* (1962, p. 290), stated that "most and perhaps all of these mammals do not seem to belong in or near lineages known from rich earlier and later faunas" and added later that they "do not know of any such markedly aberrant mammalian fauna from any other continent."

The fauna from the Divisadero Largo Formation was the basis for the Divisaderan South American Land Mammal Age (SALMA) and defined by Pascual *et al.* (1965) as an assortment of taxa, some with generalized features comparable to Casamayoran and Mustersan taxa along with others with much more modern aspect comparable to Deseadan and post-Deseadan taxa. It was precisely this peculiar concurrence of archaic and advanced taxa that prompted Pascual *et al.* (1965) to propose that this fauna could correspond to a different SALMA that could partially bridge the gap between the Mustersan and Deseadan Ages. Within this context the Divisaderan SALMA was assigned to the late Eocene.

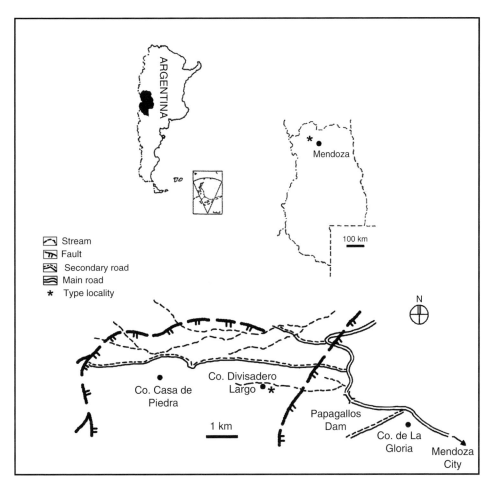

Fig. 28.1. Location map.

The type locality for the Divisadero Largo Formation is located 8 km west of the city of Mendoza in western Argentina (Fig. 28.1), an area where mainly Triassic continental sequences are overlaid by Tertiary synorogenic sediments (Simpson *et al.* 1962) that are in turn overlaid by Quaternary alluvium (Fig. 28.2) in erosive discordance.

The Divisadero Largo Formation of Chiotti (1946) comprises three members distinguished by informal designations "Conglomerado rojo," "Zona con anhidrita," and "Arcillas abigarradas" by petroleum geologists of Yacimientos Petrolíferos Fiscales. As currently understood, the Divisadero Largo Formation comprises only the two upper units of Chiotti, whereas the lower member (i.e. "Conglomerado rojo") was named the Papagallos Formation by Simpson *et al.* (1962). Thus, the lower boundary of Divisadero Largo Formation is the contact with the Papagallos Formation of possibly early Tertiary age and the upper boundary is the contact with the overlying Mariño Formation (= "Serie de las Areniscas Entrecruzadas" or "Inestratificadas"*sensu* Chiotti). At this locality, the Divisadero Largo Formation is an elongate NNE–SSW-trending

band approximately 2.2 km long and between 160 and 250 m thick, that extends from the height of Divisadero Largo to the vicinity of Papagallos. During field work at the type locality, Simpson *et al.* (1962) prospected two fossiliferous exposures, one on the east side of Cerro Divisadero Largo and the other in the cut bank of Arroyo Papagallos.

The Divisadero Largo fauna is comprised entirely of endemic taxa, a fact that has hindered comparisons with other South American faunas. The fact that no comparable faunal assemblage has ever been found in the essentially complete and fossil-rich stratigraphic sequence of Gran Barranca has always been very remarkable.

The temporal hiatus between the Mustersan and Deseadan ages in older biostratigraphic schemes was among the longest of the Cenozoic, and the recognition by Pascual *et al.* (1965, p. 175) of a Divisaderan age was an attempt to fill this gap. However, far from completing the record, the insertion of a Divisaderan age only served to emphasize further the fact that this succession of faunas was apparently still missing crucial pieces. Thus, for many years the

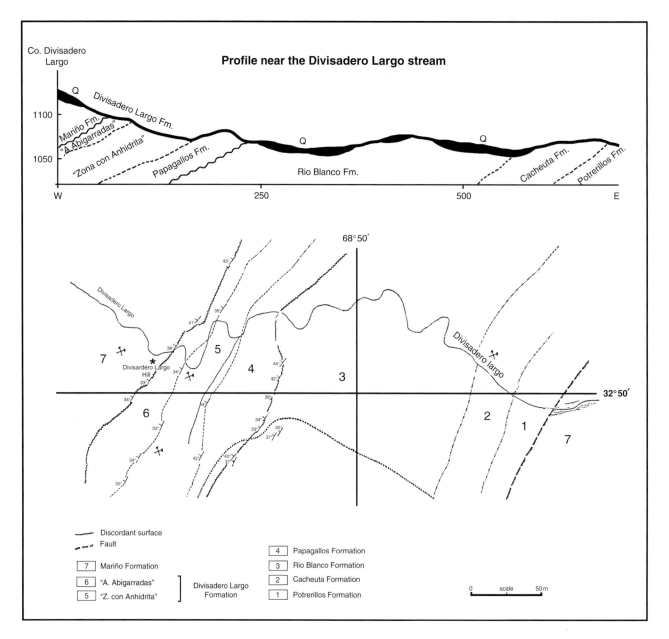

Fig. 28.2. Profile and geologic map of Divisadero Largo–Papagallos area, Mendoza Province.

standard scheme included two well-supported ages (i.e. Mustersan and Deseadan) and an intermediate but poorly understood Divisaderan age, comprising a mixture of taxa that had unclear relationships with both previous and later faunas.

During the last decade, many faunal associations have been discovered that could be referred to the post-Mustersan to pre-Deseadan interval by their intermediate composition. One is the fauna from Tinguiririca in central Chile that was the basis for the Tinguirirican Age (Flynn *et al.* 2003). Afterwards, some Patagonian local faunas (e.g. Lomas Blancas, Cañadón Blanco, La Cancha in Gran Barranca,

and Rinconada de los López) were found to be referrable to the Tinguirirican Age (Reguero *et al.* 2004). Remarkably, in spite of the fact that these assemblages were referred to this temporal interval, none was comparable with the Divisaderan fauna. Even a comparison of "stage of evolution" of their taxa did not suggest correlations between the Divisaderan and other faunas.

In addition, it seemed very striking that in such a complete sequence as the one in Gran Barranca, no fauna comparable to the Divisaderan assemblage had been discovered. Thus, Divisadero Largo fauna retained its enigmatic character as recognized by Simpson *et al.* (1962).

A review of attempts to correlate the Divisadero Largo fauna with other faunal associations reveals two clearly differing opinions: those who found a close relationship with Deseadan associations and those who considered the fauna to have closer affinities with older SALMAs (e.g. Mustersan).

Doubtless the papers by Simpson and his collaborators were the most significant contributions in the first group. The close relationship between Divisadero Largo and Deseadan faunas was supported by four phylogenetic relationships: (1) the species *Phoradiadius divortiensis* is a probable Proterotheriidae; (2) the species *Adiantoides leali* is an Adianthidae; (3) the mesotheriid "*Trachytherus? mendocensis*" is related to species of *Trachytherus* recorded at Deseadan levels in Patagonia; and (4) that *Ethegotherium carettei* displays a suite of derived characters comparable to Deseadan taxa. The species mentioned in items 3 and 4 are precisely those distinguished by Marshall *et al.* (1986, p. 946) as representing "taxonomic ties" with Deseadan associations.

Other researchers adopted the position that the Divisadero Largo fauna had a closer relationship with older associations. Ortíz-Jaureguízar (1986, 1988), applying techniques of multivariate analysis, recognized an "Infracenozoic Faunistic Episode" comprising the Itaboraian, Riochican, Casamayoran, Mustersan, and Divisaderan Ages, and a "Pre-Patagonian Faunal Cycle" comprising the last three of these ages. This criterion was followed by Pascual and Ortíz-Jaureguízar (1990) and Pascual *et al.* (1996). In addition, Bond (1991), when presenting a critique of MacFadden *et al.*'s proposal (1985, p. 243) that the lowermost levels of Salla "is either pre-Deseadan (? Divisaderan) or earliest Deseadan in age," marshaled lines of evidence to suggest that the Divisaderan fauna is older than Deseadan.

For almost 20 years, then, as other South American Paleogene faunas became better known, the endemism of the fauna from Divisadero Largo came to be viewed as a peculiar and isolated ecological enclave with no clear relationships. The discovery of mammals in lower Tertiary sediments in northwestern Argentina (e.g. Mealla and Lumbrera Formations: Pascual *et al.* 1981), in levels referred to the Eocene–Oligocene boundary in Chile (i.e. Tinguiririca age: Flynn *et al.* 2003) and the reinterpretation of several Patagonian faunas that are partly equivalent to the Tinguiririca fauna, along with a revised assessment of the temporal gap that was originally recognized by the Ameghino brothers (i.e. "Astraponotéen le plus supérieur": Bond *et al.* 1996), now provide insights into the affinities of some of the lineages occurring in this peculiar fauna.

The Divisaderan fauna

The lower and middle Divisadero Largo Formation has yielded a vertebrate fauna composed essentially of mammals and a few reptiles (i.e. turtles, crocodiles, and boas).

Minoprio (1947) mentioned the existence of a turtle shell, which he illustrated but did not describe. In their list of Cenozoic fossil vertebrates of Mendoza, Pascual and de la Fuente (1993) referred these remains to indeterminate pleurodiran turtles. The Order Crocodylia is represented by the species *Ilchunaia parca* Rusconi (1946b) on the basis of mandibular and cranial remains assigned to the Family Crocodylidae. Later, Langston (1956) tentatively assigned this species to the Sebecosuchia and Gasparini (1972) referred it with some doubt to the Sebecidae. The assignation of the original material, along with other remains found later, to Sebecosuchia is based on the high long rostrum and laterally compressed teeth with serrated edges (Gasparini 1972). *Cunampaia simplex*, described by Rusconi (1946b, 1946c) based on remains of fore and hindlimbs, was thought to be the only fossil bird in this formation, but it has recently been recognized as a crocodile by Agnolín and Pais (2006). Simpson *et al.* (1962) briefly commented on the existence of ophidian remains from the Divisadero Largo Formation but did not provide a description. Albino (1989) describes this material (vertebrae with bodies that are short, broad, high, and robust) and refers them to Family Boidae. The subfamilial assignation of this material remains doubtful; some are possibly Erycinae, others Boinae, and two fragments are indeterminate.

Mammals comprise more than 95% of the remains collected from this unit, and remarkably, almost all of them can be referred to South American native ungulates. Two marsupial species were recognized: *Groeberia minoprioi* Patterson 1952 and *Groeberia pattersoni* Simpson 1970. *Groeberia* is characterized by an unusual combination of characters (e.g., gliriform upper and lower incisors, with enamel restricted to the labial face, small canines, forward-directed orbits, a very deep mandibular ramus with a vertical symphysis. Family Groeberiidae (Patterson 1952) and Order Groeberida (Pascual *et al.* 1994) were created. Flynn and Wyss (1999) based in part on more recently described *Klohnia charrieri*, a groeberid from Tinguiririca with more derived traits than *Groeberia*, refer this family to the Argyrolagoidea (Ameghino 1904) which includes *Groeberia minoprioi, G. pattersoni, Klohnia charrieri*, and *Patagonia peregrina* Pascual and Carlini (1987), the last based on material from the Colhuehuapian at Gaiman in Patagonia.

Carlos Rusconi, the former director of the Museo de Historia Natural (now Museo de Ciencias Naturales y Antropológicas "J.C. Moyano") in Mendoza, mentioned the existence of osteoderms of "desipodinos terciarios" (*sic* Rusconi 1946a) in the Divisadero Largo Formation and in that same year (Rusconi 1946c) referred these alleged scutes to "un dasipodino del tamaño de los eutatus, pero de figura confusa." Unfortunately, none of these were illustrated by Rusconi and they are currently lost. Given that

Simpson *et al.* 1962	**In this chapter**
ORDER MARSUPIALIA Family Groeberiidae *Groeberia minoprioi* Patterson 1952.	MARSUPIALIA Order Groeberida Family Groeberiidae *Groeberia minoprioi* Patterson 1952. *Groeberia pattersoni* Simpson 1970.
ORDER LITOPTERNA Family Adianthidae *Adiantoides leali* Simpson and Minoprio, 1949.	ORDER LITOPTERNA Family Sparnotheriodonthidae *Phoradiadius divortiensis* Simpson et al. 1962.
Family Proterotheriidae? *Phoradiadius divortiensis* Simpson et al. 1962.	ORDER NOTOPTERNA Family Indaleciidae *Adiantoides leali* Simpson and Minoprio 1949.
ORDER NOTOUNGULATA Family Oldfieldthomasiidae? *Brachystephanus postremus* Simpson et al. 1962. *Xenostephanus chiottii* Simpson et al. 1962. *Allalmeia atalaensis* Rusconi 1946.	ORDER NOTOUNGULATA Unnamed family *Brachystephanus postremus* Simpson et al. 1962. *Xenostephanus chiottii* Simpson et al. 1962. *Allalmeia atalaensis* Rusconi 1946.
Family Hegetotheriidae *Ethegotherium carettei* (Minoprio 1947).	Family Henricosborniidae? *Acamana ambiguus* Simpson et al. 1962.
Family Mesotheriidae *Trachytherus? mendocensis* Simpson and Minoprio 1949.	ORDER ASTRAPOTHERIA Trigonostylopidae Gen et sp. indet. López 2006.
ORDER AND FAMILY INDET. *Acamana ambiguus* Simpson et al. 1962.	

Fig. 28.3. Comparison between taxa recognized by Simpson *et al.* (1962) and those presented in this chapter.

no armadillo scutes have ever been collected by any of the numerous later expeditions, it is possible that this material could have been chelonian plates. The lack of any record of dasypodids or any other xenarthran at Divisadero Largo is especially noteworthy because the order is very diverse and material is common in both older and younger faunal associations. Their absence at Divisadero Largo could be due to paleoenvironmental factors (López 2008).

The marked dominance of native ungulates in the Divisadero Largo Formation (Orders Notoungulata, Litopterna, Notopterna, and Astrapotheria) is striking. Although Notoungulata are the most abundant and diverse, their occurrence is restricted to two families. Simpson *et al.* (1962) reported the families Hegetotheriidae and Mesotheriidae (each represented by a single species) but referred the most abundant taxa (*Brachystephanus*, *Xenostephanus*, and *Allalmeia*) with doubt to the Family Oldfieldthomasiidae (Fig. 28.3). Bond (1981) recognized that on dental and cranial characters, these genera should be grouped with *Colbertia* (from the Itaboraian of Brazil and Eocene of northwestern Argentina) and *Maxschlosseria* (from the Patagonian Vacan) rather than with the remaining Oldfieldthomasiidae. Later, López and Bond (2003) recognized a new family as yet unnamed for these five taxa. This new family unites taxa that differ from Oldfieldthomasiidae by the possesion of:

(1) skull with shorter rostrum, proportionally larger bullae, marked sagittal and cranial ridges, and more anteriorly implanted zygomatic arch;

(2) teeth typically brachydont (whereas the molars of the Oldfieldthomasiidae have proportionally higher crowns);

(3) I1 not enlarged; P1 very small and not elongated mesiodistally; P2 and P3 with transverse-oriented major axis, simple premolar crowns with one lingual cusp and one labial cusp developed into a short ectoloph without metacone; P4 without hypocone;

(4) M1–2 with protocone and hypocone independent from each other or only basally united, ectoloph normally lacking mesostyle or with a very rudimentary one, and small and ephemeral antero- and posteroexternal fossettes; M3 normally without hypocone, when present very small and bunoid (e.g. *Allalmeia*) and with a very rudimentary mesostyle only in *Colbertia lumbrerense*; and

(5) p1 small, p2 very simple, and p3–4 not molarized.

Most of these features are relatively primitive with respect to other Oldfieldthomasiidae, yet derived with respect to Henricosborniidae (López 2008).

On the basis of a relatively complete skull and mandible preserved in two rock fragments, Minoprio (1947) described *Prohegetotherium carettei* and later Simpson *et al.* (1962) recognizing differences at the generic level, established the new combination *Ethegotherium carettei*. This species is known only by its type material, and its very high-crowned teeth were one crucial element used to refer the Divisadero Largo fauna to a Deseadan (Simpson and Minoprio 1950) or "approximately early Deseadan or latest pre-Deseadan" age (Simpson *et al.* 1962, p. 290). Although *E. carettei* is clearly a more modern species compared with the other notoungulates from Divisadero Largo, its stratigraphic provenance was never doubted. Recently, López and Manassero (2008), based on a petrographic study of formation sediments, demonstrated that it comes from younger levels of the overlying Mariño Formation, rather than from the Divisadero Largo Formation.

A similar situation concerns the other species of this fauna with derived characteristics. The description of the mesotheriid "*Trachytherus? mendocensis*" Simpson and Minoprio 1949 was based on isolated teeth from a single individual, of which the premolars (P2–3), one M1, and fragments of M2, all from the right side, are the most complete elements. These teeth, along with many other materials from the Museo de Ciencias Naturales y Antropológicas "J. C. Moyano," are regrettably lost and no illustrations were ever published except for a simple sketch. Simpson *et al.* (1962) added to the hypodigm one isolated right M, collected by Olivo Chiotti in 1945 and deposited in the Museo de La Plata (MLP 45-VII-10–2). All these teeth are high-crowned and, although their roots are not preserved, they can be considered protohypsodont, a remarkable characteristic in the general context of the Divisadero Largo fauna. Recently, Cerdeño *et al.* (2005, 2008) reported finding an indeterminate mesotheriid from the basal levels of the Mariño Formation, 100 m above the levels bearing the Divisadero Largo fauna. As this specimen is indistinguishable from the M3 referred to "*Trachytherus? mendocensis*" (the only material currently preserved), the originally supposed stratigraphic provenance of the species is brought into question. Cerdeño (2008) noted the close phylogenetic relationship of this new material with *Altitypotherium chucalensis* (Croft *et al.* 2004), a mesotherine from the early Miocene Chucal Formation in Chile.

The enigmatic species *Acamana ambiguus* Simpson *et al.* 1962 was originally of uncertain ordinal and familial affinity. When created, the peculiar combination of small I1–2, enlarged I3, reduced C, and diastema was unknown in any other South American mammal. The subsequent description of *Simpsonotus* (Pascual *et al.* 1979), an henricosborniid from the Mealla Formation, allowed Bond and Vucetich (1983) to establish possible phylogenetic relationships with *Acamana* although the type and only known material of *A. ambiguus* has been lost.

Divisadero Largo Litopterna are represented by *Phoradiadius divortiensis* (Simpson *et al.* 1962), originally referred to the Proterotheriidae. Later Soria (1980b) demonstrated its close relationship with *Sparnotheriodon epsilonoides* (Soria 1980a) and *Victorlemoinea emarginata* (Ameghino 1901), and referred these taxa to Sparnotheriodontidae. New undescribed material confirms the close phylogenetic relationship between *P. divortiensis* with species of *Victorlemoinea,* especially *V. labyrinthica* Ameghino 1901. Although the systematic position of the Sparnotheriodontidae is debated, López (1999) argued for litoptern affinity on the basis of dental characteristics.

The species *Adiantoides leali* Simpson and Minoprio 1949 was originally described as an adianthid litoptern, but the discovery of *Indalecia grandensis* Bond and Vucetich 1983 revealed affinities between the two species and justifies the establishment of the family Indaleciidae (Soria 1984). Indalecids, Amilnedwardsiidae, and Notonychopidae comprise the Order Notoptera (Soria 1989a, 1989b).

Based on fragmentary but diagnostic material, López (2006) established the first record of the Order Astrapotheria in the Divisadero Largo fauna. The material consists of a mandibular symphysis (MLP 87-II-20–74) preserving part of the crown of the left i1, the right i2 and canine, as well as the alveoli of the right i1 and i3. The roots of i2, i3, and canine can be seen on the left side. The hemimandibles are completely fused and bear evident mental foramina on both sides near the midline. The implantation of the canines and the morphology of the preserved canine crowns and incisors indicate affinities with trigonostylopid astrapotheres, in particular to *Trigonostylops* sp., a taxon characteristic of the Patagonian Casamayoran SALMA (López 2006).

Is the Divisaderan a South American Land Mammal Age?

As the two species with derived characteristics (i.e. "*Trachytherus? mendocensis*" and *Ethegotherium carettei*) do not come from the same stratigraphic level as the rest of the Divisadero Largo fauna, the mixture of primitive and modern taxa used to argue in support a different SALMA for the post-Mustersan and pre-Deseadan allocation collapses.

After excluding the elements with advanced characteristics, the age of the Divisaderan fauna still needs to be established. On the basis of its composition, the Divisadero Largo faunal assemblage shows closer affinity with Casamayoran age faunas than with faunas from the Mustersan.

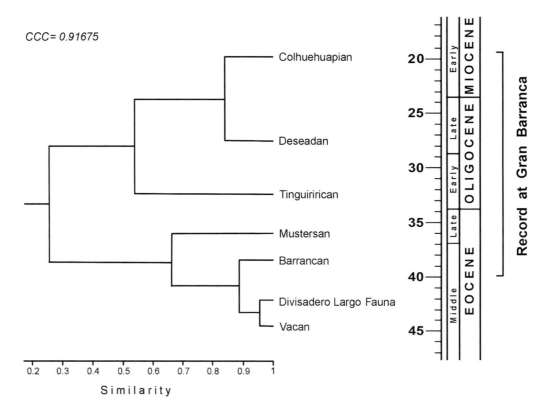

Fig. 28.4. Cluster analysis using mammal assemblages as operational units and families as characters.

The two species of *Groeberia* are not indicative of a particular age but support a correlation with faunas older than Tinguiririan.

The Sparnotheriodontidae *Phoradiadius divortiensis* is at an "evolutionary stage" comparable to *Victorlemoinea labyrinthica* Ameghino 1901 from western Río Chico, of Vacan age (Simpson 1967). The Family Sparnotheriodontidae is recorded in sediments referrable to Itaboraian, Riochican, and Casamayoran in Argentina and Brazil, and from levels assignable to Eocene in the Antarctic Peninsula.

The Indalecidae of the Divisadero Largo fauna are represented by *Adiantoides leali*, which is morphologically very similar except for its larger size to *Adiantoides magnus* Cifelli and Soria 1983 from Cañadón Vaca. Bond and Vucetich (1983) in the description of the species *Indalecia grandensis* from the Lumbrera Formation of Casamayoran age (Pascual *et al.* 1981) recognized its close relationship with *A. leali*.

Brachystephanus, Xenostephanus, and *Allalmeia* have been referred to a new unnamed family by López and Bond (2003) comprising the species of *Colbertia* from the Brazilian Itaboraian and the Lumbrera Formation in northwestern Argentina and another species from Cañadon Vaca (López 2008). The characteristics of this new family are clearly primitive with respect to those of oldfielthomasiids.

The Family Henricosborniidae characterize both Riochican and Casamayoran faunas. If the referral of *Acamana*

ambiguus to this family is confirmed, it would represent an additional element suggesting a much older age for the Divisadero Largo fauna. Likewise the astrapothere *Trigonostylops* is characteristic of the Casamayoran levels in Patagonia.

Mammals are not the only elements available for comparison. Sebecids and boas are prominent elements in Casamayoran associations and are not found in Mustersan faunas.

A multivariate cluster analysis was performed using different mammal assemblages as operational units and families as characters. The mammal-bearing units include various levels recorded in Gran Barranca (i.e. Barrancan, Mustersan, Tinguiririan, Deseadan, and Colhuehuapian) and the Vacan subage. Because the Divisadero Largo Fauna includes endemic genera, families were the hierarchical category used in the analysis.

On a data matrix of 43 taxa and 7 faunas (Appendix 28.1), cluster analysis was performed in NTSYS 2.0 software (Rohlf 1977) using the Jaccard coefficient. The Cophenetic Correlation Coefficient is elevated (CCC = 0.91675).

The cluster analysis shows two principal groups. The branch including the Divisadero Largo fauna is arranged as follows: (((Vacan subage + Divisadero Largo Fauna) Barrancan subage) Mustersan). The second branch is formed by the Tinguiririan plus the Deseadan and Colhuehuapian SALMAs (Fig. 28.4).

Based on G. G. Simpson's field notes, Cifelli (1985) recognized two subages for the Casamayoran, an older Vacan and a younger Barrancan. Remarkably, although these subages are well supported, they are seldom used in practice and the term Casamayoran is often still used. As mentioned above, the endemic nature of the taxa in the Divisadero Largo fauna hinders comparisons, and consequently any referral to either Vacan or Barrancan subage implies a degree of uncertainty. Nevertheless, the absence of a directly comparable fauna in the rich and complete stratigraphic sequence at Gran Barranca might be attributed to a still older age for the Divisadero Largo local fauna, possibly corresponding to either the Vacan subage or the temporal hiatus between the Vacan and the Barrancan. Levels of such age have not been recorded at Gran Barranca nor elsewhere in Patagonia.

ACKNOWLEDGEMENTS

I thank the editors for their kind invitation to contribute to this book. I am also very grateful to Mariano Bond, who has largely contributed to my scientific improvement. The criticisms and suggestions of the two reviewers, E. Cerdeño and D. Croft, made for a much improved manuscript. I am also grateful to M. Reguero and A. Kramarz for access to the collections under their care.

Appendix 28.1 Data matrix of families and mammal assemblages

	Vacan	Div. Largo Fauna	Barrancan	Mustersan	Tiguirirican	Deseadan	Colhuehuapian
BONAPARTHERIIDAE	1	1	1	1	0	0	0
BORHYAENIDAE	1	1	1	1	1	1	1
DIDELPHIDAE	1	1	1	1	1	1	1
GROEBERIIDAE	0	1	1	1	1	0	0
POLYDOLOPIDAE	1	1	1	1	0	0	0
PREPIDOLOPIDAE	1	1	1	1	0	0	0
DASYPODIDAE	1	1	1	1	1	1	1
GLYPTODONTIDAE	0	0	0	1	1	1	1
MEGATHERIIDAE	0	0	0	0	0	1	1
OROPHODONTIDAE	0	0	0	0	0	1	0
PALAEOPELTIDAE	0	0	0	1	1	1	1
DIDOLODONTIDAE	1	1	1	1	0	0	0
INDALECIIDAE	1	1	1	1	1	1	0
SPARNOTHERIODONTIDAE	1	1	1	0	0	0	0
ADIANTHIDAE	0	0	0	0	0	1	1
MACRAUCHENIIDAE	0	0	0	1	1	1	1
PROTEROTHERIIDAE	0	0	0	0	1	1	1
ARCHAEOHYRACIDAE	0	0	1	1	1	1	0
ARCHAEOPITHECIDAE	1	1	1	0	0	0	0
HEGETOTHERIIDAE	0	0	0	0	1	1	1
HENRICOSBORNIIDAE	1	1	0	0	0	0	0
HOMALODOTHERIIDAE	0	0	0	0	1	1	1
INTERATHERIIDAE-NOTOPIT	1	1	1	1	1	0	0
INTERATHERIIDAE-INTERAT	0	0	0	0	1	1	1
ISOTEMNIDAE	1	1	1	1	1	1	0
LEONTINIDAE	0	0	0	1	1	1	1
MESOTHERIIDAE	0	0	0	0	1	1	1
NOTOHIPPIDAE	1	1	1	1	1	1	1
NOTOSTYLOPIDAE	1	1	1	1	1	0	0
OLDFIELDTHOMASIIDAE	1	1	1	1	0	0	0
UNNAMED FAMILY	1	1	1	1	0	0	0
TOXODONTIDAE	0	0	0	0	0	1	1
ASTRAPOTHERIIDAE	1	1	1	1	1	1	1
TRIGONOSTYLOPIDAE	1	1	1	0	0	0	0
PYROTHERIIDAE	1	1	1	1	1	1	0
CEBIDAE	0	0	0	0	0	1	1
CEPHALOMYIDAE	0	0	0	0	0	1	1
DASYPROCTIDAE	0	0	0	0	1	1	1
DINOMYIDAE	0	0	0	0	0	1	1
ECHIMYIDAE	0	0	0	0	0	1	1

(*cont.*)

	Vacan	Div. Largo Fauna	Barrancan	Mustersan	Tiguirirican	Deseadan	Colhuehuapian
EOCARDIIDAE	0	0	0	0	0	1	1
ERETHIZONTIDAE	0	0	0	0	0	1	1
OCTODONTIDAE	0	0	0	0	0	1	1
CHINCHILLIDAE	0	0	0	0	1	1	1

REFERENCES

Agnolín, F. and D. F. Pais 2006. Revisión de *Cunampaia simplex* Rusconi, 1946 (Crocodylomorpha, Mesoeucrocodylia; non Aves) del terciario Inferior de Mendoza, Argentina. *Revista del Museo Argentino de Ciencias Naturales*, n.s., **8**, 35–40.

Albino, A. 1989. Los Booidea (Reptilia, Serpentes) extinguidos del territorio argentino. Ph.D. thesis, Universidad Nacional de La Plata.

Ameghino, F. 1901. Notices préliminaires sur des ongulés nouveaux des terrains crétacés de Patagonie. *Boletín de la Academia Nacional de Ciencias en Córdoba*, **16**, 349–426.

Ameghino, F. 1904. Nuevas especies de mamíferos cretáceos y terciarios de la República Argentina. *Anales de la Sociedad Científica Argentina*, **56**, 193–208; **57**, 162–175, 327–341; **58**, 35–41, 56–71, 182–192, 225–240, 241–291.

Bond, M. 1981. Un nuevo Oldfieldthomasiidae (Mammalia, Notoungulata) del Eoceno inferior (Fm. Lumbrera, Grupo Salta) del NW Argentina. *Anais II Congreso Latinoamericano de Paleontología*, **2**, 521–536.

Bond, M. 1991. Sobre las capas de supuesta Edad Divisaderense en los "Estratos de Salla", Bolivia. *I Revista Técnica de Yacimientos Petrolíferos Fiscales Bolivianos*, **12**, 701–705.

Bond, M. and M. G. Vucetich 1983. *Indalecia grandensis* gen. et sp. nov. de Eoceno temprano del Noroeste argentino, tipo de una nueva subfamilia de los Adiantidae (Mammalia, Litopterna). *Revista de la Asociación Geológica Argentina*, **38**, 107–117.

Bond, M., G. M. López, and M. A. Reguero 1996. "Astraponotéen Plus Supérieur" of Ameghino: another interval in the paleogene record of South America. *Journal of Vertebrate Paleontology*, **16** (Suppl. 3), 23 A.

Cerdeño, E. 2008. Systematic position of the Mesotheriidae (Notoungulata) from the Mariño Formation (Miocene) in Divisadero Largo, Mendoza, Argentina. *Geobios*, **40**, 767–773.

Cerdeño, E., G. M. López, and M. A. Reguero 2005. Sobre un Mesotheriidae (Mammalia, Notoungulata) de la Formación Mariño y sus implicancias sobre la identidad de la "Edad Mamífero" Divisaderense. *Ameghiniana*, **42**(4 supl.), 21R.

Cerdeño, E., G. M. López, and M. A. Reguero 2008. Biostratigraphic considerations of the Divisaderan faunal assemblage. *Journal of Vertebrate Paleontology*, **28**, 574–577.

Cifelli, R. L. 1985. Biostratigraphy of the Casamayoran, early Eocene, of Patagonia., *American Museum Novitates*, **2820**, 1–26.

Cifelli, R. L. and M. F. Soria 1983. Systematics of the Adianthidae (Litopterna, Mammalia). *American Museum Novitates*, **2771**, 1–25.

Chiotti, O. V. 1946. Estratigrafía y tectónica al Oeste de la Ciudad de Mendoza y Las Heras. Ph.D. thesis, Universidad Nacional de Córdoba, Argentina.

Croft, D. A., J. J. Flynn, and A. R. Wyss 2004. Notoungulata and Litopterna of the early Miocene Chucal Fauna, Northern Chile. *Fieldiana, Geology*, n.s., **50**, 1–52.

Flynn, J. and A. Wyss 1999. New marsupials from the Eocene-Oligocene transition of the Andean main range, Chile. *Journal of Vertebrate Paleontology*, **19**, 533–549.

Flynn, J., A. Wyss, D. Croft, and R. Charrier 2003. The Tinguiririca Fauna, Chile: biochronology, paleoecology, biogeogrphy, and a new earliest Oligocene South American Land Mammal "Age". *Palaeogeography, Palaeoclimatology, Palaeoecology*, **195**, 229–259.

Gasparini, Z. 1972. Los Sebecosuchia (Crocodilia) del territorio argentino: consideraciones sobre su "status" taxonómico. *Ameghiniana*, **9**, 23–34.

Langston, W. Jr. 1956. The Sebecosuchia: cosmopolitan crocodilians? *American Journal of Science*, **254**, 605–614.

López, G. M. 1999. *Phoradiadius divortiensis* un conflictivo Litopterna de la Formación Divisadero Largo de la provincia de Mendoza, Argentina. *Ameghiniana*, **36**(4 supl.), 14R.

López, G. M. 2002. Redescripción de *Ethegotherium carettei* (Notoungulata, Hegetotheriidae) de la Formación Divisadero Largo de la provincia de Mendoza, Argentina. *Ameghiniana*, **39**, 295–306.

López, G. M. 2006. Un posible Trigonostylopidae (Mammalia, Astrapotheria) en la Formación Divisadero Largo de Mendoza, Argentina. *Ameghiniana*, **43**(4 supl.), 44R.

López, G. M. 2008. Los ungulados de la Formación Divisadero Largo (Eoceno inferior?) de la provincia de Mendoza, *Argentina: sistemática y consideraciones bioestrati-gráficas*. Ph.D. thesis, Universidad Nacional de La Plata.

López, G. M. and M. Bond 2003. Una nueva familia de ungulados (Mammalia, Notoungulata) del Paleógeno sudamericano. *Ameghiniana*, **40**(4 supl.), 60R.

López, G. M. and M. Manassero 2008. Revision of the stratigraphic provenance of *Ethegotherium carettei* (Notoungulata, Hegetotheriidae) by sedimentary petrography. *Neues Jahrbuch für Geologie und Paläontologie Abhandlungen*, **248**, 1–9.

MacFadden, B. J., K. E. Campbell, R. L. Cifelli, O. Siles, M. N. Johnson, C. W. Naeser, and P. Zeitler 1985. Magnetic polarity stratigraphy and mammalian fauna of the Deseadan (Late Oligocene-Early Miocene) Salla beds of Northern Bolivia. *Journal of Geology*, **93**, 223–250.

Marshall, L. G., R. L. Cifelli, R. E. Drake, and G. H. Curtis 1986. Vertebrate paleontology, geology, and geochronology of the Tapera de López and Scarritt Pocket, Chubut Province, Argentina. *Journal of Palaeontology*, **60**, 920–951.

Minoprio, J. L. 1947. Fósiles de la Formación Divisadero Largo. *Anales Sociedad Científica Argentina*, **144**, 365–378.

Ortíz-Jaureguízar, E. 1986. Evolución de las comunidades de mamíferos cenozoicos sudamericanos: un estudio basado en técnicas de análisis multivariado. *Actas IV Congreso Argentino de Paleontología y Bioestratigrafía*, **2**, 191–207.

Ortíz-Jaureguízar, E. 1988. Evolución de las comunidades de mamíferos cenozoicos sudamericanos: un análisis cuali-cuantitativo basado en el registro argentino. Ph.D. thesis, Universidad Nacional de La Plata.

Pascual, R. and A. A. Carlini 1987. A new Superfamily in the extensive radiation of South American Paleogene Marsupials. *Fieldiana, Geology*, n.s., **39**, 99–110.

Pascual, R. and M. S. de la Fuente 1993. Vertebrados fósiles Cenozoicos. *Relatorio XII Congreso Geológico Argentino y II Congreso de Explotación de Hidrocarburos*, **1**, 357–363.

Pascual, R. and E. Ortíz-Jaureguízar 1990. Evolving climates and mammal fauna in Cenozoic South America. *Journal of Human Evolution*, **19**, 23–60.

Pascual, R., E. J. Ortega Hinojosa, D. Gondar, and E. P. Tonni 1965. Las edades del Cenozoico mamalífero de la Argentina, con especial atención a aquéllas del territorio bonaerense. *Anales de la Comisión de Investigaciones Científicas de la Provincia de Buenos Aires*, **6**, 165–193.

Pascual, R., M. G. Vucetich, and J. Fernández 1978. Los primeros mamíferos (Notoungulata, Henricosborniidae) de la Formación Mealla (Grupo Salta, Subgrupo Santa Bárbara): sus implicancias filogenéticas, taxonómicas y cronológicas. *Ameghiniana*, **15**, 366–390.

Pascual, R., M. Bond, and M. G. Vucetich 1981. El Subgrupo Santa Bárbara (Grupo Salta) y sus vertebrados: cronología, paleoambientes y paleobiogeografía. *Actas VIII Congreso Geológico Argentino*, **3**, 743–758.

Pascual, R., F. J. Goin, and A. A. Carlini 1994. New data on the Groeberiidae: unique Late Eocene–Early Oligocene South American Marsupials. *Journal of Vertebrate Paleontology*, **14**, 247–259.

Pascual, R., E. Ortíz-Jaureguízar, and J. J. Prado 1996. Land Mammals: paradigm for Cenozoic South American

geobiotic evolution. *Münchner Geowissenschaftliche Abhandlungen A*, **30**, 265–319.

Patterson, B. 1952. Un nuevo y extraordinario marsupial Deseadiano. *Revista del Museo Municipal de Ciencias Naturales y Tradicionales de Mar del Plata*, **1**, 39–44.

Reguero, M. A., G. M. López, M. Bond, and R. H. Madden 2004. Faunas de Edad Tinguiririquense (Eoceno tardío–Oligoceno temprano) en Patagonia. *Ameghiniana*, **41**(4 supl.), 24R.

Rohlf, F. J. 1977. *Computational Efficiency of Agglomerative Clustering Algorithms*. Yorktown Heights, NY: IBM Watson Research Center.

Rusconi, C. 1946a. Nuevo mamífero fósil de Mendoza. *Boletín Paleontológico de Buenos Aires*, **20**.

Rusconi, C. 1946b. Ave y reptil oligocenos de Mendoza. *Boletín Paleontológico de Buenos Aires*, **21**.

Rusconi, C. 1946c. Algunos mamíferos, reptiles y aves del Oligoceno de Mendoza. *Revista de la Sociedad de Historia y Geografía de Cuyo*, **2**, 1–37.

Simpson, G. G. 1967. The Ameghino's localities for Early Cenozoic mammals in Patagonia. *Bulletin of Museum of Comparative Zoology*, **136**(9), 63–76.

Simpson, G. G. 1970. Addition to knowledge of *Groeberia* (Mammalia, Marsupialia) from the Mid-Cenozoic of Argentina. *Breviora*, **362**, 1–17.

Simpson, G. G. and J. L. Minoprio 1949. A new adianthine litoptern and associated mammals from a Deseadan faunule in Mendoza, Argentina. *American Museum Novitates*, **1434**, 1–27.

Simpson, G. G. and J. L. Minoprio 1950. Fauna Deseadense de Mendoza. *Anales de la Sociedad Científica Argentina*, **149**, 245–253.

Simpson, G. G., J. L. Minoprio, and B. Patterson 1962. The mammalian fauna of the Divisadero Largo Formation. Mendoza, Argentina. *Bulletin of Museum of Comparative Zoology*, **127**(4), 239–293.

Soria, M. F. 1980a. Una nueva y problemática forma de ungulado del Casamayorense. *Actas II Congreso Argentino de Paleontología y Bioestratigrafía y I Congreso Latinoamericano de Paleontología*, **2**, 193–203.

Soria, M. F. 1980b. Las afinidades de *Phoradiadus divortiensis* Simpson, Minoprio y Patterson, 1962. *Circular Informativa de la Asociación Paleontológica Argentina*, **4**, 20.

Soria, M. F. 1984. Noticia preliminar sobre los Amilnedwardsidae, fam. nov. (Mammalia) del Eoceno temprano de Patagonia, República Argentina. *Resúmenes I Jornadas Argentinas de Paleontología de Vertebrados*, 19.

Soria, M. F. 1989a. Notopterna: un nuevo orden de mamíferos ungulados Eógenos de América del Sur. I. Los Amilnedwardsidae. *Ameghiniana*, **25**, 245–258.

Soria, M. F. 1989b. Notopterna: un nuevo orden de mamíferos ungulados Eógenos de América del Sur. II. *Notonychops powelli* gen. et sp. nov. (Notonychopidae nov.) de la Formación Río Loro (Paleoceno medio) provincia de Tucumán, Argentina. *Ameghiniana*, **25**, 259–272.

PART V SUMMARY

29 Gran Barranca: a 23-million-year record of middle Cenozoic faunal evolution in Patagonia

Richard H. Madden, Richard F. Kay, M. Guiomar Vucetich, and Alfredo A. Carlini

Abstract

Explosive Plinian volcanism provided a rich local source of fine-grained volcaniclastic sediments to central Patagonia in the middle Cenozoic. The chronology of the accumulation extends from the Middle Eocene Climate Optimum (MECO) to the Middle Miocene Climate Optimum (MMCO) and includes the only southern hemisphere continental record across the Eocene–Oligocene transition (EOT) and into the Oi-1 glaciation of Antarctica. The continuity of sedimentation is broken by hiatuses of different temporal magnitude that appear to correspond chronologically with changing sea level and the intensity of erosion. The Sarmiento Formation at Gran Barranca is subdivided into six members, Gran Barranca (41.6–38.7 Ma), Rosado (38.7 Ma), Lower Puesto Almendra (37 Ma), Vera (35–33.3 Ma), Upper Puesto Almendra (31.1–26.3 Ma), and Colhue-Huapi (20.4–18.7 Ma).

The sequence of fossil mammal faunas in the Sarmiento Formation at Gran Barranca is the standard sequence of South American Land Mammal Ages (SALMAs) and their subdivisions. The sequence includes the Barrancan SALM Subage, an intermediate level "El Nuevo," the Mustersan SALMA, the Tinguirirican SALMA, an early Deseadan level "La Cantera," the Deseadan SALMA, Colhuehuapian SALMA, and a Pinturan level. The taxonomic composition and most distinctive features of each of these levels at Gran Barranca is summarized, with special attention to the new and rich record of small mammals recovered by intensive wet washing.

Given the climate intimacy between narrow peninsular Patagonia, the southeast Pacific and South Atlantic oceans, and the vast Southern Ocean today, the terrestrial environments of middle Cenozoic Patagonia were probably subject to the influence of global climate change as recorded in the marine sea-floor record. Evidence that the terrestrial biota responded to the trends and events recorded in marine sediments of the southern oceans depends on an understanding of how the magnitude and rate of this climate change drives evolutionary and biotic change, and whether

it leaves an imprint on terrestrial sediments and their fossil archives. We summarize the fossil mammal record at Gran Barranca in light of the chronology of climate change, but are cautious in our conclusions as appropriate to the limitations imposed by deep time.

During the MECO, phytolith assemblages include graminoid elements (pooids and chloridoids) as components of savanna and/or forest communities. Climate change subsequent to the MECO is recorded at Gran Barranca by the last occurrence of crocodilians and an episode of coincident increase in hypsodonty in several clades of mammalian herbivores. Important change in paleosols, depositional environments, and ichnofacies suggest a general aridification and cooling at the close of the Barrancan and into the Mustersan. Volcanic activity intensified at this time. Pooid and chloridoid grasses become dissociated in the presence of indicators of plant stress conditions or aridification, the oldest evidence for the adaptation of chloridoid grasses to dry environments.

The EOT is preserved in the sediments of the Vera Member at Gran Barranca and the mammals of Tinguirirican SALMA. The mammals document an increase in the overall proportion of hypsodont taxa along with the continued persistence of primitive taxa within the same clades, suggesting species richness was responding to a diversification of diets in addition to the rigors of environmental abrasives. It is tempting to attribute this to the massively bedded volcaniclastic loessites, evidence of reworking by more persistent eolian processes that may have accompanied Oi-1 and an intensification of the developing circum-Antarctic flow. The first appearance of rodents in Patagonia at about 30 Ma occurs some 3 m.y. after significant change in the small marsupial fauna at the EOT.

The MMCO coincides with the presence of primates in Patagonia between 20.2 and 15.7 Ma. The mammal fauna of the Colhuehuapian is especially rich in small marsupials and rodents. Most climate indicators accord with the view that Patagonian climates at this time were warm and humid.

The South American mammal fauna evolved in geographic isolation from other continents through most of the middle Cenozoic. This geographic isolation began with the final separation of South America from Antarctica in the late Eocene prior to the opening of Drake Passage to deep-water circulation. Within Patagonia, a reduction of

The Paleontology of Gran Barranca: Evolution and Environmental Change through the Middle Cenozoic of Patagonia, eds. R. H. Madden, A. A. Carlini, M. G. Vucetich, and R. F. Kay. Published by Cambridge University Press.

geographic area and possibly subdivision by marine barriers, may have accompanied middle Cenozoic marine transgressions, especially prior to the EOT. The arrival of primates and rodents from Africa and their radiations must represent significant perturbations to continental isolation, but their impact has yet to be fully assessed.

The evolution of high-crowned teeth was a major feature of mammal evolution in South America throughout the Cenozoic, as a general faunal trend of increasing proportions of taxa with high-crowned teeth, as successive independent radiations of high-crowned taxa within clades, and as events of coincident increase involving diverse clades. The environmental context of these trends and events are explored.

Resumen

Un vulcanismo explosivo Pliniano fue la fuente de los finos sedimentos volcaniclásticos en la Patagonia central durante el Cenozoico medio. Cronológicamente, la acumulación de sedimentos se extiende desde el Óptimo Climático del Eoceno Medio (OCEM), hasta el Optimo Climático del Mioceno Medio (OCMM), e incluye el único registro continental del hemisferio austral de la Transición Eoceno-Oligoceno (TEO), y el inicio de glaciación antártica (Oi-1). La continuidad de la sedimentación se interrumpe por hiatos de diferente magnitud temporal, que parecen corresponder cronológicamente con cambios en el nivel de mar y en la intensidad de la erosión. La Formación Sarmiento se divide en seis miembros: Gran Barranca (41.6–38.7 Ma), Rosado (38.7 Ma), Puesto Almendra Inferior (37 Ma), Vera (35–33.3 Ma), Puesto Almendra Superior (31.1–26.3 Ma), y Colhue-Huapi (20.4–18.7 Ma).

La sucesión de faunas de mamíferos fósiles en Gran Barranca es la secuencia estandar de Edades Mamífero Sudamericanas y sus subdivisiones. La secuencia incluye: la Subedad Barranquense, un nivel intermedio "El Nuevo", la Mustersense, la Tinguiririquense, un nivel temprano del Deseadense "La Cantera", la Deseadense, la Colhuehuapense, y un nivel "Pinturense". Se resume la composición taxonómica y las características más sobresalientes de cada uno de estos niveles en Gran Barranca, con especial atención al novedoso registro de mamíferos pequeños colectados por un intensivo lavado y sarandeo de sedimentos en agua.

Dada la intimidad climática entre la angosta Patagonia penínsular y los mares y océanos australes circundantes, los ambientes terrestres del Cenozoico medio probablemente estuvieron sujetos a la influencia de cambio climático global, tal como es reconocido en los fondos oceánicos. La evidencia de que las biotas respondieron a los cambios y eventos preservados en el registro marino austral, depende de la comprensión de cómo la magnitud y la tasa de ese cambio climático influenció la biota y su evolución, y cuándo deja su impronta sobre los sedimentos y su registro fosilífero. Resumimos e interpretamos el registro fosilífero de Gran Barranca, de acuerdo a la cronología del cambio climático, con precaución dadas las limitaciones impuestas por el largo tiempo transcurrido.

Durante el OCEM, los conjuntos de fitolitos incluyen elementos graminoides (pooides y cloridoides) como componentes de comunidades de sabana y/o bosque. El cambio climático posterior al OCEM está marcado en Gran Barranca por el último registro de caimanes, y un evento coincidente de aumento en la altura de las coronas en varios clados de mamíferos herbívoros. Un importante cambio en los paleosuelos (ambientes de deposición) y en icnofácies, sugieren una aridificación y enfriamiento general al finalizar el Barranquense y en el Mustersense. La actividad volcánica se intensificó en este lapso. Gramíneas pooides y cloridoides se desasocian en la presencia de indicadores de condiciones de mayor estrés o aridificación, la evidencia más antigua de la adaptación de pastos cloridoides a ambientes secos. La TEO está preservada en los sedimentos del Miembro Vera en Gran Barranca y los mamíferos de la edad Tinguiririquense. Los mamíferos documentan un aumento en la proporción total de taxones hipsodontes, junto con la persistencia de taxones primitivos dentro de los mismos clados, sugiriendo que la diversidad de especies respondió a una diversificación de dietas y a una adaptación al estrés de la abrasividad ambiental creciente. Probablemente, se pueda atribuir el desarrollo de la hipsodoncia a la depositación de las loessitas masivas volcaniclásticas, retrabajadas por procesos eólicos más persistentes (que pudieron haber acompañado al Oi-1 y la intensificación de la circulación circum-antártica). El primer registro de roedores en la Patagonia, alrededor de 30 Ma, ocurre tres millones de años después del cambio significativo en la fauna de marsupiales en la TEO. El OCEM coincide con la presencia de primates en Patagonia entre 20.2 y 15.7 Ma. La fauna de mamíferos del Colhuehuapense es especialmente rica en pequeños marsupiales y roedores. La mayor parte de los indicadores climáticos son acordes con condiciones de climas cálidos y húmedos en la Patagonia.

La fauna mamalífera sudamericana evolucionó en aislamiento geográfico de otros continentes durante la mayor parte del Cenozóico medio. Este aislamiento comenzó con la separación entre América del Sur y la Antártida antes de la apertura de Drake a una circulación de aguas profundas. En la Patagonia, una reducción de superficie y posiblemente una fragmentación por barreras marinas, pudieran haber acompañado a las transgresiones marinas del Cenozoico medio, especialmente antes de la TEO. La llegada de primates y roedores de Africa y su posterior radiación, deben haber perturbado significativamete este aislamiento, pero su impacto es todavía difícil de precisar. La evolución de coronas altas fue una característica principal en la evolución de los mamíferos en América del Sur a lo largo del Cenozoico, como una tendencia de aumentar la proporción de taxones hipsodontes, como una sucesión de radiaciones independientes de taxones dentro de clados, y como eventos de aumento coincidente involucrando a distintos clados. Se discute el contexto ambiental para estas tendencias y eventos.

Introduction

Gran Barranca is a 7-km long and 200-meter high escarpment bordering the southern margin of Lake Colhue-Huapi in Patagonian Argentina at about 45.71° South latitude (Figs. 29.1, 29.2). An outline of the stratigraphy and vertebrate faunal succession at Gran Barranca was first published in the early part of the twentieth century (Ameghino 1906; see Madden and Scarano this book). Since that time, the only English-language monographic study of mammalian faunas from the sequence at Gran Barranca was that of Simpson (1948, 1967) and Cifelli (1985) based on Simpson's field notes (Simpson 1930). As important as Simpson's legacy has proved, his contribution to our knowledge about Gran Barranca was enabled by a single collection made during a brief 1-month field season in 1930, at a time when its relevance to broader questions of biotic evolution and local and global environmental change were obscure.

This volume is the latest contribution to our knowledge of the geology and paleontology of Gran Barranca based on more than 10 years of exploration, fossil collecting, and geological study. The scope of our interest in Gran Barranca has been multifaceted. Gran Barranca exposes the most complete fossiliferous continental rock record of a ~23-million-year interval of the middle Cenozoic in South America and where the superpositional sequence of many South American mammal faunas was established. Because much of the fossil-bearing rock is composed of fine-grained volcanic ash and lava, it is possible to establish a radiometric and paleomagnetic chronology for these faunas. Through study of fossils from known stratigraphic position, we have added several new levels with distinct faunas hitherto either unknown or poorly sampled. By this means we have gained a better appreciation of how mammalian faunas changed through this temporal interval. Through study of the stable element isotope geochemistry of mammalian tooth enamel and records of paleosols, trace fossils, invertebrates, and micro-plant remains, we reconstruct the climatic and biotic conditions within which the mammalian faunas lived and evolved. Finally, as a consequence of the completeness of the rock record, we can see how regional geologic events (Andean tectonism and volcanism, opening of the ocean gateway between South America and Antarctica), and global events (changing sea-surface temperatures and sea level) may have influenced mammalian evolution in this part of the world.

The rock record

Gran Barranca exposes the most complete fossiliferous continental rock record of the middle Cenozoic in South America and the southern hemisphere. The lithology of this middle Cenozoic sequence, called the Sarmiento Formation, and a revision of its stratigraphy and sedimentology is presented by Bellosi (Chapters 2 and 19, this book). A series of magnetic polarity profiles and a number of new radiometric dates allow refinement of the age of the members and units of the formation at Gran Barranca (Ré *et al.* Chapters 3 and 4, this book). The formation was deposited at Gran Barranca between about 41.6 and 18.7 million years ago from the middle through the late Eocene, across the Eocene–Oligocene transition (EOT), and into the early Miocene (Fig. 29.3).

The oldest lithologic division of the Sarmiento Formation is the Gran Barranca Member, which intergrades with the underlying Koluel-Kaike Formation. This member spans between about 41.6 Ma and 38.7 Ma. Overlying the Gran Barranca Member is the Rosado Member, consisting of a mature paleosol radiometrically dated to about 38.7 Ma. Above the Rosado Member is the Lower Puesto Almendra Member which contains a tuff dated at 37.0 Ma. To the west, the Upper Puesto Almendra Member rests directly on the Lower Puesto Almendra Member but to the east the Vera Member is interposed between the two. The Vera Member has an estimated age between 35 and 33.3 Ma. The Upper Puesto Almendra Member is as old as 31.1 Ma and several basalt flows higher up yield a range of dates from 29.2 to 26.3 Ma. Above the Upper Puesto Almendra Member is found the Colhue-Huapi Member, dated between about 20.4 and 18.7 Ma.

The faunal sequence

Since 1906, when the first stratigraphic profile of the Sarmiento Formation was published, the fossil mammal faunas at Gran Barranca have been the standard reference sequence of South American Land Mammal Ages (SALMAs). Here, one can climb through a rock sequence including the Barrancan (late middle Eocene), a transitional level known as "El Nuevo," the Mustersan (late Eocene), Tinguirirican (latest Eocene), an unnamed early Oligocene faunal level ("La Cantera"), the Deseadan (late Oligocene), and two early Miocene faunas, the Colhuehuapian and "Pinturan" SALMAs. Wherever middle Cenozoic fossil mammals are collected in South America or Antarctica, comparisons are necessarily made to the fossil content of the standard sequence in Patagonia which in turn relies on the quality and resolution of the paleontology, stratigraphy, and geochronology at Gran Barranca (Marshall *et al.* 1983; Flynn and Swisher 1995; Arratia 1996).

At the highest possible resolution, new collections of fossil mammals at Gran Barranca come from 49 discrete stratigraphic levels including levels never before sampled. Assemblages of taxa from these stratigraphic levels may eventually allow us to define mammalian biostratigraphic zones and to relate assemblages from elsewhere to these zones as their fossil content and geochronology become better known. Middle Cenozoic SALMAs will eventually

Fig. 29.1. Panorama of Gran Barranca from west (A) to east (E) showing the approximate position of the stratigraphic profiles of the Sarmiento Formation.

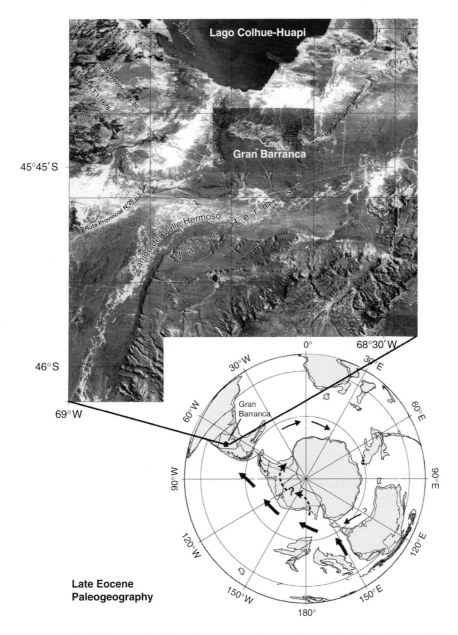

Fig. 29.2. Map of Gran Barranca. Aerial photograph of Gran Barranca superimposed on portion of a satellite map (Instituto Geográfico Militar 1998) and paleogeography of the southern oceans in the late Eocene. (After Bohaty and Zachos 2003; Lawver and Gahagan 2003.)

become substantiated as biochronological units based in large part on the age-calibrated sequence of fossil levels and zones at Gran Barranca. In some instances at Gran Barranca, SALMAs are confined to single lithostratigraphic units (members or their subdivisions) but in other instances a SALMA incorporates all or part of two rock units and in one case a member contains more than one SALMA.

Barrancan

Near the base of the Sarmiento Formation there appear faunas collectively referred to as the Barrancan SALM

Subage, a subage of the Casamayoran SALMA. Duke/ Museo de La Plata (MLP) Expeditions expended relatively little effort to collect in the 12 Barrancan levels at Gran Barranca. The composition of the Barrancan fauna at Gran Barranca is still best known through papers by Simpson (1948, 1967) and Cifelli (1985). Our understanding of the age of the Barrancan fauna is based on the geochronology at Gran Barranca (Ré *et al.* Chapters 3 and 4, this book). To date, only three localities or exposures of the Sarmiento Formation elsewhere in Patagonia can be correlated with the Gran Barranca Member on the basis of fossil content;

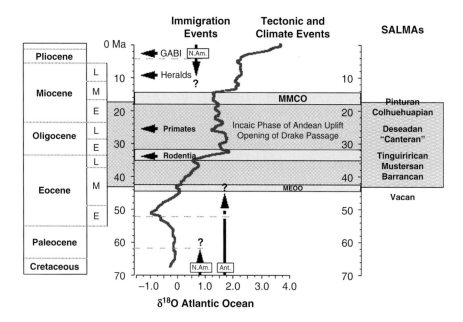

Fig. 29.3. Middle Cenozoic of South America indicating faunal immigration, faunal exchange, tectonic and climatic events, Atlantic Ocean benthic foraminiferal δ¹⁸O curve (from Miller *et al.* 1987), and superposition and approximate age of South American Land Mammal Ages recorded at Gran Barranca. N. Am., North America; Ant., Antarctica; GABI, Great American Biotic Interchange; MECO, Middle Eocene Climate Optimum; MMCO, Middle Miocene Climate Optimum; EOT = Eocene–Oligocene transition.

Colhue-Huapi Norte, Cañadón Lobo (= Cañadón Tournouër at Punta Casamayor), and Tapera de López localities I, II, III, and V (Cifelli 1985). The only emendations we offer to the composition of the Barrancan fauna at Gran Barranca is to point out that the fossil mammal assemblage from GBV-60 El Nuevo is no longer included in the Barrancan.

El Nuevo

The fossil vertebrate locality GBV-60 El Nuevo at Profile M corresponds to Level 15 of Cifelli (1985). The distinctive stratum yielding the fossil vertebrates is in the Rosado Member (Bellosi Chapter 2, this book). The composition of the mammal assemblage based on our new collections includes some noteworthy differences from typical Barrancan taxa among Notoungulata (Reguero and Prevosti this book) and Didolodontidae (Gelfo this book).

Mustersan

Above El Nuevo occur levels of the Mustersan SALMA, which at Gran Barranca is between 38.7 and 37.0 Ma. Levels with Mustersan assemblages are found in the Rosado and Lower Puesto Almendra Members (Bond and Deschamps this book; Gelfo this book; López *et al.* this book; Reguero and Prevosti this book). Bond and Deschamps (this book) proposes that the Rosado Member at Gran Barranca become the type section for the Mustersan Age, given the presence of characteristic Mustersan taxa, precise upper and lower boundaries, and intermediate position within a well-dated

mammal sequence. The composition of the Mustersan fauna at Gran Barranca is largely based on recent collections made by Duke/MLP expeditions at GBV-3 El Rosado and an American Museum of Natural History (AMNH) locality known as "Coley's Quarry," reviewed most recently by Bond and Deschamps (this book) and Gelfo (this book).

The locality GBV-3 El Rosado, just west of Profile J, is the richest Mustersan fauna at Gran Barranca and occurs in the Rosado Member. Cifelli (1985) recorded only two taxa in the AMNH collection from this locality, *Notostylops* sp. and *Antepithecus brachystephanus*, and on this basis included this level in his Barrancan Subage. Since that time, we have collected extensively at GBV-3 El Rosado and have recovered *Astraponotus*, cf. *Eopachyrucos*, in addition to *Periphragnis* and *Rhyphodon* (typical Mustersan isotemnids), and also possibly the notohippid *Puelia* (López *et al.* this book). The co-occurrence of *Notostylops* with *Astraponotus* at GBV-3 El Rosado is noteworthy, but this is not the first time *Notostylops* has been reported as part of an otherwise Mustersan assemblage (see Simpson n.d.) wherein he remarked that the Locality #5 (Cerro Blanco) fossils "en general son mamíferos típicos del Mustersense, pero no comprendo la presencia de *Notostylops*. Nunca he encontrado este género en situ junto con mamíferos parecidos" ["in general these are typical Mustersan mammals, but I don't understand the presence of *Notostylops*. I've never found this genus in situ together with such mammals"].

Coley's Quarry (GBV-64) is the *Astraponotus* quarry Coleman S. Williams worked at the base of Unit 1 of the Lower Puesto Almendra Member. The material includes *Astraponotus* and isotemnid taxa typical of Mustersan assemblages elsewhere in Patagonia. Also, AMNH material of *Periphragnis* sp. from Gran Barranca includes specimens from Coley's Quarry and from GBV-65 (the level of Simpson's #94), another possible Mustersan level.

It should be pointed out that the Rosado Member, GBV-3 El Rosado, and the Rosado Tuff are three different things. The Rosado Member is a distinctive lithostratigraphic unit comprising various strata described by Bellosi (Chapter 2, this book). GBV-3 El Rosado and GBV-60 El Nuevo are fossil vertebrate localities within the Rosado Member. The Rosado Tuff is a dated tuff sampled from the same stratum yielding the fossil mammals at GBV-3. The Rosado Member at Gran Barranca is not a synonym for the Mustersan fauna or the Mustersan SALMA. The Mustersan fossils from Coley's Quarry were collected at a stratigraphic level in the Lower Puesto Almendra Member between Profiles A and G. Thus, the Mustersan at Gran Barranca may be represented by fossil taxa from at least two levels, each in a different member. The faunule from GBV-60 El Nuevo appears to be of intermediate composition between more typical Barrancan and Mustersan assemblages at Gran Barranca, but the locality occurs within the Rosado Member.

La Cancha

Above the Mustersan is found a fauna that is similar in composition to the Chilean type fauna for the Tinguirirican SALMA (Flynn *et al.* 2003). Some characteristic elements of the fauna were known in Ameghino's time and described at the "Astraponotéen plus superior" (Bond and Deschamps this book). The Tinguirirican fauna at Gran Barranca is found in the Vera Member at a level dated to between 33.3 and 33.7 Ma. This age is similar to, but better constrained than the age of the Tinguirirican type fauna. While similar in many respects, the Chilean fauna contains the earliest record of rodents whilst the assemblage at GBV-4 La Cancha at Gran Barranca has yielded no rodents despite an intensive effort to recover small mammals. Almost all we know about the Tinguirirican at Gran Barranca is described in work reported here (Carlini *et al.* this book; Goin *et al.* this book; López this book; Reguero and Prevosti, this book).

The idea of a distinct faunal level called the "Astraponotéen plus superieur" began with Ameghino (1901) supported by evidence that accumulated piecemeal beginning with (1) the curious assemblage from Cañadón Blanco discovered by Roth in 1902, (2) the under-appreciated discovery of Profile K by Egidio Feruglio (Simpson 1936), (3) the collection of a few fossils in a similar stratigraphic context at Profile M by the Scarritt Expedition in 1930 (Simpson 1967), and (4) newer collections from diverse localities in Patagonia

(Hitz *et al.* 2000; Croft *et al.* 2003; Reguero *et al.* 2003; Bond and Deschamps this book). A Tinguirirican SALMA was finally given substance by the remarkable fauna from Chile (Flynn *et al.* 2003). Now the presence of this distinctive faunal level at Gran Barranca has been confirmed by the rediscovery of GBV-4 La Cancha.

The vertebrate locality GBV-4 La Cancha occurs in the La Cancha bed, a distinctive stratum extending nearly unbroken between Profiles K and M. The La Cancha bed occurs within the otherwise nearly monotonous and lithologically homogeneous Vera Member (Bellosi Chapter 19, this book). GBV-4 La Cancha is a flat exposure with a surface area roughly that of a "soccer pitch" near the top of the Vera Member at Profile K. The Carlini Tuff outcrops near the base of the La Cancha bed. GBV-4 La Cancha at Gran Barranca is stratigraphically correlated with the levels at Profile M where Coley Williams collected *Eomorphippus*, and in clear stratigraphic and age relationships with older and younger assemblages referred to the Mustersan and Deseadan land mammal ages. Finally, GBV-4 La Cancha has been dated by an integrated magnetic polarity and Ar/Ar geochronology (Ré *et al.* Chapter 3, this book).

The marsupial faunas from the Tinguirirican levels represent a radical departure from any other known Paleogene marsupial faunas, although marsupials recovered from the GBV-3 El Rosado level at Gran Barranca are under study and could change that picture, as noted by Goin *et al.* (this book). Goin *et al.* (this book) call attention to a radical shift in the niche structure of the smaller marsupials in the Tinguirirican fauna at Gran Barranca and carried on in the slightly younger La Cantera fauna (see below). They note that in the Paleocene and Eocene, frugivorous, omnivorous, or insectivorous marsupials dominated. By contrast, many of the Tinguirirican and later early Oligocene marsupials (and by far the most commonly occurring ones) show more herbivorous feeding habits (seeds, hard fruits, and/or abrasive food items). Particularly *Hondonadia* and *Klohnia* have molar morphologies indicative of some side-to-side or anteroposterior masticatory movements and some degree of hypsodonty. Additionally, the newly described Argyrolagoidea developed a rodent-like molar pattern. The rodent-like adaptations among South American marsupials suggest a rapid response to more cool, open, and arid environments, which opened a new series of adaptive zones by the early Oligocene. The change in marsupial niche structure from the Eocene to the Oligocene is a distinctive feature of the La Cancha fauna as compared to other (older) Paleogene associations. Given our present state of knowledge, the La Cancha marsupials represent the most dramatic marsupial turnover of the South American Cenozoic, an event termed the *Bisagra Patagónica* ("Patagonian Hinge") by Goin and colleagues.

The composition of the southern ungulate assemblage from La Cancha was discussed by López *et al.* (2005) and more recently by Reguero and Prevosti (this book) and López *et al.* (this book). GBV-4 La Cancha has yielded fragmentary remains of the pyrothere *Propyrotherium* and considerable diversity among notoungulates. The assemblage of southern ungulates from GBV-4 La Cancha has its greatest taxonomic similarity with the fauna from Tinguiririca in central Chile, and on this basis alone the GBV-4 La Cancha fauna can be assigned to the Tinguiririan SALMA.

Carlini *et al.* (this book) add substantially to knowledge of Tinguiririan dasypodids at Gran Barranca, which look markedly different from those of older faunal levels. While stegotheriins carry over from lower levels, utaetins disappear and eutatins make their first appearance and are quite diverse (three genera and five species). Euphractins decline in diversity to a single species (compared with four in Mustersan levels).

La Cantera

The next younger fauna at Gran Barranca is the La Cantera fauna (GBV-19) in Unit 3 of the Upper Puesto Almendra Member, constrained in age to between 31.1 and 29.5 Ma – at least 2.2 million years younger that the Tinguiririan at Gran Barranca, and certainly older than well-dated Deseadan faunas at Salla, Bolivia, and Scarritt Pocket and other Deseadan localities in Patagonia. La Cantera could justifiably be proposed as a SALMA of its own. Among the La Cantera marsupials, Goin *et al.* (this book) note the last record of rosendolopids and polydolopimorphians. Argyrolagoids continue in small numbers and the paucituberculate diversification continued. Large borhyaenids are represented for the first time by *Pharsophorus*. Several other characteristic elements of the Deseadan SALMA appear but many characteristically Tinguiririan elements persist, especially marsupials that do not survive in the typical Deseadan faunas of Patagonia, Brazil, and Bolivia (Carlini *et al.* this book; Goin *et al.* this book; López *et al.* this book; Reguero and Prevosti this book; Vucetich *et al.* Chapter 13, this book).

Xenarthrans are represented at La Cantera by one stegotheriin, as in the Tinguiririan, and the same three genera of eutatins are represented. On the other hand, the composition of the euphractine fauna is quite different, with just one species possibly in common: *Parutaetus chilensis*. Two other Tinguiririan euphractine genera drop out and two genera are added: *Archaeutatus* and *Prozaedyus* (Carlini *et al.* this book).

The rodent fauna in the La Cantera level is the oldest in Argentina and morphologically most primitive in South America (Vucetich *et al.* Chapter 13, this book). The represented taxa include three octodontoids, one cavioid, one dasyproctid, and one possible chinchilloid. All species have low-crowned teeth (brachydont) except the chinchilloid

which displays an incipient unilateral hypsodonty. The absence of rodents in well-documented Tinguiririan levels at Gran Barranca suggests rodents arrived from the north during the early Oligocene. Based on their similarity with African rodents of the late Eocene, rodents arrived in South America as immigrants from Africa perhaps in the middle to late Eocene.

The ungulate fauna from La Cantera contains elements of the Tinguiririan fauna along with a significant number of more evolutionarily advanced taxa that are known from Deseadan faunas (*Scarrittia*, *Henricofilholia*) including guide fossils for this SALMA. Sixty percent (60%) of the remains collected at La Cantera are clearly referable to the Leontiniidae, characteristic elements in Deseadan faunas of Patagonia. Of the leontiniid material from GBV-19 La Cantera the most abundant taxa are new species only known from this locality, e.g. *Scarrittia barranquensis* and *Henricofilholia vucetichia* (Ribeiro *et al.* this book).

The family Notohippidae is represented at GBV-19 La Cantera also by two new species, *Patagonhippus canterensis* and *Patagonhippus dukei*, closely related to well-known taxa from La Flecha and Cabeza Blanca, but more primitive morphologically (López *et al.* this book).

To date, Toxodontidae are represented only by rare upper molars generally comparable in morphology to *Proadinotherium* from Deseadan assemblages in Argentina, Chile, Bolivia, and Colombia. Toxodontidae appear abruptly at La Cantera, with as yet no plausible ancestor in older levels or faunas anywhere in Patagonia.

The typotherians include a small *Trachytherus* (Mesotheriidae) and *Prohegetotherium* (Reguero and Castro 2004; Reguero and Cerdeño 2005). Archaeohyracidae include material of either *Archaeotypotherium* or *Archaeohyrax*, and *Protarchaeohyrax gracilis*. Material of Interatheriidae from GBV-19 is similar in size and morphology to *Eopachyrucos pliciferus*.

In general, Hegetotheriidae, Mesotheriidae, and Leontiniidae are present at La Cantera but not at GBV-4 La Cancha and between Isotemnidae and Notohippidae, more primitive taxa are recorded at GBV-4 La Cancha. Among typotherians, both Notopithecinae and Interatheriinae are present at GBV-4 La Cancha, but only relatively derived Interatheriinae occur at GBV-19 La Cantera.

Two subfamilies of Astrapotheriidae are present at La Cantera; Astrapotheriinae comparable to *Parastrapotherium* (it should be noted that material of large astrapotheriine is known to occur as float down slope from both Colhuehuapian and Deseadan levels) and Albertogaudryiinae, represented by a large form comparable to *Albertogaudrya* (a genus otherwise known from Casamayoran and Mustersan levels in Patagonia) and a smaller taxon recently described by Kramarz and Bond (2009).

Some mandibular fragments from GBV-19 La Cantera can be referred to Litopterna, but generally poor preservation, makes their identification problematical, but are reminiscent of the enigmatic genus *Protheosodon* (Proterotheriidae?).

Deseadan

Until now, the Deseadan at Gran Barranca has been elusive. Marshall *et al.* (1986) complained that the precise stratigraphic context of Deseadan taxa was not recorded and Cifelli had trouble pinning it down stratigraphically based on Simpson's otherwise well-documented collection of fossil mammals at the AMNH. Since long before the time of Simpson, and as early as the work of Carlos and Florentino Ameghino, the presence of a Deseadan or "*Pyrotherium*" level at Gran Barranca has never been convincingly or adequately documented. Three localities in the Upper Puesto Almendra Member at Gran Barranca have yielded useful assemblages of fossil mammals (GBV-1, GBV-34 in Unit 3, and GBV-35) that although of limited diversity, have taxonomic compositions reminiscent of the type Deseadan fauna at La Flecha and the best-known Deseadan assemblage at Cabeza Blanca. Four new isotopic dates (26.3 to 29.2 Ma) are available for basalts in the Upper Puesto Almendra Member indicating that multiple flows are represented with relatively brief intervening periods of local erosion (Ré *et al.* Chapter 4, this book). Local unconformities occur both at the base (D7) and at the top of the basalts (D8). The Deseadan localities come from within Unit 3 of the Upper Puesto Almendra Member in association with basalts but the stratigraphic relationship between them and their associated unconformities is obscure (Ré *et al.* Chapter 3, this book).

The most abundant material from these localities are isolated specimens (mostly teeth) of large Leontiniidae and Astrapotheriidae. The leontiniid material is referred to *Scarrittia barranquensis*, and *Scarrittia* cf. *S. canquelensis* (Ribeiro *et al.* this book), and the astrapotheriid to *Parastrapotherium crassum* (Kramarz and Bond this book).

Remains of *Pyrotherium* are rare at Gran Barranca. Simpson (1930, p. 75) observed only "a few fragments, none . . . happen to have been in place." A single lower milk molar of *Pyrotherium sorondoi* (FMNH P 15068) is the only material of *Pyrotherium* collected by the Riggs expedition at Gran Barranca. The most complete material known to date are the two specimens of *Pyrotherium* collected by Alejandro Bordas in 1939, including MACN 12410, a left maxilla with M1 and M2 from Km 163 (a classic geographic place name associated with Gran Barranca that indicates nothing about its stratigraphic provenance). Bordas (1945) published a profile of a 130-m thick exposure of the Sarmiento Formation at Gran Barranca onto which he indicates the *Pyrotherium* zone just below a basalt somewhere at the west end of Gran Barranca.

No rodents or xenarthrans have been recorded from Deseadan levels at Gran Barranca.

Colhuehuapian

After a considerable temporal hiatus (in faunal representation though not necessarily in deposition) there occurs the Lower Fossil Zone of the Colhue-Huapi Member, the type fauna of the Colhuehuapian SALMA, between 20.0 and 20.2 Ma. A number of chapters in this book summarize our current state of knowledge about this faunal zone (Carlini *et al.* this book; Czaplewski this book; Kay this book; Kramarz and Bond this book; Reguero and Prevosti this book; Ribeiro *et al.* this book; Vucetich *et al.* Chapter 14, this book) and Goin *et al.* (2007) reported previously on the marsupial fauna.

Among dasypodine armadillos, astegotheriins (*Pseudostegotherium*) reappear for the first time since the Barrancan with a very tiny species (*Pseudostegotherium glangeaudi*). *Stegotherium* is present and abundant. Among Euphractinae is one species of Eutatini *Proeutatus*, and one of the Euphractin *Prozaedyus* both also known from the "Pinturan" levels (Carlini *et al.* this book).

Goin *et al.* (2007) revised the Colhuehuapian Didelphimorphia and Sparassodonta. The Colhuehuapian levels of Gran Barranca are the richest ones in marsupials of this age, including 15 species referred to six families of these two orders only. Among the most important results, they recognized the oldest record of Didelphoidea including the oldest Caluromyidae, and a great diversity of carnivores including the oldest Thylacosmilidae. They described a new species of *Necrolestes*, an unusual Santacrucian mammal of uncertain affinities, with "pretribosphenic" molar pattern. Although not revised in the context of this project, other groups of marsupials are recorded in the Colhuehuapian levels at Gran Barranca: microbiotheriids, caenolestids, and abderitids (Marshall 1976; Pascual *et al.* 1996). A wide variety of microbiotheriids, several argyrolagid taxa, and an impressive diversity of paucituberculatans (Caeonolestidae, Palaeontentidae, and Abderitidae) have been recognized (F. Goin pers. comm.).

Vucetich *et al.* (Chapter 14, this book) describe a remarkable rodent fauna from the Colhuehuapian SALMA. Strikingly, the Colhuehuapian rodents are more diverse than from any other rodent fauna in the Cenozoic. Some of the diversity is accounted for by *in situ* evolution of Deseadan Patagonian taxa whilst other lineages appear without obvious Patagonian antecedents and likely originated from northern subtropical regions after the Deseadan. The diversity of rodent cheek tooth designs including crown complexity, increased crown height, and a broader size range compared with Deseadan rodents shows that Colhuehuapian caviomorphs had already developed considerable dietary niche breadth. Especially notable are the richness of small octodontoids and the great diversity of possibly arboreal erethizontids that have their acme at this time. Many

Colhuehuapian rodent lineages had gone extinct in Patagonia after the middle Miocene; only eocardiids and chinchillids are certainly closely related to extant representatives. They note that the differences between the Gran Barranca Colhuehuapian fauna and those from other well-known Colhuehuapian sites are minor and may result largely from environmental variations.

Primates are rare elements of the Gran Barranca fauna, being found only in the Colhuehuapian (Kay this book). Kay describes a new genus, *Mazzonicebus*. The addition of this taxon brings the known diversity of Colhuehuapian primates to three genera. *Mazzonicebus* is closely related to *Soriacebus* from the younger Miocene rocks of the Pinturas Formation in Santa Cruz Province, but not recorded at Gran Barranca. The two offer the first well-documented sister-taxon relationship between a Colhuehuapian and a "Pinturan" (some times considered as early Santacrucian) primate, spanning ∼3 m.y. *Mazzonicebus* and *Soriacebus* show specializations for fruit husking and seed predation, a dietary niche similar to that seen in the more derived members of the living platyrrhine subfamily Pitheciinae (sakis and uakaries), a fact that has led some researchers to conclude that *Soriacebus* is an early representative of that clade. However, the more primitive characters of the dentition of *Mazzonicebus* compared with *Soriacebus* further document and reinforce the hypothesis that these two Argentine early Miocene taxa are an independently evolved clade that is an ecological "vicar" of the extant platyrrhine seed predation dietary niche.

Bats are another rare element of the Gran Barranca fauna known from the Colhuehuapian. *Mormopterus barrancae* is a new species of phyllostomine Molossidae. A more primitive species of this genus is found in the Deseadan of Brazil. A single tooth, 20–30% smaller than *M. barrancae*, represents a second smaller species in the Colhuehuapian of Gran Barranca. The presence of the Phyllostominae suggests a warm moist climate and the availability of lowland forest habitat in the Colhuehuapian, consistent in part with habitats inferred from early Miocene rodents and primates.

Colhuehuapian notoungulates seem poorly diversified for a fauna otherwise so rich. Only five species referable to four families occur in the Lower Fossil Zone (Reguero and Prevosti this book). Kramarz and Bond (2005) and Soria (2001) list four taxa of Proterotheriidae (Litopterna) from the Colhuehuapian of the Sarmiento Formation without specifying whether from Gran Barranca or elsewhere in Patagonia: *Lambdaconus lacerum, Paramacrauchenia inexspectata, Paramacrauchenia scamnata,* and *Prolicaphrium specillatum*. Finally, there are at least three taxa of Astrapotheriidae in the Colhuehuapian at Gran Barranca, *Parastrapotherium symmetrum, Parastrapotherium martiale,* and *Astrapotherium? ruderarium* and possibly a fourth uruguaytheriine (Kramarz and Bond this book).

"Pinturan"

Near the top of the Sarmiento Formation in the Upper Fossil Zone of the Colhue-Huapi Member between 19.7 and 18.7 Ma, there occurs a still poorly represented fauna that while resembling the Pinturan fauna from Santa Cruz Province, may be at least 1.5 m.y. older (Kramarz *et al.* this book). Although notoungulate and litoptern material from the "Pinturan" levels at Gran Barranca is fragmentary, three typotherians are present (*Interatherium, Protypotherium, Pachyrukhos*) and a fragmentary upper molar has been assigned to *Tetramerorhinus* (Proterotheriidae) (Kramarz *et al.* this book). Rodents are represented at least by seven genera of five families, including the otherwise exclusively "Pinturan" *Prostichomys*. Edentates are also diverse, represented by dasypodids, glyptodonts, and tardigrades, this latter through a single mandible fragment.

Regional and global climatic influences

Southern South America was likely influenced by worldwide and southern hemisphere climate trends and events during the middle Cenozoic. These might include (1) the period of cooling after the Mid-Eocene Climate Optimum (MECO), (2) the Eocene–Oligocene transition (EOT), and (3) the early Miocene warming trend leading to the Middle Miocene Climate Optimum (MMCO) (Lear *et al.* 2000; Zachos *et al.* 2001).

Our suspicion that biotic evidence bearing on these three events should be preserved in the terrestrial fossil record of the Sarmiento Formation at Gran Barranca (Fig. 29.3) is based on the fact that oceanic climate prevails today across Patagonia, with characteristically moderated seasonal atmospheric temperature extremes closely tied to sea-surface temperatures in the adjacent southern oceans. Proximity to Antarctica brings colder temperature extremes during the austral winter when colder polar air migrates northward, and similar such phenomena may have intensified with the onset and variation in the intensity of Antarctic glaciation. Today the Andes interrupts the westerly flow of atmospheric moisture across Patagonia, producing a pronounced orographic effect. However, the present morphology of the Patagonian Andes seems to postdate the early Miocene such that orographic influence on rainfall must have been less strongly felt when the Sarmiento Formation was deposited. The low altitude and flatness of central Patagonia makes the peninsula susceptible to sea-level change at epoch-scale and shorter durations (Miller *et al.* 2005; Pekar and Christie-Blick 2008). The surface area of the peninsula changed with sea level and parts of Patagonia may have become isolated from the rest of the continent (Bellosi Chapter 2, this book; Guerstein *et al.* this book; Kohn *et al.* this book). Beyond the reductions and increases in surface area brought by changing sea level associated with Antarctic ice volume and regional tectonism, the middle Cenozoic of South America witnessed two major biogeographic events: the progressive severance of terrestrial continuity with Antarctica and, later, the arrival of primates and rodents from Africa (Kay *et al.* 1998b; Vucetich *et al.* this book). The first event established the geographic isolation of South America and its continental mammal fauna. The immigration of rodents and primates from Africa represent important perturbations to this faunal isolation. Finally, the fossil record at Gran Barranca has yielded a record of significant evolution of high-crowned teeth in diverse clades of mammalian herbivores. Our understanding of what drives the tempo of change in tooth crown height implicates climate change and its influence on earth surface processes. Finally, our suspicion is heightened by the rich marine record of Southern Ocean climate change (Bohaty and Zachos 2003; Pagani *et al.* 2005; Thomas 2008), and the sensitivity of high latitudes to change in the atmospheric concentration of CO_2 (Emanuel *et al.* 1985). These suspicions make it all the more surprising that Kohn *et al.* (2004, this book) find such limited change in enamel oxygen isotope composition throughout the Sarmiento Formation at Gran Barranca.

Global change

Middle Eocene climate deterioration
There is little change in the composition of Barrancan mammal assemblages and phytolith assemblages through the Gran Barranca Member, a time when graminoid elements occur as components of savanna and/or forest communities (Zucol *et al.* this book). Phytolith diversity is relatively high and stable until the eruption events marked by Simpson's Y Tuff and the Mazzoni Tuff (Zucol *et al.* this book). Between 41.6 and 36.6 Ma there occurs an erratic decrease in atmospheric CO_2 concentrations and a pronounced increase in benthic $\delta^{18}O$ oxygen isotope values marking a $6\,^{\circ}C$ cooling trend after the MECO ending with late Eocene warming (Bohaty and Zachos 2003), after the first influx of Pacific water into the South Atlantic through Drake Passage between 41.3 and 39.6 Ma (Scher and Martin 2006). The interval of time marked by this pronounced deep-water cooling is represented by the Gran Barranca Member and extends through the overlying Rosado Member and into the Lower Puesto Almendra Member at Gran Barranca. During this cooling trend, at 39.85 Ma, temperature-sensitive crocodilians make their last appearance in Patagonia in Simpson's Y Tuff at Gran Barranca (Simpson 1933), about 2 m.y. earlier than the abrupt decline in crown-group crocodilian diversity between 38 and 36 Ma and contraction in their latitudinal range in the northern hemisphere (Markwick 1998a, 1998b). Crocodilian distribution is sensitive not only to cold temperatures, but to perennial rivers and the suitability of substrate.

Between 39.1 and 38.4 Ma there is a coincident increase in hypsodonty in three clades of mammalian herbivores (Notostylopidae, Interatheriidae, and Archaeohyracidae). This is the oldest evolutionary event of increasing hypsodonty in the Patagonian record (Kay *et al.* 1999; Pascual and Ortíz-Jaureguízar 2007), recalibrated by a more comprehensive geochronology (Ré *et al.* Chapter 4, this book). This period records as well important changes in paleosols, sedimentology, and ichnofacies between the Gran Barranca and Rosado Members (Bellosi Chapter 19, this book; Bellosi and González this book; Bellosi *et al.* this book) that indicate a transition from subhumid to arid and even xeric conditions. The prior co-occurrence of pooid and chloridoid grass elements is broken and indicators of plant stress conditions appear that provide evidence for aridification (Zucol *et al.* this book). At this time, an important environmental threshold into dryer conditions seems to have been crossed in terrestrial ecosystems in Patagonia.

Eocene–Oligocene transition
The Eocene–Oligocene transition (EOT), here defined as events occurring between 34.0 and 33.0 Ma, marks a more accelerated two-step event of global cooling and increasing Antarctic ice volume (Lear *et al.* 2008), the deepening of the Tasmanian Gateway between Australia and East Antarctica (Stickley *et al.* 2004), the further establishment of a global Antarctic Circumpolar Current (Huber *et al.* 2004), and the extinction of continental mammals in Antarctica (Reguero and Marenssi this book). As presently calibrated (Ré *et al.* Chapter 4, this book), the age of the mammalian fauna from GBV-4 La Cancha falls into this brief interval. The sedimentology of the Sarmiento Formation changes significantly with the Vera Member, a thick unit of massively bedded volcaniclastic loessite reworked by more persistent eolian processes representing continuous deposition and increasing sedimentation rates through the interval between Chrons C15n to C13n. Stable isotopes from tooth enamel at La Cancha do not suggest more than a 2–3 °C cooling and no detectable change in seasonality at Gran Barranca (Kohn *et al.* 2004, this book). Compared with the Mustersan, the La Cancha fauna reveals an increase in the overall proportion of hypsodont taxa through the local proliferation of high-crowned species among Notohippidae, Archaeohyracidae, and Interatheriidae. This increase in faunal hypsodonty occurred along with the continued persistence of relatively lower-crowned taxa within these same clades (Lopez *et al.* this book; Reguero *et al.* this book). This suggests that these clades were responding through both the diversification of diets and the proliferation of taxa adapting to the rigors of increased environmental abrasives. The range of mammalian taxa adjusting to changing conditions across the EOT was broadened by

the appearance of high-crowned teeth among herbivorous marsupials as well (Goin *et al.* this book).

Early to middle Miocene climate amelioration
Oxygen isotope curves for the Neogene (Zachos *et al.* 2001) indicate a prominent period of ice-free conditions extending from the late Oligocene to the MMCO, corresponding to the interval between the Colhuehuapian and the Colloncuran in Patagonia, i.e. between 20.4 and 15.7 Ma.

The oldest record of primates in the Patagonian fossil record is in the Lower Fossil Zone of the Colhue-Huapi Member at Gran Barranca at about 20.2 Ma (Kay this book). Thereafter, primates occur continuously in the Patagonian fossil record until their latest known occurrence in the Colloncuran (Kay *et al.* 1998a). Primates were present in Patagonia throughout the 4.5-million-year interval between 20.2 and 15.7 Ma. Given the modern association between primates, humid environments, and subtropical climate conditions at their southernmost occurrences today (*Alouatta caraya* at 28° S in Argentina, and 30° S in Brazil), Patagonian climate must have been warm (mean annual temperature >19 °C) and humid (mean annual precipitation >1100–1400 mm) during the interval between 20.4 and 15.7 Ma.

The fossil record for the Colhuehuapian at Gran Barranca is especially rich in small marsupials and rodents. Goin *et al.* (2007) reviewed all the published evidence for Colhuehuapian climates in Patagonia and concluded that central Patagonia supported a regionally heterogeneous vegetation that included wet forests, palm-tree associations, restricted grassy environments, and flooded or paludal areas on variable topography. Most climate indicators accord with the view that Patagonian climates at this time were warm and humid during all or most of the year at a time when the Patagonian Andes did not act as an orographic barrier to moisture-laden winds. These benign conditions are reflected especially in the richness of brachydont octodontoids and erethizontid rodents (Vucetich *et al.* Chapter 14, this book). Also, asthegotheriin dasypodine armadillos reappear again in Patagonia after a 20-Ma hiatus (Carlini *et al.* this book).

Geographic isolation
The South American mammal fauna evolved in geographic isolation from other continents through most of the middle Cenozoic (Simpson 1950, 1980; Patterson and Pascual 1968). The available evidence from age-calibrated molecular phylogenetics and paleontology suggests that continental South American land mammals became isolated from Antarctica sometime prior to about 36 Ma (Springer *et al.* 1998; Reguero *et al.* 2002; Nilsson *et al.* 2004). This biogeographic separation of South America from Antarctica occurred prior to the opening of Drake Passage to deep-water circulation

sometime between the EOT and about 28 Ma (Lawver and Gahagan 2003; Livermore *et al.* 2005).

In terms of faunal composition through the middle Cenozoic, continued geographic proximity between South America and Antarctica was less meaningful than South America's geographic proximity to Africa. Despite the progressive widening of the South and Equatorial Atlantic, waif immigration of small mammals from Africa to South America contributed rodents and primates to the South American fauna. Rodents first appear in the South American fossil record in dated context at about 31.5 Ma (Wyss *et al.* 1993) and shortly thereafter in Patagonia (Vucetich *et al.* 2004, Chapter 13, this book). The existence of caviomorphs in South America sometime prior to these dated occurrences, while logical, has yet to be adequately age-calibrated (Vucetich and Ribeiro 2003; Frailey and Campbell 2004). The geographic spread of caviomorph rodents to southern South America may not have occurred very rapidly. An example may be provided by the Sigmodontinae. Fully 6 million years elapsed between a molecular-clock divergence date for the clade Sigmodontinae between 12.3 and 13.1 Ma (Steppan *et al.* 2004) and their first appearance in the South American fossil record at 30° S latitude at about 6 Ma (Verzi and Montalvo 2008). During their accommodation to South American environments, Sigmodontinae achieved higher taxonomic level diversification (tribe) encompassing variation in locomotor and digestive anatomy (Voss 1988), cranial morphology and tooth crown shape and height (Steppan 1995), and a body size range from the smallest *Calomys musculinus* 8.2–21.2 g to largest *Holochilus magnus* 227–250 g (Carleton and Musser 1984; Patterson 1999; Williams and Kay 2001), plausibly constrained by competition with caviomorphs (Patterson and Wood 1982; Wood and Patterson 1959; Walton 1997).

The oldest known primate in South America is *Branisella* in Bolivia at tropical latitudes at about 26.0 Ma (Kay *et al.* 1998b, 2001; Takai *et al.* 2000). The impact of primates on the native South American fauna will be difficult to assess without a more continuous record from tropical latitudes. Nevertheless, their presence in Patagonia during a substantial interval of the early to middle Miocene is thought to have influenced ecological vicars, among them caenolestoid marsupials (Strait *et al.* 1990). Primates appear to have been excluded from Patagonian environments prior to the beginning of the Miocene Climatic Optimum.

Within Patagonia, a reduction of geographic area and possibly subdivision by marine barriers may have accompanied middle Cenozoic marine transgressions. Guerstein *et al.* (this book) review the history of marine deposits and sea-level variation affecting the surface area of Patagonia. A tentative geochronology for middle Cenozoic Patagonian transgressions can be established using age-calibrated sea-level curves (Miller *et al.* 2005) and their correlation with discontinuities and temporal hiatuses in sedimentation between members of the Sarmiento Formation. Not all of these highstands are preserved in marine sediments in the vicinity of Gran Barranca, nor in the dinoflagellate record described by Guerstein *et al.* (this book). Nevertheless, middle Cenozoic transgressions may have influenced mammalian evolution through their influence on the relationship between habitable area and the area of pyroclastic sediment accumulation (Ardolino *et al.* 1999; Malumián 1999).

Faunal hypsodonty and the establishment of grass-dominated ecosystems

The evolution of high-crowned teeth was a major feature of mammal evolution in South America throughout the Cenozoic, as a general faunal trend of increasing proportions of taxa with high-crowned teeth, as successive independent radiations within clades, and as events of coincident increase involving diverse clades. The evolution of high-crowned teeth in the middle Cenozoic of Patagonia occurs in diverse lineages of herbivorous mammals, including marsupials (Goin *et al.* this book), as many as six clades of native ungulates (Pascual *et al.* 1996; Reguero *et al.* this book), and several clades of rodents (Vucetich *et al.* Chapters 13 and 14, this book), all documented by the fossil record at Gran Barranca. The general trend of increasing faunal hypsodonty in the middle Cenozoic appears to have started among notoungulates in Patagonia as early as 43 Ma when lophodont ungulates begin to replace bunodont forms. By 38 Ma 23% of notoungulate genera were hypsodont (Kohn *et al.* 2004) and this trend culminated in at least 44% of genera sometime prior to 33 Ma in the early Oligocene. The first appearance of herbivorous marsupials with high-crowned teeth seems to have been at the EOT (Goin *et al.* this book). Rodents in the early Oligocene at GBV-19 La Cantera had notably more low-crowned teeth than their Deseadan (late Oligocene) and Colhuehuapian (early Miocene) successors. Notable examples of independent lineages evolving *in situ* towards higher crowns during the Deseadan to Colhuehuapian interval include Toxodontidae, Notohippidae, and Caviodea. Thus, the general faunal trend, multiple examples of the independent radiation of hypsodont taxa within clades, and several events of coincident accelerated increase in hypsodonty in Patagonia significantly antedate the initiation and early middle Cenozoic trends in faunal hypsodonty in North America (Prothero and Heaton 1996) and Europe (Collinson and Hooker 1987; Legendre 1989).

As documented at Gran Barranca, in the middle Cenozoic in Patagonia, there occurred several events of accelerated increase in crown height within the larger trend toward higher proportions of herbivore taxa with high-crowned teeth. These events occurred between 39 and 38 Ma, again between 32 and 26 Ma, and between 22 and 20 Ma. The occurrence of hypsodonty (Pascual and Ortíz-Jaureguízar 1990) in sediments containing grass phytoliths (Mazzoni

1979) has traditionally been understood to reflect the establishment of widespread grass-dominated ecosystems in South America (Jacobs *et al.* 1999). This plant–animal interaction assumes that herbivores evolve high-crowned teeth in response to the silica content of grass in the diet. However, in North America, phytoliths of open-country grasses become dominant in the Great Plains 6 million years before horses evolved hypsodonty (Strömberg 2002, 2004, 2005), and long after faunal hypsodonty became the dominant evolutionary trend among terrestrial mammalian herbivores. At Gran Barranca, grass phytoliths are abundant in the Gran Barranca Member (Zucol *et al.* this book) at least 1.5 million years before the oldest coincident increase in hypsodonty in Notohippidae, Archaeohyracidae, and Interatheriidae between 39 and 38 Ma. All this indicates that herbivores either did not exploit grasses or were adapting to some other source of dietary abrasives.

Summing up

Despite the fact that more than 100 years of research has been undertaken at Gran Barrranca, much remains to be done. The research in the present volume highlights as much about what we do not know as what we do know. The record of biotic evolution is as yet dimly perceived. Yet the collective efforts of the contributors to this book have moved forward on many questions. We now know roughly the stratigraphy of Gran Barranca and something about its mode of deposition. We have a more detailed understanding of the chronological ages of most of the vertebrate faunas. We have filled in some gaps. Most notably we have better documented the faunas clustered around the Eocene–Oligocene boundary. The work done up to now shows very clearly that it is misleading to assume that we can predict what occurred in Patagonia from global proxies of climate change, e.g. in sea-surface temperatures, or even regional effects like Andean uplift, sea-level fluctuations, the onset of Antarctic glaciation, or the opening of circumpolar oceanic circulation. What is especially clear is that different mammal groups responded in different ways to the same effects. For example, it appears that increased cheek-tooth crown height (hypsodonty) occurred over a long period of time and occurred in some taxonomic groups earlier than in others, and at different rates.

There are many gaps in our knowledge that could be addressed relatively easily. To cite a few examples, we have only begun to study the micro-plant remains despite the abundance of phytoliths through the section. Much more needs to be done with stable isotopes as proxies of climate change. And more intensive efforts must be made to recover small mammals, especially in the Gran Barranca Member. Other gains will come only with considerably more effort. Learning about the paleobiology of the mammalian species will be difficult given the fragmentary quality of the fossils. We can only hope that this book will serve as a milepost for progress in understanding the middle Cenozoic record of Patagonia not as a stop sign.

ACKNOWLEDGEMENTS

The results reported in this book have involved researchers from many institutions, most notably from the Museo de La Plata, the Universidad de Buenos Aires, and the Museo Argentino de Ciencias Naturales, enabled through the fruits of an international cooperative research program between the Museo de La Plata and Duke University and through the generous and long-term financial support of the U.S. National Science Foundation (NSF) to Richard F. Kay and Richard Madden, and the Argentine Consejo Nacional de Investigaciones Científicas y Técnicas (CONICET) and Agencia Nacional de Promoción Científica y Técnica (ANPCyT) to numerous Argentine colleagues. In our belief that the global perspective enabled by this effort is significant scientifically, the overall objective of the work at Gran Barranca has been to make recent work and enable future work to be accessible and comparable to advances in the study of the EOT and MMCO everywhere.

REFERENCES

Ameghino, F. 1901. Notices préliminaires sur des ongulés nouveaux des terrains crétacés de Patagonie. *Boletin de la Academia Nacional de Ciencias en Córdoba*, **16**, 349–426.

Ameghino, F. 1906. Les formations sédimentaires du Crétacé supérieur et du Tertiaire de Patagonie avec un parallele entre leurs faunes mammalogiques et celles de l'ancien continent. *Anales del Museo Nacional de Buenos Aires, Série 3*, **15**, 1–568.

Ardolino, A., M. Franchi, M. Remesal, and F. Salani 1999. La sedimentación y el volcanismo terciarios en la Patagonia extraandina. II. El volcanismo en la Patagonia Extraandina. *Anales del Instituto de Geología y Recursos Minerales, Buenos Aires*, **29**, 579–612.

Arratia, G. 1996. Contributions of Southern South America to Vertebrate paleontology. *Münchner Geowissenschaftliche Abhandlungen A*, **30**, 1–342.

Bohaty, S. M. and J. C. Zachos 2003. Significant Southern Ocean warming event in the late middle Eocene. *Geology*, **31**, 1017–1020.

Bordas, A. F. 1945. Notas para el conocimiento de la geología estratigráfica de algunas zonas de la Patagonia. *Anales del Museo Patagónico, Buenos Aires*, **1**, 139–184.

Carleton, M. D. and G. G. Musser 1984. Muroid rodents. In Anderson, S. and Jones, J. K. (eds.), *Orders and Families of Recent Mammals of the World*. New York: John Wiley, pp. 289–379.

Carlini, A. A., M. Ciancio, and G. J. Scillato-Yané 2005. Los Xenarthra de Gran Barranca: más de 20 Ma de historia. *Actas XVI Congreso Geológico Argentino*, **4**, 419–424.

Cifelli, R. L. 1985. Biostratigraphy of the Casamayoran, Early Eocene, of Patagonia. *American Museum Novitates*, **2820**, 1–26.

Collinson, M. E. and J. J. Hooker 1987. Vegetational and mammalian faunal changes in the early Tertiary of southern England. In Friis, E. M., Chaloner, W. G., and Crane, P. R. (eds.), *The Origins of Angiosperms and their Biological Consequences*. Cambridge, UK: Cambridge University Press, pp. 259–303.

Croft, D. A., M. Reguero, A. R. Wyss, and J. J. Flynn 2003. Large archaeohyracids (Typotheria, Notoungulata) from central Chile and Patagonia including revision of *Archaeotypotherium*. *Fieldiana, Geology*, n.s., **49**, 1–38.

Emanuel, W. R., H. H. Schugart, and M. P. Stevenson 1985. Climatic change and the broad-scale distribution of terrestrial ecosystem complexes. *Climatic Change* **7**, 29–43.

Flynn, J. J. and C. C. Swisher III 1995. Cenozoic South American Land Mammal Ages: correlation to global geochronologies. In Berggren, W. A., Kent, D. V., Aubry, M.-P., and Hardenbol, J. (eds.), *Geochronology, Time Scales, and Global Stratigraphic Correlation*, Special Publication no. 54. Tulsa, OK: Society for Sedimentary Geology, pp. 317–333.

Flynn, J. J., A. R. Wyss, D. A. Croft, and R. Charrier 2003. The Tinguiririca Fauna, Chile: biochronology, paleoecology, biogeography, and a new earliest Oligocene South American Land Mammal 'Age'. *Palaeogeography, Palaeoclimatology, Palaeoecology*, **195**, 229–259.

Frailey, C. D. and Campbell, K. E. 2004. Paleogene rodents from Amazonian Peru: the Santa Rosa Local Fauna. *Natural History Museum of Los Angeles County, Science Series*, **40**, 71–130.

Goin, F. J., A. Abello, E. Bellosi, R. F. Kay, R. H. Madden, and A. A. Carlini 2007. Los Metatheria sudamericanos de comienzos del Neógeno (Mioceno temprano, Edad-mamífero Colhuehuapense). I. Introducción, Didelphimorphia y Sparassodonta. *Ameghiniana*, **44**, 29–71.

Hitz, R., M. A. Reguero, A. R. Wyss, and J. J. Flynn 2000. New interatheriines (Interatheriidae, Notoungulata) from the Paleogene of Central Chile and Southern Argentina. *Fieldiana, Geology*, n.s., **42**, 1–26.

Huber, M., H. Brinkhuis, C. E. Stickley, K. Döös, A. Sluijs, J. Warnaar, S. A. Schellenberg, and G. L. Williams 2004. Eocene circulation of the Southern Ocean: was Antarctica kept warm by subtropical waters? *Paleoceanography*, **19**, PA4026, doi:10.1029/2004PA001014

Instituto Geográfico Militar 1998. *Carta de Imagen Satelitaria de la República Argentina, 1:250,000, Rada Tilly 4569-IV*. Available at www.ign.gob.ar

Jacobs, B. F., J. D. Kingston, and L. L. Jacobs 1999. The origin of grass-dominated ecosystems. *Annals of the Missouri Botanical Garden*, **86**, 590–643.

Kay, R. F., D. J. Johnson, and D. J. Meldrum 1998a. A new pitheciin primate from the middle Miocene of Argentina. *American Journal of Primatology*, **45**, 317–336.

Kay, R. F., B. J. MacFadden, R. H. Madden, H. Sandeman, and F. Anaya 1998b. Revised age of the Salla beds, Bolivia, and its bearing on the age of the Deseadan South American Land Mammal 'Age'. *Journal of Vertebrate Paleontology*, **18**, 189–199.

Kay, R. F., R. M. Madden, M. G. Vucetich, A. A. Carlini, M. M. Mazzoni, G. H. Ré, M. Heizler, and H. Sandeman 1999. Revised age of the Casamayoran South American Land Mammal "Age": climatic and biotic implications. *Proceedings of the National Academy of Sciences USA*, **96**, 13 235–13 240.

Kay, R. F., B. A. Williams, and F. Anaya 2001. The adaptations of *Branisella boliviana*, the earliest South American monkey. In Plavcan, J. M., van Schaik, C., Kay, R. F., and Jungers, W. L. (eds.), *Reconstructing Behavior in the Primate Fossil Record*. New York: Plenum Press, pp. 339–370.

Kohn, M. J., J. A. Josef, R. M. Madden, R. F. Kay, M. G. Vucetich, and A. Carlini 2004. Climate stability across the Eocene–Oligocene transition, southern Argentina. *Geology*, **32**, 621–624.

Kramarz, A. G. and M. Bond 2005. Los Litopterna (Mammalia) de la Formación Pinturas, Mioceno Temprano–Medio de Patagonia. *Ameghiniana*, **42**, 611–625.

Kramarz, A. G. and M. Bond 2009. A new Oligocene astrapothere (Mammalia, Meridiungulata) from Patagonia and a new appraisal of astrapothere phylogeny. *Journal of Systematic Palaeontology*, **7**, 117–128.

Lawver, L. A. and L. M. Gahagan 2003. Evolution of Cenozoic seaways in the circum-Antarctic region. *Palaeogeography, Palaeoclimatology, Palaeoecology*, **198**, 11–37.

Lear, C. H., H. Elderfield, and P. A. Wilson 2000. Cenozoic deep-sea temperatures and global ice volumes from Mg/Ca in benthic foraminiferal calcite. *Science*, **287**, 269–272.

Lear, C. H., T. R. Bailey, P. N. Pearson, H. K. Coxall, and Y. Rosenthal 2008. Cooling and ice growth across the Eocene–Oligocene transition. *Geology*, **36**, 251–254.

Legendre, S. 1989. Les communautés de mammifères du Paléogène (Eocène supérieur et Oligocène) d'Europe occidentale: structures, milieux et evolution. *Münchner Geowissenschaftliche Abhandlungen A*, **16**, 5–110.

Livermore, R., A. Nankivell, G. Eagles, and P. Morris 2005. Paleogene opening of Drake Passage. *Earth and Planetary Science Letters*, **236**, 459–470.

Malumián, N., 1999. La sedimentación y el volcanismo terciarios en la Patagonia extraandina. I. La sedimentación en la Patagonia Extraandina. *Anales del Instituto de Geología y Recursos Minerales, Buenos Aires*, **29**, 557–578.

Markwick, P. J. 1998a. Crocodilian diversity in space and time: the role of climate in paleoecology and its implication for understanding K/T extinctions. *Paleobiology*, **24**, 470–497.

Markwick, P. J. 1998b. Fossil crocodilians as indicators of Late Cretaceous and Cenozoic climates: implications for using paleontological data in reconstructing paleoclimate. *Palaeogeography, Palaeoclimatology, Palaeoecology*, **137**, 205–271.

Marshall, L. G. 1976. Revision of the South American marsupial subfamily Abderitinae (Mammalia, Caenolestidae). *Museo Municipal de Ciencias Naturales de Mar del Plata "Lorenzo Scaglia,"* **2**, 57–90.

Marshall, L. G., R. Hoffstetter, and R. Pascual 1983. Mammals and stratigraphy: geochronology of the continental mammal-bearing Tertiary of South America. *Palaeovertebrata, Mémoire Extraordinaire, Montpellier*, 1–91.

Marshall, L. G., R. L. Cifelli, R. E. Drake, and G. H. Curtis 1986. Vertebrate paleontology, geology, and geochronology of the Tapera de Lopez and Scarritt Pocket, Chubut Province, Argentina. *Journal of Paleontology*, **60**, 920–951.

Mazzoni, M. M. 1979. Contribución al conocimiento petrográfico de la Formación Sarmiento, Barranca Sur del Lago Colhué-Huapí, Provincia de Chubut. *Revista de la Asociación Argentina de Mineralogía Petrología y Sedimentología*, **10**, 33–53.

Miller, K. G., R. G. Fairbanks, and G. S. Mountain 1987. Tertiary oxygen isotope synthesis, sea level history, and continental margin erosion. *Paleoceanography*, **2**, 1–19.

Miller, K. G., M. A. Kominz, J. V. Browning, J. D. Wright, G. S. Mountain, M. E. Katz, P. J. Sugarman, B. S. Cramer, N. Christie-Blick, and S. F. Pekar 2005. The Phanerozoic record of global sea-level change. *Science*, **310**, 1293–1298.

Nilsson, M. A., U. Arnason, P. B. S. Spencer, and A. Janke 2004. Marsupial relationships and a timeline for marsupial radiation in South Gondwana. *Gene*, **340**, 189–196.

Pagani, M., J. C. Zachos, K. H. Freeman, B. Tipple, and S. Bohaty 2005. Marked decline in atmospheric carbon dioxide concentrations during the Paleogene. *Science*, **309**, 600–603.

Pascual, R. and E. Ortíz-Jaureguízar 1990. Evolving climates and mammal faunas in Cenozoic South America. *Journal of Human Evolution*, **19**, 23–60.

Pascual, R. and E. Ortíz-Jaureguízar 2007. The Gondwanan and South American episodes: two major and unrelated moments in the history of the South American mammals. *Journal of Mammalian Evolution*, **14**, 75–137.

Pascual, R., E. Ortíz-Jaureguízar, and J. L. Prado 1996. Land mammals: paradigm for Cenozoic South American geobiotic evolution. *Münchner Geowissenschaftliche Abhandlungen A*, **30**, 265–319.

Patterson, B. D. 1999. Contingency and determinism in mammalian biogeography: the role of history. *Journal of Mammalogy*, **80**, 345–360.

Patterson, B. and R. Pascual 1968. Evolution of mammals on southern continents. V. The fossil mammal fauna of South America. *Quarterly Review of Biology*, **43**, 409–451.

Patterson, B. and A. E. Wood 1982. Rodents from the Deseadan Oligocene of Bolivia and the relationships of the Caviomorpha. *Bulletin of the Museum of Comparative Zoology*, **149**, 234–271.

Pekar, S. F. and N. Christie-Blick 2008. Resolving apparent conflicts between oceanographic and Antarctic climate records and evidence for a decrease in pCO$_2$ during the Oligocene through early Miocene (34–16 Ma). *Palaeogeography, Palaeoclimatology, Palaeoecology*, **260**, 41–49.

Prothero, D. R. and T. H. Heaton 1996. Faunal stability during the Early Oligocene climatic crash. *Palaeogeography, Palaeoclimatology, Palaeoecology*, **127**, 257–283.

Reguero, M. A. and P. V. Castro 2004. Un nuevo Trachytherinae (Mammalia, †Notoungulata) del Deseadense (Oligoceno tardío) de Patagonia, Argentina; implicancias en la filogenia, biogeografía y bioestratigrafía de los Mesotheriidae. *Revista Geologica de Chile*, **31**, 45–64.

Reguero, M. A. and E. Cerdeño 2005. New late Oligocene Hegetotheriidae (Mammalia, Notoungulata) from Salla, Bolivia. *Journal of Vertebrate Paleontology*, **25**, 674–684.

Reguero, M. A., S. A. Marenssi, and S. N. Santillana 2002. Antarctic Peninsula and South America (Patagonia) Paleogene terrestrial faunas and environments: biogeographic relationships. *Palaeogeography, Palaeoclimatology, Palaeoecology*, **179**, 189–210.

Reguero, M. A., D. A. Croft, J. J. Flynn, and A. R. Wyss 2003. Small archaeohyracids from Chubut, Argentina and Central Chile: implications for trans-Andean temporal correlation. *Fieldiana, Geology*, n.s, **48**, 1–17.

Scher, H. D. and E. E. Martin 2006. Timing and climatic consequences of the opening of Drake Passage. *Science*, **312**, 428–430.

Simpson, G. G. 1930. *Scarritt-Patagonian Exped. Field Notes*. New York: American Museum of Natural History. (Unpublished.) Available at http://paleo.amnh.org/notebooks/index.html

Simpson, G. G. 1933. A new crocodilian from the *Notostylops* beds of Patagonia. *American Museum Novitates*, **623**, 1–9.

Simpson, G. G. 1936. A specimen of *Pseudostylops subquadratus* Ameghino. *Memorie dell'Istituto Geologico della Reale Università di Padova*, **11**, 1–12.

Simpson, G. G. 1948. The beginning of the Age of Mammals in South America. I. *Bulletin of the American Museum of Natural History*, **91**, 1–232.

Simpson, G. G. 1950. History of the fauna of Latin America. *American Scientist*, **38**, 361–389.

Simpson, G. G. 1967. The beginning of the Age of Mammals in South America. II. *Bulletin of the American Museum of Natural History*, **137**, 1–260.

Simpson, G. G. 1980. *Splendid Isolation: The Curious History of South American Mammals*. New Haven, CT: Yale University Press.

Simpson, G. G. n.d. *Identifications of fossils in Feruglio Collection*. Philadelphia, PA: Simpson Papers, American Philosophical Society. (Unpublished.)

Soria, M. F. 2001. Los Proterotheriidae (Litopterna, Mammalia), sistemática, origen y filogenia. *Monografias del Museo Argentino de Ciencias Naturales, Buenos Aires*, **1**, 1–167.

Springer, M. S., M. Westerman, J. R. Kavanagh, A. Burk, M. O. Woodburne, D. J. Kao, and C. Krajewski 1998. The origin of the Australasian marsupial fauna and the phylogenetic affinities of the enigmatic monito del monte and marsupial mole. *Proceedings of the Royal Society of London B*, **265**, 2381–2386.

Steppan, S. J. 1995. Revision of the Tribe Phyllotini (Rodentia: Sigmodontinae), with a phylogenetic hypothesis for the Sigmodontinae. *Fieldiana, Zoology*, n.s. **80**, 1–112.

Steppan, S. J., R. M. Adkins, and J. Anderson 2004. Phylogeny and divergence-date estimates of rapid radiations in muroid rodents based on multiple nuclear genes. *Systematic Biology*, **53**, 533–553.

Stickley, C. E., H. Brinkhuis, S. A. Schellenberg, A. Sluijs, U. Röhl, M. Fuller, M. Grauert, M. Huber, J. Warnaar, and G. L. Williams 2004. Timing and nature of the deepening of the Tasmanian Gateway, *Paleoceanography*, **19**, PA4027, doi:10.1029/2004PA001022.

Strait, S. G., J. G. Fleagle, T. M. Bown, and E. R. Dumont 1990. Diversity in body size and dietary habits of fossil caenolestid marsupials from the Miocene of Argentina. *Journal of Vertebrate Paleontology*, **10**, 44A.

Strömberg, C. A. E. 2002. The origin and spread of grass-dominated ecosystems in the late Tertiary of North America: preliminary results concerning the origin of hypsodonty. *Palaeogeography, Palaeoclimatology, Palaeoecology*, **177**, 59–75.

Strömberg, C. A. E. 2004. Using phytolith assemblages to reconstruct the origin and spread of grass-dominated habitats in the great plains of North America during the late Eocene to early Miocene. *Palaeogeography, Palaeoclimatology, Palaeoecology*, **207**, 239–275.

Strömberg, C. A. E. 2005. Decoupled taxonomic radiation and ecological expansion of open-habitat grasses in the Cenozoic of North America. *Proceedings of the National Academy of Sciences USA*, **102**, 11 980–11 984.

Takai, M., F. Anaya, N. Shigehara, and T. Setoguchi 2000. New fossil materials of the earliest new world monkey, *Branisella boliviana*, and the problem of platyrrhine origins. *American Journal of Physical Anthropology*, **111**, 263–281.

Thomas, E. 2008. Descent into the icehouse. *Geology*, **36**, 191–192.

Verzi, D. H. and C. I. Montalvo 2008. The oldest South American Cricetidae (Rodentia) and Mustelidae (Carniovora): late Miocene faunal turnover in central Argentina and the Great American Biotic Interchange. *Palaeogeography, Palaeoclimatology, Palaeoecology*, **267**, 284–291.

Voss, R. S. 1988. Systematics and ecology of ichthyomyine rodents (Muroidea): patterns of morphological evolution in a small adaptive radiation. *Bulletin of the American Museum of Natural History*, **188**, 259–493.

Vucetich, M. G., A. Carlini, R. M. Madden, and R. F. Kay 2004. New discoveries among the oldest rodents in South America: how old and how primitive? *Journal of Vertebrate Paleontology*, **24**, 125A.

Vucetich, M. G. & A. M. Ribeiro 2003. A new and primitive rodent from the Tremembé Formation (Late Oligocene) of Brazil, with comments on the morphology of the lower premolars of caviomorph rodents. *Revista Brasileira de Paleontologia, Porto Alegre*, **5**, 73–83.

Walton, A. H. 1997. Rodents. In Kay, R. F., Madden, R. H., Cifelli, R. L., and Flynn, J. J. (eds.), *Vertebrate Paleontology in the Neotropics: The Miocene Fauna of La Venta, Colombia*. Washington, D.C: Smithsonian Institution Press, pp. 392–409.

Williams, S. and R. F. Kay 2001. A comparative test of adaptive explanations for hypsodonty in ungulates and rodents. *Journal of Mammalian Evolution*, **8**, 207–229.

Wood, A. E. and B. Patterson 1959. The rodents of the Deseadan Oligocene of Patagonia and the beginnings of South American rodent evolution. *Bulletin of the Museum of Comparative Zoology, Harvard University*, **120**, 279–428.

Wyss, A. R., J. J. Flynn, M. A. Norell, C. C. Swisher, R. Charrier, M. J. Novacek, and M. C. McKenna 1993. South America earliest rodent and recognition of a new interval of mammalian evolution. *Nature*, **365**, 434–437.

Zachos, J., M. Pagani, L. Sloan, E. Thomas, and K. Billups 2001. Trends, rhythms, and aberrations in global climate 65 Ma to present. *Science*, **292**, 686–693.

Index